T0265320

Current Directions in Water Scarcity Research

WATER RESOURCE MODELING AND COMPUTATIONAL TECHNOLOGIES

VOLUME 7

Current Directions in Water Scarcity Research

VOLUME 7

Series Editors

ROBERT MCLEMAN

MICHAEL BRÜNTRUP

Volume 1: From Catchment Management to Managing River Basins
ISBN: 9780128148518

Volume 2: Drought Challenges
ISBN: 9780128148204

Volume 3: Water Productivity and Food Security
ISBN: 9780323914512

Volume 4: Indigenous Water and Drought Management in a Changing World
ISBN: 9780128245385

Volume 5: Water Resources
ISBN: 9780323853781

Volume 6: Urban Water Crisis and Management: Strategies for Sustainable Development
ISBN: 9780323918381

Current Directions in Water Scarcity Research

WATER RESOURCE MODELING AND COMPUTATIONAL TECHNOLOGIES

VOLUME 7

Edited By

MOHAMMAD ZAKWAN
School of Technology, MANUU, Hyderabad, India

ABDUL WAHID
School of Technology, MANUU, Hyderabad, India

MAJID NIAZKAR
Department of Civil and Environmental Engineering, School of Engineering, Shiraz University, Shiraz, Iran

UDAY CHATTERJEE
Department of Geography, Bhatter College, Dantan (Affiliated to Vidyasagar University), West Bengal, India

ELSEVIER

Elsevier
Radarweg 29, PO Box 211, 1000 AE Amsterdam, Netherlands
The Boulevard, Langford Lane, Kidlington, Oxford OX5 1GB, United Kingdom
50 Hampshire Street, 5th Floor, Cambridge, MA 02139, United States

ISBN: 978-0-323-91910-4
ISSN: 2542-7946

For information on all Elsevier publications
visit our website at https://www.elsevier.com/books-and-journals

Publisher: Candice Janco
Acquisitions Editor: Louisa Munro/Maria Elekidou
Developmental Editor: Sara Valentino
Production Project Manager: Sruthi Satheesh
Cover Designer: Matthew Limbert

Typeset by STRAIVE, India

Contents

Contributors xiii
About the editors xvii
Foreword xix
Preface xxi
Acknowledgments xxiii

Section I
Introduction

1. Artificial intelligence and machine learning in water resources engineering
Mohd Danish

1. Introduction 3
2. Materials and methods 4
3. Evolution of artificial intelligence and machine learning 7
4. Results and discussion 8
5. Conclusion 11
References 12

Section II
Application of artificial intelligence to water resources

2. Demystifying artificial intelligence amidst sustainable agricultural water management
Aman Srivastava, Shubham Jain, Rajib Maity, and Venkappayya R. Desai

1. Introduction 17
2. AI in agriculture 22
3. Current and future scope in AI for agriculture 29
4. Challenges of AI in agriculture 31
5. Conclusions 31
Acknowledgments 32
Conflict of interest 32
References 32
Further reading 35

3. Bidirectional long short-term memory-based empirical wavelet transform: A new hybrid artificial intelligence model for robust prediction of soil moisture content
Salim Heddam, Sungwon Kim, Ahmed Elbeltagi, and Ozgur Kisi

1. Introduction 38
2. Materials and methods 39
3. Results and discussion 45
4. Conclusions 53
5. Recommendations 53
References 54

4. Fuzzy logic modeling of groundwater potential in Marinduque, Philippines
Destiny S. Lutero, Jcob C. Malaguit, Marie Joy L. Relente, Allen L. Nazareno, and Arnold R. Salvacion

1. Introduction 57
2. Material and methods 59
3. Results 63
4. Discussion 66
5. Conclusion 67
References 67

5. Soft-computing approach to scour depth prediction under wall jets
Mohammad Aamir, Mohammad Amir Khan, and Zulfequar Ahmad

1. Introduction 71
2. Materials and methods 73
3. Results and discussion 75
4. Conclusions 80
References 80

Section III

Image processing applications in water resources

6. Assessment of water resources using remote sensing and GIS techniques

N.L. Kushwaha, Ahmed Elbeltagi, Abhishek Patel, Mohammad Zakwan, Jitendra Rajput, and Puneet Sharma

1. Introduction 86
2. Remote sensing and GIS: Tools for sustainability of water resources 87
3. Global positioning system (GPS) 92
4. Case study: Hydrological response analysis of IARI watershed using remote sensing and GIS 93
5. Future recommendations for efficient water resource management 95
6. Conclusion 96
References 96

7. Establishing spatial relationships between land use and water quality influenced by urbanization

Manish Kumar Sinha, Klaus Baier, Rafig Azzam, Preeti Rajput, and M.K. Verma

1. Introduction 100
2. Study area 102
3. Material and methods 103
4. Results and discussion 107
5. Conclusion 112
Acknowledgments 113
References 113

8. Satellite sensors, machine learning, and river channel unit types: A review

Adeyemi Olusola and Samuel Adelabu

1. Introduction 117
2. Bedrock channels 118
3. River channels and remote sensing 120
4. Materials and methods 122
5. Results and discussion 123
6. Predicting bedrock channels using machine learning algorithms 129
7. Conclusion 130
References 130

9. Geospatial modeling in the assessment of environmental resources for sustainable water resource management in a semiarid region: A GIS approach

Nanabhau Santujee Kudnar

1. Introduction 134
2. Study area 135
3. Material and methods 135
4. Results and discussion 138
5. Conclusions 148
Conflict of interest 148
References 148

10. Study of morphologic changes in the past and predicting future changes of border rivers (case study: Arvand River, Iran-Iraq Border Line)

Alireza Hajiabadi, Saeed Sakhdari, and Reza Barati

1. Introduction 153
2. Materials and methods 154
3. Results and discussion 157
4. Conclusion 162
References 162
Further reading 163

11. Rainfall-runoff modeling using GIS: A case study of Gorganrood Watershed, Iran

Mohammad Reza Goodarzi, Amir Reza R. Niknam, and Maryam Sabaghzadeh

1. Introduction 165
2. Rationale of the study 167
3. Material and methods 168
4. Results and discussion 173
5. Conclusion 180
References 180

12. A review of GIS-based hydrological models for sustainable groundwater management

Hamza Badamasi

1. Introduction 183
2. Hydrological modeling 185
3. Geographic information system (GIS) 189
4. Benefits of using GIS-based hydrological models for groundwater modeling 194
5. Drawbacks of GIS-based hydrological models for groundwater modeling 194
6. Conclusion and future prospects 195
References 196

13. Development of rainfall-runoff model using ANFIS with an integration of GIS: A case study

Sandeep Samantaray, Abinash Sahoo, Sambit Sawan Das, and Deba Prakash Satapathy

1. Introduction 202
2. Study area 203
3. Methodology 205
4. Results and discussions 208
5. Conclusion 220
References 221

14. Assessing the impact of land use and land cover changes on the water balances in an urbanized peninsular region of India

Harsh Ganapathi, Mayuri Phukan, Preethi Vasudevan, and Santosh S. Palmate

1. Introduction 226
2. Materials and methods 227
3. Results and discussion 232
4. Conclusion 239
References 239

Section IV

Advances in hydroinformatics mitigation

15. Random vector functional link network based on variational mode decomposition for predicting river water turbidity

Salim Heddam, Sungwon Kim, Ahmed Elbeltagi, and Ozgur Kisi

1. Introduction 246
2. Materials and methods 248
3. Results and discussion 255
4. Conclusions 259
5. Recommendations 261
Acknowledgments 262
References 262

16. Water quality management: Development of a fuzzy-based index in hydro informatics platform

A. Krishnakumar, S.K. Aditya, K. AnoopKrishnan, Revathy Das, and K. Anju

1. Introduction 265
2. Study area 267
3. Materials and methods 269
4. Results and discussion 274
5. Conclusion 282
Acknowledgments 282
References 282

17. Appraisal of multigene genetic programming for estimating optimal properties of lined open channels with circular shapes incorporating constant and variable roughness scenarios

Majid Niazkar

1. Introduction 285
2. Methods and materials 286

3. Results and discussion 290
4. Conclusions 293
Funding 295
Conflict of interest 295
References 297

18. Geoinformatics-based assessment of gross irrigation requirement of different crops grown in the south-western region of Haryana, India

Arvind Dhaloiya, RS Hooda, Devendra Kumar, Anurag Malik, and Ajay Kumar

1. Introduction 300
2. Study area 301
3. Material and methods 301
4. Results and discussion 304
5. Conclusions 315
Conflict of interest 315
References 315

Section V

Advances in watershed modelling

19. Theoretical background and application of numerical modeling to surface water resources

Oscar Herrera-Granados

1. Introduction 319
2. Materials and methods 321
3. Conclusions 338
References 339

20. Prophecy of groundwater fluctuation through SVM-FFA hybrid approaches in arid watershed, India

Sandeep Samantaray, Abinash Sahoo, Deba Prakash Satapathy, and Shaswati S. Mishra

1. Introduction 342
2. Study area 345
3. Material and methods 347
4. Results and discussion 352
5. Conclusions 362
References 362
Further reading 365

21. Basin-scale subsurface hydrology: Modeling of a stressed and data-scarce aquifer using hillslope-based approach

Soumyaranjan Sahoo and Suraj Jena

1. Introduction 367
2. Materials and methods 370
3. Study area and data collection 373
4. Field application 377
5. Results and discussion 379
6. Summary and conclusions 383
References 384

Section VI

Advances in numerical modelling in water resources

22. Multiphysics modeling of groundwater flow on the example of a coupled thermo-hydro-mechanical model of infiltration of water warmer or cooler than the surroundings

Anna Uciechowska-Grakowicz

1. Introduction 389
2. Mathematical models of fluid flow in porous media 391
3. Results and discussion 400
4. Conclusions 403
Acknowledgments 404
References 404

23. Hydrosalinity modeling of water and salt dynamics in irrigated soil groundwater systems

Samanpreet Kaur, Mehraj U. Din Dar, and S.K. Kamra

1. Introduction 410
2. Materials and methods 411
3. Hydrosalinity models 420
4. Results and discussions 421
5. Subsurface drainage modeling 423
6. Conclusions 427
References 427

Section VII

Optimization techniques and analytical formulations in water resource

24. Multiobjective optimization techniques for integrated urban water management: A case study of Varanasi city
Satya Prakash Maurya and Ramesh Singh

1. Introduction 433
2. Material and methods 436
3. Case study of Varanasi City 440
4. Conclusion 444
References 445

25. Hybrid extreme learning machine optimized bat algorithm based on ensemble empirical mode decomposition for modeling dissolved oxygen in river
Salim Heddam, Sungwon Kim, Ahmed Elbeltagi, and Ozgur Kisi

1. Introduction 448
2. Materials and methods 449
3. Results and discussion 456
4. Conclusions 462
5. Recommendations 463
References 463
Further reading 466

26. Application of machine learning models to side-weir discharge coefficient estimations in trapezoidal and rectangular open channels
Majid Niazkar and Mohammad Zakwan

1. Introduction 467
2. Material and methods 469
3. Results and discussion 474
4. Conclusion 477
References 478

Section VIII

Advances in sediment transport modelling and river engineering

27. The hole size analysis of bursting events around mid-channel bar using the conditional method approach
Mohammad Amir Khan, Nayan Sharma, Mohammad Aamir, Manish Pandey, Rishav Garg, and Hanif Pourshahbaz

1. Introduction 484
2. Experimental program 485
3. Bursting events 487
4. Conclusions 493
5. Limitation and scope for future research work 494
References 494

28. Magnitude frequency analysis of sediment transport: Concept, review, and application
Mohammad Zakwan, Qamar Sultana, and Gulfam Ahamad

1. Introduction 497
2. Literature review 503
3. Materials and method 508
4. Results and discussion 508
5. Applications of effective discharge in river design and monitoring 509
6. Conclusion 510
References 510

29. Last century evolution in local scour measuring techniques
Geeta Devi and Munendra Kumar

1. Introduction 513
2. Scour monitoring 516
3. Scour monitoring measurement method in the laboratory 523
4. Scour monitoring using structural dynamic 525
5. Photogrammetry 526
6. Conclusion 526
References 527
Further reading 529

Section IX

Computational intelligence in extreme hydrology: flood and droughts

30. Understanding trend and its variability of rainfall and temperature over Patna (Bihar)

Nitesh Gupta, Pradeep K. Mahato, Jitendra Patel, Padam Jee Omar, and Ravi P. Tripathi

1. Introduction 533
2. Area of research and information sources 535
3. Material and methods 536
4. Results and discussion 538
5. Conclusion 541
References 542

31. A review of climate change trends and scenarios (2011–2021)

Deborah Ayodele-Olajire and Adeyemi Olusola

1. Introduction 545
2. Methodology 547
3. Results and discussion 550
4. Conclusion 558
References 558

32. Climate change and trend analysis of precipitation and temperature: A case study of Gilan, Iran

Mohammad Reza Goodarzi, Mohammad Javad Abedi, and Mahnaz Heydari Pour

1. Introduction 562
2. Materials and methods 564
3. Performance evaluation of model used 568
4. Trend analysis 570
5. Results and discussion 570
6. Conclusions 582
References 585

33. Innovative triangular trend analysis of monthly precipitation at Shiraz Station, Iran

Mohammad Zakwan and Majid Niazkar

1. Introduction 589
2. Materials and methods 591
3. Innovative trend analysis (ITA) 593
4. Results and discussion 594
5. Conclusion 596
References 597

34. Overview of trend and homogeneity tests and their application to rainfall time series

Deepesh Machiwal, H.M. Meena, and D.V. Singh

1. Introduction 600
2. Materials and methods 604
3. Results and discussion 605
4. Limitations and future challenges 616
5. Recommendations 616
6. Concluding remarks 616
Acknowledgments 617
References 617

35. Flash floods and their impact on natural life using surface water model and GIS technique at Wadi Degla natural reserve area, Egypt

Sherif A. Abu El-Magd, Ahmed M. Masoud, Ahmed A. Abdel Moneim, and Bakr M. Bakr

1. Introduction 622
2. Materials and methods 623
3. Results and discussions 629
4. Impact on natural life 635
5. Conclusions 635
Funding 638
Availability of data and materials 638
Ethics approval and consent to participate 638
Consent for publication 638
Competing interests 638
References 638

36. GLOF Early Warning System: Computational challenges and solutions

Binay Kumar, T.S. Murugesh Prabhu, Anish Sathyan, and Arun Krishnan

1. Introduction 641
2. Materials and methods 643
3. Results and discussion 657
4. Salient features of GLOF Early Warning System 657
5. Limitations of the study 659
6. Recommendations 659
7. Conclusions 660
Acknowledgments 660
References 660

37. Flood forecasting using novel ANFIS-WOA approach in Mahanadi river basin, India

Sandeep Samantaray, Abinash Sahoo, and Shaswati S. Mishra

1. Introduction 663
2. Study area 666
3. Material and methods 668
4. Results and discussion 672
5. Conclusion 678
References 681

Index 683

Contributors

Mohammad Aamir Department of Civil Engineering, Chaitanya Bharathi Institute of Technology, Hyderabad, India

Ahmed A. Abdel Moneim Geology Department, Faculty of Science, Sohag University, Sohag, Egypt

Mohammad Javad Abedi Department of Civil Engineering, Water Resources Management Engineering, Yazd University, Yazd, Iran

Sherif A. Abu El-Magd Geology Department, Faculty of Science, Suez University, Suez, Egypt

Samuel Adelabu Department of Geography, University of the Free State, Bloemfontein, South Africa

S.K. Aditya ESSO-National Centre for Earth Science Studies (NCESS), Ministry of Earth Sciences, Government of India, Thiruvananthapuram, Kerala, India

Gulfam Ahamad Department of Computer Sciences, BGSBU, Rajouri, India

Zulfequar Ahmad Department of Civil Engineering, Indian Institute of Technology Roorkee, Roorkee, India

K. Anju ESSO-National Centre for Earth Science Studies (NCESS), Ministry of Earth Sciences, Government of India, Thiruvananthapuram, Kerala, India

K. AnoopKrishnan ESSO-National Centre for Earth Science Studies (NCESS), Ministry of Earth Sciences, Government of India, Thiruvananthapuram, Kerala, India

Deborah Ayodele-Olajire Department of Geography, University of Ibadan, Ibadan, Nigeria

Rafig Azzam Department of Engineering Geology and Hydrogeology, RWTH Aachen University, Aachen, Germany

Hamza Badamasi Department of Chemistry, Federal University Dutse, Dutse, Jigawa State, Nigeria

Klaus Baier Department of Engineering Geology and Hydrogeology, RWTH Aachen University, Aachen, Germany

Bakr M. Bakr Egyptian Environmental Affairs Agency (EEAA), Cairo, Egypt

Reza Barati Ph.D. of Civil Engineering, Head of Applied Research Group at Water Authority, Khorasan Razavi Water Authority, Mashhad, Iran

Mohd Danish Civil Engineering Section, University Polytechnic, Aligarh Muslim University, Aligarh, UP, India

Revathy Das ESSO-National Centre for Earth Science Studies (NCESS), Ministry of Earth Sciences, Government of India, Thiruvananthapuram, Kerala, India

Sambit Sawan Das Department of Civil Engineering, NIT Srinagar, Jammu & Kashmir, India

Venkappayya R. Desai Department of Civil Engineering, Indian Institute of Technology (IIT) Kharagpur, Kharagpur, West Bengal, India

Geeta Devi Delhi Technological University, New Delhi, India

Arvind Dhaloiya Haryana Space Applications Centre, Hisar, Haryana; Department of Soil and Water Engineering, Punjab Agricultural University, Ludhiana, Punjab, India

Mehraj U. Din Dar Department of Soil and Water Engineering, Punjab Agricultural University, Ludhiana, India

Ahmed Elbeltagi Agricultural Engineering Department, Faculty of Agriculture, Mansoura University, Mansoura, Egypt

Harsh Ganapathi Wetlands International South Asia, New Delhi, India

Rishav Garg Department of Civil Engineering, Galgotias College of Engineering and Technology, Greater Noida, India

Mohammad Reza Goodarzi Department of Civil Engineering, Yazd University, Yazd, Iran

Nitesh Gupta Department of Civil Engineering, Indian Institute of Technology (BHU), Varanasi, India

Alireza Hajiabadi MSc of Water Resources Engineering, Abrah GostarTadbir Consulting Engineering Co., Tehran, Iran

Salim Heddam Faculty of Science, Agronomy Department, Hydraulics Division, Laboratory of Research in Biodiversity Interaction Ecosystem and Biotechnology, University 20 Août 1955, Skikda, Algeria

Oscar Herrera-Granados Faculty of Civil Engineering, Wrocław University of Science and Technology, Wrocław, Poland

Mahnaz Heydari Pour Department of Civil Engineering, Water Resources Management Engineering, Yazd University, Yazd, Iran

RS Hooda Advisor to Agriculture and Farmer Welfare Department, Panchkula, Haryana, India

Shubham Jain Department of Civil Engineering, Indian Institute of Technology (IIT) Bombay, Mumbai, Maharashtra, India

Suraj Jena School of Infrastructure, Indian Institute of Technology Bhubaneswar, Bhubaneswar, Odisha, India

S.K. Kamra Emeritus Scientist (Hon.), ICAR-Central Soil Salinity Research Institute, Karnal, Haryana, India

Samanpreet Kaur Department of Soil and Water Engineering, Punjab Agricultural University, Ludhiana, India

Mohammad Amir Khan Department of Civil Engineering, Galgotias College of Engineering and Technology, Greater Noida, India

Sungwon Kim Department of Railroad Construction and Safety Engineering, Dongyang University, Yeongju, Republic of Korea

Ozgur Kisi Civil Engineering Department, Ilia State University, Tbilisi, Georgia; Department of Civil Engineering, Technical University of Lübeck, Lübeck, Germany

A. Krishnakumar ESSO-National Centre for Earth Science Studies (NCESS), Ministry of Earth Sciences, Government of India, Thiruvananthapuram, Kerala, India

Arun Krishnan Centre for Development of Advanced Computing (C-DAC), Pune, Maharashtra, India

Nanabhau Santujee Kudnar C. J. Patel College, Tirora, Maharashtra, India

Ajay Kumar Department of Microbiology, Chaudhary Charan Singh Haryana Agricultural University, Hisar, Haryana, India

Binay Kumar Centre for Development of Advanced Computing (C-DAC), Pune, Maharashtra, India

Devendra Kumar Haryana Space Applications Centre, Hisar, Haryana, India

Munendra Kumar Delhi Technological University, New Delhi, India

N.L. Kushwaha Division of Agricultural Engineering, ICAR-Indian Agricultural Research Institute, New Delhi, India

Destiny S. Lutero Institute of Mathematical Sciences and Physics, College of Arts and Sciences, University of the Philippines Los Baños, Los Baños, Laguna, Philippines

Deepesh Machiwal Division of Natural Resources, ICAR-Central Arid Zone Research Institute, Jodhpur, Rajasthan, India

Pradeep K. Mahato Department of Civil Engineering, Indian Institute of Technology (BHU), Varanasi, India

Rajib Maity Department of Civil Engineering, Indian Institute of Technology (IIT) Kharagpur, Kharagpur, West Bengal, India

Jcob C. Malaguit Institute of Mathematical Sciences and Physics, College of Arts and Sciences, University of the Philippines Los Baños, Los Baños, Laguna, Philippines

Anurag Malik Punjab Agricultural University, Regional Research Station, Bathinda, Punjab, India

Ahmed M. Masoud Geology Department, Faculty of Science, Sohag University, Sohag, Egypt

Satya Prakash Maurya Department of Civil Engineering, Indian Institute of Technology (BHU), Varanasi, India

H.M. Meena Division of Natural Resources, ICAR-Central Arid Zone Research Institute, Jodhpur, Rajasthan, India

Shaswati S. Mishra Department of Philosophy, Utkal University, Bhubaneswar, Odisha, India

Allen L. Nazareno Institute of Mathematical Sciences and Physics, College of Arts and Sciences, University of the Philippines Los Baños, Los Baños, Laguna, Philippines

Majid Niazkar Department of Civil and Environmental Engineering, School of Engineering, Shiraz University, Shiraz, Iran

Amir Reza R. Niknam Department of Civil Engineering, Water Resources Management Engineering, Yazd University, Yazd, Iran

Adeyemi Olusola Department of Geography, University of Ibadan, Ibadan, Nigeria; Department of Geography, University of the Free State, Bloemfontein, South Africa; Faculty of Environmental and Urban Change, York University, Toronto, ON, Canada

Padam Jee Omar Department of Civil Engineering, Motihari College of Engineering, Motihari, India

Santosh S. Palmate Texas A&M AgriLife Research, Texas A&M University, El Paso, TX, United States

Manish Pandey National Institute of Technology Warangal, Warangal, India

Abhishek Patel Regional Research Station, ICAR-Central Arid Zone Research Institute, Kukma, Bhuj, India

Jitendra Patel Department of Civil Engineering, Samrat Ashok Technological Institute, Vidisha, India

Mayuri Phukan The Energy and Resources Institute, New Delhi, India

Hanif Pourshahbaz Department of Civil and Water Engineering, Laval University, Pavillon Adrien-Pouliot, Québec, QC, Canada

T.S. Murugesh Prabhu Centre for Development of Advanced Computing (C-DAC), Pune, Maharashtra, India

Jitendra Rajput Division of Agricultural Engineering, ICAR-Indian Agricultural Research Institute, New Delhi, India

Preeti Rajput Department of Civil Engineering, Government Engineering College, Raipur, Chhattisgarh, India

Marie Joy L. Relente Institute of Statistics, College of Arts and Sciences, University of the Philippines Los Baños, Los Baños, Laguna, Philippines

Maryam Sabaghzadeh Department of Civil Engineering, Water Resources Management Engineering, Yazd University, Yazd, Iran

Abinash Sahoo Department of Civil Engineering, NIT Silchar, Silchar, Assam, India

Soumyaranjan Sahoo School of Water Resources, Indian Institute of Technology Kharagpur, Kharagpur, West Bengal, India

Saeed Sakhdari MSc of Water Resources Engineering, Department of Civil Engineering, University of Sistan and Baluchestan, Zahedan, Iran

Arnold R. Salvacion Department of Community and Environmental Resource Planning, College of Human Ecology, University of the Philippines Los Baños, Los Baños, Laguna, Philippines

Sandeep Samantaray Department of Civil Engineering, NIT Srinagar, Jammu & Kashmir, India

Deba Prakash Satapathy Department of Civil Engineering, NIT Srinagar, Jammu & Kashmir, India

Anish Sathyan Centre for Development of Advanced Computing (C-DAC), Pune, Maharashtra, India

Nayan Sharma Center for Environmental Sciences & Engineering (CESE), Shiv Nadar University, Institution of Eminence, Greater Noida, UP, India

Puneet Sharma Punjab Agricultural University, Ludhiana, India

D.V. Singh Division of Natural Resources, ICAR-Central Arid Zone Research Institute, Jodhpur, Rajasthan, India

Ramesh Singh Department of Civil Engineering, Indian Institute of Technology (BHU), Varanasi, India

Manish Kumar Sinha Department of Engineering Geology and Hydrogeology, RWTH Aachen University, Aachen, Germany Environmental and Water Resources Engineering, University Teaching Department, Chhattisgarh Swami Vivekanand Technical University Bhilai, Bhilai, Chhattisgarh, India

Aman Srivastava Department of Civil Engineering, Indian Institute of Technology (IIT) Kharagpur, Kharagpur, West Bengal, India

Qamar Sultana Department of Civil Engineering, MJCET, Hyderabad, India

Ravi P. Tripathi Department of Civil Engineering, Rajkiya Engineering College, Sonbhadra, India

Anna Uciechowska-Grakowicz Faculty of Civil Engineering, Wrocław University of Science and Technology, Wrocław, Poland

Preethi Vasudevan TERI School of Advanced Studies, New Delhi, India

M.K. Verma Environmental and Water Resources Engineering, University Teaching Department, Chhattisgarh Swami Vivekanand Technical University Bhilai, Bhilai, Chhattisgarh, India

Mohammad Zakwan School of Technology, MANUU, Hyderabad, India

About the editors

Mohammad Zakwan is presently working as an assistant professor at Maulana Azad National Urdu University, Hyderabad, India. He was awarded a PhD from the Department of Civil Engineering, Indian Institute of Technology, Roorkee. He has published numerous papers in Science Citation Index and Scopus journals with a demonstrated history of working in water resource engineering. He has been serving as editorial board member and reviewer for various Science Citation Index and Scopus journals. His research interests include modeling of infiltration, modeling river water quality, sediment rating curves, stage discharge curves, computation of effective discharge and dominant discharge, climate change, trend analysis in water resources, river engineering, hydraulics, and hydrology of rivers. He has more than 3 years of teaching and research experience.

Abdul Wahid is presently serving as a professor and the dean of the School of Technology, Maulana Azad National University, Hyderabad, India. He has also served as a visiting professor (research) at the National Defence University, Malaysia. He has more than 22 years of teaching experience. He has published numerous papers in Science Citation Index and Scopus journals with a demonstrated history of working in computational intelligence. He has been serving as editorial board member and reviewer for various Science Citation Index and Scopus journals. He has been session chair and keynote speaker in many international conferences. His research interests include computer architecture, artificial neural network, and computer security and reliability. He has worked on several national and international research projects.

Majid Niazkar received his BS in civil engineering (2012), MS in civil engineering (hydraulic structures) (2014), and PhD in civil engineering (water resources) (2019) from the Department of Civil and Environmental Engineering, Shiraz University, Shiraz, Iran. He completed a 1-year postdoctoral research, which was funded by Iran's National Elites Foundation at Shiraz University in 2021. His research interests include hydrology, open-channel hydraulics, water resources management, impact assessment of climate change, and artificial intelligence applications. He has published more than 70 papers in international peer-reviewed journals and conference proceedings. In addition, he has served as reviewer for more than 130 manuscripts for various international journals. Based on his research outputs, he was selected as one of the World's Top 2% Scientists in 2020 by Stanford University and Elsevier on October 19, 2021. Recently, he joined the editorial board of *Mathematical Problems in Engineering* (ISSN: 1024-123X).

Uday Chatterjee is an assistant professor in the Department of Geography, Bhatter College, Dantan, Paschim Medinipur, West Bengal, India, and an applied geographer with a postgraduate degree in applied geography from Utkal University and doctoral degrees in applied geography at Ravenshaw University, Cuttack, Odisha, India. He has contributed several research papers published in reputed national and international journals and has edited book volumes. He has recently jointly edited the book titled *Land Reclamation and Restoration Strategies for Sustainable Development* (November 2021; 1st Edition; Publisher: Elsevier; Editors: Dr. Gouri Sankar Bhunia, Dr. Uday Chatterjee, Dr. Anil Kashyap, Dr. Pravat Kumar Shit; ISBN: 9780128238950; https://www.elsevier.com/books/land-reclamation-and-restoration-strategies-for-sustainable-development/bhunia/978-0-12-823895-0). His research interests include urban planning, social and human geography, applied geomorphology, hazards, disasters, and environmental issues. His research work has been funded by the West Bengal Pollution Control Board (WBPCB), Government of West Bengal, India. Currently, Dr. Uday Chatterjee is the lead editor of the Special Issue on Urbanism, Smart Cities and Modelling, *GeoJournal* (Springer).

Foreword

It is a great honor to write the foreword for the book titled *Water Resource Modeling and Computational Technologies* edited by Dr. Mohammad Zakwan, Dr. Abdul Wahid, Dr. Majid Niazkar, and Dr. Uday Chatterjee to be published by Elsevier.

Water is a crucial natural resource for the entire ecosystem, so its sustainable development, protection, regulation, and management by meeting demands efficiently are all important. The complex interactions of the water resource system, human-induced stresses, and increased sectoral competition have been a challenge for its sustainability. Water resources system modeling is vital and can deliver essential information about the interaction of components within the system and support evidence-based planning and decision-making. Models are simplified replications of real-world systems with several assumptions, approximations, and limitations. Developing models is a cycle of processes that involve using knowledge of the system being modeled, a clear objective, adequate information, and some analytical and programming skills. Modeling water resource systems is complex as it includes several interdependent processes that govern the system's behavior, and computational limitations can further increase the complexity. A good model or computational technique provides better information for informed decision-making. There is a strong need to understand and consider the advances in modeling several water resource aspects (engineering, hydrological, ecological, institutional, etc.) to effectively predict the system's behavior or performance and take a significant decision.

This book provides a complete package of several tools and technologies in water resources modeling. The 37 chapters in this book are systematically categorized into 9 sections to provide a clear picture to the readers. The sections include Introduction (Section I); Application of artificial intelligence to water resources (Section II); Image processing applications in water resources (Section III); Advances in hydroinformatics mitigation (Section IV); Advances in watershed modeling (Section V); Advances in numerical modeling in water resources (Section VI); Optimization techniques and analytical formulations in water resources (Section VII); Advances in sediment transport modeling and river engineering (Section VIII); and Computational intelligence in extreme hydrology: Flood and droughts (Section IX). The tools, technologies, and approaches discussed in these sections address water resources modeling issues that are crucial for future advancement.

I take this opportunity to appreciate and congratulate the editors and authors of this book for their dedication and collaborative efforts. The book will be a valuable initiative in prompting global discussions on the challenges and opportunities in water resources modeling. Furthermore, I believe that the book will inspire new actions and advancements in this domain and will be widely acclaimed by researchers, practitioners, managers, and decision-makers in water resources management.

Sangam Shrestha
Water Engineering and Management (WEM),
Department of Civil and Infrastructure Engineering (CIE),
School of Engineering and Technology (SET),
Asian Institute of Technology (AIT),
Khlong Luang, Pathum Thani, Thailand

Preface

This book emerged out of the need to cumulate numerous applications of modern computational techniques and software in various sectors of water resources in the form of a book. The book will offer valuable guidance on exploring the use of computational methods to hydrologists, hydraulicians, environmentalists, earth scientists, and of course people involved in tackling problems in the field of water resources. As a result, it is postulated that it will be highly beneficial to practitioners, policy makers, scientists, researchers, and students.

In light of the recent advancements in computational techniques, there has been a surge in investigating their applications in various sectors of water resource engineering. In this context, this book encompasses assessment of applying numerical modeling, computational approaches, and artificial intelligence techniques to water resource engineering. An attempt has been made to include a flowchart for the methodology of various computational techniques for easy and quick understanding of future readers. The chapters in this book entail many applications in water resource management that include but are not limited to hydroinformatics, impact assessment of climate change, hydrologic modeling, river engineering, floods, droughts, image processing, GIS, water quality analysis, aquifer mapping, basin scale modeling, numerical modeling of surges and groundwater flow, optimal reservoir operation, hydraulic structures, and irrigation engineering.

Mohammad Zakwan
Abdul Wahid
Majid Niazkar
Uday Chatterjee

Acknowledgments

The preparation of this book has been guided by several civil engineering and computational intelligence pioneers. We are obliged to these experts for dedicating their time to evaluating the chapters published in this book. We thank the anonymous reviewers for their constructive comments that led to substantial improvement in the quality of this book. This work would not have been possible without the constant inspiration from our institutes, enthusiasm from our colleagues and collaborators, and support from our families. Finally, we also thank our publisher, Elsevier, and its publishing editor for their continuous support in the publication of this book.

SECTION I

Introduction

CHAPTER

1

Artificial intelligence and machine learning in water resources engineering

Mohd Danish

Civil Engineering Section, University Polytechnic, Aligarh Muslim University, Aligarh, UP, India

OUTLINE

1 Introduction 3

2 Materials and methods 4

2.1 Selection of search terms 5

2.2 Scientometric review 5

3 Evolution of artificial intelligence and machine learning 7

4 Results and discussion 8

5 Conclusion 11

References 12

1 Introduction

Water resources engineering aims to conserve water and build infrastructure to provide clean water to the people and industry. It also focuses on producing clean and efficient renewable energy through hydropower generation. The objectives of water resources engineering include conservation of aquatic ecosystem, preventing extreme flood and drought events, providing critical recreational opportunities while preserving and maintaining natural habitats and the environment, and enhancing commerce through maintenance of navigable waters (Labadie, 2014). The projects planned to fulfill these aims and objectives of water resources engineering require computational efforts. Conventionally various traditional methods based on assumptions and approximations are applied in water resources engineering. These methods involve tedious calculations. Assumptions and approximations result in sizable departures from real-world behavior, and complex problems require extra human efforts. It can also be said that human beings are slow learners and take decades to learn

Current Directions in Water Scarcity Research, Volume 7
https://doi.org/10.1016/B978-0-323-91910-4.00001-7

anything (Simon, 1983). It becomes ideal to involve machines, particularly computers, to accurately and efficiently execute water resources projects.

Recently, artificial intelligence (AI), including machine learning (ML), has gained popularity in almost every domain of science and technology. Some examples of the application of AI are speech recognition, image recognition, driverless cars, healthcare, navigation, stock market predictions, chatbots, and word processing software. The idea behind AI is to mimic the human brain to create advanced computers and machines that can perform various tasks efficiently and faster. ML, a subset of AI, makes the device use data to learn and make predictions. Due to the increase in the availability of data in water resources and the advancement in ML algorithms and computational resources, ML application in water resources engineering is rapidly increasing. ML research has been pursued using different approaches, emphasizing various aspects. The three major themes of ML focus are neural modeling and decision-theoretic techniques, symbolic concept-oriented learning, and knowledge-intensive learning systems exploring various learning tasks (Carbonell et al., 1983). Wang et al. (2009) examined the performance of autoregressive moving-average models, artificial neural networks (ANNs), adaptive neural-based fuzzy inference system (ANFIS) techniques, GP models, and support vector machines (SVM) using the long-term observations of monthly river flow discharges. They found that the ANFIS, GP, and SVM indicated best performance results based on different evaluation criteria during the training and validation. Flood forecasting models were developed using ANN-based genetic algorithms and SVM (Chau, 2006; Han et al., 2007; Yu et al., 2006). The ANN has been significantly employed by various researchers in different aspects of hydrology, viz., rainfall-runoff modeling, flood forecasting, estimating evapotranspiration, sediment load concentration (Chau et al., 2005; Goovaerts, 2000; Govindaraju, 2000; Kisi, 2007; Nagy et al., 2002; Nayak et al., 2004; Tokar and Johnson, 1999; Toth et al., 2000; Wu et al., 2009; Wu and Chau, 2011).

Additionally, ANN has been applied in groundwater level forecasting (Adamowski and Chan, 2011; Daliakopoulos et al., 2005; Yoon et al., 2011). The use of wavelet AI models in hydrology has increased gradually over the years and has attracted research interest given the robustness and accuracy of the approach (Nourani et al., 2014). Multiple researchers in water resources have focused on combining different algorithms and their applications to solve their problems. However, the analysis of various strategies for selecting definite algorithms for different scenarios has not been extensively conducted. Furthermore, there is an immediate need to create a road map for future research to excel the proper use of ML in water resources.

Therefore, to provide an organized and inclusive direction for future research, a detailed literature review on the application of AI and ML in water resources engineering is presented here. This chapter aims to enable the researchers working in this field to quickly identify the current spectrum of study and find the research areas yet to be explored. The present chapter is organized as follows: Section 2 presents the materials and methods used in this study. In contrast, Section 3 provides an overview of AI and ML evolution, while Section 4 discusses the results.

2 Materials and methods

ML algorithms are broadly classified as shallow learning and deep learning. Most ML models are referred to as shallow learning, including ANNs. In contrast, deep learning is the branch of ML that constitutes the study of ANNs with more than one hidden layer. A deep learning algorithm requires a large amount of data comparable to various shallow learning algorithms. The present chapter focuses on the different shallow ML algorithms applied in water resources engineering. Since the Web of Science (WoS) database accounts for the indexing of high-quality publications (Shukla et al., 2019; Zavadskas et al., 2014), this database was used to collect the literature presented in this chapter. To ensure the relevance and quality of the collected citations, the subject, keywords, language, and document types were filtered accordingly.

2.1 Selection of search terms

Some of the shallow learning algorithms that are often applied in water resources engineering are namely linear regression, logistic regression, ANN, decision trees (DT), gene expression programming (GEP), genetic programming (GP), support vector machines (SVM), *k*-nearest neighbor (KNN), *k*-means clustering algorithm, AdaBoost, random forest (RF), hidden Markov model (HMM), spectral clustering (SC), and group method of data handling (GMDH). This chapter analyzes the articles related to the shallow learning algorithm mentioned above and their water resource engineering applications. Additionally, a selection of papers from 1989 to 2022 which have been frequently cited was collected and then analyzed.

The query used in the search engine of WoS was the web of science category, WC=" Water resources" and "Engineering, Civil." The algorithms of shallow learning, namely linear regression, logistic regression, ANN, DT, GEP, GP, multigene genetic programming (MGGP), SVM, KNN, *k*-means clustering algorithm, Adaboost, RF, HMM, SC, and GMDH was put in all search fields including title, keywords, and abstract of the articles. A manual review was conducted to weed out any irrelevant articles obtained by using the query string (WC=("Water Resources" AND "Engineering, Civil")) AND ALL=("Linear Regression" OR "Logistic Regression" OR "Support Vector Machines*" OR "SVM" OR "Decision Trees*" OR "Artificial Neural Network" OR "ANN" OR "Genetic Programming" OR "Gene Expression Programming" OR "Group Method of Data Handling" OR "GMDH" OR "GEP" OR "*k*-Nearest Neighbor*" OR "KNN" OR "*K*-Means*" OR "AdaBoost" OR "Random Forest*" OR "Hidden Markov Model" OR "Spectral Clustering" OR "Machine Learning" OR "Artificial Intelligence" OR "Multigene Genetic Programming"). An asterisk symbol was added after the keyword to include the variation in the usage of keywords by different authors (i.e., singular or plural form of a keyword). Further, the results were refined by filtering out the articles in a language other than English and excluding retracted articles, corrections, discussions, or book chapters. Through utilizing the processes mentioned above, articles were obtained from 1989 till 2022 and the search query was performed on February 21st, 2022. During this period of 34 years, 2890 publications were obtained.

2.2 Scientometric review

Fig. 1 shows the number of papers published from 1989 to the beginning of 2022. A single article was published in 1989, followed by 6, 8, and 11 in the subsequent years. After that, a minor dent in the number of publications was observed. It is evident from Fig. 1 that since 2001 the uptrend in the number of papers has increased significantly. This uptrend tends

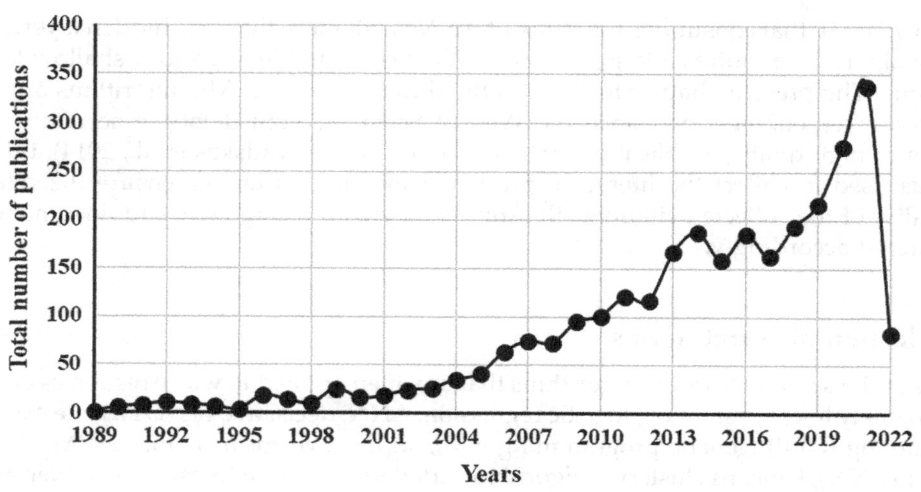

FIG. 1 Number of papers over the years.

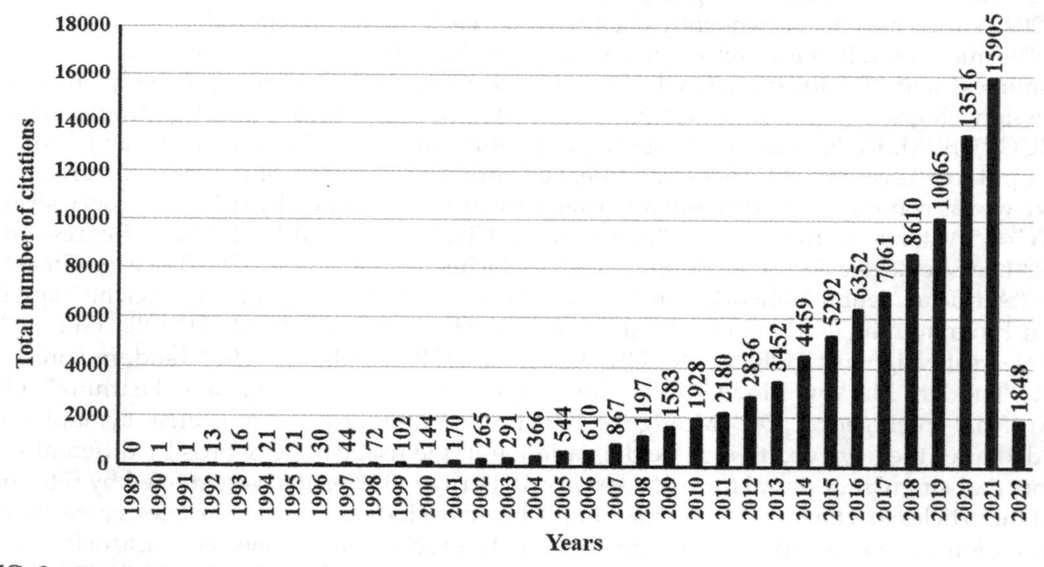

FIG. 2 Distribution of citations over the years.

to follow the enhancing interest in AI and ML in water resources. The maximum number of publications of 338 was in 2021. The total number of publications till February 2022 is 83, and if this trend continues, it will reach a new milestone at the end of this year.

However, the total number of publications witnessed a few ups and downs, but the growth in citation per year has been exponential since 2001, as shown in Fig. 2. Two thousand eight

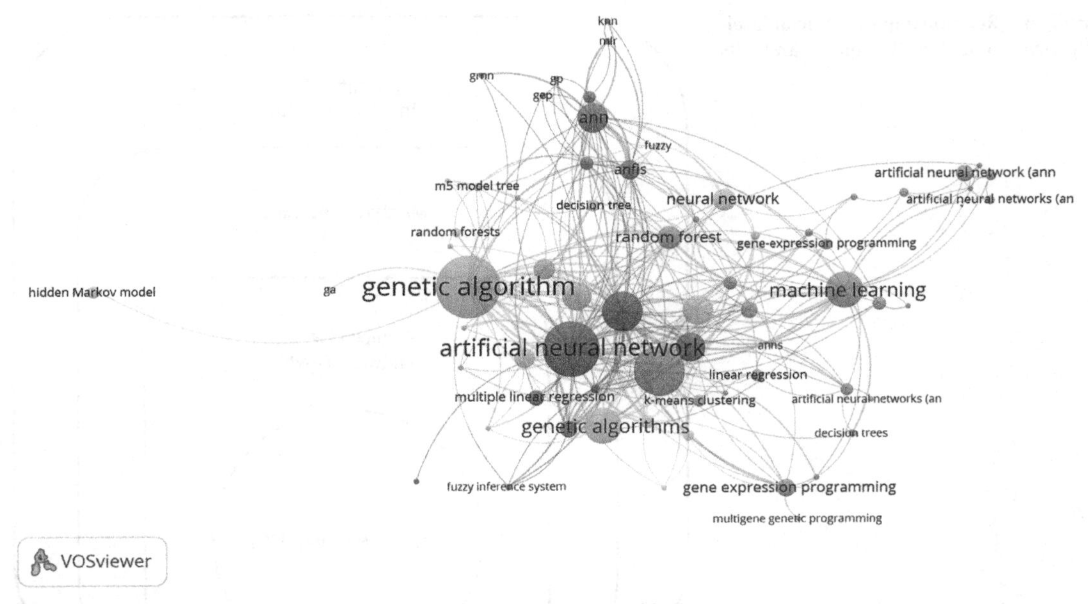

FIG. 3 Co-occurrence of author keywords of publications (1989–2022).

hundred and ninety publications have received 89,862 citations from 1989 to 2022, with an average citation count of 31.09 and an h-index of 126. During the period, the highest number of citations received was 15,905 in 2021. Govindaraju (2000) presented a review on ANNs in the hydrologic application while discussing essential aspects such as physical interpretation of ANN architecture, optimal training data set, adaptive learning, and extrapolation. It received the highest of 1024 citations, with an average of 44.52 citations per year.

The keywords obtained from the collected articles were analyzed to establish the connections and research interest between the various ML algorithms. Fig. 3 shows multiple authors' trends of keyword usage from 1989 to 2022. The primary keywords in the top 3 are the ANN, SVM, and ML, with total occurrences of 696, 178, and 106, respectively.

3 Evolution of artificial intelligence and machine learning

According to Google trends, AI and ML are among the most searched terms on Google web search. Often these two terms are used interchangeably, but they are entirely different fundamentally. AI is a branch of science dedicated to developing intelligent computer programs and machines that can solve various problems creatively. On the other hand, ML is an application of AI in which the device has access to data so that the system will be able to learn and improve without being explicitly programmed. There are various ML algorithms. Among them, ANNs are mainly used. The latest inclusion in ML algorithms is deep learning, which uses complex multilayered neural networks inspired by the human brain. The relationship between AI and ML algorithms is shown in Fig. 4.

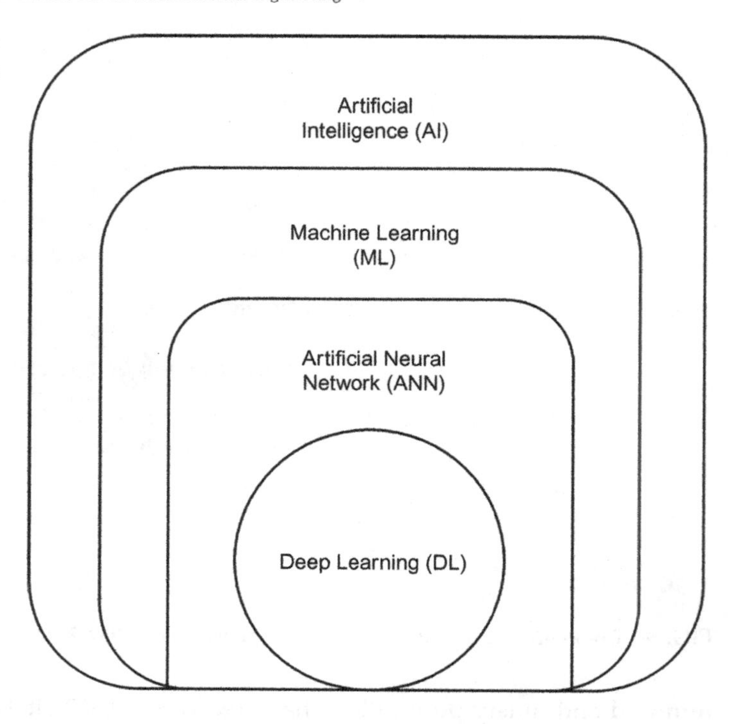

FIG. 4 Relationship of artificial intelligence, machine learning and its algorithms.

4 Results and discussion

The analysis of the applications of shallow ML algorithms in water resources engineering is presented in this section. Research content presented in Table 1 was obtained by filtering the research papers where shallow learning algorithms were initially applied in water resources engineering. The comparison of time intervals of origin of new algorithms and their applications in water resources have been presented in Fig. 4. The ANN was first proposed in 1943. But in water resources engineering, it was used to assess the effects of ozone on water quality in 1989 (Grasso et al., 1989). Cui et al. (2022) proposed an ANN model for groundwater level prediction in an intensively irrigated region of Northwest Bangladesh while considering the rainfall, evapotranspiration, groundwater abstraction, and irrigation return flow as the input variables. They have found that the ANN model was able to predict the groundwater level better than that of ANFIS and ANFIS-particle swarm optimization techniques. Gieske and De Vries (1990) published the singular value decomposition algorithm using linear regression. They compared the solutions obtained to estimate the chloride balance of the Nnywane/Pitsanyane basins with positive semidefinite singular matrices using quadratic programming and the single value decomposition algorithm. Their analysis observed that the quadratic programming led to solutions that were not unique and could ignore linear dependencies in the design matrix. In contrast, the SVD algorithm, on the other hand, provided an adequate solution in such circumstances. The accuracy of both the linear and logistic

TABLE 1 Research content of various shallow learning algorithms.

S. no.	Machine learning algorithms	Initial application in water resources	Year of creation of the algorithm	Application year	Research content
1.	Artificial neural network (ANN)	(Grasso et al., 1989)	1943	1989	Assessment of the effects of ozone on water quality
2.	Linear regression	(Gieske and De Vries, 1990)	1958	1990	The singular value decomposition algorithm is suggested
3.	Multigene genetic programming (MGGP)	(Cobaner et al., 2016)	1992	2016	Estimation of groundwater levels with surface observations
4.	Decision trees (DT)	(Mnrphy and Olson, 1996)	1986	1996	Quantitative occurrence of sulfate-reducing bacteria on a groundwater basis
5.	Logistic regression	(James and Havens, 1996)	1958	1996	Algal bloom probability in a large subtropical lake
6.	Genetic programming (GP)	(Savic et al., 1999)	1992	1999	Rainfall–runoff modeling
7.	Bayesian network (BN)	(Stow et al., 2003)	1985	2003	Estuarine water quality models
8.	Hidden Markov model (HMM)	(Thyer and Kuczera, 2003)	1960	2003	Calibrating an HMM to the long-term rainfall time-series data obtained from various sites using the Bayesian approach
9.	Support vector machines (SVM)	(Ibbitt and Woods, 2004)	1995	2004	SVM applied in hydrological modeling
10.	*k*-Means clustering	(Maruyama et al., 2005)	1967	2005	Potential water resources availability (PWRA) in an area
11.	*k*-Nearest neighbor	(Ostfeld and Salomons, 2005)	1951	2005	Hydrodynamic and water quality model for surface water bodies
12.	Gene expression programming (GEP)	(Guven and Aytek, 2009)	1999	2009	Stage-discharge relationship
13.	Random forest (RF)	(Herrera et al., 2010)	2001	2010	Predicting water consumption in urban areas
14.	Adaboost	(Pai et al., 2014)	1995	2014	Predicting debris flow disaster
15.	Group method of data handling (GMDH)	(Baydaroglu and Kocak, 2014)	1968	2014	Prediction of evaporation amount
16.	Spectral clustering (SC)	(Candelieri et al., 2015)	2000	2015	Identify leaks within a water distribution network

regressions is relatively low while handling complex variables (Adamowski and Chan, 2011; Muzzammil et al., 2015a; Tehrany et al., 2015).

A specialized form of a genetic algorithm that can mimic natural evolutionary processes such as mutation and crossover is known as GP (Niazkar and Zakwan, 2021; Noh et al., 2020). Savic et al. (1999) used GP in rainfall-runoff modeling for the Kirkton Catchment and compared the results with the ANN model. Recently, GP has been applied to propose rainfall-runoff models (Herath et al., 2021) and prediction of wind speed (Dong et al., 2021). Prediction of the discharge coefficient of gabion weir (Farzin et al., 2021) and optimization of properties of lined canals (Niazkar, 2020) has also been carried out recently using GP. The accuracy of the GP model can be enhanced by taking different objective functions. However, GP cannot be implemented for evolving complex models that include more than one gene. Thus, MGGP, a modified version of GP, can be employed. This model combines a linearly weighted model consisting of individual gene trees that use one or more gene trees and calibrates coefficients of the gene trees using statistical regression methods such as the least-square method (Niazkar and Niazkar, 2020; Niazkar and Zakwan, 2021; Noh et al., 2020). Both GA and GP are robust evolutionary methods for effectively modeling complex problems. They can be more efficient if combined while removing their limitations.

GEP is yet another evolutionary method invented by Candida Ferreira in 1999. In GEP, a gene represents a set of linear chromosomes of fixed length as in a genetic algorithm and then expressed as a phenotype in the form of expression trees (Danish, 2014; Ferreira, 2002; Muzzammil et al., 2015b). The initial application of GEP in water resources engineering was to develop a stage-discharge relationship (Guven and Aytek, 2009). Gharehbaghi and Ghasemlounia (2022) carried out an experimental investigation of the discharge coefficient of sharp-crested V-notch weir and then used the dataset to develop a prediction model using GEP.

DT were developed in 1986, and a decade later, Mnrphy and Olson (1996) applied this technique in the quantitative occurrence of sulfate-reducing bacteria on groundwater. Estuarine water quality models using a Bayesian network were developed (Stow et al., 2003). The DT are prone to overfitting and ignore the problems caused by inter-data correlation (Han et al., 2002, 2007). The HMM was created in 1960, but it found its application in long-term multisite rainfall time series in 2003 (Thyer and Kuczera, 2003). Ibbitt and Woods (2004) used SVM in developing hydrological modeling. It is difficult to find a suitable kernel function to deal with missing data while applying SVM, and its efficiency in some cases is relatively low (Tehrany et al., 2015). Therefore, shallow learning applications are used primarily on dealing with structured data. Deep learning is a better solution for processing unstructured data such as images or videos (Xu et al., 2021). Like the HMM, KNN, and *k*-means clustering were introduced in 1951 and 1967, but it was not applied in water resources engineering until 2005. Maruyama et al. (2005) studied potential water resources availability (PWRA) in an area using a k-means clustering algorithm.

In contrast, Ostfeld and Salomons (2005) developed a hydrodynamic and water quality model for surface water bodies using the KNN algorithm. Prediction of water consumption in an urban area using RF was published by Herrera et al. (2010). Adaboost was applied in predicting debris flow disaster by Pai et al. (2014). Ivakhenko proposed the GMDH (Hussain et al., 2021; Ivakhenko et al., 2003). Several decades later, Baydaroglu and Kocak (2014) used GMDH to predict the amount of evaporation. Identification of leaks within a water

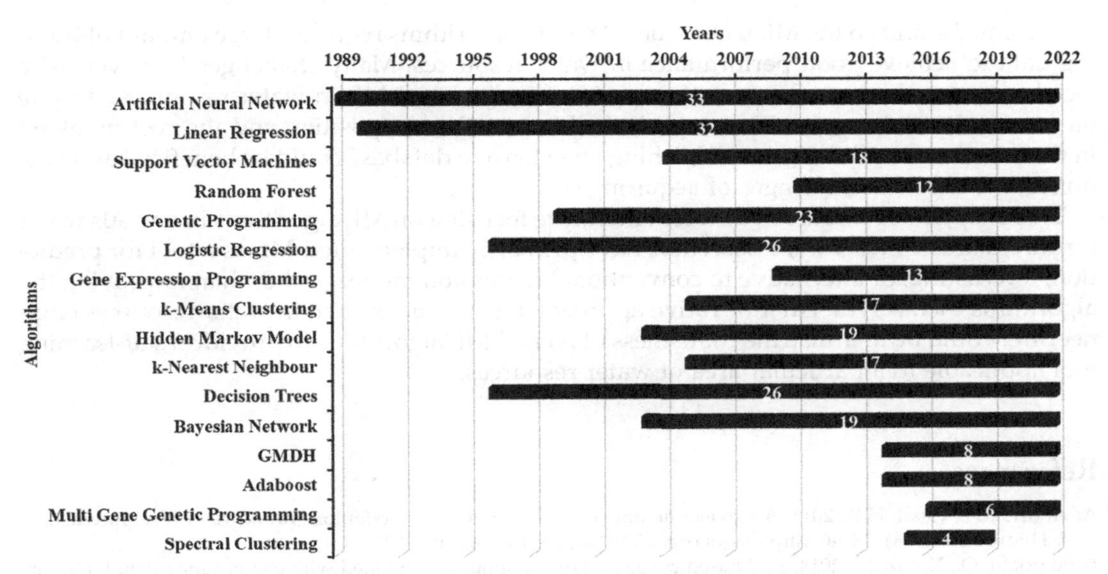

FIG. 5 Comparison of origin of machine learning algorithms and their application in water resources engineering.

distribution network using SC was conducted by Candelieri et al. (2015). It can be observed from Fig. 5 that all the algorithms have continuously been used till 2022 since their first application year in water resources engineering except SC. The initial application of SC was first published in 2015, but there has been a time gap in its application since 2019.

While the articles reviewed in this chapter on applications of AI and ML in water resources engineering are not exhaustive. These techniques have made a significant impact in this area. Some algorithms, such as Markov random field, conditional random field, mean shift, Q-learning etc., have yet to be applied in water resources engineering, and these techniques must be explored for future research in water resources engineering.

5 Conclusion

An ANN is a robust tool. It can solve complex problems, making it the most practical ML tool in water resources. Thus, its application has significantly increased in the past 20 years. Generally, these techniques have played a significant role in developing automated technologies in water resources. Due to more direct applicability and lesser computational or storage resources requirement, logistic regression is a favorable tool among various researchers. On the other hand, the DT-based approach is highly interpretable. It can process samples with unrelated features and missing attributes, yet this approach is prone to overfitting and ignoring the problems caused by inter-data correlation.

Moreover, support vector machine utilizes little data to solve nonlinear problems. Still, it is difficult to find a suitable kernel function to deal with missing data while applying SVM, and its efficiency in some cases is relatively low. Almost all techniques in the world have some shortcomings, and so the ML techniques. The ML algorithms require a large amount of training data to achieve good performance in water resources. Many challenges have yet to be faced by researchers working on the application of AI and ML in water resources, despite its various breakthroughs. The main difficulties are data acquisition and the cost involved in these processes. Therefore, establishing an extensive database could be beneficial and economical to ease the challenges of acquiring data.

This chapter has discussed different literature focusing on ML algorithms in various water resources areas. Presently, researchers have primarily implemented ML as a tool for prediction, forecasting, or alternative to conventional regression models rather than improving the algorithms themselves. An innovative approach for future research in water resources engineering would be to utilize the robustness of AI and ML algorithms to provide a self-learning tool applicable to a particular area of water resources.

References

Adamowski, J., Chan, H.F., 2011. A wavelet neural network conjunction model for groundwater level forecasting. J. Hydrol. 407 (1–4), 28–40. https://doi.org/10.1016/j.jhydrol.2011.06.013.

Baydaroglu, O., Kocak, K., 2014. SVR-based prediction of evaporation combined with chaotic approach. J. Hydrol. 508, 356–363. https://doi.org/10.1016/j.jhydrol.2013.11.008.

Candelieri, A., Soldi, D., Archetti, F., 2015. Cost-effective sensors placement and leak localization the Neptun pilot of the ICeWater project. J. Water Supply Res. Technol. Aqua 64 (5), 567–582. https://doi.org/10.2166/aqua.2015.037.

Carbonell, J.G., Michalski, R.S., Mitchell, T.M., 1983. An overview of machine learning. In: Machine Learning. Springer Berlin Heidelberg, pp. 3–23, https://doi.org/10.1007/978-3-662-12405-5_1.

Chau, K.W., 2006. Particle swarm optimization training algorithm for ANNs in stage prediction of Shing Mun River. J. Hydrol. 329 (3–4), 363–367. https://doi.org/10.1016/j.jhydrol.2006.02.025.

Chau, K.W., Wu, C.L., Li, Y.S., 2005. Comparison of several flood forecasting models in Yangtze river. J. Hydrol. Eng. 10 (6), 485–491. https://doi.org/10.1061/(ASCE)1084-0699(2005)10:6(485).

Cobaner, M., Babayigit, B., Dogan, A., 2016. Estimation of groundwater levels with surface observations via genetic programming. J. Am. Water Works Assoc. 108 (6), E335–E348. https://doi.org/10.5942/JAWWA.2016.108.0078.

Cui, F., Al-Sudani, Z.A., Hassan, G.S., Afan, H.A., Ahammed, S.J., Yaseen, Z., 2022. Boosted artificial intelligence model using improved alpha-guided grey wolf optimizer for groundwater level prediction: comparative study and insight for federated learning technology. J. Hydrol. 606. https://doi.org/10.1016/J.JHYDROL.2021.127384.

Daliakopoulos, I.N., Coulibaly, P., Tsanis, I.K., 2005. Groundwater level forecasting using artificial neural networks. J. Hydrol. 309 (1–4), 229–240. https://doi.org/10.1016/j.jhydrol.2004.12.001.

Danish, M., 2014. Prediction of scour depth at bridge abutments in cohesive bed using gene expression. Int. J. Civil Eng. Technol. 5 (11), 25–32.

Dong, Y., Niu, J., Liu, Q., Sivakumar, B., Du, T., 2021. A hybrid prediction model for wind speed using support vector machine and genetic programming in conjunction with error compensation. Stochastic Environ. Res. Risk Assess. 35 (12), 2411–2424. https://doi.org/10.1007/s00477-021-01996-0.

Farzin, S., Nastaran, S., John, A., 2021. Discharge coefficients for rectangular broad-crested gabion weirs: experimental study. J. Irrig. Drain. Eng. 147 (3). https://doi.org/10.1061/(ASCE)IR.1943-4774.0001535.

Ferreira, C., 2002. Gene expression programming in problem solving. In: Soft Computing and Industry, pp. 635–653, https://doi.org/10.1007/978-1-4471-0123-9_54.

Gharehbaghi, A., Ghasemlounia, R., 2022. Application of AI approaches to estimate discharge coefficient of novel kind of sharp-crested V-notch weirs. J. Irrig. Drain. Eng. 148 (3). https://doi.org/10.1061/(ASCE)IR.1943-4774.0001646.

Gieske, A., De Vries, J.J., 1990. Conceptual and computational aspects of the mixing cell method to determine groundwater recharge components. J. Hydrol. 121 (1–4), 277–292. https://doi.org/10.1016/0022-1694(90)90236-Q.

Goovaerts, P., 2000. Geostatistical approaches for incorporating elevation into the spatial interpolation of rainfall. J. Hydrol. 228 (1–2), 113–129. https://doi.org/10.1016/S0022-1694(00)00144-X.

Govindaraju, R.S., 2000. Artificial neural networks in hydrology. II. Hydrologic applications. J. Hydrol. Eng. 5 (2), 124–137. https://doi.org/10.1061/(ASCE)1084-0699(2000)5:2(124).

Grasso, D., Walter, J.W., DeKam, J.A., 1989. Effects of preoxidation with ozone on water quality: a case study. J. Am. Water Works Assoc. 81 (6), 85–92. https://doi.org/10.1002/j.1551-8833.1989.tb03221.x.

Guven, A., Aytek, A., 2009. New approach for stage-discharge relationship: gene-expression programming. J. Hydrol. Eng. 14 (8), 812–820. https://doi.org/10.1061/(ASCE)HE.1943-5584.0000044.

Han, D., Chan, L., Zhu, N., 2007. Flood forecasting using support vector machines. J. Hydroinf. 9 (4), 267–276. https://doi.org/10.2166/hydro.2007.027.

Han, D., Cluckie, I.D., Karbassioun, D., Lawry, J., Krauskopf, B., 2002. River flow modelling using fuzzy decision trees. Water Resour. Manag. 16 (6), 431–445. https://doi.org/10.1023/A:1022251422280.

Herath, H.M.V.V., Chadalawada, J., Babovic, V., 2021. Genetic programming for hydrological applications: to model or to forecast that is the question. J. Hydroinf. 23 (4), 740–763. https://doi.org/10.2166/hydro.2021.179.

Herrera, M., Torgo, L., Izquierdo, J., Perez-Garcia, R., 2010. Predictive models for forecasting hourly urban water demand. J. Hydrol. 387 (1–2), 141–150. https://doi.org/10.1016/j.jhydrol.2010.04.005.

Hussain, A., Shariq, A., Danish, M., Ansari, M.A., 2021. Discharge coefficient estimation for rectangular side weir using GEP and GMDH methods. Adv. Comput. Des. 6 (2), 135–151. https://doi.org/10.12989/acd.2021.6.2.135.

Ibbitt, R., Woods, R., 2004. Re-scaling the topographic index to improve the representation of physical processes in catchment models. J. Hydrol. 293 (1–4), 205–218. https://doi.org/10.1016/j.hydrol.2004.01.016.

Ivakhenko, A.G., Savchenko, E.A., Ivakhenko, G.A., 2003. Problems of future GMDH algorithms development. Syst. Anal. Model. Simul. 43 (10), 1301–1309. https://doi.org/10.1080/0232929032000115029.

James, R.T., Havens, K.E., 1996. Algal bloom probability in a large subtropical lake. Water Resour. Bull. 32 (5), 995–1006.

Kisi, O., 2007. Streamflow forecasting using different artificial neural network algorithms. J. Hydrol. Eng. 12 (5), 532–539. https://doi.org/10.1061/(ASCE)1084-0699(2007)12:5(532).

Labadie, J.W., 2014. Advances in water resources systems engineering: applications of machine learning. In: Modern Water Resources Engineering. Humana Press, pp. 467–523, https://doi.org/10.1007/978-1-62703-595-8_10.

Maruyama, T., Kawachi, T., Singh, V.P., 2005. Entropy-based assessment and clustering of potential water resources availability. J. Hydrol. 309 (1–4), 104–113. https://doi.org/10.1016/j.jhydrol.2004.11.020.

Mnrphy, P., Olson, B.H., 1996. Decision-tree construction and analysis. J. Am. Water Works Assoc. 88 (2), 59–67.

Muzzammil, M., Alam, J., Danish, M., 2015a. Application of gene expression programming in flood frequency analysis. J. Water Resour. Soc. 35 (2), 1–6.

Muzzammil, M., Alam, J., Danish, M., 2015b. Scour prediction at bridge piers in cohesive bed using gene expression programming. Aquat. Procedia 4, 789–796. https://doi.org/10.1016/j.aqpro.2015.02.098. Icwrcoe.

Nagy, H.M., Watanabe, K., Hirano, M., 2002. Prediction of sediment load concentration in rivers using artificial neural network model. J. Hydraul. Eng. ASCE 128 (6), 588–595. https://doi.org/10.1061/(ASCE)0733-9429(2002)128:6(588).

Nayak, P.C., Sudheer, K.P., Rangan, D.M., Ramasastri, K.S., 2004. A neuro-fuzzy computing technique for modeling hydrological time series. J. Hydrol. 291 (1–2), 52–66. https://doi.org/10.1016/j.jhydrol.2003.12.010.

Niazkar, M., 2020. Assessment of artificial intelligence models for calculating optimum properties of lined channels. J. Hydroinf. 22 (5), 1410–1423. https://doi.org/10.2166/hydro.2020.050.

Niazkar, M., Niazkar, H.R., 2020. COVID-19 outbreak: application of multi-gene genetic programming to country-based prediction models. Electron. J. Gen. Med. 17 (5), em247. https://doi.org/10.29333/EJGM/8232.

Niazkar, M., Zakwan, M., 2021. Application of MGGP, ANN, MHBMO, GRG, and linear regression for developing daily sediment rating curves. Math. Probl. Eng. 2021. https://doi.org/10.1155/2021/8574063.

Noh, H., Kwon, S., Seo, I.W., Baek, D., Jung, S.H., 2020. Multi-gene genetic programming regression model for prediction of transient storage model parameters in natural rivers. Water 13 (1), 76. https://doi.org/10.3390/W13010076.

Nourani, V., Baghanam, A.H., Adamowski, J., Kisi, O., 2014. Applications of hybrid wavelet-artificial intelligence models in hydrology: a review. J. Hydrol. 514, 358–377. https://doi.org/10.1016/j.jhydrol.2014.03.057.

Ostfeld, A., Salomons, S., 2005. A hybrid genetic—instance based learning algorithm for CE-QUAL-W2 calibration. J. Hydrol. 310 (1–4), 122–142. https://doi.org/10.1016/j.jhydrol.2004.12.004.

Pai, P.F., Li, L.L., Hung, W.Z., Lin, K.P., 2014. Using ADABOOST and rough set theory for predicting debris flow disaster. Water Resour. Manag. 28 (4), 1143–1155. https://doi.org/10.1007/s11269-014-0548-8.

Savic, D.A., Walters, G.A., Davidson, J.W., 1999. A genetic programming approach to rainfall-runoff modelling. Water Resour. Manag. 13 (3), 219–231. https://doi.org/10.1023/A:1008132509589.

Shukla, A.K., Janmaijaya, M., Abraham, A., Muhuri, P.K., 2019. Engineering applications of artificial intelligence: a bibliometric analysis of 30 years (1988–2018). Eng. Appl. Artif. Intel. 85, 517–532. https://doi.org/10.1016/j.engappai.2019.06.010.

Simon, H.A., 1983. Why should machines learn? In: Machine Learning. Springer Berlin Heidelberg, pp. 25–37, https://doi.org/10.1007/978-3-662-12405-5_2.

Stow, C.A., Roessler, C., Borsuk, M.E., Bowen, J.D., Reckhow, K.H., 2003. Comparison of estuarine water quality models for total maximum daily load development in Neuse River Estuary. J. Water Resour. Plan. Manag. 129 (4), 307–314. https://doi.org/10.1061/(ASCE)0733-9496(2003)129:4(307).

Tehrany, M.S., Pradhan, B., Jebur, M.N., 2015. Flood susceptibility analysis and its verification using a novel ensemble support vector machine and frequency ratio method. Stochastic Environ. Res. Risk Assess. 29 (4), 1149–1165. https://doi.org/10.1007/s00477-015-1021-9.

Thyer, M., Kuczera, G., 2003. A hidden Markov model for modelling long-term persistence in multi-site rainfall time series 1. Model calibration using a Bayesian approach. J. Hydrol. 275 (1–2), 12–26. https://doi.org/10.1016/S0022-1694(02)00412-2.

Tokar, A.S., Johnson, P.A., 1999. Rainfall-runoff modeling using artificial neural networks. J. Hydrol. Eng. 4 (3), 232–239. https://doi.org/10.1061/(ASCE)1084-0699(1999)4:3(232).

Toth, E., Brath, A., Montanari, A., 2000. Comparison of short-term rainfall prediction models for real-time flood forecasting. J. Hydrol. 239 (1–4), 132–147. https://doi.org/10.1016/S0022-1694(00)00344-9.

Wang, W.-C., Chau, K.-W., Cheng, C.-T., Qiu, L., 2009. A comparison of performance of several artificial intelligence methods for forecasting monthly discharge time series. J. Hydrol. 374 (3–4), 294–306. https://doi.org/10.1016/j.jhydrol.2009.06.019.

Wu, C.L., Chau, K.W., 2011. Rainfall-runoff modeling using artificial neural network coupled with singular spectrum analysis. J. Hydrol. 399 (3–4), 394–409. https://doi.org/10.1016/j.jhydrol.2011.01.017.

Wu, C.L., Chau, K.W., Li, Y.S., 2009. Methods to improve neural network performance in daily flows prediction. J. Hydrol. 372 (1–4), 80–93. https://doi.org/10.1016/j.jhydrol.2009.03.038.

Xu, Y., Zhou, Y., Sekula, P., Ding, L., 2021. Machine learning in construction: from shallow to deep learning. Dev. Built Environ. 6, 100045. https://doi.org/10.1016/j.dibe.2021.100045.

Yoon, H., Jun, S.-C., Hyun, Y., Bae, G.-O., Lee, K.-K., 2011. A comparative study of artificial neural networks and support vector machines for predicting groundwater levels in a coastal aquifer. J. Hydrol. 396 (1–2), 128–138. https://doi.org/10.1016/j.jhydrol.2010.11.002.

Yu, P.-S., Chen, S.-T., Chang, I.-F., 2006. Support vector regression for real-time flood stage forecasting. J. Hydrol. 328 (3–4, SI), 704–716. https://doi.org/10.1016/j.jhydrol.2006.01.021.

Zavadskas, E.K., Skibniewski, M.J., Antucheviciene, J., 2014. Performance analysis of Civil Engineering Journals based on the Web of Science® database. Arch. Civil Mech. Eng. 14 (4), 519–527. https://doi.org/10.1016/J.ACME.2014.05.008.

SECTION II

Application of artificial intelligence to water resources

CHAPTER

2

Demystifying artificial intelligence amidst sustainable agricultural water management

Aman Srivastava[a], *Shubham Jain*[b], *Rajib Maity*[a], *and Venkappayya R. Desai*[a]

[a]Department of Civil Engineering, Indian Institute of Technology (IIT) Kharagpur, Kharagpur, West Bengal, India [b]Department of Civil Engineering, Indian Institute of Technology (IIT) Bombay, Mumbai, Maharashtra, India

OUTLINE

1 Introduction 17
  1.1 Review objectives and chapter organization 22
2 AI in agriculture 22
  2.1 AI in preagricultural (preparatory) activities 23
  2.2 AI during agricultural activities 26
  2.3 AI in postagricultural activities 27
3 Current and future scope in AI for agriculture 29
4 Challenges of AI in agriculture 31
5 Conclusions 31
Acknowledgments 32
Conflict of interest 32
References 32
Further reading 35

1 Introduction

Agriculture is an important livelihood activity for feeding the population of human beings and animals globally and thereby sustaining the growth of living beings. In developing countries, the well-being of millions of people from across the countries is, directly and indirectly,

https://doi.org/10.1016/B978-0-323-91910-4.00002-9

dependent on agriculture for their livelihood (Ramankutty et al., 2018). Worldwide, over one billion people are working in the agricultural industry generating USD 3.2 trillion for the global economy (Raman, 2017). Developing countries like India, where over 60% of the population is dependent on agrarian livelihoods, are aiming for making the entire economy including the agricultural sector a USD 5 trillion economies (Singh et al., 2020). Thus, it becomes an imperative industry to reconsider, given the advancement of technologies over conventional approaches in agriculture.

According to United Nations Food and Agriculture Organization (UN-FAO), it is expected that the world population will increase by 2 billion from today by 2050; however, only an additional 4% land cover would be available under cultivation (Bongaarts, 2007; Hunter et al., 2017; Mindtree, 2018). This rapid surge in population can have a tremendous impact on the already scarce resources (such as land and water) required for agricultural activities. Given that agriculture accounts for over 70% of global water use (World Bank, 2021) alongside ~60% water wastage, issues such as water scarcity during peak irrigation requirements may further aggravate the water stress (Mancosu et al., 2015). Furthermore, rain-fed agriculture is a dominant form of practice in the arid, semiarid, and drylands across the globe, comprising ~1.3 billion ha (80%) of cultivated areas contributing ~60% (~2.4 billion tons) of global food production by using ~4000 km^3 of water (Ondrasek, 2014; Tian et al., 2021; Elbeltagi et al., 2022). This indicates the dominance of conventional practices amidst modernizing agriculture and demands the scope of introducing recent computational techniques for sustainable water resources management (Kumar et al., 2022). Also, under rapidly rising population scenarios and considering the ongoing practices, 57% more water is likely required to sustain the population through agriculture by 2050 (Boretti and Rosa, 2019). Past researches have estimated that agricultural production would be required to step up almost 70% to fulfill these growing demands (Hunter et al., 2017; World Bank, 2021). Therefore, to meet this growing food demand, more crops are required to be cultivated on decreasing fertile land cover and water availability amidst urbanization and encroachment. This necessitates reducing water consumption and enhancing water-use efficiency.

According to an estimate, ~10% of additional barren land is available that can be converted as agricultural ground, meaning that a mere 10% of additional land compensates for the 70% increase in agricultural demand (Scanlon et al., 2007). However, to combat this challenge, there is a need of increasing the productivity of the existing agriculture system. To achieve this, seven broad steps have been followed since traditional times to bring systematicity in agrarian activities. This included (1) preparation of soil; (2) followed by sowing of seeds; (3) thereby adding manures and fertilizers; (4) followed by irrigation; (5) then providing weed protection; (6) followed by harvesting, and (7) finally the storage of the harvested crops (Hsiao et al., 2007) (Fig. 1). However, challenges are being faced by farmers in each of these aforesaid steps. They include:

1. The climatic factors, such as temperature, rainfall, and humidity, are unpredictable and variable in space and time. Hence, it becomes difficult for the farmers to make rational decisions for preparing the soil, sowing seeds, harvesting produce, as well as difficult for the meteorologist to precisely forecast the weather conditions in view of rainfall and temperature that could help farmers prepare better for agricultural produce;
2. As humans and other living beings require proper nutrition to stay healthy, in a similar manner quality agricultural product is required to be properly cultivated and nourished

7. Storage
Packaging and transportation of crops after harvest, followed by storage to guarantee food security during non agriculture period.
6. Harvesting
To gather ripe crops from the fields, and post harvest handling such as cleaning, sorting, packing, etc. Quite labour intensive activity.
5. Weed Protection
To prevent the loss of soil nutrients, crop yield, weeds are required to e plucked from near the crops.
06
07
05
01
04
02
03
1. Preparation of Soil
First step of farming where farmers prepares soil, consisting of removing debris such as rocks and adding fertilizers & organic matter.
2. Sowing of Seeds
This stage requires attention on the distance between two seeds, depth of planting seeds. Climatic conditions also need to be evaluated.
3. Adding Fertilizers
To maintain soil fertility, and providing plants with healthy nutrition such as nitrogen, phosphorus and potassium, fertilizers are req.
4. Irrigation
To keep the soil moist and maintain humidity, precise irrigation is required

FIG. 1 Key steps in a typical agriculture lifecycle.

which is primarily driven by the nutrition content of the soil. Nutrition deficiency of soil in view of Nitrogen, Potassium, and Phosphorous can lead to poor quality of crops. Providing adequate nutrition to a crop during production is a challenging task for the farmers;

3. Productive agriculture requires the removal of unwanted plants (called weeds) from the field, which tends to consume the nutrients or compete for the natural resources (such as water and fertilizers) that are meant for the crops. As a result of this, the productivity and quality of the crops get compromised. Thus, weed protection plays an important role in the agriculture lifecycle which is a laborious task to attain, especially in developing countries;
4. Besides, lack of irrigation and drainage facilities and lack of storage management can further result in agricultural distress in terms of lower productivity and loss of/damages to harvested crops. Hence, given the challenges in the conventional practice of agricultural activities, it becomes imperative to discuss the emerging scope of modern technologies, such as artificial intelligence (AI), in overall agricultural management in the context of optimal natural resources consumption.

Agriculture, in the 21st century, is experiencing a major shift toward the development of agricultural technologies largely due to the focus on increasing productivity for fulfilling the rising food demands of the current and future generations. In addition to the existing constraints, agricultural systems are enormously variable in space and time, thus making it difficult to generalize a solution and apply it. With the advent of technology amidst digitalization, the thinking process is pushed beyond the imagination wherein attempts are made to coalesce a normal brain with an artificial one resulting in a whole new field called "artificial intelligence" (AI) (Jha et al., 2019). AI could be called an intelligent machine tool that mimics how the human brain thinks, learn, make decisions, and mind works while solving a problem. In fact, government, non-government, and agricultural organizations from across the globe are deploying AI in agriculture so as to exploit the benefits of AI in yielding healthier crops, monitoring soil and environmental conditions, controlling pests, increasing productivity, improving efficiency, addressing labor shortages, and addressing sustainability issues amidst low water availability (Yahya, 2018). Bannerjee et al. (2018) established the advancements in AI according to the different roles in the agriculture lifecycle and have given a brief overview of various AI techniques.

Artificial neural networks (ANNs) have been one of the most opted algorithms in the agricultural sector due to its benefit of predicting and forecasting based on parallel reasoning (refer to Fig. 2). Due to its ease of programming with increased capacity to compute and remarkable ability of self-organization, and adaptive learning, ANN has discovered ways into almost all the (agrarian) activities. Instead of writing every logic, and taking every factor into account during programming, neural networks, in contrast, are required to be fed the right input and they could train themselves (like the human brain learns) to carry out a particular task, that is required for the system (Fig. 3) (Jain et al., 1996). Similarly, machine learning (ML) among other applications is defined as the scientific field that gives machines the ability to learn without being strictly programmed (Samuel, 2000) (Fig. 4).

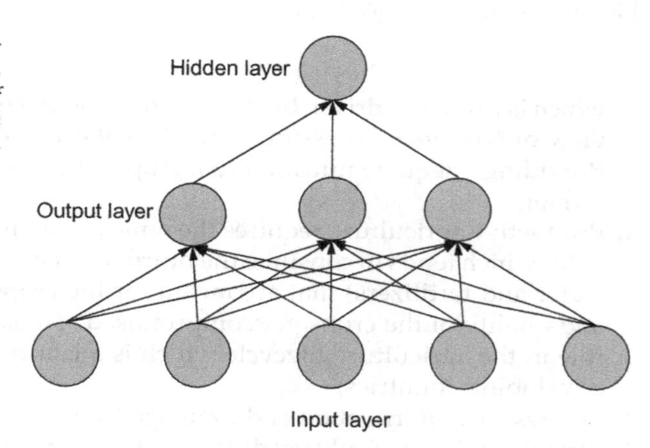

FIG. 2 A representative model of artificial neural network (ANN). *Adapted from Hallinan, J.S., 2013. Computational intelligence in the design of synthetic microbial genetic systems. Methods Microbiol. 40, 1–37, https:/doi.org/10.1016/B978-0-12-417029-2.00001-7.*

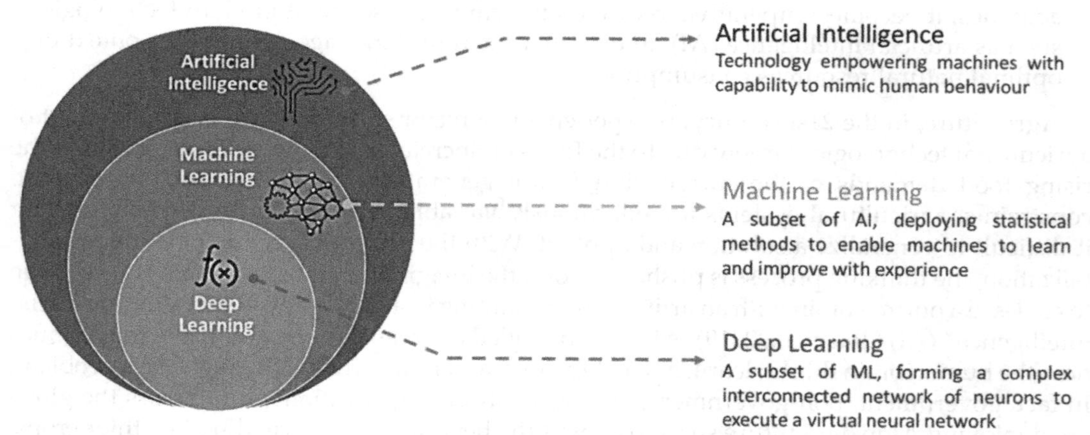

FIG. 3 Venn diagram showing the difference between artificial intelligence, machine learning, and deep learning.

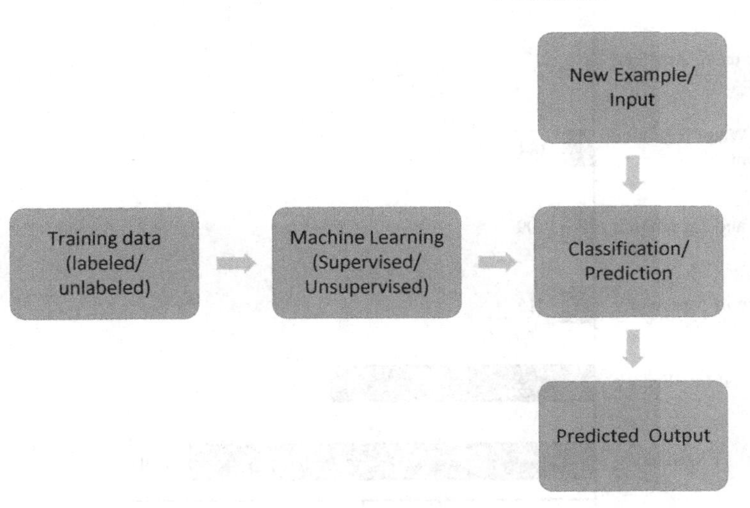

FIG. 4 Flowchart depicting the different processes involved in a machine learning model.

There is a substantial need for improvisation in agricultural water management owing to disruptions caused by climate change uncertainties and unpredictability, greenhouse gas emissions, water scarcity and insecurity, and natural disasters. To combat these risks and vulnerabilities and in order to bridge the gap between conventional agricultural practices and contemporary sustainable farming methodologies, AI-based developments and prospects are paving pathways to reducing the negative economic, socioeconomic, and environmental impacts (Bonab et al., 2021; Araújo et al., 2021). The technologies which are based on AI have shown improved efficiency in fields and provide scalable solutions in order to cope with growing challenges in the agricultural sectors, such as irrigation and water management, soil management, crop yield, crop monitoring, weeding, and crop establishment (Kim et al., 2008). For example, to ensure Sustainable Development Goal (SDG)-6 (sustainable management of water and sanitation for all) and SDG-13 (urgent actions to promote climate-related investments to combat climate change impacts), AI is playing a pioneering role in the digital transformation of water supply (Pigola et al. 2021). The development of numerical tools based on AI applications has resulted in addressing key global water management-based challenges, such as controlling water tariff increase; preventing aging and deteriorating water assets; recognizing water resources as a growth-limiting factor in water-food-energy nexus, both in terms of quality and scarcity under the compounded impact of climate change and increased urbanization; and integrating soil, water, and vegetation at a watershed level to aid natural resource management for agriculture (Nourani et al., 2014).

Besides, while conducting a literature survey for this chapter, it was observed that for different keywords on AI, the number of publications available is systematically varying (Fig. 5). For example, the number of publications on AI in the agricultural sector is recorded higher than AI in water resources management. Whereas publications explicitly on automation in the agricultural sector are recorded far lower than both. Even the publication trends in AI are further stressing the objective of bringing more attention to AI-based techniques in the agricultural industry so as to develop strategies for overcoming the ongoing and future

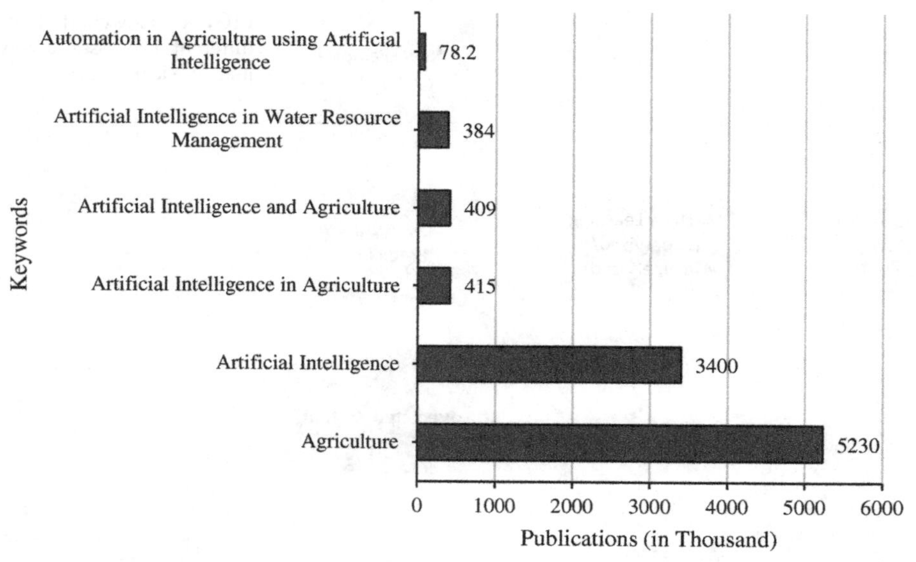

FIG. 5 Trends in publications related to artificial intelligence on Google Scholar (last updated on 31 October 2021).

challenges. This chapter thus reviews the advancements in AI technologies related to the agricultural lifecycle. The review objectives and classification of this chapter into various sections are described in the following section.

1.1 Review objectives and chapter organization

This chapter aims to review the scope of integrating advancements in AI in agricultural water management. Although the chapter largely focuses on entire agricultural processes, understanding of deploying AI techniques to optimally manage natural resources, such as land, water, nutrients, fertilizers, etc., are also considered, given the various stages of agrarian processes. These stages have been categorized as preagricultural activities, during agricultural activities, and postagricultural activities and are covered under Section 2. Given the emerging trends and future developmental scope in AI, reviews further describe the ongoing and future scope in water focused AI-based applications for agriculture in Sections 3 and 4. The chapter finally concludes by highlighting the strength and limitations of employing AI in the agricultural industry in Section 5.

2 AI in agriculture

The preagricultural activities comprise those stages whereby the farmers prepare their farms for sowing and cultivation. Several environmental, meteorological, and soil conditions, such as temperature, season, humidity, rainfall, wind speed, topography, soil type, terrain, soil moisture content, etc., influence preagricultural activities (Attia et al., 2015). Besides, issues such as flash floods, recurrent droughts, and extreme weather conditions negatively influence farmers' livelihood alongside increasing water shortages, thereby threatening long-term viability and profitability. For instance, a single season of no rain can drastically affect agricultural productivity, thereby interlinked livelihood, economy, and food supply-demand nexus (Hasegawa et al., 2018; Wang et al., 2018). Conserving and better managing water supply majorly contributes to securing sustainable crop production at a lower environmental burden. AI can help combat such difficulties by carefully analyzing past weather conditions and satellite information and by providing possible ranges of future scenarios. AI can be employed to measure and monitor parameters like solar radiation, precipitation, temperature speculations, wind speed, moisture content, crop health, etc. These operations result in obtaining accurate data concerning potential crop growth and yields, which would help in planning the type of crop that can be grown, time of sowing the seeds, time of harvesting the field, and quantity of resources (water, fertilizers, etc.) required for healthy growth of crops (Javaid, 2021).

Efficient and scalable management of natural resources, such as fresh water, for irrigation purposes, is a major concern in many cropping systems, especially in the semiarid and arid areas located across underdeveloped and developing nations (Kim et al., 2008). In the context of delivering efficient water supply, AI learning tools are being applied in decision-making tasks. This includes scope related to maximizing information and data availability for making better decisions and enhancing service deliveries using water utilities. Furthermore, such extended operations can generate a scope in optimizing capital investment (CAPEX), reducing operating costs (OPEX), and including social and environmental externalities (Gilbert, 2018).

2.1 AI in preagricultural (preparatory) activities

In general, the first step toward agriculture is the selection of appropriate quality and type of seed to be sown in a particular environment concerning weather conditions, soil type, and water availability. Attia et al. (2015) showed that seed quality and variety selection set the maximum performance level for all plants. The emergence of technologies like AI has empowered farming communities in providing the best selection of crops with an improved selection of hybrid seed choices based on the field conditions and as per the suitability of the individual farmer (Nie et al., 2019). AI-based technologies are being developed aiming to deliver water-smart solutions so as to achieve better harvests with less water thereby managing agricultural water sustainably. For example, a seed variety named "Arize hybrid rice seeds" has been developed which is specifically designed to be tolerant to submergence and dry conditions. As a result, the seeds have survived unfavorable conditions such as flash floods and recurrent droughts. Such technologies have substantially increased productivity, efficiency, and profits, and reduced natural resources consumption, more especially the irrigational requirements (Choudhary and Suri, 2018).

Soil contains water (as moisture), nutrients, and proteins in appropriate ratios that make it vital for crop survival. However, if the ratio of crop health and productivity drivers gets misbalanced, water relations in the soil-plant continuum get affected in the first place. The resultant consequences endanger soil characteristics such as organic richness, fertility, moisture level, etc. (Lenka and Mohapatra, 2015). One of the major advancements of AI has been observed in monitoring soil and crop health. The images resulting from aerial photogrammetry through are being used for seed and soil health analysis. Aerial images captured by means of UAVs (unmanned aerial vehicles) are being used to create 3-D maps for early soil analysis, which are being used to map the terrain of the field, identify the areas of nutrient deficiency, and create heat maps with different soil properties like moisture content, organic carbon, and total nitrogen of the scanned soil. The implementation of such approaches is then achieved by imaging foliage which is then run through software that can distinguish between normal and unhealthy growth patterns.

Various test cases have been in practice where the UAVs, mostly consisting of quadcopters, are being utilized by farmers to sow the seeds in the field. UAV-based drones are flying robots that use GPS navigation and software-controlled flight, (also installed in their flight controller systems) and are capable of performing autonomous flight operations (Zeng et al., 2016). The utility consists of two separate chambers which store the seeds in pods and nutrients (fertilizers), which are linked to a feeder system, equipped with a motor control mechanism (Jiang et al., 2019) and this utility is then mounted on a quadcopter for action. As the quadcopter hovers over the field, a camera records the terrain of the field, and prefeed GPS coordinates are marked in the system where the quadcopter shoots (pushes) the pods with seeds along with the nutrients through the feeder into the soil providing all the nutrients necessary for the growth of the crop. These test cases have proved to be 85% more efficient both in terms of cost and time for sowing seeds (Misra et al., 2020). Further, an automated system has been introduced to determine the actual position of seeds, establishing the ultra-high precision placement of the seeds.

2.1.1 Case studies: "Agri-e-calculator and sowing app"

Several applications have been developed in the 21st century in view of precisely governing the preagricultural activities. One such application is called "Agri-e-calculator," which is one of the smart applications being available in the market that helps the farmers choose the most affordable and most suitable crop based on the dependent factors, such as the weather conditions, soil type, terrain, water availability in the region, etc. (Misra et al., 2020). Once the farmer decides to cultivate the desired crop, the application would direct the user (here the farmer) to take all the required inputs related to the field and crops. In addition, the application provides significant estimates, as an output, such as the quantity of water required for preparing the field or for irrigation, seeds type and quality, cultivation equipment cost (if any), fertilizers cost and quantity, the amount of effort in terms of labor days and its corresponding cost, and the crop yield along with an extrapolated market price at the harvest time, and the subsequent profitability from the whole agricultural cycle. All the aforementioned data is received by the farmer at just a few clicks of the button. Such AI-based applications have enabled farmers to make the most appropriate decisions concerning their farm conditions.

Besides, under the collaborative efforts, International Crops Research Institute for the Semi-Arid Tropics (ICRISAT) initiated a pilot project on providing advisory services to the farmers of Andhra Pradesh state of India (ICRISAT, 2021). The services are based on the application of the Cortana Intelligence Suite including ML and Power BI, together with

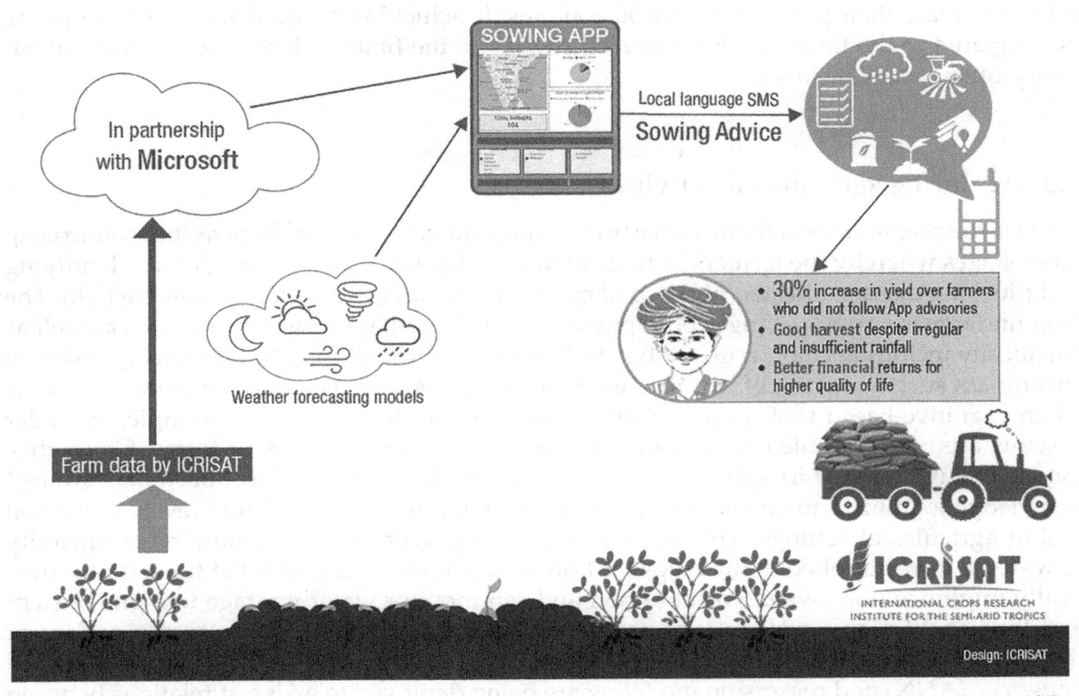

FIG. 6 Working of sowing application. *Source: ICRISAT—International Crops Research Institute for the Semi-Arid Tropics, 2021. Sowing App Infographic. https://i0.wp.com/www.icrisat.org/wp-content/uploads/2017/05/Sowing-App-Infographgic.jpg?ssl=1 (Accessed 31 October 2021).*

enabling the transformation of the data into intelligent actions. In this service, farmers receive around 10 sowing advisories through SMS in their local language. Messages contained essential information regarding recommendations on sowing date, land preparation, soil test-based fertilization, farmyard manure application, seed treatment, optimum sowing depth; and directions on preventive weed management, maintaining proper plant density, applying nutrients, harvesting, and correct storage practices. To acquire all these details, farmers are required to carry any mobile phone capable of receiving text messages (refer to Fig. 6). To test the practical applicability of such applications, a case study was conducted by the ICRISAT at the Devanakonda administrative area in the Kurnool district of Andhra Pradesh. Selected farmers were delivered the message (as instructions) for sowing groundnut in the last week of June and the first week of July 2016. This sowing period was estimated by application based on analyzing historic climate data (1986–2015) for the study area. Here, daily rainfall data for this period was analyzed to estimate the moisture adequacy index (MAI). Simulations were performed to predict MAI for the next five days. The findings provided instructions on the optimum sowing period (as mentioned above). This is followed by generating further advisories to farmers which continue right from appropriate sowing techniques to correct selection of harvesting methods. As a part of the validation of the sowing recommendation and advisories delivered to the selected farmers, it was found that they were able to increase their productivity by 30% alongside achieving reduced water consumption, as compared to the farmers who preferred sowing in the first week of June (against the advisory of last week of June).

2.2 AI during agricultural activities

Once the preparatory activities have been conducted, it is followed by activities comprising those stages whereby the farmers indulge in day-to-day tasks, such as irrigation, identifying and plucking disease-infected portions of crops, and removing the weeds from the field. The crop undergoes various changes both physically, such as shape, size, texture, etc., as well as chemically including color, reflectivity, etc. It becomes difficult for a traditional algorithm to encompass such a vast set of features and form a relation between every parameter. This is when cognitive-based tools powered by AI can come to the rescue. For example, sprinkler systems are often installed for carefully supplying water in gardens or farms. Given this, modern AI methods of irrigation have gone a step further, wherein the "machine learning" knowledge is utilized to enable the system to track the real-time moisture levels in the soil within agricultural settings. This allows determining, with precision, how much quantity of water is to be supplied to the crops for how much duration and at what time. It can drastically minimize huge water consumption and can monitor water wastage to a great extent and thus could play a vital role in managing scarce water resources across dry regions of the world (Jha et al., 2019). With the emergence of AI, techniques such as "artificial neural network" (ANN) and regression modeling are being deployed to arrive at relations between evapotranspiration, precipitation, and soil that eventually help in providing remote agricultural management decisions (Coopersmith et al., 2014). Data from various sensors, such as a visible near-infrared spectrophotometer, are being used to collect soil spectra, and apply regression models to predict soil properties in view of moisture content, organic carbon, and total nitrogen (Morellos et al., 2016).

Agriculture consists of a month-long lifecycle, where the crop grows under changing weather conditions. The regular change in the surrounding environment may lead to dramatic losses due to the diseases. These diseases can be both natural (due to the change in weather) and more often or so, due to humming pests and insects breeding in the vicinity of these agricultural fields, which ultimately leads to the reduction in the productivity of the crops and poses a grave concern to a farmer. AI sensors are being developed these days which primarily uses image sensing techniques to determine characteristics of diseases developed in crops, especially on their leaves. In this process, color imaging is conducted via AI machines that are able to distinguish between healthy and diseased leaves, and thereby remove them through integration with robotics (Microsoft Stories India, 2017). With technologies like photogrammetry and remote sensing (Everaerts, 2008; Colomina and Molina, 2014), the massive task of identifying and then eliminating and diagnosing of the diseases the diseased portions of the crop has turned into a color puzzle game.

Controlling the weed growth in the farms is of high importance to farmers. Globally, close to 250 species have been identified which are resistant to modern herbicides, and thus more than 40 billion USD is lost annually for eradicating weeds alone from the farms of soybean and corn (WSSA, 2016). As per World Health Organization (WHO) estimates, on an annual basis, over one million cases of people (especially farmers or farm laborers) require medical aid due to illness caused during spraying the pesticides in the field crop manually.

The use of UAVs, which are also used to identify crop issues can provide rescue to farmers by remotely spraying the pesticides on the crops and thereby preventing the direct contact of pesticides with farmers (Giles and Billing, 2015). Thus AI-enabled UAVs not only can mitigate health-associated risks and problems with manual spraying of pesticides but can provide instructions to farmers as to how much and where exactly to spray within the agricultural field. In addition, UAVs can be easily used and have the potential to reach the spaces and corners where the equipment and labors are difficult to operate. Researches involving UAV technologies, their methods, systems, and limitations are examined in (Huang et al., 2013). A multispectral camera mounted on a UAV can scan the whole crop field and generates a spatial map that could help in manifesting the crop condition through normalized difference vegetation index (NDVI). This step allows farmers to evaluate which type of fertilizers and pesticides to be applied to the crop (Mogili and Deepak, 2018). An overall management summary for weed control using AI is described in Eli-Chukwu (2019).

2.3 AI in postagricultural activities

The emergence of technologies, such as AI, satellite-based remote sensing, and advanced analytics have created an ecosystem for smart farming. By integrating appropriate technologies and tools, it enables farmers to achieve higher average yields and even better price control. Predictive analytics is one of the upcoming research areas that is having the potential to bring revolution in the agricultural market. Having a prediction regarding the crop yield could be beneficial for marketing strategies by the farmer and play a pivotal role in crop cost estimation for the policymakers and the market. Analyzing market demand, forecasting prices, and determining the optimal time to harvest a crop are some of the key aspects that can be effectively estimated using AI applications. Further, robots are being deployed nowadays to harvest fruits and vegetables. As each type of produce has its own set of unique requirements, robotics-based approaches thus demand rigorous research and mechanical expertise.

AI in agriculture have developed applications and tools which provide farmers appropriate guidance regarding water management, crop rotation, timely harvesting, type of crop to be grown, optimum planting, pest attacks, and nutrition management. Images of crops taken under white or ultraviolet (UV) light are being fed into a machine-learning algorithm to determine how ripe a particular fruit/crop is, and also to come up with the remaining time for the fruit to ripe and harvest. This also enables the farmers to create different levels of ripeness (readiness) based on the crop category and to segregate them into different stacks before dispatching them to the market.

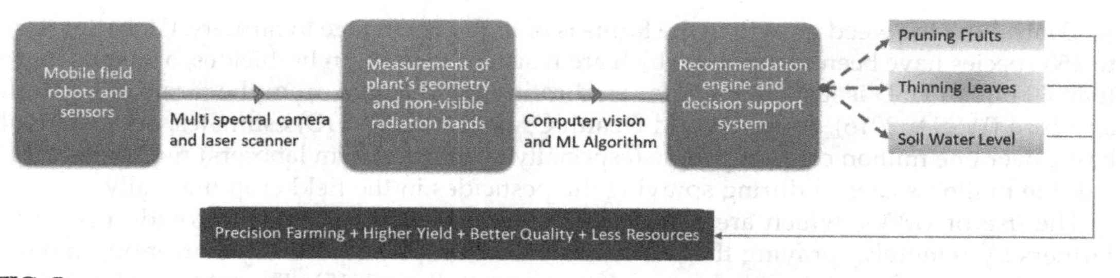

FIG. 7 A summary of how robotics is helping in Digital Farming. *Adapted from Mindtree, 2018. Artificial Intelligence in Agriculture. https://www.mindtree.com/sites/default/files/2018-04/Artificial%20Intelligence%20in%20Agriculture.pdf (Accessed 31 October 2021).*

The 2020 global market valued the robotics-based applications at USD 27.73 billion which is expected to reach USD 74.1 billion by 2026 (https://www.mordorintelligence.com/industry-reports/robotics-market). A considerable budget has been planned to be dedicated to agricultural robots. Agricultural robots are defined as specialized articles of technology that are capable of assisting farmers with a wide range of operations (Bechar and Vigneault, 2017). In this regard, AI-based companies are developing robots that can easily perform multiple tasks in farming fields (Fig. 7). The primary role of a robot is to provide automotive solutions for tasks that are labor-intensive, repetitive, and physically demanding. Given the effectiveness, robots are being developed with specialized tools to equip them with the ability to detect and pick out sensitive fruits and vegetables. For example, to combat manual dexterity required in fields to handle fruits and vegetables, a unique robot has been developed called "Vegebot," which relies on a computer vision-based technology to achieve high accuracy when harvesting crops (for more information: https://stories.pinduoduo-global.com/agritech-hub/robots-in-agriculture-and-farming). These types of robots are trained to check the quality of crops and detect weeds with picking and packing of crops at the same time. Further, robots are being trained to harvest crops at a faster pace with higher volumes compared to manual efforts by humans and animals. Thus, agricultural robotics is proving to be proficient at capturing valuable data for fine-tuning AI and ML-based algorithms, thereby further improving crop yields (Saiz-Rubio and Rovira-Más, 2020).

In the agricultural lifecycle, considering the harvesting process as the final step is misleading, as farmers are still required to ensure that the yielded crops should reach the market and profitable revenue should get generated in view of achieving high economic efficiency. Besides, there have been various developments in the field of food monitoring and quality control, logistics, storage mechanisms that have been developed with help of AI systems in place, discussed in detail by (Bannerjee et al., 2018). One such system, called "Agri Reach," has been developed by Delhi-based Sohan Lal Commodity Management (SLCM), India, which is an AI-based system allowing the company to monitor more than 700+ agricommodities and micromanage their storage environment remotely (Dey, 2018). The system could effectively predict storage conditions and suggest preventive actions to reduce rotting damage and could also predict curative processes if storage conditions had deteriorated. The system helped in reducing the storage losses at the warehouses from 10% to 0.5%.

3 Current and future scope in AI for agriculture

Farm Automation is one of the buzzwords recently in the arena of agriculture. It is the process of utilizing various technological innovations to upgrade and automate the many processes of farming. AI has been one of the technologies being deployed at the forefront to achieve farm automation. These techniques are meant to ease the labor and time-intensive process of agriculture that challenges farmers around the world. While using the machine-learning algorithms in connection with images captured by satellites and drones, AI-enabled technologies could predict weather conditions, analyze crop sustainability and evaluate farms for the presence of diseases or pests and poor plant nutrition on farms with data such as temperature, precipitation, wind speed, and solar radiation. Given the wide-ranging scope of inclusion of modern AI-driven technologies in conventional farming practices, contemporary researches on AI development are gradually shifting to agriculture amidst depleting water resources, increasing human and livestock population, rising climate change vulnerabilities, and declining food production due to human-induced rapid land-use land cover transformations.

Crop Monitoring and Health assessment remain one of the most significant areas in agriculture to provide drone-based solutions in collaboration with AI and computer vision technology. Drone technology has emerged as one of the leading applications that have enabled technological evolution in the field of agriculture. Drones are being tested throughout the world for helping farmers for enabling them to make better decisions and help them in making efficient, productive, and scalable in their fields. There are several applications of drone technology within various stages of the agricultural cycle such as for soil and field analysis, planting seeds, weed identification, pesticide spraying, crop monitoring, irrigation management, health assessment, irrigation equipment monitoring, herd and wildlife monitoring, and disaster management (Ahirwar et al., 2019; Natu and Kulkarni, 2016; Veroustraete, 2015). Some of the recent researches highlighted the possible future applications of the same (Krishna, 2017). While other researchers indicated potential difficulties and limitations, in particular the limited load capacity of drones (Vroegindeweij et al., 2014). There are hopes for the use of drones in harvesting fruits from trees and concepts of experimental drones that can collect coconuts from coconut trees (Rajan et al., 2020). For a detailed study of the application of drones for agriculture automation from planting to harvest, readers can refer to Kulbacki et al. (2018). A summary of the various applications of drones in agriculture can also be found in Talaviya et al. (2020).

Besides, studies have been conducted on the investigation for determining the precise crop water requirements using free satellite imageries. Scaling-up of such advancements may enhance efficiency in water applications for agriculture alongside facilitation of the works being done by irrigation advisors and technicians. Coherent to this, Aguilar et al. (2018) evaluated and validated the MOD16 algorithm, which is acquiring satellite-based information on evapotranspiration losses from agricultural fields. The study compared the data on evapotranspiration, as obtained from satellite, with ground-based eddy covariance measurements, as a case study approach across five Northwestern Mexico locations. The study site, as reported, was water-deficient given arid and semiarid climatic conditions such that agrarian activities remained limited to wheat cultivation and natural vegetation (shrubs). Indicators found that the results being obtained from MOD16 underestimated evapotranspiration which lead to the conclusion that MOD16 can provide promising findings to the emerging farmers concerning crop water requirements. Nevertheless, one of the limitations being raised was the limited grounded investigation to assess the accuracy of the satellite information. In general, the development and scaling-up of such emerging, low-cost, and advanced technologies can provide a substantial input to revise current decision support systems for crop water demand and irrigation scheduling.

AI has proved itself as a dependable source of technology for various interdisciplinary roles, which previously remained human-centric, such as call centers, suggestions, expert advice, insurance, loan, market price evaluation, etc. With the advent of AI, farmers can leverage the ever-increasing capability provided by AI systems including, but not limited to, chatbot services which are 24×7 online, crop care services, expert advice on the go, loan, and insurance services, best price to sell in the market, etc. With the help of AI-powered virtual assistants, farmers can get answers and recommendations to their specific problems from their homes themselves. Additionally, the sensing abilities of AI-based applications for analyzing fluctuations in market price can provide early warnings to the farmers, thereby could direct for appropriate planting and harvesting in view of achieving improved productivity (Chukkapalli et al., 2020). Machine-learning algorithms are being trained to analyze past records of statistical data for market price and the corresponding factors (demand, supply, international market trend, fuel price, etc.), to come up with a prediction on the most probable prices that a particular crop could be sold for in the market. This service could help the farmer get a safeguard from market fluctuations and mitigate the risk of price loss, by providing a plan for better release of their commodity in the market (Nayal et al., 2021). Given achieving agricultural water management in contemporary times, chatbots are pioneering in providing interactive sessions with farmers on best crop selection, hybrid seed choices, seed choices as per weather conditions, soil types, plant diseases, and increase the return-on-crops, etc. (Talaviya et al., 2020). Coherent to this, the deep learning-based "DeHaat" is a growing startup in the AgriTech sector in India having the potential to provide 24×7 support to marginalized farmers on the aforementioned parameters (Nuthalapati and Nuthalapati, 2021).

The lack of information about the market conditions is a concern for smallholder farmers to price their agricultural produce. Farmers are often compelled to sell their produce to middlemen who exploit this knowledge asymmetrically to their advantage. With this context of pricing issue, the Government of Karnataka State in India has initiated the use of price forecasting for agricultural produce to keep the farmers informed not only about the current but to keep them aware about the possible future rates for their commodities depending upon the market attributes (Microsoft Stories India, 2017). In this regard, the Government took the initiative of informing farmers of the predicted commodity price of tur (a type of pulse native to Karnataka State which is also the second-largest producer of the same). The model considers datasets on historical sowing areas, production yields, weather patterns, and other relevant information. In addition, it uses remote sensing data from geostationary satellite images to predict crop yields at every stage of the farming process. The output of the model includes predictions about arrival dates and crop volumes, enabling local governments and farmers to predict commodity prices 3 months in advance for major crop markets. This allows State Government to accurately plan the minimum support price for the agricultural produce.

4 Challenges of AI in agriculture

Though in trials and testing of AI in agriculture, the system seems to perform satisfactorily on the technical parameters and offers vast opportunities for its wide-ranging applications. However, there exists a lack of familiarity with these machine-learning-based technical solutions in these fields across the globe. Farming is dependent on several factors, some internal (crop characteristics) and a few external (weather and climate). The external factors, such as weather conditions, soil conditions, pest movements, etc., can change with time and have an important role in governing the growth and health of any crop. Therefore, what might look like an appropriate solution at the starting phase of the agricultural cycle may not be observed optimal in the subsequent phases, which can be attributed to the inherent uncertainties in the external parameters. Though contemporary AI systems showcase a need for time robustness to the changes in these external factors; however, it is yet to reach a stage where accurate determination and prediction of the future courses of these external events can be made to minimize the uncertainty in planning.

Furthermore, the limitation with image-based AI training models is that they require a tremendous amount of data to be fed as input for the system to train itself for making precise predictions as per the user requirement. Raw (or primary) data collection is a tedious, costly, and time-consuming affair. For the vast agricultural lands spread across the globe, the spatial data is relatively easy to gather and organize given the advancements in the remote sensing technology and the increasing use of satellite-based data acquisition; however, the temporal data over the field, which is one of the key parameters for the machine to learn over time and develop the patterns and acquire robustness for the system, is quite difficult to acquire from the existing sources. Even after the collection of a high amount of raw data in a database, it requires the data to be organized and filtered for the ML models to train and learn, which, in turn, requires quality hardware and resources for data processing.

The future of farming depends largely on the adoption of cognitive solutions. While large-scale research is still in progress and some applications are already available in the market, the research in this industry is still highly underserved. When it comes to handling realistic challenges faced by farmers and using autonomous decision-making and predictive solutions to solve them, farming is still at a nascent stage. Another important aspect is the exorbitant cost of different cognitive solutions available in the farming market. The solutions need to become more affordable to ensure the technology reaches the masses. An open-source platform would make the solutions more affordable, resulting in rapid adoption and higher penetration among the farmers (Mindtree, 2018).

5 Conclusions

AI can be applied cross-disciplinarily and it can also bring a paradigm shift in the conventional farming practices leading to substantial water conservation. Further, AI-based developments can control wastage of natural resources, such as soil, water, and land, and increase the competitive edge for agricultural firms while ensuring sustainability. In addition,

AI could direct toward improving the crop yield per unit area of land. This is empowered by algorithms mimicking human cognition, bringing in technologies, such as ML, to analyze big data and come up with models capable of making effective decisions. These mathematically AI-powered solutions will not only enable farmers to do more (say productivity-wise) with less (say natural resources consumption-wise), it will also improve quality and ensure a faster go-to-market for crops. AI technologies can help farmers to rapidly analyze land, water, and soil characteristics, the health of crops, etc., and save time, thereby allowing farmers to choose the appropriate crop for cultivation corresponding to each season in view of achieving the best yield. AI-based predictions enable suggesting appropriate pesticides and crops that too at the suitable place and at right time. Such information can be critical for farming communities that suffer due to unprecedented disruptions caused by the large-scale incidence of diseases, lack of irrigation-based technologies, and frequent encounters with natural hazards (such as floods or droughts) under changing climatic conditions. Amidst the advancement in AI-based technologies in agriculture, there remains huge space still untouched, given introducing further advancement in automatic response systems and leveraging emergent technologies of catboats for supporting farmers from drylands who are prone to water scarcity. Additionally, technologies are still lacking which can answer farmers' queries and can provide relevant advice and recommendations to their specific farm-related concerns under limiting irrigation conditions. Such technologies, if developed and scaled-up, can further propel the growth of AI-powered technologies for the benefit of the global agricultural industry under changing climate and stressing water availability.

Acknowledgments

The authors appreciate the Indian Institute of Technology (IIT) Kharagpur authorities for providing technical conveniences and a research environment for conducting this review research. The authors would like to thank Mr. Sandeep R. Mahajan (Coordinator, Lead India Jalgaon Group, India) for providing deep insights regarding various aspects of rural-based farming infrastructure and issues faced by marginalized farmers.

Conflict of interest

The authors declare that they have no known competing financial interests or personal relationships that could have appeared to influence the work reported in this chapter.

References

Aguilar, A.L., Flores, H., Crespo, G., Marín, M.I., Campos, I., Calera, A., 2018. Performance assessment of MOD16 in evapotranspiration evaluation in Northwestern Mexico. Water 10 (7), 901. https://doi.org/10.3390/w10070901.

Ahirwar, S., Swarnkar, R., Bhukya, S., Namwade, G., 2019. Application of drone in agriculture. Int. J. Curr. Microbiol. Appl. Sci. 8 (01), 2500–2505. https://doi.org/10.20546/ijcmas.2019.801.264.

Araújo, S.O., Peres, R.S., Barata, J., Lidon, F., Ramalho, J.C., 2021. Characterising the agriculture 4.0 landscape—emerging trends, challenges and opportunities. Agronomy 11 (4), 667. https://doi.org/10.3390/agronomy11040667.

Attia, A., Shapiro, C., Kranz, W., Mamo, M., Mainz, M., 2015. Improved yield and nitrogen use efficiency of corn following soybean in irrigated sandy loams. Soil Sci. Soc. Am. J. 79 (6), 1693–1703. https://doi.org/10.2136/sssaj2015.05.0200.

Bannerjee, G., Sarkar, U., Das, S., Ghosh, I., 2018. Artificial intelligence in agriculture: a literature survey. Int. J. Sci. Res. Comput. Sci. Appl. Manag. Stud. 7 (3), 1–6.

Bechar, A., Vigneault, C., 2017. Agricultural robots for field operations. Part 2. Operations and systems. Biosyst. Eng. 153, 110–128. https://doi.org/10.1016/j.biosystemseng.2016.11.004.

Bonab, A.B., Rudko, I., Bellini, F., 2021. A review and a proposal about socio-economic impacts of artificial intelligence. In: Dima, A.M., D'Ascenzo, F. (Eds.), Business Revolution in a Digital Era. Springer Proceedings in Business and Economics, Springer, Cham, https://doi.org/10.1007/978-3-030-59972-0_18.

Bongaarts, J., 2007. Food and agriculture organization of the United Nations: the state of food and agriculture: agricultural trade and poverty: can trade work for the poor? Popul. Dev. Rev. 33 (1), 197. https://link.gale.com/apps/doc/A162101584/AONE?u=anon~96cae425&sid=googleScholar&xid=3363c0bc.

Boretti, A., Rosa, L., 2019. Reassessing the projections of the world water development report. NPJ Clean Water 2 (1), 1–6. https://doi.org/10.1038/s41545-019-0039-9.

Choudhary, A.K., Suri, V.K., 2018. System of rice intensification in promising rice hybrids in North-Western Himalayas: crop and water productivity, quality, and economic profitability. J. Plant Nutr. 41 (8), 1020–1034. https://doi.org/10.1080/01904167.2018.1431675.

Chukkapalli, S.S.L., Mittal, S., Gupta, M., Abdelsalam, M., Joshi, A., Sandhu, R., Joshi, K., 2020. Ontologies and artificial intelligence systems for the cooperative smart farming ecosystem. IEEE Access 8, 164045–164064. https://doi.org/10.1109/ACCESS.2020.3022763.

Colomina, I., Molina, P., 2014. Unmanned aerial systems for photogrammetry and remote sensing: a review. ISPRS J. Photogramm. Remote Sens. 92, 79–97. https://doi.org/10.1016/j.isprsjprs.2014.02.013.

Coopersmith, E.J., Minsker, B.S., Wenzel, C.E., Gilmore, B.J., 2014. Machine learning assessments of soil drying for agricultural planning. Comput. Electron. Agric. 104, 93–104. https://doi.org/10.1016/j.compag.2014.04.004.

Dey, S., 2018. Now, AI to Help Reduce Wastage of Crops and Post-Harvest Loss. https://www.theweek.in/news/sci-tech/2018/03/05/ai-reduce-wastage-crops-and-post-harvest-loss.html. (Accessed 31 October 2021).

Elbeltagi, A, Raza, A., Hu, Y., Al-Ansari, N., Kushwaha, N.L., Srivastava, A., et al., 2022. Data intelligence and hybrid metaheuristic algorithms-based estimation of reference evapotranspiration. Appl. Water Sci. 12 (7), 1–18. https://doi.org/10.1007/s13201-022-01667-7.

Eli-Chukwu, N.C., 2019. Applications of artificial intelligence in agriculture: a review. Eng. Appl. Sci. Res. 9 (4), 4377–4383. https://doi.org/10.48084/etasr.2756.

Everaerts, J., 2008. The use of unmanned aerial vehicles (UAVs) for remote sensing and mapping. Int. Arch. Photogramm. Remote. Sens. Spat. Inf. Sci. 37, 1187–1192. 2008.

Gilbert, M., 2018. The role of artificial intelligence for network automation and security. In: Gilbert, M. (Ed.), Artificial Intelligence for Autonomous Networks. Chapman and Hall/CRC, pp. 1–23.

Giles, D., Billing, R., 2015. Deployment and performance of a UAV for crop spraying. Chem. Eng. Trans. 44, 307–312. https://doi.org/10.3303/CET1544052.

Hasegawa, T., Fujimori, S., Havlík, P., Valin, H., Bodirsky, B.L., Doelman, J.C., et al., 2018. Risk of increased food insecurity under stringent global climate change mitigation policy. Nat. Clim. Chang. 8 (8), 699–703. https://doi.org/10.1038/s41558-018-0230-x.

Hsiao, T.C., Steduto, P., Fereres, E., 2007. A systematic and quantitative approach to improve water use efficiency in agriculture. Irrig. Sci. 25 (3), 209–231. https://doi.org/10.1007/s00271-007-0063-2.

Huang, Y., Thomson, S.J., Hoffmann, W.C., Lan, Y., Fritz, B.K., 2013. Development and prospect of unmanned aerial vehicle technologies for agricultural production management. Int. J. Agric. Biol. Eng. 6 (3), 1–10. http://www.ijabe.org/index.php/ijabe/article/view/900/pdf.

Hunter, M.C., Smith, R.G., Schipanski, M.E., Atwood, L.W., Mortensen, D.A., 2017. Agriculture in 2050: recalibrating targets for sustainable intensification. Bioscience 67 (4), 386–391. https://doi.org/10.1093/biosci/bix010.

ICRISAT—International Crops Research Institute for the Semi-Arid Tropics, 2021. Sowing App Infographic. https://i0.wp.com/www.icrisat.org/wp-content/uploads/2017/05/Sowing-App-Infographgic.jpg?ssl=1. (Accessed 31 October 2021).

Jain, A.K., Mao, J., Mohiuddin, K.M., 1996. Artificial neural networks: a tutorial. Computer 29 (3), 31–44. https://doi.org/10.1109/2.485891.

Javaid, N., 2021. Integration of context awareness in internet of agricultural things. ICT Express. https://doi.org/10.1016/j.icte.2021.09.004 (in press).

Jha, K., Doshi, A., Patel, P., Shah, M., 2019. A comprehensive review on automation in agriculture using artificial intelligence. Artif. Intell. Agric. 2, 1–12. https://doi.org/10.1016/j.aiia.2019.05.004.

Jiang, F., Pourpanah, F., Hao, Q., 2019. Design, implementation, and evaluation of a neural-network-based quadcopter UAV system. IEEE Trans. Ind. Electron. 67 (3), 2076–2085. https://doi.org/10.1109/TIE.2019.2905808.

Kim, Y., Evans, R.G., Iversen, W.M., 2008. Remote sensing and control of an irrigation system using a distributed wireless sensor network. IEEE Trans. Instrum. Meas. 57 (7), 1379–1387. https://doi.org/10.1109/TIM.2008.917198.

Krishna, K.R., 2017. Push Button Agriculture: Robotics, Drones, Satellite-Guided Soil and Crop Management, first ed. CRC Press.

Kulbacki, M., Segen, J., Knieć, W., Klempous, R., Kluwak, K., Nikodem, J., et al., 2018. Survey of drones for agriculture automation from planting to harvest. In: 2018 IEEE 22nd International Conference on Intelligent Engineering Systems (INES). IEEE, pp. 353–358, https://doi.org/10.1109/INES.2018.8523943.

Kumar, P., Vishwakarma, D.K., Markuna, S., Ali, R., Kumar, D., Jadhav, N., et al., 2022. Evaluation of Catboost method for predicting weekly pan evaporation: case study of subtropical and subhumid regions of India. Res. Sq. https://doi.org/10.21203/rs.3.rs-1538970/v1.

Lenka, S.K., Mohapatra, A.G., 2015. Gradient descent with momentum based neural network pattern classification for the prediction of soil moisture content in precision agriculture. In: 2015 IEEE International Symposium on Nanoelectronic and Information Systems. IEEE, pp. 63–66, https://doi.org/10.1109/iNIS.2015.56.

Mancosu, N., Snyder, R.L., Kyriakakis, G., Spano, D., 2015. Water scarcity and future challenges for food production. Water 7 (3), 975–992. https://doi.org/10.3390/w7030975.

Microsoft Stories India, 2017. Digital Agriculture: Farmers in India are using AI to Increase Crop Yields. https://news.microsoft.com/en-in/features/ai-agriculture-icrisat-upl-india/. (Accessed 31 October 2021).

Mindtree, 2018. Artificial Intelligence in Agriculture. https://www.mindtree.com/sites/default/files/2018-04/Artificial%20Intelligence%20in%20Agriculture.pdf. (Accessed 31 October 2021).

Misra, N.N., Dixit, Y., Al-Mallahi, A., Bhullar, M.S., Upadhyay, R., Martynenko, A., 2020. IoT, big data and artificial intelligence in agriculture and food industry. IEEE Internet Things J. https://doi.org/10.1109/JIOT.2020.2998584.

Mogili, U.R., Deepak, B.B.V.L., 2018. Review on application of drone systems in precision agriculture. Procedia Comput. Sci. 133, 502–509. https://doi.org/10.1016/j.procs.2018.07.063.

Morellos, A., Pantazi, X.E., Moshou, D., Alexandridis, T., Whetton, R., Tziotzios, G., et al., 2016. Machine learning based prediction of soil total nitrogen, organic carbon and moisture content by using VIS-NIR spectroscopy. Biosyst. Eng. 152, 104–116. https://doi.org/10.1016/j.biosystemseng.2016.04.018.

Natu, A.S., Kulkarni, S.C., 2016. Adoption and utilization of drones for advanced precision farming: a review. Int. J. Recent Innov. Trends Comput. Commun. 4 (5), 563–565.

Nayal, K., Raut, R.D., Queiroz, M.M., Yadav, V.S., Narkhede, B.E., 2021. Are artificial intelligence and machine learning suitable to tackle the COVID-19 impacts? An agriculture supply chain perspective. Int. J. Logist. Manag. https://doi.org/10.1108/IJLM-01-2021-0002.

Nie, P., Zhang, J., Feng, X., Yu, C., He, Y., 2019. Classification of hybrid seeds using near-infrared hyperspectral imaging technology combined with deep learning. Sens. Actuators B 296, 126630. https://doi.org/10.1016/j.snb.2019.126630.

Nourani, V., Baghanam, A.H., Adamowski, J., Kisi, O., 2014. Applications of hybrid wavelet-artificial intelligence models in hydrology: a review. J. Hydrol. 514, 358–377. https://doi.org/10.1016/j.jhydrol.2014.03.057.

Nuthalapati, C.S., Nuthalapati, C., 2021. Has open innovation taken root in India? Evidence from startups working in food value chains. Circ. Econ. Sustain. 1 (4), 1207–1230. https://doi.org/10.1007/s43615-021-00074-5.

Ondrasek, G., 2014. Water scarcity and water stress in agriculture. In: Ahmad, P., Wani, M. (Eds.), Physiological Mechanisms and Adaptation Strategies in Plants Under Changing Environment. Springer, New York, NY, pp. 75–96. In this issue. https://doi.org/10.1007/978-1-4614-8591-9_4.

Pigola, A., da Costa, P.R., Carvalho, L.C., Silva, L.F.D., Kniess, C.T., Maccari, E.A., 2021. Artificial intelligence-driven digital technologies to the implementation of the sustainable development goals: a perspective from Brazil and Portugal. Sustainability 13 (24), 13669. https://doi.org/10.3390/su132413669.

Rajan, D., Kalathil, G.T., Jacob, B., Abraham, A.J., Thankachan, B., 2020. Coconut Harvesting Drone. http://du2016-grp063-05.blogspot.com/. (Accessed 31 October 2021).

Raman, R., 2017. The impact of genetically modified (GM) crops in modern agriculture: a review. GM Crops Food 8 (4), 195–208. https://doi.org/10.1080/21645698.2017.1413522.

Ramankutty, N., Mehrabi, Z., Waha, K., Jarvis, L., Kremen, C., Herrero, M., Rieseberg, L.H., 2018. Trends in global agricultural land use: implications for environmental health and food security. Annu. Rev. Plant Biol. 69, 789–815. https://www.annualreviews.org/doi/abs/10.1146/annurev-arplant-042817-040256.

Saiz-Rubio, V., Rovira-Más, F., 2020. From smart farming towards agriculture 5.0: a review on crop data management. Agronomy 10 (2), 207. https://doi.org/10.3390/agronomy10020207.

Samuel, A.L., 2000. Some studies in machine learning using the game of checkers. IBM J. Res. Dev. 44 (1.2), 206–226. https://doi.org/10.1147/rd.441.0206.

Scanlon, B.R., Jolly, I., Sophocleous, M., Zhang, L., 2007. Global impacts of conversions from natural to agricultural ecosystems on water resources: quantity versus quality. Water Resour. Res. 43 (3). https://doi.org/10.1029/2006WR005486.

Singh, A.K., Upadhyaya, A., Kumari, S., Sundaram, P.K., Jeet, P., 2020. Role of agriculture in making India $5 trillion economy under Corona pandemic circumstance: role of agriculture in Indian economy. J. AgriSearch 7 (2), 54–58.

Talaviya, T., Shah, D., Patel, N., Yagnik, H., Shah, M., 2020. Implementation of artificial intelligence in agriculture for optimisation of irrigation and application of pesticides and herbicides. Artif. Intell. Agric. 4, 58–73. https://doi.org/10.1016/j.aiia.2020.04.002.

Tian, X., Engel, B.A., Qian, H., Hua, E., Sun, S., Wang, Y., 2021. Will reaching the maximum achievable yield potential meet future global food demand? J. Clean. Prod. 294, 126285. https://doi.org/10.1016/j.jclepro.2021.126285.

Veroustraete, F., 2015. The rise of the drones in agriculture. EC Agric. 2 (2), 325–327. https://www.researchgate.net/profile/Frank-Veroustraete/publication/282093589_The_Rise_of_the_Drones_in_Agriculture/links/5d49be4092851cd046a6aaae/The-Rise-of-the-Drones-in-Agriculture.pdf.

Vroegindeweij, B.A., Van Wijk, S.W., Van Henten, E., 2014. Autonomous Unmanned Aerial Vehicles for Agricultural Applications. https://edepot.wur.nl/329497.

Wang, J., Vanga, S.K., Saxena, R., Orsat, V., Raghavan, V., 2018. Effect of climate change on the yield of cereal crops: a review. Climate 6 (2), 41. https://doi.org/10.3390/cli6020041.

World Bank, 2021. Water in Agriculture. https://www.worldbank.org/en/topic/water-in-agriculture#1. (Accessed 31 October 2021).

WSSA—Weed Science Society of America, 2016. WSSA Calculates Billions in Potential Economic Losses from Uncontrolled Weeds. https://wssa.net/2016/05/wssa-calculates-billions-in-potential-economic-losses-from-uncontrolled-weeds/. (Accessed 31 October 2021).

Yahya, N., 2018. Agricultural 4.0: its implementation toward future sustainability. In: Green Urea. Green Energy and Technology, Springer, Singapore, https://doi.org/10.1007/978-981-10-7578-0_5.

Zeng, Y., Zhang, R., Lim, T.J., 2016. Wireless communications with unmanned aerial vehicles: opportunities and challenges. IEEE Commun. Mag. 54 (5), 36–42. https://doi.org/10.1109/MCOM.2016.7470933.

Further reading

Bendig, J., Bolten, A., Bareth, G., 2012. Introducing a low-cost mini-UAV for thermal-and multispectral-imaging. Int. Arch. Photogramm. Remote. Sens. Spat. Inf. Sci. 39, 345–349.

Burger, W., Burge, M.J., 2013. Principles of Digital Image Processing. Undergraduate Topics in Computer Science, Springer, London, https://doi.org/10.1007/978-1-84882-919-0_1.

Plant, R.E., Pettygrove, G.S., Reinert, W.R., 2000. Precision agriculture can increase profits and limit environmental impacts. Calif. Agric. 54 (4), 66–71. https://pdfs.semanticscholar.org/7691/1bb3182e178845178188e9cd681a03abbf58.pdf.

CHAPTER

3

Bidirectional long short-term memory-based empirical wavelet transform: A new hybrid artificial intelligence model for robust prediction of soil moisture content

Salim Heddam[a], *Sungwon Kim*[b], *Ahmed Elbeltagi*[c], *and Ozgur Kisi*[d,e]

[a]Faculty of Science, Agronomy Department, Hydraulics Division, Laboratory of Research in Biodiversity Interaction Ecosystem and Biotechnology, University 20 Août 1955, Skikda, Algeria [b]Department of Railroad Construction and Safety Engineering, Dongyang University, Yeongju, Republic of Korea [c]Agricultural Engineering Department, Faculty of Agriculture, Mansoura University, Mansoura, Egypt [d]Civil Engineering Department, Ilia State University, Tbilisi, Georgia [e]Department of Civil Engineering, Technical University of Lübeck, Lübeck, Germany

OUTLINE

1 Introduction 38
2 Materials and methods 39
2.1 Study site and data used 39
2.2 Performance assessment of the models 42
2.3 Methodology 42
3 Results and discussion 45
4 Conclusions 53
5 Recommendations 53
References 54

Current Directions in Water Scarcity Research, Volume 7
https://doi.org/10.1016/B978-0-323-91910-4.00003-0

1 Introduction

Soil moisture plays an important role and is involved in the formation of the continuous process of infiltration, transpiration, water evaporation, and agriculture water management, and understanding its variation over space and time is crucial for several agricultural, irrigation, and hydrological applications (Babaeian et al., 2021; Judge et al., 2021; Gururaj et al., 2021; Zhu et al., 2021). Monitoring and quantifying soil moisture content (SM) is of significant importance and can help in better management and planning of irrigation, especially in the arid and semiarid regions characterized by natural limitations of the water supply (Babaeian et al., 2021). Soil moisture can be directly and accurately quantified with high degree of precision using in situ measurements; however, obtaining a large and continuous SM data is hard to be obtained with regards to the growing cost of instruments, equipment and infrastructure needed under the safety conditions required (Zhao et al., 2021). Over the past decades, machines learning models (ML) were successfully applied for soil moisture retrieval and the merits and limitations of each one have been deeply highlighted.

Chai et al. (2021) used the multilayer perceptron neural *network* (MLPNN) model for modeling SM using three input variables, i.e., the (H-) and V-polarized brightness temperature (TbH, TbV), and surface temperature (T_S), and they obtained a root mean square error (RMSE) of 3.67% and coefficient of determination (R^2) of 0.89. Ahmed et al. (2021a) compared between long-short term memory (LSTM), support vector regression (SVR), multivariate adaptive regression *splines* (*MARS*), and three hybrid Boruta-random forest (BRF) coupled with LSTM, SVR and MARS, namely, LSTM-BRF, SVR-BRF and MARS-BRF models in predicting SM. The BRF was used a robust tool for better selection of input variables and it was found that the hybrid LSTM-BRF and single LSTM were more accurate than the other models and exhibiting high accuracies of approximately $\approx$0.994, $\approx$0.995, $\approx$0.995, $\approx$0.154 and $\approx$0.120 and $\approx$0.877, $\approx$0.933, $\approx$0.739, $\approx$2.047 and $\approx$1.587 in terms of R, Willmott index of agreement (d), Nash-Sutcliffe efficiency (NSE), RMSE, and mean absolute error (MAE), respectively. Ahmed et al. (2021b) have proposed a new modeling strategy for accurately predicting surface SM using high number of predictors collected by the satellite-derived global land data assimilation program. They proposed the use of signal decomposition algorithm, i.e., the complete ensemble empirical mode decomposition with adaptive noise (CEEMDAN) for improving the performances of the gated recurrent unit (GRU) and convolutional neural network (CNN) models. In total, four models were compared, i.e., CEEMDAN-CNN-GRU, CNN-GRU, CEEMDAN-GRU, and GRU, and it was found that: (i) the performances of the models significantly decreased from one day ahead forecasting horizon to 30 days ahead forecasting horizon, and (ii) the best accuracies were obtained by the CEEMDAN-CNN-GRU with R NSE RMSE and MAE exhibiting the values of 0.996, 0.995, 0.021 and 0.013, respectively. Bartels et al. (2021) used the MLPNN model for predicting gravimetric SM using four categories of input variables namely, topography, soil properties, climatic and rainfall variables, showing good predictive accuracies with NSE and RMSE of approximately $\approx$0.93 and $\approx$0.017, respectively. Chen et al. (2021) compared between CNN and partial least squares regression (PLS) for predicting SM content and showing that the CNN ($R^2 \approx 0.989$ and RMSE$\approx$0.016) was slightly more accurate compared to the PLS ($R^2 \approx 0.983$ and RMSE$\approx$0.020). In a recently published paper, Chaudhary et al. (2021) compared between several machine leaning models in predicting SM from dual polarimetric

Sentinel-1 radar backscatter data. The proposed models were: SVR with linear, polynomial, radial and sigmoid kernel, i.e., SVR-Li, SVR-Po, SVR-RB, SVR-SG), random forest regression (RFR), MLPNN, radial basis function (RBFNN), Wang and Mendel's (WM), subtractive clustering (SC), adaptive neuro fuzzy inference system (ANFIS), hybrid fuzzy interference system (HyFIS), and dynamic evolving neural fuzzy inference system (DENFIS). Obtained results showed that subtractive clustering (SC) was more accurate and exhibited the best accuracies.

Several other machines learning models for soil moisture retrieval can be found in the literature for example, MLPNN, regression tree (RT), and Gaussian process regression (GPR) (Senanayake et al., 2021), MLPNN and RFR (Zhang et al., 2021a), extreme gradient boosting (XGBoost) (Karthikeyan and Mishra, 2021), CNN, LSTM, and convolutional LSTM (ConvLSTM) models (Li et al., 2021), and RF regression (Carranza et al., 2021). Consequently, we propose in the present study new modeling strategy for predicting soil moisture measured at different depths using the empirical wavelet transform (EWT) preprocessing signal decomposition for improving the performances of three machine learning models namely, the SVR, the GPR and the bidirectional LSTM (BiLSTM). The novelty of our study is that, the prediction of SM is based only on soil temperature as a sole predictor. Thus, we compared between the SVR, GPR and BiLSTM and the hybrid SVR_EWT, GPR_EWT, and BiLSTM_EWT for better prediction of soil moisture measured at 20, 30, 50, 75, and 100 cm depths.

2 Materials and methods

2.1 Study site and data used

This study was performed using data for the USGS web site (https://nwis.waterdata.usgs.gov/nd/nwis). Fig. 1 presents map of the selected station. The selected station was: USGS 475646097372201 152-054-31BBB, Grand Forks County, North Dakota, USA (Latitude 47°

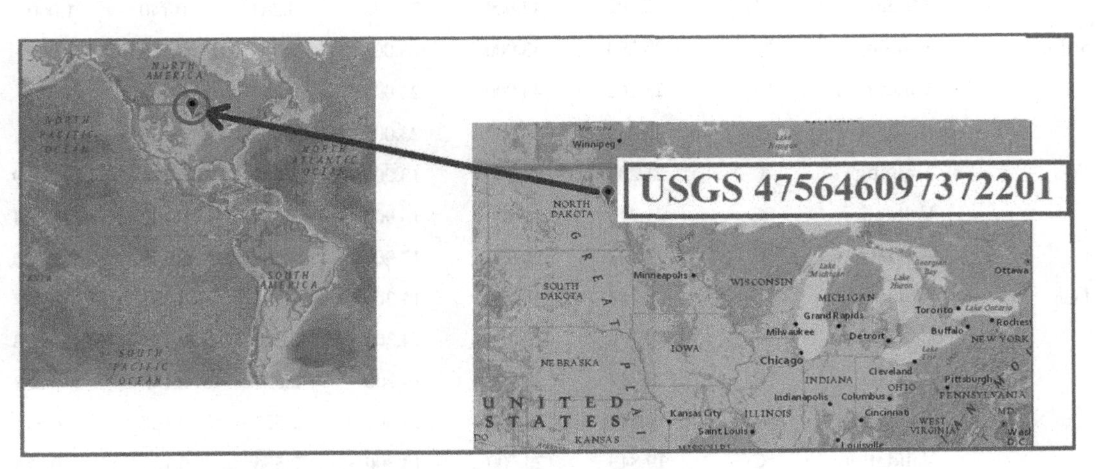

FIG. 1 Map showing the location of the USGS station North Dakota, USA.

56′48.24″, Longitude 97°37′21.69″). For modeling soil moisture (moisture content, soil, volumetric, percent of total volume *SM*: %), we use only soil temperature (T_S: °C) as a single predictor. We used data measured at *fifteen minute intervals of time*. Data was recorded during the period ranging from 01 Juan 2021 at 00:00 to 21 September 2021 at 15:00 with a total of 11,545 patterns divided into training (70%: 8082) and validation (30%: 3463). Soil moisture and soil temperature were measured at several depths namely, 20 cm, 30 cm, 50 cm, 75 cm, and 100 cm. The statistical parameters were calculated and reported in Table 1. Fig. 2 shows the Graph changes of T_S (°C) and SM (%) at different depths.

TABLE 1 Summary statistics of soil moisture content and soil temperature.

Variables	Subset	Unit	X_{mean}	X_{max}	X_{min}	S_x	C_v	R
USGS 475646097372201 152-054-31BBB, Grand Forks County, North Dakota, USA								
SM_{20}	Training	%	15.173	40.000	9.800	4.414	0.291	1.000
	Validation	%	15.063	40.000	9.800	4.433	0.294	1.000
	All data	%	15.140	40.000	9.800	4.425	0.292	1.000
SM_{30}	Training	%	22.007	40.000	15.000	6.011	0.273	1.000
	Validation	%	22.009	40.000	15.000	5.996	0.272	1.000
	All data	%	22.008	40.000	15.000	6.008	0.273	1.000
SM_{50}	Training	%	17.716	34.000	14.000	5.645	0.319	1.000
	Validation	%	17.737	34.000	14.000	5.660	0.319	1.000
	All data	°%	17.668	34.000	14.000	5.604	0.317	1.000
SM_{75}	Training	%	23.425	42.000	21.000	3.314	0.141	1.000
	Validation	%	23.349	41.000	21.000	3.198	0.137	1.000
	All data	°%	23.402	42.000	21.000	3.284	0.140	1.000
SM_{100}	Training	%	25.253	43.000	23.000	3.980	0.158	1.000
	Validation	%	25.262	43.000	23.000	4.048	0.160	1.000
	All data	°%	25.256	43.000	23.000	4.003	0.159	1.000
T_{S20}	Training	°C	21.234	26.800	13.900	3.142	0.148	−0.329
	Validation	°C	21.245	26.800	13.900	3.174	0.149	−0.334
	All data	°C	21.237	26.800	13.900	3.152	0.148	−0.330
T_{S30}	Training	°C	20.738	25.400	14.300	2.830	0.136	−0.377
	Validation	°C	20.770	25.400	14.300	2.850	0.137	−0.398
	All data	°C	20.747	25.400	14.300	2.837	0.137	−0.383
T_{S50}	Training	°C	19.820	23.500	13.300	2.525	0.127	−0.511
	Validation	°C	19.843	23.500	13.300	2.556	0.129	−0.526
	All data	°C	19.827	23.500	13.300	2.535	0.128	−0.515

TABLE 1 Summary statistics of soil moisture content and soil temperature.—Cont'd

Variables	Subset	Unit	X_{mean}	X_{max}	X_{min}	S_x	C_v	R
T_{S75}	Training	°C	18.516	21.500	11.600	2.311	0.125	−0.104
	Validation	°C	18.540	21.500	11.600	2.326	0.125	−0.109
	All data	°C	18.523	21.500	11.600	2.316	0.125	−0.106
T_{S100}	Training	°C	17.289	20.000	10.400	2.360	0.136	0.103
	Validation	°C	17.248	20.000	10.300	2.438	0.141	0.113
	All data	°C	17.277	20.000	10.300	2.384	0.138	0.106

Abbreviations: X_{mean}, *mean;* X_{max}, *maximum;* X_{min}, *minimum;* S_x, *standard deviation;* C_v, *coefficient of variation;* R, *coefficient of correlation,* SM, *moisture content, soil, volumetric, percent of total volume;* T_S, *soil temperature;* R, *correlation coefficient between soil temperature and soil moisture at the same depth.*

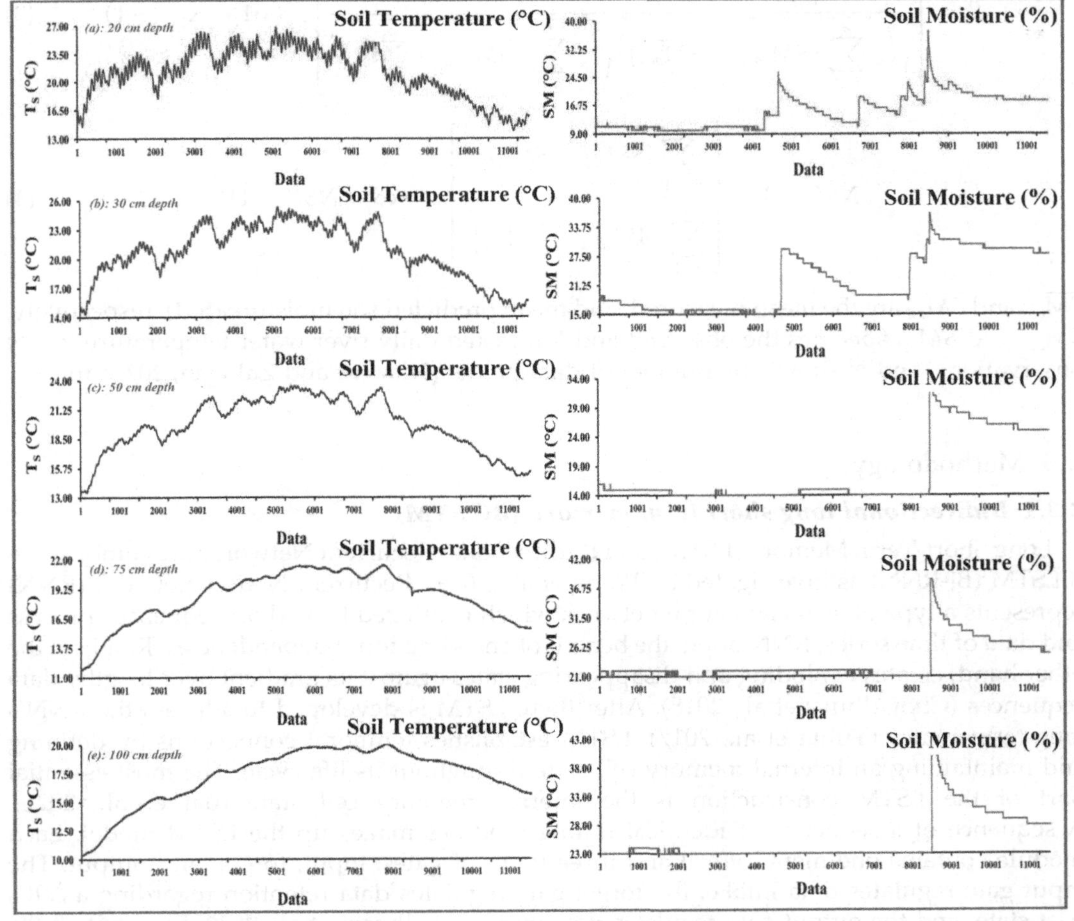

FIG. 2 Graph showing changes in soil water temperature (°C) and soil moisture (%) measured at the USGS 475646097372201 Grand Forks County, North Dakota, USA. The values are plotted at *fifteen minute intervals of time.*

2.2 Performance assessment of the models

All models were evaluated using root mean square error (RMSE), mean absolute error (MAE), correlation coefficient (R), and Nash-Sutcliffe efficiency (NSE). Expressions are given as:

$$RMSE = \sqrt{\frac{1}{N}\sum_{i=1}^{N}\left[(SM_{obs,i}) - (SM_{est,i})_i\right]^2}, \quad (0 \leq RMSE < +\infty) \tag{1}$$

$$MAE = \frac{1}{N}\sum_{i=1}^{N}|SM_{obs,i} - SM_{est,i}|, \ (0 \leq MAE < +\infty) \tag{2}$$

$$R = \left[\frac{\frac{1}{N}\sum_{i=1}^{N}(SM_{obs,i} - \overline{SM_{obs}})\ (SM_{est,i} - \overline{SM_{est}})}{\sqrt{\frac{1}{N}\sum_{i=1}^{n}(SM_{obs,i} - \overline{SM_{obs}})^2}\sqrt{\frac{1}{N}\sum_{i=1}^{n}(SM_{est,i} - \overline{SM_{est}})^2}}\right], \ (-1 < R \leq +1) \tag{3}$$

$$NSE = 1 - \left[\frac{\sum_{i=1}^{N}[SM_{obs} - SM_{est}]^2}{\sum_{i=1}^{N}\left[SM_{obs,i} - \overline{SM_{obs}}\right]^2}\right], (-\infty < NSE \leq 1) \tag{4}$$

$\overline{SM}_{obs}$ and $\overline{SM}_{est}$ are the mean measured, and mean predicted soil moisture (SM), respectively, SM_{obs} and SM_{est} specifies the observed and forecasted daily river water temperature for *ith* observations, and N shows the number of data points (Niazkar and Zakwan, 2021a,b).

2.3 Methodology

2.3.1 *Bidirectional long short term memory (BiLSTM)*

Long Short-Term Memory (LSTM) and Bi-directional Recurrent Networks are combined in BiLSTM (Bi-RNN) as investigated by Wang et al. (2019). Recurrent Neural Network (RNN) represents a type of artificial neural network which is utilized to analyze sequences process and data of time series. RNN offers the benefit of encoding input dependencies. RNN, on the other hand, creates exploding and disappearing states against its gradient over lengthy data sequences (Chowdhury et al., 2018). After then, LSTM is developed to address the RNN's long-term issues (Yulita et al., 2017). LSTM establishes temporal connections by defining and maintaining an internal memory cell state throughout its life cycle. The most essential part of the LSTM construction is the internal memory cell state (Bai et al., 2021). A sequence of a sequence of identical timing modules makes up the LSTM model. Each module contains one memory cell and three types of gates: input, forget, and output. The input gate regulates data intake, the forget gate regulates data retention regarding a cell's past state, and the output gate regulates data output (Aldhyani et al., 2020; Fan et al., 2020;

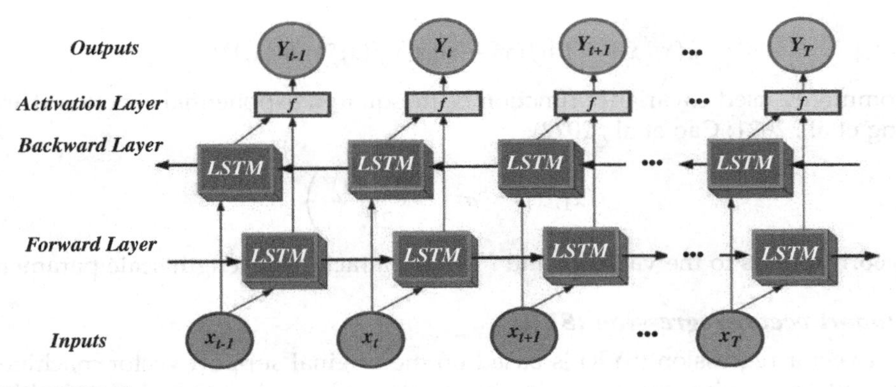

FIG. 3 The BiLSTM layout.

Huang et al., 2020; Ouma et al., 2021). As various gate constructions exist, LSTM may selectively influence the state of each instant by providing information. However, because both the LSTM and the RNN can only receive input from the previous context, the Bidirectional Recurrent Neural Network (Bi-RNN) is utilized to make further progress. Bi-RNN is capable of handling data from both the front and back ends (Zhang et al., 2019). When Bi-RNN and LSTM are combined, BiLSTM is formed, as illustrated in Fig. 3. This method combines the advantages of the LSTM in the storage form in cell memory and Bi-RNN with access information from context. It allows the BiLSTM to benefit from the LSTM's feedback for the following layer (Yulita et al., 2017). BiLSTM, on the other hand, can manage data that is dependent on a long term. BiLSTM can consider both past and future data information. The principle is: the same output connects two LSTM networks with opposite timings. The forward LSTM can obtain the past data information of the input sequence and the backward LSTM can obtain the future data information of the input sequence (Minh-Tuan and Kim, 2019).

2.3.2 Gaussian process regression (GPR)

The main idea of the Gaussian process regression (GPR) (Rasmussen and Nickisch, 2010) is to construct a nonparametric regression model based on the Bayesian inference based on supervised learning (Cao et al., 2022), and it is highly recommended for building a model with only fewer available training dataset (Yuan et al., 2009). For any approximation function used for mapping a set of input variables x to an output variable y, the Gaussian process $f(x)$ is parameterized by its mean function $m(x)$ and covariance function $k(x_i, x_j)$, that is (Fang et al., 2020; Jiang et al., 2021):

$$y = f(x) \tag{5}$$

$$f(x) \sim G\tilde{}P(m(x), k(x_i, x_j)) = f(x) \sim G\tilde{}P(m(x), K) \tag{6}$$

where $m(x)$ represents the mean function, and K the covariance function based on kernel functions k, which can be expressed as follow (Ouyang and Zou, 2021):

$$m(x) = E[f(x)] \tag{7}$$

$$k(x_i, x_j) = E\left[(f(x) - m(x))(f(x_j) - m(x_j))\right] \tag{8}$$

One commonly used covariance function is the squared exponential function (Fang et al., 2020; Jiang et al., 2021; Cao et al., 2022):

$$k(x_1, x_1) = \sigma_f^2\left(\frac{-(x_1 - x_2)^2}{2l^2}\right) \tag{9}$$

where σ_f corresponds to the variance and l is the characteristic length scale parameter.

2.3.3 *Support vector regression (SVR)*

Support vector regression (SVR) is based on the original support vector machine (SVM) and proposed for solving regression problems (Vapnik et al., 1997). The transition from the SVM to the SVR was achieved by adding a ε-insensitive error function (Abbasi et al., 2021). The most significant particularity of the SVR is that it uses the structural risk minimization instead of empirical risk minimization (Okkan et al., 2021). Using a set of training data with N patterns which could take the following form:

$$Tr = \{(x_1, y_1), (x_2, y_2), \cdots, (x_l, y_l)\}, x_i \in R^n \tag{10}$$

The SVR regression function is used for linking the input to the output as follow:

$$f(x) = w^T\varphi(x) + b \tag{11}$$

where w and b are corresponds to the models parameters, i.e., weight and bias, to be optimized during the training process (Hafeez et al., 2021). Therefore, the SVR should solve the following optimization problem:

$$\text{Minimize } \frac{1}{2}\|w\|^2 + C\sum_{i=0}^{t}(\xi_i + \xi_i^*) \tag{12}$$

where $C > 0$ is a regularization factor, $\|.\|$ is a 2-norm, $\xi, \xi^* \in R$ are the slack variables (Abbasi et al., 2021; Chen et al., 2021; Alrashidi et al., 2021).

2.3.4 *Empirical wavelet transform (EWT)*

The empirical wavelet transform (EWT) is a preprocessing signal decomposition algorithm proposed by Gilles (2013), and it is used for decomposing a nonlinear and nonstationary signal in several multiresolution analysis (MRA) components. Great attention has been attributed to the application of EWT During the last few years and several applications can be found in the literature, i.e., estimation of vibration powers of a high-speed train (Wang et al., 2022), forecasting air pollutant concentration (Kim et al., 2021a), COVID-19 disease identification (Gaur et al., 2022), bearing fault diagnosis (Zhang et al., 2021b), energy consumption forecasting (Peng et al., 2022), and short-term wind speed forecasting (Yang et al., 2021). From a mathematical point of view, the EWT is a combination of the empirical mode decomposition (EMD) and wavelet analysis (Liang et al., 2021). Using the EWT, the normalized frequency domain $\omega \in [0,\pi]$ should be divided into suite of segments with $N+1$

boundaries, and the empirical scaling function ($\varphi(\omega)$) and empirical wavelets ($\Psi(\omega)$) can be defined as follows, for which the $\beta(x)$ is a supplementary function and (r) is a ratio for the transition phase of the boundary (Liang et al., 2021; Kim et al., 2021b; Gilles, 2013):

$$\widehat{\Psi}_{(n)}(\omega) = \begin{cases} 1, (1+r)\omega_n \le |\omega| < (1-r)\omega_{n+1} \\ \cos\left[\frac{\pi}{2}\beta\left(\frac{1}{2r\omega_{n+1}}(|\omega|) - (1-r)\omega_{n+1}\right)\right], (1-r)\omega_{n+1} \le |\omega| < (1+r)\omega_{n+1} \\ \sin\left[\frac{\pi}{2}\beta\left(\frac{1}{2r\omega_n}(|\omega|) - (1-r)\omega_n\right)\right], (1-r)\omega_n \le |\omega| < (1+r)\omega_n \\ 0, \text{others} \end{cases} \tag{13}$$

$$\widehat{\varphi}_{(n)}(\omega) = \begin{cases} 1, |\omega| < (1-r)\omega_n \\ \cos\left[\frac{\pi}{2}\beta\left(\frac{1}{2r\omega_n}(|\omega|) - (1-r)\omega_n\right)\right], [(1-r)\omega_n] \le |\omega| < [(1+r)\omega_n] \\ 1, \text{otherwise} \end{cases} \tag{14}$$

$$\beta(x) = x^4(35 - 84x + 70x^2 - 20x^3) \tag{15}$$

$$r < \min_n \left(\frac{\omega_{n+1} - \omega_n}{\omega_{n+1} + \omega_n}\right) \tag{16}$$

The EWT does not use predefined basis function, and the transition phase is centered around ω_n which has a width of $2\lambda\omega_n$ where $0 < \lambda < 1$ (Gaur et al., 2022; Liang et al., 2021; Kim et al., 2021b; Gilles, 2013).

3 Results and discussion

We applied in the present chapter three machines learning models for modeling soil moisture (*SM*) using only soil temperature (T_S) namely, support vector regression (SVR), Gaussian Process Regression (GPR), and the bidirectional long short term memory (BiLSTM) deep learning model. Two scenarios are analyzed in the present work: (i) using only soil temperature (T_S), i.e., single models and (ii) using the empirical wavelet transform (EWT) as preprocessing signal decomposition for soil T_S, and the obtained multiresolution analysis (MRA) components were used as new input variables, i.e., hybrid models (Fig. 4). In the present study, the T_s was decomposed into fifteen MRA, i.e., MRA1, MRA2,..., MRA15. For the first scenario, the models were designated as SVR, GPR, and BiLSTM. Similarly, for the second scenario the models were designated as BiLSTM_EWT, GPR_EWT, and SVR_EWT. For models evaluation, we calculated four performances metrics namely, NSE, R, RMSE, and MAE. In addition, the soil moisture was modeled at four different depths namely, 20 cm depth (SM_{20}), 30 cm depth (SM_{30}), 50 cm depth (SM_{50}), 75 cm depth (SM_{20}), and 100 cm depth (SM_{100}). The flowchart of the proposed modeling approach is shown in Fig. 5.

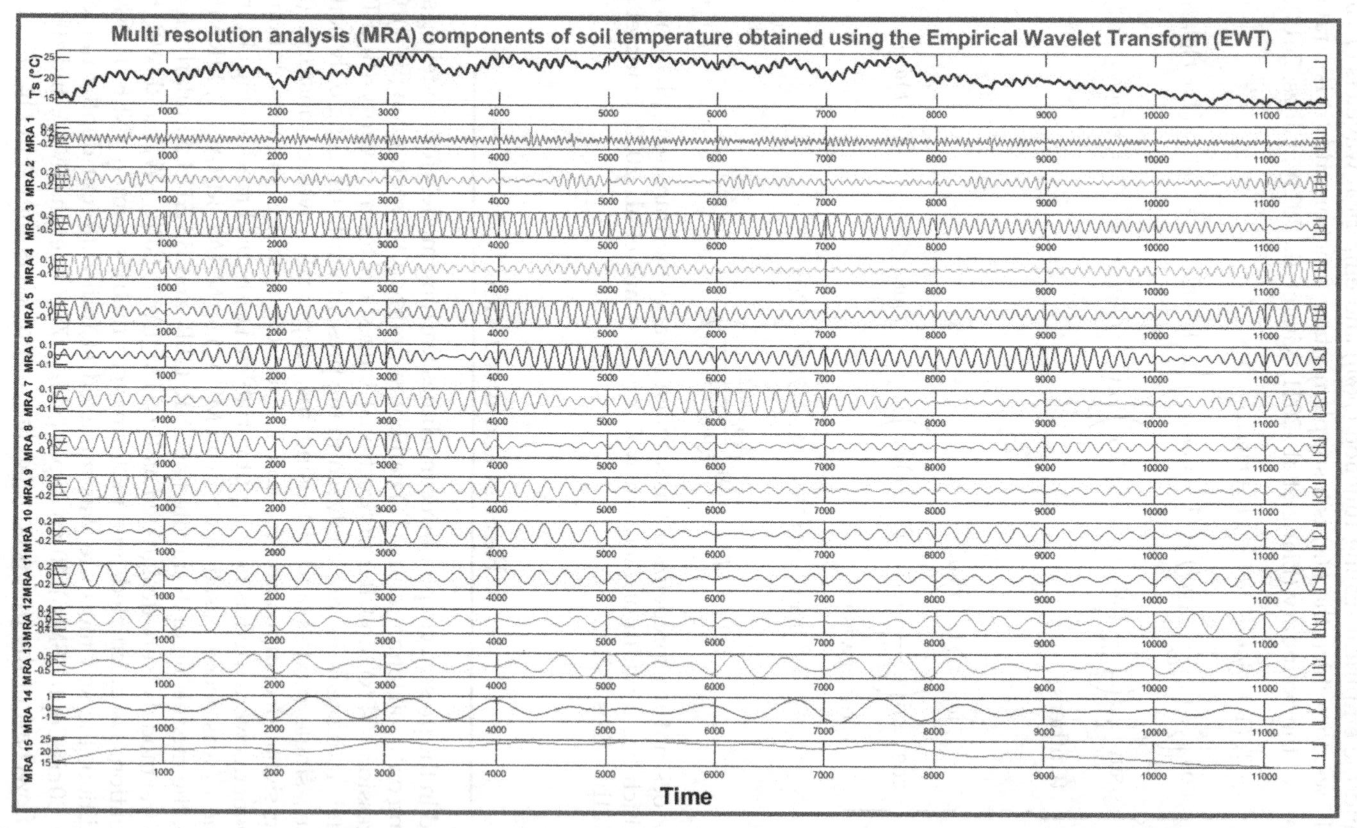

FIG. 4 Multi resolution analysis components (MRA) of soil temperature (T_s) obtained using the empirical wavelet transform (EWT).

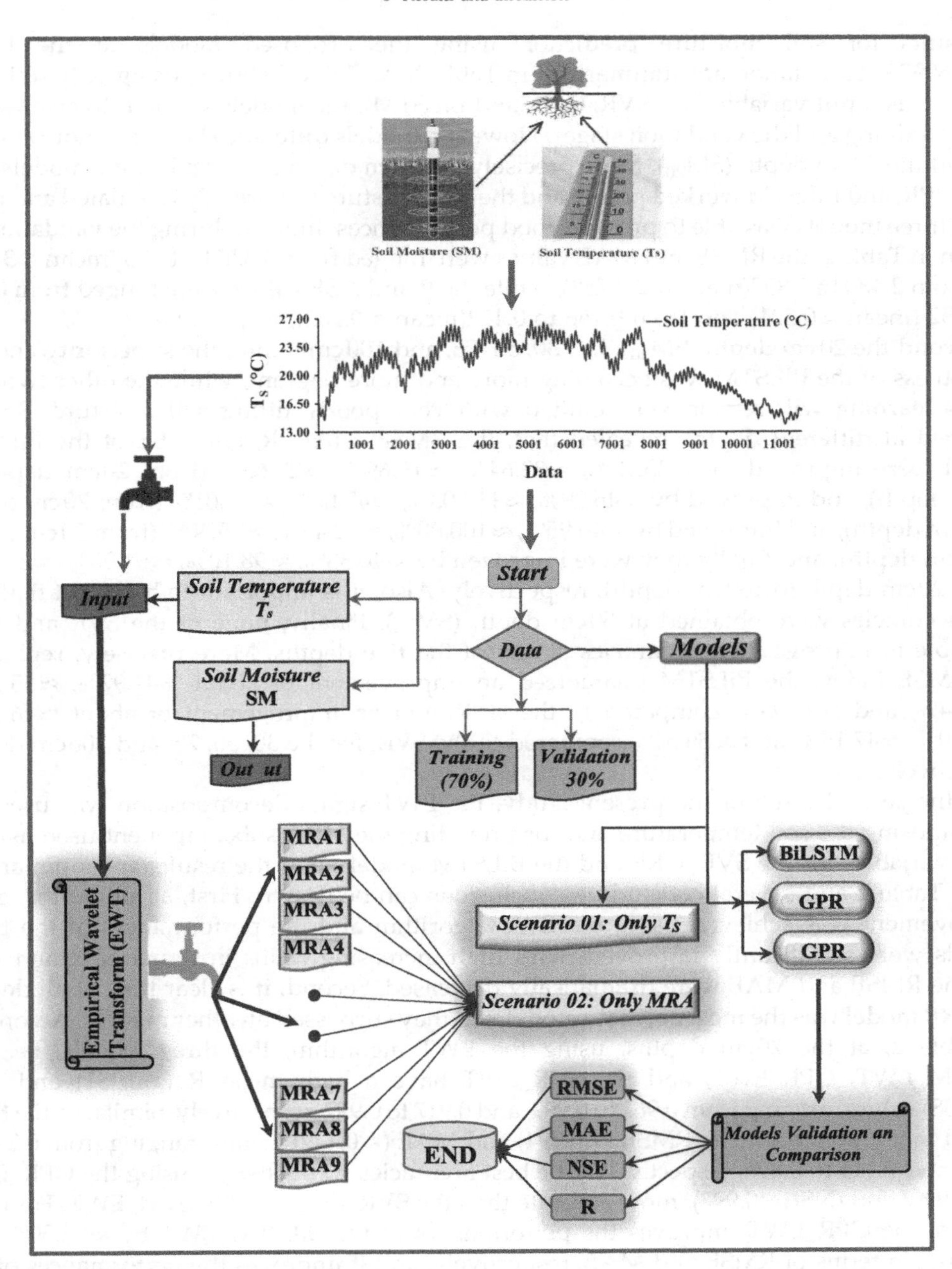

FIG. 5 Flowchart of the proposed soil moisture modeling strategy.

Results for soil moisture prediction using the proposed models at the USGS 475646097372201 station are summarized in Table 2. As Table 2 shows, using only soil temperature as input variable (i.e., SVR, GPR, and BiLSTM), the models yielded diverse results for the training and the validation stages. However, models difference becomes more evident beyond the 20 cm depth (SM_{20}). More precisely, at 20 cm depth (SM_{20}) all single models, i.e., SVR, GPR, and BiLSTM worked poorly and the soil moisture was weakly calculated and none of the three models was able to provide good performances. Indeed, during the validation, as shown in Table 2, the RMSE and MAE values were ranged from 3.370 to 4.073 (mean $\approx$3.795) and from 2.480 to 2.838(mean $\approx$2.2.788), while the R and NSE values were ranged from 0.453 to 0.652 (mean $\approx$0.521) and from 0.156 to 0.422(mean $\approx$0.262), respectively.

Beyond the 20 cm depth (SM_{20}), i.e., 30, 50, 75, and 100 cm depth, the superiority and the robustness of the BiLSTM was becoming more and more flagrant, while the other two machines learning will remain very limited with very poorly fitting soil moisture. Results obtained at different depths revealed that, the RMSE, MAE, R, and NSE of the BiLSTM model were improved by $\approx$35.12%, $\approx$83.64%, $\approx$15.69%, $\approx$22.66% (from 20 cm depth to 30 cm depth), and improved by $\approx$45.39%, $\approx$113.03%, $\approx$47.06%, $\approx$72.01% (from 20 cm depth to 50 cm depth), and improved by $\approx$40.95%, $\approx$100.00%, $\approx$62.49%, $\approx$75.40% (from 20 cm depth to 75 cm depth), and finally they were improved by $\approx$40.33%, $\approx$98.10%, $\approx$51.30%, $\approx$64.67% (from 20 cm depth to 100 cm depth), respectively. Also, it is important to highlight that, the best accuracies were obtained at 50 cm depth (SM_{50}). Finally, none of the SVR and GRP was able to improve their accuracies at any of the five depths. More precisely, regarding the RMSE index, the BiLSTM guaranteed an improvement of about $\approx$41.92%, $\approx$55.66%, $\approx$37.64%, and $\approx$47.93%, compared to the GPR, and an improvement of about $\approx$46.51%, $\approx$65.70%, $\approx$47.15%, and $\approx$59.51% compared to the SVR, for the 30, 50, 75, and 100 cm depth, respectively.

In the second stage of the present study, the EWT signal decomposition was used for decomposing the soil temperature and for providing the MRA subcomponent used as new input variables for the SVR, GRP and the BiLSTM models, and the results are summarized at the Table 2 and several concluding conclusions can be drawn. First, at all depths, great improvement was achieved using the EWT algorithm and the performances of the three models were significantly improved, with high increasing ratios in terms of R and NSE and the RMSE and MAE were dramatically decreased. Second, it is clear that, at all depths the GRP model was the most accurate model and they surpassed all other models. According to Table 2, at the 20 cm depths, using the EWT algorithm, the three models, i.e., the BiLSTM_EWT, GPR_EWT, and the SVR_EWT have a high mean R ($\approx$0.981) and NSE ($\approx$0.958) values, ranging from 0.961 to 0.999, and 0.917 to 0.997, respectively. Similarly, the three hybrid models have a mean RMSE ($\approx$0.794) and MAE ($\approx$0.449) values, ranging from 0.224 to 1.280, and 0.145 to 0.890, respectively. The best accuracies was achieved using the GPR_EWT (R$\approx$0.999) and (NSE$\approx$0.997), more accurate than the SVR_EWT and BiLSTM_EWT. For comparison, the GPR_EWT improves the performances of the BiLSTM_EWT by $\approx$82.50% and $\approx$83.70% in terms of RMSE and MAE, respectively, and it improves the performances of the SVR_EWT by $\approx$74.48% and $\approx$53.37% in terms of RMSE and MAE, respectively. Overall, comparison between the models with and without the EWT show that the means R, NSE, RMSE, and MAE of the single models (i.e., BiLSTM, SVR, and GPR) were improved by $\approx$88.10%, $\approx$265.31%, $\approx$79.07%, and $\approx$83.70% using the hybrid models (i.e., BiLSTM_EWT,

TABLE 2 Performances of different models at different depths.

Models	Training				Validation			
	R	NSE	RMSE	MAE	*R*	NSE	RMSE	MAE
Results at 20 cm depth (SM_{20})								
BiLSTM	0.612	0.375	3.489	2.577	0.652	0.422	3.370	2.480
BiLSTM_EWT	0.994	0.988	0.481	0.317	0.961	0.917	1.280	0.890
GPR	0.493	0.243	3.840	2.976	0.459	0.209	3.942	3.046
GPR_EWT	0.999	0.999	0.092	0.035	0.999	0.997	0.224	0.145
SVR	0.486	0.198	3.952	2.741	0.453	0.156	4.073	2.838
SVR_EWT	0.998	0.996	0.288	0.066	0.982	0.961	0.878	0.311
Results at 30 cm depth (SM_{30})								
BiLSTM	0.873	0.762	2.931	1.951	0.881	0.775	2.841	1.918
BiLSTM_EWT	0.997	0.994	0.473	0.354	0.974	0.944	1.419	1.050
GPR	0.564	0.318	4.965	4.048	0.578	0.334	4.892	3.984
GPR_EWT	0.999	0.999	0.073	0.026	0.998	0.995	0.430	0.161
SVR	0.495	0.176	5.456	3.875	0.520	0.215	5.312	3.749
SVR_EWT	0.999	0.999	0.220	0.065	0.990	0.978	0.879	0.403
Results at 50 cm depth (SM_{50})								
BiLSTM	0.959	0.920	1.602	0.651	0.948	0.899	1.784	0.694
BiLSTM_EWT	0.997	0.995	0.417	0.290	0.981	0.961	1.108	0.589
GPR	0.705	0.497	4.013	2.505	0.696	0.484	4.024	2.509
GPR_EWT	0.999	0.999	0.051	0.012	0.999	0.998	0.277	0.186
SVR	0.470	0.120	5.310	2.620	0.486	0.138	5.202	2.523
SVR_EWT	1.000	0.999	0.167	0.058	0.993	0.986	0.672	0.233
Results at 75 cm depth (SM_{75})								
BiLSTM	0.942	0.888	1.109	0.584	0.919	0.844	1.264	0.610
BiLSTM_EWT	0.989	0.977	0.500	0.363	0.937	0.823	1.345	0.813
GPR	0.787	0.620	2.043	1.237	0.774	0.598	2.027	1.238
GPR_EWT	0.999	0.999	0.107	0.033	0.997	0.993	0.272	0.169
SVR	0.668	0.445	2.469	1.595	0.664	0.440	2.392	1.562
SVR_EWT	0.996	0.993	0.285	0.055	0.990	0.975	0.505	0.212

Continued

TABLE 2 Performances of different models at different depths.—Cont'd

Models	Training				Validation			
	R	NSE	RMSE	MAE	*R*	NSE	RMSE	MAE
Results at 100 cm depth (SM_{100})								
BiLSTM	0.940	0.884	1.354	0.844	0.915	0.836	1.641	0.876
BiLSTM_EWT	0.975	0.950	0.894	0.536	0.947	0.891	1.337	0.688
GPR	0.640	0.409	3.060	1.884	0.628	0.394	3.152	1.936
GPR_EWT	0.999	0.999	0.043	0.008	0.987	0.974	0.653	0.353
SVR	0.268	0.0718	3.985	2.128	0.267	0.07	4.053	2.115
SVR_EWT	0.955	0.905	1.229	0.267	0.940	0.879	1.409	0.516

SVR_EWT, and GPR_EWT). The scatterplots of measured against calculated soil moisture (SM: %) at 20 cm depth (SM_{20}) are depicted in Fig. 6.

According to Table 2, at the 30 cm depths, using the EWT algorithm, the BiLSTM_EWT improve the BiLSTM by decreasing the RMSE and MAE by ≈50.05% and ≈45.25%, and by increasing the R and NSE by ≈10.55% and ≈21.80%, respectively. The GPR_EWT improves the GPR by decreasing the RMSE and MAE by ≈91.21% and ≈95.95%, and by increasing the R and NSE by ≈72.66% and ≈197.90%, respectively. Finally, The SVR_EWT improves the SVR by decreasing the RMSE and MAE by ≈83.45% and ≈89.25%, and by increasing the R and NSE by ≈90.38% and ≈354.88%, respectively. It is clear that the best accuracies were achieved using the GPR_EWT model. The scatterplots of measured against calculated soil moisture (SM: %) at 30 cm depth (SM_{30}) are depicted in Fig. 7.

At the 50 cm depths, the GPR_EWT model performed slightly better than the SVR_EWT tacking into account the *R* and NSE with only negligible difference; however, when the models were compared tacking into account the RMSE and MAE, the superiority of the GPR_EWT was more evident showing an improvement of about ≈58.78% and ≈20.17%, respectively. However, the superiority of the GPR_EWT and the SVR_EWT in comparison to the BiLSTM_EWT is significant and large with an improvement in terms of RMSE and MAE of about ≈75.00% and ≈68.42%, and ≈39.35% and ≈60.44%, respectively. Consequently, When EWT is used, GPR_EWT model performed better, with larger R and NSE values, and smaller RMSE and MAE values. The scatterplots of measured against calculated soil moisture (SM: %) at 50 cm depth (SM_{50}) are depicted in Fig. 8.

At the 75 cm depths, the GPR_EWT was always the best model exhibiting the high R (≈0.997) and NSE (≈0.993) and the lowest RMSE (≈0.272) and MAE (≈0.169). The GPR_EWT guaranteed an improvement of about ≈46.139%, and ≈20.28% in terms of RMSE and MAE compared to the SVR_EWT. In addition, the GPR_EWT improves the accuracies of the BiLSTM_EWT by approximately ≈79.77% and ≈79.21% in terms of RMSE and MAE. The scatterplots of measured against calculated soil moisture (SM: %) at 75 cm depth (SM_{75}) are depicted in Fig. 9.

Finally, at the 100 cm depths, it is clear that the hybrid models using the EWT were also more accurate than the single models, and the GPR_EWT was more accurate although, their

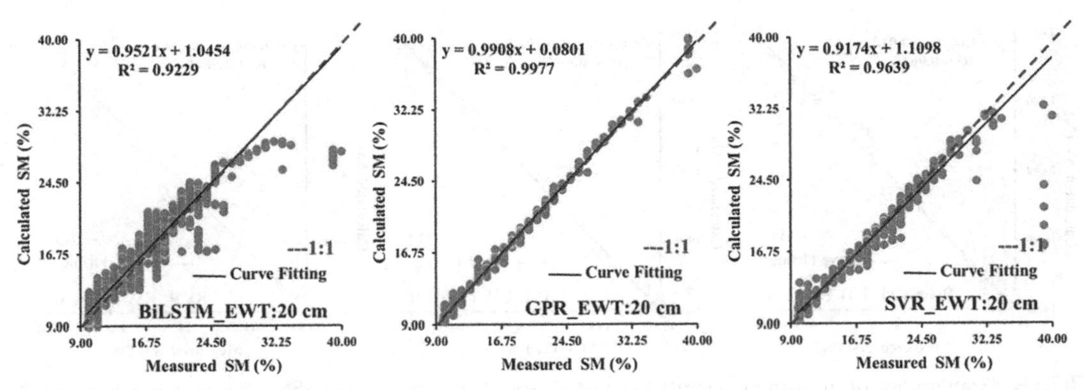

FIG. 6 Scatterplots of measured against calculated soil moisture content (SM: %) at the 20 cm depth: validation stage.

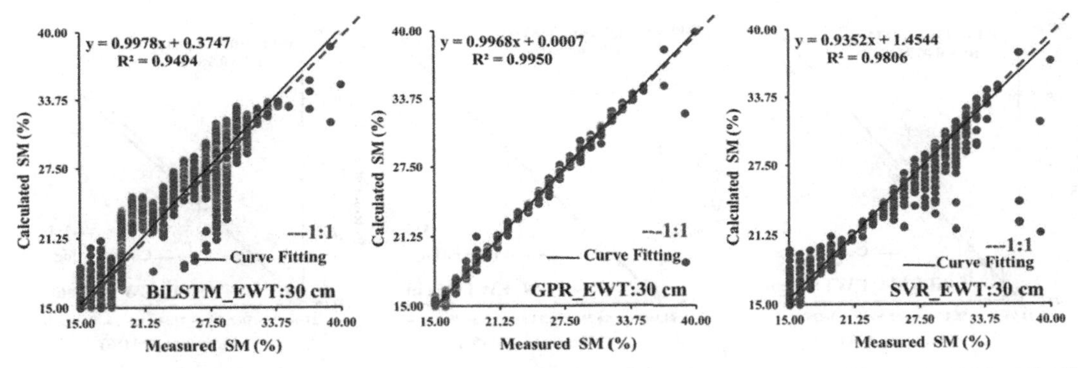

FIG. 7 Scatterplots of measured against calculated soil moisture content (SM: %) at the 30 cm depth: validation stage.

performances were slightly decreased compared to the 75 cm depths. Indeed, the most significant loss in models performances has been observed for the SVR_EWT model for which the R and NSE were dropped from ≈0.990 and ≈0.975 to ≈0.940 and ≈0.879, respectively, while the performances of the BiLSTM_EWT were slightly improved compared to the 75 cm depths for which the R and NSE rose to ≈0.947 and ≈0.891, respectively. In overall, the means R, NSE, RMSE, and MAE calculated using the single models were improved by the hybrid models with ratios of approximately ≈58.87%, ≈111.077%, 61.57%, and 68.39%, respectively. The scatterplots of measured against calculated soil moisture (SM: %) at 100 cm depth (SM_{100}) are depicted in Fig. 10.

At the end of the present discussion, it is important to mention the advantages and disadvantages of the proposed approach. According to the obtained results it clear that, the most significant advantage is the capability of the machines learning models in improving their performances using the EWT as a preprocessing algorithm, this is at least partly because the calculated MAR subcomponents were more easily captured by the models compared to the original complicated input signal. In another hand, the disadvantage of the proposed

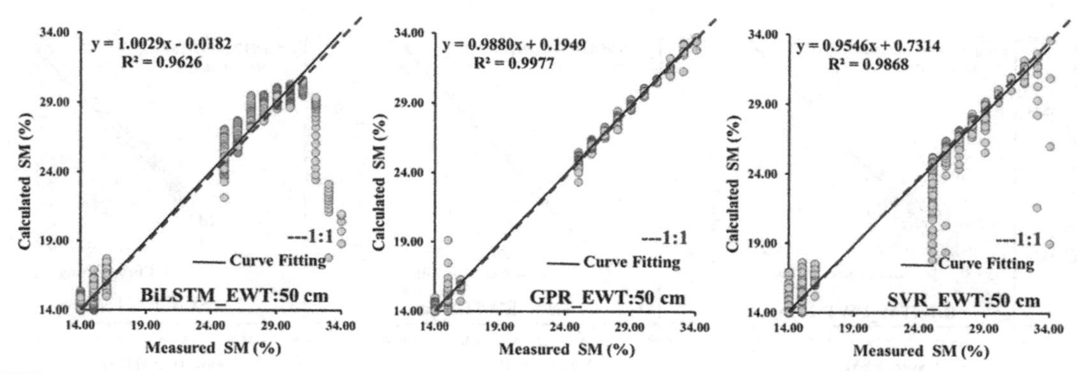

FIG. 8 Scatterplots of measured against calculated soil moisture content (SM: %) at the 50cm depth: validation stage.

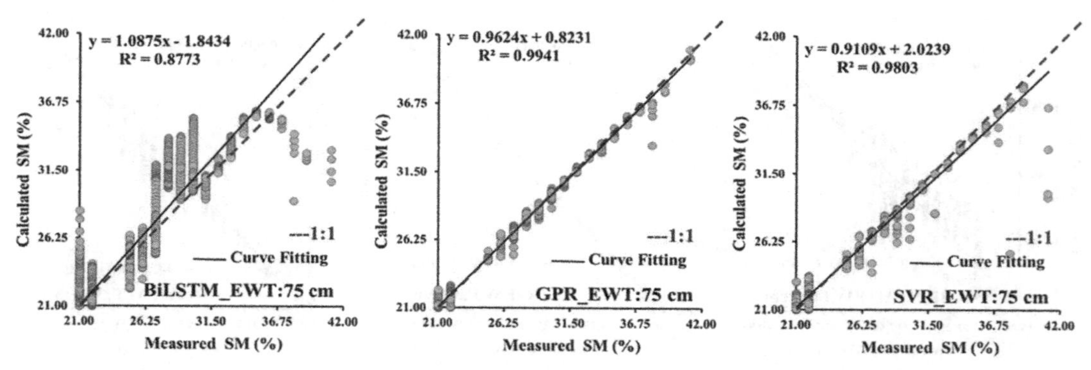

FIG. 9 Scatterplots of measured against calculated soil moisture content (SM: %) at the 75cm depth: validation stage.

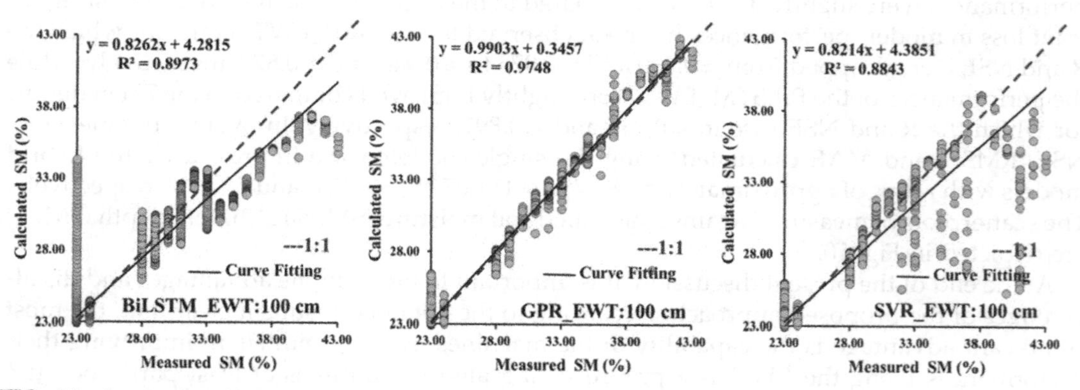

FIG. 10 Scatterplots of measured against calculated soil moisture content (SM: %) at the 100cm depth: validation stage.

method is that the structure of the models become more complicated by increasing the number of the inputs, i.e., the number of MRA, making the training process difficult and time consuming.

4 Conclusions

By this study, soil moisture contents at different depths were modeled by a new conjunction deep learning method, bidirectional long short-term memory-based empirical wavelet transform and outcomes were compared with the two conjunctions GPR-EWT and SVR-EWT methods and standalone BiLSTM, GPR and SVR methods. The methods are assessed with respect to *R*, NSE, RMSE and MAE statistics. All the statistics agree that the BiLSTM method can be successfully used in mapping soil moisture using only soil temperature data; its RMSE/MAE are 15/19%, 42/52%, 56/72%, 38/51%, and 48/55% lower than those of the GPR in cases of 20 cm, 30 cm, 50 cm, 75 cm and 100 cm depths while the corresponding values for the SVR are 17/13%, 47/49%, 66/73%, 47/61%, and 60/59%, respectively. It is evident from the relative differences provided; the GPR is more successful than the SVR in modeling soil moisture content in all depths. It is also clear that the difference between GPR or SVR and BiLSTM is much lower in case of modeling soil moisture at the depth of 20 cm. The reason of this might be the fact that the nonlinearity between SM and soil temperature (T_S) increases by increasing depth as expected and deep learning method provides better performance in modeling SM compared to GPR and SVR methods.

The outcomes indicated that combining EWT with the BiLSTM, GPR and SVR considerably improved their accuracy in simulating (training stage) and modeling (validation stage) SM for different depths; for example, for the of 20 cm depth, the improvements in R, NSE, RMSE, MAE of BiLSTM are 62%, 164%, 86%, 88% and 47%, 117%, 62%, 64% for the training and validation stages, respectively. The corresponding improvements respectively are 103%, 311%, 98%, 99% and 118%, 377%, 94%, 95% for the GPR, 105%, 403%, 93%, 98% and 117%, 516%, 78%, 89% for the SVR. The outcomes also reveal that applying EWT improves the accuracy of GPR much more compared to other two methods; the improvements in R, NSE, RMSE, MAE are 103%, 311%, 98%, 99% and 118%, 377%, 94%, 95% for the training and validation stages of the 20 cm depth case, respectively. It is interesting that the BiLSTM provides high efficiency in modeling SM in all depths using only soil temperature data as input among the single methods. It provides NSE more than 0.800 (very good fit) for the SM_{50}, SM_{75} and SM_{100} predictions while the NSE of the GPR and SVR ranges very low from 0.394 to 0.598 and from 0.070 to 0.440, respectively. This shows the superiority of the BiLSTM over other methods in modeling SM at five different depths.

5 Recommendations

This work presented the application of BiLSTM, GPR and SVR methods in modeling SM at different depths using soil temperature as input. The effect of EWT on models' accuracy was assessed. As future studies, we recommend comparison of BiLSTM method with other deep learning methods such as convolutional neural networks or gene regulatory networks in modeling soil moisture by also considering more affective input variables. The empirical mode decomposition-based data preprocessing methods can be combined with the selected single machine learning methods so as to see their effects on models' accuracies.

References

Abbasi, M., Farokhnia, A., Bahreinimotlagh, M., Roozbahani, R., 2021. A hybrid of random forest and deep auto-encoder with support vector regression methods for accuracy improvement and uncertainty reduction of long-term streamflow prediction. J. Hydrol. 597, 125717. https://doi.org/10.1016/j.jhydrol.2020.125717.

Ahmed, A.M., Deo, R.C., Ghahramani, A., Raj, N., Feng, Q., Yin, Z., Yang, L., 2021a. LSTM integrated with Boruta-random forest optimiser for soil moisture estimation under RCP4. 5 and RCP8. 5 global warming scenarios. Stochastic Environ. Res. Risk Assess., 1–31. https://doi.org/10.1007/s00477-021-01969-3.

Ahmed, A.M., Deo, R.C., Raj, N., Ghahramani, A., Feng, Q., Yin, Z., Yang, L., 2021b. Deep learning forecasts of soil moisture: convolutional neural network and gated recurrent unit models coupled with satellite-derived MODIS, observations and synoptic-scale climate index data. Remote Sens. (Basel) 13 (4), 554. https://doi.org/10.3390/rs13040554.

Aldhyani, T.H., Al-Yaari, M., Alkahtani, H., Maashi, M., 2020. Water quality prediction using artificial intelligence algorithms. Appl. Bionics Biomech. 2020. https://doi.org/10.1155/2020/6659314.

Alrashidi, M., Alrashidi, M., Rahman, S., 2021. Global solar radiation prediction: application of novel hybrid data-driven model. Appl. Soft Comput. 112 (107768). https://doi.org/10.1016/j.asoc.2021.107768.

Babaeian, E., Paheding, S., Siddique, N., Devabhaktuni, V.K., Tuller, M., 2021. Estimation of root zone soil moisture from ground and remotely sensed soil information with multisensor data fusion and automated machine learning. Remote Sens. Environ. 260, 112434. https://doi.org/10.1016/j.rse.2021.112434.

Bai, Y., Bezak, N., Zeng, B., Li, C., Sapač, K., Zhang, J., 2021. Daily runoff forecasting using a cascade long short-term memory model that considers different variables. Water Resour. Manag. 35 (4), 1167–1181. https://doi.org/10.1007/s11269-020-02759-2.

Bartels, G.K., dos Reis Castro, N.M., Pedrollo, O., Collares, G.L., 2021. Soil moisture estimation in two layers for a small watershed with neural network models: assessment of the main factors that affect the results. Catena 207, 105631. https://doi.org/10.1016/j.catena.2021.105631.

Cao, Q.D., Miles, S.B., Choe, Y., 2022. Infrastructure recovery curve estimation using Gaussian process regression on expert elicited data. Reliab. Eng. Syst. Saf. 217, 108054. https://doi.org/10.1016/j.ress.2021.108054.

Carranza, C., Nolet, C., Pezij, M., Van Der Ploeg, M., 2021. Root zone soil moisture estimation with random forest. J. Hydrol. 593, 125840. https://doi.org/10.1016/j.jhydrol.2020.125840.

Chai, S.S., Goh, K.L., Chang, Y.H.R., Sim, K.Y., 2021. Coupling normalization with moving window in backpropagation neural network (BNN) for passive microwave soil moisture retrieval. Int. J. Comput. Intell. Syst. 14 (1), 1–11. https://doi.org/10.1007/s44196-021-00034-9.

Chaudhary, S.K., Srivastava, P.K., Gupta, D.K., Kumar, P., Prasad, R., Pandey, D.K., et al., 2021. Machine learning algorithms for soil moisture estimation using Sentinel-1: model development and implementation. Adv. Space Res. https://doi.org/10.1016/j.asr.2021.08.022.

Chen, W., Xu, H., Chen, Z., Jiang, M., 2021. A novel method for time series prediction based on error decomposition and nonlinear combination of forecasters. Neurocomputing 426, 85–103. https://doi.org/10.1016/j.neucom.2020.10.048.

Chowdhury, S., Dong, X., Qian, L., Li, X., Guan, Y., Yang, J., Yu, Q., 2018. A multitask bi-directional RNN model for named entity recognition on Chinese electronic medical records. BMC Bioinf. 19 (17), 75–84. https://doi.org/10.1186/s12859-018-2467-9.

Fan, H., Jiang, M., Xu, L., Zhu, H., Cheng, J., Jiang, J., 2020. Comparison of long short term memory networks and the hydrological model in runoff simulation. Water 12 (1), 175. https://doi.org/10.3390/w12010175.

Fang, S., Hu, R., Yuan, X., Liu, S., Zhang, Y., 2020. Resolution enhancement for lung 4D-CT based on transversal structures by using multiple Gaussian process regression learning. Phys. Med. 78, 187–194. https://doi.org/10.1016/j.ejmp.2020.09.011.

Gaur, P., Malaviya, V., Gupta, A., Bhatia, G., Pachori, R.B., Sharma, D., 2022. COVID-19 disease identification from chest CT images using empirical wavelet transformation and transfer learning. Biomed. Signal Process. Control 71, 103076. https://doi.org/10.1016/j.bspc.2021.103076.

Gilles, J., 2013. Empirical wavelet transform. IEEE Trans. Signal Process. 61 (16), 3999–4010. https://doi.org/10.1109/TSP.2013.2265222.

Gururaj, P., Umesh, P., Shetty, A., 2021. Assessment of surface soil moisture from ALOS PALSAR-2 in small-scale maize fields using polarimetric decomposition technique. Acta Geophys. 69 (2), 579–588. https://doi.org/10.1007/s11600-021-00557-x.

Hafeez, G., Khan, I., Jan, S., Shah, I.A., Khan, F.A., Derhab, A., 2021. A novel hybrid load forecasting framework with intelligent feature engineering and optimization algorithm in smart grid. Appl. Energy 299, 117178. https://doi.org/10.1016/j.apenergy.2021.117178.

Huang, L., Cai, T., Zhu, Y., Zhu, Y., Wang, W., Sun, K., 2020. LSTM-based forecasting for urban construction waste generation. Sustainability 12 (20), 8555. https://doi.org/10.3390/su12208555.

Jiang, Y., Jia, J., Li, Y., Kou, Y., Sun, S., 2021. Prediction of gas-liquid two-phase choke flow using Gaussian process regression. Flow Meas. Instrum. 81, 102044. https://doi.org/10.1016/j.flowmeasinst.2021.102044.

Judge, J., Liu, P.W., Monsiváis-Huertero, A., Bongiovanni, T., Chakrabarti, S., Steele-Dunne, S.C., et al., 2021. Impact of vegetation water content information on soil moisture retrievals in agricultural regions: an analysis based on the SMAPVEX16-MicroWEX dataset. Remote Sens. Environ. 265, 112623. https://doi.org/10.1016/j.rse.2021.112623.

Karthikeyan, L., Mishra, A.K., 2021. Multi-layer high-resolution soil moisture estimation using machine learning over the United States. Remote Sens. Environ. 266, 112706. https://doi.org/10.1016/j.rse.2021.112706.

Kim, J., Wang, X., Kang, C., Yu, J., Li, P., 2021a. Forecasting air pollutant concentration using a novel spatiotemporal deep learning model based on clustering, feature selection and empirical wavelet transform. Sci. Total Environ. 801, 149654. https://doi.org/10.1016/j.scitotenv.2021.149654.

Kim, Y., Ha, J.M., Na, K., Park, J., Youn, B.D., 2021b. Cepstrum-assisted empirical wavelet transform (CEWT)-based improved demodulation analysis for fault diagnostics of planetary gearboxes. Measurement 183, 109796. https://doi.org/10.1016/j.measurement.2021.109796.

Li, Q., Wang, Z., Shangguan, W., Li, L., Yao, Y., Yu, F., 2021. Improved daily SMAP satellite soil moisture prediction over China using deep learning model with transfer learning. J. Hydrol. 600, 126698. https://doi.org/10.1016/j.jhydrol.2021.126698.

Liang, R., Zhang, Z., Li, H., Chi, P., Li, G., Tao, Y., 2021. Partial discharge location of power cables based on an improved single-terminal method. Electr. Pow. Syst. Res. 193, 107013. https://doi.org/10.1016/j.epsr.2020.107013.

Minh-Tuan, N., Kim, Y.H., 2019. Bidirectional long short-term memory neural networks for linear sum assignment problems. Appl. Sci. 9 (17), 3470. https://doi.org/10.3390/app9173470.

Niazkar, M., Zakwan, M., 2021a. Assessment of artificial intelligence models for developing single-value and loop rating curves. Complexity 2021. https://doi.org/10.1155/2021/6627011.

Niazkar, M., Zakwan, M., 2021b. Application of MGGP, ANN, MHBMO, GRG, and linear regression for developing daily sediment rating curves. Math. Probl. Eng. 2021. https://doi.org/10.1155/2021/8574063.

Okkan, U., Ersoy, Z.B., Kumanlioglu, A.A., Fistikoglu, O., 2021. Embedding machine learning techniques into a conceptual model to improve monthly runoff simulation: a nested hybrid rainfall-runoff modeling. J. Hydrol. 598, 126433. https://doi.org/10.1016/j.jhydrol.2021.126433.

Ouma, Y.O., Cheruyot, R., Wachera, A.N., 2021. Rainfall and runoff time-series trend analysis using LSTM recurrent neural network and wavelet neural network with satellite-based meteorological data: case study of Nzoia hydrologic basin. Complex Intell. Syst., 1–24. https://doi.org/10.1007/s40747-021-00365-2.

Ouyang, Z.L., Zou, Z.J., 2021. Nonparametric modeling of ship maneuvering motion based on Gaussian process regression optimized by genetic algorithm. Ocean Eng. 238, 109699. https://doi.org/10.1016/j.oceaneng.2021.109699.

Peng, L., Wang, L., Xia, D., Gao, Q., 2022. Effective energy consumption forecasting using empirical wavelet transform and long short-term memory. Energy 238, 121756. https://doi.org/10.1016/j.energy.2021.121756.

Rasmussen, C.E., Nickisch, H., 2010. Gaussian processes for machine learning (GPML) toolbox. J. Mach. Learn. Res. 11, 3011–3015.

Senanayake, I.P., Yeo, I.Y., Walker, J.P., Willgoose, G.R., 2021. Estimating catchment scale soil moisture at a high spatial resolution: integrating remote sensing and machine learning. Sci. Total Environ. 776, 145924. https://doi.org/10.1016/j.scitotenv.2021.145924.

Vapnik, V., Golowich, S.E., Smola, A., 1997. Support vector method for function approximation, regression estimation, and signal processing. Adv. Neural Inf. Proces. Syst., 281–287.

Wang, M., Sheng, X., Li, M., Li, Y., 2022. Estimation of vibration powers flowing to and out of a high-speed train bogie frame assisted by time-domain response reconstruction. Appl. Acoust. 185, 108390. https://doi.org/10.1016/j.apacoust.2021.108390.

Wang, S., Wang, X., Wang, S., Wang, D., 2019. Bi-directional long short-term memory method based on attention mechanism and rolling update for short-term load forecasting. Int. J. Electr. Power Energy Syst. 109, 470–479. https://doi.org/10.1016/j.ijepes.2019.02.022.

Yang, R., Liu, H., Nikitas, N., Duan, Z., Li, Y., Li, Y., 2021. Short-term wind speed forecasting using deep reinforcement learning with improved multiple error correction approach. Energy, 122128. https://doi.org/10.1016/j.energy.2021.122128.

Yuan, J., Liu, C.L., Liu, X., Wang, K., Yu, T., 2009. Incorporating prior model into Gaussian processes regression for WEDM process modeling. Expert Syst. Appl. 36 (4), 8084–8092. https://doi.org/10.1016/j.eswa.2008.10.048.

Yulita, I.N., Fanany, M.I., Arymuthy, A.M., 2017. Bi-directional long short-term memory using quantized data of deep belief networks for sleep stage classification. Procedia Comput. Sci. 116, 530–538. https://doi.org/10.1016/j.procs.2017.10.042.

Zhang, K., Ma, C., Xu, Y., Chen, P., Du, J., 2021b. Feature extraction method based on adaptive and concise empirical wavelet transform and its applications in bearing fault diagnosis. Measurement 172, 108976. https://doi.org/10.1016/j.measurement.2021.108976.

Zhang, L., Liu, Y., Ren, L., Teuling, A.J., Zhang, X., Jiang, S., et al., 2021a. Reconstruction of ESA CCI satellite-derived soil moisture using an artificial neural network technology. Sci. Total Environ. 782, 146602. https://doi.org/10.1016/j.scitotenv.2021.146602.

Zhang, S., Bi, K., Qiu, T., 2019. Bidirectional recurrent neural network-based chemical process fault diagnosis. Ind. Eng. Chem. Res. 59 (2), 824–834. https://doi.org/10.1021/acs.iecr.9b05885.

Zhao, B., Dai, Q., Zhuo, L., Zhu, S., Shen, Q., Han, D., 2021. Assessing the potential of different satellite soil moisture products in landslide hazard assessment. Remote Sens. Environ. 264, 112583. https://doi.org/10.1016/j.rse.2021.112583.

Zhu, P., Zhang, G., Wang, H., Zhang, B., Liu, Y., 2021. Soil moisture variations in response to precipitation properties and plant communities on steep gully slope on the Loess Plateau. Agric Water Manag 256, 107086. https://doi.org/10.1016/j.agwat.2021.107086.

CHAPTER

4

Fuzzy logic modeling of groundwater potential in Marinduque, Philippines

Destiny S. Lutero[a], Jcob C. Malaguit[a], Marie Joy L. Relente[b], Allen L. Nazareno[a], and Arnold R. Salvacion[c]

[a]Institute of Mathematical Sciences and Physics, College of Arts and Sciences, University of the Philippines Los Baños, Los Baños, Laguna, Philippines [b]Institute of Statistics, College of Arts and Sciences, University of the Philippines Los Baños, Los Baños, Laguna, Philippines [c]Department of Community and Environmental Resource Planning, College of Human Ecology, University of the Philippines Los Baños, Los Baños, Laguna, Philippines

OUTLINE

1 Introduction	57
2 Material and methods	59
2.1 Study site	59
2.2 Data	59
2.3 Groundwater potential mapping using fuzzy logic	61
3 Results	63
4 Discussion	66
5 Conclusion	67
References	67

1 Introduction

Groundwater is one of the most important natural resources on Earth and is also a critical part of the hydrologic cycle that is hidden in soil pore spaces, fractures of rock formations, natural voids, and aquifers (Apostolaki et al., 2020; Beckie, 2013; Flores, 2014; Tang et al., 2017; Van der Gun, 2021). It is an important part of hydrological systems that support

https://doi.org/10.1016/B978-0-323-91910-4.00004-2

approximately 50% and 30% of global drinking and total water supply, respectively (Alley, 2009; Beckie, 2013; Flores, 2014). Apart from provisioning, groundwater also provides different ecosystem services that include nutrient cycling, purification and long-term storage of water, and drought and flood mitigation (Griebler and Avramov, 2015). However, despite its importance, there has been limited information (abundance and quality) on groundwater (Ajami, 2021; Shrestha and Pandey, 2016).

Delineating groundwater potential is critical for sustainable groundwater management (Chen et al., 2019; Miraki et al., 2019). A groundwater potential map is an important input for groundwater identification, management, and protection programs (Chen et al., 2018; Jha et al., 2010; Naghibi and Pourghasemi, 2015). Aside from being costly and time consuming, the traditional approaches for groundwater mapping exploration, such as test drilling, geophysical techniques, and stratigraphy analysis factors do not also account for different factors that influence groundwater occurrence (Jha et al., 2010; Oh et al., 2011; Rahmati et al., 2016). There are several statistical and data mining techniques that have been developed for mapping groundwater potential. These techniques include weights-of-evidence, index entropy, frequency ratio, classification regression tree, boosted regression tree, and random forest (Al-Abadi et al., 2016; Kim et al., 2019; Khoshtinat et al., 2019; Kordestani et al., 2019; Naghibi et al., 2015). However, these methods require data on location of springs, wells, or qanat along with different geo-environmental variables that influence groundwater potential (Salvacion, 2022). Hence, application of these techniques to map groundwater potential in areas with limited or no available data on well locations is considered as a great limitation.

Several environmental factors influence groundwater potential in an area which includes climate, topography, slope, landforms, land use, drainage pattern, secondary porosity, lithology, geology, fracture density, and fracture connectivity and aperture (Mukherjee, 1996; Oh et al., 2011; Rahmati et al., 2016). Over the past several years, numerous studies have explored the use of other environmental indicators to delineate groundwater potential of a region. These indicators include distance to fault, drainage density, distance to river/stream, topographic wetness index (TWI), and soil texture (Al-Abadi et al., 2016; Al-Fugara et al., 2020; Chen et al., 2018, 2019; Golkarian and Rahmati, 2018; Guru et al., 2017; Ibrahim-Bathis and Ahmed, 2016; Kim et al., 2019; Lee et al., 2019; Kordestani et al., 2019; Khoshtinat et al., 2019; Naghibi et al., 2015; Nampak et al., 2014; Oh et al., 2011; Rahmati et al., 2016; Razandi et al., 2015; Thapa et al., 2017). According to Khoshtinat et al. (2019), closer distance to fault can result in higher groundwater potential because water can easily penetrate cracks and fractures on rocks and soil. On the contrary, higher drainage density can increase surface run-off resulting in lower infiltration, thus lower groundwater potential (Kumar et al., 2007; Razandi et al., 2015). Meanwhile, higher groundwater potential is greater in areas closer to the river with a permanent or longer period of flow (Golkarian and Rahmati, 2018; Oh et al., 2011). Higher TWI indicates higher groundwater potential due to its effect on the zoning and size of saturated areas hence affecting occurrence of springs (Chen et al., 2019; Nampak et al., 2014; Pourtaghi and Pourghasemi, 2014). Lastly, soil texture influences the surface water infiltration and groundwater recharge and storage (Al-Abadi et al., 2016; Chen et al., 2018; Ibrahim-Bathis and Ahmed, 2016).

Creating a groundwater potential map while considering several environmental factors can be done using the fuzzy logic approach. Fuzzy logic is a soft-computing approach that is based on the fuzzy set theory introduced by Zadeh (1965), which is an extension of the classical set theory where the membership of an object to a set is binary. Fuzzy set theory provides a mathematical description for both quantitative and qualitative data that involve imprecision, ambiguity, and vagueness and the membership of an element can take on any value between 0 and 1 (Zadeh, 1965; Aouragh et al., 2017). Adhering to this logic, specialists can develop models of complex nonlinear functions in a simple, quick, and efficient way while allowing vagueness, impressions, and suboptimality (Chen and Paydar, 2012; Dubey et al., 2013). This approach also allows the development of models using both qualitative and quantitative information (Kim and Beresford, 2011; Salvacion et al., 2015). Lastly, fuzzy logic enables model development in the absence of precise quantitative measurements (Kim and Beresford, 2011; Salvacion et al., 2015). Several studies have applied this technique in groundwater mapping (Aouragh et al., 2017; Arumaikkani et al., 2017; Rafati and Nikeghbal, 2017; Sadeghfam et al., 2016). Aouragh et al. (2017) used the analytic hierarchy process to determine the relative importance (weights) of the spatial data on lithology, lineaments, karsts domains, drainage, land cover, and slope which is important for the weighted aggregation function. Arumaikkani et al. (2017) used fuzzy product, fuzzy sum, and fuzzy gamma functions as its aggregators for their spatial data on geomorphology, geology, lineament, slope, soil and land use. Rafati and Nikeghbal (2017) used fuzzy gamma function only as its aggregator for the spatial data on geology, slope, elevation, drainage, fault, and joint. Sadeghfam et al. (2016) used catastrophe fuzzy membership functions and Jenks optimization method to address the discrepancies in the layers of data.

This chapter discusses the use of fuzzy logic approach to analyze groundwater potential using spatial data on elevation, slope, TWI, distance to river, drainage density, and annual rainfall of Marinduque, Philippines. Each spatial datum corresponds to a membership function that is defined according to the nature of the spatial datum. Due to limited data points, different fuzzy aggregators were used to compare groundwater potential maps.

2 Material and methods

2.1 Study site

Marinduque (Fig. 1) is an island province in the heart of the Philippine archipelago (Salvacion, 2017). The province is dominated by rolling to moderately steep mountains in the inner portion and low-lying coastal areas in the outer region (Salvacion, 2018). Fishing and farming are the main sources of livelihood in Marinduque (Salvacion, 2019). Although groundwater is the primary source of potable water in the province, there is a scarcity of public data on locations of deep-water wells and official groundwater potential map of the province (Salvacion, 2022).

2.2 Data

Data on different environmental factors that influence groundwater potential for Marinduque were collected from different sources. Elevation data of the province were extracted from the advanced space-borne thermal emission and reflection radiometer

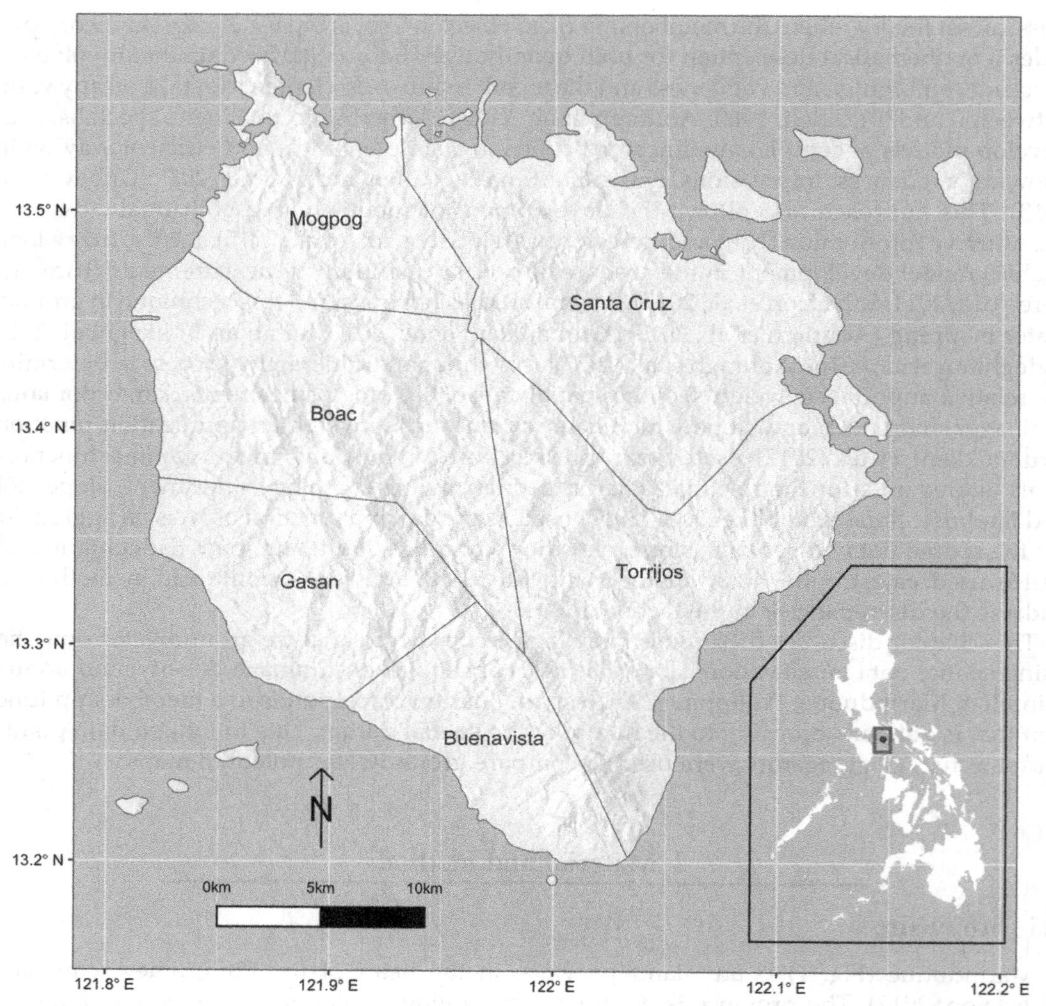

FIG. 1 Location map of Marinduque showing the six municipalities of the province namely: Boac, Mogpog, Santa Cruz, Torrijos, Buenavista, and Gasan.

(ASTER) global digital elevation model (GDEM) (Abrams et al., 2015). This elevation data was used to generate slope, topographic wetness index, and drainage density map of Marinduque (Salvacion, 2016). Distance from the river was calculated from the river network data of Marinduque that was obtained from the Philippine GIS data clearinghouse (http://philgis.org/). Lastly, monthly rainfall data of Marinduque were obtained from the WorldClim database (Fick and Hijmans, 2017). To have a uniform spatial resolution, all input data were resampled to 30m spatial resolution following the spatial resolution of the elevation data using the nearest neighbor approach (Zhu, 2017).

2.3 Groundwater potential mapping using fuzzy logic

The groundwater potential map of Marinduque was derived using fuzzy logic. First, membership functions were defined for each variable using parameters from literature (Phong et al., 2021; Tella and Balogun, 2020; Thapa et al., 2017). Since groundwater mapping is a multivariable problem, an aggregation function was determined to define a single membership value for each data point.

2.3.1 *Identification of membership function*

In fuzzy logic, the process of translating both qualitative and quantitative information to mathematical expressions requires the definition of a function that will describe the degree of belongingness or membership of an element to a (fuzzy) set. The membership of an element can take on any value between 0 and 1 inclusive. Oftentimes, choosing the appropriate membership function for a particular data set depends on the expert's choice or the nature of the data. The commonly used membership functions are triangular and trapezoidal functions. In this study, we considered membership functions that follow one of the forms described below:

$$\mu_{variable} = \begin{cases} 1 & \text{if } 0 \leq x < a \\ \dfrac{b-x}{b-a} & \text{if } a \leq x \leq b \\ 0 & \text{if } x > b \end{cases} \quad (1)$$

$$\mu_{variable} = \begin{cases} 1 & \text{if } x > d \\ \dfrac{x-c}{d-c} & \text{if } c \leq x \leq d \\ 0 & \text{if } 0 \leq x < c \end{cases} \quad (2)$$

2.3.2 *Determination of aggregation function*

Multivariable problems such as groundwater mapping need an aggregation function to combine the multiple membership values of each variable. There are different aggregation functions, but the most commonly used are the arithmetic mean, fuzzy "and," fuzzy "or," fuzzy algebraic product, fuzzy algebraic sum, and fuzzy gamma operator (Bonham-Carter, 1994). The specification of each of these aggregation functions, given n variables, is summarized in Table 1.

The average aggregation function can be viewed as a special case of the weighted aggregation function. For the average aggregation function, the weights $\omega_i = \frac{1}{n}$ for all $i=1, \ldots, n$. The weights ω_i for the weighted aggregation function were determined by identifying the frequency of use of the variables in the literature cited in Salvacion (2022) whereas $\lambda=0.9$ was the value used for the gamma operator.

For each aggregation function, commonly used metrics were computed to determine its performance for the given data.

TABLE 1 Table of specifications for aggregator functions given n variables.

Operation	Equation
Fuzzy and (minimum)	$\mu = min\{\mu_1, \mu_2, \ldots, \mu_n\}$
Fuzzy or (maximum)	$\mu = max\{\mu_1, \mu_2, \ldots, \mu_n\}$
Fuzzy algebraic product	$\mu = \prod_{i=1}^{n} \mu_i$
Fuzzy algebraic sum	$\mu = 1 - \prod_{i=1}^{n}(1 - \mu_i)$
Gamma operator	$\mu = \left(1 - \prod_{i=1}^{n}(1 - \mu_i)\right)^{\lambda} \left(\prod_{i=1}^{n} \mu_i\right)^{1-\lambda}$ where $\lambda \in (0, 1)$
Weighted	$\mu = \sum_{i=1}^{n} \omega_i \mu_i$ where $\sum_{i=1}^{n} \omega_i = 1$

2.3.3 *Calculation of the performance metrics of the fuzzy aggregation functions*

It is also necessary to determine how the results of the different fuzzy aggregation functions will be able to depict the actual source of groundwater. Several metrics were calculated to evaluate the performance of each aggregation function. These metrics are as follows:

$$Accuracy = \frac{TP + TN}{TP + FP + FN + TN}$$

$$Sensitivity = \frac{TP}{TP + FN}$$

$$Sensitivity = \frac{TN}{TN + FP}$$

where

	Predicted (present)	Predicted (not present)
Actual (Present)	**TP**	**FN**
Actual (Not Present)	**FP**	**TN**

True Positives (TP): The cases in which the membership function specifies that there is a potential source of groundwater, and the actual class indicates presence of groundwater.

True Negatives (TN): The cases in which the membership function specifies that there is no potential source of groundwater, and the actual class indicates absence of groundwater.

False Positives (FP): The cases in which the membership function specifies that there is a potential source of groundwater, and the actual class indicates absence of groundwater.

False Negatives (FN): The cases in which the membership function specifies that there is no potential source of groundwater, and the actual class indicates presence of groundwater.

3 Results

The membership function value for each parcel of land was calculated using the membership function specified for each variable (see Fig. 2). The membership functions for elevation (E), slope (S), distance to river (DR), and drainage density (DD) follow the form of Eq. (1) whereas the membership function for topographic wetness index (TWI) and annual rainfall (AR) follow Eq. (2).

The maps of the six variables considered for groundwater potential in Marinduque are shown in Fig. 3. The figure depicts that the groundwater potential is moderate to high everywhere if annual rainfall alone is considered. When distance to the river alone is factored in, groundwater potential was high along areas that appear to branch from the sides of the island going inward. Drainage density showed high potential at the far north, middle, and far south areas of the island in contrast to elevation where high potential is found along the eastern and western banks of the island. Slope showed a dispersed moderate to high distribution of groundwater potential throughout the island while in terms of TWI, groundwater potential throughout the island is low.

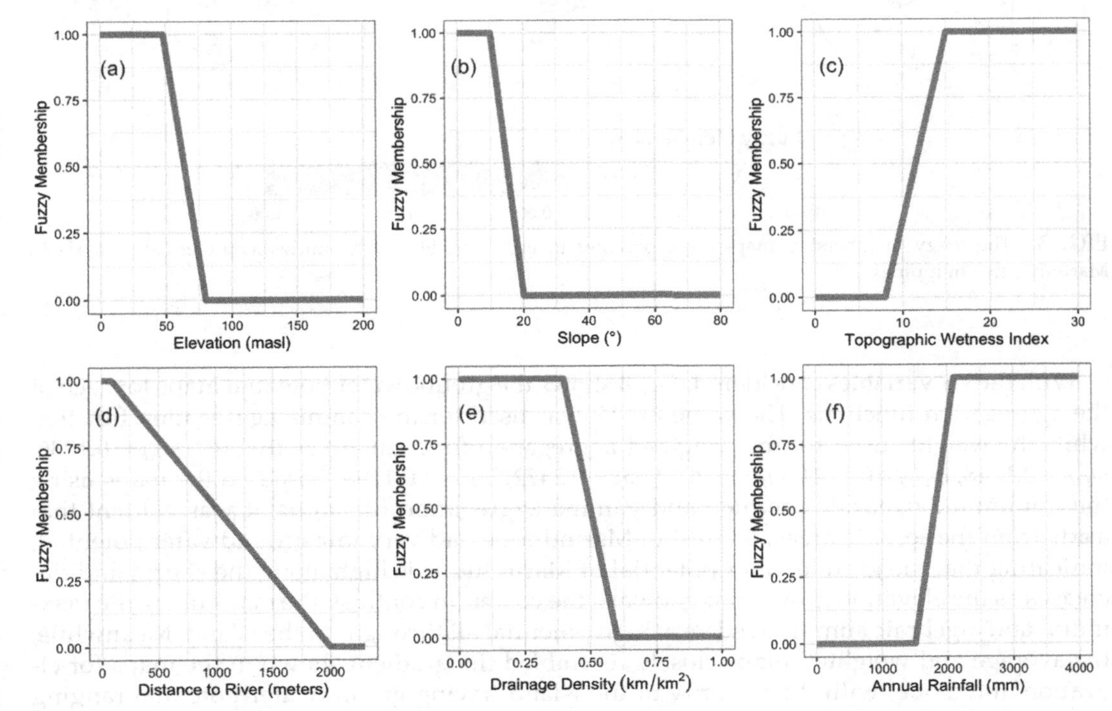

FIG. 2 Graph of membership functions employed in each variable: (A) elevation; (B) slope; (C) topographic wetness index (TWI); (D) distance to river; (E) drainage density; and (F) annual rainfall.

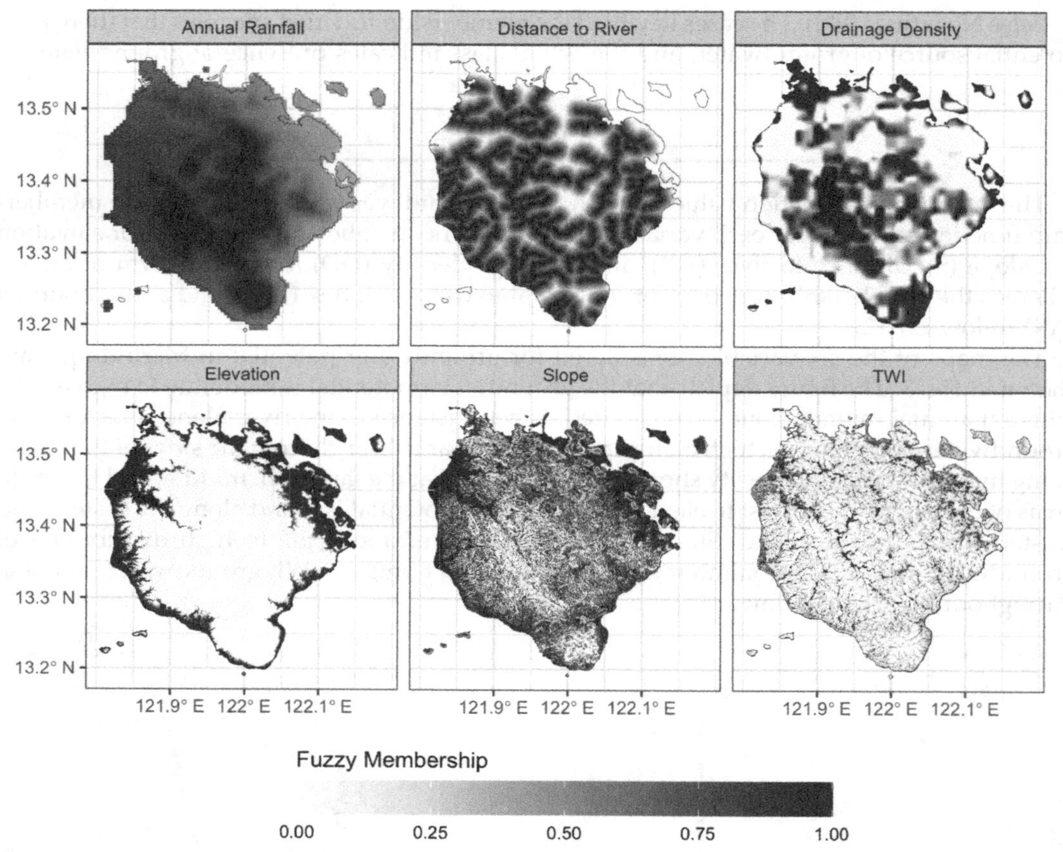

FIG. 3 The fuzzy membership maps corresponding to each variable for groundwater potential analysis in Marinduque, Philippines.

With the six variables considered, Fig. 4 shows the groundwater potential maps for each of the aggregation functions. The value $\lambda=0.9$ was used for the gamma aggregation function while the weights used for the weighted aggregation function are as follows: $\omega_{AR}=0.1236$, $\omega_{DR}=0.1348$, $\omega_{DD}=0.1124$, $\omega_{E}=0.2022$, $\omega_{S}=0.2472$, $\omega_{TWI}=0.1798$. Based on the maps using the minimum, algebraic product, and gamma aggregation functions, it was evident that aside from the seaside areas, the rest of Marinduque had very low groundwater potential, indicating that the groundwater potential of Marinduque is high along the shores and decreases as the elevation grows going toward the center. In contrast, the maps using the maximum and algebraic sum showed very high potential all throughout the island. Meanwhile, the average and weighted maps closely resembled the gradient shown in the maps for elevation and slope, with the majority of the island having groundwater potential ranging from low to high.

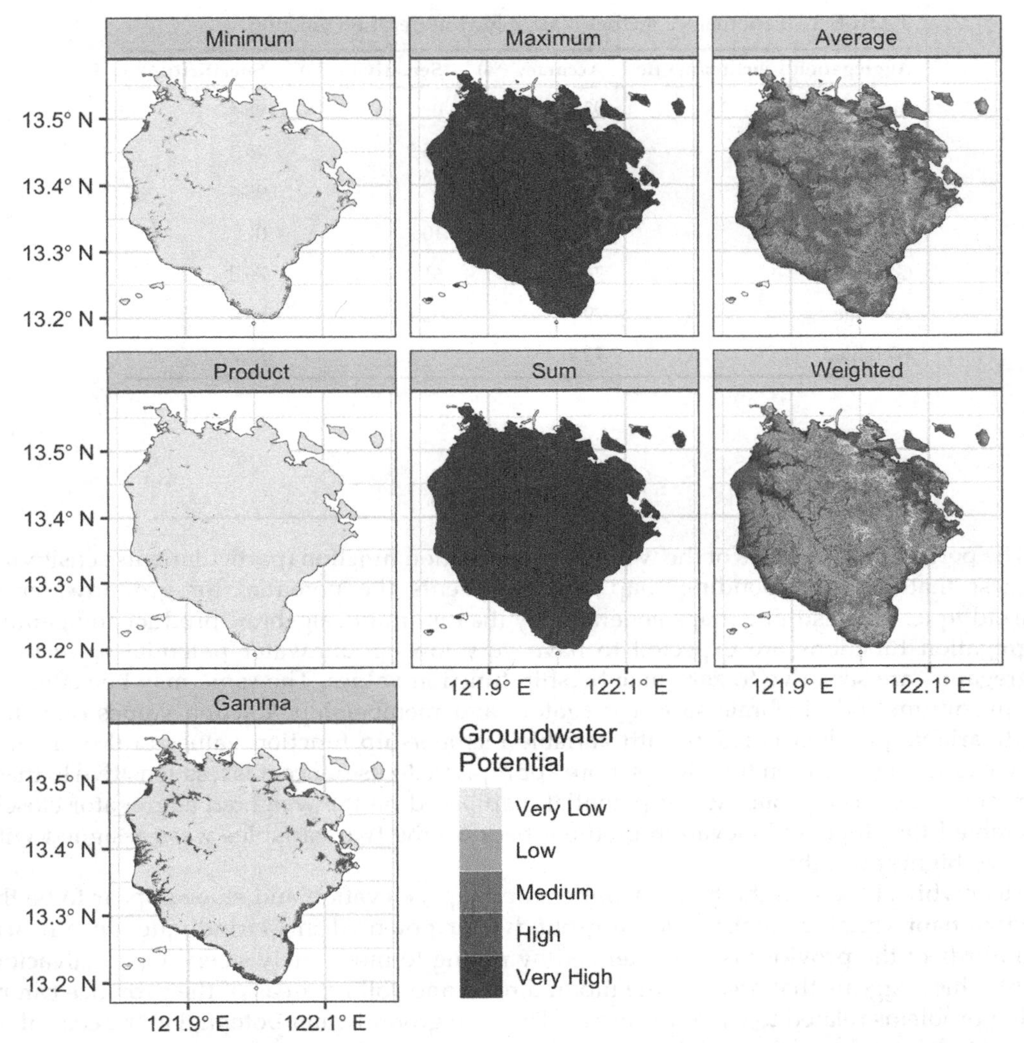

FIG. 4 The groundwater potential maps corresponding to each aggregation function for Marinduque, Philippines.

The performance metrics for each fuzzy aggregation function examined are summarized in Table 2. Both the minimum and algebraic product aggregation functions had the highest accuracy of 93.0%, while algebraic sum had the lowest accuracy of 2.6%. In terms of sensitivity, the algebraic sum aggregation function obtained a perfect score (100%), whereas the minimum and algebraic product both received zero. For the specificity, again, both the minimum and algebraic product aggregation functions scored the highest (95.4%). Furthermore, the algebraic sum had the smallest specificity, which was zero. Overall, the weighted aggregation function performed fairly across all metrics.

TABLE 2 Performance metrics of each fuzzy aggregation function.

Aggregation function/metric	Accuracy (%)	Sensitivity (%)	Specificity (%)
Minimum	93.0	0	95.4
Maximum	47.5	95.2	46.3
Algebraic product	93.0	0	95.4
Algebraic sum	2.6	100	0
Gamma	76.4	61.9	76.8
Average	79.9	52.4	80.6
Weighted	72.8	71.4	72.9

4 Discussion

The performance metrics of the weighted aggregation function (particularly its sensitivity) suggest that its corresponding map best represents the potential for groundwater in Marinduque. The resulting maps generated by the minimum, algebraic product and gamma aggregation functions are expected to have very low groundwater potential since these aggregators are sensitive to zero membership function values. The same may be stated for the maximum and algebraic sum aggregators and membership function values of 1. If a multivariable problem is laden with variable membership function values of 0 or 1, then the weighted aggregation function is more appropriate to use since it is less sensitive to these extreme values. The groundwater potential map based on the weighted aggregator closely resembled the slope and elevation gradient because the two variables were assigned with the two highest weights.

Meanwhile, based on the fuzzy membership maps, elevation and slope appear to be the predominant variables that influence groundwater potential in Marinduque since nearly two-thirds of the province is characterized by rolling to moderately steep slope (Salvacion, 2016). This suggests that areas in the mountainous and rolling areas of the province can be facing problems related to water resources. The high groundwater potential in the coastal region of the province means that there is available water supply to support the agricultural and domestic activities in the province, since most of the province's population is concentrated near the coastal region (Salvacion and Macandog, 2015). However, this might also indicate that the province's groundwater resource is extremely vulnerable to excessive extraction, saltwater intrusion, and domestic pollution. According to Roy et al. (2008), in areas with growing populations, unrestrained groundwater withdrawal to support water requirements for agriculture, domestic, and other purposes can later result in a water crisis. Similarly, over extraction of groundwater due to increasing population demand in coastal regions can trigger saltwater intrusion (Moore and Joye, 2021). Furthermore, biological and chemical contaminants associated with human activities can degrade the quality of the groundwater (Fienen and Arshad, 2016).

5 Conclusion

This chapter on groundwater potential analysis in Marinduque highlights the use of fuzzy logic approach to generate maps given publicly available data on annual rainfall, distance to river, drainage density, elevation, slope, and topographic wetness index. Results show that elevation and slope are the predominant variables that influence groundwater potential in Marinduque due to its topography. The performance metrics imply that among the seven aggregation functions, the weighted aggregator provided the best groundwater potential map for the island based on performance measures. This means that the island has high to very high groundwater potential along the shores and has generally low to high potential elsewhere. This suggests further that areas in the mountainous and rolling areas of the province can be facing problems related to water resources. Additionally, the general framework developed in this study may be adapted to assess groundwater potential in other locales and settings.

References

Abrams, M., Tsu, H., Hulley, G., et al., 2015. The advanced spaceborne thermal emission and reflection radiometer (ASTER) after fifteen years: review of global products. Int. J. Appl. Earth Obs. Geoinf. 38 (1), 292–301. https://doi.org/10.1016/j.jag.2015.01.013.

Ajami, H., 2021. Geohydrology: groundwater. In: Alderton, D., Elias, S.A. (Eds.), Encyclopedia of Geology, second ed. Academic Press, Oxford, pp. 408–415, https://doi.org/10.1016/B978-0-12-409548-9.12388-7.

Al-Abadi, A.M., Al-Temmeme, A.A., Al-Ghanimy, M.A., 2016. A GISbased combining of frequency ratio and index of entropy approaches for mapping groundwater availability zones at Badra-Al Al-Gharbi-Teeb areas, Iraq. Sustain. Water Resour. Manag. 2, 265–283 (2016). https://doi.org/10.1007/s40899-016-0056-5.

Al-Fugara, A., Pourghasemi, H.R., Al-Shabeeb, A.R., Habib, M., Al-Adamat, R., Al-Amoush, H., Collins, A.L., 2020. A comparison of machine learning models for the mapping of groundwater spring potential. Environ. Earth Sci. 79, 206. https://doi.org/10.1007/s12665-020-08944-1.

Alley, W.M., 2009. Ground water. In: Likens, G.E. (Ed.), Encyclopedia of Inland Waters. Academic Press, Oxford, pp. 684–690, https://doi.org/10.1016/B978-012370626-3.00015-6.

Apostolaki, S., Akinsete, E., Koundouri, P., Samartzis, P., 2020. Freshwater: the importance of freshwater for providing ecosystem services. In: Goldstein, M.I., DellaSala, D.A. (Eds.), Encyclopedia of the World's Biomes. Elsevier, Oxford, pp. 71–79, https://doi.org/10.1016/B978-0-12-409548-9.12117-7.

Aouragh, M.H., Essahlaoui, A., El Ouali, A., El Hmaidi, A., Kamel, S., 2017. Groundwater potential of middle atlas plateaus, Morocco, using fuzzy logic approach, GIS and remote sensing. Geomat. Nat. Haz. Risk 8 (2), 194–206. https://doi.org/10.1080/19475705.2016.1181676.

Arumaikkani, G.S., Chelliah, S., Gopalan, M., 2017. Revelation of Groundwater Possible Region Using Fuzzy Logic Based GIS Modeling. https://www.semanticscholar.org/paper/Revelation-of-Groundwater-Possible-Region-Using-GIS-Arumaikkani-Chelliah/3bddea95a883e54d1d83fbb3c310685880508a72.

Beckie, R.D., 2013. Groundwater. In: Reference Module in Earth Systems and Environmental Sciences. Elsevier, https://doi.org/10.1016/B978-0-12-409548-9.05923-6.

Bonham-Carter, G.F., 1994. Tools for map analysis: multiple maps. In: Bonham-Carter, G.F. (Ed.), Geographic Information Systems for Geoscientists. Pergamon, pp. 267–337, https://doi.org/10.1016/B978-0-08-041867-4.50014-X (Chapter 9).

Chen, W., Li, H., Hou, E., Wang, S., Wang, G., Panahi, M., Li, T., Peng, T., Guo, C., Niu, C., Xiao, L., Wang, J., Xie, X., Ahmad, B.B., 2018. GIS-based groundwater potential analysis using novel ensemble weights-of-evidence with logistic regression and functional tree models. Sci. Total Environ. 634, 853–867. https://doi.org/10.1016/j.scitotenv.2018.04.055.

Chen, W., Tsangaratos, P., Ilia, I., Duan, Z., Chen, X., 2019. Groundwater spring potential mapping using population-based evolutionary algorithms and data mining methods. Sci. Total Environ. 684, 31–49. https://doi.org/10.1016/j.scitotenv.2019.05.312.

Chen, Y., Paydar, Z., 2012. Evaluation of potential irrigation expansion using a spatial fuzzy multi-criteria decision framework. Environ. Model. Softw. 38, 147–157. https://doi.org/10.1016/j.envsoft.2012.05.010.

Dubey, S., Pandey, R.K., Gautam, S.S., 2013. Literature review on fuzzy expert system in agriculture. Int. J. Soft Comput. Eng. 2 (6), 289–291.

Fick, S.E., Hijmans, R.J., 2017. WorldClim 2: new 1-km spatial resolution climate surfaces for global land areas. Int. J. Climatol. 37, 4302–4315. https://doi.org/10.1002/joc.5086.

Fienen, M.N., Arshad, M., 2016. The international scale of the groundwater issue. In: Jakeman, A.J., Barreteau, O., Hunt, R.J., Rinaudo, J.-D., Ross, A. (Eds.), Integrated Groundwater Management: Concepts, Approaches and Challenges. Springer International Publishing, Cham, pp. 21–48, https://doi.org/10.1007/978-3-319-23576-9_2.

Flores, R.M., 2014. Co-produced water management and environmental impacts. In: Flores, R.M. (Ed.), Coal and Coalbed Gas. Elsevier, Boston, pp. 437–508, https://doi.org/10.1016/B978-0-12-396972-9.00008-2 (Chapter 8).

Griebler, C., Avramov, M., 2015. Groundwater ecosystem services: a review. Freshw. Sci. 34, 355–367. https://doi.org/10.1086/679903.

Golkarian, A., Rahmati, O., 2018. Use of a maximum entropy model to identify the key factors that influence groundwater availability on the Gonabad Plain, Iran. Environ. Earth Sci. 77, 369. https://doi.org/10.1007/s12665-018-7551-y.

Guru, B., Seshan, K., Bera, S., 2017. Frequency ratio model for groundwater potential mapping and its sustainable management in cold desert, India. J. King Saud Univ. Sci. 29, 333–347. https://doi.org/10.1016/j.jksus.2016.08.003.

Ibrahim-Bathis, K., Ahmed, S.A., 2016. Geospatial technology for delineating groundwater potential zones in Doddahalla watershed of Chitradurga district, India. Egypt. J. Remote Sens. Space Sci. 19, 223–234. https://doi.org/10.1016/j.ejrs.2016.06.002.

Jha, M.K., Chowdary, V.M., Chowdhury, A., 2010. Groundwater assessment in Salboni Block, West Bengal (India) using remote sensing, geographical information system and multi-criteria decision analysis techniques. Hydrgeol. J. 18, 1713–1728. https://doi.org/10.1007/s10040-010-0631-z.

Khoshtinat, S., Aminnejad, B., Hassanzadeh, Y., Ahmadi, H., 2019. Application of GIS-based models of weights of evidence, weighting factor, and statistical index in spatial modeling of groundwater. J. Hydroinf. 21, 745–760. https://doi.org/10.2166/hydro.2019.127.

Kim, K.S., Beresford, R.M., 2011. Use of a climatic rule and fuzzy sets to model geographic distribution of climatic risk for European canker (*Neonectria galligena*) of apple. Phytopathology 102 (2), 147–157. https://doi.org/10.1094/PHYTO-01-11-0018.

Kim, J.-C., Jung, H.-S., Lee, S., 2019. Spatial mapping of the groundwater potential of the Geum River basin using ensemble models based on remote sensing images. Remote Sens. (Basel) 11, 2285. https://doi.org/10.3390/rs11192285.

Kordestani, M.D., Naghibi, S.A., Hashemi, H., Ahmadi, K., Kalantar, B., Pradhan, B., 2019. Groundwater potential mapping using novel data-mining ensemble model. Hydrgeol. J. 27, 211–224. https://doi.org/10.1007/s10040-018-1848-5.

Kumar, P.K.D., Gopinath, G., Seralathan, P., 2007. Application of remote sensing and GIS for the demarcation of groundwater potential zones of a river basin in Kerala, southwest coast of India. Int. J. Remote Sens. 28, 5583–5601. https://doi.org/10.1080/01431160601086050.

Lee, S., Hyun, Y., Lee, M.-J., 2019. Groundwater potential mapping using data mining models of big data analysis in Goyang-si, South Korea. Sustainability 11, 1678. https://doi.org/10.3390/su1106167.

Miraki, S., Zanganeh, S.H., Chapi, K., Singh, V.P., Shirzadi, A., Shahabi, H., Pham, B.T., 2019. Mapping groundwater potential using a novel hybrid intelligence approach. Water Resour. Manag. 33, 281–302. https://doi.org/10.1007/s11269-018-2102-6.

Moore, W.S., Joye, S.B., 2021. Saltwater intrusion and submarine groundwater discharge: acceleration of biogeochemical reactions in changing coastal aquifers. Front. Earth Sci. 9, 231. https://doi.org/10.3389/feart.2021.600710.

Mukherjee, S., 1996. Targeting saline aquifer by remote sensing and geophysical methods in a part of Hamirpur-Kanpur, India. Hydrol. J. 19, 1867–1884 (1996).

Naghibi, S.A., Pourghasemi, H.R., 2015. A comparative assessment between three machine learning models and their performance comparison by bivariate and multivariate statistical methods in groundwater potential mapping. Water Resour. Manag. 29, 5217–5236. https://doi.org/10.1007/s11269-015-1114-8.

Naghibi, S.A., Pourghasemi, H.R., Dixon, B., 2015. GIS-based groundwater potential mapping using boosted regression tree, classification and regression tree, and random forest machine learning models in Iran. Environ. Monit. Assess. 188, 44. https://doi.org/10.1007/s10661-015-5049-6.

Nampak, H., Pradhan, B., Manap, M.A., 2014. Application of GIS based data driven evidential belief function model to predict groundwater potential zonation. J. Hydrol. 513, 283–300. https://doi.org/10.1016/j.jhydrol.2014.02.053.

Oh, H.-J., Kim, Y.-S., Choi, J.-K., Park, E., Lee, S., 2011. GIS mapping of regional probabilistic groundwater potential in the area of Pohang City, Korea. J. Hydrol. 399, 158–172. https://doi.org/10.1016/j.jhydrol.2010.12.027.

Phong, T.V., Pham, B.T., Trinh, P.T., Ly, H.-B., Vu, Q.H., Ho, L.S., Le, H.V., Phong, L.H., Avand, M., Prakash, I., 2021. Groundwater potential mapping using GIS-based hybrid artificial intelligence methods. Ground Water 59 (5), 745–760. https://doi.org/10.1111/gwat.13094.

Pourtaghi, Z.S., Pourghasemi, H.R., 2014. GIS-based groundwater spring potential assessment and mapping in the Birjand Township, southern Khorasan Province, Iran. Hydrgeol. J. 22, 643–662. https://doi.org/10.1007/s10040-013-1089-6.

Rafati, S., Nikeghbal, M., 2017. Groundwater exploration using fuzzy logic approach in GIS for an area around an anticline, Fars province. ISPRS Int. Arch. Photogramm. Remote Sens. Spat. Inf. Sci. 42W4, 441–445. https://doi.org/10.5194/isprs-archives-XLII-4-W4-441-2017.

Rahmati, O., Pourghasemi, H.R., Melesse, A.M., 2016. Application of GIS-based data driven random forest and maximum entropy models for groundwater potential mapping: a case study at Mehran Region, Iran. Catena 137, 360–372. https://doi.org/10.1016/j.catena.2015.10.010.

Razandi, Y., Pourghasemi, H.R., Neisani, N.S., Rahmati, O., 2015. Application of analytical hierarchy process, frequency ratio, and certainty factor models for groundwater potential mapping using GIS. Earth Sci. Inform. 8, 867–883. https://doi.org/10.1007/s12145-015-0220-8.

Roy, M., Nilson, L., Pal, P., 2008. Development of groundwater resources in a region with high population density: a study of environmental sustainability. Environ. Sci. 5, 251–267. https://doi.org/10.1080/15693430802358605.

Sadeghfam, S., Hassanzadeh, Y., Nadiri, A.A., Khatibi, R., 2016. Mapping groundwater potential field using catastrophe fuzzy membership functions and Jenks optimization method: a case study of Maragheh-Bonab plain, Iran. Environ. Earth Sci. 75 (7), 545. https://doi.org/10.1007/s12665-015-5107-y.

Salvacion, A., Macandog, D., 2015. Spatial analysis of human population distribution and growth in Marinduque Island, Philippines. J. Mar. Island Cult. 4 (1), 27–33. https://doi.org/10.1016/J.IMIC.2015.06.003.

Salvacion, A.R., Pangga, I.B., Cumagun, C.J.R., 2015. Assessment of mycotoxin risk on corn in the Philippines under current and future climate change conditions. Rev. Environ. Health 30 (3), 135–142. https://doi.org/10.1515/reveh-2015-0019.

Salvacion, A.R., 2016. Terrain characterization of small island using publicly available data and open-source software: a case study of Marinduque, Philippines. Model. Earth Syst. Environ. 2 (1), 1–9. https://doi.org/10.1007/s40808-016-0085-y.

Salvacion, A.R., 2017. Exploring determinants of child malnutrition in Marinduque Island, Philippines. Hum. Ecol. 45 (6), 853–863. https://doi.org/10.1007/s10745-017-9951-0.

Salvacion, A.R., 2018. Spatial pattern and determinants of village level poverty in Marinduque Island. GeoJournal. https://doi.org/10.1007/s10708-018-9944-6.

Salvacion, A.R., 2019. Mapping land limitations for agricultural land use planning using fuzzy logic approach: a case study for Marinduque Island, Philippines. GeoJournal. https://doi.org/10.1007/s10708-019-10103-4.

Salvacion, A.R., 2022. Groundwater potential mapping using maximum entropy. In: Kumar, P., Nigam, G.K., Sinha, M.K., Singh, A. (Eds.), Water Resources Management and Sustainability. Advances in Geographical and Environmental Sciences, Springer Singapore, pp. 239–256, https://doi.org/10.1007/978-981-16-6573-8_13.

Shrestha, S., Pandey, V.P., 2016. In: Shrestha, S., Pandey, V.P., Shivakoti, B.R., Thatikonda, S. (Eds.), Groundwater as an environmental issue in Asian cities. Butterworth-Heinemann, Groundwater Environment in Asian Cities, pp. 1–13, https://doi.org/10.1016/B978-0-12-803166-7.00001-5 (Chapter 1).

Tang, Y., Zhou, J., Yang, P., Yan, J., Zhou, N., 2017. Groundwater. In: Tang, Y., Zhou, J., Yang, P., Yan, J., Zhou, N. (Eds.), Groundwater Engineering. Springer Natural Hazards, Springer, Singapore, pp. 1–34, https://doi.org/10.1007/978-981-10-0669-2_1.

Tella, A., Balogun, A., 2020. Ensemble fuzzy MCDM for spatial assessment of flood susceptibility in Ibadan, Nigeria. Nat. Hazards. https://doi.org/10.1007/s11069-020-04272-6.

Thapa, R., Gupta, S., Guin, S., Kaur, H., 2017. Assessment of groundwater potential zones using multi-influencing factor (MIF) and GIS: a case study from Birbhum district, West Bengal. Appl. Water Sci. 7, 4117–4131. https://doi.org/10.1007/s13201-017-0571-z.

Van der Gun, J., 2021. Chapter 24—Groundwater resources sustainability. In: Mukherjee, A., Scanlon, B.R., Aureli, A., Langan, S., Guo, H., McKenzie, A.A. (Eds.), Global Groundwater. Elsevier, pp. 331–345, https://doi.org/10.1016/B978-0-12-818172-0.00024-4.

Zadeh, L.A., 1965. Fuzzy sets. Inf. Control 8 (3), 338–353. https://doi.org/10.1016/S0019-9958(65)90241-X.

Zhu, A.-X., 2017. Resampling, Raster. In: International Encyclopedia of Geography. American Cancer Society, pp. 1–5, https://doi.org/10.1002/9781118786352.wbieg0878.

CHAPTER

5

Soft-computing approach to scour depth prediction under wall jets

Mohammad Aamir[a], *Mohammad Amir Khan*[b], *and Zulfequar Ahmad*[c]

[a]Department of Civil Engineering, Chaitanya Bharathi Institute of Technology, Hyderabad, India [b]Department of Civil Engineering, Galgotias College of Engineering and Technology, Greater Noida, India [c]Department of Civil Engineering, Indian Institute of Technology Roorkee, Roorkee, India

OUTLINE

1 Introduction 71
2 Materials and methods 73
2.1 *Effect of various parameters on equilibrium depth of scour* 73
2.2 *Existing prediction equations for maximum scour depth* 74
3 Results and discussion 75
3.1 *Statistical error analysis* 76
3.2 *Artificial neural network (ANN) model* 77
3.3 *Adaptive neuro-fuzzy interference system (ANFIS) model* 78
4 Conclusions 80
References 80

1 Introduction

The phenomenon of scouring in the areas closely surrounding the hydraulic structures has always been an area of utmost importance for hydraulic investigators because of its significance in influencing the safety and stability of hydraulic structures. Incessant scouring reveals the foundation of those structures, thus triggering a hazard to their strength and stability. Almost all of the failures of hydraulic structures can be attributed directly or

https://doi.org/10.1016/B978-0-323-91910-4.00005-4

indirectly to the scouring activity around their foundations. Local scour under two-dimensional plane jets initiates when the scouring power of the incoming jet transcends the minimum threshold value of bed shear stress and the condition for incipient motion of sediment is exceeded. Jet dispensing from opening of a sluice progresses as a wall jet while traversing the rigid apron. These jets possess a high value of width to thickness ratio. Once it reaches the erodible bed, the scouring process is instigated, which begins to progress with time until the equilibrium condition is reached. In the further analyses, d_s = equilibrium depth of scour, x_s = streamwise horizontal length from apron end to equilibrium depth of scour, a = aperture of sluice gate, V = incoming velocity of jet, d_t = depth of tailwater level, L = streamwise length of the solid apron.

Many investigators like Chatterjee et al. (1994), Aderibigbe and Rajaratnam (1998), Lim and Yu (2002), Dey and Sarkar (2006), and Aamir and Ahmad (2017) proposed equations for maximum scour depth prediction. Recently, Aamir and Ahmad (2015, 2021) have investigated the characteristics of submerged horizontal two-dimensional jets and the consequent scour initiating toward the end of the protective apron, and also developed an empirical relationship to predict the equilibrium depth of scour. Aamir and Ahmad (2016) put forth a comprehensive review on scour under the influence of turbulent two-dimensional jets. Aamir and Ahmad (2019) compared the results of soft-computing methods used for the estimation of maximum scour depth under the influence of two-dimensional horizontal jets, with those obtained using the existing published equations.

Lately, investigators have demonstrated profound interest in applying soft-computing methods for the prediction of equilibrium depth of scour around several hydraulic structures. Investigations have been undertaken on the prediction of the equilibrium depth of scour under pile groups and bridge piers using the methodology of genetic programming and artificial neural networks (ANNs) by Kambekar and Deo (2003), Bateni et al. (2007a, b), Lee et al. (2007), Firat and Gungor (2009), Azamathulla et al. (2010), Kaya (2010), Hashemi et al. (2011), and Ismail et al. (2013). Farhoudi et al. (2010) applied neuro-fuzzy model to predict scour characteristics downstream of stilling basins. Ebtehaj et al. (2017) predicted equilibrium depth of scour around bridge piers utilizing the technique of self-adaptive extreme learning machine. Pourzangbar et al. (2017) also used GP and ANN for the prediction of equilibrium depth of scour at seawalls. Results disclosed that the soft-computing models were more precise as matched with empirical equations obtained experimentally. Pandey et al. (2020a,b) used a genetic algorithm to estimate the time-dependent scour depth around a bridge pier. Pandey et al. (2021) predicted scour depth near spur dikes using two novel tree-based ensemble models, namely, stacked boosting regression tree (SBRT) and stacked bagging regression tree (SBGT). Also, Pandey et al. (2022) presented three robust AI-based techniques including gradient boosting decision tree (GBDT), cascaded forward neural network (CFNN), and kernel ridge regression (KRR) to predict the scour depth around spur dike in cohesive sediment mixtures.

In this chapter, an analysis of experimental data for local scour under the influence of wall jets is presented. Data have been adapted from Aamir and Ahmad (2017), Dey and Sarkar (2006), Aderibigbe and Rajaratnam (1998), and Chatterjee et al. (1994). A set of 290 published experimental data in respect of local scouring under the action of two-dimensional horizontal plane jets is used for the present analysis. Range of values for different parameters is provided in Table 1, where F = incoming Froude number of jet ($= V/(ga)^{0.5}$), and g = gravitational

TABLE 1 Range of parameters.

Investigator	Number of data	Range of values for different variables							Dimensions of flume		
		d_s/a	F	d_t/a	D_{50}/a	L/a	σ_g	Hours	Length (m)	Width (cm)	Depth (cm)
Aamir and Ahmad (2017)	27	0.73–9.8	2.7–25.2	6.67–30	0.02–1.34	33–100	–	6–8	10	60	54
Dey and Sarkar (2006)	205	2.27–8.16	2.37–4.87	6.57–13.85	0.02–0.44	26.67–55	1.1–3.9	12–24	10	60	71
Aderibigbe and Rajaratnam (1998)	30	1.32–24.4	1.21–21.54	12–60	0.05–1.35	0	1.3–3.1	5–52	5	32	65
Chatterjee et al. (1994)	28	0.9–4.1	1.02–5.46	5.82–15.5	0.02–0.22	13.2–33	1.2–1.4	0.2–4.7	9	60	69
Total	290	0.9–24.4	1.02–21.54	5.82–60	0.02–1.35	0–55	1.1–3.9	0.2–52	5–10	32–60	65–71

acceleration. Scour prediction equations from literature are evaluated for their performance, when subjected to application of the available data. Although, soft-computing techniques have been applied by many researchers owing to their accuracy in scour prediction under different hydraulic structures, for example, abutments, piers, spur dikes etc., but there is a lacking to acquire and implement these techniques for scour prediction under two-dimensional plane wall jets. Published studies have merely applied a limited spectrum of published laboratory data to acquire ANN model. Contemplating the significance of the problem of scour under wall jets, this study has been undertaken for the development of ANFIS- and ANN-based models utilizing a broader spectrum of data as input variables, to enable improved and more precise scour depth estimation, having broader application to practical problems.

2 Materials and methods

2.1 Effect of various parameters on equilibrium depth of scour

A comprehensive analysis of the available literature in respect of turbulent jets reveals that the equilibrium depth of scour under the impact of these jets is dependent primarily on the level of tailwater, approaching Froude number of jet, aperture of the sluice gate, length of the solid apron, and median particle size of sediment. A matrix of correlation is constructed to evaluate the dependency of different variables on equilibrium depth of scour, as shown in Table 2, indicating whether there exists a positive or negative correlation between different independent variables and d_s/a. Nondimensional depth of scour, d_s/a is evaluated to be correlated positively with F and d_t/a, implying that as there occurs an increase in F or d_t/a, the value of d_s/a increases; and negatively with L/a and D_{50}/a, implying that there is a decrease in the value of d_s/a as L/a or D_{50}/a increases.

F-test is also a statistical tool to determine the importance of each independent variable on the asymptotic scour depth. In this chapter, F-test was carried out and the outcome is

TABLE 2 Correlation matrix.

	d_s/a	L/a	F	d_t/a	D_{50}/a
d_s/a	1				
L/a	−0.30	1			
F	0.91	−0.16	1		
d_t/a	0.42	−0.53	0.79	1	
D_{50}/a	−0.45	−0.22	0.58	0.59	1

TABLE 3 Results of F-test.

Parameter	F-value
L/a	24
F	114
d_t/a	65
D_{50}/a	72

TABLE 4 Scour depth prediction equations.

Investigator	Equation
Chatterjee et al. (1994)	$\frac{d_s}{a} = 0.775F$
Aderibigbe and Rajaratnam (1998)	$\frac{d_s}{a} = 3.35F_{d(95)} - 6.11$; where $F_{d(95)}$ = densimetric Froude number based on D_{95}
Lim and Yu (2002)	$\frac{d_s}{a} = 1.04F_d^{1.47}\sigma_g^{-0.69}\left(\frac{D_{50}}{a}\right)^{0.33}K'_L$; where K_L' = factor given by $K'_L = e^{-0.004F_d^{-0.35}\sigma_g^{-0.5}\left(\frac{D_{50}}{a}\right)^{-0.5}\left(\frac{L}{a}\right)^{1.4}}$; σ_g = geometric standard deviation
Dey and Sarkar (2006)	$\frac{d_s}{a} = 2.59F_d^{0.94}\left(\frac{L}{a}\right)^{-0.37}\left(\frac{d_t}{a}\right)^{0.16}\left(\frac{D_{50}}{a}\right)^{0.25}$
Aamir and Ahmad (2017)	$\frac{d_s}{a} = 0.238{F_d}^{1.343}\left(\frac{D_{50}}{a}\right)^{0.255}\left(\frac{d_t}{a}\right)^{0.248}$

presented in Table 3. It is clear from the figure that the Froude number is the most significant variable that affects the asymptotic scour depth, followed by median particle size, tailwater depth, and length of solid apron.

2.2 Existing prediction equations for maximum scour depth

Various prediction equations exist for the estimation of the equilibrium depth of scour under two-dimensional wall jets. Almost all of these equations are based on laboratory experiments. Table 4 shows the equations proposed by various researchers for prediction of the equilibrium depth of scour. It can be noted that Chatterjee et al. (1994), and Aderibigbe and Rajaratnam (1998) equations are formulated in such a manner that the estimated scour depth is dependent only on the approaching Froude number of jet or densimetric Froude number. On the other hand, rest of the given equations are formulated taking into account the influence of various other variables affecting the equilibrium depth of scour. All of the listed equations are evaluated in this study for their functioning and implementation against the available experimental data, as listed in Table 1.

3 Results and discussion

The data set adapted in this chapter is used to evaluate the applicability to the existing equations, and observed scour depth is compared with predicted scour depth by the existing equations. The results are presented in Fig. 1A–E. The straight line in each of these figures

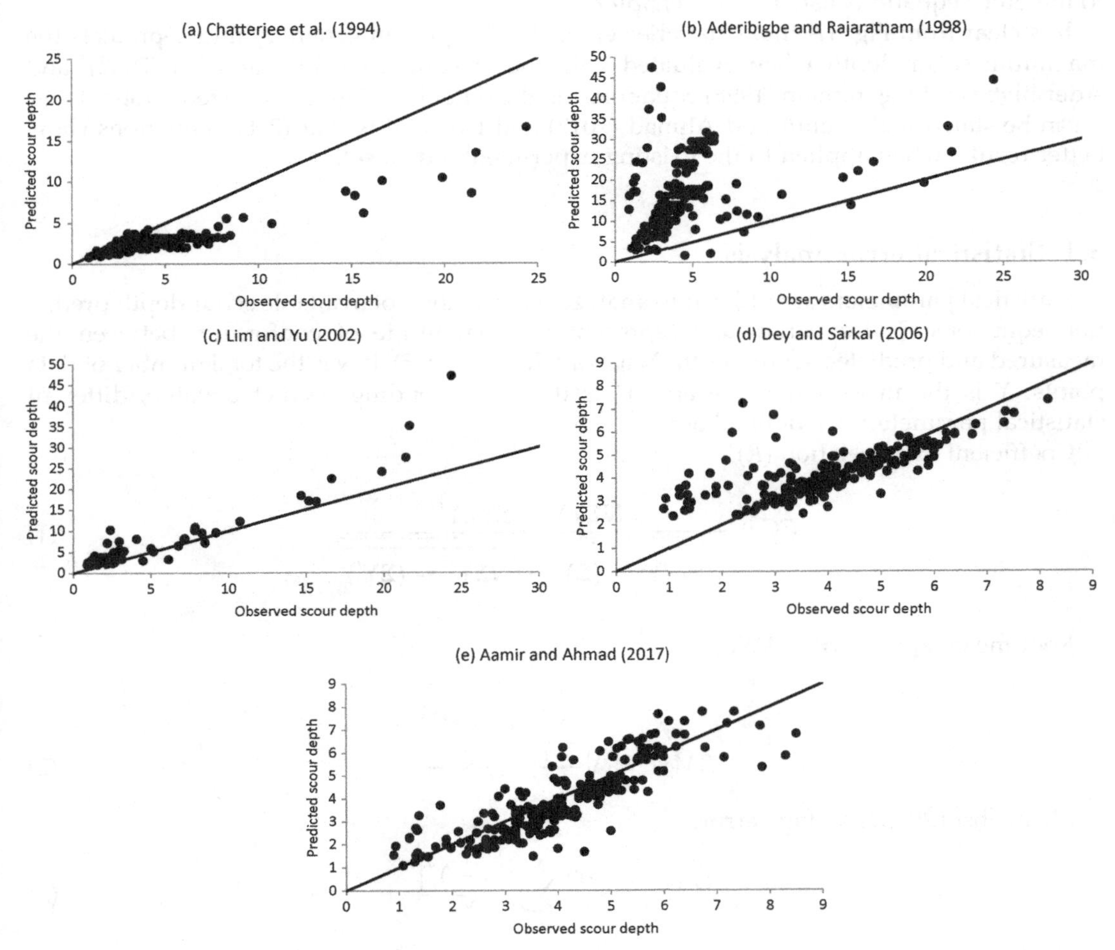

FIG. 1 Contrast of experimental scour depth with predicted scour depth using (A) Chatterjee et al. (1994), (B) Aderibigbe and Rajaratnam (1998), (C) Lim and Yu (2002), (D) Dey and Sarkar (2006), and (E) Aamir and Ahmad (2017) equations.

represents the line on which the predicted value of scour depth is equal to the experimental value. Each equation has a particular range of application, and the data points which lie within that range for a particular equation only are applied to evaluate that equation. It is realized that the maximum asymptotic scour depth estimated by Aderibigbe and Rajaratnam (1998), and Lim and Yu (2002) equations scatter and depart in the positive direction from the perfect agreement line, which indicates that these two equations mostly over-predict the asymptotic scour depth. There is a deviancy in the negative direction in the case of Chatterjee et al. (1994) equation, indicating that this equation tends to under-predict the equilibrium depth of scour. Maximum asymptotic scour depth predicted by Aamir and Ahmad (2017), and Dey and Sarkar (2006) equations are observed to present a lesser degree of deviancy from the line of perfect agreement. Hence, these two equations prove to be more reliable and accurate in predicting the asymptotic scour depth under two-dimensional jets, as compared to the other equations listed in this chapter.

It is clear from Fig. 1A that Chatterjee et al. (1994) equation generally under-predicts the maximum scour depth when evaluated with the existing data. Lim and Yu (2002), and Aderibigbe and Rajaratnam (1998) equations, on the other hand, give over-predictions. Thus, it can be stated that, Aamir and Ahmad (2017) and Dey and Sarkar (2006) equations show better results when applied to the existing experimental data set.

3.1 Statistical error analysis

Statistical parameters are utilized to analyze the precision of available scour depth prediction equations. These parameters represent the magnitude of conformity between the measured and predicted scour depth (Najafzadeh et al., 2018). If N is the total number of data points, Y is the measured value and Y' is the corresponding predicted value, different statistical parameters are defined as:

Coefficient of correlation (R),

$$CC = \frac{N\Sigma YY' - \Sigma Y\Sigma Y'}{\sqrt{N\Sigma Y^2 - (\Sigma Y)^2}\sqrt{N\Sigma Y'^2 - (\Sigma Y')^2}} \tag{1}$$

Root mean square error (MSE),

$$RMSE = \sqrt{\frac{\sum_{i=1}^{N}(Y_i - Y'_i)^2}{N}} \tag{2}$$

Mean absolute percentage error,

$$MAPE = \frac{100}{N}\sum_{i=1}^{N}\frac{|Y_i - Y'_i|}{|Y_i|} \tag{3}$$

$$BIAS = \frac{1}{N}\sum_{i=1}^{N}(Y_i - Y'_i) \tag{4}$$

TABLE 5 Values of statistical parameters for the published equations.

Investigator	*R*	*RMSE*	*MAPE*	*BIAS*	*SI*
Aamir and Ahmad (2017)	0.63	0.29	0.51	−0.003	0.25
Dey and Sarkar (2006)	0.59	0.31	0.56	−0.005	0.28
Lim and Yu (2002)	0.38	0.58	0.92	0.027	2.90
Aderibigbe and Rajaratnam (1998)	0.29	1.31	1.88	−0.035	3.11
Chatterjee et al. (1994)	0.43	0.49	0.62	−0.012	1.74

Scatter index,

$$SI = \frac{\sqrt{\frac{1}{N}\sum_{i=1}^{N}\left[\left(Y_i - \overline{Y_i}\right) - \left(Y'_i - \overline{Y'_i}\right)\right]^2}}{\frac{1}{N}\sum_{i=1}^{N} Y_i} \tag{5}$$

Statistical parameters were calculated for each of the existing equations to assess their implementation against the data set used in this study. Table 5 presents the values of each of these parameters for the existing equations.

The value of coefficient of correlation (*R*) is found to be maximum in case of Aamir and Ahmad (2017), and the values for all mean errors (*RMSE, MAPE, BIAS* and *SI*) in this case are minimum. Thus, it can be concluded from the statistical error analysis that Aamir and Ahmad (2017) equation outperforms others when applied and evaluated with the existing data. However, for practical applications, these values of statistical errors are still high and need to be further improved. Hence, it was proposed to develop soft-computing-based scour depth prediction models, which motivated subsequent sections of this study.

3.2 Artificial neural network (ANN) model

ANN is a data-driven computer model which is utilized for the mapping of data among a given set of output and input variables by replicating the natural reasoning activity of the brain of a human-being (Azamathulla et al. 2005). A characteristic neural network comprises three layers of neurons. Foremost layer is known as the input layer, subsequent one is the hidden layer and the ultimate layer is known as the output layer. Many investigations have been done in the past involving application of ANN in calculation of local scour depth around hydraulic structures (Azamathulla et al., 2006; Guven and Gunal, 2008; Guven and Azamathulla, 2012). The books of Kosko (1992) and Wassermann (1993) can be referenced back for a detailed operation of the neural networks.

In the current investigation, the FFBP (feed-forward back-propagation) neural network model was used. The network consisted of three layers, which implies that only one hidden layer was used. In a FFBP model, the entered data is served into the first layer and the target configurations are connected with output unit. The error (target output) is promulgated back to the neural network for modification of weights.

The network was optimized by employing the enumeration technique, in which different network topologies are tested to achieve the optimized number of hidden layers, and consequent number of neurons in each hidden layer. Different values of learning rate were also tested. Based on the MSE criterion, a typical network topology of 4-9-1 with a learning rate of 0.06 was observed to be optimum, with MSE = 0.0025. Seventy-five percent of the available data were utilized for training and validation of the ANN model, whereas the other 25% data were used for the purpose of testing.

Fig. 2 shows an assessment between the predicted and observed maximum scour depth values using ANN model for (A) training, and (B) testing data set. The value of R is evaluated to be 0.95 in case of training, and 0.96 in case of testing data sets. This demonstrates that the model is efficient and can be utilized for estimation of the equilibrium scour depth under two-dimensional horizontal jets.

3.3 Adaptive neuro-fuzzy interference system (ANFIS) model

ANFIS is a fuzzy Sugeno model applied to the structure of adaptive systems to enable learning and adaptation. ANFIS combines the advantages of both neural networks and fuzzy inference systems (FIS). In neuro-fuzzy models, the problem of adjustment of the membership functions of principle variables is undertaken by a multilayer feed forward neural network. Fuzzy logic facilitates the interaction between the output and input space with a set of If-Then statements. In this chapter, the hybrid learning algorithm was used for training and adaptation of the FIS. This algorithm syndicates the least-squares method with the back-propagation.

Fig. 3 shows an assessment between the predicted and observed maximum scour depth values using ANFIS model for (A) training, and (B) testing data set. The value of R is evaluated to be 0.93 for training and 0.92 for testing. This shows that the performance of

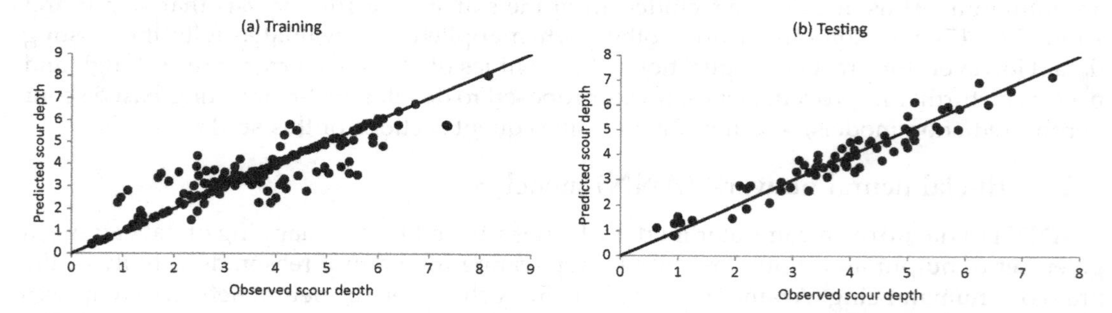

FIG. 2 Contrast between observed and predicted depth of scour using ANN model for (A) training, and (B) testing.

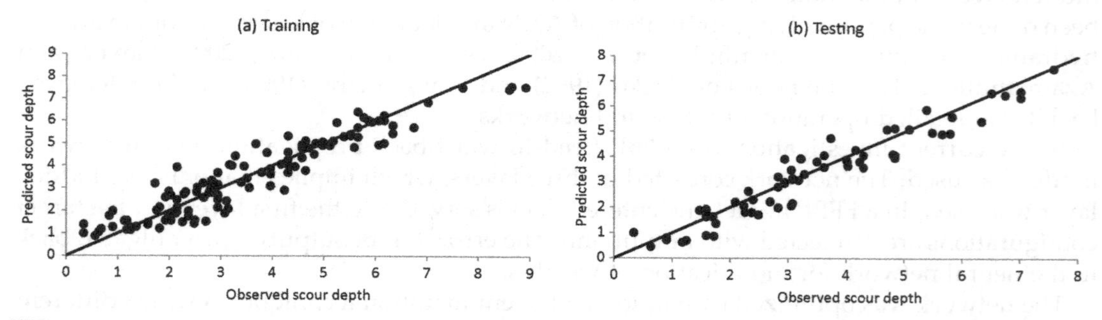

FIG. 3 Contrast between observed and predicted scour depth using ANFIS model for (A) training, and (B) testing.

the ANFIS model is very good for the prediction of the equilibrium depth of scour under two-dimensional wall jets.

Table 6 presents the values of statistical parameters for ANFIS and ANN models developed in the present study. The proposed models outperform the existing prediction equations. On comparison, the ANN and ANFIS models are observed to have a considerably superior value of *R* and significantly lesser values of *MAPE* and *RMSE*. This shows that the proposed models are highly advantageous over the existing equations. Over and above better prediction of the equilibrium depth of scour, the ANFIS and ANN models also possess a much broader range of pertinence, since they were formulated using an extensive range of laboratory data. Hence, the models proposed in this study can be effectively utilized as a scour depth prediction method for two-dimensional wall jets.

Fig. 4 shows a percent error graph, which depicts a contrast in between the percent errors and the percent of data evaluated, for the ANFIS and ANN models, and Aamir and Ahmad

TABLE 6 Values of statistical parameters for ANN and ANFIS models.

Model	Stage	*R*	*RMSE*	*MAPE*	*BIAS*	*SI*
ANN	Training	0.95	0.003	0.031	0.173	−0.004
	Testing	0.96	0.003	0.035	0.171	0.002
ANFIS	Training	0.93	0.011	0.042	0.002	0.234
	Testing	0.92	0.013	0.051	0.003	0.125

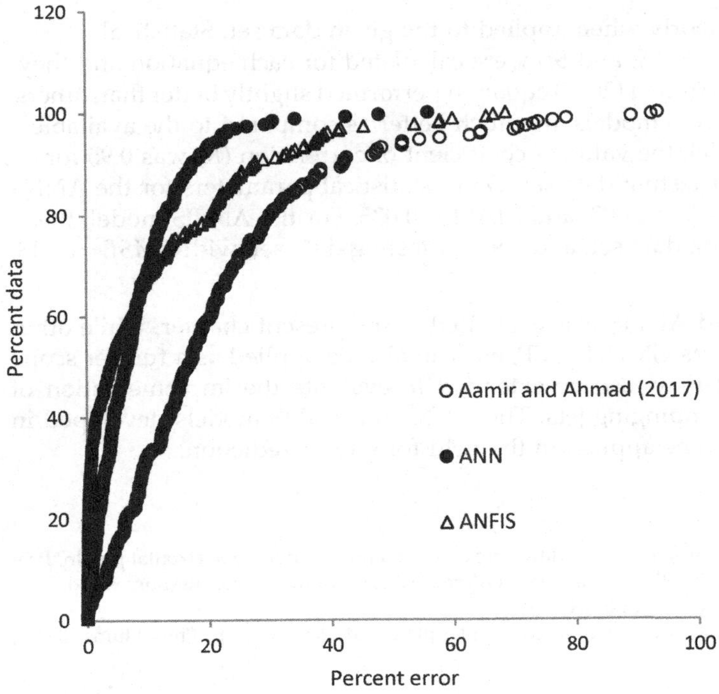

FIG. 4 Percent error graph.

(2017) equation, since this equation is superior to others. It is observed from Fig. 4 that the percent error in case of the ANFIS and ANN models is significantly lesser than that of Aamir and Ahmad (2017) equation. This proves that the given models can be used effectively for practical purposes, in determining the maximum scour depth under the influence of two-dimensional horizontal jets.

4 Conclusions

Published scour depth prediction equations pertaining to wall jets were evaluated by implementing those equations to a sizeable set of existing laboratory data. The existing equations perform poorly when applied to the given data set. Hence, ANFIS and ANN models were developed as precise and easily applicable tools to predict equilibrium scour depth. Following are the conclusions of this study:

1. Equilibrium scour depth is positively correlated to F and d_t/a, while it gives negative correlation with L/a and D_{50}/a. This is because the scouring capacity of the jet increases with increasing jet Froude number and tailwater depth. On the other hand, scouring capacity of the jet decreases with increasing length of solid apron, as it gets exhausted while traversing longer length of the apron. Similarly, scour depth decreases with increasing median particle size of sediment, since higher energy is required to transport larger size of sediment particles.
2. F-test revealed that the Froude number is the most significant variable that affects the asymptotic scour depth, followed by median particle size, tailwater depth, and length of solid apron.
3. Existing equations perform poorly when applied to the given data set. Statistical parameters (*R*, *RMSE*, *MAPE*, *BIAS*, and *SI*) were calculated for each equation and they gave poor values. Aamir and Ahmad (2017) equation performed slightly better than others.
4. The developed ANN and ANFIS models are much better as compared to the available equations. For the ANN model, the value of coefficient of correlation (R) was 0.95 for training data set, and 0.96 for testing data set. Other statistical parameters for the ANN model were calculated as $RMSE=0.003$; and $MAPE=0.035$. For the ANFIS model, the value of R was 0.93 for training data set, and 0.92 for testing data set, with $RMSE=0.013$; and $MAPE=0.051$.

The performance of ANN and ANFIS was evaluated in the present chapter, while other soft-computing techniques such as GMDH, GEP, etc. can also be applied as a further scope of this work. These techniques may also be extended to evaluate the implementation of prediction equations in case of impinging jets. The ANN and ANFIS models developed in this chapter are recommended to be applied in the field for scour prediction.

References

Aamir, M., Ahmad, Z., 2015. Estimation of scour depth downstream of an apron under 2D horizontal jets. In: Proceedings of HYDRO 2015 International, 20th International Conference on Hydraulics, Water Resources and River Engineering. Indian Institute of Technology Roorkee, India.

Aamir, M., Ahmad, Z., 2016. Review of literature on local scour under plane turbulent wall jets. Phys. Fluids 28 (10), 105102.

Aamir, M., Ahmad, Z., 2017. Prediction of local scour depth downstream of an apron under wall jets. In: Garg, V., Singh, V., Raj, V. (Eds.), Development of Water Resources in India. Water Science and Technology Library, vol. 75. Springer, Cham, pp. 375–385. 32.

Aamir, M., Ahmad, Z., 2019. Estimation of maximum scour depth downstream of an apron under submerged wall jets. J. Hydroinf. 21 (4), 523–540.

Aamir, M., Ahmad, Z., 2021. Effect of apron roughness on flow characteristics and scour depth under submerged wall jets. Acta Geophys. https://doi.org/10.1007/s11600-021-00672-9. in press.

Aderibigbe, O., Rajaratnam, N., 1998. Effect of sediment gradation on erosion by plane turbulent wall jets. J. Hydraul. Eng. 124 (10), 1034–1042.

Azamathulla, H.M., Deo, M.C., Deolalikar, P.B., 2005. Neural networks for estimation of scour downstream of a ski-jump bucket. J. Hydraul. Eng. 131 (10), 898–908.

Azamathulla, H.M., Deo, M.C., Deolalikar, P.B., 2006. Estimation of scour below spillways using neural networks. J. Hydraul. Res. 44 (1), 61–69.

Azamathulla, H.M., Ghani, A.A., Zakaria, N.A., Guven, A., 2010. Genetic programming to predict bridge pier scour. J. Hydraul. Eng. 136 (3), 165–169.

Bateni, S.M., Borghei, S.M., Jeng, D.S., 2007a. Neural network and neuro-fuzzy assessments for scour depth around bridge piers. Eng. Appl. Artif. Intel. 20 (3), 401–414.

Bateni, S.M., Jeng, D.S., Melville, B.W., 2007b. Bayesian neural networks for prediction of equilibrium and time-dependent scour depth around bridge piers. Adv. Eng. Softw. 38 (2), 102–111.

Chatterjee, S.S., Ghosh, S.N., Chatterjee, M., 1994. Local scour due to submerged horizontal jet. J. Hydraul. Eng. 120 (8), 973–992.

Dey, S., Sarkar, A., 2006. Scour downstream of an apron due to submerged horizontal jets. J. Hydraul. Eng. 132 (3), 246–257.

Ebtehaj, I., Sattar, A.M.A., Bonakdari, H., Zaji, A.H., 2017. Prediction of scour depth around bridge piers using self-adaptive extreme learning machine. J. Hydroinf. 19 (2), 207–224.

Farhoudi, J., Hosseini, S.M., Sedghi-Asl, M., 2010. Application of neuro-fuzzy model to estimate the characteristics of local scour downstream of stilling basins. J. Hydroinf. 12 (2), 201–211.

Firat, M., Gungor, M., 2009. Generalized regression neural networks and feed forward neural networks for prediction of scour depth around bridge piers. Adv. Eng. Softw. 40 (8), 731–737.

Guven, A., Azamathulla, H.M., 2012. Gene-expression programming for flip-bucket spillway scour. Water Sci. Technol. 65 (11), 1982–1987.

Guven, A., Gunal, M., 2008. Prediction of scour downstream of grade-control structures using neural networks. J. Hydraul. Eng. 134 (11), 1656–1660.

Hashemi, S.G., Shahidi, A.E., Kazeminezhad, M.H., Mansoori, A.R., 2011. Prediction of pile group scour in waves using support vector machines and ANN. J. Hydroinf. 13 (4), 609–620.

Ismail, A., Jeng, D.S., Zhang, L.L., Zhang, J.S., 2013. Predictions of bridge scour: application of a feed-forward neural network with an adaptive activation function. Eng. Appl. Artif. Intel. 26 (5–6), 1540–1549.

Kambekar, A.R., Deo, M.C., 2003. Estimation of pile group scour using neural networks. Appl. Ocean Res. 25 (4), 225–234.

Kaya, A., 2010. Artificial neural network study of observed pattern of scour depth around bridge piers. Comput. Geotech. 37 (3), 413–418.

Kosko, B., 1992. Neural Networks and Fuzzy Systems. Prentice Hall, Englewood Cliffs, NJ.

Lee, T.L., Jeng, D.S., Zhang, G.H., Hong, J.H., 2007. Neural network modeling for estimation of scour depth around bridge piers. J. Hydrodyn. 19 (3), 378–386.

Lim, S.Y., Yu, G., 2002. Scouring downstream of sluice gate. In: Proceedings of the 1st International Conference on Scour of Foundations. vol. 1. Texas Transportation Institute, College Station, TX, pp. 395–409.

Najafzadeh, M., Movahed, F.S., Sarkamaryan, S., 2018. NF-GMDH-based self-organized systems to predict bridge pier scour depth under debris flow effects. Mar. Georesour. Geotechnol. 36 (5), 589–602.

Pandey, M., Jamei, M., Ahmadianfar, I., Karbasi, M., Lodhi, A.S., Chu, X., 2022. Assessment of scouring around submerged spur dike in cohesive sediment mixtures: a comparative study on three rigorous machine learning models. J. Hydrol. 606, 127330.

Pandey, M., Jamei, M., Karbasi, M., Ahmadianfar, I., Chu, X., 2021. Prediction of maximum scour depth near spur dikes in uniform bed sediment using stacked generalization ensemble tree-based frameworks. J. Irrig. Drain. Eng. 147 (11), 4021050.

Pandey, M., Zakwan, M., Khan, M.A., Bhave, S., 2020a. Development of scour around a circular pier and its modelling using genetic algorithm. Water Supply 20 (8), 3358–3367.

Pandey, M., Zakwan, M., Sharma, P.K., Ahmad, Z., 2020b. Multiple linear regression and genetic algorithm approaches to predict temporal scour depth near circular pier in non-cohesive sediment. ISH J. Hydraul. Eng. 26 (1), 96–103.

Pourzangbar, A., Saber, A., Yeganeh-Bakhtiary, A., Ahari, L.R., 2017. Predicting scour depth at seawalls using GP and ANNs. J. Hydroinf. 19 (3), 349–363.

Wassermann, P.D., 1993. Advanced Methods in Neural Computing. Van Nostrand Reinhold, New York.

SECTION III

Image processing applications in water resources

CHAPTER

6

Assessment of water resources using remote sensing and GIS techniques

N.L. Kushwaha[a], Ahmed Elbeltagi[b], Abhishek Patel[c], Mohammad Zakwan[d], Jitendra Rajput[a], and Puneet Sharma[e]

[a]Division of Agricultural Engineering, ICAR-Indian Agricultural Research Institute, New Delhi, India [b]Agricultural Engineering Department, Faculty of Agriculture, Mansoura University, Mansoura, Egypt [c]Regional Research Station, ICAR-Central Arid Zone Research Institute, Kukma, Bhuj, India [d]School of Technology, MANUU, Hyderabad, India [e]Punjab Agricultural University, Ludhiana, India

OUTLINE

1 Introduction 86

2 Remote sensing and GIS: Tools for sustainability of water resources 87
- 2.1 Hydrological management 87
- 2.2 Watershed management 87
- 2.3 Precision irrigation 89
- 2.4 Flood disaster management 90
- 2.5 Salinity management 90
- 2.6 Groundwater management 91

3 Global positioning system (GPS) 92

4 Case study: Hydrological response analysis of IARI watershed using remote sensing and GIS 93
- 4.1 Background and methodology 93
- 4.2 Results and discussion 93

5 Future recommendations for efficient water resource management 95

6 Conclusion 96

References 96

Current Directions in Water Scarcity Research, Volume 7
https://doi.org/10.1016/B978-0-323-91910-4.00006-6

1 Introduction

Nowadays, remote sensing (RS) and geospatial technology (geographical information system—GIS) are applied in the assessment, development, and planning of available water resources. Effective consumption of water resources is important to meet future demand. Water as a natural resource supports economic and human resources upliftment of a nation. Presently, many countries are not capable to meet the water demand for domestic as well as agricultural uses (Kushwaha et al., 2021). The reasons for the nonavailability of water may include climate change, inappropriate or substandard infrastructure, the excessive draft of groundwater and river bodies, water contamination from industrial and agricultural actions, eutrophication resulting due to water quality changes, the salinity of soils from return irrigation flows, infestations of exotic plant and animals, uncontrolled fish farming, demographic changes including land-use changes due to ongoing developments, and alterations of water quality from sediment inflow (Loucks and van Beek 2017).

RS technology gathers information of study object remotely. The EMR (electromagnetic radiation) emitted from the object is captured by the sensors and converted into the desired information. The EMR spectrum comprises specific wavelengths intervals of interest. Table 1 illustrates the bands of interest in satellite RS and GIS. GIS with its advanced data storage, management, analysis, and display capabilities, are important and useful techniques in the development of environmental models. The potential to produce information in both the spatial and temporal domains is the most significant benefit of RS data for hydrological studies (Burrough et al., 2015; Kushwaha et al., 2016).

A vast diversity of studies on water resource planning and management have been carried out using RS and GIS in different continents (Kushwaha et al., 2022). In this sense, Senthilkumar et al. (2019) assessed the groundwater recharge potential zone in the Amaravathi aquifer system, southern India using RS and GIS techniques. In this study, 248 wells were constructed and used for monitoring water table behavior. Another attempt was made by Yousaf et al. (2021) to monitor the real-time crop water requirement for efficient irrigation water management. They have used open-source data (Landsat-8) and Moderate Resolution Imaging Spectroradiometer (MODIS) in their study. Kushwaha et al. (2016) assessed the hydrological behavior of Takarla-Ballowal watershed, in Shivalik hills using

TABLE 1 Most commonly used spectrum band in satellite remote sensing.

Sr. no.	Spectrum band		Wavelength (μm)
1.	Visible (VIS)	Blue	0.4–0.5
		Green	0.5–0.6
		Red	0.6–0.7
2.	Infrared (IR)	Near-infrared (NIR)	0.7–1.3
		Mid infrared (MIR)	1.3–3.0
		Thermal infrared (NIR)	3.0–5.0 and 8.0–14.0
3.	Microwave regions	–	10^4–10^6

RS and GIS for sustainable water resource management. The result indicated that watershed have medium to high relief with moderate permeable soil, resulting in moderate runoff volume causing the problem of soil erosion in the watershed. Sahana and Sajjad (2019) applied RS and GIS in the Sundarban region, India to identify and mapped the susceptible areas of storm surge flood. They found that the villages and lowland areas in the vicinity of the river were most susceptible to surge floods. Previous research has demonstrated that researchers worldwide are utilizing remotely acquired spatial data for water accounting, watershed management, groundwater monitoring and development, irrigation management, monitoring of seawater intrusion, flood disaster management, and green and blue water use, among other purposes.

The incorporation of RS and GIS technology in hydrological modeling has been effectively used in recent times as an advanced tool to monitor, assess and development of water resources sustainably. Brief introductions of RS, GIS, and GPS with their adaptability to water resources management are discussed in the present chapter.

2 Remote sensing and GIS: Tools for sustainability of water resources

2.1 Hydrological management

Hydrological processes are dynamic and varied in time as well as spatial scale. In the past times, traditional approaches like in situ or point measurements were employed for studying the hydrological phenomenon. The findings of such studies are then used to analyze and synthesize to find the aerial estimates. The limitations of these primitive approaches of hydrological variable estimation include labor-intensive, time-consuming, and applicability to the small area/at a specific location. The advancement of RS and GIS technologies has opened the window to investigate various aspects of the hydrological cycle, such as spatial and temporal distribution of meteorological parameters, quantification of water balance components, etc. (Kushwaha et al., 2016). Three crucial aspects of RS in the area of water resource management are as follows:

Qualitative assessment: In the first class, general observations, which relate qualitative aspects of the study (e.g., discharge of industrial waste changes the color of the river, indicating a location for observation point).

Geometric assessment: It deals with the collection of the data related to area, shape, pattern, dimension, and distribution of features of study for example information of land cover in a particular area that affects runoff, infiltration, evapotranspiration (ET), and moisture content of the soil.

Correlation development: Finding relation between RS observations and ground truth data site to study a hydrologic variable like rainfall measurement, soil wetness, etc.

2.2 Watershed management

Proper planning and its execution are necessary for conserving input resources, i.e., land-water for desirable production. RS, such as through aerial and space-borne sensors, is used successfully to estimate watershed priorities, potentials, and management requirements, as well as for periodic supervision (Khan et al., 2001; Kushwaha et al., 2016; Kushwaha and Yousuf, 2017; Pandey and Sharma, 2017). Numerous geomorphologic assessments can be produced from RS data, including area, dimensions, shape and terrain of the watershed, drainage density, and rock formations. Wavelength ranges of 0.6–0.7 μm and 0.8–1.1 μm were found applicable for the morphometric analysis of micro-watersheds (El Baroudy and Moghanm, 2014; Kushwaha and Bhardwaj, 2017). The drainage network is primarily a function of the basin's lithological behavior and stream structure (stream orders, length, sinuosity, bifurcation ratio). Drainage networks and linear aspects of channels can easily, be acquired employing RS techniques (Kudnar, 2020; Kushwaha and Bhardwaj, 2016, 2017). When delineating watersheds or stream networks, we proceed through a series of steps in ArcGIS environment. Few steps are mandatory, while others are optional depending on the characteristics of the input data. Flow across a surface will always be in the steepest downslope direction. Once the direction of flow out of each cell is known, it is possible to determine which and how many cells flow into any given cell. This information can be used to define watershed boundaries and stream networks. The following flowchart (Fig. 1) shows the process of extracting hydrologic information, such as watershed boundaries and stream networks, from a digital elevation model (DEM).

Water balance models are among the most intelligent hydrological models because they simulate all process hydrological parameters for the ground atmosphere system up to a definite extent. Different hydrological processes which are extensively being modeled, use the RS and GIS-based processed data at large spatial and temporal scale model include: the large scale (regional or watershed level) studies on hydrological modeling concerning different hydrological processes such as (i) precipitation (Wang et al., 2018), which usually serves as model input, (ii) ET (Pan et al., 2018), (iii) runoff (including subsurface) (Fok et al., 2021), and (iv) groundwater flow (Shu et al., 2018) intensively use the remotely sensed and GIS processed data.

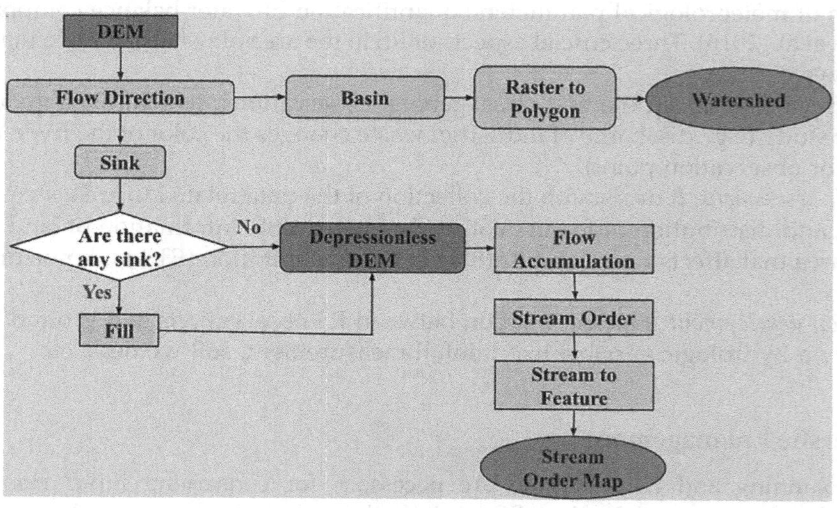

FIG. 1 Methodology for watershed delineation using ArcGIS.

A decision support system integrated with RS and GIS enables effective watershed management. Aher et al. (2014) developed "WATMIS: Watershed Management Information System," a comprehensive tool based on the DSS principle developed for the watershed of Pimpalgaon Ujjaini in Maharashtra, India. WATMIS can prioritize conservation structures, water, and land management based on queries asked and model the crop yield. Additionally, it evaluates crop yields, delineates priority zones for resources conservation, and implements information repository solutions through information to decision backing, including database management, visual analysis, query, and customizing provision as per user's requirement.

2.3 Precision irrigation

Water shortage is a big challenge and a limiting factor that has impacted many country's economies. Consequently, good management of water supplies and water loss minimization (Elbeltagi et al., 2020a,b). The agricultural sector is the prime water consuming sector, which uses a high percentage of the total water worldwide (Elbeltagi et al., 2020a,b; Farg et al., 2012; Pôças et al., 2015). Irrigation scheduling includes deciding when and how much irrigation water to be applied and preventing the resulting water deficit from decreasing yields. Efficient irrigation scheduling helps maximize benefits while reducing inputs such as water for irrigation and the cost of electricity. Factors influencing irrigation schedules include crop size, growth stages, soil resources, and weather conditions (Kushwaha et al., 2022). RS plays a crucial role as a method for estimating crop water needs accurately. It integrates sensors, information systems, and knowledge management to maximize performance by considering volatility and uncertainty in farming systems (Elsayed et al., 2015). Also, the thermal imaging technique uses high-resolution thermal images to detect the plant water content and its status based on the temperature of the plant (Jerbi et al., 2015). In this technique, infrared radiation is measured to determine the plant temperature and spatial distribution is created. The higher plant temperature attributes to the lower plant water content as the stomatal activities are restricted resulting in reduced energy dissipation and higher plant temperature (Patel et al., 2001).

Spectral reflectance properties rely on light assimilation at a particular wavelength range that is related to crop features. Vegetation indices were developed using the red and NIR bands and were useful for monitoring potato growth (Elsayed et al., 2017). Also, a combination of RS data with artificial intelligence methods has produced good results because it depends on estimating optimal climate parameters for predicting crop water productivity (Elbeltagi et al., 2020a, b).

Precision irrigation dramatically helps in managing the irrigation water effectively. This management deals with the idea of water availability, supply, and demand in a particular agricultural area. In this direction, Elbeltagi et al. (2021) assessed the blue and green water ET in the Nile Delta basin from 1997 to 2017 in Egypt. The study applied the Landsat satellite image to detect ET's temporal and spatial variation in the arc GIS environment. The findings from this study suggested the best saving both water use components of ET. They recommended the relative irrigation supply is an effective indicator for estimating green and blue water ET with an R^2 of 0.99 and 0.76, respectively.

These GIS-based maps were used to identify the efficient and unsustainable water management areas at various irrigation districts. This left the water managers, farmers, and administrators to rethink groundwater resource utilization, spreading knowledge and awareness of sustainable groundwater use for better and efficient irrigation water management among users at district levels.

2.4 Flood disaster management

Flooding is an extreme natural event that occurs in the hydrological cycle. Sometimes, it is essential to regain soil fertility by introducing nutrients and fine-grained sediments frequently; however, it may consequence in the fatal injury of humans or other animals, and their habitat, and unrecoverable disruption to infrastructure facilities. Real-time flood distribution mapping and continuous monitoring were achieved through RS and GIS methodology. The accessibility of sensors capable of detecting and estimating flooding across a broad range of the EMR spectrum facilitates us with a cost-effective solution at a large extent and magnitude of the flooding (Sanyal and Lu, 2004). RS and GIS techniques have been successfully illustrated as an integrated framework for flood risk mapping (Sinha et al., 2008). Some researchers took a more quantitative research approach, utilizing multiple criteria decision analysis (MCDA), analytic hierarchy process (AHP), frequency ratio (FR) model, normal standardized vegetation index (NDVI) model, and practical, robust GIS techniques to provide accurate information and interpretation for flood-prone areas.

At crisis, most users need "near-real-time turnaround" information to cope with the situation more effectively. Response time is less and challenging for those involved in hydrological modeling, calibration/testing tests, damage evaluation, and flood mitigation preparation. Flooding conditions generally occurring under cloudy conditions challenges the optical sensors to capture short-term flood events. For these reasons, the active SAR is a way out for flood monitoring and mapping. When combined with a preflood map, the SAR data is most beneficial for identifying flood-affected areas and then presenting in a GIS with details about the cadastral and road network. The Hydrological Engineering Center-River Analysis System (HEC-RAS) model (Brunner, 2016) is the most widely used flood modeling tool across the globe (Schumann et al., 2009), and it can simulate both stable and turbulent flow conditions. Precise calculation of ET, as it is the second most significant component of the water balance after rainfall, so it is crucial for monitoring and managing floods, conservation of water in areas with limited water supplies, also where accurate ground-level measurements are scarce due to low population densities (Mohammadi et al., 2015).

2.5 Salinity management

Salinity is undesirable under the current context of the food demand of the increasing world population. However, these are widely distributed across the world on varying scales. The arid and semiarid regions are highly affected by salinity. About 20% of irrigated farmland worldwide is affected by salinity (Shahid et al., 2018), resulting in soil erosion, loss of arable land, and climate deterioration. Thus, daily monitoring and mapping of salt-affected areas are of great importance in providing appropriate information for salinity management and reclamation. Conventionally quantitative determination of soil salinity was carried out by calculating the electrical conductivity (EC) of soil solution extracts or extracts with a higher-than-average water content (Kumar and Sharma, 2020). Since the extraction of soil water from samples at specific field water content is impractical, EC saturation extracts were generated at 1:1, 1:2, and 1:5 soils:water ratios, referred to as EC1:1, EC1:2, and EC1:5, were commonly applied for predicting the soil salinity. However, using such a conventional method requires significant time and financial and human resources, typically leading to a high cost of soil salinity analysis.

Integration of RS and GIS techniques together is beneficial for the assessment of degraded land in terms of excessive salt accumulation. Asfaw et al. (2018) introduced a statistical model (regression model) for mapping salinity extent using the RS and GIS in the arid regions of the study area in Ethiopia. They identified the several indexes, i.e., SI, NDSI, BI, NDVI, vegetation soil salinity index (VSSI), and soil-adjusted vegetation index (SAVI), which can be derived from the optical RS data as LANDSAT imageries. The SI was the best to develop the model that helps generate the salinity maps using RS data and GIS technique through image enhancement. These salinity maps were generated through a developed model (having $R^2=0.78$). These may be used for salinity assessment at a large scale and thus will help integrated management of arid soils.

2.6 Groundwater management

Groundwater research is yet another pivotal use of RS in the management of subsurface water. Numerous researchers have conducted extensive evaluations of the RS data utility in groundwater management. RS applications in groundwater studies are applied in three fields namely: geomorphological analysis for groundwater modeling, groundwater storage, and groundwater recharge potential. Extraction of geological and surface information using different sensors (e.g., Landsat TM, IRS-LISS) helps to analyze the status of groundwater studies. RS methods can give water level information to a usable accuracy and precision. With modern available spatial data like radar interferometry and Lidar altimetry, fine resolution DEM is very useful in improving groundwater simulations.

Identifying groundwater priority regions is a crucial multicriteria evaluation concern in which thematic maps of hydro-geological variables are combined in a GIS tool to assess groundwater availability. Thematic maps based on RS were created for lithology, landforms, lineaments, and surface water. These were coupled with additional data, including drainage density, slope, and soil type. These thematic maps were used to assess the logical conditions characterizing the groundwater potential. The recognized groundwater potential zones are relatively consistent with borehole data obtained in the field. The potential for classifying groundwater/recharge potential regions using RS data from IRS-LISS and Landsat TM sensors is well reported in the literature. The general procedure for groundwater potential zone mapping is represented in Fig. 2.

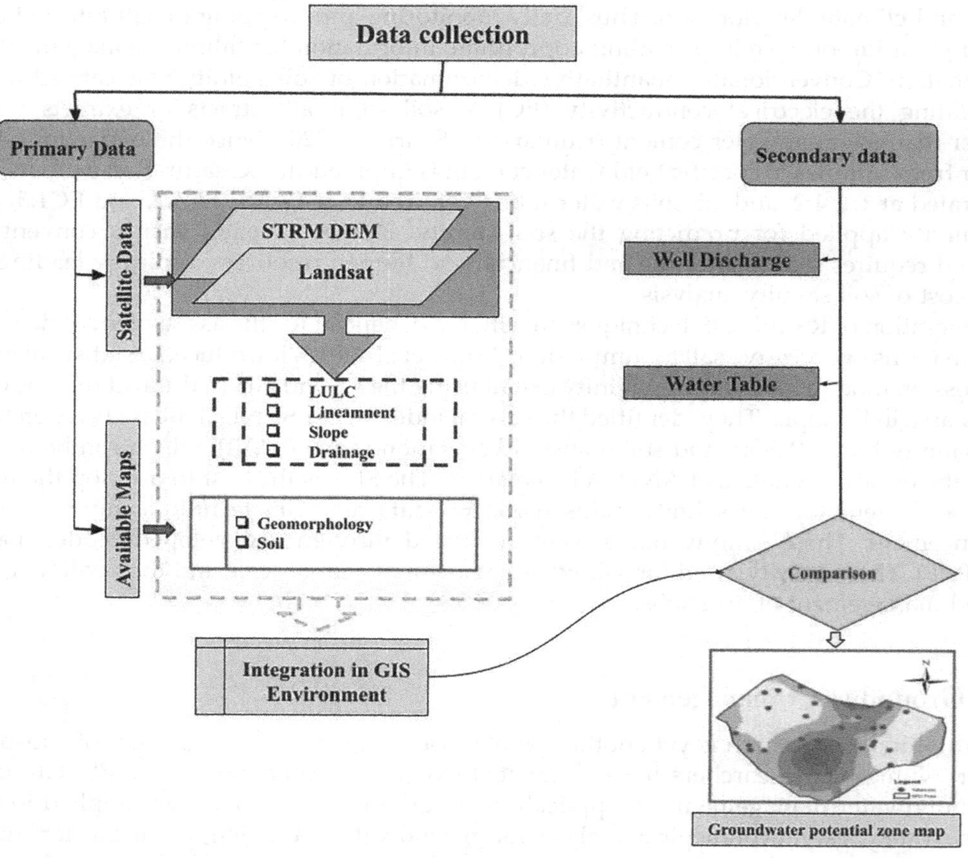

FIG. 2 General methodology for groundwater potential zone mapping.

3 Global positioning system (GPS)

For most application technologies in precision agriculture (PA), GPS systems are crucial as these enable users to act in real-time to schedule a particular production activity at a specific location under investigation. There are numerous applications of GPS systems in farming such as soil mapping, field contouring, crop monitoring, and production tracking. Commonly a GPS contains a GPS receiver, or differential global positioning system (DGPS), mounted on the vehicle traveling through the field, and information retrieval unit, and a software package responsible for mapping and visualization. The farmer moves around the field to contour an area and collects the data using both GPS devices use the same equipment. During the plant season, the farmer can also traverse the field and record data on pest and disease-stricken areas. Consequently, the farmer should apply the correct cultivation care to know the areas with those problems. Concerning soil mapping, GPS is used to monitor the position of the soil samples that are collected to produce the corresponding maps. As regards production mapping, GPS is used to gather the location of each area of which production status is known using other sensors. Subsequently, derived information is used to create the respective maps. Lastly, GPS systems are used to drive farm vehicles in specific field areas during crop surveying. A farm vehicle assists in applying varied rate agrichemicals depending on each zone's soil and production system features.

4 Case study: Hydrological response analysis of IARI watershed using remote sensing and GIS

4.1 Background and methodology

Water as a natural resource supports the economic and human resources upliftment of a nation. Assessment of water resources is an important step for efficient and sustainable planning of existing water resources (Elbeltagi et al., 2022). Water balance analysis of a region provides vital information about the water resources availability and aids in effective planning for their utilization. The present study was carried out for the Indian Agricultural Research Institute (IARI) watershed, New Delhi employing the SWAT model in the ArcGIS environment. The study aims to assess the various water balance components including runoff (Q), ET, and groundwater recharge. Daily weather records of the previous 31 years (1990–2020) were obtained from the observatory located in the IARI, New Delhi. The Advanced Spaceborne Thermal Emission and Reflection Radiometer (ASTER) DEM (Fig. 3) was used to delineate the watershed. The land use and land cover (LULC) map was prepared using Sentinel-2 satellite imagery. Soil texture analysis was carried out of the IARI watershed to prepare a soil map. All the input data tables were prepared in an ArcGIS environment and the SWAT model was run after performing a SWAT check. Then various water balance components were determined. The water balance equation for the various components is as follows:

$$P - Q - \text{ET} - \text{Base flow} \pm \Delta\text{TWS} - (\text{other components}) = 0 \tag{1}$$

where P = precipitation, Q = runoff, ET = evapotranspiration, and ΔTWS = change in terrestrial water storage, and other components consist of soil moisture, shallow and deep groundwater storage, glacier and soil moisture.

4.2 Results and discussion

The LULC classification revealed a mixed type of land use pattern in the IARI watershed with an average curve number value of 76.27. The SWAT model formed 80 hydrological response units (HRU) of the IARI watershed. The water balance components have been generated from the SWAT model. The annual average precipitation during the period 1990–2020

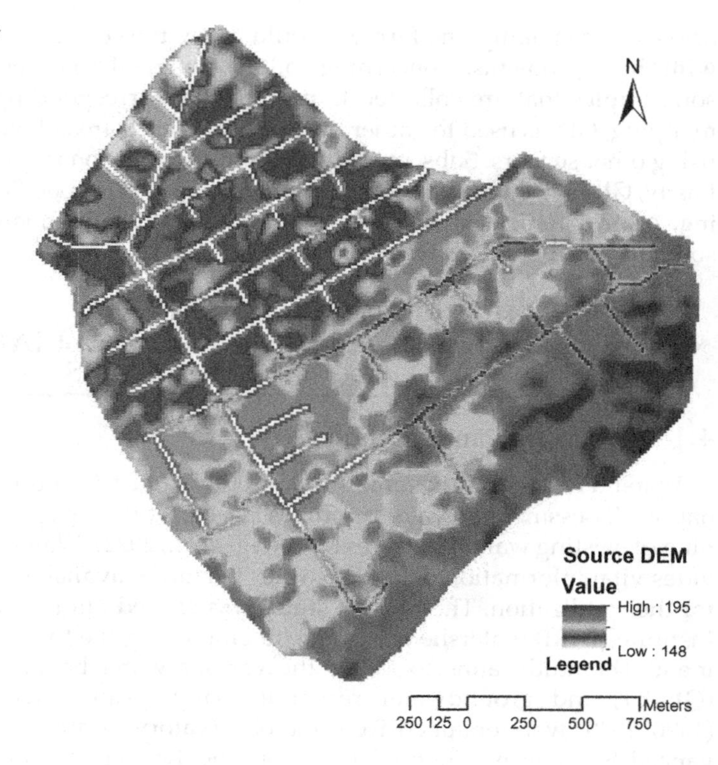

FIG. 3 The DEM of the study watershed.

was obtained as 795.8mm. The output from the model for the IARI watershed is shown in Fig. 4, which represents the various water balance components values. Annual average evaporation and transpiration combinedly was 377mm while the average potential evapotranspiration (PET) for the study area was 1783.9mm. Surface runoff was generated from the watershed as 241.57mm and the ratio of average runoff to the average rainfall was found as 0.30. According to output, the average precipitation = 795.8mm; average total runoff (surface + lateral + return) = 383.71mm, average groundwater recharge = 5.7mm, and average evapotranspiration = 377mm.

So, from Eq. (1);

$$795.8 - 383.71 - 377 - 5.7 = 29.39 \tag{2}$$

The value 29.39 includes other components such as interception, shallow and deep groundwater storage, soil moisture and snow, and glaciers. Therefore, 795.8 – 383.71 – 377 – 5.7 – 29.39 = 0. Results showed that the study watershed has the potential for groundwater recharge by utilizing surface runoff through artificial recharge techniques to recharge deep aquifers. The study findings could be useful to the water resource engineers for sustainable management and planning of existing water resources.

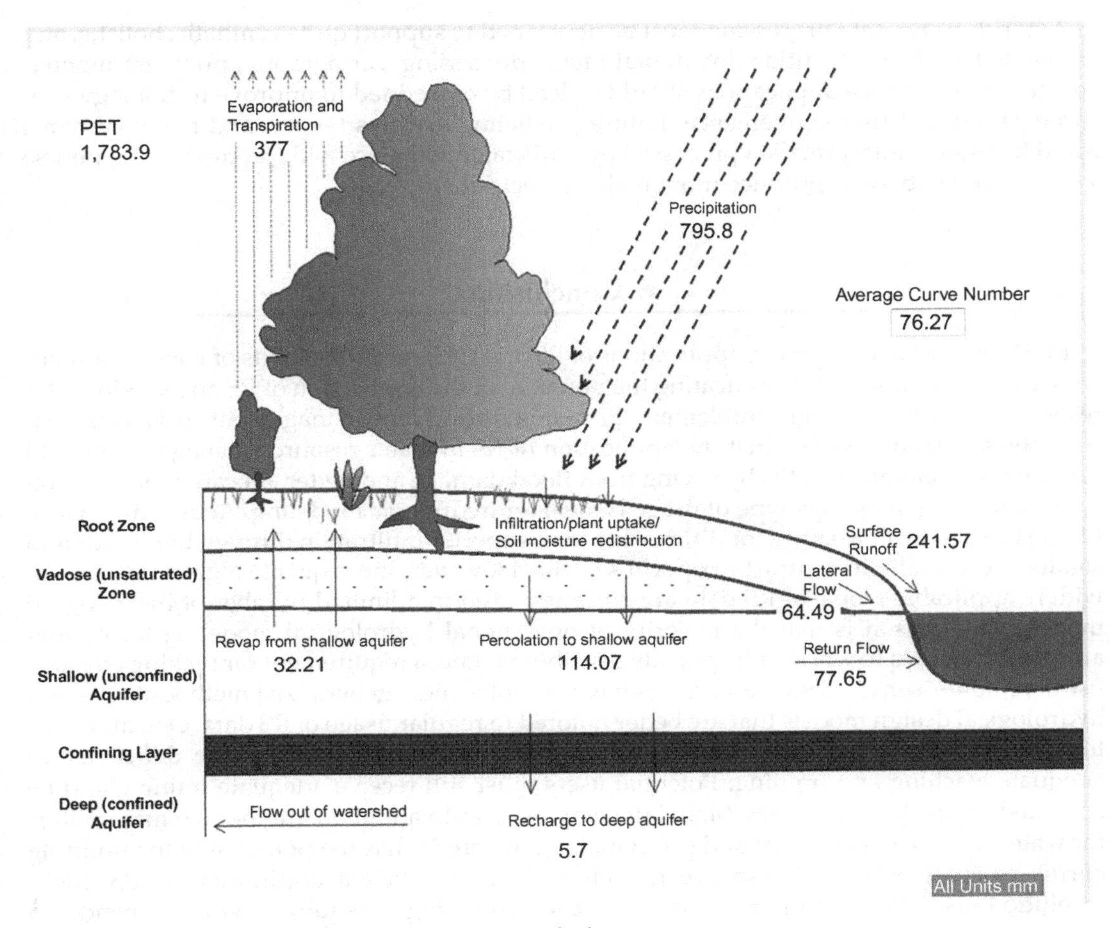

FIG. 4 Water balance components of the study watershed.

5 Future recommendations for efficient water resource management

Most new emerging information demands regarding the ground level situation recent advances in RS have occurred mainly due to hydrological analysis where existing methods have limitations. These include the general circulation model (GCM) application, land parameterizations approach, concepts of snow hydrology, and precise estimation of soil moisture. With time it is expected that more multitemporal, multispectral, and multivariant data available from several satellite platforms. It can be clearly stated that long-term monitoring could effectively derive results if data is remotely sensed at high temporal and moderate spatial resolution. The imaging spectrometer is likely to improve the quality of data collection. Comparing global data sets where satellite sensing is of prime importance requires standardization of procedures for geo-coding, calibration, normalization, and atmosphere correction.

Algorithms for data processing must be developed to support quite synthetic, holistic, and automated analysis. Additionally, digital image processing, cartography, photogrammetry, and computer science applications should indeed be embedded to improve technology synchronization and user convenience. Future predicting systems (expert and fuzzy systems) could be used to interpret data processed by artificial intelligence with professional expertise stored in a database to produce more realistic decisions.

6 Conclusion

In the present book chapter, applications of RS and GIS in various fields of water resources were discussed thoroughly, indicating the potential of the application of RS and GIS in water resources planning, design, implementation, monitoring and management. It is also clear from the above discussions that the application of RS in water resource management could earn better economic benefits by saving from flood damage and better allocation of water resources for irrigation. This type of usage is ideal for hydrologic modeling. Active microwave RS can be used in all weather conditions due to its superior infiltration through the clouds and shadows especially in contrast to optical sensors. However, the requisite algorithms are not widely applicable globally. RS data are currently used in a limited number of hydrological models. One reason is that the majority of operational hydrological modeling techniques are not engineered to work with spatially distributed data, a requirement for making effective use of remotely sensed data. Researchers have to evolve more generalized methodologies and hydrological design models that are better tailored to regular usage of RS data. One more factor contributing to the low adoption of RS in hydrologic processes is the deficiency of adequate teaching and training. Potential users must still receive adequate training and be educated on the benefits of RS. Model development and validation are the two main pillars for water resource assessment and policymaking, where RS has the potential to the uplifting current scenario of water resource modeling. For the efficient application of RS, high-resolution spatial and temporal data for image processing is required, which depends on the quality of sensors to give desired information regarding the object under observation.

References

Aher, P.D., Adinarayana, J., Gorantiwar, S.D., Sawant, S.A., 2014. Information system for integrated watershed management using remote sensing and GIS. In: Srivastava, P.K., Mukherjee, S., Gupta, M., Islam, T. (Eds.), Remote Sensing Applications in Environmental Research. Springer International Publishing, Cham, pp. 17–34, https://doi.org/10.1007/978-3-319-05906-8_2.

Asfaw, E., Suryabhagavan, K.V., Argaw, M., 2018. Soil salinity modeling and mapping using remote sensing and GIS: the case of Wonji sugar cane irrigation farm, Ethiopia. J. Saudi Soc. Agric. Sci. 17 (3), 250–258. https://doi.org/10.1016/j.jssas.2016.05.003.

Brunner, G., 2016. HEC-RAS River Analysis System Hydraulic Reference Manual. USACE CEC, Davis, CA, USA.

Burrough, T., Late, P.P.A., McDonnell, R.A., 2015. Principles of Geographical Information Systems, third ed. Oxford University Press, Oxford, New York.

El Baroudy, A.A., Moghanm, F.S., 2014. Combined use of remote sensing and GIS for degradation risk assessment in some soils of the northern Nile Delta, Egypt. Egypt. J. Remote Sens. Space Sci. 17 (1), 77–85. https://doi.org/10.1016/j.ejrs.2014.01.001.

Elbeltagi, A., Aslam, M.R., Mokhtar, A., Deb, P., Abubakar, G.A., Kushwaha, N.L., et al., 2021. Spatial and temporal variability analysis of green and blue evapotranspiration of wheat in the Egyptian Nile Delta from 1997 to 2017. J. Hydrol. 594, 125662. https://doi.org/10.1016/j.jhydrol.2020.125662.

Elbeltagi, A., Deng, J., Wang, K., Hong, Y., 2020a. Crop water footprint estimation and modeling using an artificial neural network approach in the Nile Delta, Egypt. Agric. Water Manag. 235 (February), 106080. https://doi.org/10.1016/j.agwat.2020.106080.

Elbeltagi, A., Kushwaha, N.L., Srivastava, A., Zoof, A.T., 2022. Artificial intelligent-based water and soil management. In: Poonia, R.C., Singh, V., Nayak, S.R. (Eds.), Deep Learning for Sustainable Agriculture. Academic Press, pp. 129–142, https://doi.org/10.1016/B978-0-323-85214-2.00008-2 (Chapter 5).

Elbeltagi, A., Zhang, L., Deng, J., Juma, A., Wang, K., 2020b. Modeling monthly crop coefficients of maize based on limited meteorological data: a case study in Nile Delta, Egypt. Comput. Electron. Agric. 173 (August 2019), 105368. https://doi.org/10.1016/j.compag.2020.105368.

Elsayed, S., Elhoweity, M., Ibrahim, H., Hassan Dewirc, Y., Migdadic, H., Schmidhaltere, U., 2017. Thermal imaging and passive reflectance sensing to estimate the water status and grain yield of wheat under different irrigation. Agric. Water Manag. 189 (May), 98–110. https://doi.org/10.1016/j.agwat.2017.05.001.

Elsayed, S., Rischbeck, P., Schmidhalter, U., 2015. Comparing the performance of active and passive reflectance sensors to assess the normalized relative canopy temperature and grain yield of drought-stressed barley cultivars. Field Crop Res. 177, 148–160. https://doi.org/10.1016/j.fcr.2015.03.010.

Farg, E., Arafat, S.M., Abd El-Wahed, M.S., El-Gindy, A.M., 2012. Estimation of evapotranspiration ETcand crop coefficient Kcof wheat, in South Nile Delta of Egypt using integrated FAO-56 approach and remote sensing data. Egypt. J. Remote Sens. Space Sci. 15 (1), 83–89. https://doi.org/10.1016/j.ejrs.2012.02.001.

Fok, H.S., Chen, Y., Wang, L., Tenzer, R., He, Q., 2021. Improved Mekong Basin runoff estimate and its error characteristics using pure remotely sensed data products. Remote Sens. 13 (5), 996. https://doi.org/10.3390/rs13050996.

Jerbi, T., Wuyts, N., Angela Cane, M., Faux, A., 2015. High resolution imaging of maize (*Zea maize*) leaf temperature in the field: the key role of the regions of interest. Funct. Plant Biol.

Khan, M.A., Gupta, V.P., Moharana, P.C., 2001. Watershed prioritization using remote sensing and geographical information system: a case study from Guhiya, India. J. Arid Environ. 49 (3), 465–475. https://doi.org/10.1006/jare.2001.0797.

Kudnar, N.S., 2020. GIS-based assessment of morphological and hydrological parameters of Wainganga River basin, Central India. Model. Earth Syst. Environ. 6 (3), 1933–1950. https://doi.org/10.1007/s40808-020-00804-y.

Kumar, P., Sharma, P.K., 2020. Soil salinity and food security in India. Front. Sustain. Food Syst. 4 (174), 1–15. https://doi.org/10.3389/fsufs.2020.533781. 533781.

Kushwaha, N.L., Bhardwaj, A., 2016. Micro-watershed prioritization using RUSLE, Remote Sensing and GIS. Ecoscan 10 (3–4), 585–590. https://scholar.google.com/citations?view_op=view_citation&hl=en&user=qRkt9BoAAAAJ&citation_for_view=qRkt9BoAAAAJ:qjMakFHDy7sC. (Accessed 3 September 2021).

Kushwaha, N.L., Bhardwaj, A., 2017. Remote sensing and GIS based morphometric analysis for micro-watershed prioritization in Takarla-Ballowal watershed. J. Agric. Eng. 54 (3), 48–56. https://www.indianjournals.com/ijor.aspx?target=ijor:joae&volume=54&issue=3&article=006. (Accessed 3 September 2021).

Kushwaha, N.L., Bhardwaj, A., Verma, V.K., 2016. Hydrologic response of Takarla-Ballowal watershed in Shivalik foot-hills based on morphometric analysis using remote sensing and GIS. J. Indian Water Resour. Soc. 36 (1), 17–25.

Kushwaha, N.L., Elbeltagi, A., Mehan, S., Malik, A., Yousuf, A., 2022. Comparative study on morphometric analysis and RUSLE-based approaches for micro-watershed prioritization using remote sensing and GIS. Arab. J. Geosci. 15 (7), 564. https://doi.org/10.1007/s12517-022-09837-2.

Kushwaha, N.L., Rajput, J., Elbeltagi, A., Elnaggar, A.Y., Sena, D.R., Vishwakarma, D.K., et al., 2021. Data intelligence model and meta-heuristic algorithms-based pan evaporation modelling in two different agro-climatic zones: a case study from northern India. Atmosphere 12 (12), 1654. https://doi.org/10.3390/atmos12121654.

Kushwaha, N.L., Yousuf, A., 2017. Soil erosion risk mapping of watersheds using RUSLE, remote sensing and GIS: a review. Res. J. Agric. Sci. 8 (2), 269–277.

Loucks, D.P., van Beek, E., 2017. Water resources planning and management: an overview. In: Loucks, D.P., van Beek, E. (Eds.), Water Resource Systems Planning and Management: An Introduction to Methods, Models, and Applications. Springer International Publishing, Cham, pp. 1–49, https://doi.org/10.1007/978-3-319-44234-1_1.

Mohammadi, A., Costelloe, J.F., Ryu, D., 2015. Evaluation of remotely sensed evapotranspiration products in a large scale Australian arid region: Cooper Creek, Queensland. In: 21st International Congress on Modelling and Simulation (Modsim2015), December, pp. 2346–2352.

Pan, S., Liu, L., Bai, Z., Xu, Y.-P., 2018. Integration of remote sensing evapotranspiration into multi-objective calibration of distributed hydrology-soil-vegetation model (DHSVM) in a humid region of China. Water 10 (12), 1841. https://doi.org/10.3390/w10121841.

Pandey, M., Sharma, P.K., 2017. Remote sensing and GIS based watershed prioritization. In: Presented at the 2017 IEEE International Geoscience and Remote Sensing Symposium (IGARSS), pp. 6182–6185, https://doi.org/10.1109/IGARSS.2017.8128420.

Patel, N.R., Mehta, A.N., Shekh, A.M., 2001. Canopy temperature and water stress quantificaiton in rainfed pigeonpea (*Cajanus cajan* (L.) Millsp.). Agric. For. Meteorol. 109, 223–232.

Pôças, I., Paço, T.A., Paredes, P., Cunha, M., Pereira, L.S., 2015. Estimation of actual crop coefficients using remotely sensed vegetation indices and soil water balance modelled data. Remote Sens. 7 (3), 2373–2400. https://doi.org/10.3390/rs70302373.

Sahana, M., Sajjad, H., 2019. Vulnerability to storm surge flood using remote sensing and GIS techniques: a study on Sundarban biosphere reserve, India. Remote Sens. Appl.: Soc. Environ. 13, 106–120. https://doi.org/10.1016/j.rsase.2018.10.008.

Sanyal, J., Lu, X.X., 2004. Application of remote sensing in flood management with special reference to monsoon asia: a review. Nat. Hazards 33 (2), 283–301. https://doi.org/10.1023/B:NHAZ.0000037035.65105.95.

Schumann, G., Bates, P.D., Horritt, M.S., Matgen, P., Pappenberger, F., 2009. Progress in integration of remote sensing–derived flood extent and stage data and hydraulic models. Rev. Geophys. 47 (4). https://doi.org/10.1029/2008RG000274.

Senthilkumar, M., Gnanasundar, D., Arumugam, R., 2019. Identifying groundwater recharge zones using remote sensing & GIS techniques in Amaravathi aquifer system, Tamil Nadu, South India. Sustain. Environ. Res. 29 (1), 15. https://doi.org/10.1186/s42834-019-0014-7.

Shahid, S.A., Zaman, M., Heng, L., 2018. Soil salinity: historical perspectives and a world overview of the problem. In: Zaman, M., Shahid, S.A., Heng, L. (Eds.), Guideline for Salinity Assessment, Mitigation and Adaptation Using Nuclear and Related Techniques. Springer International Publishing, pp. 43–53.

Shu, Y., Li, H., Lei, Y., 2018. Modelling groundwater flow with MIKE SHE using conventional climate data and satellite data as model forcing in Haihe plain, China. Water 10 (10), 1295. https://doi.org/10.3390/w10101295.

Sinha, R., Bapalu, G.V., Singh, L.K., Rath, B., 2008. Flood risk analysis in the Kosi river basin, north Bihar using multi-parametric approach of Analytical Hierarchy Process (AHP). J. Indian Soc. Remote Sens. 36 (4), 335–349. https://doi.org/10.1007/s12524-008-0034-y.

Wang, J., Zhang, Y., Cheng, Y., Zhang, X., Feng, X., Huang, W., Zhou, H., 2018. Detecting snowfall events over mountainous areas using optical imagery. Water 10 (11), 1514. https://doi.org/10.3390/w10111514.

Yousaf, W., Awan, W.K., Kamran, M., Ahmad, S.R., Bodla, H.U., Riaz, M., et al., 2021. A paradigm of GIS and remote sensing for crop water deficit assessment in near real time to improve irrigation distribution plan. Agric. Water Manag. 243, 106443. https://doi.org/10.1016/j.agwat.2020.106443.

CHAPTER

7

Establishing spatial relationships between land use and water quality influenced by urbanization

Manish Kumar Sinha[a,c], *Klaus Baier*[a], *Rafig Azzam*[a], *Preeti Rajput*[b], and *M.K. Verma*[c]

[a]Department of Engineering Geology and Hydrogeology, RWTH Aachen University, Aachen, Germany [b]Department of Civil Engineering, Government Engineering College, Raipur, Chhattisgarh, India [c]Environmental and Water Resources Engineering, University Teaching Department, Chhattisgarh Swami Vivekanand Technical University Bhilai, Bhilai, Chhattisgarh, India

OUTLINE

1 Introduction 100

2 Study area 102

3 Material and methods 103

3.1 Summary of data 103

3.2 Data processing 104

3.3 Correlation between water quality and land use and statistical significance testing 105

4 Results and discussion 107

4.1 Results of spatial assessment 108

4.2 Interpretation of correlations coefficients 112

5 Conclusion 112

Acknowledgments 113

References 113

Current Directions in Water Scarcity Research, Volume 7
https://doi.org/10.1016/B978-0-323-91910-4.00007-8

1 Introduction

Davis and Golden (1954) described urbanization as a "switch from spread out pattern of human settlements to one of concentration in urban centers." Definitions for the term "urban" vary from country to country, but urban areas are oftentimes defined by either administrative or political boundaries or by a minimal threshold population size (UNICEF, 2012).

In the 2011 revision of world urbanization prospects by the United Nations (UN-DESA, 2015), Indian urban areas are defined as a place satisfying the following three criteria simultaneously, (i) a minimum population of 5000, (ii) at least 75% of male working population engaged in nonagricultural pursuits, and (iii) a density of population of at least 400 per sq. km. The Reserve Bank of India classifies cities into six different tiers, according to their population size and defines that city of more than 100,000 inhabitants are classified as Urban Centers (Planning_Commission (GOI), 2008). India has experienced a continuous concentration of population in tier 1 cities over the years. The number of tier 1 cities has increased from 24 in 1901 to 393 in 2001. Fig. 1 shows the trend of rural and urban population, urban-rural ratio and urban population growth in India for the years 1960–2015. Note that for better visibility the graph for urban population growth was plotted per mill. The figure shows that rural population has decreased from over 80% in 1960 to below 70% in 2015. Despite a decline of urban population growth since the 1980s, urban population is still increasing, coupled with an increase in urban-rural ratio. Between 1960 and 2015, the urban population percentage has increased from below 20% to over 30% (United_Nations, 2017).

On the November 1, 2000, the state Chhattisgarh was formed and simultaneously, Raipur city was declared capital of Chhattisgarh (Census-of-India, 2011). While Raipur had already been a growing city before becoming capital, a large population growth was predicted as a consequence Raipur becoming a capital city (Agrawal, 2013). Table 1 shows actual and predicted population numbers of Raipur for the years 2000 up until 2025. The data reveal that

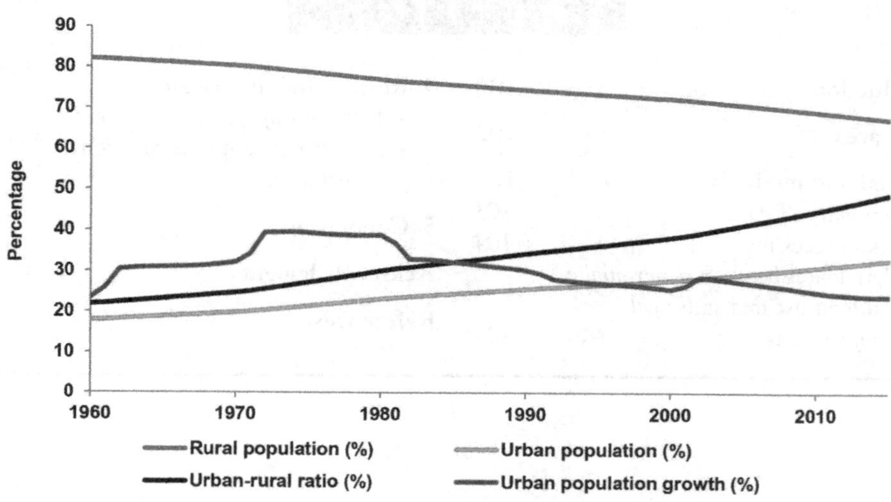

FIG. 1 Trends of Indian rural and urban population, 1960–2015 (Fay and Opal, 2000).

TABLE 1 Raipur city population, 2000–2025 (projected) (UN-DESA, 2015).

Year	2000	2005	2010	2015	2020	2025
Raipur population (million)	0.680	0.858	1.088	1.361	1.621	1.974

the population of Raipur roughly doubled between the years 2000 and 2015. With increasing population in Raipur, the numbers and percentages of slum population increased as well. In 1991, 21% of the total population of Raipur were slum dwellers. This increased slightly to 24% in 2001 and then dramatically to an estimated 43% in 2005 (GIZ, ASEM, and RMC, 2011). The heavy increase in population in the last few decades has caused a multitude of problems concerning a wide range of areas. One major problem is the outdated and nonfunctioning sanitation system of Raipur (GIZ, ASEM, and RMC, 2011; Suman et al., 2011). To address this issue, a City Sanitation Plan (CSP) for Raipur has been developed by the Raipur Municipal Corporation (RMC) in collaboration with the Deutsche Gesellschaft für Internationale Zusammenarbeit (GIZ). The CSP explains the current sanitation situation in detail, giving information on slums, water supply, access to toilets, waste water treatment, and solid waste management.

As first key issue, the CSP names poorly designed, operated and managed toilets in urban poor areas. 20% have no access to any toilets at all and on average, there is only one toilet seat for 463.5 urban poor users. Due to this limited access and due to the poor conditions of community toilets, a total of 57% of the urban poor population resorts to open defecation. Suman et al. (2011) Inappropriate handling of solid waste is named as the second key issue. Untreated solid waste dumps are a typical sight in all parts of Raipur. Waste gets dumped in open areas and storm water drains where pollutants can easily be mobilized through contact with water, ultimately polluting water bodies. SUDA & RMC (2017) and Suman et al. (2011) there is currently no fully connected sewage network, more than half of Raipur's properties are connected to unscientifically designed septic tanks. These septic tanks contribute to the contamination of water bodies and water supply systems because they overflow directly into open drainage channels. The storm water drainage network consists of both natural drains and man-made drains that were constructed on the side of Raipur's roads. The majority of the drains (locally known as "nallahs") are open and therefore prone to being clogged with untreated solid waste. The drains are cleaned only infrequently and clogged drains are a very common sight. The blockages cause a multitude of problems, e.g., serving as breeding grounds for mosquitoes, the fouling wastes produce unpleasant odors that are noticeable all over the city. Most importantly, messy and clogged drains are where solid waste and sewage comes into contact with water, and therefore, pollutants are mobilized which sooner or later may lead to contamination of groundwater. It causes deterioration of water quality in several areas of the city (GIZ, ASEM, and RMC, 2011; Suman et al., 2011).

Raipur clearly suffers from severe consequences of rapid urbanization. Slum population has increased dramatically since Raipur became capital of Chhattisgarh in 2000 (UN-DESA, 2015). The public service lacks organization, staff and financial means and the especially the poor state of the sanitation system causes a variety of problems, including endangering the health of Raipur's inhabitants (SUDA & RMC, 2017).

In order to reduce existing and prevent future water quality deterioration, research must focus on the causes and sources of pollution. Point sources such as municipal wastewater treatment plants or industrial plants are relatively simple to identify, but it has become obvious that a large proportion of water pollution is caused by nonpoint sources whose identification and containment is much more complex (Bourgeois-Calvin, 2008). The impacts of different land use types on water quality have become an important object of research with the aim to improve water management (Foley et al., 2005). Previous research has shown that large-scale land use types like urban, agricultural or forest land use can be linked to specific pollutants and nutrients (e.g., Bourgeois-Calvin, 2008; Tong and Chen, 2002).

While relationships between surface water quality and land use have been explored for larger scales, little is known about the impacts of specific urban land use types on surface water. Thus, the objective of this work is to contribute to the investigation of relationships between urban land use types and surface water quality. Raipur city is a suitable study area because the municipal area of Raipur is dominated by urban land use and various surface water bodies are spread out across the city area. The method that is applied in this work is to evaluate correlation coefficients between land use and surface water quality data. These correlation coefficients are evaluated to explore if there are any noticeable relationships and if so, whether or not they are meaningful.

2 Study area

Raipur is the capital and largest city of the Indian state Chhattisgarh (Fig. 2). It was formerly part of the state Madhya Pradesh, until the November 1st, 2000, when 16 districts of

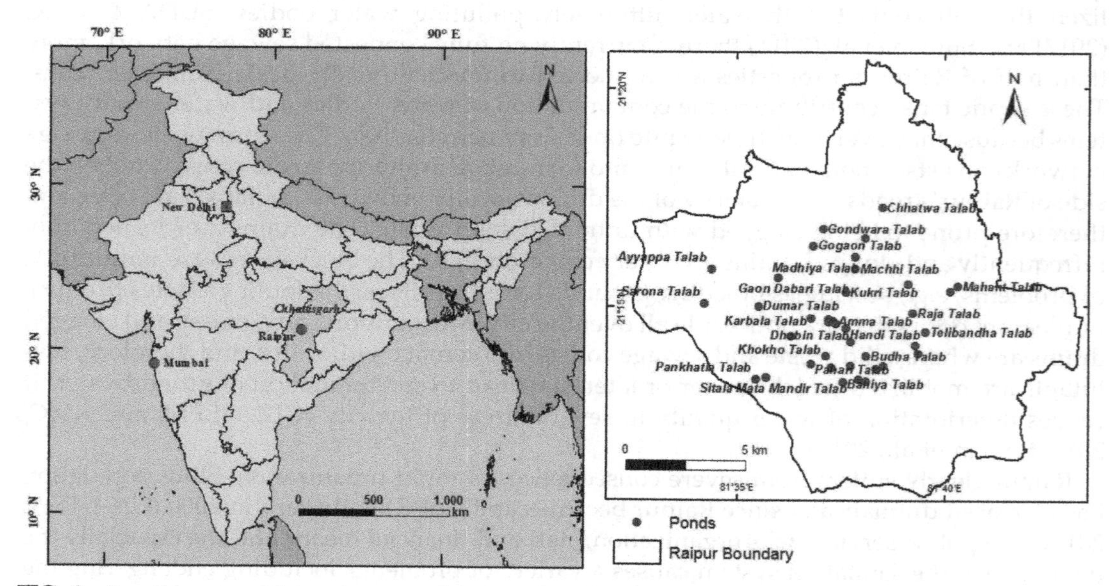

FIG. 2 Location map of the study area and the sampling locations in Raipur city.

Madhya Pradesh were formed into the new state Chhattisgarh (Sinha et al., 2022; Vishwakarma et al., 2014). As of the official census data, the population of Raipur was 1,010,087 in 2011. In 2015, Raipur population is estimated to be around 1.361 million. The city is divided into 67 wards and spreads across an area of over 193 km^2 (Sinha et al., 2019a). Raipur is located at 21.25°N, 81.63°E and its elevation varies from 268 to 304 m above mean sea level. In the Köppen climate classification scheme, Raipur is classified as Aw (tropical wet climate), as it experiences less than 60 mm of precipitation and less than 4% of the total annual precipitation in the driest month. The annual mean temperature is 26.8 °C and annual mean precipitation is 1276 mm (Sinha et al., 2019b, 2021a, 2022). Evapotranspiration is high (average 4.9 mm/day) and the relative humidity averages 33% to 85% (Chandrakar and Kumar Sinha, 2021; Rajput and Sinha, 2020; Tiwari and Sinha, 2020). Raipur experiences a hot summer that lasts from March to mid-June (Mukherjee et al., 2017) where temperatures often rise above 45°C and. The mean temperature in May is 35.4°C (and mean maximum temperature is 45.2°C).

Ponds in Raipur are suffering from bad water quality and pollutant contamination, leading to a variety of problems as the pond water is used for irrigation, washing, bathing and other activities where people have direct contact to the contaminated water (e.g., Parashar and Parashar, 2012; Shobhnath et al., 2000; Sinha et al., 2021b; Swarnakar and Choubey, 2016). On account of this, it is crucial to understand why the pond water is below desired quality. Examining the influence of land use is one potential way to find causes and sources of surface water pollution (e.g., Bourgeois-Calvin, 2008; Brett et al., 2005; Omernik, 1976; Tufford et al., 2003).

3 Material and methods

Analyzing relationships between land use and water quality is a complicated field of research, because there is no optimal methodology that will yield 100% accurate and realistic results. One major reason that complicates the procedure of linking land use and water quality data is that these data have different dimensions. To link water quality and land use data, the land use data has to be transformed into a one-dimensional numerical parameter. By doing that, spatial information is lost, which can degrade the quality of results. The impact that these methodical limitations have on the resulting Pearson correlation coefficients is difficult to estimate, but it might be very significant. Because of the use of suboptimal methodology and the lack of superior alternatives, the field of research has had trouble to identify strong and significant relationships between land use and water quality {Rajput, 2022 #2259}.

3.1 Summary of data

To create a comprehensive data base, a variety of different sources of surface water quality data was used. 38 location water quality data were utilized to conduct this study. All gathered water quality data consist of compilations of water quality parameter values that were determined by analyzing pond water samples. The statics of the sample are as shown in Table 2. The land use data is based on existing land use in Raipur and includes projected land-use changes as planned by the Raipur Development Authority. The most practical solution

TABLE 2 Statistics of water quality parameters used in the study.

Parameters	Unit	Min	Max	Median	Parameters	Unit	Min	Max	Median
Alkalinity (total)	mg/L	85.1	484	200	Hardness (total)	mg/L	2.4	580	136.5
BOD	mg/L	0.1	30	3.7	HCO_3	mg/L	146	732	221
Ca	mg/L	9	170	30.7	K	mg/L	2.4	905	37
Cl	mg/L	5.9	696	150	Mg	mg/L	7	260.5	17
COD	mg/L	0.2	87	31	Na	mg/L	12.4	430	115
Coli (fecal)	MPN/100mL	4	1305	21.5	NO_3	mg/L	0	21.6	4
Coli (total)	MPN/100mL	15	3200	102.5	pH	–	7.1	8.94	8.06
DO	mg/L	1.2	10.5	6.4	PO_4	mg/L	0	3.3	0.6
EC	μS/cm	130	2330	884	SO_4	mg/L	0.3	680	28
F	mg/L	0.1	1	0.4	TDS	mg/L	28.1	1491	573
Fe	mg/L	0	1.4	0.3	Turbidity	NTU	3	232	14.3

was to use GIS land-use data that was created according to the Raipur Master Plan 2021. The land use is classified into six different types: Commercial, Industrial, Public Utilities/Facilities/Transportation, Public/Semipublic, Recreational and Residential. Further GIS data that was used in this work include a Shapefile of Raipur city drainage that was created within the framework of a survey on topographic sheets of India and a dataset of Raipur slum locations which was created by the RMC in the context of the city sanitation plan.

3.2 Data processing

Before linking the data on surface water quality and the land use, they were structured in a way to make further steps possible. The surface water quality data base consisted of sample values for 22 different water quality parameters in 38 sampled ponds. In this work, the method that was used to quantify land use is to analyze an area in regards to the percentages that each land use class covers in this area. As the goal of this work was to find relationships between land use and water quality, it was important to choose areas that resemble catchment areas of the ponds. To create meaningful catchment areas, exhaustive data on the open drainage network and on the pond inlets and outlets is required. This was carried out by creating buffer areas surrounding the ponds in ArcGIS. As a simple approximation, buffers A were created as circular areas with a diameter of 1 km. Buffers B are 1 km wide and 2.5 km long and are directed toward the ponds' approximate catchment areas. These directions were estimated using Raipur city drainage data. Each buffer area's land use distribution was analyzed using ArcGIS. It was determined, what percentage of each buffer is classified as the land use types Commercial, Industrial, Public Utilities/Facilities/Transportation, Public/Semipublic, Recreational or Residential. Additionally, it was determined what percentage of each buffer is classified as slum area.

3.3 Correlation between water quality and land use and statistical significance testing

In order to improve water quality in areas that suffer from polluted ground or surface water, it is inevitable to know the sources and the transport mechanisms of the pollutants. Correlations of land use and water quality have been studied for several decades with a large part of these studies focusing on water quality of streams in the United States of America (Brett et al., 2005; Omernik, 1976; Tufford et al., 2003).

Land use is a tool to describe human activities and socioeconomic functions on the Earth's surface that are directly related to the land (Harrison, 2006; Siderelis and Nagy, 1994). Over time, numerous land use classifications have been developed, because for different needs, different classifications are required. The lack of a standardized approach can cause a number of complications. Fisher et al. (2005) Remote sensing data like satellite imageries or aerial photography are commonly used to gather land use information. But the interpretation of remote sensing data is subjective and mistakes during the assignation of land use classes to area segments are inevitable. Anderson (1976) for research purposes related to urbanization, working with land use is a powerful tool. Because urbanization and land use changes go hand in hand (Carlson and Traci Arthur, 2000).

The quality of water is quantified by the content of chemical components (e.g., nitrate, dissolved oxygen), physical properties (turbidity, temperature, etc.), biological components (algae, bacteria, etc.) and radiological characteristics (WHO, 2011). Several institutions like the Bureau of Indian Standards and the World Health Organization investigate the effects that these characteristics have on humans. Both the BIS and the WHO have published reports which examine water quality parameters and establish drinking water limits for these parameters (IS10500, 2012; WHO, 2011). Table 3 lists the parameters that are considered in this work and specifies the drinking water limits established by the BIS and the WHO.

TABLE 3 Relevant water quality parameters and drinking water limits as established by the BIS and the WHO (BIS, 2012; WHO, 2011).

Parameter	Unit	BIS acceptable/permissible limit	WHO limit
Alkalinity (total, as $CaCO_3$)	mg/L	200/600	–
Biochemical oxygen demand (BOD)	mg/L	–	–
Calcium (as Ca)	mg/L	75/200	–
Chemical oxygen demand (COD)	mg/L	–	–
Chloride (as Cl)	mg/L	250/1000	250
Coliform (fecal)	MPN/100mL	0	–
Coliform (total)	MPN/100mL	0	–
Dissolved oxygen (DO)	mg/L	–	–
Electric conductivity (EC)	μS/cm	–	–
Fluoride (as F)	mg/L	1.0/1.5	1.5

Continued

TABLE 3 Relevant water quality parameters and drinking water limits as established by the BIS and the WHO (BIS, 2012; WHO, 2011).—Cont'd

Parameter	Unit	BIS acceptable/permissible limit	WHO limit
Hardness (total, as $CaCO_3$)	mg/L	200/600	500
Hydrogen carbonate (as HCO_3)	mg/L	–	–
Iron (as Fe)	mg/L	0.3	0.3
Magnesium (as Mg)	mg/L	30/100	–
Nitrate (as NO_3)	mg/L	45	50
Ph	–	6.5–8.5	6.5–8.5
Phosphate (as PO_4)	mg/L	–	–
Potassium (as K)	mg/L	–	–
Sodium (as Na)	mg/L	–	200
Sulfate (as SO_4)	mg/L	200/400	–
Total dissolved solids (TDS)	mg/L	500/2000	500
Turbidity	NTU	1/5	–

Pearson's correlation coefficient r is an index that is commonly used to describe relationships between two variables (Lane et al., 2014). The index ranges from −1 to 1. A Pearson correlation coefficient of −1 indicates a perfect negative linear relationship between the tested variables, whereas an r of 1 indicates a perfect positive linear relationship between the variables (e.g., Devore, 2015; Lane et al., 2014). The Pearson correlation coefficient is defined as the ratio of the covariance of two variables to the square root of their variances. For a set of N two-dimensional data points $[x_1, x_2, \ldots, x_N]$ and $[y_1, y_2, \ldots, y_N]$, this is expressed as

$$r = \frac{C_{xy}}{\sqrt{C_{xx}C_{yy}}} = \frac{C_{xy}}{\sigma_x \sigma_y} \tag{1}$$

with

$$C_{xx} = \sigma_x^2 = \frac{1}{N-1}\sum_i (x_i - \overline{x})^2 \tag{2}$$

$$C_{yy} = \sigma_y^2 = \frac{1}{N-1}\sum_i (y_i - \overline{y})^2 \tag{3}$$

$$\overline{x} = \frac{1}{N}\sum_i x_i \; \overline{y} = \frac{1}{N}\sum_i y_i \tag{4}$$

Evans (1996) provided a guide to describe the strength of linear relationships based on Pearson correlation coefficient ranges (Table 4).

TABLE 4 Strength of r based on (Evans, 1996).

Range of $\|r\|$	Strength
0–0.19	Very weak
0.20–0.39	Weak
0.40–0.59	Moderate
0.60–0.79	Strong
0.80–1	Very strong

The number of pairs of values has a strong influence on the resulting correlation coefficients. Larger sample size produce correlation coefficients that give a more realistic estimation of the linear relationship between two variables than small sample sizes (Bonett and Wright, 2000; Onwuegbuzie and Daniel, 1999). Therefore, correlation coefficients that are calculated using only small sample sizes can be misleading and might result in misinterpretation of the analyzed data (Onwuegbuzie and Daniel, 1999).

The *t*-test is a statistical significance test that can be used to analyze the informative value of correlation coefficients (Artusi et al., 2002). With this test, significance levels and confidence levels are determined to quantify the significance of correlations.

The test states that a correlation between two variables is significant with a probability P if t_{N-2} is larger than $t_{N-2,\,P}$ (Onwuegbuzie and Daniel, 1999). P is then specified as the confidence level and $\alpha = 1 - P$ is called significance level.

t_{N-2} is calculated using formula 10 (N being the sample size):

$$t_{N-2} = \frac{r}{\sqrt{\frac{1-r^2}{N-2}}} \tag{5}$$

$t_{N-2,\,P}$ is determined by using a table for the critical values of Student's t distribution with N degrees of freedom (Anderson and Bancroft, 1952). Because of the small sample sizes that were used to calculate correlation coefficients in this work, the application of significance tests was of critical importance. Due to the large number of calculated Pearson correlation coefficients in this work, the *t*-test was only performed for correlations larger than $r = 0.40$.

4 Results and discussion

It is evident that most of the analyzed ponds accumulate in the central part of Raipur city. Fig. 3 shows the land use map of Raipur according to the Raipur Master Plan 2021. The locations of slums are depicted as well. The map shows that most of the industrial area is located in the north of Raipur. Aside from roads that are distributed evenly throughout the city, most of the public utilities, facilities, and transportation land use are situated in the northeast of Raipur. Public and semipublic areas are mainly found in the center and in the western part of the city whereas recreational land use takes up most of the outer areas. Residential land use takes up a large part of Raipur's area and is encountered in most parts of the city.

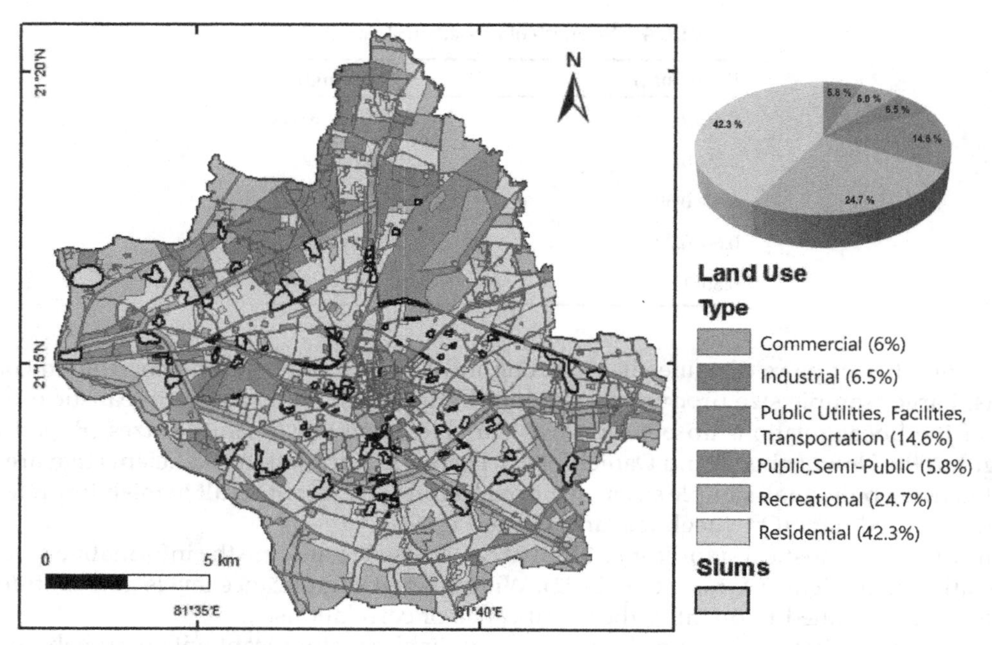

FIG. 3 Land use of Raipur area in 2021, location of slums.

As indicated above, residential land use is the largest contributor in terms of area, taking up over 40% of the total area in Raipur. About one fourth of the total area is associated with recreational land use. Fifteen percent of the total area belongs to public utilities, facilities and transportation. Both public and semipublic, commercial, and industrial areas take up around 6%, respectively.

4.1 Results of spatial assessment

Two types of buffers (buffers A and buffers B) were created surrounding the relevant ponds (Fig. 4). Buffers A are circular areas with a diameter of 1 km. Buffers B are 1 km wide, 2.5 km long and are directed toward the ponds' approximate catchment areas. The blue lines indicate drainage pathways and are radially directed outward from the center of Raipur outward.

In order to find and quantify relationships between land use and water quality, Pearson correlation coefficients were calculated using the surface water quality data from and the area percentages of land use types in each buffer. Statistical confidence levels of the correlations were also determined, allowing an assessment about the informative value of the correlation coefficients. The calculated Pearson correlation coefficients are compiled in Table 5 (for buffers A) and Table 6 (for buffers B).

From total number of 153 calculated correlation coefficients in Table 5, 127 are of very weak or weak strength (green cells). In four different cases, correlation coefficients of $r = 0.00$ were found. The land use class "Residential" and "Slums" produced only very weak or weak

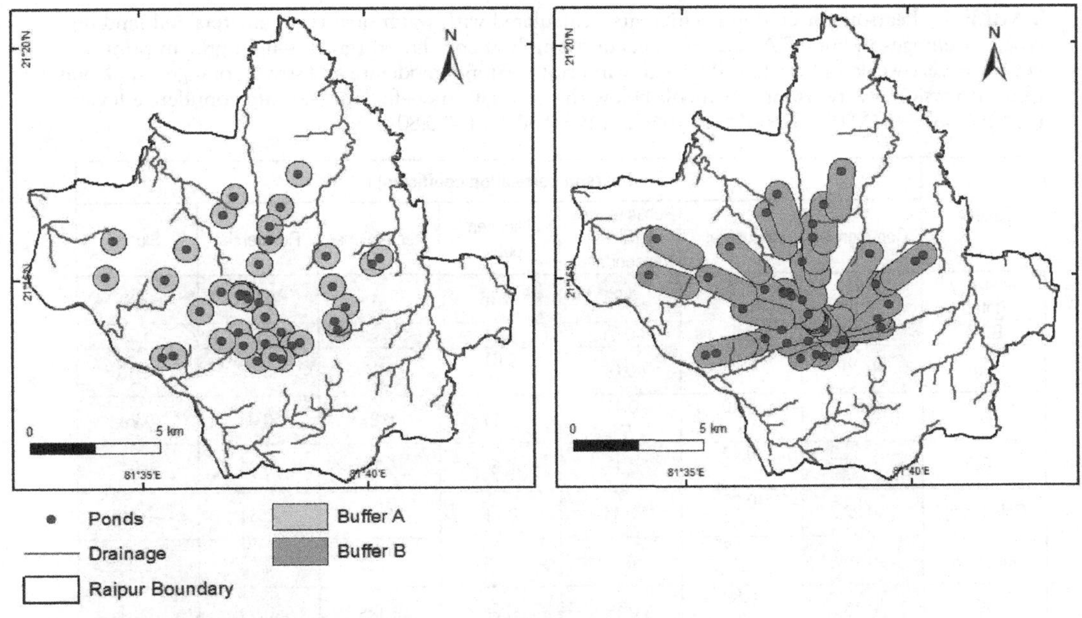

FIG. 4 Buffer zones A and B around the ponds and drainage pathways.

correlation coefficients with any water quality parameters. 21 moderate and strong correlations were found (yellow cells). They are mainly associated with the land use class "Industrial," which shows six moderate and six strong correlations. All five very strong correlations are also associated with the land use class "Industrial." The strongest correlations were found between industrial land use and the water quality parameters total alkalinity ($r = -0.86$), dissolved oxygen ($r = -0.89$), hydrogen carbonate ($r = -0.98$), sodium ($r = -0.93$) and turbidity ($r = 0.94$).

For all correlation coefficients larger than $r = 0.40$, confidence levels were calculated. It is noticeable that most correlation coefficients for the industrial land use have low confidence levels. 14 of these correlations' coefficients have a confidence level of $P < 0.80$, whereas only 2 have a confidence level of $P = 0.80$ and 1 has a confidence level of 0.90. For the correlations of the other land use classes, the calculated confidence levels are generally higher. Especially for recreational land use, the examined correlations all have high confidence levels of $P = 0.99$.

Table 6 also shows 153 correlation coefficients, of which 130 are of very weak or weak strength, compared to 127 very weak or weak correlations in Table 5. Similar to the results of using buffers A, the strongest correlations using buffers B were found with industrial land use. Six water quality parameters correlate moderately with industrial land use and another six water quality parameters exhibit strong correlations with industrial land use. The two strongest correlations are found between industrial land use and dissolved oxygen ($r = -0.82$) and between industrial land use and turbidity ($r = 0.83$). Both of these correlations were among the strongest correlations for using buffers A as well. However, compared to the

TABLE 5 Pearson correlation coefficients r calculated with water quality parameters and land use type percentages in buffers A. Colors indicate strength of correlation (green—light gray in print version = very weak and weak; yellow—gray in print version = moderate and strong; orange—dark gray in print version = very strong). Symbols below the correlation coefficient indicate confidence levels (****: $P=0.99$; ***: $P=0.95$;**: $P=0.90$; *: $P=0.80$; +: $P<0.80$).

Buffer A	Pearson correlation coefficient *r*						
	Commercial	Industrial	Public utilities, facilities, transportation	Public, semi-public	Recreational	Residential	Slums
Alkalinity (total)	0.48 ***	-0.86 +	0.23	-0.36	-0.14	0.00	0.06
BOD	-0.03	-0.26	-0.21	0.11	-0.26	0.09	0.09
Ca	-0.38	-0.41 +	-0.04	0.61 ****	0.30	-0.17	-0.10
Cl	0.56 ***	-0.70 +	0.41 ***	-0.27	-0.27	-0.08	0.29
COD	-0.04	-0.47 +	0.17	-0.30	0.25	-0.06	0.04
Coli (fecal)	-0.22	-0.80 +	-0.14	0.21	0.52 ****	-0.34	-0.10
Coli (total)	-0.10	-0.78 +	-0.09	-0.14	0.46 ****	-0.15	-0.03
DO	-0.16	-0.89 *	0.19	-0.04	0.09	-0.09	-0.34
EC	-0.13	-0.59 +	0.16	0.10	-0.22	-0.01	0.10
F	-0.35	-0.24	0.24	-0.09	-0.15	0.21	-0.08
Fe	-0.29	-	-0.20	-0.02	0.04	0.04	-0.24
Hardness (total)	-0.37	-0.54 +	-0.09	0.11	0.20	-0.06	-0.36
HCO_3	0.36	-0.98 *	0.19	-0.30	-0.08	0.01	0.19
K	-0.37	-0.21	0.23	-0.08	-0.27	0.11	-0.06
Mg	-0.32	-0.54 +	0.25	-0.01	-0.23	0.10	-0.13
Na	-0.01	-0.93 **	0.13	-0.14	0.11	-0.24	0.05
NO_3	-0.48 **	0.67 +	-0.17	0.39	-0.01	-0.05	0.09
pH	0.22	0.65 +	0.00	-0.50 ***	-0.27	0.31	-0.07
PO_4	0.36	-0.69 +	-0.27	-0.15	0.54 ****	-0.30	-0.18
SO_4	-0.17	-0.21	0.04	0.00	-0.03	0.00	-0.21
TDS	-0.12	-0.58 +	0.07	0.05	-0.11	-0.02	-0.02
Turbidity	-0.34	0.94 +	-0.15	-0.16	0.22	0.03	-0.15

results in Table 5, where three correlation coefficients were higher than 0.9 or lower than −0.9, no correlation coefficients of the magnitudes are found using buffers B. Another evident difference between the results of Table 5 and Table 6 is that Table 5 shows four correlation coefficients of moderate strength for slums (compared to zero in Table 6). The confidence

TABLE 6 Pearson correlation coefficients r calculated with water quality parameters and land use type percentages in buffers B. Colors indicate strength of correlation (green—light gray in print version = very weak and weak; yellow—gray in print version = moderate and strong; orange—dark gray in print version = very strong). Symbols below the correlation coefficient indicate confidence levels (****: $P=0.99$; ***: $P=0.95$; **: $P=0.90$; *: $P=0.80$; +: $P<0.80$).

Buffer B	Pearson correlation coefficient *r*						
	Commercial	Industrial	Public utilities, facilities, transportation	Public, semi-public	Recreational	Residential	Slums
Alkalinity (total)	0.48 ****	-0.61 +	0.06	-0.39	-0.10	0.27	0.13
BOD	0.13	-0.18	-0.08	0.68 ****	-0.28	-0.31	-0.37
Ca	-0.03	-0.41 +	0.19	0.23	0.12	-0.42 ***	-0.32
Cl	0.56 ****	-0.48 +	0.14	-0.18	-0.18	0.05	0.41 **
COD	-0.14	-0.43 +	0.21	-0.16	0.30	-0.15	-0.01
Coli (fecal)	-0.23	-0.73 +	-0.01	0.10	0.20	-0.20	0.33
Coli (total)	-0.26	-0.65 +	0.02	-0.07	0.33	-0.24	0.43 *
DO	-0.02	-0.82 **	0.15	0.11	0.03	-0.16	-0.56 ***
EC	0.26	-0.40	-0.01	0.00	-0.11	0.07	0.34
F	-0.20	-0.27	0.23	-0.30	-0.04	0.28	0.07
Fe	-0.14	-	-0.30	-0.07	0.20	0.00	-0.12
Hardness (total)	-0.24	-0.57 +	0.03	0.13	0.22	-0.25	-0.38
HCO_3	0.27	-0.74 +	-0.08	-0.20	-0.03	0.17	0.38
K	0.06	-0.22	0.00	0.09	-0.26	0.20	-0.13
Mg	0.18	-0.53 +	0.34	0.14	-0.27	-0.13	-0.33
Na	0.10	-0.56 +	0.01	-0.13	0.19	-0.09	0.12
NO_3	-0.18	0.64 +	-0.22	0.25	-0.06	0.01	-0.59 ***
pH	0.36	0.38	0.05	-0.07	-0.20	0.12	-0.06
PO_4	-0.16	-0.63 +	-0.25	-0.06	0.47 ****	-0.16	0.13
SO_4	0.06	-0.22	0.13	0.21	-0.02	-0.27	-0.30
TDS	0.15	-0.35	0.00	-0.04	-0.09	0.09	0.12
Turbidity	-0.26	0.83 *	-0.21	-0.16	0.19	0.13	-0.04

levels for the correlation coefficients associated with industrial land use are generally low, with 12 confidence levels of $P<0.80$ and only two confidence levels higher than that. In contrast, the calculated confidence levels associated with the other land uses are predominantly higher: The correlations with the land uses "Commercial, "Public Utilities,

Facilities, Transportation," "Public, Semipublic," and "Recreational," all exhibit confidence levels of $P=0.99$. Confidence levels for correlations associated with slums vary between $P=0.8$ and $P=0.95$.

4.2 Interpretation of correlations coefficients

To summarize these findings, several significant moderate and few significant strong linear relationships between land use types and water quality parameters were found. Of the high number of strong and very strong linear relationships that were found between industrial land use and water quality parameters, most are of low ($P=0.80$) or very low ($P<0.80$) significance; therefore, they do not hold much informative value. For most land use types, the results using buffers A and buffers B featured noticeable differences with regard to the strength and significance of the calculated correlation coefficients, indicating that the results are greatly influenced by the choice of analyzed areas. Several significant moderate and few strong linear relationships between land use and water quality parameters were found in this study. There are many factors that limit the informative value of the findings, such as a suboptimal data base and flaws in the methodology. Concerning the field of research, more work has to be put into the development of a methodology that enables to link land use and water quality data while respecting the fact that land use data is not numerical. The previous approach of linking the data through correlation coefficients may not prove to be the most useful.

5 Conclusion

Within the context of urbanization and its far-reaching impacts on infrastructure in general and water resources in particular, the goal of this work was to identify and quantify relationships between land use and water quality on an urban scale. Due to its numerous water bodies that are spread throughout the city area and due to its predominant urban land use, Raipur city, the vastly growing capital of Chhattisgarh, India, served as an ideal study area. Water quality data of 38 ponds, all located within the municipal area of Raipur, were obtained from previous studies on pond water quality in Raipur. The land use data was transformed into numerical values by creating buffer zones around the ponds and analyzing these buffer zones in terms of their land use type percentages. In order to find relationships between land use and water quality, the water quality and land use data were linked through Pearson correlation coefficients. Twenty-two different water quality parameters and six different land use types (and slum areas in addition) were analyzed. Industrial land use produced the strongest correlation coefficients. Applying circular buffer zones with a diameter of 1 km (buffer type A), industrial land use showed very strong negative correlations with total alkalinity ($r=-0.86$), dissolved oxygen ($r=-0.89$), hydrogen carbonate ($r=-0.98$), sodium ($r=-0.93$), and a very strong positive correlation with turbidity ($r=0.94$). Buffer zones with a length of 2.5 km and a width of 1 km, directed upstream toward drainage pathways (buffer type B) produced somewhat weaker correlations. Because only few data pairs were used in the calculation of correlation with industrial land use, the confidence levels are mainly below 0.80.

To verify the relationships between industrial land use and various water quality parameters, a larger data base is necessary. The other land use types produced weaker correlation coefficients than industrial land use in general. Strong correlations were found between public/semipublic land use and calcium ($r=0.61$) and between public/semipublic land use and biochemical oxygen demand ($r=0.68$). Both correlations are of high significance with a confidence level of 0.99. The rest of the correlation coefficients are distributed into few moderate correlations and predominantly weak and very weak correlations. These results may be utilized for planning purpose of the study area. The results are also valuable to local authorities and stakeholder who manage water quality status in the city.

Acknowledgments

The authors would like to thank the Deutscher Akademischer Austauschdienst- German Academic Exchange Service (DAAD), as the research material is based upon work within the program "A New Passage to India." One of the authors would like to acknowledge the help in field data collection provided by LIH workgroup.

References

Agrawal, V., 2013. Raipur as a new capital: impact on population. Int. J. Eng. Res. Sci. Technol. 2 (1), 41–46.
Anderson, J.R., 1976. A Land Use and Land Cover Classification System for use With Remote Sensor Data. vol. 964 US Government Printing Office.
Anderson, R.L., Bancroft, T.A., 1952. Statistical Theory in Research. McGraw-Hill Book Company, Inc., New York.
Artusi, R., Verderio, P., Marubini, E., 2002. Bravais-Pearson and Spearman correlation coefficients: meaning, test of hypothesis and confidence interval. Int. J. Biol. Markers 17, 148–151. 12113584.
Bonett, D.G., Wright, T.A., 2000. Sample size requirements for estimating Pearson, Kendall and Spearman correlations. Psychometrika 65 (1), 23–28. https://doi.org/10.1007/bf02294183.
Bourgeois-Calvin, A., 2008. Relationship between Land Use and Surface Water Quality in a Rapidly Developing Watershed in Southeast Louisiana. University of New Orleans.
Brett, M.T., Arhonditsis, G.B., Mueller, S.E., Hartley, D.M., Frodge, J.D., Funke, D.E., 2005. Non-point-source impacts on stream nutrient concentrations along a forest to urban gradient. Environ. Manag. 35 (3), 330–342. https://doi.org/10.1007/s00267-003-0311-z.
Carlson, T.N., Traci Arthur, S., 2000. The impact of land use—land cover changes due to urbanization on surface microclimate and hydrology: a satellite perspective. Glob. Planet. Chang. 25 (1), 49–65. https://doi.org/10.1016/S0921-8181(00)00021-7.
Census-of-India, 2011. Census of India 2011, District Census Handbook. Census-of-India, Raipur. Retrieved from: https://censusindia.gov.in/2011census/dchb/2211_PART_B_DCHB_RAIPUR.pdf.
Chandrakar, S., Kumar Sinha, M., 2021. Comparative analysis of NDVI and LST to identify urban Heat Island effect using remote sensing and GIS. In: Vikas, D., Sri, R.K.M., Marta, M.-D., Vaibhav, D. (Eds.), Water Resource Technology. De Gruyter, pp. 99–110.
Davis, K., Golden, H.H., 1954. Urbanization and the development of pre-industrial areas. Econ. Dev. Cult. Chang. 3 (1), 6–26. https://doi.org/10.1086/449673.
Devore, J.L., 2015. Probability and Statistics for Engineering and the Sciences. Cengage Learning.
Evans, J.D., 1996. Straightforward Statistics for the Behavioral Sciences. Brooks/Cole.
Fay, M., Opal, C., 2000. Urbanization Without Growth: A not so Uncommon Phenomenon. vol. 2412, World Bank Publications.
Fisher, P., Comber, A.J., Wadsworth, R., 2005. Land use and land cover: contradiction or complement. In: Re-presenting GIS, pp. 85–98.
Foley, J.A., DeFries, R., Asner, G.P., Barford, C., Bonan, G., Carpenter, S.R., et al., 2005. Global consequences of land use. Science 309 (5734), 570–574. https://doi.org/10.1126/science.1111772.

GIZ, ASEM, & RMC, 2011. City Level Strategy: City Sanitation Plan for Raipur. Retrieved from Alchemy Urban Systems (P) Ltd.: http://www.urbansanitation.org/e31169/e49836/e49828/.

Harrison, A., 2006. National Land Use Database: Land Use and Land Cover Classification. Office of the Deputy Prime Minister, Ed.

IS10500, B, 2012. Indian Standard Drinking Water–Specification (Second Revision). Bureau of Indian Standards (BIS), New Delhi.

Lane, D.M., Scott, D., Hebl, M., Guerra, R., Osherson, D., Zimmer, H., 2014. Introduction to Statistics. Rice University, Houston, TX, pp. 474–476.

Mukherjee, A., Gupta, A., Ray, R.K., Tewari, D., 2017. Aquifer response to recharge–discharge phenomenon: inference from well hydrographs for genetic classification. Appl. Water Sci. 7 (2), 801–812. https://doi.org/10.1007/s13201-015-0293-z.

Omernik, J.M., 1976. The Influence of Land Use on Stream Nutrient Levels. vol. 76 US Environmental Protection Agency, Office of Research and Development, Corvallis Environmental Research Laboratory, Eutrophication Survey Branch.

Onwuegbuzie, A.J., Daniel, L.G., 1999. Uses and Misuses of the Correlation Coefficient.

Parashar, A.K., Parashar, R., 2012. Environmental Impact Assessment of Water Quality of Major Ponds of Raipur City.

Planning_Commission (GOI), 2008. Eleventh Five Year Plan (2007–2012). Retrieved from: Oxford University Press, New Delhi.

Rajput, P., Sinha, M.K., 2020. Geospatial evaluation of drought resilience in sub-basins of Mahanadi river in India. Water Supply 20 (7), 2826–2844. https://doi.org/10.2166/ws.2020.178.

Shobhnath, Mishra, A.K., Shrivastava, M., Pati, G.C., Tewari, D., Mukerjee, A., 2000, 29 March. Impact of urbanization on ground water resource of Raipur urban agglomerate. In: Paper Presented at the Workshop on Strategy for Ground Water Development, Madhya Pradesh, India.

Siderelis, K., Nagy, Z., 1994. A Standard Classification System for the Mapping of Land Use and Land Cover. State of North Carolina, Governor's Office of State Planning.

Sinha, M.K., Baghel, T., Baier, K., Verma, M.K., Jha, R., Azzam, R., 2019a. Impact of urbanization on surface runoff characteristics at catchment scale. In: Water Resources and Environmental Engineering I. Springer, pp. 31–42.

Sinha, M.K., Baier, K., Azzam, R., Baghel, T., Verma, M.K., 2019b. Semi-distributed modelling of stormwater drains using integrated hydrodynamic EPA-SWM model. In: Harmony Search and Nature Inspired Optimization Algorithms. Springer, pp. 557–567.

Sinha, M.K., Baier, K., Azzam, R., Verma, M.K., Jha, R., 2021a. Analysis of intensity-duration-frequency and depth-duration-frequency curve projections under climate variability. In: Climate Change Impacts on Water Resources: Hydraulics, Water Resources and Coastal Engineering. Springer International Publishing, pp. 407–421, https://doi.org/10.1007/978-3-030-64202-0_35.

Sinha, M.K., Baier, K., Azzam, R., Verma, M.K., Kumar, S., 2022. Impacts of climate variability on urban rainfall extremes using statistical analysis of climatic variables for change detection and trend analysis. In: Kumar, P., Nigam, G.K., Sinha, M.K., Singh, A. (Eds.), Water Resources Management and Sustainability. Springer Singapore, Singapore, pp. 333–387.

Sinha, M.K., Rajput, P., Baier, K., Azzam, R., 2021b. GIS-based assessment of urban groundwater pollution potential using water quality indices. In: Groundwater Resources Development and Planning in the Semi-Arid Region. Springer, pp. 293–313, https://doi.org/10.1007/978-3-030-68124-1_15.

SUDA & RMC, 2017. Sewage Master Plan of Raipur City: Inception Report. Retrieved from.

Suman, A., RMC, PHE, 2011. Raipur City: Storm Water Drainage Report. Retrieved from: Raipur Muncipcal Cooperaion, Chhattisgarh.

Swarnakar, A.K., Choubey, S., 2016. Testing and analysis of pond water in Raipur City, Chhattisgarh, India. Int. J. Sci. Res. 5 (4), 1962–1965. https://doi.org/10.21275/23197064.

Tiwari, A., Sinha, M.K., 2020. Comparing station based and gridded rainfall data for hydrological modelling. CSVTU Res. J. Eng. Technol. 9 (01), 62–74.

Tong, S.T., Chen, W., 2002. Modeling the relationship between land use and surface water quality. J. Environ. Manag. 66 (4), 377–393.

Tufford, D.L., Samarghitan, C.L., McKellar, H.N., Porter, D.E., Hussey, J.R., 2003. Impacts of urbanization on nutrient concentrations in small southeastern coastal streams. J. Am. Water Resour. Assoc. 39 (2), 301–312. https://doi.org/10.1111/j.1752-1688.2003.tb04385.x.

UN-DESA, 2015. World Urbanization Prospects: The 2014 Revision. United Nations Department of Economics and Social Affairs, Population Division, New York, NY, USA.

UNICEF, 2012. The State of the World's Children 2012: Children in an Urban World. United Nations Publications.

United_Nations, 2017. World Economic Situation and Prospects 2017. Retrieved from: United Nations, New York.

Vishwakarma, J., Sinha, M.K., Verma, M.K., Ahmad, I., 2014. Application of remote sensing and GIS in groundwater prospect mapping. Int. J. Eng. Res. Technol. 3 (10), 549–555.

WHO, G, 2011. Guidelines for Drinking-Water Quality. 216 World Health Organization, pp. 303–304.

CHAPTER

8

Satellite sensors, machine learning, and river channel unit types: A review

Adeyemi Olusola[a,b] *and Samuel Adelabu*[b]

[a]Faculty of Environmental and Urban Change, York University, Toronto, ON, Canada
[b]Department of Geography, University of the Free State, Bloemfontein, South Africa

OUTLINE

1 Introduction 117
2 Bedrock channels 118
  2.1 *Reach level classification* 119
3 River channels and remote sensing 120
  3.1 *Wavelengths, sensors and river science* 121
4 Materials and methods 122
  4.1 *Review* 122
  4.2 *Case study* 123
5 Results and discussion 123
  5.1 *Knowledge production* 123
6 Predicting bedrock channels using machine learning algorithms 129
7 Conclusion 130
References 130

1 Introduction

The interpretation of landscapes has been a major focus of geomorphology since the emergence of the discipline. Earlier attempts in interpreting landscapes by Abraham Werner and James Hutton laid the foundation of modern geomorphology (Faniran et al., 2006). Of great importance is the work of James Hutton who is associated with the doctrine of uniformitarianism. This doctrine rests on the principle that the explanation of the earth's formations and features, including the underlying strata, should be sought not in pre-supposed cataclysmic agents of destruction but in agents which have always existed and could be seen in operation

Current Directions in Water Scarcity Research, Volume 7
https://doi.org/10.1016/B978-0-323-91910-4.00008-X

in running water, wind, glaciers, gravity, etc. (Faniran et al., 2006; Olusola, 2019). The most profound paradigm on landscape interpretation comes from the work of the American physical geographer, William Morris Davis. The Davisian cycle (also variously called geographical cycle, geomorphic cycle, fluvial cycle, etc.) provides the temporal framework and the recognition that all landscapes are palimpsests, overwriting by numerous tectonic and climatic processes (Bloom, 2002).

Furthermore, based on the principle of a process-response system, i.e., a geomorphic process will elicit a coherent response (Schumm and Lichty, 1965; Chorley and Kennedy, 1971) given an amount of change in its input. Under this view, landscape elicits a response by rearrangement or alteration of landscape attributes like drainage area, land use aspects, etc. However, the complexity of the landscape belies the simplicity of the process-response model. Complexity results from the interaction of an array of geomorphic processes acting on a heterogeneous landscape, which already bears the imprint of previous events, at a variety of spatial and temporal scales. At the basin scale, there exist distinct geomorphic domains characterizing the fluvial system—hillslope, colluvial, and alluvial (Adeyemi et al., 2020; Olusola et al., 2020). These domains commonly referred to as process domains can be defined as a scale delimited landscape unit within which the landscape attributes bear the signal of a particular process or assemblage of processes (White, 2002; Olusola, 2019).

A process domain is here defined as a scale delimited landscape unit within which the landscape attributes bear the signal of a particular process or assemblage of processes (Adeyemi et al., 2020). Numerous river classification systems have stemmed from the process domain concept (Leopold and Wolman, 1957; Rosgen, 1994; Schumm, 1963) across different scales from the continental to the reach scale. The approach of a process domain is very promising as it offers the opportunity to study the fluvial legacy left by underlying geology, anthropogenic forcings, surficial deposits on one hand, and also helps in identifying discriminants as distinct process domains are traversed by the rivers. Understanding the spatial arrangement and linkages of distinct morphological groups (hillslope-colluvial-alluvial) within a larger landscape mosaic will lead to a more generalized conceptual framework for interpreting fluvial systems (Olusola et al., 2020). The focus of this chapter is on the mapping and interpretation of river channel unit types using satellite sensors and machine learning algorithms. To achieve the aim, this chapter will provide a background to river channel unit types (domains), remote sensing and sensors as applied to river sensing, a review of some existing studies and a case study.

2 Bedrock channels

Bedrock channels occur mainly but not exclusively, in former to actively incising portions of landscapes and where channels are cut to or into resistant rock units, most often in actively uplifting areas (Whipple, 2004). Bedrock channels on the other hand cannot substantially widen, lower, or shift their bed without eroding bedrock (Turowski, 2017). Bedrock channels are not restricted to headwater regions as often expected and can extend to large drainage areas depending on river network geometry and underlying geology. A known fact is that bedrock channels erode through a suite of processes that are poorly understood. Bedrock is removed by plucking along pre-existing joints or fractures, by the forceful collapse of bubbles in turbulent flows known as cavitation, abrasion, and dissolution (Whipple et al., 2000).

2.1 Reach level classification

Channel reach has been described as a span of a river course showing similar and consistent bedforms or channel units. It is determined using the dimension of the channel widths (usually 10–15 channel widths). Broadly, channel reach level can be broken down into colluvial, bedrock and alluvial. These channels and their roughness properties are assumed to show some association with gradient and position within the channel network. It is expected that the change from one morphology to another may be gradual in the field, as this classification imposes order on a continuum of natural morphologies. Channel classification broadly covers a range of scales over which different covariates influence channel properties. Specifically, at the reach scale, the river channel form and process provide a framework for comparing channels at finer spatial scales. Although across sites and regions, variations are bound to exist in channel classification. However, river channel classification as described by Montgomery and Buffington (1997) gives a robust picture of a general channel classification.

1. Cascade reaches

These types of reach classification occupy steep slopes, dissipating high rates of energy and showing longitudinal and lateral disorganized bed material (cobbles and boulders) confined by valley walls (see Montgomery and Buffington, 1997; Grant et al., 1990).

2. Step-pool reaches

These reaches are described as huge clasts morphed into distinct river channel spanning accumulations that create a series of steps separating pools containing finer material. The result of the stepped morphology creates alternating turbulent flow over steps and tranquil flow in pools (see Whittaker and Jaeggi, 1982; Chin, 1989; Montgomery and Buffington, 1993).

3. Plane-bed reaches

The plane-bed reaches are typically, relatively featureless gravel/cobble bed channel units often termed glides, riffles, and rapids (Ikeda, 1975, 1977; Bisson, 1982; Dietrich et al., 1979).

4. Pool-rifle reaches

These reaches are most often than not found within unconfined valley settings where the valley walls are not constricting and they contain features such as bars, pools and riffles as a result of an oscillating cross-channel flow that causes flow convergence and scour on alternating banks of the channel (Church and Jones, 1982; Montgomery and Buffington, 1997).

5. Bedrock reaches

Bedrock reaches have been described as being void of any alluvial bed material and most often than not they are largely confined and lack floodplains. In terms of slope, they occur on steeper slopes (Montgomery and Buffington, 1997). In low-gradient reaches, bedrock channels show a high transport capacity relative to sediment supply, whereas those in steep debris-flow-prone channels also reflect recent debris flow scour. In terms of its response potential to disturbance(s), bedrock reaches are generally insensitive to short-term changes in sediment supply/discharge. It is expected that changes in bedrock channel geometry do not occur over short time scales because bedrock channels are confined; channel width and depth will increase in response to a greater discharge not by incision but by a simple expansion of flow.

Identification of a priori prediction of channel types (domains) has potential for various ecological and fluvial geomorphological studies. In addition, it provides an important benchmark for assessing the impact of large woody debris loss accompanying intensive timber harvesting. Classical channel unit type studies are based on intensive fieldwork or what can be called intrusive studies using various tools and modeling techniques (Olusola 2019; Olusola et al., 2020; Adeyemi et al., 2022). In recent times, there has been renewed interest in the use of non-intrusive methods to aid intrusive methods in the mapping and understanding of channel unit types across domains. These studies (Walia and Mipun, 2010; Salam and Alam, 2014; Heim et al., 2019; Langat et al., 2019; Legleiter and Fosness, 2019; Kuenzer et al., 2019; Mahala, 2020; Fu et al., 2020; Betz et al., 2020; Hanif et al., 2020) have leveraged on the increasing earth observation products and geo-computational techniques.

3 River channels and remote sensing

Traditional river studies fall within the domain of fluvial geomorphologists and ecologists. These river scientists have provided over the years, the inherent dynamics across various river systems and their interactions with the living and nonliving material of the ecosystem. However, the evolution in remote sensing techniques especially in the last three decades has warranted the need to embrace this emerging technology in river sciences. The abundance of spatiotemporal datasets available and archived at the global level across various resolutions and the improvement in geo-computational tools and platforms have opened up a new vista for fluvial geomorphologists to produce continuous data across scales with greater precision. Also, remote sensing offers a shift in paradigm by providing an enabling platform to study and understand river systems as a holistic system (Carbonneau and Piégay, 2012). The science behind remote sensing has to do with imaging as a result of the reflection of various objects. This is made possible as a result of the physics of the electromagnetic spectrum (Carbonneau and Piégay, 2012). When electromagnetic energy is incident on any given earth surface feature, three fundamental energy interactions with the feature are possible, these are reflection, absorption and or transmission. The proportions of these interactions will vary for different earth surface features, depending on their material type and condition. It is this difference that helps in distinguishing features on an image as captured by the sensor. Hence, objects on the earth's surface imprint differently on acquired images across different sensors. The electromagnetic spectrum is thus divided into different waves according to their wavelengths. These are X-rays, ultra-violet, visible light, infrared, microwaves and radio waves. Our focus here will only be on the visible, infrared and microwaves.

The visible light, 0.4–0.7 μm, is called visible because the human eyes can detect it. The range of 0.4–0.7 μm of reflected sunlight contains information on the object. The basic color bands here are red, green and blue. For the color bands, their radiation can be captured on the sensor, a technique referred to as reflectance spectroscopy. The reflectance spectroscopy has allowed for the identification of various objects across the face of the earth using bands. The infrared band is divided into near, middle, thermal and far and it ranges between 0.7 μm and 1 mm. The infrared band acquisition is the same way as the visible light; however, the acquired radiation is the energy radiated from the object in form of heat (thermal). The microwaves and radio waves, 1 mm to >1 m, hold a variety of applications beyond fluvial geomorphology. Different bands exist here such as but are not limited to L, S, Xu, K, and C. The micro and radio waves have longer wavelengths and can penetrate surfaces others cannot.

3.1 Wavelengths, sensors and river science

Various satellite sensors have accompanying wavelengths and this has implications for their applications in river sensing (Marcus and Fonstad, 2010). The spectral reflectance of objects determined by their properties as captured by various sensors aboard different satellites holds great potential for image analysis. In broad terms, sensors can be divided into passive and active sensors. Passive sensors make use of the sun when it is illuminating the earth, while active sensors have their source for illumination. Most often than not passive sensors are of short wavelengths than active sensors (Table 1).

Despite the plethora of satellites and their sensors, their applicability is not only limited by either being active or passive but also by the spatial and spectral resolutions. The spatial resolution refers to the pixel size of a single image as captured by the sensor. In simple terms, a sensor with a spatial resolution of 30 m can conveniently capture an object of size ($900 m^2$) in area. While spectral resolution refers to the ability of the sensor to capture at specific wavelengths. The more wavelengths a sensor can capture, the more capability it can offer.

The finer the wavelength or spectral resolution of a sensor, the more their potentials. Hence, we have high-resolution images (10–100 cm/pixel) that can provide many details of the earth's surface albeit some of these images are costly such as GeoEye, Worldview

TABLE 1 Examples of satellite and their sensors.

Satellite	Sensors
Landsat 4, 5, 7, 8	TM, ETM, OLI, TIRS
Terra	ASTER & MODIS
Aqua	MODIS
NOAA	VIIRS, AVHRR
GOES	Imager and Sounder
Sentinel-2	MSI
Ocean Surface Topography Mission/Jason-2	Altimeters—Radar and Laser
Sentinel-1	Altimeters—Radar and Laser (lidar)
Advanced Land Observing Satellite (ALOS)	Phased Array L-band Synthetic Aperture Radar (PALSAR)

TM, thematic mapper; ETM, enhanced thematic mapper; OLI, operational land imager; TIRS, thermal infrared sensor; ASTER, advanced spaceborne thermal emission reflection radiometer; MODIS, MODerate-resolution imaging spectroradiometer; AVHRR, advanced very high-resolution radiometer; VIIRS, visible infrared imaging radiometer suite; MSI, multispectral instrument.

1–4, Pleidas Neo, QucikBird, Spot, and IKONOS. These images provide scenes of the earth at a much finer scale and can be used in a variety of applications. On the other hand, low-resolution images (5 m–>1 km) which some classifications have divided into low and medium-resolution images capture details at a broader level and are mostly useful for large medium to large scale studies. Examples include but are not limited to the Landsat, Sentinel, Moderate Resolution Imaging Spectroradiometer (MODIS) Terra, and MODIS Aqua. Irrespective of the resolution, images could be multispectral or hyperspectral. However, Radio Detection and Ranging (RADAR) together with its counterpart Light Detection and Ranging (LiDAR) have emerged with great usage and capability in geomorphology. Reviews about these have been provided in several texts (see Marcus and Fonstad, 2010; Farr, 2011). These two sensors (RADAR and LiDAR) can penetrate clouds and fogs and provide information outside the realm of other sensors. They both provide information on topography and change detection using amplitude and phase information. Examples include but are not limited to ASTER, shuttle radar topographic mission (SRTM), and synthetic aperture radar (SAR). The interesting thing is that these images aboard different platforms have been operating for long, except for recent missions such as the Sentinels and some others, and they provide a temporal archive of images with very short revisit days from 5 to 30 days.

One of the drawbacks of space-borne satellites and their sensors asides from spectral and spatial resolution is their revisit days. A growing technology that is addressing the issue of revisit days among other things is the airborne satellites (drones). Airborne satellites are gradually gaining ground in landform studies. Drone technologies are providing capabilities in terms of resolution at a much finer scale and with improved spectral resolutions. Drones have increased the chances of targeted studies focusing on areas of interest with controlled temporal intervals. The fact that airborne platforms are controlled and can be targeted provides an avenue for fluvial geomorphologists to embrace river sensing with minimal field mapping using point clouds. A point cloud is a collection of data points mapped in three dimensions. Each point has its X, Y, and Z values according to where it is in space. A drone using a photogrammetry camera system can assemble a point cloud as one of the outputs from the resulting three-dimensional image. In either case, the resulting cloud is a detailed and accurate picture of the scanned area.

4 Materials and methods

4.1 Review

Various disciplines in the natural sciences have embraced the use of Earth Observation and their products. Especially in vegetation studies, hydrology, environmental monitoring and assessments, etc. In the last decade, there has been an increase as a result of an increase in software that could process these images across all levels and computational tools such as it is available in Python and R. In the last 5 years or thereabout, the emergence of Google Earth Engine and Amazon Web Services, cloud computing platforms, has provided an immense opportunity to run analysis and computations by leveraging on a cloud-based architecture.

To avoid duplication of reviews already provided in extant literature (see Marcus and Fonstad, 2010; Carbonneau and Piégay, 2012), this study streamlined the search string to focus only on "river channels and remote sensing" using the SCOPUS database, R Ecosystem

(R Core Team, 2020) and Bibliometrix package (Aria and Cuccurullo, 2017). Exclusions were made in the following disciplines medicine, arts and humanities, immunology, and business.

4.2 Case study

Channel morphological variables were measured in the field while others were derived. Measured variables are the channel width, depth, wetted perimeter, Bank Height Left and Right, while Slope, Cross-sectional Area, Hydraulic Radius, Velocity, Discharge, Stream power, Specific Stream Power were derived using standard hydraulic equations (Olusola et al., 2020). Spatial data sets for each river point were extracted using the Google Earth Engine platform. Two satellite datasets Landsat 8 (TIRS/OLI) and Sentinel-1 (SAR) bands were extracted for each point. The RGB for the Landsat was pan-sharpened with the panchromatic band (15 m) using the Gram-Schmidt algorithm (see Zhang and Roy, 2016; Agapiou, 2020). The extracted bands (1–11) for Landsat 8 and HV and HH for Sentinel-1 for each of the bedrock were exported in *.csv*.

4.2.1 Data analysis

The spatial data were combined with the morphological data and analyzed using Orange. The Orange software is an easy-to-use open-source machine learning and data visualization. It builds data analysis workflows visually, with a large, diverse toolbox. Four basic learners were used to predict bedrock channel unit types. These are *k*-nearest neighbor (KNN), logistic regression (LR), random forest (RF) and support vector machines (SVM). Asides from Logistic regression, other learners are regression and classification-based machine learning algorithms. The best classifier was determined using root mean square error.

kNN was performed using the Euclidean metric with uniform weights, for the logistic regression the ridge L1 was the regularization type preferred due to its feature selection capabilities. The number of trees for the random forest was set at 10 while splits smaller than 5 as subsets were restricted. The kernel for the SVM was set at linear while the cost and regression loss epsilon was set at 1 and 0.10, respectively. Cross-validation with 20 folds was used.

5 Results and discussion

5.1 Knowledge production

A total of 1646 documents fulfilled the search string and these were used to provide a review on river sensing and river channels. The period for these documents ranges from 1970 to 2021. The entire documents comprise 1190 articles, 2 books, 24 book chapters, 371 conference papers, 32 reviews, etc. The entire documents (1646) were produced by over 4000 authors across different countries (Fig. 1). The darker shades of blue depict areas where most of the studies are coming from. The major areas of scientific production are largely in the United States of America and China (Fig. 1). However, there is a fair representation across the continents.

The spread in knowledge production is expected as these two countries (United States of America and China) are leading in space-based technologies. In recent times, European Union has also emerged as a leading giant in remote sensing technologies. In terms of

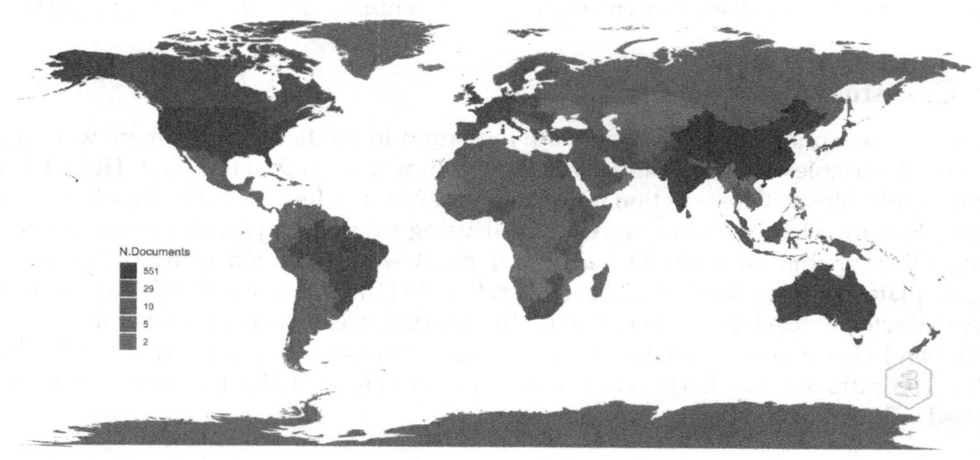

FIG. 1 Country scientific production.

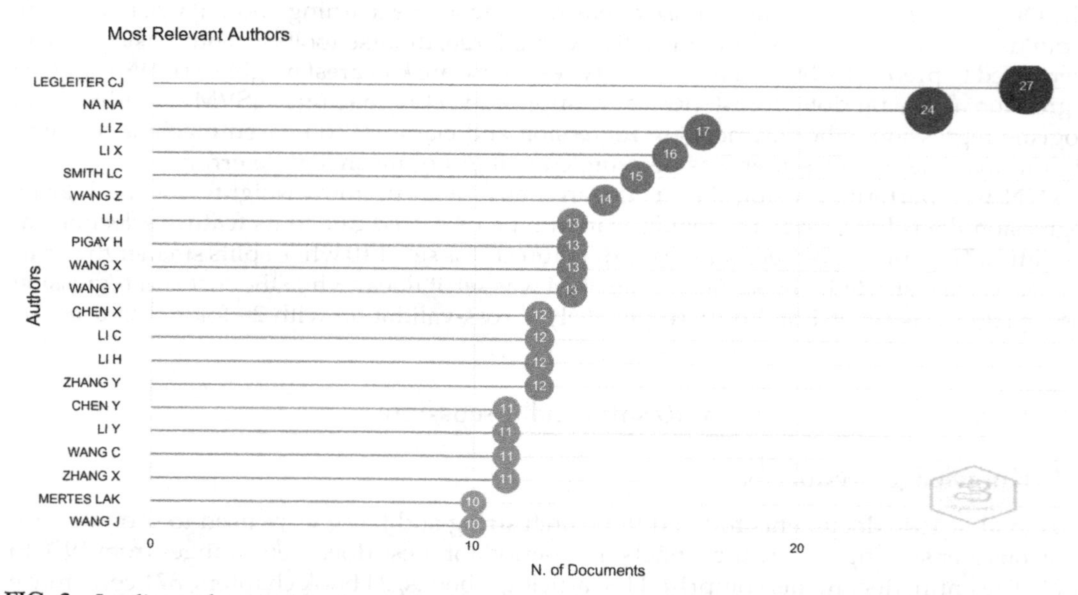

FIG. 2 Leading authors in river channels and remote sensing.

authorship (Fig. 2), Carl Legleiter who works with the United States Geological Survey (USGS) is one of the prominent authors in river sensing and river channels with 27 documents.

Authors who have contributed to papers in river channels using remote sensing technologies have published in some leading journals in the field (Fig. 3). These studies have set the tone,

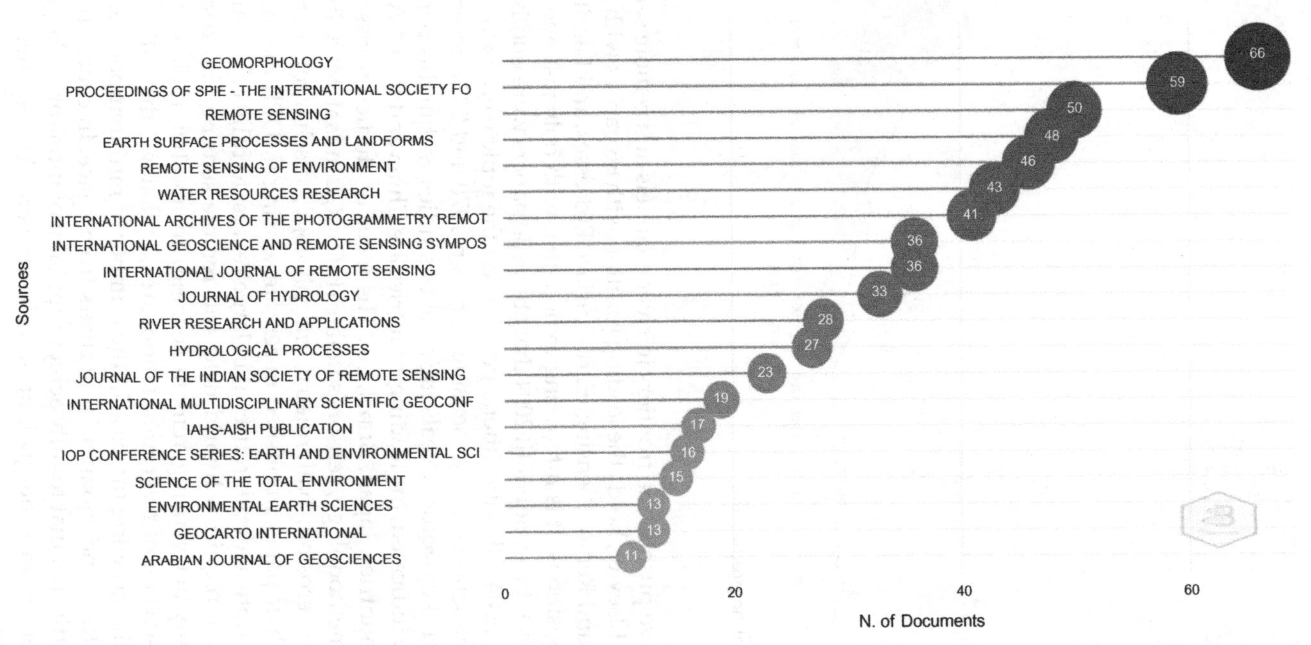

FIG. 3 Leading journals in river channels and remote sensing.

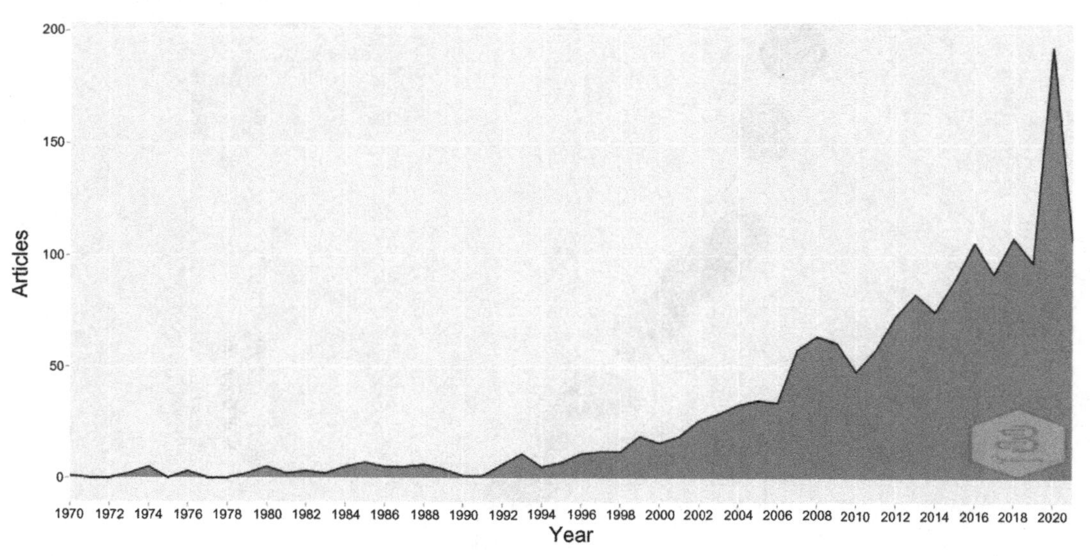

FIG. 4 Annual scientific production.

trajectories and have pushed the frontiers in river channels and remote sensing. Geomorphology, published by Elsevier, holds the leading position with 66 papers with proceedings of SPIE closely following, and Remote Sensing, Earth Surface Processes, and Landforms occupying the third and fourth positions. It is interesting to note that with only just one paper published on this topic in 1970 and it was not until 2004 that above 30 papers were published in a single year. Between 1970 and 2004, all publications per year on the topic were below 30 (Fig. 4).

Knowledge production on river sensing between 1970 and 2021 across countries and authors based on their keywords reveals the span of studies within this period at a broad scale (Fig. 5). As already indicated, the leading countries are China and the USA, in the mix we have other European countries, Japan, Brazil, etc. The focus of studies across these countries and authors suggests methodology, sensor/satellite (LiDAR, Landsat, MODIS), and river channel studies (migration, erosion, bathymetry, morphology, discharge). Legleiter and Fosness (2019) posited in their paper published in remote sensing (MDPI) that as a result of the limitations in passive optical depth retrieval in large revers, to fully carry out a complete bathymetric coverage will require a hybrid field and remote sensing approach. The focus of the study was on bathymetry using remote sensing. In another review published by Smith (1997) on remote sensing of river inundation, area, and stage, the author alluded to the fact that as a result of the growing increase in the number of multitemporal satellite data, it is becoming easier to study and monitor large rivers from space. In concluding the study, Smith (1997), however, affirmed that to fully achieve this aim of monitoring large rivers from space, there is the need for a more synergistic approach between SAR together with either the visible or the infrared data.

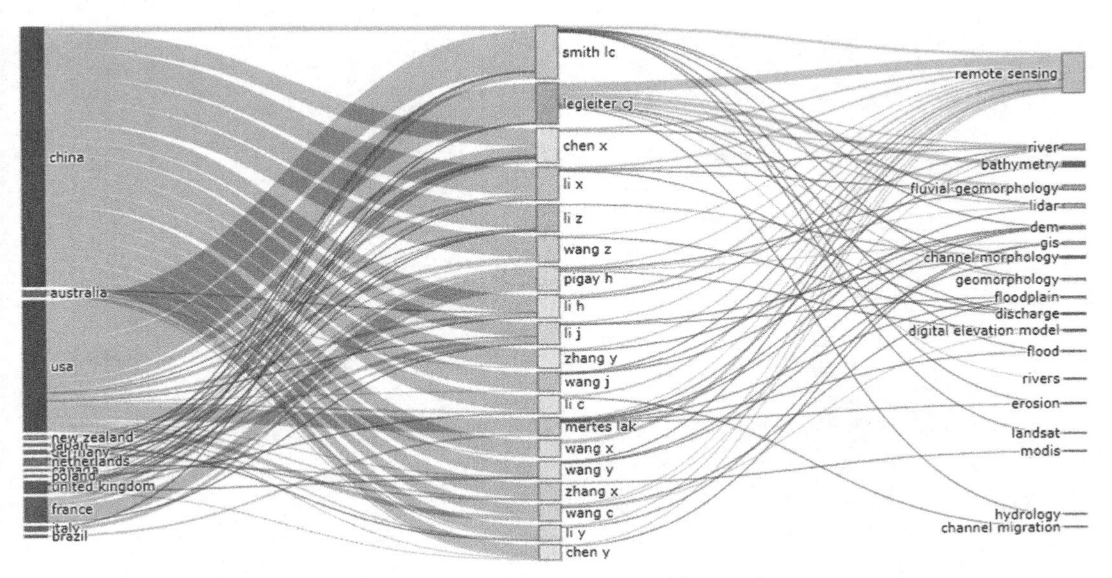

FIG. 5 Three-fieldplot showing first 20 authors, their countries and keywords.

Other authors have focused on various aspects of the river channel using various tools and techniques. Betz et al. (2020) in their study on analyzing the Corridor of the Naryn River in Kyrgyzstan Based on OpenAccess Data published in remote sensing (MDPI) alluded to the fact that remote sensing-based approaches have become increasingly popular in river science due to the increase in the spatial scale of analysis. They concluded in their study using SRTM and Landsat OLI in the automated mapping of various riverscape parameters such as riparian zone extent, distribution of riparian vegetation, active channel width and confinement, as well as stream power that their result strengthens the claim that remote sensing is capable of providing information on the measure parameters even in data-scarce regions. In addition, Huang and Jensen (1997) in their conference paper (A machine learning approach to automated construction of knowledge bases for expert systems for remote sensing image analysis with GIS data) presented at the annual meeting of the American Society for Photogrammetry and Remote Sensing, Baltimore, MD (United States) developed a knowledge base for wetland classification using SPOT image and GIS data. They posited that their observations suggest that the method could provide an effective approach to integrating remote sensing and GIS in wetland studies. However, asides from these studies, several other studies as captured within the period have also considered planform changes, sediment monitoring and changing land use within drainage basins (Fig. 6).

The central theme between 1970 and 2021 has been on dams, channel morphology, and sediment transport (Fig. 6). Basic themes revolve around flood, erosion, bathymetry, LiDAR, hydrology, Landsat, water quality, channel migration, erosion. While niche themes are studies around soil moisture, groundwater, tectonics, SAR, meander, etc. In terms of the use of satellite and sensors (Fig. 7) LiDAR has been applied in 34 of the publications, while Landsat in 29, MODIS in 24, UAVs together with drones in 18, SAR in 12.

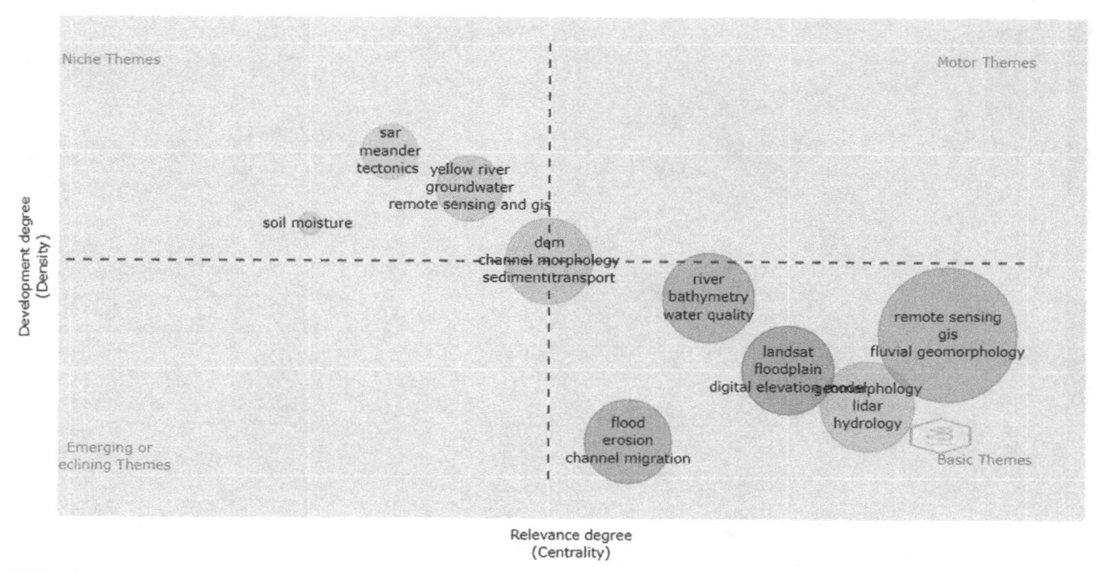

FIG. 6 Thematic clusters of keywords on river channels and remote sensing between 1970 and 2021.

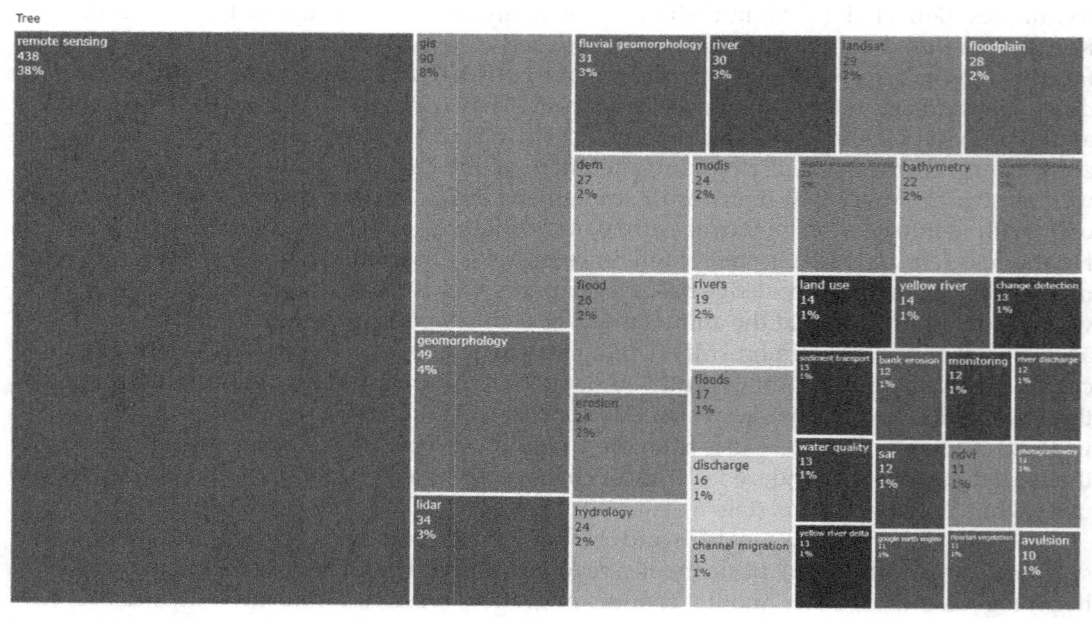

FIG. 7 Treemap showing keywords and their frequencies.

Studies focusing on rive channels using remote sensing are beginning to emerge; however, very few to no studies have focused on medium- to small-scale basins. This is understandable due to the spatial resolutions of freely available Earth Observation products. Sentinel provides 10m, while Landsat provides 30m. However, with various pan sharpening techniques issues around spatial resolution are being resolved albeit not without care and caution. In addition, the use of machine learning algorithms in domain process mapping and prediction is missing.

6 Predicting bedrock channels using machine learning algorithms

Domain units such as bedrock channels are largely impressive in appearance and form. These channel types are underlain by Basement Complex formations. The result of the model evaluation shows that in terms of accuracy and precision kNN performs better than other models, although this is closely followed by the random forest (Table 2).

The top five features selected for the random forest for channel unit type predictions are cross-sectional area, total stream power, wetted perimeter, band 5 (near-infrared), and discharge, while kNN top five features are slope, discharge, hydraulic radius, cross-sectional area and depth. Some studies have identified discriminating factors such as slope, discharge and cross-sectional areas as important variables in channel classification using only morphological variables (Wohl and Merritt, 2005; Olusola et al., 2018). The variations observed between the two learners could be due to their strength in classifying bedrock channels. Bedrock channels are very different in terms of their morphology and dynamics (Olusola et al., 2020). Their mode of formation and dynamics are fundamentally distinct, hence, morphological variables that could aid in their prediction are bound to be different. The choice of the Near-Infrared Band (NIR) in random forest suggests the possibility of the combination of bands with morphological variables in discriminating channel unit types. Indices have been developed in environmental studies using such as the normalized difference vegetation index, soil moisture index, normalized pigment chlorophyll ratio index, etc. Even though most of these indices are in combination with other bands, there is the possibility to explore combining band 5 with other bands or with channel morphological variables in the prediction of bedrock channels.

TABLE 2 Model comparisons for channel unit type prediction.

Model	Area under curve	Classification accuracy	F1	Precision
kNN	0.72	0.89	0.89	0.89
SVM	0.53	0.78	0.79	0.79
Random forest	0.72	0.86	0.83	0.88
Logistic regression	0.69	0.71	0.66	0.60

7 Conclusion

This study has provided a review of the application of remote sensing to river channels and existing gaps in knowledge. Several studies have been carried out as regards identifying controls, predicting and mapping of channel unit types across the landscape using only morphological variables. A major gap has been to add to the already growing literature in dynamic geomorphology as regards channel morphology, types, classification and prediction using Earth Observation products. Hence, the relevance of this study and its position in tropical climes is providing a baseline study upon which several studies can emerge.

Channel unit types are distinct geomorphic history and assemblage of geomorphic processes, which in a way produce distinctive forms. Concerning river systems, these unit types can be used to understand and predict the relationship between supply limited and transport capacity continuum as well as ecological zones and structures along and within stream segments. The ability to correctly classify channel types based on morphological variables and spectral bands as shown in this study analysis provides a way to predict domains a priori. The success achieved in classification will be of great help to various stakeholders within an interest in river channel studies. Furthermore, the relevance of predicting channel types is useful in other studies involving predicting channel response to disturbance, modeling river processes and landscape evolution, aspects of natural resources management, mapping habitat and explaining the spatial distributions of aquatic and riparian species or community types. The ability to increase discrimination among channel units underscores potential benefits for existing classification systems.

However, care must be taken in the use of spectral bands at a much finer scale even with pan-sharpening. Spectral reflectances can help decipher and clean datasets to ensure the extracted bands are a reflection of the river channels. The discriminating variables identified either for the kNN or the random forest suggest that in attempting to predict channel types, variables either morphological or spatial are sensitive to the model/learner and also to individual channel types and specific stream reaches.

References

Adeyemi, O., Olutoyin, F., Olumide, O., 2020. Downstream hydraulic geometry across headwater channels in upper Ogun River basin, Southwestern Nigeria. Afr. Geogr. Rev. 39 (4), 345–360.

Adeyemi, O.O., Onafeso, O., Olutoyin, A.F., Samuel, A, 2022. River sensing: the inclusion of red band in predicting reachscale types using machine learning algorithms. Hydrol. Sci. J. https://doi.org/10.1080/02626667.2022.2098752.

Agapiou, A., 2020. Evaluation of Landsat 8 OLI/TIRS level-2 and sentinel 2 level-1C fusion techniques intended for image segmentation of archaeological landscapes and proxies. Remote Sens. 12 (3), 579.

Aria, M., Cuccurullo, C., 2017. bibliometrix: An R-tool for comprehensive science mapping analysis. In: J. Informetr., fourth ed. 11. Elsevier, pp. 959–975.

Betz, F., Lauermann, M., Cyffka, B., 2020. Open source riverscapes: analyzing the corridor of the Naryn River in Kyrgyzstan based on open access data. Remote Sens. 12 (16), 2533.

Bisson, P.A., 1982. A system of naming habitat types in small streams, with examples of habitat utilization by salmonids during low stream flow. In: Acquisition and Utilization of Aquatic Habitat Inventory Information.

Bloom, A., 2002. Teaching about relict, no-analog landscapes. Geomorphology 47, 303–311.

Carbonneau, P., Piégay, H. (Eds.), 2012. Fluvial Remote Sensing for Science and Management. John Wiley & Sons.

Chin, A., 1989. Step pools in stream channels. Prog. Phys. Geogr. 13 (3), 391–407. https://doi.org/10.1177/030913338901300304.

Chorley, R.J., Kennedy, B.A., 1971. Physical Geography: A Systems Approach. Prentice-Hall.

Church, M., Jones, D., 1982. Channel bars in gravel-bed rivers. In: Hey, R.D., Bathurst, J.C., Thorne, C.R. (Eds.), Gravel-Bed Rivers. John Wiley and Sons, Chichester, UK, pp. 291–324.

Dietrich, W.E., Smith, J.D., Dunne, T., 1979. Flow and sediment transport in a sand-bedded meander. J. Geol. 87, 305–315.

Faniran, A., Jeje, L.K., Ebisemiju, F.S., 2006. Essentials of Geomorphology. Penthouse Publishers.

Farr, T.G., 2011. Remote sensing in geomorphology. In: The SAGE Handbook of Geomorphology. 210. SAGE Publishing.

Fu, Y., Dong, Y., Xie, Y., Xu, Z., Wang, L., 2020. Impacts of regional groundwater flow and river fluctuation on floodplain wetlands in the middle reach of the Yellow River. Water 12 (7), 1922.

Grant, G.E., Swanson, F.J., Wolman, M.G., 1990. Pattern and origin of stepped-bed morphology in high-gradient streams, Western Cascades, Oregon. Geol. Soc. Am. Bull. 102 (3), 340–352. https://doi.org/10.1130/0016-7606(1990)102<0340:paoosb>2.3.CO;2.

Hanif, M., Hidayat, R.A., Triyatno, H.R., 2020. The high resolution imagery to mapping the dynamics of river flow patterns, and deposition on fluvial landform of Sianok canyon. J. Remote Sens. GIS Technol. 6, 8–14.

Heim, B., Juhls, B., Abramova, E., Bracher, A., Doerffer, R., Gonçalves-Araujo, R., et al., 2019. Ocean colour remote sensing in the Laptev Sea. In: Remote Sensing of the Asian Seas. Springer, Cham, pp. 123–138.

Huang, X., Jensen, J.R., 1997. A machine-learning approach to automated knowledge-base building for remote sensing image analysis with GIS data. Photogramm. Eng. Remote Sens. 63 (10), 1185–1193.

Ikeda, H., 1975. On the bed configuration in alluvial channels: their types and condition of formation with reference to bars. Geogr. Rev. Jpn. 48, 712–730.

Ikeda, H., 1977. On the origin of bars in the meandering channels. Bull. Environ. Res. Center Univ. Tsukuba 1, 17–31.

Kuenzer, C., Heimhuber, V., Huth, J., Dech, S., 2019. Remote sensing for the quantification of land surface dynamics in large river delta regions—a review. Remote Sens. 11 (17), 1985.

Langat, P.K., Kumar, L., Koech, R., 2019. Monitoring river channel dynamics using remote sensing and GIS techniques. Geomorphology 325, 92–102.

Legleiter, C.J., Fosness, R.L., 2019. Defining the limits of spectrally based bathymetric mapping on a large river. Remote Sens. 11 (6), 665.

Leopold, L.B., Wolman, M.G., 1957. River channel patterns: braided, meandering, and straight. US Government Printing Office.

Mahala, A., 2020. The significance of morphometric analysis to understand the hydrological and morphological characteristics in two different morpho-climatic settings. Appl. Water Sci. 10 (1), 1–16.

Marcus, W.A., Fonstad, M.A., 2010. Remote sensing of rivers: the emergence of a subdiscipline in the river sciences. Earth Surf. Process. Landf. 35 (15), 1867–1872.

Montgomery, D.R., Buffington, J.M., 1993. Channel classification, prediction of channel response, and assessment of channel condition. University of Washington, Seattle, p. 84.

Montgomery, D.R., Buffington, J.M., 1997. Channel-reach morphology in mountain drainage basins. Geol. Soc. Am. Bull. 109, 596–611.

Olusola, A.O., 2019. Process-Form Dynamics of Upper Ogun River basin, Southwestern Nigeria (Unpublished Ph.D. Thesis). Department of Geography, University of Ibadan, Ibadan.

Olusola, A.O., Fashae, O.A., Faniran, A., 2018. Classification and prediction of channel morphology within selected third-order basins (Southwestern Nigeria). In: Conference of the Arabian Journal of Geosciences. Springer, Cham, pp. 323–326.

Olusola, A., Fashae, O., Onafeso, O., Adelabu, S., 2020. Morphologic and hydraulic variability of small bedrock and alluvial channels in relation to lithological controls, upper Ogun River basin, Southwestern Nigeria. Phys. Geogr. 41 (6), 537–557.

R Core Team (2020). R: A language and environment for statistical computing. R Foundation for Statistical Computing, Vienna, Austria. URL https://www.R-project.org/.

Rosgen, D.L., 1994. A classification of natural rivers. Catena 22 (3), 169–199.

Salam, M.A., Alam, M.S., 2014. Identification and delineation of Turag River basin boundary using remote sensing techniques. J. Environ. Sci. Nat. Resour. 7 (1), 169–175.

Schumm, S.A., 1963. In: A tentative classification of alluvial river channels: an examination of similarities and differences among some Great Plains rivers. Vol. 477. US Department of the Interior, Geological Survey.
Schumm, S.A., Lichty, R.W., 1965. Time, space and causality in geomorphology. Am. J. Sci. 263, 110–119.
Smith, L.C., 1997. Satellite remote sensing of river inundation area, stage, and discharge: a review. Hydrol. Process. 11 (10), 1427–1439.
Turowski, J.M., 2017. Alluvial cover controlling the width, slope and sinuosity of bedrock channels. Earth Surf. Dyn. 6, 29–48. https://doi.org/10.5194/esurf-2017-46.
Walia, D., Mipun, B.S., 2010. Hydrological behavior of Umshing River, East Khasi Hills, Meghalaya. Acta Geophys. 58 (5), 908–921.
Whipple, K.X., 2004. Bedrock rivers and the geomorphology of active orogens. Annu. Rev. Earth Planet. Sci. 32, 151–185.
Whipple, K.X., Hancock, G.S., Anderson, R.S., 2000. River incision into bedrock: mechanics and relative efficacy of plucking, abrasion, and cavitation. Geol. Soc. Am. Bull. 112 (3), 490–503.
White, R., 2002. *Geomorphic process domains in a mountain basin*. Doctoral dissertation, University of British Columbia.
Whittaker, J.G., Jaeggi, M.N., 1982. Origin of step-pool systems in mountain streams. J. Hydraul. Eng. 108 (6), 758–773.
Wohl, E., Merritt, D., 2005. Prediction of mountain stream morphology. Water Resour. Res. 41 (8). https://doi.org/10.1029/2004WR003779.
Zhang, H.K., Roy, D.P., 2016. Computationally inexpensive Landsat 8 operational land imager (OLI) pansharpening. Remote Sens. 8 (3), 180.

CHAPTER

9

Geospatial modeling in the assessment of environmental resources for sustainable water resource management in a semiarid region: A GIS approach

Nanabhau Santujee Kudnar
C. J. Patel College, Tirora, Maharashtra, India

OUTLINE

1 Introduction 134
2 Study area 135
3 Material and methods 135
4 Results and discussion 138
4.1 Geomorphology and soil types 138
4.2 Geology 138
4.3 Slope 138
4.4 Landforms 138
4.5 Climate and rainfall 139
4.6 Hydrology 140
4.7 Water availability analysis 140
4.8 Water inflow 140
4.9 Measurement of water discharge 141
4.10 Water balance study 141
4.11 Water source sustainability 142
4.12 Land use 142
4.13 Irrigation projects 142
4.14 Environmental resources for sustainable water resource management 142
4.15 Water quality analysis (WQA) 143
4.16 Water quality analysis (WQA) and analytic hierarchy process (AHP) 145
4.17 Priority 147
5 Conclusions 148
Conflict of interest 148
References 148

Current Directions in Water Scarcity Research, Volume 7
https://doi.org/10.1016/B978-0-323-91910-4.00009-1

Abbreviations

AHP analytical hierarchical process
APML Adani Power Maharashtra Limited
CWC Central Water Commission
IMD Indian Meteorological Department
MWRRA Maharashtra Water Resources Regulatory Authority
NEERI National Environmental Engineering Research Institute
SOI Survey of India
WGS World Geodetic System
WQI water quality index

1 Introduction

Water is a great source of economic development. If equitable, community-based, and practical development of water resources has to be ensured, then traditional systems shall have to be revitalized and developed (Aslam et al., 2020; Tsihrintzis, 2017; Soutter et al., 2009; Archibald and Marshall, 2018). Per capita water availability is declining due to increasing population. Also, increasing urbanization and industrialization have increased the stress on water management. As a result, water management is one of the major challenges of the 21st century. Currently, there is a growing imbalance between the demand and supply of water in front of the watershed, the uncertainty of water availability, and limitations on its use, the problem of floods and droughts, the current low efficiency, the gap between the constructed irrigation capacity and actual use, declining groundwater (Dongare et al., 2013; Pathare and Pathare, 2020; Khan et al., 2001; Todmal, 2020).

Today we see many problems and challenges such as wastage in urban distribution systems, deteriorating water quality, natural reservoirs and encroachments on rivers and streams. While looking for solutions to many of these problems on earth today, there are some challenges that need to be addressed, ensuring clean water and sanitation, planning measures for water scarcity, planning for drought and flood conditions and territorial allocation of water in a fair and strategic manner among various water users. Much research has been done to protect the ecosystem, to protect and increase the quality of groundwater and surface water, to increase the productivity and efficiency of water use, and thus to manage water (Cannata et al., 2018; Bouteraa et al., 2019; Chattaraj et al., 2017; Mahadevan et al., 2020; Laurent et al., 1998). Today, despite the huge investments and improvements made in the water sector, there are still some serious problems and challenges in the world regarding the current state of water resources and their management. Due to the rapid growth of water due to population growth, urbanization and changing lifestyles, there is a huge gap between demand and supply. This poses a serious challenge to water security (Karande et al., 2020; Ahamed et al., 2020; Ouro-Sama et al., 2020; Ahmad and Pandey, 2018; Andualem et al., 2020; Jothimani et al., 2020; Redvan and Mustafa, 2021).

Conflict is growing among water users in inter-regional, inter-regional and upstream and downstream areas of the river basin. The availability of water for drinking and other household uses in rural areas is still a challenge and hence the need of the hour to manage surface and ground water (Kadam et al., 2020; Xu et al., 2020; Ramadan et al., 2019; Rahaman et al., 2016; Salami and Ehteshami, 2016).The work plan of watershed management can be prepared with the help of GIS and RS techniques. Many environmental disasters have been created during the last century because of human activities. Man has affected water quality (Cieszynska et al., 2012; Guettaf et al., 2017; Lkr et al., 2020; Majumdar and Chatterjee 2021; Walega et al., 2016; Sharma and Kansal, 2011) every-where by pursuing fast economic development and by adopting a physical life and a materialistic culture. Even though water is a renewable resource it is possible to develop sustainable practices for its use provided that the quality of water is not compromised. The more critical and important principle enunciated by the (Ksiazek et al., 2019; Sharma and Dutta, 2020; Dutta et al., 2017; Kudnar, 2020a,b; Teshome and Halefom, 2020; Majumdar et al., 2021; Khademalrasoul and Amerikhah, 2021) is that environmental flows must be maintained in case of all perennial rivers, and if there are impounding structures like dams, then sufficient water must be reserved for releasing it into the main river.

In the study, geomorphology, soil types, geology, slope, relief, landforms, climate and rainfall, hydrology, water availability analysis, water inflow, water balance, water quality analysis (WQA), measurement of water discharge, water resource management, resource units, inventory of existing medium/minor/lift irrigation schemes constructed the areas through geospatial modeling for sustainable development in Tirora tehsil, central India.

2 Study area

Tirora tehsil, north latitudes 21°13′05″ to 21°33′30″ and East longitudes 79°47′50″ to 80° 05′00″ and is located in Gondia district, Maharashtra (Fig. 1). Tiroda tehsil has a geographical area of 617.10 sq. km and has 114 villages and five administrative divisions. It has an average annual temperature of 26.40°C, an average maximum temperature of 29.30°C and an average minimum temperature of 24.12°C. The average rainfall in this place is 1400 mm and it receives maximum rainfall from monsoon. Eighty-five percent of the rainfall from monsoon rains falls in this region. Tiroda tehsil is bounded by four talukas of Gondia district. It is bounded on the north by Gondia, on the south by Sadak-Arjuni, on the east by Goregaon and on the west by Bhandara district and on the north by Tiroda tehsil (Kudnar, 2022). Total 2065 hector irrigated area in study area 12,750 hector area irrigated by tanks. In this area, total number of tanks is 152; out of which 2 medium projects, 27 small irrigation projects, and 123 Malgujari tanks.

3 Material and methods

The present study is carried out of topographical maps on a 1:50,000 scales using the SOI. The geomorphic unit, geology, land use, soil, and water analysis, Collection of rainfall and runoff data, maps, reports. River behavior for all the river flow condition, long term rainfall data for the entire Wainganga basin as well as small tributaries, particularly for tehsil area have been collected from IMD, the stream flow and sediment load data at Rajegaon as well as water availability and discharge data of near Dhapewada dam have been collected being monitored by CWC and MWRRA. Water sample has been collected from the tehsil area and also at river Wainganga and physiochemical water has been analyzed, Satellite data products multispectral imageries have been acquired for time series analysis of various hydrological as well

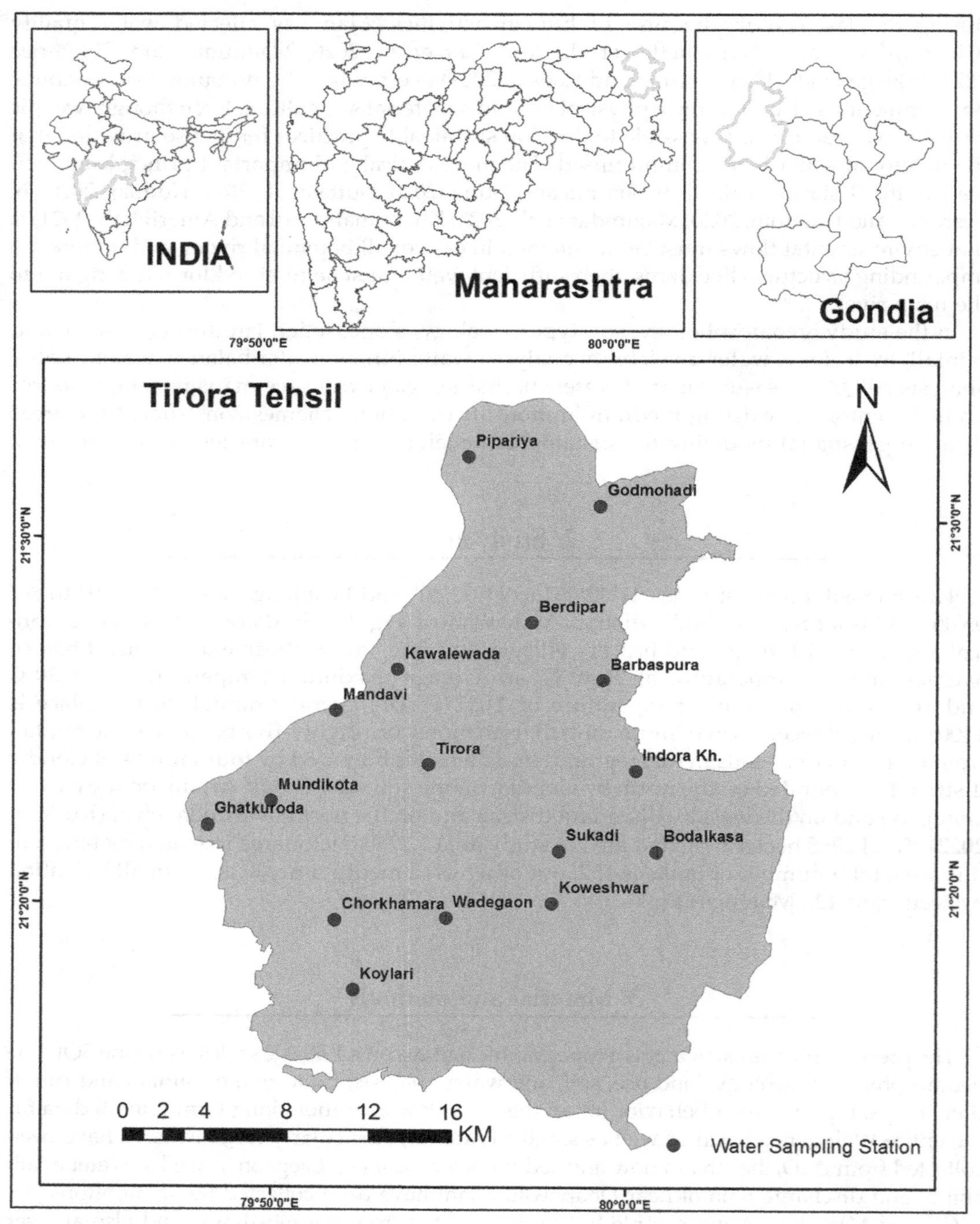

FIG. 1 Study area, Tirora Tehsil.

geomorphological features of in study area. The following steps have opted for the study environmental network of the tehsil was analyzed. Using SOI topographic maps and Universal Transverse Mercator (UTM) zone 44 N projection was georeferenced using WGS 84 datum, in ArcGIS desktop 9.3. Water sample has been collected from the downstream as well as upstream of water at river Wainganga (Kudnar and Rajasekhar, 2020) and environmental as well as geomorphology, soil types, geology, slope, relief, landforms, climate and rainfall, hydrology, water availability analysis water inflow water balance, measurement of water discharge, demographics, prioritization of land resource management has been analyzed (Fig. 2). To attempt this 16 sample WQA, water samples were analyzed for various parameters correlation, ranking prioritization and priority mappings such as TSS, TDS, pH, Total Alkalinity, COD, BOD, Total Coliform, Chloride, Iron, Fecal, Mg, and Nitrate applied AHP method and GIS techniques.

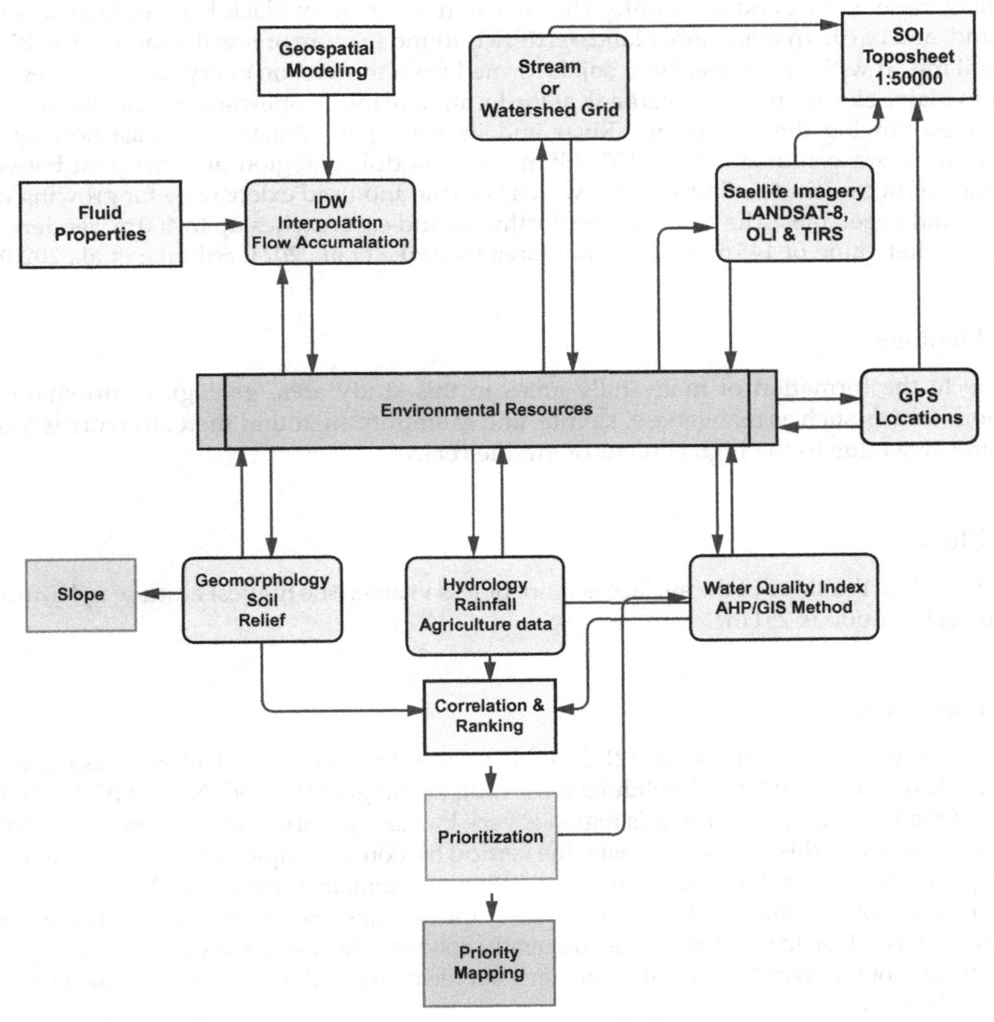

FIG. 2 Flowchart of the methodology.

4 Results and discussion

4.1 Geomorphology and soil types

These Tiroda tehsil have higher mountain ranges to the southeast and southeast, and to a lesser extent to the northwest and northwest. Accordingly, the highest height of the mountain range is 420 m and the lowest height is 251 m. (Mundikota area 21°36′96″ N 79°83′60″ E) and southeastern parts Chorkhamara area 21°26′66″ N, 79°95′57″ E of the tehsil, while the denudational that including the Gangazari (21°44′86″ N 80°05′72″ E), Bodalkasa (21°35′06″ N 80°01′93″ E), and Sukadi (21°35′19″ N 79°97′95″ E). The mountain ranges formed by the Satpuda mountain range are found in this place. Some of the mountain ranges are found in this area. The area also contains geological formations containing precious metals such as manganese, kinite, and saliminite. The soils here are mainly black kanhar, khardi, sihar, morand, and bardi. In this Kanhar land, fertility is found in more or less the same form. Birdie is a soil mixed with limestone. Sihar soil is formed from the friction of crystalline stones. The river in plains elevation of the riverbank at the location of the Dhapewada dam at 267 m. of the southwest-flowing the Wainganga River and its principal tributary, the east-flowing the Bawanthari, are confined within 155–285 m. The floodplain region and the area between 255 and 265 m constitute a flattish plain which is fertile and used extensively for growing rice, pulses, and vegetables. The height at the northwest and east reaches up to 400 m, rendering a relative relief value of 145 m for the study area (Salunke et al., 2021; Salunke et al., 2020).

4.2 Geology

Due to the formation of many hilly areas in this study area, geological formations of precious metals such as manganese, kainite, and saliminite are found there. Iron ore is found in some areas due to the high content of granite rocks.

4.3 Slope

In this place the slope of the surface is more or less visible. The highest altitude is 420 m and the lowest altitude is 251 m.

4.4 Landforms

Dissected hilltops Nagzira ranges (21°28′44″ N to 79°94′68″ E), dissected hills Bodalkasaranges area (21°32′62″ N to 80°02′60″ E), subdued hills Nimgaon ranges (21°39′36″ N to 80°04′02″ E). The speed of the Wainganga River in this region is very low and it starts flowing slowly. Therefore, the friction action in this region, especially the vertical friction, is completely stopped. As a result, the depth of the river bed decreases and the width of the container increases. The silt carried by the river does not flow into the stream, so it stays in the container. So the character of this river has become shallow. Due to this, the river is frequently flooded. The floods have created silt deposits on both sides of the river for several kilometers. On the banks of the river, silt is formed on both sides of the river.

4.5 Climate and rainfall

The study area has an average elevation of 300 m above sea level and the climate is hot and dry as it is far away from the sea. Summers and winters are just as intense. In Tiroda tehsil, although the temperature is lower in winter due to hills, the temperature is higher in summer and relative humidity is 62%. Summer hot winds in May have a major impact on agricultural production here (Table 1). Most of the rainfall is in July and August. It starts raining from the second week of June to the first week of October with south-westerly monsoon winds. The average rainfall in this place is between 1300 and 1500 mm. Tiroda tehsil has minimum rainfall of 1357.8 mm and maximum rainfall of 1415.8 mm (Table 2). To ascertain the certainty of rainfall that may occur (dependable rainfall), the probability analysis of the rainfall data for the period from 1917 to 2016 is also carried out for different probability levels of 50%, 80%, and 90%. For doing the analysis, daily rainfall data has been converted to the average monthly data. The monthly rainfall data has thereafter been analyzed for different probability levels. The analyzed rainfall results will help in the estimation of river discharge during the lean season, or during various months and also during the year (Geremew et al., 2020; Sobhani and Zengir, 2020; Hussen et al., 2018). Lower probability indicates, less assurance to the occurrence of rainfall, while higher probability indicates, rainfall has more chance to occur. As the 75% and 90% dependability are of great importance in the field of hydrology the same were estimated and found to be 1194 mm and 1023 mm. Correlation of Bodalkasa medium tank capacity and amount of rainfall is $r=0.728$ which is positive and high; it indicates that there is a high relation between water capacity of tanks and amount of rainfall Bodalkasa tank. Correlation of Chorkhamara medium tank capacity and amount of rainfall is $r=0.43$

TABLE 1 Temperature.

Month	Jan.	Feb.	Mar.	Apr.	May	June	July	Aug.	Sept.	Oct.	Nov.	Dec.
Average annual maximum tem. (O°C)	26.6	31.6	35.2	39	42.1	38.1	30.5	29.9	30.8	31	29.3	27.9
Average annual minimum tem. (O°C)	13.3	15.4	19.6	24.6	28.9	27.9	24.3	24.1	23.9	21.2	15.2	12.9

Source: IMD, Pune.

TABLE 2 Rainfall.

Station	Jan.	Feb.	Mar.	Apr.	May	June	July	Aug.	Sept.	Oct.	Nov.	Dec.
Tirora (mm)	13.2	27.7	14.2	12.2	12.7	204.7	458.5	411	200.1	52.1	14.2	8.6
Bodhal Kasa (mm)	15.2	23.1	13.7	11.4	8.1	204.2	494.8	359.9	183.4	48.5	12.7	4.3
Chorkha Mara (mm)	13.2	25.4	12.7	17	7.4	196.6	532.4	388.1	210.3	51.1	7.4	4.6

Source: IMD, Pune.

TABLE 3 Long term trend of rainfall (mm) Gondia and Tirora (2013–18).

Sl. no.	Year	Premonsoon (Gondia)	Premonsoon (Tirora)	Monsoon (Gondia)	Monsoon (Tirora)	Postmonsoon (Gondia)	Postmonsoon (Tirora)	Winter (Gondia)	Winter (Tirora)	Annual (Gondia)	Annual (Tirora)
1	2013	16.6	55.5	1677.9	1530.5	226.1	165	35	34	1955.6	1787
2	2014	40.8	106.5	930.3	980.6	188.9	174.8	35.7	61.7	1195.7	1323.6
3	2015	114.2	207	930.1	963.6	222.9	208.8	19	19.5	1286.2	1398.9
4	2016	73.6	73.5	919.9	960.5	386.7	191	3.4	10	1383.6	1235
5	2017	24.5	38	682.6	835.5	146.5	113.5	2.6	0	856.2	987
6	2018	25.1	26	680.1	1131.6	110	111.5	20.1	20.3	1418	1289.4

Source: IMD, Pune.

which is positive and moderate; it indicates that there is moderate relation between water capacity of tanks and the amount of rainfall of Chorkhama tank (Table 3).

4.6 Hydrology

The temporal variation of the groundwater level conveys the characteristics like its recharge potential due to precipitation, the withdrawal by different stresses like domestic use, agricultural use, and industrial use. In the present study, groundwater level data of NEERI are used to study groundwater fluctuation. The water level (bgl) varied from 3.28 to 22.06m premonsoon and 1.47 to 10.71m postmonsoon. The majority of the sources are having a shallow water table (<10 m below ground level). The water level measured below ground level (Rajasekhar et al., 2020; Datta et al., 2020), has been converted to water level (above mean sea level) by subtracting the water level (bgl) from the reduced level of the source.

4.7 Water availability analysis

As Gondia district is completely plateau, lakes have been created naturally in some places. Out of these, Bodhankasa, Dhapewada, and Chorkhamara medium projects have higher capacity and higher water holding capacity is coming. All available water resources of the study area are utilized for full agribusiness. The rest of the reservoir is used for the Adani power project, as well as a large number of water-based fishing businesses. Fishing ponds are constructed by digging and closing dams on shallow waste water bodies which have ample water availability and water retention and in which they have become unusable for agriculture. There are 22 lakes in Tiroda tehsil on 250ha and 100–250ha size. The former cattle ponds are important for agribusiness, fisheries, Shingada and vegetable trade. The number of former cattle ponds is 123. The Water Resources Department, Government of Maharashtra has permitted APML to withdraw a total of 70 Mm^3/annum (about 1,95,000 m^3/day) of raw water for the near Dhapewada dam, on the river Wainganga near Kawalewada village.

4.8 Water inflow

The water inflow analysis has been carried out using water storage and discharge data from 2016 through 2018 being recorded by Dhapewada Irrigation Project in Tirora. Table 4

TABLE 4 Status of monthly inflow at Dhapewada dam.

Parameter	Monthly inflow of water (mcm/m)												
	Jan.	Feb.	Mar.	Apr.	May	Jun.	July	Aug.	Sep.	Oct.	Nov.	Dec.	Annual
Average	77.44	56.61	32.83	21.16	14.99	202.19	1250.49	2579.32	1723.76	453.97	160.36	95.92	555.75
Maximum	438.92	227.94	107.37	105.10	59.48	1143.28	5340.35	4819.30	5645.30	1711.42	785.37	789.76	1496.54
Minimum	17.88	2.65	0.19	0.11	0.03	18.72	208.50	539.23	211.14	88.72	39.67	4.22	134.99
Standard deviation	76.90	49.15	27.83	25.50	16.89	236.45	994.11	1180.14	1371.41	407.45	144.06	131.98	259.71

showed the long term average monthly the Wainganga River runoff data of the Pauni site (1975–2010) and the Rajegaon site after applying a necessary correction factor to convert the flows near Dhapewada dam. The average monthly flows and their statistical variations, viz., maximum, minimum, standard deviation in Table 4, in the average monthly inflow of water vary from 14.99 to 2579.32 mcm/m.

4.9 Measurement of water discharge

The data was stored and transferred to the computer separately according to the sections. The cross-sections were divided into no. of sections and the average depth of each section was used for computation of discharge for each section in all respective cross-sections. Finally, to compute the total discharge, the value of each section has been added. Two cross-sections have been surveyed at upstream, i.e., Bapera-Chandori is 10 km u/s and Kumli-Sawra is 15 km u/s, to assess the inflow of water at near Dhaphewada dam. Similarly, at down-stream, i.e., Mandavi (Approx 3 km d/s of APML intake well) to assess the discharge rate of water from Dhaphewada dam, the Wainganga near Bapera-Chandori is 10 km u/s.

4.10 Water balance study

The water balance analysis has been carried out based on the probable monthly flows for all the 12 months with flows at 75%, 90%, and 95% of exceedance, and by assuming that the barrage is normally full at the beginning. The capacity of the Dhapewada dam is 44.05 Mm^3 and the losses due to evaporation and groundwater seepage corresponding to 20% of the barrage capacity that worked out to be 8.81 Mm^3/year is divided proportionately to all months. The daily losses are worked out to be 0.024 Mm3/day. Similarly, irrigation and other daily demands (15% of the barrage storage capacity) and industrial requirements of 70 Mm^3/year have been calculated with daily demands as 0.191781 Mm^3/day for industrial purposes. The water balance has been carried out making use of all inflows and outflows, the storage available at the end of each month have been determined. If the storage has been found more than the storage capacity of the barrage then the downward flow has been calculated using the equation as mentioned in earlier sections. The calculated downward flows have thereafter been compared with the estimated environmental flows as per working group recommendation and the failure months representing the downward flows less than the estimated environmental flows have been identified for different exceedance of probabilities (Mishra and Rai, 2016; Rajasekhar et al., 2021).

4.11 Water source sustainability

In this study area, daily water requirements of 1,91,760 m^3 are being sourced from the Dhapewada dam at Kawalewada on the Wainganga River, Bodalkasa dam is located near Bodalkasa village, in Tirora tehsil, gross storage capacity 17.392 MCM, length of the dam is 510-m and 16.454 MCM live storage capacity. The Dhapewada dam storage capacity is 44 Mm^3, which has been designed, and constructed for having an annual water requirement of about 70 Mm^3 supply of water to the Tiroda Thermal Power Plant and remaining water to canal command area for irrigation. Since the Dhapewada, Bodalkasa, and Chorkhambara dam supply of water from its storage for industrial and irrigation use, drinking and sustainable development.

4.12 Land use

The land deployment in Tiroda tehsil is built-up land is 0.55%, agricultural land is 54.25%, and water bodies are 6.05%. The forest cover is 20.2%, dense forest is 17.75%, open scrap is 1.15%, and space vegetation is 1.30%. This is an important factor affecting the natural structure. At present we see a large scale tree planting program in this tehsil. In this place, dense forest fallows occupy 7.85% of the eastern region as well as there is fallow land in which scrub land 2%, without scrub 9.10% of the land you can see (Mishra et al., 2017; Shailaja et al., 2019; Bisen and Kudnar, 2013).

4.13 Irrigation projects

Gondia District is "land of Tanks"; in Tirora tehsil having 152 total tanks. Whereas 02 medium, 05 small, 22 very small, and 123 tanks are under the category of Mama Talav (Village Tanks). The medium irrigation project capacity in this study area is 10,246 ha out of which 10,043 hectares are irrigated. The capacity of small irrigation project is 2300 ha out of which 2201 ha are irrigated. Tank Irrigation Projects having 489 ha. area potential; whereas 450 ha. the area is irrigated; which is merely 1.30% to the entire potential of all irrigation projects.

4.14 Environmental resources for sustainable water resource management

In study area, 65%–72% area under agriculture is irrigated by a tank, and farmers have used channels and little unit's distribution facility for transportation of water from the tank to farm. Water capacity of small irrigation tanks (100–250 hector irrigation capacity) within the study area is 4.11 cubic meter, out of which 3.78 cubic meter water is often used for irrigation purposes and remaining 0.32 cubic meter dead water are going to be available for Fishery. The Malgujari tanks (MM tank) use small irrigation tanks having more water capacity for fishery occupation after the most occupation, i.e., agriculture, the entire geographic area of the tehsil is 51,616 hector (617.1 sq. km.); out of which 33,588 hector areas are under agriculture, 2974 hector area may be a barren land and 35,347 hector areas aren't useful for agriculture. Out of total area under agriculture, 20,650 hector area irrigated. Total 2065 hector irrigated area in tehsil 12,750 hector area irrigated by tanks. Within the study reason majority of the tanks having fewer amounts of water capacity thanks to ignorance of state and local community water capacity of tanks is reducing day by day. For the event of agriculture sector irrigation is the most vital tool. Because the agriculture sector develops it automatically affects the economic level of farmers during a particular area. Tanks of the study area aren't equally distributed in Tirora tehsil. To uplift, the fishery occupation in these area Government of Maharashtra realizes notification that no individual can collect fish from different tanks. For fishing occupation, a gaggle of fishermen may constitute a Co-operative Society. Water capacity of small irrigation tanks (above 250 hector irrigation capacity) within the study area is 11.169 kL, out of which 10.6278 kL water are often used for irrigation purposes and remaining 0.5412 kL dead water are going to be available for Fishery. The tiny irrigation tank (0–100 hector irrigation capacity) in Tirora tehsil having a water capacity of is 11.26 cubic meters and it's completely used for irrigation purposes.

4.15 Water quality analysis (WQA)

Groundwater is a community resource utilized through wells, springs, canals, etc. Different inventions began to meet the growing demand for water. Modern mechanization, manufacturing, agriculture, industries, chemical fertilizers and pesticides have adversely affected the ground water and the quality of water is deteriorating day by day. Therefore, different results have been studied to study water quality Table 5 shows the natural, chemical and

TABLE 5 Status of water quality in the Tirora Tehsil (river water).

Sl. no.	Parameter	Method of analysis	Unit	Permissible limits	Intake well (Kawalewada Village)	Karti	Bapera	Swara	Mandavi
1	Total suspended solids (TSS)	2540D	mg/L		21	23	20	23	22
2	Total dissolved solids (TDS)	2540C	mg/L	2000	350	320	310	340	330
3	pH	4500H+B		6.5–8.5	8.1	8.1	8	7.9	7.9
4	Total alkalinity	2320B	mg/L	600	109	117	113	108	121
5	Chemical oxygen demand (COD)	5220B	mg/L		14	11.4	13.3	18.1	52.2
6	BOD	5210B	mg/L		5	3.5	4.5	6.1	5.1
7	Total coliform	9221B	MPN/100 mL	Shall not be detectable in any 100 mL sample	16,000			2800	9200
8	Chloride (as Cl-)	4500Cl-B	mg/L	1000	14	14	14	19	14
9	Iron (as Fe)	3500FeB	mg/L	0.3	0.24	0.1	<MRL>	0.14	<MRL>
10	Fecal coliform	9221E	MPN/100 mL	Shall not be detectable in any 100 mL sample	2200			330	1700
11	Magnesium (as Mg)	3500 MgB	mg/L	100	22	12	10	11	10
12	Nitrate	4500NO3-E	mg/L	45	0.28	0.28	0.68	0.71	0.14

MRL = minimum reporting limit (MRL for CN = 0.02 mg/L, for Mn = 1.0 mg/L, for residual chlorine = 1 mg/L, for Ni = 0.5 mg/L, for As = 2.0, for Hg = 2.0.
Source: Environmental Laboratory—APML.

bacteriological tests separation of water. As it turns out, 5 of the above 12 parameters have been taken. From Table 6, you can see the difference in the properties of water depending on the season in this place. You can see the difference in the pH of water in the 3 years from 2016 to 2018. In the post monsoon, we see changes in the pH such as 7.6 in 2016, 7.85 in 2017,

TABLE 6 Seasonal changes in water quality in Tirora Tehsil.

Sl. no.	Parameter	Year	Premonsoon (mg/L)	Monsoon (mg/L)	Postmonsoon (mg/L)	Winter (mg/L)
1	pH	2016	7.6	7.35	6.9	7.05
		2017	7.85	7.75	7.7	7.8
		2018	7.7	8	8.1	8
2	Fluoride	2016	0.35	0.3	0.25	0.35
		2017	0.35	0.5	0.3	0.25
		2018	0.35	0.4	0.3	0.3
3	Nitrates	2016	4.6	3.5	2.15	2.3
		2017	2.45	2.85	2.1	2.05
		2018	2.3	2.45	2.2	3.1
4	Sulfate	2016	9.8	7.4	4.2	6.5
		2017	8.3	12.5	8.3	6.5
		2018	8.1	9.3	7.4	8.2
5	Iron	2016	0.07	0.06	0.06	0.07
		2017	0.075	0.08	0.065	0.06
		2018	0.075	0.07	0.07	0.065
6	Chlorides	2016	11.2	10.3	8	9.7
		2017	11.2	17.1	9.7	9.1
		2018	10.4	10.7	10.1	10.7
7	Total dissolved solids (TDS)	2016	276	152	106	146
		2017	170	236	168	122
		2018	170	186	152	146
8	Total hardness	2016	124	86	74.2	83.5
		2017	92.5	114	84.5	74.2
		2018	88.5	95	88	90
9	Electric conductivity	2016	428	240	168	224
		2017	266	366	278	196
		2018	262	290	236	230

Source: Environmental Laboratory—APML.

and 7.7 in 2018. But the same change you see in post men has changed drastically. You see a difference of 6.9 in 2016, 7.7 in 2017, and 8.1 in 2018 (Ramadan et al., 2019; Sharip et al., 2016).

4.16 Water quality analysis (WQA) and analytic hierarchy process (AHP)

Applied data from 2011 before monsoon (mg/L) calculate 16 parameters of ground water qualities (Table 7) and AHP method, weight was found to be: EC < TDC < TH < Cl^- < SO_4^{2-} - < pH < NO_3^- < F^- < Fe. The consistency value of the AHP model was within acceptable limits (CR = 0.01 < 0.1). Which shows that the correlation is in plus and the relativity of electronic conductivity is in minus. Of electric conductivity, the lowest 540 and the highest 3139 are shown in Fig. 3A. The TDS of water in Fig. 3B is found between 107 and 511. Fig. 3C shows total hardness with a range of 547 with a minimum of 133 and a maximum of 680. The pH range of 1.69 with minimum 7.21 and maximum 8.90 (Fig. 3D), standard ration 0.41, variation 0.173, scavenge 0.545, and kurtosis 1.518. Fluoride is shown in table with a range of 1.71 with a minimum of 0.06 and a maximum of 1.77. The appearance of Nitrate, with a value of minimum 0.11 and a maximum of 279. In addition, many parameters have been studied and mentioned in this study area.

TABLE 7 Statistical analysis of ground water quality in Tirora Tehsil.

Parameter	Range	Minimum	Maximum	Mean	Std. deviation	Variance	Skewness	Kurtosis
EC	2599.0	540.0	3139.0	1091.513	594.1086	352,965.083	3.030	10.473
TDS	1701	310	2011	604.75	436.041	190,131.933	2.451	7.254
TH	547	133	680	244.44	158.372	25,081.729	1.691	2.569
pH	1.69	7.21	8.90	8.0113	0.41635	0.173	0.545	1.518
F^-	1.71	0.06	1.77	0.5556	0.60928	0.371	0.995	−0.756
NO_3^-	278.89	0.11	279.00	19.3100	69.28936	4801.015	3.992	15.957
TA	404	107	511	174.44	133.679	17,869.996	2.295	4.116
Na^+	490	190	680	377.69	160.968	25,910.763	0.517	−0.667
Do	2.58	6.32	8.90	7.2613	0.74218	0.551	0.828	0.165
K^+	399	12	411	48.63	98.780	9757.450	3.733	14.354
COD	16.70	1.40	18.10	11.8500	4.60486	21.205	−1.428	2.128
Ca^{2+}	219.960	20.040	240.000	86.47163	66.449737	4415.568	1.086	0.149
Mg^{+2}	101.00	10.00	111.00	44.1844	33.36655	1113.327	0.760	−0.746
HCO_3	412	239	651	320.69	90.080	8114.363	3.682	14.380
Cl^-	3001.80	13.00	3014.80	690.4219	882.47737	778,766.316	1.541	1.978
SO_4^{2-}	469.00	12.00	481.00	52.7981	118.89455	14,135.913	3.565	13.109

Source: NEERI and Central Ground Water Board (CGWB).

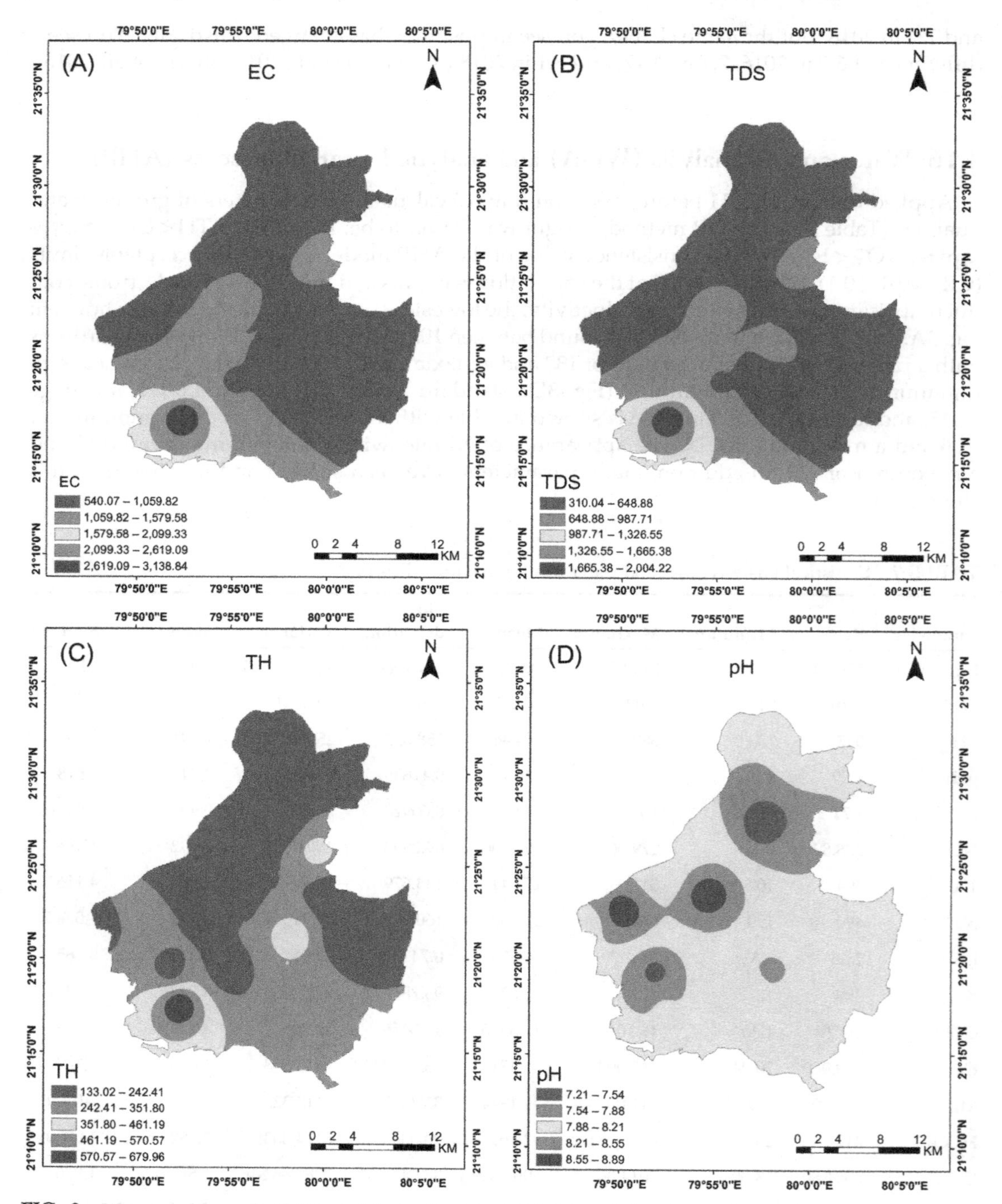

FIG. 3 Water quality analysis (A) electric conductivity (EC), (B) total dissolved solids (TDS), (C) total hardness (TH), and (D) potential of hydrogen (pH).

4.17 Priority

Table 8 and Fig. 4 show that form of salinity water is shown by the electric conductivity. It consists of four water sample in lotion low salinity and it is 25% of its appearance. Medium salinity water also contains five samples and their percentage is 31.25% and high salinity

TABLE 8 Classification of ground water for irrigation based on EC.

Type	EC (μS/cm)	No. of samples	% of samples
Low salinity water	<250	4	25
Medium salinity water	250–750	5	31.25
High salinity water	750–2250	7	43.75
Total		**16**	**100.0**

Residual sodium carbonate (RSC).

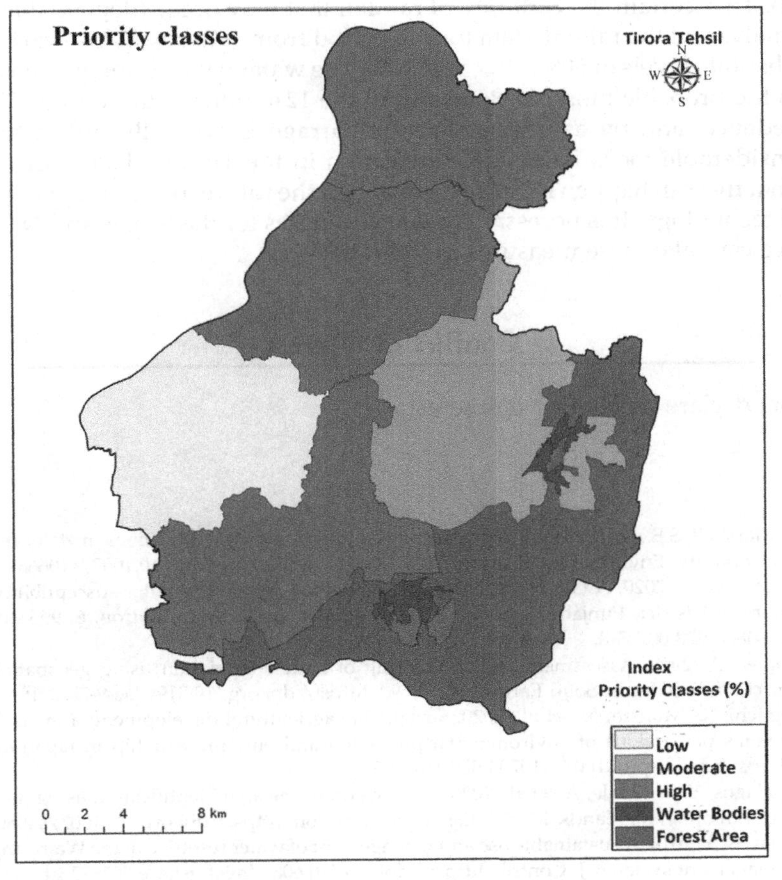

FIG. 4 Priority classes.

water consists of seven stations with a percentage of 43.75%. The quality of water in these study areas is better in riverine areas but water quality in these areas seems to be declining due to the use of chemical fertilizers and pesticides. The water quality in the northern part is also very good. From the above play, we can see that the nature of this groundwater has changed a lot in modern times. The growing population and the water required for it, the scarcity of water in the reservoirs and the insensitivity to the use of groundwater have created a dire situation in the form of groundwater. In groundwater today, its properties have changed from time to time.

5 Conclusions

To evaluate the geomorphology, soil types, geology, slope, relief, landforms, climate and rainfall, hydrology, water availability analysis water inflow water balance, measurement of water discharge, demographics, can be sustained by judiciously using both the surface and subsurface waters. The majority of the sources are having a shallow water table (<10 m below ground level). To ascertain the certainty of rainfall that may occur (dependable rainfall), the probability analysis of the rainfall data for the period from 1971 to 2018 is also carried out for different probability levels of 50%, 80%, and 90%. The water balance analysis has been carried out based on the probable monthly flows for all the 12 months with flows at 75%, 90%, and 95% of exceedance, and by assuming that the barrage is normally full at the beginning. Presently considerable focus is on major irrigation in the Tirora tehsil. From this research you can see that this can happen in the present and in the future to use and plan a region using geographical technology. It is necessary to take measures for the tribals and Nasalizes in this region and we can take these measures in the future.

Conflict of interest

The authors declare no conflict of interest.

References

Archibald, T.W., Marshall, S.E., 2018. Review of mathematical programming applications in water resource management under uncertainty. Environ. Model. Assess. 23, 753–777. https://doi.org/10.1007/s10666-018-9628-0.

Aslam, B., Ismail, S., Ali, I., 2020. A GIS-based DRASTIC model for assessing aquifer susceptibility of Safdarabad tehsil, Sheikhupura District, Punjab Province, Pakistan. Model. Earth Syst. Environ. 6, 995–1005. https://doi.org/10.1007/s40808-020-00735-8.

Ahmad, N., Pandey, P., 2018. Assessment and monitoring of land degradation using geospatial technology in Bathinda district, Punjab, India. Solid Earth 9 (1), 75–90. https://doi.org/10.5194/se-9-75-2018.

Ahamed, T., Noguchi, R., Muhsin, N., et al., 2020. Sustainable agricultural development: a micro-level GIS-based study on women's perceptions of environmental protection and entrepreneurship in Japan and Bangladesh. GeoJournal. https://doi.org/10.1007/s10708-020-10169-5.

Andualem, T.G., Hagos, Y.G., Kefale, A., et al., 2020. Soil erosion-prone area identification using multi-criteria decision analysis in Ethiopian highlands. Model. Earth Syst. Environ. https://doi.org/10.1007/s40808-020-00757-2.

Bisen, D.K., Kudnar, N.S., 2013. A sustainable use and management of water resource of the Wainganga river basin: a traditional management systems. J. Contrib. https://doi.org/10.6084/m9.figshare.663573.v1.

Bouteraa, O., Mebarki, A., Bouaicha, F., Nouaceur, Z., Laignel, B., 2019. Groundwater quality assessment using multivariate analysis, geostatistical modeling, and water quality index (WQI): a case of study in the Boumerzoug-El Khroub valley of Northeast Algeria. Acta Geochim. 38, 796. https://doi.org/10.1007/s11631-019-00329-x.

Cannata, M., Neumann, J., Rossetto, R., 2018. Open source GIS platform for water resource modelling: FREEWAT approach in the Lugano Lake. Spat. Inf. Res. 26, 241–251. https://doi.org/10.1007/s41324-017-0140-4.

Chattaraj, S., Srivastava, R., et al., 2017. Semi-automated object-based landform classification modelling in a part of the Deccan plateau of Central India. Int. J. Remote Sens. 38 (17), 4855–4867. https://doi.org/10.1080/01431161.2017.1333652.

Cieszynska, M., Wesolowski, M., Bartoszewicz, M., et al., 2012. Application of physicochemical data for water-quality assessment of watercourses in the Gdansk municipality (South Baltic coast). Environ. Monit. Assess. 184, 2017–2029. https://doi.org/10.1007/s10661-011-2096-5.

Datta, A., Gaikwad, H., Kadam, A., et al., 2020. Evaluation of groundwater prolific zones in the unconfined basaltic aquifers of Western India using geospatial modeling and MIF technique. Model. Earth Syst. Environ. https://doi.org/10.1007/s40808-020-00791-0.

Dongare, V.T., Reddy, G.P.O., Maji, A.K., et al., 2013. Characterization of landforms and soils in complex geological formations—a remote sensing and GIS approach. J. Indian Soc. Remote Sens. 41, 91–104. https://doi.org/10.1007/s12524-011-0195-y.

Dutta, G., Vinay Kumar, P., Mohammad, S., 2017. Retrieving characteristics of inertia gravity wave parameters with least uncertainties using the hodograph method. Atmos. Chem. Phys. 17, 14811–14819. https://doi.org/10.5194/acp-17-14811-2017.

Geremew, G.M., Mini, S., Abegaz, A., 2020. Spatiotemporal variability and trends in rainfall extremes in EnebsieSarMidir district, Northwest Ethiopia. Model. Earth Syst. Environ. 6, 1177–1187. https://doi.org/10.1007/s40808-020-00749-2.

Guettaf, M., Maoui, A., Ihdene, Z., 2017. Assessment of water quality: a case study of the Seybouse River (north east of Algeria). Appl Water Sci 7, 295–307. https://doi.org/10.1007/s13201-014-0245-z.

Hussen, B., Mekonnen, A., Pingale, S.M., 2018. Integrated water resources management under climate change scenarios in the sub-basin of Abaya-Chamo, Ethiopia. Model. Earth Syst. Environ. 4, 221–240. https://doi.org/10.1007/s40808-018-0438-9.

Jothimani, M., Abebe, A., Dawit, Z., 2020. Mapping of soil erosion-prone sub-watersheds through drainage morphometric analysis and weighted sum approach: a case study of the Kulfo River basin, rift valley, Arba Minch, Southern Ethiopia. Model. Earth Syst. Environ. 6, 2377–2389. https://doi.org/10.1007/s40808-020-00820-y.

Kadam, A.K., Umrikar, B.N., Sankhua, R.N., 2020. Assessment of recharge potential zones for groundwater development and management using geospatial and MCDA technologies in semiarid region of Western India. SN Appl. Sci. 2, 312. https://doi.org/10.1007/s42452-020-2079-7.

Karande, U.B., Kadam, A., Umrikar, B.N., et al., 2020. Environmental modelling of soil quality, heavy-metal enrichment and human health risk in sub-urbanized semiarid watershed of western India. Model. Earth Syst. Environ. 6, 545–556. https://doi.org/10.1007/s40808-019-00701-z.

Khan, M.A., Gupta, V.P., Moharana, P.C., 2001. Watershed prioritization using remote sensing and geographical information system: a case study from Guhiya, India. J. Arid Environ. 49 (3), 465–475. https://doi.org/10.1006/jare.2001.0797.

Khademalrasoul, A., Amerikhah, H., 2021. Assessment of soil erosion patterns using RUSLE model and GIS tools (case study: the border of Khuzestan and Chaharmahal Province, Iran). Model. Earth Syst. Environ. 7, 885–895. https://doi.org/10.1007/s40808-020-00931-6.

Kudnar, N.S., 2020a. GIS-based assessment of morphological and hydrological parameters of Wainganga river basin, Central India. Model. Earth Syst. Environ. 6, 1933–1950. https://doi.org/10.1007/s40808-02000804-y.

Kudnar, N.S., 2020b. GIS-based investigation of topography, watershed, and hydrological parameters of Wainganga River Basin, Central India. In: Sustainable Development Practices Using Geoinformatics. Scrivener Publishing LLC, pp. 301–318, https://doi.org/10.1002/9781119687160.ch19.

Kudnar, N.S., Rajasekhar, M., 2020. A study of the morphometric analysis and cycle of erosion in Waingangā Basin, India. Model. Earth Syst. Environ. 6, 311–327. https://doi.org/10.1007/s40808-019-00680-1.

Kudnar, N.S., 2022. Geospatial modeling in the assessment of environmental resources for sustainable water resource management in a Gondia District, India. In: Rai, P.K., Mishra, V.N., Singh, P. (Eds.), Geospatial Technology for Landscape and Environmental Management. Advances in Geographical and Environmental Sciences. Springer, Singapore, https://doi.org/10.1007/978-981-16-7373-3_4.

Laurent, R., Anker, W., Graillot, D., 1998. Spatial modeling with geographic information system for determination of water resources vulnerability application to an area in massif central (France). J. Am. Water Resour. Assoc. 34 (1), 123–134. https://doi.org/10.1111/j.1752-1688.1998.tb05965.x.

Lkr, A., Singh, M.R., Puro, N., 2020. Assessment of water quality status of Doyang River, Nagaland, India, using water quality index. Appl. Water Sci. 10, 46. https://doi.org/10.1007/s13201-019-1133-3.

Ksiazek, L., Wos, A., Florek, J., et al., 2019. Combined use of the hydraulic and hydrological methods to calculate the environmental flow: Wislokariver, Poland: case study. Environ. Monit. Assess. 191, 254. https://doi.org/10.1007/s10661-019-7402-7.

Mahadevan, H., Krishnan, K.A., Pillai, R.R., et al., 2020. Assessment of urban river water quality and developing strategies for phosphate removal from water and wastewaters: integrated monitoring and mitigation studies. SN Appl. Sci. 2, 772. https://doi.org/10.1007/s42452-020-2571-0.

Majumdar, S., Chatterjee, U., 2021. Modelling urban growth using urban growth deterministic model in Kolkata metropolitan area: a geo-statistical approach. Model. Earth Syst. Environ. 7, 2241–2249. https://doi.org/10.1007/s40808-020-00985-6.

Majumdar, S., Kose, M., Chatterjee, U., 2021. Gully erosion mapping by multi-criteria decision analysis techniques and geoinformatics in Adana Province, Turkey. Earth Syst. Environ. https://doi.org/10.1007/s41748-020-00198-y.

Mishra, V.N., Rai, P.K., 2016. A remote sensing aided multi-layer perceptron-Markov chain analysis for land use and land cover change prediction in Patna district (Bihar), India. Arab. J. Geosci. 9, 249. https://doi.org/10.1007/s12517-015-2138-3.

Mishra, V.N., Prasad, R., Kumar, P., et al., 2017. Dual-polarimetric C-band SAR data for land use/land cover classification by incorporating textural information. Environ. Earth Sci. 76, 26. https://doi.org/10.1007/s12665-016-6341-7.

Ouro-Sama, K., Solitoke, HD., Tanouayi, G., et al., 2020. Spatial and seasonal variation of trace elements contamination level of the waters from the hydrosystem Lake Togo-Lagoon of Aného (South of Togo). SN Appl. Sci. 2, 811. https://doi.org/10.1007/s42452-020-2593-7.

Pathare, J.A., Pathare, A.R., 2020. Prioritization of micro-watershed based on morphometric analysis and runoff studies in Upper Darna Basin, Maharashtra, India. Model. Earth Syst. Environ. https://doi.org/10.1007/s40808-020-00745-6.

Rahaman, K.M., Ahmed, F.R.S., Nazrul Islam, M., 2016. Modeling on climate induced drought of north western region, Bangladesh. Model. Earth Syst. Environ. 2, 45. https://doi.org/10.1007/s40808-016-0089-7.

Rajasekhar, M., Gadhiraju, S.R., Kadam, A., et al., 2020. Identification of groundwater recharge-based potential rainwater harvesting sites for sustainable development of a semiarid region of southern India using geospatial, AHP, and SCS-CN approach. Arab. J. Geosci., 13–24. https://doi.org/10.1007/s12517-019-4996-6.

Rajasekhar, M., Sudarsana Raju, et al., 2021. Multi-criteria land suitability analysis for agriculture in semi arid region of Kadapa District, southern India: geospatial approaches. Remote Sens. Land 5 (2), 59–72. https://doi.org/10.21523/gcj1.2021050201.

Ramadan, E.M., Fahmy, M.R., Nosair, A.M., et al., 2019. Using geographic information system (GIS) modeling in evaluation of canals water quality in Sharkia governorate, East Nile Delta, Egypt. Model. Earth Syst. Environ. 5, 1925–1939. https://doi.org/10.1007/s40808-019-00618-7.

Redvan, G., Mustafa, U., 2021. Flood prioritization of basins based on geomorphometric properties using principal component analysis, morphometric analysis and Redvan's priority methods: a case study of Harşit River basin. J. Hydrol. 603 (Part C). https://doi.org/10.1016/j.jhydrol.2021.127061.

Salami, E.S., Ehteshami, M., 2016. Application of neural networks modeling to environmentally global climate change at San Joaquin Old River station. Model. Earth Syst. Environ. 2, 38. https://doi.org/10.1007/s40808-016-0094-x.

Salunke, V.S., Lagad, S.J., et al., 2021. A geospatial approach to enhance point of the interest and tourism potential centers in Parner tehsil in Maharashtra, India. Int. J. Sci. Res. Sci. Eng. Technol. 8 (1), 186–196. https://doi.org/10.32628/IJSRSET218136.

Salunke, V.S., Bhagat, R.S., et al., 2020. Geography of Maharashtra. Prashant Publication, Jalgaon, pp. 1–229.

Shailaja, G., Kadam, A.K., Gupta, G., et al., 2019. Integrated geophysical, geospatial and multiple-criteria decision analysis techniques for delineation of groundwater potential zones in a semi-arid hard-rock aquifer in Maharashtra, India. Hydrogeol. J. 27, 639–654. https://doi.org/10.1007/s10040-018-1883-2.

Sharip, Z., Saman, J.M., Noordin, N., et al., 2016. Assessing the spatial water quality dynamics in Putrajaya Lake: a modelling approach. Model. Earth Syst. Environ. 2, 46. https://doi.org/10.1007/s40808-016-0104-z.

Sharma, U., Dutta, V., 2020. Establishing environmental flows for intermittent tropical rivers: why hydrological methods are not adequate? Int. J. Environ. Sci. Technol. 17, 2949–2966. https://doi.org/10.1007/s13762-020-02680-6.

Sharma, D., Kansal, A., 2011. Water quality analysis of River Yamuna using water quality index in the national capital territory, India (2000–2009). Appl. Water Sci. 1, 147–157. https://doi.org/10.1007/s13201-011-0011-4.

Sobhani, B., Zengir, V.S., 2020. Modeling, monitoring and forecasting of drought in south and southwestern Iran, Iran. Model. Earth Syst. Environ. 6, 63–71. https://doi.org/10.1007/s40808-019-00655-2.

Soutter, M., Alexandrescu, M., Schenk, C., et al., 2009. Adapting a geographical information system-based water resource management to the needs of the Romanian water authorities. Environ. Sci. Pollut. Res. 16, 33–41. https://doi.org/10.1007/s11356-008-0065-5.

Teshome, A., Halefom, A., 2020. Potential land suitability identification for surface irrigation: in case of Gumara watershed, Blue Nile basin, Ethiopia. Model. Earth Syst. Environ. 6, 929–942. https://doi.org/10.1007/s40808-020-00729-6.

Todmal, R.S., 2020. Understanding the hydrometeorological characteristics and relationships in the semiarid region of Maharashtra (western India): implications for water management. Acta Geophys. 68, 189–206. https://doi.org/10.1007/s11600-019-00386-z.

Tsihrintzis, V.A., 2017. Integrated water resources management, efficient and sustainable water systems, protection and restoration of the environment. Environ. Process. 4, 1–7. https://doi.org/10.1007/s40710-017-0271-6.

Walega, A., Młyński, D., Bogdał, A., Kowalik, T., 2016. Analysis of the course and frequency of high water stages in selected catchments of the upper Vistula Basin in the south of Poland. Water 8, 394. https://doi.org/10.3390/w8090394.

Xu, H., Cai, C., Du, H., et al., 2020. Responses of water quality to land use in riparian buffers: a case study of Huangpu River, China. GeoJournal. https://doi.org/10.1007/s10708-020-10150-2.

CHAPTER

10

Study of morphologic changes in the past and predicting future changes of border rivers (case study: Arvand River, Iran-Iraq Border Line)

Alireza Hajiabadi[a], *Saeed Sakhdari*[b], *and Reza Barati*[c]

[a]MSc of Water Resources Engineering, Abrah GostarTadbir Consulting Engineering Co., Tehran, Iran [b]MSc of Water Resources Engineering, Department of Civil Engineering, University of Sistan and Baluchestan, Zahedan, Iran [c]Ph.D. of Civil Engineering, Head of Applied Research Group at Water Authority, Khorasan Razavi Water Authority, Mashhad, Iran

OUTLINE

1 Introduction 153
2 Materials and methods 154
2.1 *Landsat image of the study area* 154
2.2 *1:5000 topographic maps* 155
3 Results and discussion 157
3.1 *River width changes* 157
3.2 *River meanders properties* 157
3.3 *Predicting future morphologic changes of the river* 160
4 Conclusion 162
References 162
Further reading 163

1 Introduction

River channel migration is a phenomenon that can take place due to erosion of the river banks and deposition over the time, and there is a lot of phenomena that relate this important

https://doi.org/10.1016/B978-0-323-91910-4.00010-8

issue in the field of water resources management (Akbari et al., 2012; Barati et al., 2014, 2018; Ghandehary and Barati, 2018; Barati and Salehi Neyshabouri, 2019; Tajnesaie et al., 2020; Nienhuis et al., 2020; Bray and Kellerhals, 2020; Ghosh et al., 2020; Scorpio et al., 2020; Hiemstra et al., 2020; Grams et al., 2020; Alam et al., 2021; Moghadam et al., 2021; Badfar et al., 2021). This can have a great impact on the land use in the vicinity of the river and can cause damage to the structures built along it (Munasinghe et al., 2021; Wang et al., 2021; Wu et al., 2021; Xu et al., 2021). In the case of Arvand River, since it lies on the Iran-Iraq border line, the problem could be even more urgent (Maghrebi et al., 2020, 2021a,b; Noori et al., 2021). Arvand is a large river in south west of Iran, located on Iran-Iraq border. This River is formed by joining Tigris, Euphrates, Karoon, and Karkhe rivers. Length of this river from Iran-Iraq border to its estuary on the Persian Gulf is 90 km. Because of its mild slope, it seems completely still and the waves higher than 10 cm are rarely seen unless in the case of storms or ship crossings. Arvand River has some specific properties such as: being located on the border, having some drowned ships, being a tidal river, possibility of navigation, salinity due to proximity to the sea, low and reverse speed of flow at tides and great width of the river. Based on these reasons, morphologic changes, erosion and sedimentation in this river are different from other rivers.

In the present study, the analysis and prediction of changes in Arvand River has been considered. There are several sources of data and information such as large-scale topographic maps, land use maps, Landsat images of four different times, Google Earth images, site visits, and talks with locals and authorities, and results of other studies have been collected to do this examination.

2 Materials and methods

Available information of Arvand River, used in this study include: large-scale topographic maps, land use maps, Landsat images of four different times, Google Earth images, site visits, and talks with locals and authorities and results of other studies.

2.1 Landsat image of the study area

Landsat images are the satellite images that can be accessed easily in different time periods and with acceptable resolution for the considered issue. In this study these images were used from four time periods as listed in Table 1. Available images cover a 34 years period and

TABLE 1 Specifications of Landsat images used in the present study.

Specifications of the images	1973	1990	2002	2006
Date of producing the image	6/3/1973	6/12/1990	7/31/2002	3/20/2006
Type of sensor	MSS	TM	ETM+	ETM+
Image route	178	165	165	165
Row of image	39	39	39	39
Resolution (m)	57	28.5	14.25	14.25

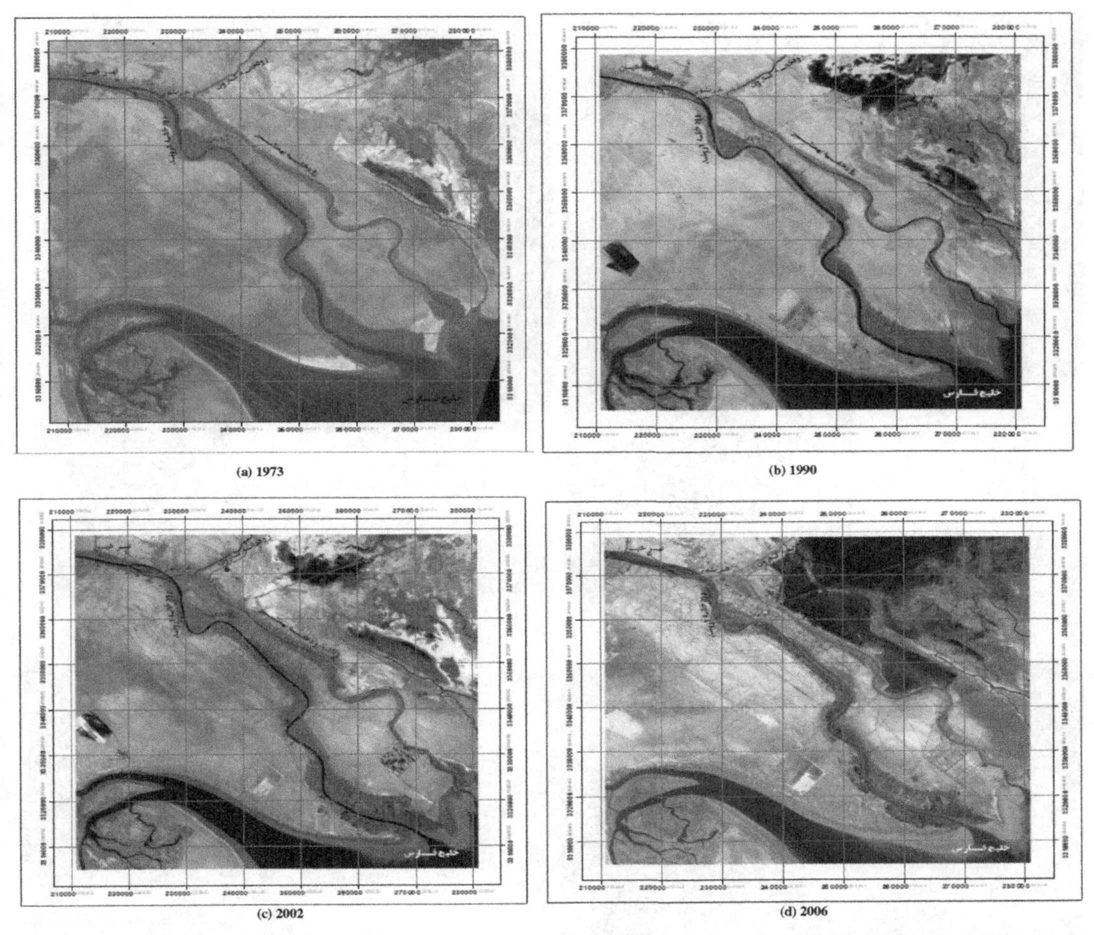

FIG. 1 Landsat images of Arvand River and study area for different time periods (A) 1973, (B) 1990, (C) 2002, and (D) 2006.

mostly are related to rainy months. Fig. 1 illustrates Arvand River and study area for different time periods from 1973 to 2006.

2.2 1:5000 topographic maps

Large-scale topographic maps of the study river are scale of 1:5000 and have been produced from aerial images, taken by the air forces of the army in 2004. Since the date of production of aerial images and Landsat images are nearly the same (2006), correspondence of two images were studied and they were proven to be of good correspondence as depicted in Fig. 2. In the figure, red lines indicate the river's route based on 1:5000 topographic maps.

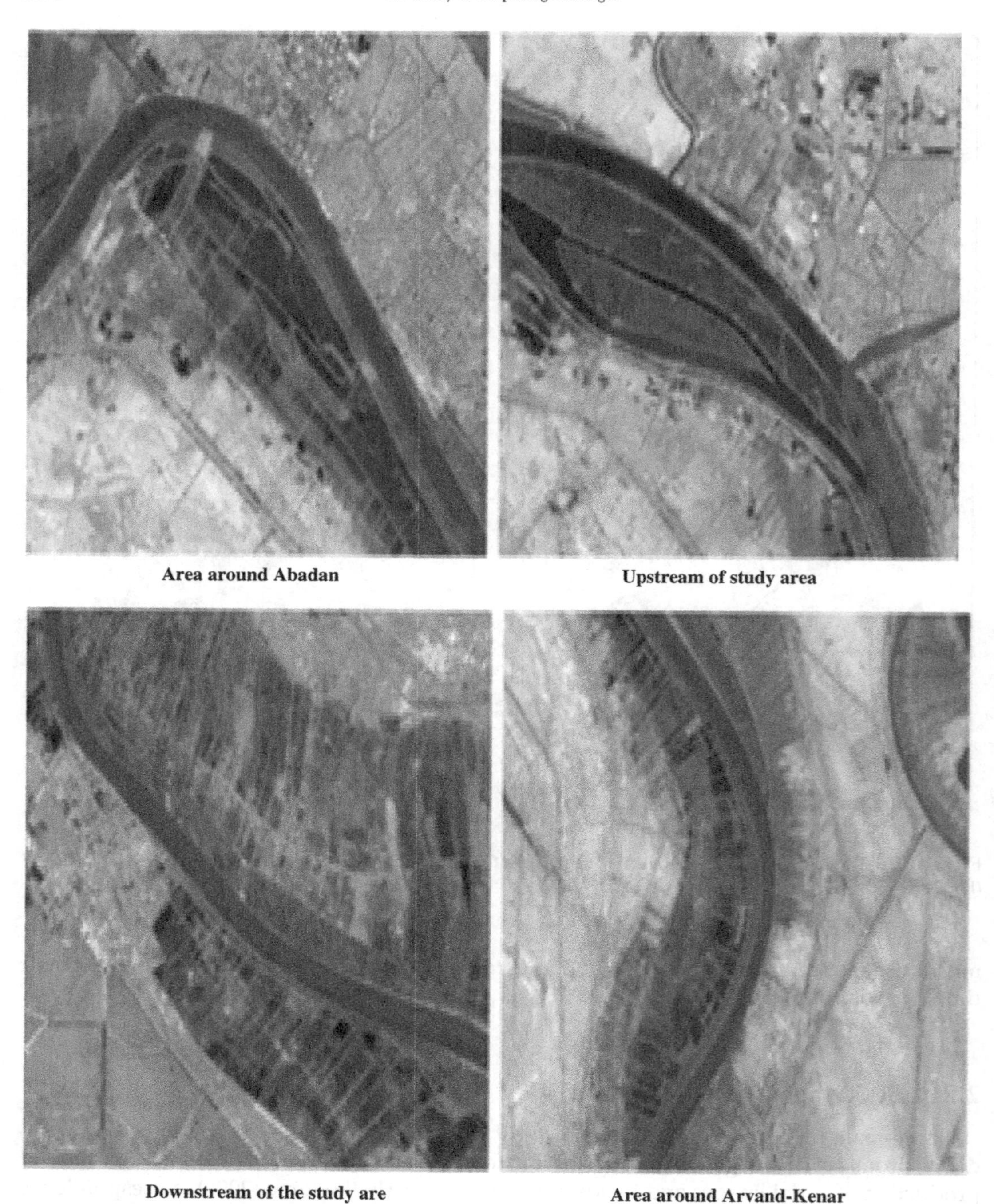

FIG. 2 Arvand River in 1:5000 topographic maps and Landsat images 2006.

3 Results and discussion

Using available 1:5000 topographic maps, river plan and centerline of flow were extracted and based on it, the morphologic specifications of the river were analyzed as following:

3.1 River width changes

To analysis changes of river width along its route, width of Arvand was measured in 184 sections along the study reach and the mean, minimum, and maximum widths respectively were 456, 243, and 702 m. The results show that the changes of the river width along its route don't have a specific relationship, but from KM 75 that the river goes toward the estuary and Persian Gulf, the river width starts to increase. Based on results of field visits, there are a lot of factors that effects on the changes of width of the Arvand River. These effective factors are erosion and sedimentation, joining other tributaries, withdrawal of water, meanders, growth of canebrakes on the banks and drowned objects.

3.2 River meanders properties

For a detailed analysis of morphologic specifications of the river and estimating the related parameters, the study reach was divided into 11 subdivisions.

The map of variations and conditions of the subdivisions was prepared based on available information such as (1) Landsat images of 1973, 1990 and 2006, which cover a 34 years period; (2) the river route in 1:5000 topographic maps using 2006 Landsat images; (3) satellite images of 2002 and 2006 for a short time span; and (4) Google Earth images of the area during 2004 to 2013, which are nearly identical to Landsat images of 2006.

By considering the meander ratio it can be said that about 70% of the study reach is in stable condition and the remaining 30% is instable in form of two local stable pits. By considering the relative curvature, it can be said that about 30.8% of the study reach is in stable condition, and in remaining 69.2%, the stability is in form of distribution of shear tension at the end of outer bank. By considering the central angle, it can be said that about 30.8% of the study reach has a straight form, 12.2% of the study reach is meander like, and the remaining 57% is composed of developed and undeveloped meanders.

It can be noted that Landsat images used at the present study are of an accuracy of 15 to 57 m. Also, these images may have been taken in tidal situations, so the results may have some inaccuracy. To study the changes of river, firstly the main channel were extracted from the images. Secondly, using functions available in Arc GIS-Arc Map, flow centerline of the river was drawn, and this way the centerline and main channel were extracted for years 1973, 1990 and 2006. To analysis the river changes, layers of flow centerline and river limits, were separately overlaid, in three time periods above. An example is presented in Fig. 3.

By comparing river plan and flow central line of different years, trend of changes of Arvand River could be studied. Approximately, mean width of river in 1973, 1990 and 2006 respectively are 700, 550 and 455 m which indicate a decline in width of Arvand River over the time. This is due to some important factors such as reduction of discharge of river from upstream, intensification of sedimentation due to drowned ships in the river, and

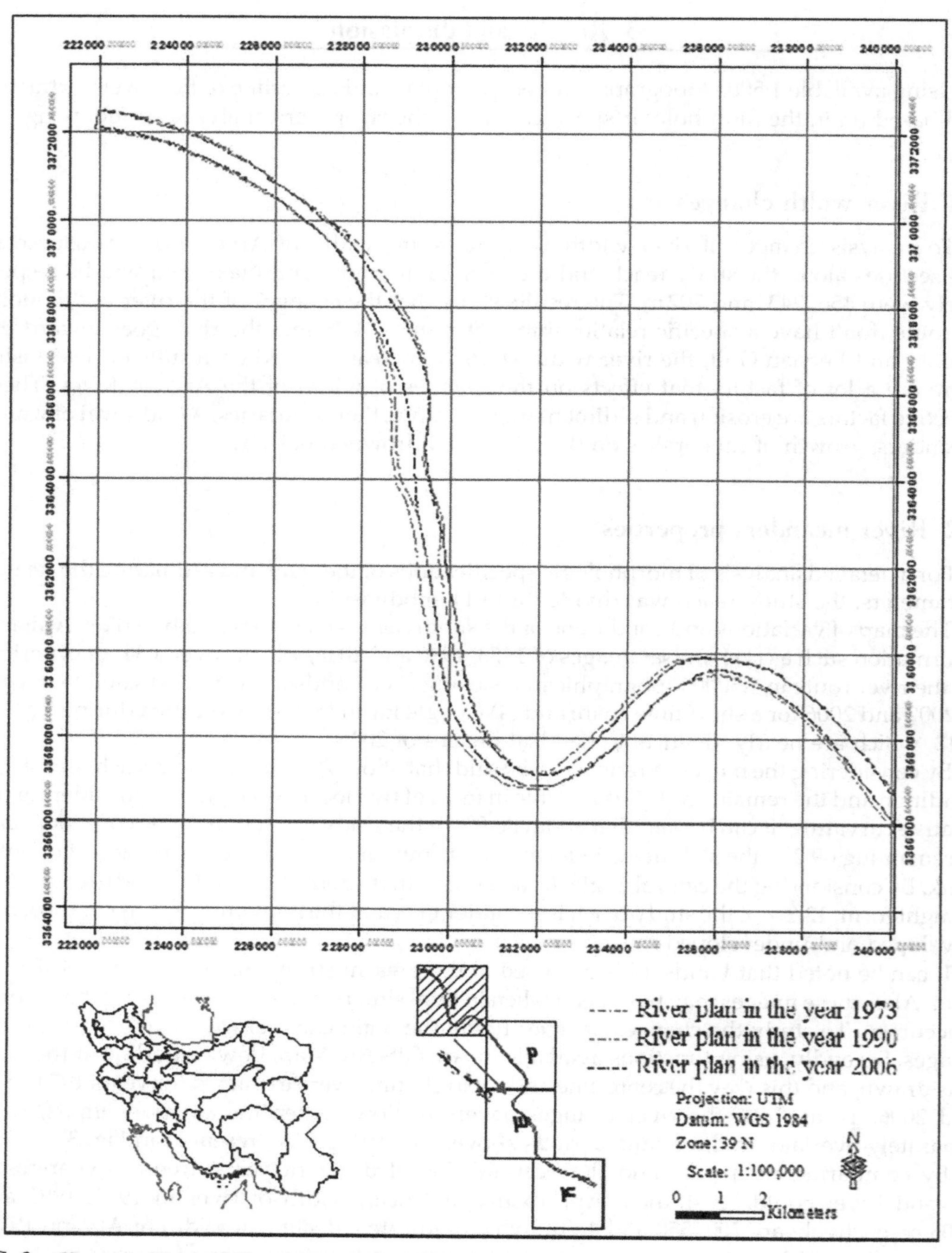

FIG. 3 Changes of Arvand River bed over the 1973 to 2006 period.

TABLE 2 Variations in subdivisions of Arvand River.

Subdivision no.	Start *y*	Start *x*	End *y*	End *x*	Type of changes
1	3,372,335	221,949	3,369,522	227,389	Decline of river width in both banks
2	3,369,522	227,389	3,363,889	229,306	Intense decline of river width in both banks
3	3,363,889	229,306	3,341,029	251,280	Sedimentation in Iraqi bank and erosion in Iranian bank
4	3,341,029	251,280	3,337,080	248,419	Erosion and sedimentation in both banks
5	3,337,080	248,419	3,320,185	255,612	Decline of river width in both banks
6	3,320,185	255,612	3,357,959	238,821	Sedimentation in Iraqi bank and erosion in Iranian bank
7	3,357,959	238,821	3,358,322	230,297	Erosion and sedimentation in both banks
8	3,358,322	230,297	3,313,274	270,156	Intense decline of river width in both banks

TABLE 3 Area and mean width of each subdivision in different periods.

Subdivision no.	Length	Specification of bed in 1973 Area (Ha)	Specification of bed in 1973 Mean width (m)	Specification of bed in 2006 Area (Ha)	Specification of bed in 2006 Mean width (m)	Area changes (Ha)	Width changes (m)
1	6254	262.54	420	206.58	330	−55.96	−89
2	6173	459.50	744	274.78	445	−184.73	−299
3	5597	311.06	556	244.83	437	−66.23	−118
4	10,780	511.41	474	385.97	358	−125.43	−116
5	21,353	1430.11	670	920.54	431	−509.58	−239
6	4920	276.03	561	202.12	411	−73.91	−150
7	20,185	1460.67	724	968.03	480	−492.64	−244
8	16,283	1835.72	1127	953.45	586	−882.27	−542

limiting the river banks by embankments. For a more detailed study of changes, the study reach is divided into eight subdivisions and type of changes of the river plan was analyzed with the results given in Table 2. Type of changes were divided as decline of river width in both banks, intense decline of river width in both banks, sedimentation in Iraqi bank and erosion in Iranian bank and erosion and sedimentation in both banks. By calculating the area and mean width of each of subdivisions related to 1973 and 2006, change of bed width and area of Arvand River were studied and results are given in Table 3. For both area and mean width, a downward trend can be observed in all subdivisions. Most of declines in area and river width are related to subdivision 8 with a magnitude of 882 (Ha) and 542 (m), respectively. The decrease of the river width is in the range of 116 and 542 (m) and the decline of area is in the range of 55 and 882 (Ha). A linear correlation with $R^2=0.77$ can be observed between the area and mean width of each of subdivisions.

Study of central line showed that the centerline of the river reach changed as the centerline has deviated toward Iran in subdivisions 2, 3, and 6, while the changes were small in subdivision 1, and the changes take place on both Iranian and Iraqi banks in other subdivisions. Based on results of field visits, important factors that effects on the centerline of the Arvand River are erosion and sedimentation, joining other tributaries, withdrawal of water, meanders, growth of canebrakes on the banks and drowned objects.

3.3 Predicting future morphologic changes of the river

Studying statistical results of geometric parameters of the river and variation of these parameters along the river, especially the radius of curvature of plan and ratio of river width to radius of curvature of plan shows that development of meander loops of Arvand River is of a normal trend, on the other hand, erosion of outer bank can be seen in most of subdivisions. In Arvand River, erosion in banks is much less than sedimentation. Based on this, morphologic changes of the river in future, in any of subdivisions are as: (1) in subdivision 1, declining bed width would continue at a low rate, until it reaches equilibrium. It could be noted that, due to future water resources management plans, inflow from tributaries would decrease, so the tidal flows would have a decisive role in equilibrium in the river; (2) in subdivision 2, there is a curve that in outer bank (Iran), erosion and in inner bank (Iraq) sedimentation takes place. If this erosion is not controlled, it would continue in the future; (3) in subdivision 3, intensification of erosion in Iran and intense sedimentation In Iraq is observed which will continue in the future; (4) in subdivision 4 there are two curves, that considering the past trends shows a continuation of erosion in outer bank and sedimentation in inner bank in the future; (5) subdivision 5 encompasses a straight route, in which mostly sedimentation and reduction of width of cross sections have taken place. Reduction of width of cross-section at a low rate is predictable in the future; (6) subdivision 6, encompasses a curve, in which erosion in outer bank and sedimentation in inner bank can be seen. The same trend will continue in the future. The important point is that Iran is located on the outer bank, so erosion would take place and the river would be deviated toward Iran; (7) study of changes in subdivision 7, indicates erosion in outer bank and sedimentation in inner bank and reduction of width along the river. The same trend would continue; (8) subdivision 8 is a relatively straight route in which sedimentation has taken place and the same trend would continue in the future. Considering what was discussed above, future morphologic changes of Arvand River was predicated and an example is presented in Fig. 4. The land use on banks of Arvand River includes harbors, oil facilities, farms, channels, embankments, residential areas, etc. Based on this analysis, erosion and sedimentation along this river, in addition to affecting Iran-Iraq border, causes destruction of farms and harbors or reduction of their efficiency, and damage to residential areas.

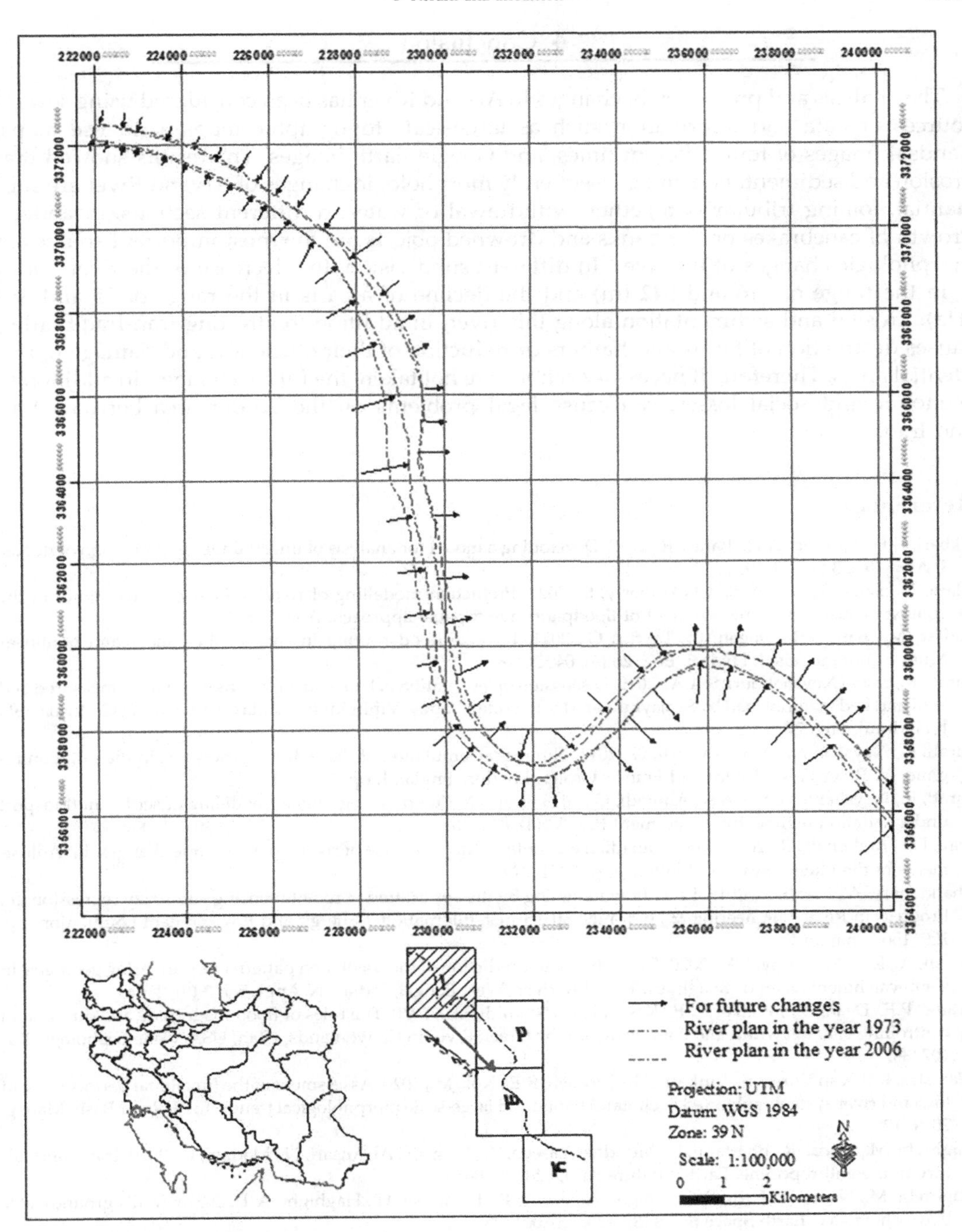

FIG. 4 Predicting morphologic changes of Arvand River in the future.

4 Conclusion

The analysis and prediction of changes in Arvand River has been considered using several sources of data and information such as large-scale topographic maps, land use maps, Landsat images of four different times, and Google Earth images. The results showed that erosion and sedimentation and consequently morphologic changes of Arvand River are substantial. Joining tributaries together, withdrawal of water in different sections, meanders, growth of canebrakes on the banks and drowned objects are the most important sources of morphologic changes of the river. In different subdivision, the decrease of the river width is in the range of 116 and 542 (m) and the decline of area is in the range of 55 and 882 (Ha). Erosion and sedimentation along this river, in addition to affecting Iran-Iraq border, causes destruction of farms and harbors or reduction of their efficiency, and damage to residential areas. Therefore, if necessary actions are not taken, the future changes, in addition to economic and social losses, can cause legal problems on the border area between Iran and Iraq.

References

Akbari, G.H., Nezhad, A.H., Barati, R., 2012. Developing a model for analysis of uncertainties in prediction of floods. J. Adv. Res. 3 (1), 73–79.

Alam, N., Saha, S., Gupta, S., Chakraborty, S., 2021. Prediction modelling of riverine landscape dynamics in the context of sustainable management of floodplain: a geospatial approach. Ann. GIS 27, 1–16.

Badfar, M., Barati, R., Dogan, E., Tayfur, G., 2021. Reverse flood routing in rivers using linear and nonlinear Muskingum models. J. Hydrol. Eng. 26 (6), 04021018.

Barati, R., Salehi Neyshabouri, S.A.A., 2019. Discussion of 'evaluation of bed load equations using field-measured bed load and bed material load by Sanjaykumar Madhusudan Yadav, Vipin Kumar Yadav, and Anurag Gilitwala. ISH J. Hydraul. Eng. 27, 1–3.

Barati, R., Neyshabouri, S.S., Ahmadi, G., 2014. Numerical simulation of the sediment transport in the saltation regime. In: River Flow. Taylor and Francis Group, London, England, pp. 833–841.

Barati, R., Neyshabouri, S.A.A.S., Ahmadi, G., 2018. Issues in Eulerian–Lagrangian modeling of sediment transport under saltation regime. Int. J. Sediment Res. 33 (4), 441–461.

Bray, D.I., Kellerhals, R., 2020. Some Canadian examples of the response of rivers to man-made changes. In: Adjustments of the Fluvial System. Routledge, pp. 351–372.

Ghandehary, A., Barati, R., 2018. River flow modeling by the application of remote sensing and fuzzy regression. In: Progress in River Engineering & Hydraulic Structures. International Energy and Environment Foundation, pp. 129–150 (Chapter 7).

Ghosh, A., Roy, M.B., Roy, P.K., 2020. Estimation and prediction of the oscillation pattern of meandering geometry in a sub-catchment basin of Bhagirathi-Hooghly river, West Bengal, India. SN Appl. Sci. 2 (9), 1–24.

Grams, P.E., Dean, D.J., Walker, A.E., Kasprak, A., Schmidt, J.C., 2020. The roles of flood magnitude and duration in controlling channel width and complexity on the Green River in Canyonlands, Utah, USA. Geomorphology 371, 107438.

Hiemstra, K.S., van Vuren, S., Vinke, F.S.R., Jorissen, R.E., Kok, M., 2020. Assessment of the functional performance of lowland river systems subjected to climate change and large-scale morphological trends. Int. J. River Basin Manag. 20, 1–12.

Maghrebi, M., Noori, R., Bhattarai, R., Mundher Yaseen, Z., Tang, Q., Al-Ansari, N., Madani, K., 2020. Iran's agriculture in the anthropocene. Earth's Future 8 (9), e2020EF001547.

Maghrebi, M., Noori, R., Partani, S., Araghi, A., Barati, R., Farnoush, H., Haghighi, A.T., 2021a. Iran's groundwater hydrochemistry. Earth Space Sci. 8 (8), e2021EA001793.

Maghrebi, M., Noori, R., Darougheh, F., Razmgir, R., Farnoush, H., Taherpour, H., Kløve, B., 2021b. Decline in Iran's River Flows.

Moghadam, H.M., Karami, G.H., Bagheri, R., Barati, R., 2021. Death time estimation of water heritages in Gonabad plain, Iran. Environ. Earth Sci. 80 (4), 1–10.

Munasinghe, D., Cohen, S., Gadiraju, K., 2021. A review of satellite remote sensing techniques of River Delta morphology change. Remote Sens. Earth Syst. Sci. 4 (1), 44–75.

Nienhuis, J.H., Ashton, A.D., Edmonds, D.A., Hoitink, A.J.F., Kettner, A.J., Rowland, J.C., Törnqvist, T.E., 2020. Global-scale human impact on delta morphology has led to net land area gain. Nature 577 (7791), 514–518.

Noori, R., Maghrebi, M., Mirchi, A., Tang, Q., Bhattarai, R., Sadegh, M., Madani, K., 2021. Anthropogenic depletion of Iran's aquifers. Proc. Natl. Acad. Sci. 118 (25), e2024221118.

Scorpio, V., Andreoli, A., Zaramella, M., Moritsch, S., Theule, J., Dell'Agnese, A., Comiti, F., 2020. Restoring a glacier-fed river: past and present morphodynamics of a degraded channel in the Italian Alps. Earth Surf. Process. Landf. 45 (12), 2804–2823.

Tajnesaie, M., Jafari Nodoushan, E., Barati, R., Azhdary Moghadam, M., 2020. Performance comparison of four turbulence models for modeling of secondary flow cells in simple trapezoidal channels. ISH J. Hydraul. Eng. 26 (2), 187–197.

Wang, Y.H., Cai, S.L., Yang, Y.D., Zhong, Z.Y., Liu, F., 2021. Morphological consequences of upstream water and sediment changes and estuarine engineering activities in Pearl River estuary channels over the last 50 years. Sci. Total Environ. 765, 144172.

Wu, S., Xu, Y.J., Wang, B., Cheng, H., 2021. Riverbed dune morphology of the lowermost Mississippi River–implications of leeside slope, flow resistance and bedload transport in a large alluvial river. Geomorphology 385, 107733.

Xu, Y.J., Wang, B., Xu, W., Tang, M., Tsai, F.T.C., Smith, L.C., 2021. Four-decades of bed elevation changes in the heavily regulated upper Atchafalaya River, Louisiana, USA. Geomorphology 386, 107748.

Further reading

Alavipanah, S.K., 2004. Application of Remote Sensing in Earth Sciences. Tehran University Publications.

River Training Studies of Arvand River Neighboring Abadan and Khosroabad Refinery, Sabzab Arvand Consulting Engineering Co, 1389.

Sadri Nasab, M., Fayyaz Mohammadi, M., 2010. Simulation of flow and oscillation of water surface in estuary of Arvand river. J. Mar. Sci. Iran. Spring and Summer.

CHAPTER

11

Rainfall-runoff modeling using GIS: A case study of Gorganrood Watershed, Iran

Mohammad Reza Goodarzi[a], *Amir Reza R. Niknam*[b], *and Maryam Sabaghzadeh*[b]

[a]Department of Civil Engineering, Yazd University, Yazd, Iran [b]Department of Civil Engineering, Water Resources Management Engineering, Yazd University, Yazd, Iran

OUTLINE

1 Introduction 165
2 Rationale of the study 167
3 Material and methods 168
3.1 Data and software 168
3.2 Methodology 168
4 Results and discussion 173
4.1 LULC 173
4.2 Soil map 173
4.3 Curve number 175
4.4 Computation of runoff 175
4.5 Calibration and validation of the SCS-CN model 179
5 Conclusion 180
References 180

1 Introduction

Hydrological modeling is one of the most important processes in water resource applications. In particular, analysis or modeling of the rainfall runoff is the main challenging factor in hydrological modeling that is essential for the development, planning, and management of water resources (Karunanidhi et al., 2020). Rainfall-runoff models are used to describe the relationship between rainfall and runoff in a drainage basin. By applying

https://doi.org/10.1016/B978-0-323-91910-4.00011-X

the model, the rate of rainfall transformed into runoff will be calculated. Due to climate change, human disturbance in the natural environment, and unprincipled construction in vulnerable areas, flooding occurs more frequently and intensely than in the past (Nkwunonwo et al., 2015).

In the hydrological cycle, rainfall can be considered as one of the important and main sources in creating surface runoff (Bansode and Patil, 2014; Karunanidhi et al., 2020; Thilagavathi et al., 2014). Furthermore, the rainfall-runoff transformation is a nonlinear process and varies according to different times and locations. The rainfall runoff is crucial in planning water resources development (Mishra et al., 2006). Spatial ground techniques could increase conventional methods in the studies concerning rainfall runoff (Karunanidhi et al., 2020).

The soil conservation service-curve number (SCS-CN) model is a method that can be used in the absence of runoff measurements. This method was developed by the US Department of Agriculture. The SCS-CN method is mainly influenced by important features of the runoff-related watershed such as land use/land cover (LULC), hydrological soil group (HSG), slope and climatic conditions (Hawkins et al., 2009; Huang et al., 2006). The combination of watershed parameters with climatic factors in a single entity performed by the SCS-CN model is called a curve number (CN). Therefore, the higher CN indicates high runoff and low surface infiltration, while the lower CN indicates low runoff and high surface infiltration (Al-Ghobari et al., 2020). The SCS-CN model was first developed for the basins covering an area of less than 15 km^2, it was then modified for greater basins by using the weighting curve method and the spatial inputs including LULC and soil properties (Ramakrishnan et al., 2009).

Remote sensing (RS) and geographic information system (GIS) are tools that are widely used in various studies to evaluate the hydrological reactions of basins. They also play a very important role in estimating the various parameters required by the SCS-CN method. Integrated use of RS with GIS helps in restoring, collecting, analyzing, manipulating, retrieving, and interpreting the spatially referenced and nonspatial data on large scale and is very useful for classifying satellite images and analyzing parameters such as landuse/landcover (LULC), etc. (Sudhakar et al., 2010; Thilagavathi et al., 2014; Tiwari et al., 2014). CN values can be calculated using existing tables and curves (traditional method), but one of the advantages of using RS and GIS with hydrological models over the traditional method is that it reduces costs and time, and also provides high-accuracy results (Cheng et al., 2006).

Different studies are carried out in recent years concerning this subject, for example, Nayak and Jaiswal (2003) founded that there is a good correlation between the measured and estimated runoff depth using the SCS-CN integrated with GIS. Du et al. (2012) studied the impacts of urban expansion on the runoff and urban flooding in the Qinhuai river, China, and stated that by the expansion of impenetrable urban areas, 2.3% flood events are increased to 13.9%. Phetprayoon (2015) used the SCS-CN hydrological model integrated with GIS to model the upstream runoff of Lam Ta Kong basin in Thailand. The comparison of the modeled and measured results in this study indicates the acceptable accuracy of the model used for runoff simulation. Rajbanshi (2016) used the NRCS-CN method to investigate the depth of surface runoff and average runoff. The results revealed that the average surface runoff of the Konar basin is 71,510.76 m^3 from 2000 to 2009, and it only shows 6.3% of the annual rainfall. Satheeshkumar et al. (2017) investigated the runoff modeling process in

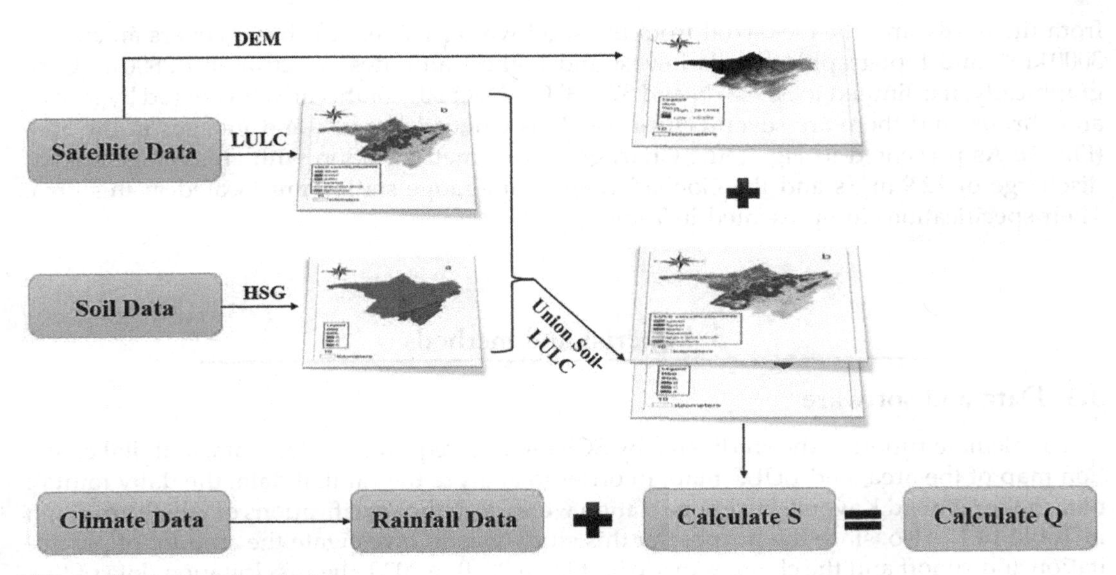

FIG. 1 Flow chart of methodology.

Northern India and considered the accuracy of runoff production potential based on the SCS-CN model as acceptable. Odiji et al. (2020) used the SCS-CN method to estimate the surface runoff of the Upper Benue basin, Nigeria. Their research showed that there is a high correlation between surface runoff and rainfall. Al-Ghobari et al. (2020) implemented the SCS-CN approach integrated with GIS and RS to estimate the runoff of Western Saudi Arabia. They reported that the SCS-CN in combination with RS and GIS deserved more consideration for better basin management and conservation. Kumar et al. (2021) used an integrated combination of SCS-CN and GIS to estimate the surface runoff of the Sind river in a 10-years period. Results indicated that the average annual surface runoff of the Sind river is calculated to be 133.7 mm. The average volume of total runoffs was 35.04*10.8 m^2 that is 17.21% of the annual precipitation.

The purpose of this study is to obtain the runoff rate of the study area by RS and GIS. Thus, the SCS-CN was adopted for this research. SCS-CN is an experimental method for obtaining the runoff of the basins lacking adequate data. Then, the land-use and land cover (LULC) maps produced by RS were combined with the area's hydrologic soil groups' maps and the CN map to obtain the area's run-off depth. In the next step, the rainfall-runoff of 2010 and 2020 were compared in order to determine the impact of land use on runoff. Further, the impact of urban and agricultural expansion on the area's rainfall-runoff was investigated under two scenarios (Fig. 1).

2 Rationale of the study

As a part of the main Gorganrood basin located in Golestan province, Northeastern Iran, the study area is called the middle Gorganrood (Gorganrood-e Miani). This area is limited to the Atrak-Sofla basin from the north, the Namak Lake basin from the south, the Caspian Sea from the north, and the Nekarrod from the southwestern side. This basin covers an area of 3000 km^2, and topographically, its lowest and highest altitudes are 20 m and 2800 m. Geographically, it is limited to 36 °43′ N and 37° 23′ E. About 50% of this area is covered by forests and shrubs, and there are several cities—such as Gonbad Kavus—and villages in the area (Fig. 2). As presented in Fig. 2, the Ghazaghli hydrometric station with an average annual discharge of 12.9 m^3/s and the Gonbad Kavus rain-gauge station are located in this area. Their specifications are presented in Table 1.

3 Material and methods

3.1 Data and software

To calculate runoff in the study area by SCS method requires rainfall data, a digital elevation map of the area, and LULC map. In order to analyze the rainfall data, the daily rainfall data of the Gonbad Kavous rain gauge station were used, the specifications of which are given in Table 14.1. Also, since the purpose of this study was to investigate the amount of precipitation and runoff and the changes made in it from 2010 to 2020, the precipitation data of this station in this period were obtained from the Meteorological Organization. In order to validate the runoff values obtained from the model, the daily discharge data of Ghazaghli station, whose location is shown in Fig. 2, were obtained from Iran Water Resources Management Company. Landsat satellite imagery was used to obtain the LULC map. The Landsat satellite has been continuously imaging the earth's surface since 1972 and has provided data that can be used as a valuable resource for research on land-use change. These images can be downloaded for free from the website of the United States Geological Survey (USGS). In this study, Landsat 5 and 8 images for 2010 and 2020 have been used, the specifications of which are given in Table 2.

Furthermore, as presented in Fig. 2C, a digital elevation model (DEM) with a local accuracy of 30 m was downloaded from the USGS website. As shown in Fig. 2C, the lowest height of this region is 13.292 m and the highest point of the region is 2813.94 m, the eastern part of the region has a lower height and the western part of the region has a higher height. In order to obtain the LULC map, satellite images were processed in ENVI software, which was combined with the DEM map, precipitation data, tables, and related calculations, which will be described below, in the Arc GIS environment and using the HEC-GeoHMS toolbox. Finally, the amount of runoff in the area was calculated.

3.2 Methodology

The SCS-CN method, developed by the United States Department of Agriculture (USDA), is a common method used for rainfall-runoff modeling (Al-Ghobari et al., 2020). This method is used for basins where there is no measured data on runoff values. By this method, the runoff of the basin could be computed according to the precipitation data and CN that indicates the characteristics of the basin in terms of infiltration (Bansode and Patil, 2014).

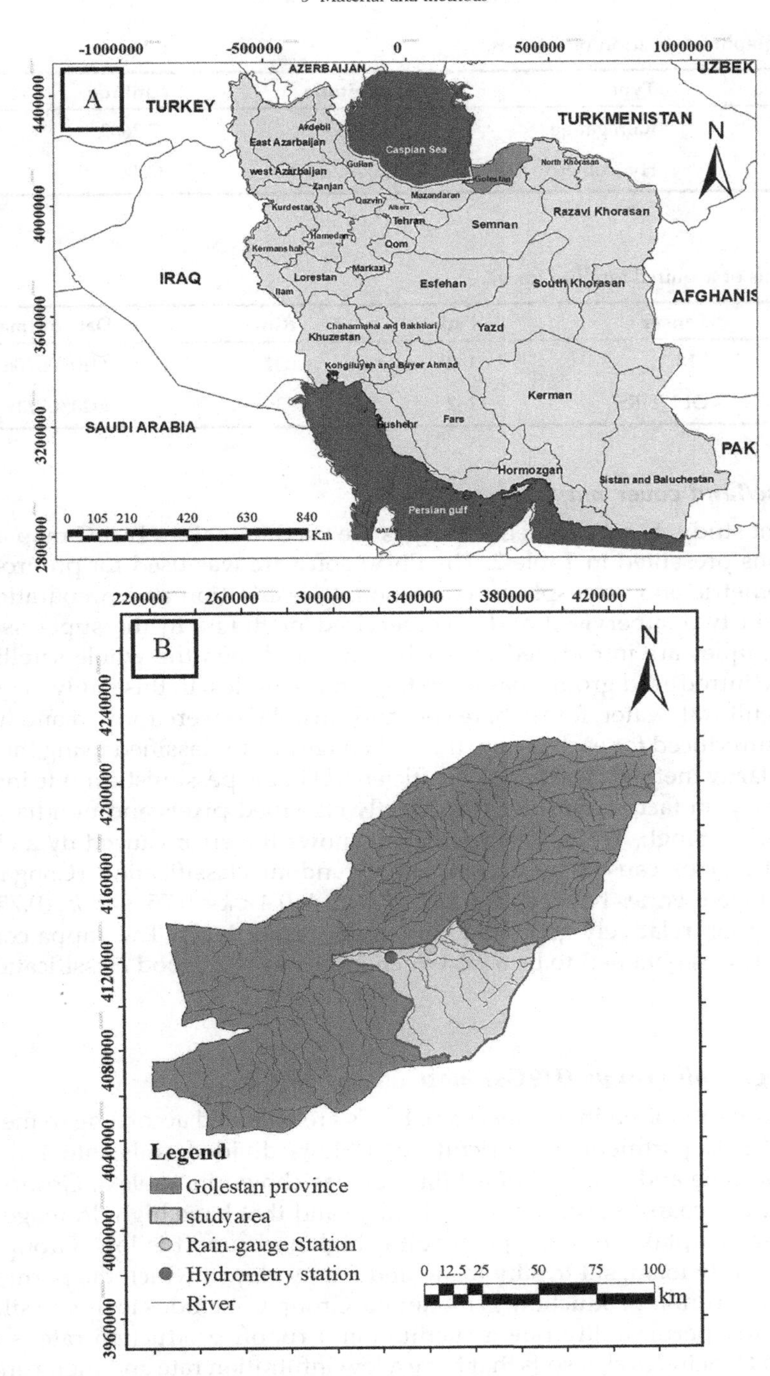

FIG. 2 Location of study are (B), DEM (C) and Iran (A).

TABLE 1 Geographical location of stations.

Name	Type	Longitude	Latitude	Elevation
Gonbad kavous	Rain gauge	55.2128	37.2669	37.5
Ghazaghli	Hydrometry	55.047	37.22	25

TABLE 2 Details of acquired satellite image.

Satellite	Sensor	Path	Row	Date of image acquisition
Landsat_5	TM	162	034	2010/06/06
Landsat_8	OLI_TIRS	162	034	2020/07/19

3.2.1 *Land use/land cover map*

In the present study, Landsat satellite images are used to make a LULC map according to the specifications presented in Table 2. The ENVI software was used for preprocessing and then, the radiometric and atmospheric correction. Classification and preparation of LULC map are done in two supervised and unsupervised methods. In the supervised method, instructional samples are introduced for each group and then the whole satellite image is divided into the introduced groups based on the same samples. In this study, 6 groups of urban lands, agricultural, water, forest, bare soil, and shrubland were formed and instructional samples were introduced for each group, then all images were classified using the supervised maximum similarity method. The kappa coefficient (k) in the pessimistic mode indicates classification accuracy; in fact, it considers incorrectly classified pixels and nondiagonal values of the error matrix (Singh, 1989). This coefficient shows the error caused by a classification compared to the error caused by a completely random classification (Congalton, 1991). The kappa coefficient varies between 0 and 1. If $k<0.4$, $0.4<k<0.75$, and $k>0.75$, the classification will be poor, relatively good and very good, respectively. The kappa coefficient for this classification was obtained to be 0.98, which indicates very good classification accuracy (Foody, 2020).

3.2.2 *Hydrologic soil groups (HSGs) map*

The HSGs map is required in the next step. HSGs are obtained according to the soil type of the area. The US Department of Agriculture (USDA) divided soils into four categories according to soil type and water permeability rate, as shown in Table 3. Group A includes soils such as gravel, coarse sand, sand, and loamy sand that have high drainage properties, so the rate of water uptake in this type of soil is high and runoff is low. Group B includes Loamy sand to sandy loam, silt to silty loam, and loamy clay in which the permeability rate of water, as well as runoff production, is moderate. Group C includes loamy to silty clay soils in which the water permeability rate is medium and runoff production rate is medium to high, and group D includes clay soils that have a low infiltration rate and high runoff production rate (Werchan et al., 1974). As presented in Table 3, Group A has the most infiltration and

TABLE 3 USDA-SCS Soil classification (Werchan et al., 1974).

Hydrologic soil group	Runoff potential	Infiltration rate (mm/h)
A	Low	>7.5
B	Moderate	3.8–7.5
C	Moderate	1.3–3.8
D	High	<1.3

least potential for runoff, and Group D has the least infiltration and the most potential for runoff (Neilsen and Hjelmfelt, 1998).

3.2.3 *Runoff computation according to the SCS method*

One method of estimating runoff in watersheds without flow measurements is to use the US Department of Agriculture CN method. In the SCS method, it is necessary to determine the CNs, which is a function of soil characteristics and land use. Due to its simplicity, this method has quickly become one of the most common methods among engineers and experts and is used for small urban and agricultural basins, medium natural basins and basins for which there is no flow data. (Mishra et al., 2006). In the SCS method, the runoff height is computed by experimental expression (1), where Ia is early precipitation losses including interception, depression storage, and infiltration. $Ia = \lambda S$, where λ is the experimental and regional coefficient that is usually 0.2. The runoff height multiplied by the basin's area equals the volume of the area (Al-Ghobari et al., 2020).

$$P > 0.2S \qquad Q = \frac{(P - Ia)^2}{(P - Ia + S)} \tag{1}$$

$$P \leq 0.2S \qquad Q = 0$$

where R is the runoff height (mm), P is rainfall (mm), and S is the factor affecting the soil water retention and it is computed according to expression (2).

$$S = \frac{25,400}{\mathrm{CN}} \tag{2}$$

where CN is the curve number indicating the amount of infiltrated water into the basin. The value of CN is between 0 and 100. CN equal to zero means that no runoff is obtained from rainfall and CN equals to 100 means that all rainfall is converted to runoff and the height of the runoff will be equal to the height of rainfall. The higher curve number indicates more runoff, and the lower curve number indicates less runoff (Lian et al., 2020). The curve number is obtained according to the land cover and HSGs based on Table 4 (Division, 1986). Vegetation plays an essential role in soil permeability and runoff in the region. Urban areas have the highest runoff production due to the presence of asphalt and lack of vegetation. Forest areas also have less potential for runoff due to vegetation and soils with high permeability. After obtaining the curve number by using Eq. (3), the average curve number of the whole area will be obtained.

TABLE 4 CN for various LULC and soil groups for normal antecedent moisture conditions.

Sr. no.	Cover type	CNs for hydrologic soil group			
		A	B	C	D
1	Urban	77	85	90	92
2	Water	100	100	100	100
3	Forest	36	60	73	79
4	Bare soil	77	86	91	94
5	Agriculture	65	76	84	88
6	Grass and shrub	49	69	79	84

$$\sum\left(\frac{CNi^{*}\ Ai}{A}\right) = \overline{CN} \tag{3}$$

where $\overline{CN}$ denotes average weight (CN) in the basin, and CNi is the value of CN at area Ai.

3.2.4 Model validation

To assess the validity of simulated data, the two indicators of determination coefficient (R^2) and Nash-Sutcliffe Model Efficiency (NSE) coefficient were used according to expressions (4) and (5).

$$R^2 = \frac{\left[\sum_{i=1}^{n}(Oi - \overline{O}) \cdot (Si - \overline{O})\right]^2}{\sum_{i=1}^{n}(Oi - \overline{O})^2 \cdot \sum_{i=1}^{n}(Si - \overline{S})^2} \tag{4}$$

$$\text{NSE} = 1 - \frac{\sum_{i=1}^{n}(Si - Oi)^2}{\sum_{i=1}^{n}(Oi - \overline{O})^2} \tag{5}$$

where $\overline{O}$is average observational data, $\overline{S}$is average simulated data, Oi is observational data, Si is simulated data, and R^2 denotes the linear relationship between simulated and observational data and its value varies from 0 to 1 and the more closer it is to 1, the better the simulation would be. The NSE coefficient is also between 0 and 1, and if it is more than 0.75, the simulation would be better; the values between 0.36 and 0.75 indicate acceptable simulation, and the values less than 0.36 indicate unacceptable simulation. In the present study, λ is changed until the best assessment in R^2 and NSE is obtained (Motovilov et al., 1999). For this purpose, the value of λ is changed from 0.02 to 0.4 to assess the operation of the model according to the above criteria. The best operation coincides with the best value of λ.

3.2.5 Creating scenarios for the future

As the population grows, cities are getting larger every day, and the urban and agricultural lands which are contributed to by humans are growing. As the cities grow, the penetrable lands turn into impenetrable lands such as roads. As the soil penetrability decreases, the rainfall runoff increases, and the risk of flooding in the area will raise. Therefore, in this study, two scenarios were proposed in order to investigate the effects. In the first scenario, urban cells were resampled to 2 pixels in GIS so that they can bear the effects of urbanization. It was presupposed that the existing urban areas are expanded, and its effects on the curve number of the area, soil retention coefficient, and at last, depth and volume of runoff were investigated. In the second scenario, it was presupposed that the agricultural lands—which are created by humans—are expanded, and similar to the first scenario, the agricultural lands of the modeled places were resampled in order to investigate their impact on the depth and volume of runoff in the area.

4 Results and discussion

4.1 LULC

The methods mentioned in the second section were adopted and the LULC map of the study area was made. Six land-use groups were defined for this map: 1. Urban 2. Forest 3. Water 4. Bare soil 5. Shrubland 6. Agricultural.

Fig. 3 shows the LULC map of 2010 and 2020. As Fig. 3 indicates, urban and agricultural lands are mostly in the northern part of the area, and the southern part is covered with forests, grasslands, and shrublands.

According to this map, the area of each level is presented in Table 5. According to Table 5, the largest part of this area is covered with low vegetation in the form of shrubs encompassing 26% of the whole area. The area of forests that play important roles in the absorption of water and preventing flooding in the area is reduced to 26 km^2 in these 10 years. Most changes in these 10 years concern the urban and agricultural lands. Their area is increased to 108 and 62 km^2, respectively. Therefore, according to impenetrable areas such as asphalt streets and roads in the urban areas, it is expected that the runoff will increase in 2020 in comparison with 2010.

Furthermore, as stated in Section 3.2.4, two scenarios were created. In the first scenario, the expansion of the urban area was investigated in GIS by resampling each urban cell to two pixels. In the second scenario, the expansion of agricultural land was investigated by resampling each agricultural cell to two pixels. Two new maps were made for LULC, and as shown in Table 5, in the first scenario, urban lands cover the largest area (28%), and in the second scenario, agricultural lands cover the largest area (33%).

Fig. 3C and D shows the LULC maps under the two scenarios. As presented in Fig. 3C, urban use covers a large part of the northern side, and according to Fig. 3D, agricultural use covers the northern and the middle part of the area.

4.2 Soil map

Due to the role of soil hydrological groups in showing how and to what extent water infiltration, HSGs are among the determining factors in calculating the runoff of the region. Therefore, they were used to investigate the characteristics of the catchment in the ability to

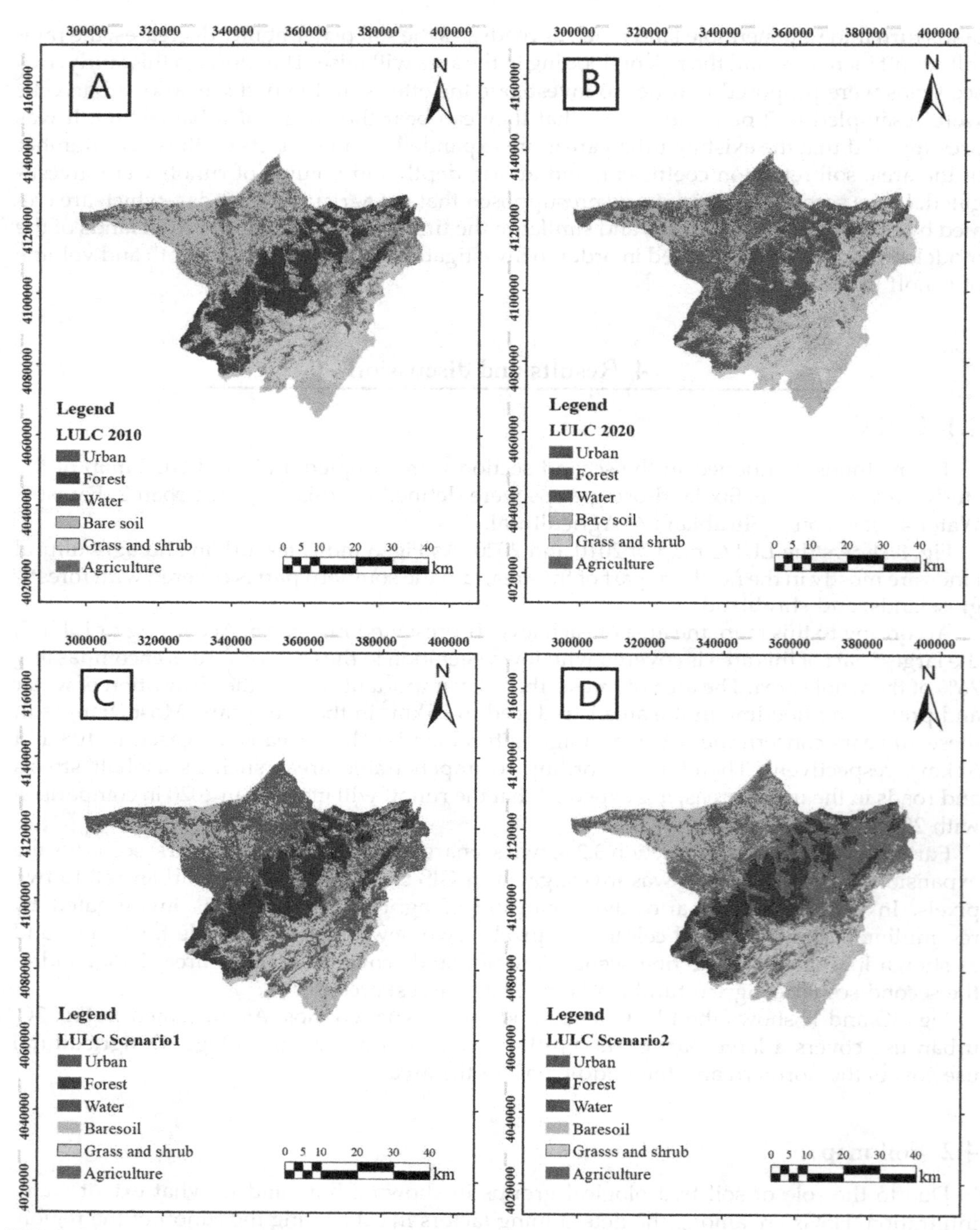

FIG. 3 LULC map (A) 2010, (B) 2020, (C) Scenario1, and (D) Scenario2.

TABLE 5 Area related to each landcover in percentage.

	Urban	Forest	Water	Bare soil	Grass and shrub	Agriculture
2010	7.198	25.51	0.524	26.808	31.259	8.701
2020	10.602	24.687	0.49	24.672	29.109	10.44
Senario1	28.912	25.383	0.426	16.259	23.506	5.514
Senario2	3.726	20.508	0.165	23.465	18.414	33.722

generate runoff according to the amount of rainfall. The US Department of Agriculture has divided all soils into four groups in terms of runoff production capacity: A, B, C, and D. Table 3 presented in Section 11.3.2.2 shows different soil types and classifications of HSGs. The HSGs map of the study area was made as presented in Fig. 4. According to the map, most of the soil in the study area is grouped as the HSG C. Fig. 4B presents the percentages of HSGs of the study area. Group C constitutes the greatest part of the area's soil (70%). As explained before, this group has a low infiltration rate, and consequently, it has a great potential for runoff. Having a higher infiltration rate and less potential for runoff, Groups A and B constitute 26% and 2% of the whole area, respectively, and they mostly cover the northern part of the area.

4.3 Curve number

By integrating the LULC map with the HSG map and by using the expressions presented in Section 2, the curve number of the study area in 2010 and 2020 under the two scenarios were obtained.

The curve number in the study area is between 49 and 100. The curve number of the northern part is higher than other parts of the area, a reason for which could be the urban and barren lands covering this part. The average curve number in 2010 and 2020 are 79.32 and 79.66, respectively, and for the first and second scenarios, 80.5 and 80.07, respectively. As it is evident, the average curve number of the first scenario is a greater number, revealing that the expansion of urban lands relative to agricultural lands will have a greater impact on the runoff of the area. The increase in curve number in 2020 relative to 2010 is because of urbanization.

4.4 Computation of runoff

In the next step, using the existing experimental relationships and in the GIS environment, the soil retention rate (s) was obtained in millimeters. The infiltration of the basin varies from 0 to 264 mm. The lowest infiltration in the northern parts of the basin is about 25 mm, which is related to urban areas, and the highest infiltration is 264 mm, which is related to shrubland and grasslands in the south of the region and with type A soil. After that, the highest amount of infiltration is related to forest areas with an infiltration rate of 93 mm. As it is known, in addition to land cover, soil type also affects the rate of water infiltration. Forest areas were expected to have more influence due to dense vegetation. However, due to the type of soil in this area, which was in hydrological group C (with low permeability), the infiltration rate was lower than the areas with hydrological group A. Finally, using daily rainfall data and basin intrusion values and using Eq. (1), the runoff depth of the area in terms of millimeters was obtained in four modes, as shown in Fig. 5.

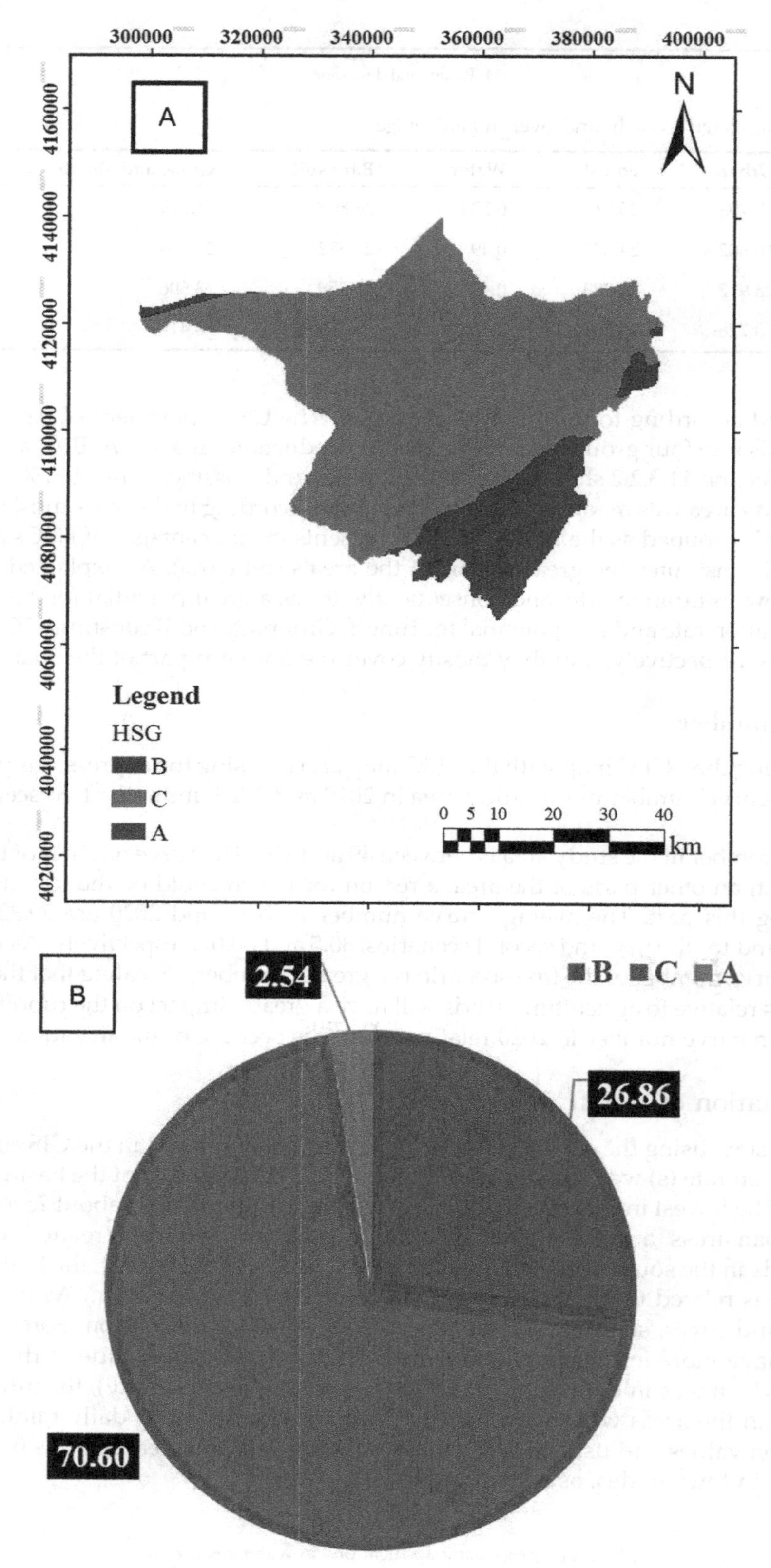

FIG. 4 (A) HSG map. (B) Percentage of HSG in the study area.

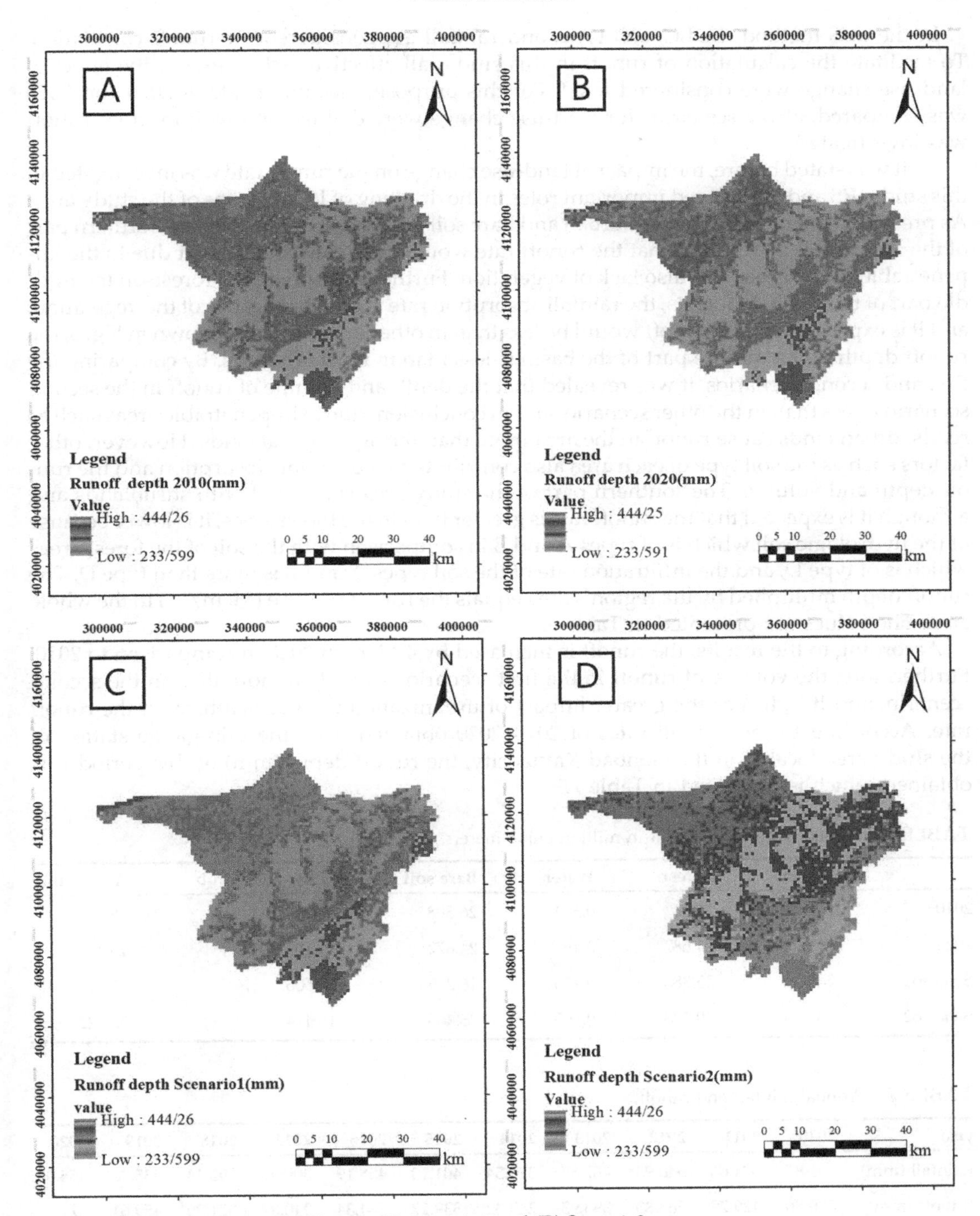

FIG. 5 Runoff depth (A) 2010, (B) 2020, (C) Scenario1, and (D) Scenario2.

In the SCS method, LULC, soil type, and rainfall are considered for runoff calculation. To facilitate the calculation of runoff in this study, all effective factors except the effect of land-use change were considered fixed. For this purpose, the runoff rate in 2010 and 2020 was compared, also 2 scenarios for land-use change were defined and its impact on runoff was investigated.

As it was stated before, the impact of land-use change on the runoff rate was investigated in this study. RS and GIS played important roles in the drawing of LULC maps of the study area. As presented in Fig. 3, most of the urban and bare soil areas are expanded in the northern part of the area and it is expected that the runoff rate would be greater in this part due to the impenetrable urban areas and also lack of vegetation. Furthermore, there are forests in the middle part of the basin. In forests, the rainfall absorption rate is greater because of the vegetation, and it is expected that the runoff would be less than in other areas; thus, as shown in Fig. 5, the runoff depth of the middle part of the basin is less than in northern parts. By comparing the first and second scenarios, it was revealed that the depth and volume of runoff in the second scenario is less than in the other scenario, and in conclusion, due to impenetrable areas such as roads, urban lands cause runoff in the area more than the agricultural lands. However, other factors such as the soil type of each area also contribute to the rainfall absorption and the runoff depth and volume. The southern part of the study area is covered with shrublands and although it is expected that the runoff rate is greater than in the forest areas, it is lower because of the hydrologic soil which is of types A and B in comparison with the soil of the forest areas which is of type D, and the infiltration rate of the soil types A and B is more than type D. The runoff depth multiplied by the region's area equals the runoff volume (Mcm/yr) in the whole area. The values are presented in Table 6.

According to the results, the runoff is increased by 4 Mcm in 2020 in comparison to 2010. Furthermore, the volume of runoff in the first scenario is 4.5 Mcm more than in the second scenario, and it indicates the greater impact of urbanization than agriculture on the runoff rate. According to the rainfall rates of 2010–2020 obtained from the rain-gauge station of the study area located in the Gonbad Kavus city, the runoff depth (mm) of this period was obtained which is presented in Table 7.

TABLE 6 Annual volume of runoff in million cubic meters.

	Urban	Forest	Water	Bare soil	Grass and shrub	Agriculture
2010	7.198	25.51	0.524	26.808	31.259	8.701
2020	10.602	24.687	0.49	24.672	29.109	10.44
Senario1	28.912	25.383	0.426	16.259	23.506	5.514
Senario2	3.726	20.508	0.165	23.465	18.414	33.722

TABLE 7 Annual rainfall and runoff.

year	2010	2011	2012	2013	2014	2015	2016	2017	2018	2019	2020
Rainfall (mm)	401.87	504.66	640.93	467.87	318.54	401.13	458.19	309.91	393.44	535.23	455.08
Runoff (mm)	370.86	429.79	563.83	393.93	249.32	339.27	384.34	240.87	321.39	459.61	372.1

Fig. 6 shows the relationship between rainfall and the runoff computed by the SCS-CN method. The coefficient of determination for rainfall and runoff rate was 0.98. This coefficient indicates that the obtained values of runoff have an acceptable accuracy, and this relationship could be used to find the response of this basin to rainfall. The slope of this line also shows the runoff coefficient that is 0.98 according to the graph. According to the results achieved by Peng and You (2006), the SCS method yields more accurate results in the case of the basins with a runoff coefficient higher than 0.5.

4.5 Calibration and validation of the SCS-CN model

The model calibration was implemented by considering λ as from 0.02 to 0.2 with 0.01 steps. The 2010–15 period was considered as the assessment period, and the 2016–18 period, as the validation period. Coefficient of determination (R^2) and Nash-Sutcliffe Model Efficiency (NSE) coefficient were obtained and compared in each stage, as a result of which, $\lambda=0.2$ is considered as the best value for the middle Gorganrood basin. A summary of the calibration and validation results.

Considering $\lambda=0.2$ is presented in Table 8.

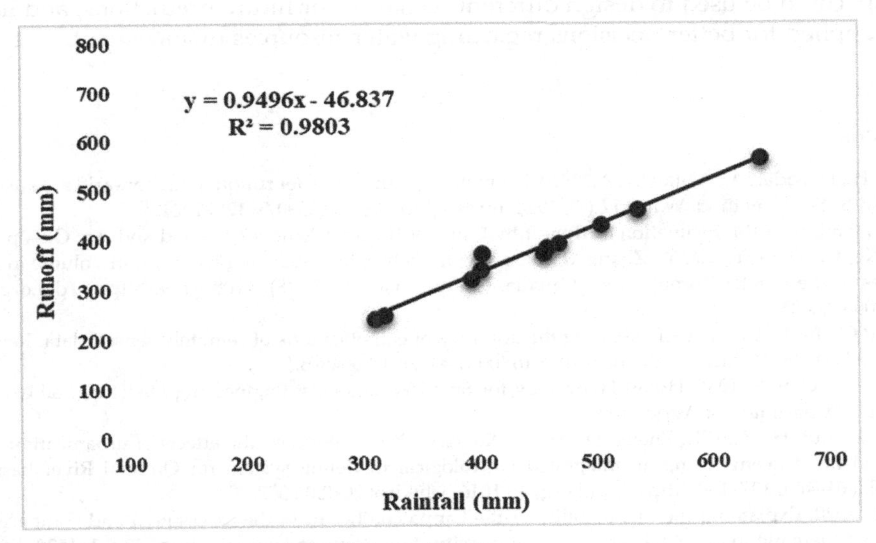

FIG. 6 Correlation coefficient for yearly rainfall-runoff series.

TABLE 8 Results of calibration and validation SCS-model.

	Date	R^2	NSE
Calibration	2010–15	0.7	0.56
Validation	2016–18	0.93	0.76

5 Conclusion

Analysis or modeling of the rainfall-runoff is the main challenging factor in hydrological modeling that is essential for the development, planning, and management of water resources. In this chapter, the SCS-CN method—that is an experimental method to estimate the runoff rate of the areas lacking sufficient data—was used to estimate the runoff rate of the study area in 2010 and 2020. For this purpose, the Landsat images were first preprocessed and the LULC maps were made in a supervised classification form. By comparing the two maps in the 2 years, it was recognized that in this period, the urban area is expanded more than other uses. This map was then integrated with HSGs map of the area, DEM map, and curve number map to compute the runoff depth by experimental expressions. The results indicated that the runoff volume in 2020 is 4 Mcm more than in 2010. Further, two scenarios were created to investigate the impact of land-use change on the runoff rate, thus, in the first scenario, each urban pixel was resampled to two pixels, and in the second scenario, each agricultural pixel was resampled to two pixels. The comparison of the estimated runoff for these scenarios showed that the runoff volume of the first scenario is 4.4 Mcm/yr more than the second scenario's, revealing that urbanization contributes to the runoff rate more than the expansion of agriculture. Also, according to the calibration and validation results that took place in the two periods 2010–15 and 2016–18, respectively, the best value of X was determined. GIS could be used to design different scenarios for future predictions, and its results could be applied for better decisions regarding water resources management.

References

Al-Ghobari, H., Dewidar, A., Alataway, A., 2020. Estimation of surface water runoff for a semi-arid area using RS and GIS-based SCS-CN method. Water 12 (7), 1924. https://doi.org/10.3390/w12071924.

Bansode, A., Patil, K., 2014. Estimation of Runoff by Using SCS Curve Number Method and Arc GIS., p. 5.

Cheng, Q., Ko, C., Yuan, Y., Ge, Y., Zhang, S., 2006. GIS modeling for predicting river runoff volume in ungauged drainages in the greater Toronto area, Canada. Comput. Geosci. 32 (8), 1108–1119. https://doi.org/10.1016/j.cageo.2006.02.005.

Congalton, R.G., 1991. A review of assessing the accuracy of classifications of remotely sensed data. Remote Sens. Environ. 37 (1), 35–46. https://doi.org/10.1016/0034-4257(91)90048-B.

Division, U. S. S. C. S. E, 1986. Urban Hydrology for Small Watersheds: Engineering Division, Soil Conservation Service. US Department of Agriculture.

Du, J., Qian, L., Rui, H., Zuo, T., Zheng, D., Xu, Y., Xu, C.Y., 2012. Assessing the effects of urbanization on annual runoff and flood events using an integrated hydrological modeling system for Qinhuai River basin, China. J. Hydrol. 464–465, 127–139. https://doi.org/10.1016/j.jhydrol.2012.06.057.

Foody, G.M., 2020. Explaining the unsuitability of the kappa coefficient in the assessment and comparison of the accuracy of thematic maps obtained by image classification. Remote Sens. Environ. 239, 111630. https://doi.org/10.1016/j.rse.2019.111630.

Hawkins, R.H., Ward, T., Woodward, D.E., Van Mullem, J., 2009. Curve Number Hydrology: State of the Practice Reston. ASCE, Washington, DC, USA, p. 106.

Huang, M., Gallichand, J., Wang, Z., Goulet, M., 2006. A modification to the soil conservation service curve number method for steep slopes in the loess plateau of China. Hydrol. Process. 20 (3), 579–589. https://doi.org/10.1002/hyp.5925.

Karunanidhi, D., Anand, B., Subramani, T., Srinivasamoorthy, K., 2020. Rainfall-surface runoff estimation for the lower Bhavani basin in South India using SCS-CN model and geospatial techniques. Environ. Earth Sci. 79 (13), 335. https://doi.org/10.1007/s12665-020-09079-z.

Kumar, A., Kanga, S., Taloor, A.K., Singh, S.K., Đurin, B., 2021. Surface runoff estimation of Sind river basin using integrated SCS-CN and GIS techniques. HydroResearch 4, 61–74. https://doi.org/10.1016/j.hydres.2021.08.001.

Lian, H., Yen, H., Huang, J.-C., Feng, Q., Qin, L., Bashir, M.A., et al., 2020. CN-China: revised runoff curve number by using rainfall-runoff events data in China. Water Res. 177, 115767. https://doi.org/10.1016/j.watres.2020.115767.

Mishra, S., Tyagi, J., Singh, V., Singh, R., 2006. SCS-CN-based modeling of sediment yield. J. Hydrol. 324 (1–4), 301–322. https://doi.org/10.1016/j.jhydrol.2005.10.006.

Motovilov, Y.G., Gottschalk, L., Engeland, K., Rodhe, A., 1999. Validation of a distributed hydrological model against spatial observations. Agric. For. Meteorol. 98, 257–277.

Nayak, T., Jaiswal, R., 2003. Rainfall-runoff modelling using satellite data and GIS for Bebas river in Madhya Pradesh. J. Inst. Eng. India Civil Eng. Div. 84, 47–50.

Neilsen, R.D., Hjelmfelt, A.T., 1998. Hydrologic soil group assignment. In: Proceedings of Water Resources Engineering, pp. 1297–1302.

Nkwunonwo, U., Malcolm, W., Baily, B., 2015. Relevance of social vulnerability assessment to flood risk reduction in the Lagos metropolis of Nigeria. Br. J. Appl. Sci. Technol. 8, 366–382. https://doi.org/10.9734/BJAST/2015/17518.

Odiji, C.A., Aderoju, O.M., Ekwe, M.C., Oje, D.T., Imhanfidon, J.O., 2020. CASE STUDY: surface runoff estimation in an upper watershed using geo-spatial based soil conservation service-curve number method. Glob. J. Environ. Sci. Manag. 6 (3). https://doi.org/10.22034/gjesm.2020.03.10.

Peng, D., You, J., 2006. Application of modified SCS model into runoff simulation. Int. J. Water Res. Environ. Eng. 17, 20–24.

Phetprayoon, T., 2015. Application of GIS-based curve number method for runoff estimation in agricultural-forest watershed, Thailand. KKU Res. J. 20, 155–167. https://doi.org/10.14456/kkurj.2015.13.

Rajbanshi, J., 2016. Estimation of runoff depth and volume using NRCS-CN method in Konar catchment (Jharkhand, India). J. Civ. Environ. Eng. 6. https://doi.org/10.4172/2165-784X.1000236.

Ramakrishnan, D., Bandyopadhyay, A., Kusuma, K.N., 2009. SCS-CN and GIS-based approach for identifying potential water harvesting sites in the Kali watershed, Mahi River basin, India. J. Earth Syst. Sci. 118 (4), 355–368. https://doi.org/10.1007/s12040-009-0034-5.

Satheeshkumar, S., Venkateswaran, S., Kannan, R., 2017. Rainfall-runoff estimation using SCS-CN and GIS approach in the Pappiredipatti watershed of the Vaniyar sub basin, South India. Model. Earth Syst. Environ. 3 (1), 24. https://doi.org/10.1007/s40808-017-0301-4.

Singh, A., 1989. Review article digital change detection techniques using remotely-sensed data. Int. J. Remote Sens. 10 (6), 989–1003.

Sudhakar, H., Subramani, T., Lakshmanan, E., 2010. GIS study on vulnerability assessment of water quality in a part of Cauvery River. Int. J. Environ. Sci. 1, 1–17.

Thilagavathi, N., Subramani, T., Suresh, M., Ganapathy, C., 2014. Rainfall Variation and Groundwater Fluctuation in Salem Chalk Hills Area, Tamil Nadu, India.

Tiwari, M.K., Gaur, M., Siyag, P., Kumar, A., 2014. Impact Assessment of Land Use Change on Runoff Generation Using Remote Sensing & Geographical Information System.

Werchan, L.E., Lowther, A., Ramsey, R.N., 1974. Soil Survey of Travis County. US Government Printing Office, Texas.

CHAPTER

12

A review of GIS-based hydrological models for sustainable groundwater management

Hamza Badamasi

Department of Chemistry, Federal University Dutse, Dutse, Jigawa State, Nigeria

OUTLINE

1 Introduction 183

2 Hydrological modeling 185

2.1 Brief history of hydrological modeling 185

2.2 Hydrological models classifications 185

2.3 Hydrological model's calibration, validation and sensitivity analyses 188

3 Geographic information system (GIS) 189

3.1 Integrating GIS with hydrological models 190

3.2 Some case studies to demonstrate the application of GIS in hydrological model 192

4 Benefits of using GIS-based hydrological models for groundwater modeling 194

5 Drawbacks of GIS-based hydrological models for groundwater modeling 194

6 Conclusion and future prospects 195

References 196

1 Introduction

Groundwater is among the most crucial and indispensable resources on Earth. Nearly two billion people rely on it for drinking and agricultural practices (Li and Merchant, 2013; Dar et al., 2020). Groundwater is an important part of the hydrological processes, contributing significantly to the hydrological cycle at both local and regional levels (Fauzia et al., 2021). Industrialization, urbanization, agricultural and activities have contributed to groundwater

Current Directions in Water Scarcity Research, Volume 7
https://doi.org/10.1016/B978-0-323-91910-4.00012-1

overuse, putting immense stresses on global groundwater resources supply (Nampak et al., 2014; Aladejana et al., 2016; De Filippis et al., 2020). Groundwater potential evaluation is the key aspect of the policy frameworks needed for a sustainable, effective, and judicious exploitation of groundwater resources of a given country (Aladejana et al., 2016). Environmental impact study and evaluation of groundwater resources include characterization and constant monitoring of a considerable number of various chemical and physical parameters such as pH, electrical conductivity, temperature, nitrates, and many more (Li and Merchant, 2013). Such an environmental conceptual framework is critical in the development of mathematical models (Refsgaard et al., 2010; Criollo et al., 2019). It was established that an efficient and realistic way of discovering water resources potential is through the development of groundwater potential projecting models (Singh et al., 2011; Mogaji et al., 2014). A model could be seen as a prototype of the actual earth system (Devi et al., 2015). It is commonly designed for predicting the efficiency and development of various hydrologic processes (Abdulkareem et al., 2018). Hydrological modeling is a vital component for a proper understanding of the hydrological cycle since it involves the transport of water and its elements across the land's surface and sub-surface systems (Bajracharya et al., 2018). Hydrological modeling generates simple theoretical models for hydrological processes (Devkota and Gyawali, 2015). Over time, several hydrologic models are being designed to evaluate the hydrological patterns of groundwater (Thakur et al., 2017; Pandey et al., 2020). SWAT, HEC-HMS, BTOPMC, TOPMODEL, MODFLOW-2005, and CWatM are among these models (Takeuchi et al., 1999; Feldman, 2000; Boyce et al., 2020; Burek et al., 2020). The use of hydrological modeling for a groundwater resources management is usually constrained by insufficient data, making the process labor-intensive and sluggish. Geographic Information System (GIS) is a state-of-the-art technology that plays a substantial role in addressing those problems (Gogu et al., 2001; Jha and Chowdary, 2007; Elbeih, 2015). In recent times, research centered on GIS has gained considerable popularity in groundwater resource management due to the spatial characteristic of the groundwater information (De Filippis et al., 2020). Among numerous methods and techniques for groundwater modeling, GIS is recognized as the most effective and cost-effective method, requiring relatively little physicomechanical activities (Termeh et al., 2019; Jhariya et al., 2021). GIS is capable of recording, storing, processing, analyzing, and visualizing vast arrays of spatial and non-spatial groundwater information in a number of formats, and is, therefore, an appropriate tool for addressing groundwater resource issues at local and global scales (Khatami and Khazaei, 2014; Criollo et al., 2019). Being autonomous tools, GIS and hydrological model could work separately; however, their functions can be more effective and fully realized in the management of groundwater resources by integration rather than working independently (Thakur et al., 2017). Hydrologic model and GIS integration have been proven to be very successful and efficient in the collection, handling, analyzing, and visualization of groundwater resource data (Singh and Prakash, 2002; Jha et al., 2007; Thakur et al., 2017). Integration had been established to reduce model runtime, avoid data segregation, and accelerate data output production (Thakur et al., 2017). Therefore, this chapter aims to present an overview of the use of GIS-based hydrologic models for groundwater resource management. The chapter would serve as a guideline for hydrologists, engineers, environmentalists, and policymakers in the proper understanding of the role and current state of GIS-based hydrological models for sustainable groundwater management.

2 Hydrological modeling

Hydrology, as a component of water resources, describes the distribution, transportation, and groundwater quality on Earth. As a fundamental feature of hydrology, the hydrological cycle describes the continuous flow of water beneath and above the Earth's surface (Heywood et al., 2006). A model is a concept that depicts an aspect of the natural or man-made environment that may be in analog, physical states, or mathematical expression. A hydrologic model is the numerical expression of the responsiveness of the catchment area to hydrological events for a given period. Hydrological modeling produces streamlined conceptual models mostly empirical mathematical expressions that define events or physical processes (Fares, 2008).

2.1 Brief history of hydrological modeling

Hydrology has a longstanding heritage that stretches back to many centuries (Biswas, 1970). The origin of hydrological modeling could be dated directly to the 19th century when Mulvany (1850) established the idea of measuring concentration-time and logical system for estimating peak flows used for municipal drainage development. Theis (1935) established the relationship between the removal of the head of the piezometric and the well pump discharge. Muskat (1937) presented an essay on homogeneous fluid flow through porous media. Since then the groundwater field of study has grown dramatically. Plenty of articles have been written to describe the hydrogeological, mathematical, numerical, and engineering aspects of groundwater processes. Decades of the 1960s and beyond marked the emergence of a digital era and hydrological modeling took a tremendous breakthrough (Singh and Fiorentino, 1996; Singh, 2018). Crawford and Linsley (1966) were researchers that developed the first hydrologic model called Stanford watershed; the development of other models then followed (Duan et al., 2003). Over the years, computational capability increased intensely, and hydrology started to mature and grow both in width and depth (Singh et al., 2007). The last half-century had acknowledged an exceptional expansion of newly developed tools and methods for the evaluation of hydrological information. The tools consist of fussy logic, artificial neural networks, geostatistical methods, calibration, and optimization techniques (Bogardi, 2017). The emergence and advancement of hydrological modeling through innovative technology has continued to grow, discovering consistently innovative applications, meeting ever-increasing demands, and attracting more researchers to work in this domain (Chalkias et al., 2016).

2.2 Hydrological models classifications

Hydrological models are important components and instruments for groundwater issues and environmental strategic planning (Abdulkareem et al., 2018). Accurate classification of the hydrological model is of particular relevance to hydrologists, engineers, and other researchers since it will help in understanding the characteristic features of the models before making decisions on using them (Gupta et al., 2015). Two categories of hydrological models had been reported: physical models and mathematical models (Abdulkareem et al., 2018). The physical models represent the actual system. It is categorized into a scale and analog model. The mathematical model describes the behavior of a system through a series of equations, along with probabilistic reasoning that expresses the relations between variables and parameters. In addition, mathematical models are categorized into theoretical, empirical, and conceptual models. The theoretical, empirical, and conceptual models can be further grouped into linear, non-linear, steady, non-steady, lumped, distributed, deterministic, and stochastic models (Gupta et al., 2015). According to Gautam (2013), the models may be categorized based on the process description (empirical, physical, and conceptual), spatial presentation (Distributed and lumped), and aspect of randomness (stochastic and deterministic). Chow et al. (1988) also classify hydrological models based on randomness, space, and time. Fig. 1 depicts how hydrological models are classified.

Numerous hydrological models were developed and are now being applied for groundwater resource management research (Tigabu et al., 2020). Some of the models used for groundwater resource management are presented below:

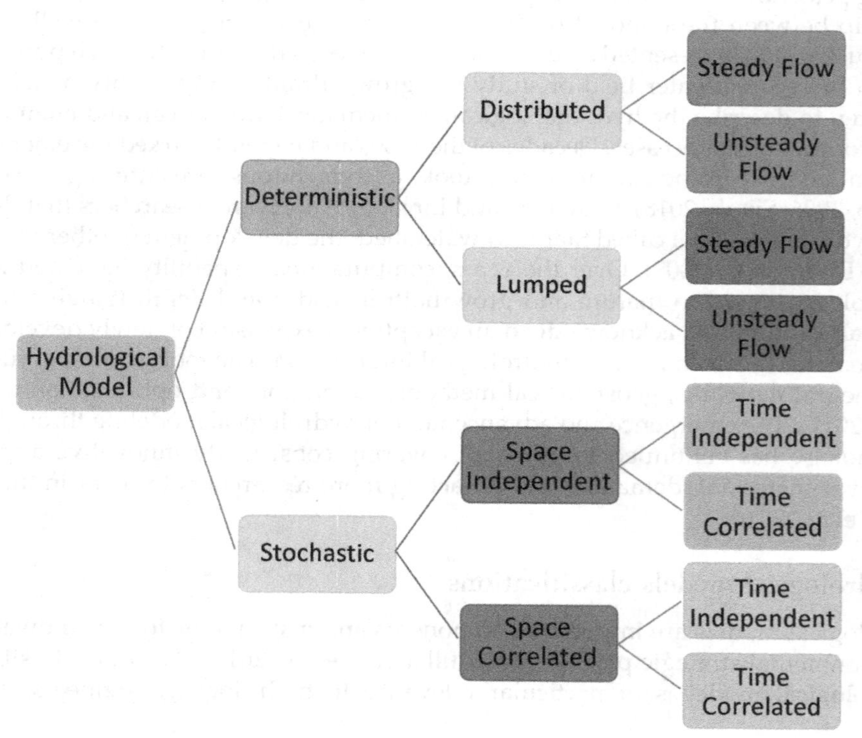

FIG. 1 Classification of hydrological models.

2.2.1 *Modular groundwater flow model (MODFLOW-2005)*

MODFLOW-2005 is a software package employed for modeling of the groundwater recharge using fixed differences. The model algorithm employs a modular plan that encourages the creation and introduction of new frameworks known as processes and packages which operate with or change responses to the numerical equations of the groundwater recharge (Bedekar et al., 2016). With the introduction that new features, the MODFLOW-2005 base system was differentiated into many separate varieties tailored to satisfy unique simulation requirements (Harbaugh, 2005; Hanson et al., 2014). The divergence restricted each separate MODFLOW developed to its individual purpose, such that there was no more single, complete, and overall hydraulic simulation structure (Boyce et al., 2020).

2.2.2 *Soil and water assessment tool (SWAT)*

SWAT is a day-to-day watershed model built to forecast the effects of land use on groundwater, agrochemicals, and sediments outputs in broad hydrologic watersheds. It is a semi-distributed, continuous eco-hydrological that has been successfully applied for groundwater resources management throughout the world (Tigabu et al., 2020). The model's main components include climate conditions, hydrogeology, erosion, temperature, plant development, nutrients, pesticides, land development, quantification of groundwater resources, and stream and reservoir mapping. It separates a catchment into sub-catchments. A stream channel connects each sub-catchment, which is additionally subdivided into a Hydrologic Response Unit, HRU (Adeogun et al., 2014). HRU is simulated as lumped and non-geolocated within each sub-basin, making SWAT computationally effective for long-standing modeling (Liu et al., 2020).

2.2.3 *Dynamic watershed simulation model (DWSM)*

A DWSM is a physically-based, storm event, distributed parameter model. It was effectively applied for simulation of groundwater-surface and sub-surface, flood prediction, and soil erosion. The model was designed to be applied for remote areas, and it describes the comprehensive hydrological cycle like rainfall, evaporation, groundwater flow, channel flow, and much more (Gupta et al., 2015).

2.2.4 *MIKE-SHE model (European hydrological model)*

MIKE-SHE is a completely distributed, physically-based, and deterministic model developed in the 1990s to simulate fully coupled ground and surface water flows, recharge, evapotranspiration processes, and water quality evaluation among others (Refsgaard and Storm, 1996). The model can also be applied to various hydrological cycle processes like rainfall, evaporation, saturated and unsaturated groundwater recharge, river flow, and others (Sandu and Virsta, 2015).

2.2.5 *HBV model (hydrological bureau water department model)*

HBV is a continuous and semi-distributed model that works on daily and monthly precipitation data, temperature, and evapotranspiration (Bergström, 1976). In HBV, the whole catchment area is separated into more catchment areas, which are additionally separated into various altitude and vegetation areas.

2.2.6 *Groundwater loading effects of agricultural management systems model (GLEAMS)*

GLEAMS is a theoretical, continuous, and field model for modeling runoff from chemicals and erosion from agricultural practices (Svetlosanov and Knisel, 1982). GLEAMS used the assumption that land use, soils, and rainfall are homogeneous for a given region. It operations involve four stages: hydrogeology, sediment loadings, transportation of pesticides, and nutrient. It also measures surface runoff and sediments lost. GLEAMS could potentially be applied to assess the influences of agricultural system management methods on the quality of water (Knisel, 1980).

2.2.7 *Community water model (CWatM)*

A CWatM is an open-source model developed using Python program design. It utilizes global freely accessible 15 dataset in the netCDF4 file set-up for interpretation, storing, and processing of information in a convenient manner. The model considers surface and groundwater flows, as well as the influence of human practices on them. It can simulate hydrological processes at various resolutions (Burek et al., 2020).

2.3 Hydrological model's calibration, validation and sensitivity analyses

Model calibration is an integral aspect of every modeling activity. To ensure the effective functioning of modeling techniques, it is critical that they are thoroughly calibrated and validated before any simulation exercise (Todini and Biondi, 2017). Model calibration and measurement efforts are established to achieve fair communications between field records and the model's performance, as well as to increase the integrity and validity of the model (Döll et al., 2003; Fares, 2008; Doherty, 2015). Calibration is the method of changing the hydrological model's parameters to correspond with the real field information. It is a mechanism by which the model's parameters are systematically modified, and the model is recurrently run till the simulated outcomes agree with the field data at an appropriate degree of precision and accuracy (Al-Abed et al., 2005). There are two main ways of calibration of models (Schaake, 2003). The first way is to establish whether one parameter is preferable over the other parameter, and the second way is to select the most preferred parameters (Beven and Binley, 1992). The calibration process in the hydrological analysis is a procedure or technique by which the model's input parameters are positioned manner that the model's outcome in the pattern of simulation model flow can match the aspect of the observed flow (Kumara and Bhattacharjyab, 2020). Calibration of hydrological models can also be achieved by the methods of trial and error or by automation. Calibration by automation may be achieved by defining the function's objectives (Schaake, 2003). Validation of hydrological models is a method of comparing the simulated outcomes with the observed values without modifying the calibrated parameters. According to Refsgaard (1997), general methods for calibrating and validating hydrological models have been thoroughly discussed. Hassanizadeh and Carrera (1992) had, however, reported that there was no agreement on the approach to follow. Refsgaard and Storm (1996) stressed that a robust parameterization protocol is important in preventing procedural issues in successive stages of model calibration and validation activities.

Sensitivity analysis is a method of optimization that studies how changes in the outcomes of hydrological models could be differentiated between different sources of variance inputs. The goal of sensitivity investigation is to analyze the behavior of the model according to the various model parameters and input values (McCuen, 1973; Al-Abed et al., 2005). If appropriately used, sensitivity analysis can offer an improved knowledge of the communication between hydrological models and physical parameters to be modeled (Fares, 2008). It is extremely suggested that the sensitivity investigation of various hydrologic model parameters should be initiated before the beginning of the calibration exercises (Beven and Binley, 1992). Presently, there are different techniques for sensitivity analysis (Hamby, 1994; Beven, 2001). Morris's method has been proven to be a good method, and it provides estimates of the two-factor interface effects (Campolongo and Braddock, 1999).

3 Geographic information system (GIS)

Information technologies have evolved rapidly over the last past decade and had influenced contemporary approaches to solving numerous problems. Among the information technology gadgets, GIS has attracted growing interest. It is a tool employed for capturing, storing, exploring, managing, and eventually displaying all forms of geographical information (Foote and Lynch, 2009; Khatami and Khazaei, 2014). In GIS, the reality is portrayed using digital information, which describes positions in space and attributes information that made up of letters and numbers lists containing temporal information that describes when the other information is valid in time (Gogu et al., 2001). Because of its tremendous capacity to identify and manipulate spatial data, GIS has diverse applications and is therefore of considerable significance in different scientific disciplines (Chen et al., 2004). GIS is considered to be a veritable tool for solving different issues related to groundwater management (ESRI, 2012; Kresic and Mikszewski, 2012; Khatami and Khazaei, 2014). It is recognized to be an effective tool for the integration of data in one format, providing a reliable structure for analyzing the spatial variability, and permitting proper management of geographic information data (Moore et al., 1991; Shrestha et al., 2017). GIS is having a remarkable capability to handle information from different sources provided that both of them are having similar spatial references. For example, information from wells and boreholes, contour maps, geological maps, and images from the satellite could be integrated together. This feature enables each of the aforesaid information to be utilized concurrently, creating a more robust model. The models might help researchers to develop a better knowledge of ground and surface water flows, and their interconnections (Jordan, 2004). In groundwater resource studies, GIS is used for four different purposes which include: (1) database management and hydrological analysis, (2) hydrogeological map elucidation, (3) vulnerability evaluation, and (4) hydrogeological databank for hydrological models. The first three are a hydrogeological continuation of conventional GIS technology. The last one is concerned with developing interrelationships between GIS and hydrologic models used in groundwater studies.

3.1 Integrating GIS with hydrological models

About twenty years through the late 1960s and early 1970s, hydrological models and GIS have evolved independently with very few connections. Significant innovations to integrate GIS into hydrological models had not started till the end of the 1980s (Goodchild et al., 1992; Singh and Fiorentino, 1996). Today, GIS experts, hydrologists, and environmentalists have gradually identified the shared advantages of that integration considering the gigantic achievements of the last 10 years (Sui and Maggio, 1999). Different hydrological modeling methods have provided the opportunity for users of GIS to work above data storage and processing level, but to execute highly developed modeling and simulation. GIS through its unique advantages in handling Digital Elevation Models (DEM) data for hydrological modeling, has supported numerous researchers with innovative information for proper information and visualization. The widespread proliferation of GIS in communities has the capacity in making different hydrological models more accessible and to encourage a broad number of users to communicate their activities and outcomes (Sui and Maggio, 1999). Many GIS systems can effectively perform overlay and index operational activities, but they cannot conduct process-based groundwater modeling activities relating to the flow and movement of groundwater. Integrating GIS with hydrological models, on the other hand, could provide an effective method for handling, storage, controlling, and displaying hydrogeological information. Integration not only facilitates their widespread capabilities in managing and control of groundwater resources but also broadens their spectrum of applications (Gao, 2002; Thakur et al., 2011). Modelers can enjoy the benefits of the integration of advanced hydrological modeling codes into the GIS environment, thereby lowering model setup and investigation time (Marchant et al., 2013). Integration also helps in reducing information isolation and improves data integrity (Bhatt et al., 2014; Alcaraz, 2016). According to Singh and Fiorentino (1996) and Gogu et al. (2001), the integration encompasses three major parts: (a) spatial data creation, (b) spatial coupling of the model's layers, and (c) the interface between GIS and hydrologic model. GIS aids in the design, calibration, and modification of hydrological models, and can interpret the hydrological modeling results into a groundwater resources program (Richards et al., 1993). In the literature, three methods of GIS and hydrological model integration, namely, loose, tight, and embedded coupling were reported (Goodchild et al., 1992; Dwivedi et al., 2017). In loose coupling, the GIS and hydrological model emanate from various software tools, and information is transferred via input/output model exchange files. A loose coupling involves treating the two modules individually and permits communication through manually-enabled file exchange only (Fig. 2A). It is the common interaction between the hydrological model and GIS (Rumbaugh and Rumbaugh, 2011). Examples include visual MODFLOW and Groundwater VISTAS (Guiger and Franz, 1996). In tight coupling (Fig. 2B), the hydrologic model and GIS operate independently, and the first one functions as a platform where the information is pre-processed, analyzed, and then displayed. Thus, a direct flow of information exists between two components. The commonly utilized GIS tools for tight coupling are ESRI, ARGUS One, QGIS, and Map Window GIS (Wang et al., 2016). Embedding coupling can occur when a hydrological model is embedded in a GIS environment (Fig. 2C) or when a GIS is embedded in a hydrological model (Fig. 2D). The coupling of ArcGIS with Archydro is an illustration of the hydrological model embedded in GIS's environment (Simões, 2013). Other examples of hydrological models and their coupling approaches are presented in Table 1.

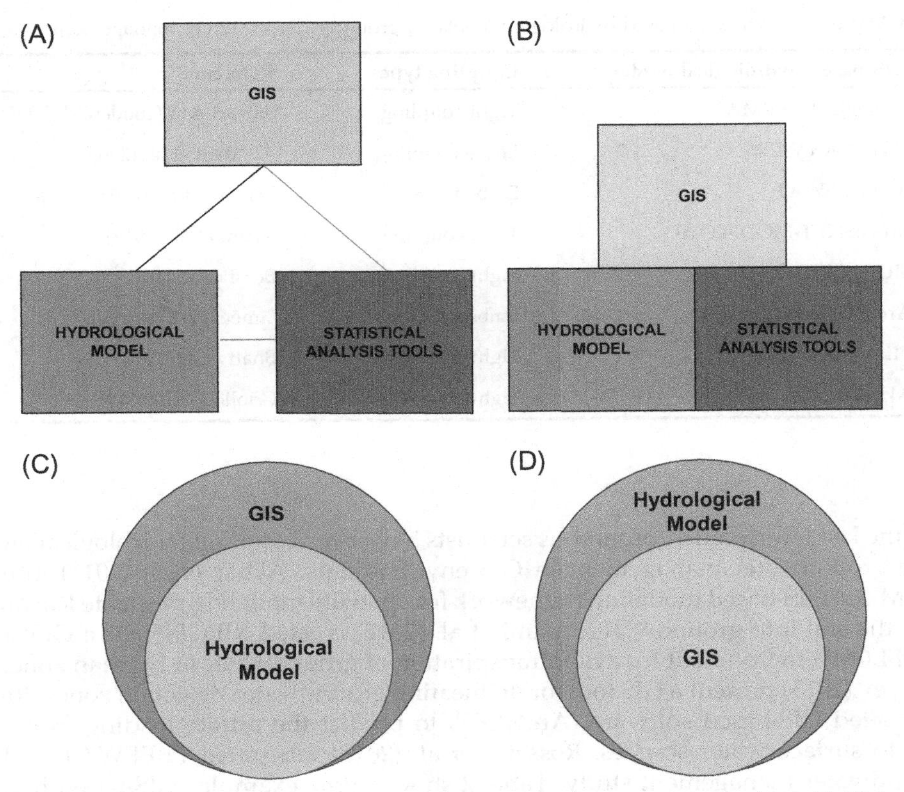

FIG. 2 Methods of Integration of GIS with a hydrological model: (A) Loose coupling, (B) Tight coupling; (C) Hydrological model embedded in GIS, (D) GIS embedded in hydrological Model.

TABLE 1 Comparison of some hydrologic models.

Hydrological models	Year of development	Spatial representation	Energy balance
TOPMODEL	1979	Semi-distributed	No
MOBIDIC	2009	Distributed	Yes
MIKE-SHE	1995	Distributed	No
PIHM	2007	Distributed	Yes
USDAHL	1978	Lumped	No
TRIBS	2004	Distributed	Yes
ParFLow	1996	Distributed	Yes
SWAT	1990	Semi-distributed	Yes
MODFLOW	2005	Distributed	Yes

TABLE 2 Some GIS-based hydrologic models for groundwater resources management.

GIS-based hydrological model	Coupling types	Reference
MODFLOW-ERMA	Tight coupling	Steyart and Goodchild (1994)
SWBM-ArcVIEW	Loose coupling	Al-Abed et al. (2005)
ArcGIS-SWAT	Embedded	Olivera et al. (2006)
RIPGISNET-MODFLOW	Tight coupling	Ajami et al. (2012)
SID & GRID	Tight coupling	Borsi et al. (2012)
ArcGIS-ArcHydro	Embedded	Simões (2013)
PIHMgis	Tight coupling	Bhatt et al. (2014)
AkvaGIS	Tight coupling	Criollo et al. (2019)

Over the last few decades, numerous scientists have been coupling hydrological models to facilitate groundwater management in GIS environments. Akbar et al. (2011) introduced ArcPRZM-3, a GIS-based modeling framework for spatially modeling pesticide leaching ability from the soil into groundwater. Ajami et al. (2012) created RIPGISNET, a GIS software in MODFLOW environment for evapotranspiration of groundwater in riparian zone. Toews and Gusyev (2013) present a GIS tool for delineating groundwater detection zones. Rios et al. (2013) created GIS-based software (ArcNLET) to predict the nitrate loading from sewage systems to surface water sources. Rossetto et al. (2018) integrated FREEWAT with QGIS for groundwater management study. Table 2 shows some examples GIS-based hydrologic models for groundwater resources management.

3.2 Some case studies to demonstrate the application of GIS in hydrological model

3.2.1 GIS-based hydrological modeling with swat: A case study of Jebba reservoir's upstream watershed in Nigeria

Adeogun et al. (2014) investigated the effectiveness and practicability of a SWAT model coupled with GIS software (MapWindow) in the prediction of stream flow in a watershed upstream of Nigeria's Jebba reservoir. SWAT flow predictions were calibrated (from January 1990 to December 1992) and validated (from January 1993 to December 1995) against measured flow data, and the model's performance was evaluated using the coefficient of determination R^2 and the Nasch-Sutcliffe Efficiency (NSE). During the calibration period, the model results showed a good correlation with the observed data, with values of 0.76 for R^2 and 0.72 for NSE. R^2 and NSE values for validation periods were 0.71 and 0.78, respectively. This fascinating SWAT result suggests that a properly calibrated SWAT can be an ideal modeling tool for assisting water resource management policies and decisions at the watershed level in the study area and other places in Nigeria.

3.2.2 *Hydrological modeling based on GIS in Kayu Ara river basin, Malaysia*

To demonstrate the application of a GIS-based hydrological model, a study was conducted in the Sungai Kayu Ara river basin in Kuala Lumpur, Malaysia (Alaghmand et al., 2019). A HEC-HMS connected to the GIS environment via the HEC-GeoHMS GIS extension was used in the study. The effects of rainfall and land-use development conditions on river basin hydrological response, different rainfall durations, and ARIs in existing, intermediate, and ultimate river basin land-use development conditions were considered. The results validate GIS's reasonable capability as an efficient tool in this process.

3.2.3 *GIS-hydrological models for management of water resources in Zarqa river catchment, Jordan*

Al-Abed et al. (2005) investigated the benefits of using GIS-based hydrological models as a water management tool to study Jordan's largest river basin, the Zarqa River basin. The Spatial Water Budget Model (SWBM) and the HEC-HMS/HEC-GeoHMS extension model were used in their research. Calibration and validation of the models were performed. Both models produced satisfactory results, with R^2 values of 0.90 and 0.85 for the HEC-HMS and SWBM models, respectively, and R^2 values of 0.75 and 0.80 for the SWBM and HEC-HMS models for calibration and validation. After calibration and validation, models can be used to verify cases that involve climate change and land-use change at the catchment scale.

3.2.4 *A comparative study of HEC-HMS and the Xinanjiang model in GIS-based hydrological modeling*

Bakir and Xingnan (2008) compared the performance of HEC-HMS to that of the Xinanjiang conceptual model using historical flood data from the Wanjiabu catchment in China (2008). Their findings show that HEC-HMS is more convenient for flood stimulation, particularly in terms of parameter optimization, but it is not as accurate as the Xinanjiang model. The reasonable explanation for this might be that the Xinanjiang model seems to have more parameters, flexible and highly to fit a flood event. Based on the data obtained, it is recommended that with improvements in data conditions, runoff output might be measured on a grid format and the outcomes for both models adequately improved.

3.2.5 *GIS-based hydrological modeling in the Sandusky watershed with SWAT*

Qi and Grunwald (2005) used the Soil and Water Assessment Tool to perform a spatially distributed calibration and validation of water flow (SWAT). They ran surface, groundwater, and total flow simulations in the Sandusky watershed in Ohio, which is part of the Lake Erie watershed and the Great Lakes basin. The validation of total water flow revealed a mean error range of 0.03 to 4.00 m^3/s, a root mean square error range of 0.06 to 2.56 m^3/s, correlation coefficients ranging from 0.70 to 0.90, and Nash–Sutcliffe coefficients ranging from 0.40 to 0.73. Except for winter rainfall runoff events, the simulation results of water flow in the Sandusky watershed and sub-watersheds were satisfactory. The significance of spatially distributed calibration and validation was demonstrated in this study.

4 Benefits of using GIS-based hydrological models for groundwater modeling

Hydrology is among the expanding area which lately engaged GIS to confront various concerns in the area. The main purpose of integrating GIS with hydrological model revolves around the fact that the hydrologic cycle is a dynamic process with temporal and spatial variability (Khatami and Khazaei, 2014). Previously, hydrograph separation had been applied for groundwater recharge research. However, because of its in-built challenges, it is nowadays substituted with GIS-based hydrologic models (Melati et al., 2019). GIS alone could simply be applied for overlay and index processes, but could not be used for groundwater flow and discharge operations. However, integrating GIS with hydrological models can offer efficient tools for handling, storage, manipulation, visualization, and presentation of groundwater resources data (Pathak et al., 2018). GIS-based hydrological models could offer a spatial component that other hydrological models did not have (Heywood et al., 2006). GIS and hydrologic modeling were also coupled to address broader local and international water resource management issues, such as the influence of global climate change on groundwater (Shannon, 2011). An additional merit of GIS-based methods in hydrologic modeling is that it could integrate various layers of geographical information and generate newly assembled data which is relatively convenient for forming hydrological variables that are dependent or independent with each other; for instance, producing evaporation from the relative humidity and temperature. Temperature and relative humidity might be kept in various layers as dependent variables and combining these two layers may produce the evaporation layer. Combining these layers requires that all their sources should be projected into similar coordinates and scales (Foote and Lynch, 2009). GIS-based hydrologic modeling is rapid and consistent for identifying the appropriate groundwater recharge zones, especially in a hard rock landscape. GIS is demonstrated to be valuable in facilitating the calibration of the models (Al-Abed et al., 2005). Numerous studies had confirmed the benefit of coupling GIS with the hydrological model. For example, Xu et al. (2014) utilized Arc/View GIS tool to generate spatially linked coverage for the PDTank mode. The integration had been proven to be an excellent tool for managing watershed-based water resources. Similarly, Kalogeropoulos et al. (2020) concluded that integrating GIS with a hydrological model provides a potentially viable framework for increasing our capacity to use a mathematical description of our natural environment. Kumara and Bhattacharjyab (2020) confirmed the valuable applications of coupling GIS with the hydrological models when they studied two GIS-based models in the Bhagirathi-Alaknanda River catchment area in Northern India.

5 Drawbacks of GIS-based hydrological models for groundwater modeling

Several authors have reported the technical problems linked to the integration of GIS with the hydrological models (Adam and Gangopadhyay, 1997). However, quite a few publications have documented the broad conceptual problems related to GIS integration with the hydrologic model (Howari et al., 2007). For instance, the ability to process the HEC-RAS or HEC-HMS models in Arc Info or a CAD platform does not improve the theoretic groundwork neither does the application of the model. Thus, the integration of GIS with hydrologic models has led to remarkable representational inconsistencies (Gan et al., 1997; Su and Troch, 2003). Nowadays, GIS-based hydrological modeling is dominated by the use of lumped deterministic models (Singh and Frevert, 2002). Numerous scientific investigations have questioned the future of these deterministic models (Thakur et al., 2017; Marsh et al., 2020). Other weaknesses of GIS-based hydrologic models include: (i) uncertainty of the parameters of the model, and initial conditions (ii) inadequate process representation, choices, and interconnections (Fatichi et al., 2016), and (iii) complexity problems such as the degree of physics-like numerical expressions, parameter numbers, and spatial isolation requirements (Hrachowitz and Clark, 2017).

6 Conclusion and future prospects

Groundwater is an essential resource for sustainable growth. Increased industrialization and urbanization have threatened groundwater potential worldwide. It had been acknowledged that an efficient and realistic approach to groundwater resources management is through the development of groundwater potential modeling techniques. Hydrological modeling has come on leaps and bounds from its humble beginning in the 1850s. Developments of hydrological modeling have taken place at a growing speed, motivated largely by the ease of access to nearly unlimited information technology prospects, advanced equipment, and GIS prowess. Over the years, the use of GIS has gained relevance as computer systems have evolved at an astounding pace, due to their capability to not just manage, evaluate and handle geospatial information, but also contribute to the development of different hydrological models for managing the groundwater. GIS is perfectly suited for groundwater resource projects that encourage the use of spatiotemporal information from various sources. GIS is primarily used in various spatial exploration methods to characterize spatial changes relevant to the fundamentals of the hydrological cycle such as rainfall, evapotranspiration, water quality, etc. In the future, GIS and hydrological modeling would continue to be used together to understand and analyze the dynamic flow of water through the environment GIS and hydrologic modeling would continue to be used in tandem in the future to study and fully understand the complex movement of water through the environment. These two powerful resources would be used to address extremely complex and critical groundwater resource issues on local and regional scales. Coupling GIS with the hydrological model could not only empower us with innovative simulation tools and operational techniques that are fully compatible and configurable across various programs but also free us from the limitations of the present models and the restricted spatiotemporal structure of the existing GIS. This new pattern of the coupling GIS and hydrologic models would help us to reflect beyond the technical challenges that dominated the last decade. Most of the GIS-based hydrological models are built from commercial GIS frameworks that are scarcely accessible to communities and institutions with scarce resources. Open sources that allow scientists to access freely and enable users to expand, change and reorganize the program while retaining the actual program should therefore be developed and made available.

References

Abdulkareem, J.H., Pradhan, B., Sulaiman, W.N.A., Jamil, N.R., 2018. Review of studies on hydrological modelling in Malaysia. Model. Earth Syst. Environ. 4, 1577–1605.

Adam, N.R., Gangopadhyay, A., 1997. Database Issues in Geographic Information Systems. Kluwer Academic Publishers, Boston, MA, pp. 23–43.

Adeogun, A.G., Sule, B.F., Salami, A.W., Okeola, O.G., 2014. GIS-based hydrological modeling using SWAT: case study of watershed of Jebba reservoir in Nigeria. Niger. J. Technol. 33 (3), 351–358.

Ajami, H., Maddock, T., Meixner, T., Hogan, J.F., Guertin, D.P., 2012. RIPGIS-NET: a GIS tool for riparian groundwater evapotranspiration in MODFLOW. Ground Water 50 (1), 154–158.

Akbar, T.A., Lin, H., DeGroote, J., 2011. Development and evaluation of GIS-based ArcPRZM-3 system for spatial modeling of groundwater vulnerability to pesticide contamination. Comput. Geosci. 37 (7), 822–830.

Al-Abed, N., Abdulla, F., Abu Khyarah, A., 2005. GIS-hydrological models for managing water resources in the Zarqa River basin. Environ. Geol. 47, 405–411.

Aladejana, O.O., Anifowose, A.Y.B., Fagbohun, B.J., 2016. Testing the ability of an empirical hydrological model to verify a knowledge-based groundwater potential zone mapping methodology. Model. Earth Syst. Environ. 2, 174.

Alaghmand, S., Abdullah, R.B., Abustan, I., Kordi, E., Vosoogh, B., 2019. GIS-based hydrological modelling in Kayu Ara river basin, Malaysia. J. Flood Eng. 10 (2), 81–97.

Alcaraz, M., 2016. GIS Platform for Management of Shallow Geothermal Resources (Ph.D. thesis). Polytechnic University of Catalunya, UPC, Spain.

Bajracharya, A.R., Bajracharya, S.R., Shrestha, A.B., Maharjan, S.B., 2018. Climate change impact assessment on the hydrological regime of the Kaligandaki Basin, Nepal. Sci. Total Environ. 625, 837–848.

Bakir, M., Xingnan, Z., 2008. GIS-based hydrological modeling: a comparative study of HEC-HMS and the Xinanjiang model. In: Twelfth International Water Technology Conference, IWTC12 2008 Alexandria, Egypt, pp. 855–865.

Bedekar, V., Morway, E.D., Langevin, C.D., Tonkin, M., 2016. MT3D-USGS Version 1: A U.S. Geological Survey Release of MT3DMS Updated with New and Expanded Transport Capabilities for Use with MODFLOW. In: Techniques and Methods 6-A53. U.S. Geological Survey, p. 69.

Bergström, S., 1976. Development and Application of a Conceptual Runoff Model for Scandinavian Catchments. SMHI RHO 7. Norrköping, p. 134.

Beven, K.J., 2001. Rainfall–Runoff Modeling. Wiley & Sons, Chichester, pp. 217–225.

Beven, K., Binley, A., 1992. The future of distributed models: model calibration and uncertainty prediction. Hydrol. Process. 6 (3), 279–298.

Bhatt, G., Kumar, M., Duffy, C.J., 2014. A tightly coupled GIS and distributed hydrologic modelling framework. Environ. Model. Softw. 62, 70–84.

Biswas, A.K., 1970. History of Hydrology. North Holland Publishing Company, p. 336.

Bogardi, I., 2017. Fuzzy logic. In: Singh, V.P. (Ed.), Handbook of Applied Hydrology. McGraw-Hill Education, New York, pp. 1–5 (Chapter 12).

Borsi, I., Rossetto, R., Schifani, C., 2012. The SID & GRID project: developing GIS embedded watershed medeling. In: 21st Century Watershed Technology: Improving Water Quality and Environment Conference Proceedings, Bari, Italy.

Boyce, S.E., Hanson, R.T., Ferguson, I., Schmid, W., Henson, W., Reimann, T., et al., 2020. One-Water Hydrologic Flow Model: A MODFLOW Based Conjunctive-use Simulation Software. Techniques and Methods 6-A60, U.S. Geological Survey, p. 435.

Burek, P., Satoh, Y., Kahil, T., Tang, T., Greve, P., Smilovic, M., et al., 2020. Development of the community water model (CWatM v1.04) a high-resolution hydrological model for global and regional assessment of integrated water resources management. Geosci. Model Dev. 13, 3267–3298.

Campolongo, F., Braddock, R.D., 1999. The use of graph theory in the sensitivity analysis of the model output: a second order screening method. Reliab. Eng. Syst. Saf. 64 (1), 1–12.

Chalkias, C., Stathopoulos, N., Kalogeropoulos, K., Karymbalis, E., 2016. Applied hydrological modeling with the use of geoinformatics: theory and practice. In: Empirical Modeling and Its Applications. IntechOpen, pp. 62–86.

Chen, Y., Takara, K., Cluckie, I.D., De Smedt, F.H., 2004. GIS and Remote Sensing in Hydrology, Water Resources and Environment. IAHS, Wallingford, Oxfordshire.

Chow, V.T., Maidment, D., Mays, L., 1988. Applied Hydrology. McGraw-Hill, New York.

Crawford, N.H., Linsley, R.K., 1966. Digital Simulation in Hydrology: The Stanford Watershed, Technical Report 39.

Criollo, R., Velascob, V., Nardib, A., de Vriesb, L.M., Rierab, C., Scheiberb, L., et al., 2019. AkvaGIS: an open source tool for water quantity and quality management. Comput. Geosci. 127, 123–132.

Dar, T., Rai, N., Bhat, A., 2020. Delineation of potential groundwater recharge zones using analytical hierarchy process (AHP). Geol. Ecol. Landsc. 5, 1–16.

De Filippis, G., Pouliaris, C., Kahuda, D., Vasile, T.A., Manea, V.A., Zaun, F., et al., 2020. Spatial data management and numerical modelling: demonstrating the application of the QGIS-integrated FREEWAT platform at 13 case studies for tackling groundwater resource management. Water 12, 41.

Devi, G.K., Ganasri, B.P., Dwarakish, G.S., 2015. A review on hydrological models. In: International Conference on Water Resources, Coastal and Ocean Engineering (ICWRCOE), pp. 1001–1007.

Devkota, L.P., Gyawali, D.R., 2015. Impacts of climate change on hydrological regime and water resources management of the Koshi River Basin, Nepal. J. Hydrol. Reg. Stud. 4, 502–515.

Doherty, J., 2015. Calibration and Uncertainty Analysis for Complex Environmental Models – PEST: Complete Theory and What it Means for Modelling the Real World. Watermark Numerical Computing.

Döll, P., Kaspar, F., Lehner, B., 2003. Mint: a global hydrological model for deriving water availability indicators: model tuning and validation. J. Hydrol. 270 (1–2), 105–134.

Duan, Q., Gupta, H.V., Sorooshian, S., Rousseau, A.N., Turcotte, R., 2003. Calibration of Watershed Models. AGU, Washington, p. 345.

Dwivedi, D., Dafflon, B., Arora, B., Wainwright, H.M., Finsterle, S., 2017. Spatial analysis and geostatistical methods. In: Singh, V.P. (Ed.), Handbook of Applied Hydrology. McGraw-Hill Education, New York, pp. 1–9 (Chapter 20).

Elbeih, S.F., 2015. An overview of integrated remote sensing and GIS for groundwater mapping in Egypt. Ain Shams Eng. J. 6, 1–15.

ESRI, 2012. ArcGIS 10. Environmental Systems Research Institute, Redlands, United States of America. http://desktop.arcgis.com/en/. (Accessed 15 June 2021).

Fares, A., 2008. Overview of the hydrological modeling of small coastal watersheds on tropical islands. WIT Trans. State Art Sci. Eng. 33, 1–34.

Fatichi, S., Vivoni, E.R., Ogden, F.L., Ivanov, V.Y., Mirus, B., Gochis, D., et al., 2016. An overview of current applications, challenges, and future trends in distributed process-based models in hydrology. J. Hydrol. 537, 45–60.

Fauzia, Surinaidu, L., Rahman, A., Ahmed, S., 2021. Distributed groundwater recharge potentials assessment based on GIS model and its dynamics in the crystalline rocks of South India. Sci. Rep. 11, 11772.

Feldman, A.D., 2000. Hydrologic Modeling System HEC-HMS, Technical Reference Manual. U.S. Army Corps of Engineers. Hydrologic Engineering Center, HEC, Davis.

Foote, K.E., Lynch, M., 2009. Geographic Information Systems as an Integrating Technology: Context, Concepts, and Definitions. The Geographer's Craft Project, Department of Geography, the University of Colorado at Boulder. (Online). Available from: https://foote.geography.uconn.edu/gcraft/notes/notes.html. (Accessed 10 June 2021).

Gan, T.Y., Dlamini, E.M., Biftu, G.F., 1997. Effects of model complexity and structure, data quality, and objective functions on hydrologic modeling. J. Hydrol. 192, 81–92.

Gao, J., 2002. Integration of GPS with remote sensing and GIS: reality and prospect. Photogramm. Eng. Remote Sens. 68, 447–453.

Gautam, D.K., 2013. Assessing costs and benefits of adaptation: methods and data. In: UNDP/ADAPT Asia-Pacific First Regional Training Work Shop. Thailand.

Gogu, R.C., Carabin, G., Hallet, V., Peters, V., Dassargues, A., 2001. GIS-based hydrogeological databases and groundwater modeling. Hydrogeol. J. 9, 555–569.

Goodchild, M., Haining, R., Wise, S., 1992. Integrating GIS and spatial data analysis: problems and possibilities. Int. J. Geogr. Inf. Syst. 6 (5), 407–423.

Guiger, N., Franz, T., 1996. Visual MODFLOW Version 2.0. Waterloo Hydrologic Software, Ontario, Canada, p. 567.

Gupta, S.K., Sharma, G., Jethoo, A.S., Tyagi, J., Gupta, N.K., 2015. A critical review of hydrological models. In: Hydro 2015 International IIT. 20th International Conference on Hydraulics, Water Resources and River Engineering. IIT Roorkee, India.

Hamby, D.M., 1994. A review of techniques for parameter sensitivity analysis of environmental models. Environ. Monit. Assess. 32, 135–154.

Hanson, R.T., Boyce, S.E., Schmid, W., Hughes, J.D., Mehl, S.M., Leake, S.A., Maddock, T., III, Niswonger, R.G., 2014. One-water hydrologic flow model (MODFLOW-OWHM). In: Techniques and Methods 6-A51. U.S. Geological Survey, p. 120.

Harbaugh, A.W., 2005. MODFLOW-2005, the U.S. Geological Survey modular ground-water model – the ground-water flow process. In: Techniques and Methods 6-A16. U.S. Geological Survey.

Hassanizadeh, S.M., Carrera, J., 1992. Editorial. Validation of geo-hydrological models (special issue). Adv. Water Resour. 15, 1–3.

Heywood, I., Cornelius, S., Carver, S., 2006. An Introduction to Geographical Information Systems. Harlow, Pearson Education.

Howari, M.F., Sherif, M.M., Singh, P.V., Al-Asam, S.M., 2007. Application of GIS and remote sensing techniques in identification, assessment and development of groundwater resources. In: Thangarajan, M. (Ed.), Groundwater Resource Evaluation, Augmentation, Contamination, Restoration, Modeling and Management. Springer, Netherlands, pp. 1–25.

Hrachowitz, M., Clark, M.P., 2017. HESS opinions: the complementary merits of competing modelling philosophies in hydrology. Hydrol. Earth Syst. Sci. 21, 3953–3973.

Jha, M.K., Chowdary, V.M., 2007. Challenges of using remote sensing and GIS in developing nations. Hydrogeol. J. 15, 197–200.

Jha, M.K., Chowdhury, A., Chowdary, V.M., Peiffer, S., 2007. Groundwater management and development by integrated remote sensing and geographic information systems: prospects and constraints. Water Resour. Manag. 21 (2), 427–467.

Jhariya, D.C., Khan, R., Mondal, K.C., Kumar, T., Indhulekha, K., Singh, V.K., 2021. Assessment of groundwater potential zone using GIS-based multi-influencing factor (MIF), multi-criteria decision analysis (MCDA) and electrical resistivity survey techniques in Raipur city, Chhattisgarh, India. J. Water Supply Res. Technol. AQUA 70 (3), 375–400.

Jordan, D.L., 2004. An introduction to GIS applications in hydrology. Southwest Hydrol. 3, 14–16.

Kalogeropoulos, K., Stathopoulos, N., Psarogiannis, A., Pissias, E., Louka, P., Petropoulos, G.P., Chalkias, C., 2020. An integrated GIS-hydro modeling methodology for surface runoff exploitation via small-scale reservoirs. Water 12, 3182.

Khatami, S., Khazaei, B., 2014. Benefits of GIS application in hydrological modeling: a brief summary. J. Water Manag. Res. 70 (1), 41–49.

Knisel, W.G., 1980. CREAMS: A Field Scale Model for Chemicals, Runoff, and Erosion from Agricultural Management Systems. USDA, Conservation Research Report no. 26., p. 643.

Kresic, N., Mikszewski, A., 2012. Hydrogeological Conceptual Site Models: Data Analysis. CRC Press, Boca Raton.

Kumara, D., Bhattacharjyab, R.K., 2020. Evaluating two GIS-based semi-distributed hydrological models in the Bhagirathi-Alkhnanda River catchment in India. Water Policy 22, 991–1014.

Li, R., Merchant, J.W., 2013. Modeling vulnerability of groundwater to pollution under future scenarios of climate change and biofuels-related land use change: a case study in North Dakota, USA. Sci. Total Environ. 447, 32–45.

Liu, W., Park, S., Bailey, R.T., Molina-Navarro, E., Andersen, E.H., Thodsen, H., et al., 2020. Quantifying the streamflow response to groundwater abstractions for irrigation or drinking water at catchment scale using SWAT and SWAT–MODFLOW. Environ. Sci. Eur. 32, 113.

Marchant, A.P., Banks, V.J., Royse, K.R., Quigley, S.P., 2013. The development of a GIS methodology to assess the potential for water resource contamination due to new development in the 2012 Olympic Park site, London. Comput. Geosci. 51, 206–215.

Marsh, C.B., Pomeroy, J.W., Wheater, H.S., 2020. The Canadian hydrological model (CHM) v1.0: a multi-scale, multi-extent, variable-complexity hydrological model–design and overview. Geosci. Model Dev. 13, 225–247.

McCuen, R.H., 1973. The role of sensitivity analysis in hydrologic modeling. J. Hydrol. 18, 37–53.

Melati, M.D., Fan, F.M., Athayde, G.B., 2019. Groundwater recharge study based on hydrological data and hydrological modelling in a South American volcanic aquifer. Compt. Rendus Geosci. 351, 441–450.

Mogaji, K.A., Lim, H.S., Abdullah, K., 2014. Regional prediction of groundwater potential mapping in a multifaceted geology terrain using GIS-based Demspter-Shafer model. Arab. J. Geosci. 8 (5), 3235–3258.

Moore, I.D., Grayson, R.B., Ladson, A.R., 1991. Digital terrain modelling: a review of hydrological, geomorphological and biological applications. In: Terrain Analysis and Distributed Modelling in Hydrology. John Wiley and Sons, UK, pp. 7–34.

Mulvany, T.J., 1850. On the use of self-registering rain and flood gauges. Proc. Inst. Civ. Eng. 4 (2), 1–8 (Dublin, Ireland).

Muskat, M., 1937. The Flow of Homogeneous Fluids through Porous Media. McGraw-Hill, New York.

Nampak, H., Pradhan, B., Manap, M.A., 2014. Application of GIS based data driven evidential belief function model to predict groundwater potential zonation. J. Hydrol. 513, 283–300.

Olivera, F., Valenzuela, M., Srinivasan, R., Choi, J., Cho, H., Koka, S., Agrawal, A., 2006. ARCGIS-SWAT: a geodata model amd GIS interface for SWAT. J. Am. Water Resour. Assoc. 42 (2), 295–309.

Pandey, V.P., Dhaubanjar, S., Bharati, L., Thapa, B.R., 2020. Spatio-temporal distribution of water availability in Karnali-Mohana Basin, Western Nepal: hydrological model development using multi-site calibration approach (part-A). J. Hydrol. Reg. Stud. 29, 100690.

Pathak, R., Awasthi, M.K., Sharma, S.K., Hardaha, M.K., Nema, R.K., 2018. Ground water flow modelling using MODFLOW – a review. Int. J. Curr. Microbiol. App. Sci. 7 (2), 83–88.

Qi, C., Grunwald, S., 2005. GIS-based hydrologic modeling in the Sandusky watershed using SWAT. Trans. ASAE 48 (1), 169–180.

Refsgaard, J.C., 1997. Parameterization, calibration and validation of distributed hydrological models. J. Hydrol. 198, 69–97.

Refsgaard, J.C., Storm, B., 1996. Construction, calibration and validation of hydrological models. In: Abbott, M.B., Refsgaard, J.C. (Eds.), Distributed Hydrological Modeling. Kluwer Academic, Dordrecht, the Netherlands, pp. 41–54.

Refsgaard, J.C., Hojberg, A.L., Moller, I., Hansen, M., Sondergaard, V., 2010. Groundwater modeling in integrated water resources management - visions for 2020. Ground Water 48, 633–648.

Richards, C.J., Raaza, H., Raaza, R.M., 1993. Integrating geographic information systems and MODEFLOW for groundwater resources assessments. Water Resour. Bull. 29 (5), 847–853.

Rios, J.F., Ye, M., Wang, L., Lee, P.Z., Davis, H., Hicks, R., 2013. ArcNLET: a GIS-based software to simulate groundwater nitrate load from septic systems to surface water bodies. Comput. Geosci. 52, 108–116.

Rossetto, R., De Filippisa, G., Borsib, I., Fogliac, L., Cannatae, M., Criollof, R., et al., 2018. Integrating free and open source tools and distributed modelling codes in GIS environment for data-based groundwater management. Environ. Model. Softw. 107, 210–230.

Rumbaugh, J.O., Rumbaugh, D.B., 2011. Tutorial Manual for Groundwater Vistas, Version 6. Environmental Simulations.

Sandu, A.-M., Virsta, A., 2015. Applicability of MIKE-SHE to simulate hydrology in Argesel rive catchment. Agric. Agric. Sci. Procedia 6, 517–524.

Schaake, J.C., 2003. Introduction. In: Duan, Q., Gupta, H.V., Sorooshian, S., Rousseau, A.N., Turcotte, R. (Eds.), Calibration of Watershed Models. Water Science and Applications Series, Wiley.

Shannon, J., 2011. Application of GIS in Hydrologic Modeling. http://www.edc.uri.edu/nrs/classes/nrs409509/509_2011/shannon.pdf. (Accessed 5 July 2021).

Shrestha, S., Kafle, R., Pandey, V.P., 2017. Evaluation of index-overlay methods for groundwater vulnerability and risk assessment in Kathmandu Valley, Nepal. Sci. Total Environ. 575, 779–790.

Simões, S.J.C., 2013. Interaction between GIS and hydrologic model: a preliminary approach using ArcHydro framework data model. Rev. Ambient. Água 8 (3), 84–92. https://doi.org/10.4136/ambi-agua.1251.

Singh, V.P., 2018. Hydrologic modeling: progress and future directions. Geosci. Lett. 5, 15.

Singh, V.P., Fiorentino, M., 1996. Geographical information systems in hydrology. In: Hydrologic Modeling with GIS. Kluwer Academic Publishers, pp. 1–13.

Singh, V.P., Frevert, D.K., 2002. Mathematical Models of Small Watershed Hydrology and Applications. Water Resources Publications, Highlands Ranch.

Singh, A.K., Prakash, S.R., 2002. An integrated approach of remote sensing, geophysics and GIS to evaluation of groundwater potentiality of Ojhala subwatershed, Mirjapur District, UP, India. In: Asian Conference on GIS, GPS, Aerial Photography and Remote Sensing, Bangkok-Thailand.

Singh, V.P., Jain, S.K., Tyagi, A., 2007. Risk and Reliability Analysis. ASCE Press, Reston, p. 783.

Singh, C., Shashtri, S., Singh, A., Mukherjee, S., 2011. Quantitative modelling of groundwater in Satluj River basin of Rupnagar district of Punjab using remote sensing and geographic information system. Environ. Earth Sci. 62, 871–881.

Steyart, L.T., Goodchild, M.F., 1994. Integration geographic information system and environmental simulation models: a status review. In: Michenerm, W.K., Brunt, J.W., Stafford, S.G. (Eds.), Environmental Information Management and Analysis: Ecosystem to Global Scale. Taylor and Francis, London, UK, pp. 333–355.

Su, Z.B., Troch, P.A., 2003. Applications of quantitative remote sensing to hydrology. Phys. Chem. Earth 28, 1–2.

Sui, D., Maggio, R., 1999. Integrating GIS with hydrological modeling: practices, problems and prospects. Comput. Environ. Urban. Syst. 23, 33–51.

Svetlosanov, V., Knisel, W.G., 1982. European and U. S. Case Studies in Application of the CREAMS Model. Int. Institute for Applied Systems Analysis, Laxenburg, Austria, CP-82-S11., p. 148.

Takeuchi, K., Ao, T.Q., Ishidaira, H., 1999. Introduction of block-wise use of TOPMODEL and Muskingum-Cunge method for the hydroenvironmental simulation of a large ungauged basin. Hydrol. Sci. J. 44 (4), 633–646.

Termeh, S.V.R., Khosravi, K., Sartaj, M., Keesstra, S.D., Tsai, F.T.C., Dijksma., et al., 2019. Optimization of an adaptive neuro-fuzzy inference system for groundwater potential mapping. Hydrogeol. J. 27 (7), 2511–2534.

Thakur, J.K., Srivastava, P.K., Pratihast, A.K., Singh, S.K., 2011. Estimation of evapotranspiration from wetlands using geospatial and hydrometeorological data. In: Geospatial Techniques for Managing Environmental Resources. Springer, Netherlands, pp. 53–67.

Thakur, J.K., Singh, S.K., Ekanthalu, V.S., 2017. Integrating remote sensing, geographic information systems and global positioning system techniques with hydrological modeling. Appl Water Sci 7, 1595–1608.

Theis, C.V., 1935. The relation between the lowering of the piezometric surface and the rate and duration of discharge of a well using ground-water storage. Trans. Am. Geophys. Union 16, 519–524.

Tigabu, T.B., Wagner, P.D., Hörmann, G., Kiesel, J., Fohrer, N., 2020. Climate change impacts on the water and groundwater resources of the Lake Tana Basin, Ethiopia. J. Water Clim. Change, jwc2020126. https://doi.org/10.2166/wcc.2020.126.

Todini, E., Biondi, D., 2017. Calibration, parameter estimation, uncertainty, data assimilation, sensitivity analysis, and validation. In: Singh, V.P. (Ed.), Handbook of Applied Hydrology. McGraw-Hill Education, New York, pp. 1–19 (Chapter 22).

Toews, M.W., Gusyev, M.A., 2013. GIS tool to delineate groundwater capture zones. In: GNS Science Report. GNS Science, Lower Hutt, NZ, p. 19.

Wang, L., Jackson, C.R., Pachocka, M., Kingdon, A., 2016. A seamlessly coupled GIS and distributed groundwater flow model. Environ. Model. Softw. 82, 1–6.

Xu, Z.X., Ito, K., Schultz, G.A., Li, J.Y., 2014. Integrated hydrologic modeling and GIS in water resources management. J. Comput. Civ. Eng. 15 (3), 217–223.

CHAPTER

13

Development of rainfall-runoff model using ANFIS with an integration of GIS: A case study

Sandeep Samantaray[a], *Abinash Sahoo*[b], *Sambit Sawan Das*[a], and *Deba Prakash Satapathy*[a]

[a]Department of Civil Engineering, NIT Srinagar, Jammu & Kashmir, India [b]Department of Civil Engineering, NIT Silchar, Silchar, Assam, India

OUTLINE

1 Introduction 202

2 Study area 203

3 Methodology 205

3.1 Preparation of thematic maps 205

3.2 AHP 205

3.3 ANFIS 207

3.4 Random forest 207

3.5 Model performance evaluation 208

4 Results and discussions 208

4.1 Elevation 208

4.2 Slope 209

4.3 Rainfall 210

4.4 Distance from river 212

4.5 Flow length 212

4.6 LULC 212

4.7 Drainage density 212

5 Conclusion 220

References 221

Current Directions in Water Scarcity Research, Volume 7
https://doi.org/10.1016/B978-0-323-91910-4.00013-3

1 Introduction

Besides runoff, there is a rise in demand for information on other hydrological parameters. Report on the amount and spatial extension of snow is essential for flood forecasting, tourism, and hydropower. Moreover, information on groundwater and soil is vital for water supply, agriculture, hydropower planning, and forecasting landslides. However, overparameterization of several rainfall-runoff models makes their behavior less reliant on their structures (Kirchner, 2006). This certainly restricts a number of added understandings regarding hydrological processes in a watershed and decreases the efficiency of predictions (Beven, 2001). Typically, flood occurrences are considered the most common natural hazards around the world (Stefanidis and Stathis, 2013). Therefore in many cities, flood hazard management is a significant challenge. Population growth, rapid urbanization, climate change, and economic development will intensify severity of this challenge. Irreparable and considerable destruction to bridges, transport, farmlands, and several other urban infrastructures prove to be an urgent necessity for flood prevention and control (Agnihotri et al., 2022; Pradhan et al., 2014; Tehrany et al., 2014a, b). Two constituents of flood risk management are flood risk mitigation and analysis. Flood risk analysis intends to assess where flood occurrence risk is excessively high, and hazard mitigation activities are necessary. Hence, complete analysis of flood risk by identifying risky and hazardous areas is a vital measure in risk management for estimating number of destructions that can occur due to flood (Meyer et al., 2009).

One of the definitive studies in hydrology is the estimation of surface runoff based on distribution of rainfall in a catchment. With watershed's actual situation because of insufficient data and on the other hand, the complexity of hydrological systems causes certain usage of rainfall-runoff recreation models. As the measurement of all constraints affecting runoff flow in a watershed is difficult, selecting a proper model with least input data necessities, a simple architecture, and reasonable accurateness is vital (Yao et al., 2018). For mapping flood sensitivity in a catchment, various techniques are commonly utilized. Among them recently used techniques are multicriteria evaluation (Balogun et al., 2015), artificial neural network (ANN) (Campolo et al., 2003; Tiwari and Chatterjee, 2010; Samantaray et al., 2019, 2020), decision tree analysis (Tehrany et al., 2013), weight of evidence (Tehrany et al., 2014a, b), fuzzy theory (Pulvirenti et al., 2011), logistic regression (LR) approaches (Nandi et al., 2016) and frequency ratio (Rahmati et al., 2016).

Alaghmand et al. (2010) utilized HEC-HMS (Hydrologic Modeling System) and HEC-RAS (Hydrological Engineering Centre-River Analysis System) for developing flood vulnerability maps across River Sungai Kayu Ara near Kuala Lumpur in Malaysia using river depth and flow velocity maps. They observed that magnitude of rainfall event had more influence on the flood hazard map compared to land-use development conditions for the basin. Soulis and Valiantzas (2013) proposed a method for determining SCS-CN parameter values in heterogeneous watersheds from rainfall-runoff data. Majidi and Shahedi (2012) used HEC-HMS hydrological model for simulating the process of rainfall-runoff in Abnama catchment situated in Southern Iran. They concluded that the proposed model could be applied with equitable approximations in hydrological simulations in Abnama watershed. Mahmoud (2014) estimated potential runoff coefficient utilizing GIS based on land use, slope, and hydrologic soil group of Egypt. Skaugen and Onof (2014) derived runoff dynamics from the spread of distance from points in watersheds to adjacent streams, and the distribution is determined from GIS for each catchment. A comparison between the applied model's performances is made with Swedish Hydrologiska Byråns Vattenbalansavedlning (HBV) model, and it is found that the developed model performed equally well. Basarudin et al. (2014) intended to enumerate the effect of rain on the hydrological model in Kelantan River watershed through extreme rain events using semidistributed HEC-HMS. Their study demonstrated that changes in rainfall patterns have a substantial effect in determining runoff depth and peak discharge for specified study areas. Dottori et al. (2017) used daily river discharge data and a GIS based model to prepare flood hazard maps of several rivers. They compared them with the satellite derived maps and further investigated the sensitivity of the flood modeling. Ahmadisharaf et al. (2017) used a flood simulation model and GIS-based multiple-criteria decision-making (MCDM) technique for sustainability-based flood hazard mapping of the Swannanoa River basin, providing a more sustainable perspective of flood management. HEC-RAS, HEC-HMS, HEC-GeoHMS, and HEC-GeoRAS were applied for developing rainfall-runoff model and spatial analysis of the watershed (Thakur et al., 2017; Ibrahim-Bathis and Ahmed, 2016). Yao et al. (2018) used the SCS-CN model for developing rainfall-runoff model and studying risk features of urbanized function regions in Beijing. Results showed that during urban rainwater management, geospatial clustering characteristics of urban runoff risk should be considered. Kim et al. (2019) used ANN and remote sensing technology to improve the flood hazard mapping analysis in Kendal Regency in Indonesia and observed that the use of ANN overcame the lack of data. Mojaddadi et al. (2017) used frequency ration approach and SVM for flood hazard mapping and to estimate flood probability using 13 flood conditioning parameters as input. Rahmati et al. (2019) used SVM, BRT, and GAM to model a multi hazard map for avalanches, rockfalls, and floods in Asara basin located in Iran using the topo-hydrological and geo-environmental factors.

The objective of the study is to study the effect of various conditional parameters on the rainfall-runoff model using GIS. The next part of the study involves the application of ANFIS and RF models to find out the most influential parameter on the rainfall-runoff process based on various statistical measures.

2 Study area

The Basin of River Brahmani is an inter-state basin situated within geographical co-ordinate of latitudes 20°28′ to 23°35′ N and longitudes 83°52′ to 87°03′ E. The river basin consists of a total drainage area of 39,033 km^2, out of which 15,769 km^2 in Jharkhand state, 900 km^2 in Chhattisgarh state, and 22,364 km^2 in Odisha state (Fig. 1). In Odisha, the districts covered by the river basin are Sundargarh, Keonjhar, Sambalpur, Deogarh, Angul, Dhenkanal, Jajpur, and Kendrapada. In Chhattisgarh, it covers the districts of Raigarh, Jashpur, and Sarguja, and in Jharkhand, it covers the districts of Ranchi, Lohardega, Simdega, Gumla, and West Singhbhum. The river merges at the confluence point near Vedvyasa, in Odisha, at an elevation of 200 m above mean sea level. After the confluence, the river courses through central Odisha till it interlaces in deltaic plain.

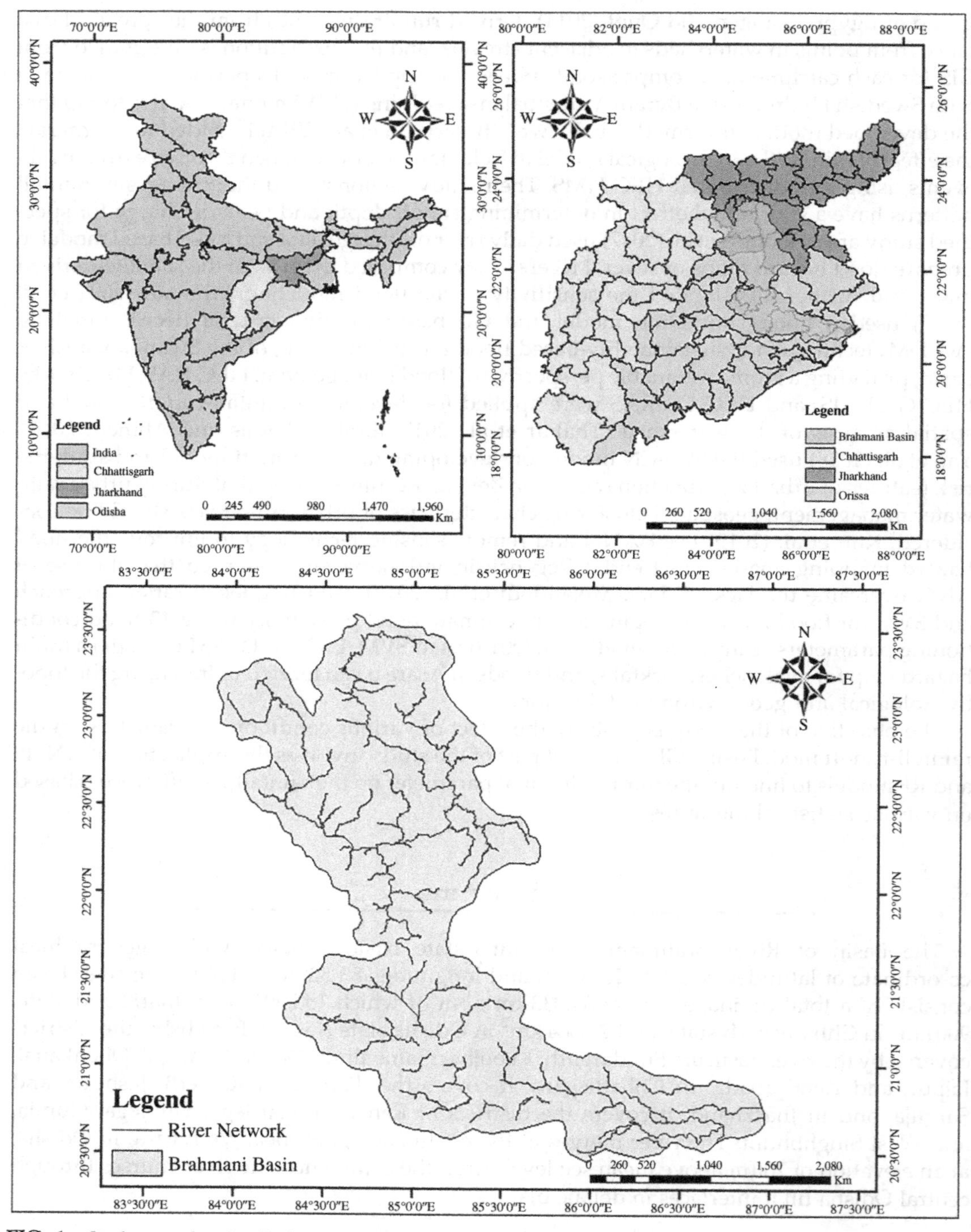

FIG. 1 Study area showing Brahmani river basin.

3 Methodology

Conventional secondary data (i.e., rainfall, LULC) remotely detected raster datasets, like DEM and primary field survey data, were collected. From USGS earth explorer website, DEM data with 30 m resolution consisting of satellite images and Landsat-8 data for LULC were collected. Monthly average rainfall data of the gauging sites were collected from CWC, Bhubaneswar, from 1990 to 2020.

3.1 Preparation of thematic maps

Here, seven most common flood influencing aspects, namely slope, DEM, rainfall, drainage density, distance from the river, flow length, and LULC, are used to prepare seven individual thematic maps. Elevation and slope maps were prepared from DEM, whereas rainfall map was prepared using IDW (inverse distance weighting) interpolation technique. For stream network delineation, hydrological analysis was performed in ArcGIS. Using the Euclidean distance tool, distance from river thematic layer was estimated. Map of LULC vector data was developed utilizing the data collected from USGS Landsat-8 data and supervised classification.

3.2 AHP

The factors utilized to access flood hazard zones don't have an equivalent grade of effect. Assessing the influence of individual factors doesn't give the entire required depiction; hence, a combination of all elements was essential. Certainly, the effect of factors on each other is found utilizing a schematic plan. This technique permits to find weight of each component on all other elements. For conducting a complete evaluation of each factor's influence on flood occurrence in the selected study basin, a weight value was allocated to each factor equal to its comparative significance. Using pairwise comparison procedure, this weight was determined.

The primary step of the AHP method is to explain the decision problem. The next step involves preparation of a pairwise comparison matrix. At this step, an $n*n$ dimension matrix of influencing factors is prepared. Scale of ranking value proposed by Saaty (1980) is represented in Table 1. Here, each factor is provided with an arithmetic value ranging from

TABLE 1 Considered parameters and its purpose of use.

Parameter	Source	Purpose of uses
Digital elevation model (DEM)	USGS Earth Explorer (SRTM data)	Land elevation, slope angle, flow network delineation, flow length and distance from river
Rainfall	CWC, Bhubaneswar	Estimate rainfall occurrence data of different gauging stations and prepare rainfall distribution map
Land use and land cover	USGS Earth Explorer (Landsat-8 data)	Land use and land cover classification

TABLE 2 Significance scale.

Scale	Numerical rating
Equal significance	1
Moderate significance	3
Strong significance	5
Very strong significance	7
Extreme significance	9
Intermediate values	2,4,6,8
Reciprocals	½, 1/3, 1/4, 1/5, 1/6, 1/7, 1/8, 1/8

TABLE 3 Models showing different input combination.

Models	Model^1	Model^2	Model^3	Model^4
Input combination	Slope, elevation, rainfall, DD, distance from river, flow length, LULC	Slope, elevation, rainfall, distance from river, flow length, LULC	Slope, rainfall, DD, distance from river, flow length, LULC	Slope, elevation, rainfall, DD, distance from river, flow length
Model no	Model^5	Model^6	Model^7	Model^8
Input combination	Elevation, rainfall, DD, distance from river, flow length, LULC	Slope, elevation, rainfall, DD, flow length, LULC	Slope, elevation, rainfall, DD, distance from river, LULC	Slope, elevation, DD, distance from river, flow length, LULC

1 to 9, subject to its importance compared to other combined factors. A value of 1 signifies that both aspects are equally important. In contrast, the value of 9 specifies that a row factor is much more important than an equivalent column factor.

Present study uses a 7*7 pairwise comparison matrix for studying influence of factors on each other. Subsequently, a CR (consistency ratio) numerical index was utilized for examining consistency of the pairwise comparison matrix (Table 2).

The selected criteria are arranged in a hierarchical order given in Table 3. Value of each row characterizes significance amid two parameters. The weights are normalized based on the AHP method. This was incorporated into the GIS tool for generating a flood hazard map.

After that, CR was obtained. Based on CR value theory, the findings may be too unreliable if CR is superior to 0.1. On the other hand, the findings seem to be completely reliable if CR equals to 0. For calculating CR value, CI (consistency index) is divided by RI (random inconsistency index). Number of criteria determines the RI value.

$$CR = CI/RI \tag{1}$$

$$CI = \frac{(\lambda - n)}{(n - 1)} \tag{2}$$

where n is quantity of aspects; λ is average value of consistency vectors.

3.3 ANFIS

ANFIS is a mixture of fuzzy logic ("if-then") and ANN principles, aimed at capturing advantages of both the techniques for estimating nonlinear functions in a solitary framework (Jang, 1993; Mohanta et al., 2021; Samantaray and Sahoo, 2021). A multilayer feed-forward network is incorporated by this estimator, where on input data, each node performs a specific function and trains a FIS (fuzzy interface system). Performance of ANFIS is governed by an assortment of shape, and a number of membership functions (MFs) since these two features affect accuracy and computational complexity of specified ANFIS model. Commonly three kinds of FIS are utilized: the Takagi-Sugeno, Tsukamoto, and Mamdani. Between these FISs, the major difference is how each finds the resultant (or "then") part of ANFIS model. Based on the literature study, we applied Takagi-Sugeno FIS for determining the resulting portion of ANFIS model.

Layer 1: Adaptive nodes are present in each node

$$O_{1,i} = \mu A_i(x)$$

$$O_{1,i} = \mu B_i(y)$$

where A and B are linguistic variables; x and y are input nodes; $\mu A_i(x)$ and $\mu B_i(y)$ are MFs for that node.

Layer 2: comprises static nodes signified as π. Product of all input signals to that node gives an output of each node:

$$O_{1,i} = W_i = \mu A_i(x)\mu B_i(y), \quad i = 1, 2$$

where W_i is each node's output.

Layer 3: incorporates static node signified as N. In this layer, nodes are normalized outputs of layer 2, which are denoted as standard firepower:

$$O_{3,i} = \overline{w_l} = \frac{w_i}{w1 + w2}, \quad i = 1, 2$$

Layer 4: Each node is in association with a node function

$$O_{4,i} = \overline{w_l} f_i = \overline{w_l}(p_i x + q_i x + r_i)$$

where $\overline{w_l}$ is normalized firepower of layer 3; p_i, q_i, and r_i are parameters of node. This layer's parameters can be inferred as resultant parameters.

Layer 5: comprises a solitary node signified as $\sum$ where all input signals are summed up to output layer:

$$O_{5,i} = \sum \overline{w_l} f_i = \frac{\sum w_i f_i}{\sum w_i}, \quad i = 1, 2$$

For implementing mathematical processes as above, we utilized ANFIS and fuzzy logic toolboxes in MATLAB.

3.4 Random forest

Ho (1995) first introduced RF, which is broadly employed in land cover classification (Pal, 2005) in addition to in medical and electronic problems (Cooper and Jacob, 1946). It also can function as a spatial modeling tool for water resources and environmental concerns (Booker and Snelder, 2012; Lee et al., 2017; Rahmati and Pourghasemi, 2017). RF works on the basis of modern-day techniques, which combine information where many decision trees are prepared, and then all trees are assimilated for prediction (Cutler et al., 2007). To develop an RT, recursive fragmentation is utilized to divide each node into two smaller nodes, and the process continues accordingly (Therneau et al., 2013). None of the designated data would be removed from input samples in this stage for making the following subset, reducing variance. Specific data might be utilized more than once during training which increases consistency of the model. This subclass can be utilized for evaluating the performance of the model. In brief, the RF technique used for classification is such that first "*t*" samples are extracted arbitrarily from training dataset. Eventually, the average outcome from predictions of all solitary trained trees determines classification output. Because of internal validation of each tree's classification outcomes within it and by weighing outcomes of each tree, the RF model can yield accurate results (Adelabu et al., 2015).

3.5 Model performance evaluation

The applied models are evaluated based on SSE, RMSE, R^2, and MAE statistical measures. Related equations of the statistical measures are given by

$$\text{SSE} = \sum_{k=1}^{N} (P_i - Q_i)^2 \tag{3}$$

$$\text{RMSE} = \sqrt{\frac{1}{N}\sum_{k=1}^{N} (P_i - Q_i)^2} \tag{4}$$

$$R^2 = \left(\frac{\sum_{i=1}^{n} P_i - \overline{P})(Q_i - \overline{Q})}{\sqrt{\sum_{i=1}^{n} P_i - \overline{P})^2 (Q_i - \overline{Q})^2}} \right)^2 \tag{5}$$

$$\text{MAE} = \frac{1}{N}\sum_{i=1}^{N} |P_i - Q_i| \tag{6}$$

where Q_i, P_i and $\overline{Q}$, $\overline{P}$ are monthly values of observed, predicted, and average of observed, predicted flood level; N is number of observed data in train or test set.

4 Results and discussions

4.1 Elevation

Elevation has a significant effect on runoff propagation in the selected study zone. Additionally, this constraint has a significant part in controlling flow direction movement and

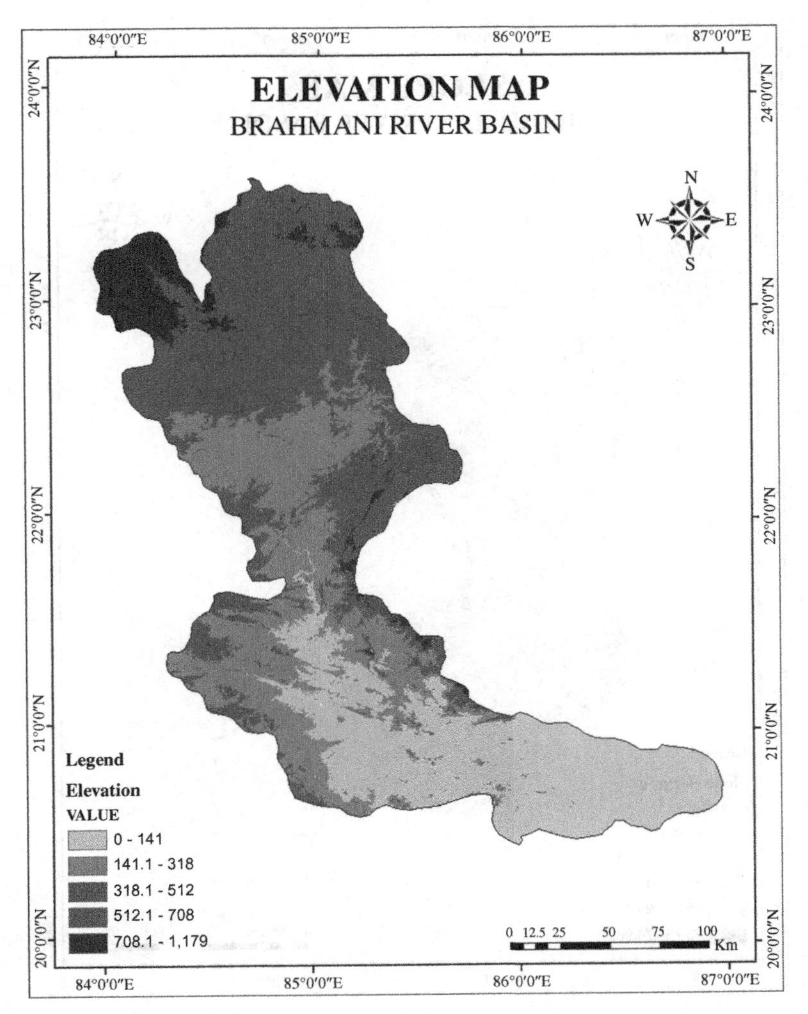

FIG. 2 Elevation map.

depth of runoff. Water normally flows from the end having higher elevation to lower elevation and flat low-lying regions may flood quicker than locations having greater elevation (Das, 2019; Dahri and Abida, 2017; Fernández and Lutz, 2010). Classification is done in ArcGIS based on natural breaking. The resultant map is grouped into five categories, as shown in Fig. 2.

4.2 Slope

Elevation contours dictate slopes directly related to the lithology, flow velocity, drainage, and soil type. In addition, gradient partly controls infiltration procedure. Surface runoff surges considerably as gradient rises; accordingly, infiltration decreases (Das, 2019; Das

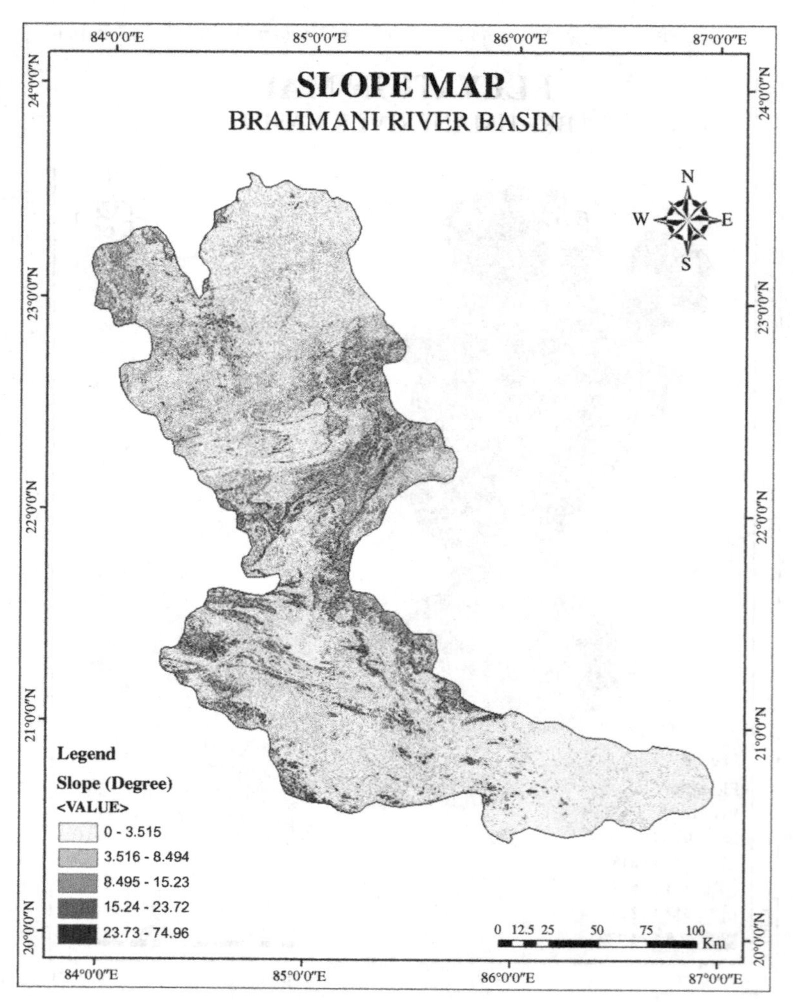

FIG. 3 Slope map.

and Pardeshi, 2018). Because of this, areas with a sudden decline of a slope have a higher flood probability since a massive volume of water becomes stationary causing a severe flood situation (Li et al., 2012; Pradhan, 2010). To prepare a slope map, a DEM (digital elevation model) of 30-m spatial resolution is used in Arc/GIS environment. As shown in Fig. 3, the slope map is divided into five categories using quantile classification technique (Tehrany et al., 2015).

4.3 Rainfall

A large number of previous literature established a relationship between rainfall and flood occurrence of a region (Zhao et al., 2018; Hong et al., 2018a, b; Rozalis et al., 2010; Zhang and

Smith, 2003; Goel et al., 2000). It cannot precisely be found to what magnitude an increase of rainfall will lead to flood (Kay et al., 2006). Instead, it can be said that rainfall is the principal factor for the occurrence of flood. When natural river networks cannot carry surplus water due to excess rain, flooding occurs. The average annual rainfall was computed based on monthly rainfall data collected from four rain gauging stations. Thirty years of rainfall data (1991–2019) was used for developing a rainfall map, where the average annual rainfall is about 3.546 m. Subsequently, by employing interpolation techniques of kriging, a mean annual rainfall map was generated (Fig. 4). The rainfall scale is divided into five classes from 3.612 to 4.346 m.

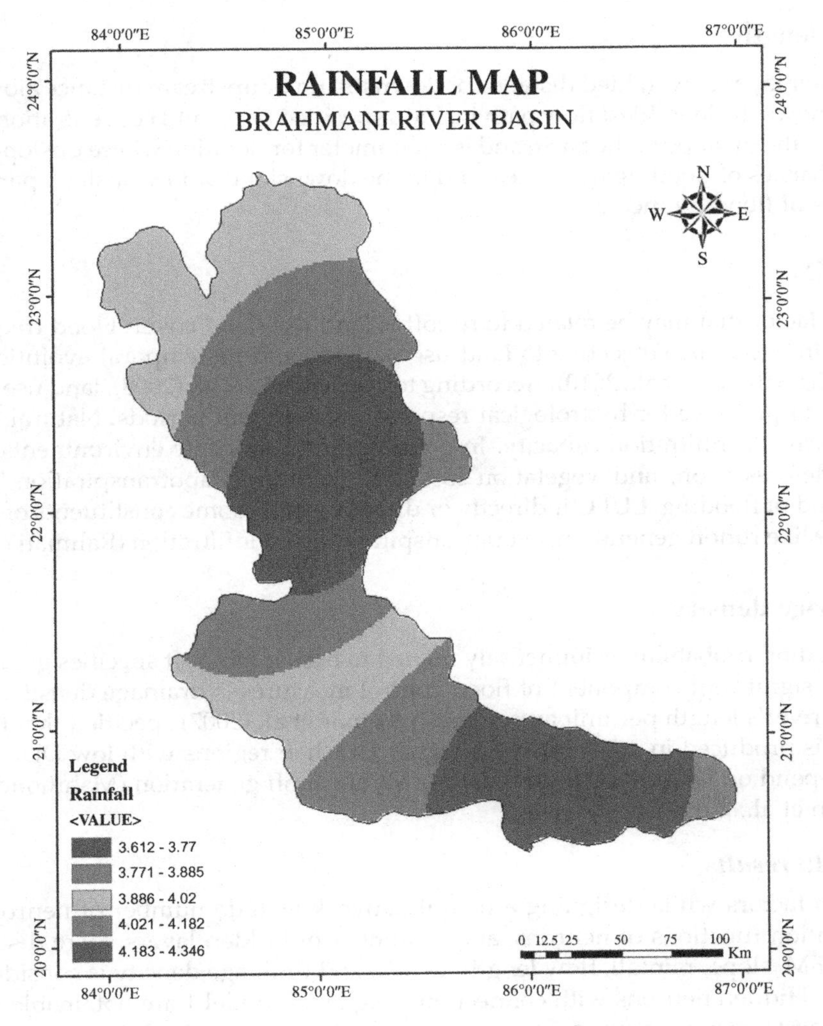

FIG. 4 Rainfall map.

4.4 Distance from river

Distance of a region from drainage network determines expansion of a flood event (Predick and Turner, 2008). Areas situated close to drainage networks, generally suffer from floods more than regions that are at a far distance as the nearby locations are within flow path (Mahmoud and Gan, 2018). Different researchers considered different distance from the drainage network that is vulnerable to the flood. Several studies specified that terrestrial water storage like dams, ponds, lakes are also related to flooding at a local level (Reager et al., 2014; Antonelli et al., 2008). The elevation and slope become higher as distance increases. Velocity and extent of runoff significantly depend upon distance from rivers (Glenn et al., 2012). Distance from river map was made and categorized into five classes (Darabi et al., 2019).

4.5 Flow length

It is a raster layer of weighted distance or downstream or upstream distance along the path of flow. It shows the lengthiest flow path inside a specified basin and concentration of a basin. It also shows the upslope of the basin and is a parameter for flooding where upslope is high; at that part, chances of flooding decreases, and in the downslope or low upslope part of the basin, chances of flooding increases.

4.6 LULC

Another factor that may be related to runoff is land use/land cover. Flood frequency of a region can intensely be subjective to land-use patterns and its temporal evolution (García-Ruiz et al., 2008; Benito et al., 2010). According to García-Ruiz et al. (2008), land use of a region has highest importance for hydrological responses at different periods. Natural land cover varies in terms of infiltration capacity. In contrast, anthropogenic environments like built-up areas, deforestation, and vegetation loss can decrease evapotranspiration, increasing the likelihood of flooding. LULC indirectly or directly affects some constituents of the hydrological cycle like runoff generation, evapotranspiration, and infiltration (Rahmati et al., 2016).

4.7 Drainage density

High flooding probability is immensely related to higher DD as it specifies greater surface runoff. So a significant component of flood control measures is drainage density, which expresses the river's length per unit area. Dinesh Kumar et al. (2007) specified that higher surface runoff is produced in areas having a higher DD than regions with low DD. Thus flood risk may depend on DD, which is a severe aspect of runoff generation (Mahmoud and Gan, 2018; Ogden et al., 2011).

4.7.1 ANFIS results

The main factors while designing a neural network include number of neurons in each layer, activation functions of neurons, and a number of hidden layers. Here distance from the river, DEM, slope, rainfall, flow length, LULC, and drainage density is considered input for Model 1. Hidden neurons with connection weights of Model 1 are obtainable in Table 4, and model performance comparisons are accessible in Table 5. Fig. 5 represents the scatter plot between the observed and predicted runoff values of Model 1.

TABLE 4 Hidden neuron with connection weights (Model 1).

Sl. no.	I_1 Distance from river	I_2 DEM	I_3 Slope	I_4 LULC	I_5 Drainage density	I_6 Flow length	I_7 Rainfall
HN#1	0.006584	0.0005	0.00628	−0.00632	0.0038	0.0032	0.00634
HN#2	0.01795	0.009241	−0.006941	0.18245	0.06324	0.041368	−0.042358
HN#3	−0.004785	0.00062	0.0344	0.069456	−0.003854	0.00521	0.00382
HN#4	0.06854	0.008264	0.063275	0.00624	0.03941	−0.0548	0.004168
HN#5	0.0036187	−0.006354	0.00418	0.036856	0.0924	0.006349	0.00369
HN#6	0.083217	0.08541	0.0395	0.008249	0.0031426	0.00762	0.0098547
HN#7	0.00369	0.03168	−0800415	0.025	0.00899	0.00318	0.00365
HN#8	−0.008002	−0.00652	0.00369	0.0038	−0.005521	0.0056	−0.09524
HN#9	0.00345	−0.000953	0.082453	0.06874	−0.0325	−0.0963	0.06349
HN#10	0.00832	0.0037	0.000816	0.063259	0.00642	0.008367	0.005267
HN#11	0.003269	0.008214	0.097154	0.0824	0.03954	0.005681	−0.082654
HN#12	0.00321	0.00389	0.00351	0.00421	0.0563278	0.0596	0.00985
HN#13	0.00065	0.002467	−0.0532	−0.09825	0.1006	0.0256	0.0634
HN#14	0.002389	0.0029	0.0063	0.00524	0.0035891	−0.0082	0.08256
HN#15	0.003718	−0.0391	0.00095	0.06389	0.004125	−0.0634	0.0063147

TABLE 5 Performance of ANFIS and RF models.

Performance indicator	Train	Test	Validation
ANFIS			
R^2	0.96725	0.94909	0.95492
MAE	10.3597	13.9624	12.694
RMSE	6.1963	9.2351	8.3498
RF			
R^2	0.99839	0.9809	0.98426
MAE	3.6972	8.90216	6.09521
RMSE	1.0524	4.5972	2.687

4.7.2 Sensitivity analysis

Table 6 presents the performance of eight proposed models for sensitivity analysis. Here number of inputs used for the model formulation is seven, and the number of scenarios is eight. This indicates that we left one input parameter on a rotating basis for different input

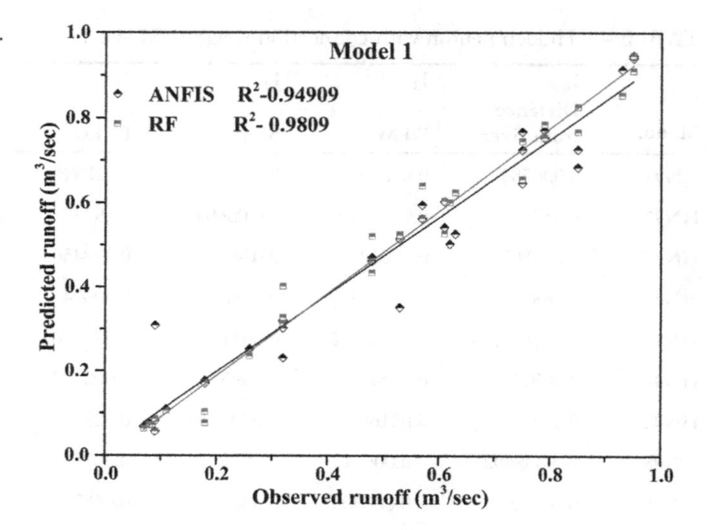

FIG. 5 R^2 for testing phase of model 1.

TABLE 6 Consequences of input factors.

Factors	SSE	RMSE	R^2	Error (%)
ANFIS				
M#1	7.14	9.354	0.94909	0.05091
M#2	8.89	12.642	0.93887	0.06113
M#3	9.06	14.69	0.93452	0.06548
M#4	8.32	11.035	0.94596	0.05404
M#5	9.35	16.083	0.93166	0.06834
M#6	8.61	11.9386	0.94011	0.05989
M#7	7.7	10.387	0.9441	0.0559
M#8	8.93	13.386	0.93684	0.06316
RF				
M#1	1.63	0.835	0.9809	0.0191
M#2	3.42	2.996	0.97204	0.02796
M#3	4.69	4.638	0.96948	0.03052
M#4	1.96	1.3054	0.9799	0.0201
M#5	5.57	5.967	0.96794	0.03206
M#6	2.8	2.345	0.97587	0.02413
M#7	2.38	1.93	0.9786	0.0214
M#8	4.21	3.642	0.97052	0.02948

scenarios of considered models (Models 2–8). Monthly monsoon (June–October) data for each input parameter from 1981 to 2020, 240 data, are used to develop the model. We used 70% of data for training persistence, 15% of data for validation resolves, and 15% of data for testing purposes to develop the model. To develop ANFIS model network, the Levenberg-Marquardt training algorithm is used. Similarly, the bagging or bootstrap aggregating is used to train "forest" generated by the RF algorithm. Bagging is a collective metaalgorithm which enhances the precision of ML algorithms. Fig. 6 illustrates comparison of actual vs. predicted runoff for all models (Models 2–8).

4.7.3 Comparison results of runoff prediction in sensitivity analysis models

The model accuracy assessment is defined in error terms of forecasting or difference amid observed and predicted values. In the present research, R^2, SSE, and RMSE methods are broadly used to examine the ANN and RF performance. Every technique is assessed from predicted ANN values and simulated targets. Performance of proposed models for runoff prediction is evaluated depending on training and testing data (Table 6). The forecasting results of the RF model gave tremendous similitude with the actual data based on R^2. The outcomes obtained presented less SSE and RMSE for the RF model. Sensitivity analysis executed in this study validates that slope is the most significant factor for runoff. The standardized value using ANFIS displays that model without slope has the lowest value ($R^2 = 0.93166$, Model 5) followed by elevation ($R^2 = 0.93452$, Model 3) and then rainfall ($R^2 = 0.93684$, Model 8). Similarly for RF, Model 5 is least efficient with $R^2 = 0.96794$ followed by Model 3, Model 8, Model 1, Model 6, Model 7, and Model 4. Detailed values are provided in Table 6. In general, the factor with minimum error is the most influencing factor toward formulation of the model.

Weights obtained from ANFIS and RF models are considered as system inputs and, after normalization, were moved to MATLAB. After analyzing the results obtained from present study, we can conclude that ANFIS overestimated a "very high" impact compared to the RF model and therefore has the lowest precision.

4.7.4 The interrelationship and pairwise comparison of influencing constituents

The seven conditioning factors, such as slope, elevation, rainfall, DD, distance from river, flow length, and LULC, documented in the preceding segment, were utilized for studying its impact on rainfall-runoff and its prediction results. By separately considering each factor's effect, runoff cannot be assessed as all aspects do not have an equivalent degree of impact on the runoff process.

4.7.5 Determined CR

RI depends on number of conditions. In present work, the inputs are eight, and the resultant RI = 1.32.

Finally, calculated CR value is 0.08045, and since the CR value is less to 0.1 and consistency of the weight is accepted.

Here, $\lambda = 7.63718$

CI = 0.1062.

RI = 1.32.

CR = $0.08045 < 0.1$.

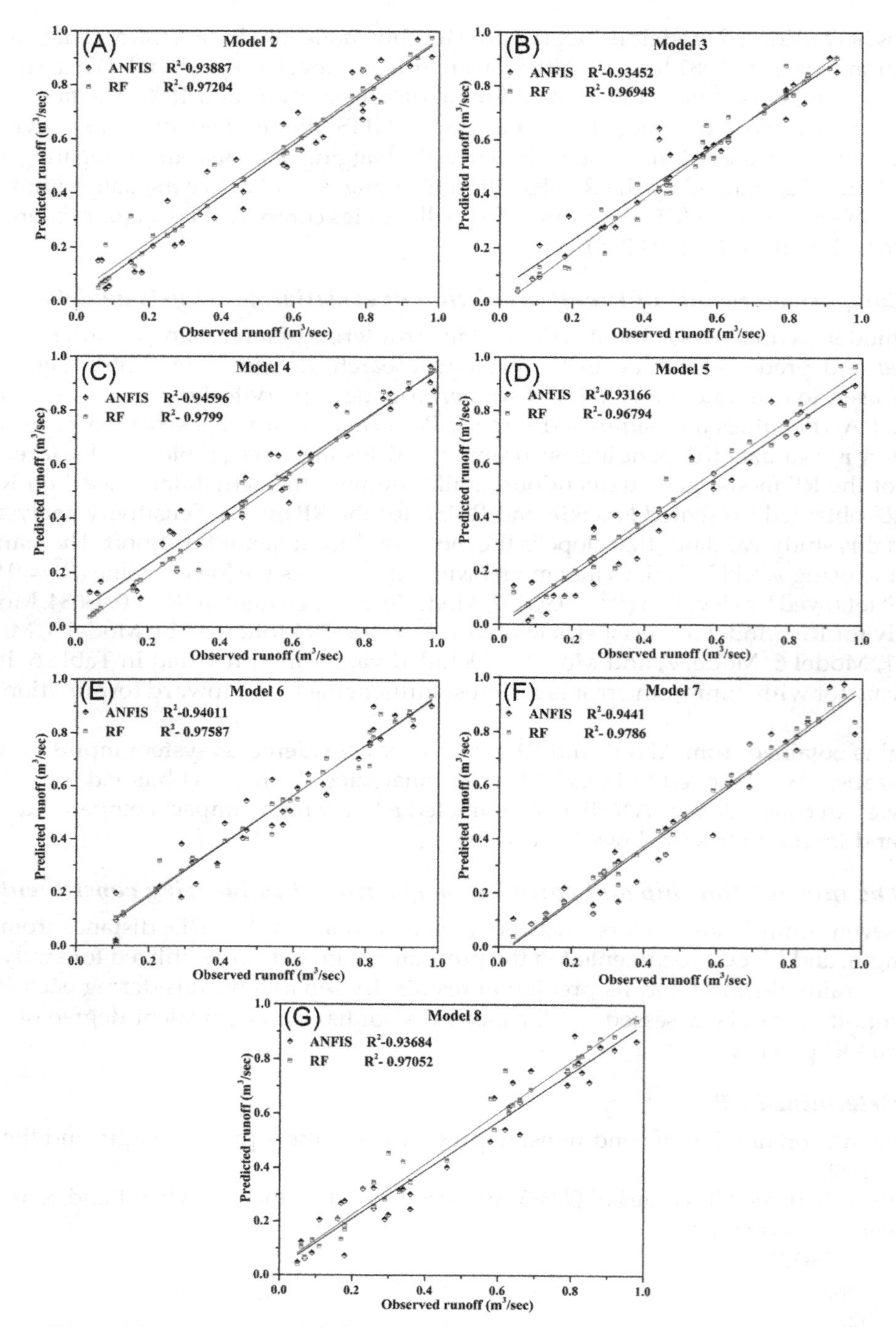

FIG. 6 Coefficient of determination for (A) Model 2, (B) Model 3, (C) Model 4, (D) Model 5, (E) Model 6, (F) Model 7, (G) Model 8, in testing phase.

Classes of factors based on weights

For model 1 using ANFIS technique, we found better accuracy in runoff prediction only when independent variables contain the current time step values. If not so, the model doesn't predict well. The overall input factors (catchment characteristics factors) for RF indicated that RMSE of predicted values and observed values are between 0.835 and 5.967. From sensitivity analysis, the value of R^2 varies within the range of 0.96794–0.9809. The above range directs that the entire input factors have an impact on river flow. The above results indicate that between the various input factors, slope is the most influential, and LULC is a negligible influential factor on the flow of river and flood. Table 7 provides classes of factors according to weights, and Table 8 gives classes of runoff influencing factors and weights utilized in AHP.

TABLE 7 Classes of the factors according to weights.

Rank	Factors	Classes	Flood hazard level	Pixel no.	Pixel (%)	Area (sq. km)	Area (%)	Weight (%)
1	Slope (degree)	0–3.515	Very high	24,299,645	54.16	21,872.14	54.16	33
		3.516–8.494	High	11,893,803	26.51	10,705.63	26.51	
		8.195–15.23	Moderate	4,680,306	10.43	4212.75	10.43	
		15.24–23.72	Low	2,880,474	6.42	2592.72	6.42	
		23.73–74.69	Very low	1,114,509	2.48	1003.17	2.48	
2	Elevation (m)	0–141	Very high	12,139,339	27.06	10,926.64	27.06	23
		141.1–318	High	11,779,459	26.25	10,602.71	26.25	
		318.1–512	Moderate	8,731,012	19.46	7858.80	19.46	
		512.1–708	Low	9,283,170	20.69	8355.79	20.69	
		708.1–1179	Very low	2,935,757	6.54	2642.48	6.54	
3	Rainfall (cm)	4.183–4.346	Very high	3484	15.17	6136.34	15.17	16
		4.021–4.182	High	2689	11.71	4736.12	11.71	
		3.886–4.02	Moderate	6447	28.07	11,355.06	28.07	
		3.771–3.885	Low	5494	23.92	9676.54	23.92	
		3.612–3.77	Very low	4856	21.14	8552.84	21.14	
4	Drainage density (km/sq. km)	0.09884–0.1125	Very high	2524	9.90	3994.36	9.90	12
		0.08585–0.09883	High	1350	5.30	2136.44	5.30	
		0.07705–0.08584	Moderate	5788	22.71	9159.81	22.71	

Continued

TABLE 7 Classes of the factors according to weights.—Cont'd

Rank	Factors	Classes	Flood hazard level	Pixel no.	Pixel (%)	Area (sq. km)	Area (%)	Weight (%)
		0.06909–0.07704	Low	12,499	49.05	19,780.31	49.05	
		0.05923–0.06908	Very low	3323	13.04	5258.82	13.04	
5	Distance from river (m)	0–2790.7	Very high	9646	37.77	15,265.29	37.77	8
		2790.71–5581.41	High	7443	29.14	11,778.93	29.14	
		5581.42–8372.11	Moderate	4586	17.96	7257.58	17.96	
		8372.12–11,581.4	Low	2806	10.99	4440.64	10.99	
		11,581.5–17,790.7	Very low	1057	4.14	1672.76	4.14	
6	Flow length (m)	0–77,118.38	Very high	15,453,273	34.44	13,909.51	34.44	5
		77,118.39–164,997.5	High	9,783,172	21.80	8805.85	21.80	
		164,997.6–254,670	Moderate	9,373,821	20.89	8437.39	20.89	
		254,670.1–342,549.1	Low	7,149,948	15.94	6435.68	15.94	
		342,549.2–455,536.5	Very low	3,108,523	6.93	2797.99	6.93	
7	Land use and land cover	Waterbodies	Very high	9,771,905	2.42	977.19	2.42	3
		Built-up area	High	20,815,467	5.15	2081.55	5.15	
		Barren land	Moderate	3,023,638	0.75	302.36	0.75	
		Vegetation	Low	206,493,707	51.13	20,649.34	51.13	
		Forest area	Very low	163,793,522	40.55	16,379.35	40.55	

TABLE 8 Classes of runoff influencing factors and weights utilized in AHP.

Rank	Factors	Classes	A	B	C	D	E	CR	Class weight	Class weight (%)
1	Slope (degree)	0–3.515	1.00					0.0171	0.470	47
		3.516–8.494	0.50	1.00					0.284	28
		8.195–15.23	0.25	0.33	1.00				0.131	13
		15.24–23.72	0.14	0.25	0.50	1.00			0.072	7
		23.73–74.69	0.13	0.17	0.25	0.50	1.00		0.043	4

TABLE 8 Classes of runoff influencing factors and weights utilized in AHP.—Cont'd

Rank	Factors	Classes	A	B	C	D	E	CR	Class weight	Class weight (%)
2	Elevation (m)	0–141	1.00					0.0142	0.420	42
		141.1–318	0.50	1.00					0.266	27
		318.1–512	0.33	0.50	1.00				0.167	17
		512.1–708	0.25	0.33	0.50	1.00			0.095	9
		708.1–1179	0.17	0.20	0.25	0.50	1.00		0.053	5
3	Rainfall (cm)	4.183–4.346	1.00					0.0139	0.459	46
		4.021–4.182	0.50	1.00					0.254	25
		3.886–4.02	0.25	0.50	1.00				0.151	15
		3.771–3.885	0.20	0.33	0.50	1.00			0.087	9
		3.612–3.77	0.14	0.20	0.25	0.50	1.00		0.049	5
4	Drainage density (km/sq. km)	0.09884–0.1125	1.00					0.0422	0.445	45
		0.08585–0.09883	0.50	1.00					0.297	30
		0.07705–0.08584	0.25	0.33	1.00				0.147	15
		0.06909–0.07704	0.17	0.20	0.33	1.00			0.073	7
		0.05923–0.06908	0.13	0.14	0.20	0.33	1.00		0.037	4
5	Distance from river (m)	0–2790.7	1.00					0.0103	0.468	47
		2790.71–5581.41	0.50	1.00					0.268	27
		5581.42–8372.11	0.25	0.50	1.00				0.144	14
		8372.12–11,581.4	0.17	0.25	0.50	1.00			0.076	8
		11,581.5–17,790.7	0.13	0.17	0.25	0.50	1.00		0.044	4
6	Flow length (m)	0–77,118.38	1.00					0.0201	0.411	41
		77,118.39–164,997.5	0.50	1.00					0.269	27
		164,997.6–254,670	0.33	0.50	1.00				0.169	17
		254,670.1–342,549.1	0.25	0.33	0.50	1.00			0.096	10
		342,549.2–455,536.5	0.20	0.20	0.25	0.50	1.00		0.056	6
7	Land use and land cover	Waterbodies	1.00					0.0172	0.436	44
		Built-up area	0.50	1.00					0.257	26
		Barren land	0.33	0.50	1.00				0.176	18
		Vegetation	0.20	0.33	0.33	1.00			0.082	8
		Forest Area	0.14	0.20	0.25	0.50	1.00		0.049	5

4.7.6 Discussion

Each year due to monsoon rainfall, India is mostly affected by floods. By identifying most vital aspects that influences flood events and by development of flood hazard maps through artificial intelligence techniques flood mitigation can be conducted. It is well-known that climatic conditions significantly affect flood events. Based on authors' knowledge, there have been no systematic studies on extreme events in Brahmani river basin. The objective of this study is to cover this knowledge gap by utilizing commonly applied artificial intelligence models. Studies related to development of flood hazard maps have been implemented in various locations worldwide. Yet, impact of change in climatic conditions on flood remains ambiguous, particularly more in-depth research is required on the seasonality effect of climatic factors. Even though some studies consider rainfall as the most vital climatic aspect for causing flood in some areas (Çelik et al., 2012; Khosravi et al., 2016), slope, distance from river, elevation are found to be the most influencing factors of flood events (Bui et al., 2016; Termeh et al., 2018).

In many studies, the proper understanding of flood events is not detailed enough. There are several factors related to flood hazards. For example, the change in land-use cover is likely to have an impact on future flood hazards (Beckers et al., 2013). As a result, to obtain accurate flood hazard maps, it is very important to understand the conditioning factors in the study area properly. According to obtained results, both RF and ANFIS have their benefits. In addition, the convergence of RF is faster compared to ANFIS. In addition, experimental outcomes that are not stated here revealed that the performance of ANFIS, ANN, and RF is considerably different. For improving efficiency of proposed methods and for more rapid convergence, it is essential to modulate parameters.

Regarding the percentage of high and very high flood vulnerable zones, resultant maps indicated a similar spatial distribution by three techniques. However, RF obtained a smaller percentage of 35.39% than ANFIS (39.29%). This denotes that RF can generate more practical outcomes for flood mitigation which can efficiently decrease the cost and time of the LULC plan in the study basin.

5 Conclusion

In this study, it can be observed that GIS can provide a suitable platform for the convergence of large volumes of multidisciplinary datasets. Several areas of India and other developing nations do not have adequate historical recorded data and comprehensive runoff information for physical-based distributed models. The present research provides better solutions for flood management to overcome such drawbacks. The applied methods for determining runoff are instrumental in the forecast/prediction of temporal variations of surface runoff that are helpful in environmental/hydrologic engineering applications. Also, the developed techniques are economical and have high accurateness. This research concludes that simulated results can be beneficial in planning and management practices of water and land resources inside the Baitarani river watershed. These models can be best utilized in water-scarce regions and ungauged watersheds where recorded data are inadequate, and estimation of runoff is compulsory for sustaining the water resources.

References

Adelabu, S., Mutanga, O., Adam, E., 2015. Testing the reliability and stability of the internal accuracy assessment of random forest for classifying tree defoliation levels using different validation methods. Geocarto Int. 30 (7), 810–821.

Agnihotri, A., Sahoo, A., Diwakar, M.K., 2022. Flood prediction using hybrid ANFIS-ACO model: a case study. In: Inventive Computation and Information Technologies. Springer, Singapore, pp. 169–180.

Ahmadisharaf, E., Kalyanapu, A.J., Chung, E.S., 2017. Sustainability-based flood hazard mapping of the Swannanoa River watershed. Sustainability 9 (10), 1735.

Alaghmand, S., Abdullah, R.B., Abustan, I., Vosoogh, B., 2010. GIS-based river flood hazard mapping in urban area (a case study in Kayu Ara River Basin, Malaysia). Int. J. Eng. Technol. 2 (6), 488–500.

Antonelli, C., Eyrolle, F., Rolland, B., Provansal, M., Sabatier, F., 2008. Suspended sediment and 137Cs fluxes during the exceptional December 2003 flood in the Rhone River, Southeast France. Geomorphology 95 (3–4), 350–360.

Balogun, A.-L., Matori, A.-N., Hamid-Mosaku, A.I., 2015. A fuzzy multi-criteria decision support system for evaluating subsea oil pipeline routing criteria in East Malaysia. Environ. Earth Sci. 74, 4875–4884.

Basarudin, Z., Adnan, N.A., Latif, A.R.A., Tahir, W., Syafiqah, N., 2014. Event-based rainfall-runoff modelling of the Kelantan River Basin. IOP Conf. Ser.: Earth Environ. Sci. 18 (1), 012084 (IOP Publishing).

Beckers, A., Dewals, B., Erpicum, S., Dujardin, S., Detrembleur, S., Teller, J., Pirotton, M., Archambeau, P., 2013. Contribution of land use changes to future flood damage along the river Meuse in the Walloon region. Nat. Hazards Earth Syst. Sci. 13 (9), 2301–2318.

Benito, G., Rico, M., Sánchez-Moya, Y., Sopeña, A., Thorndycraft, V.R., Barriendos, M., 2010. The impact of late Holocene climatic variability and land use change on the flood hydrology of the Guadalentín River, Southeast Spain. Glob. Planet. Chang. 70 (1–4), 53–63.

Beven, K., 2001. On explanatory depth and predictive power. Hydrol. Process. 15 (15), 3069–3072.

Booker, D.J., Snelder, T.H., 2012. Comparing methods for estimating flow duration curves at ungauged sites. J. Hydrol. 434, 78–94.

Bui, D.T., Pradhan, B., Nampak, H., Bui, Q.T., Tran, Q.A., Nguyen, Q.P., 2016. Hybrid artificial intelligence approach based on neural fuzzy inference model and metaheuristic optimization for flood susceptibilitgy modeling in a high-frequency tropical cyclone area using GIS. J. Hydrol. 540, 317–330.

Campolo, M., Soldati, A., Andreussi, P., 2003. Artificial neural network approach to flood forecasting in the river Arno. Hydrol. Sci. J. 48, 381–398.

Çelik, H.E., Coskun, G., Cigizoglu, H.K., Ağıralioğlu, N., Aydın, A., Esin, A.I., 2012. The analysis of 2004 flood on Kozdere Stream in Istanbul. Nat. Hazards 63 (2), 461–477.

Cooper Jr., H.H., Jacob, C.E., 1946. A generalized graphical method for evaluating formation constants and summarizing well-field history. Eos Trans. Am. Geophys. Union 27 (4), 526–534.

Cutler, D.R., Edwards Jr., T.C., Beard, K.H., Cutler, A., Hess, K.T., Gibson, J., Lawler, J.J., 2007. Random forests for classification in ecology. Ecology 88 (11), 2783–2792.

Dahri, N., Abida, H., 2017. Monte Carlo simulation-aided analytical hierarchy process (AHP) for flood susceptibility mapping in Gabes Basin (southeastern Tunisia). Environ. Earth Sci. 76 (7), 1–14.

Darabi, H., Choubin, B., Rahmati, O., Haghighi, A.T., Pradhan, B., Kløve, B., 2019. Urban flood risk mapping using the GARP and QUEST models: a comparative study of machine learning techniques. J. Hydrol. 569, 142–154.

Das, S., 2019. Geospatial mapping of flood susceptibility and hydro-geomorphic response to the floods in Ulhas basin, India. Remote Sens. Appl.: Soc. Environ. 14, 60–74.

Das, S., Pardeshi, S.D., 2018. Integration of different influencing factors in GIS to delineate groundwater potential areas using IF and FR techniques: a study of Pravara basin, Maharashtra, India. Appl. Water Sci. 8 (7), 1–16.

Dinesh Kumar, P.K., Gopinath, G., Seralathan, P., 2007. Application of remote sensing and GIS for the demarcation of groundwater potential zones of a river basin in Kerala, southwest coast of India. Int. J. Remote Sens. 28 (24), 5583–5601.

Dottori, F., Kalas, M., Salamon, P., Bianchi, A., Alfieri, L., Feyen, L., 2017. An operational procedure for rapid flood risk assessment in Europe. Nat. Hazards Earth Syst. Sci. 17 (7), 1111–1126.

Fernández, D.S., Lutz, M.A., 2010. Urban flood hazard zoning in Tucumán Province, Argentina, using GIS and multicriteria decision analysis. Eng. Geol. 111 (1–4), 90–98.

García-Ruiz, J.M., Regüés, D., Alvera, B., Lana-Renault, N., Serrano-Muela, P., Nadal-Romero, E., Navas, A., Latron, J., Martí-Bono, C., Arnáez, J., 2008. Flood generation and sediment transport in experimental catchments affected by land use changes in the Central Pyrenees. J. Hydrol. 356 (1–2), 245–260.

Glenn, E.P., Morino, K., Nagler, P.L., Murray, R.S., Pearlstein, S., Hultine, K.R., 2012. Roles of saltcedar (Tamarix spp.) and capillary rise in salinizing a non-flooding terrace on a flow-regulated desert river. J. Arid Environ. 79, 56–65.

Goel, N.K., Kurothe, R.S., Mathur, B.S., Vogel, R.M., 2000. A derived flood frequency distribution for correlated rainfall intensity and duration. J. Hydrol. 228 (1–2), 56–67.

Ho, T.K., 1995, August. Random decision forests, (Vol. 1, IEEE, pp. 278–282.

Hong, H., Panahi, M., Shirzadi, A., Ma, T., Liu, J., Zhu, A.X., Chen, W., Kougias, I., Kazakis, N., 2018a. Flood susceptibility assessment in Hengfeng area coupling adaptive neuro-fuzzy inference system with genetic algorithm and differential evolution. Sci. Total Environ. 621, 1124–1141.

Hong, H., Tsangaratos, P., Ilia, I., Liu, J., Zhu, A.X., Chen, W., 2018b. Application of fuzzy weight of evidence and data mining techniques in construction of flood susceptibility map of Poyang County, China. Sci. Total Environ. 625, 575–588.

Ibrahim-Bathis, K., Ahmed, S.A., 2016. Rainfall-runoff modelling of Doddahalla watershed—an application of HEC-HMS and SCN-CN in ungauged agricultural watershed. Arab. J. Geosci. 9 (3), 170.

Jang, J.S., 1993. ANFIS: adaptive-network-based fuzzy inference system. IEEE Trans. Syst. Man Cybern. Syst. 23 (3), 665–685.

Kay, A.L., Jones, R.G., Reynard, N.S., 2006. RCM rainfall for UK flood frequency estimation. II. Climate change results. J. Hydrol. 318 (1–4), 163–172.

Khosravi, K., Nohani, E., Maroufinia, E., Pourghasemi, H.R., 2016. A GIS-based flood susceptibility assessment and its mapping in Iran: a comparison between frequency ratio and weights-of-evidence bivariate statistical models with multi-criteria decision-making technique. Nat. Hazards 83 (2), 947–987.

Kim, D.E., Gourbesville, P., Liong, S.Y., 2019. Overcoming data scarcity in flood hazard assessment using remote sensing and artificial neural network. Smart Water 4 (1), 1–15.

Kirchner, J.W., 2006. Getting the right answers for the right reasons: linking measurements, analyses, and models to advance the science of hydrology. Water Resour. Res. 42 (3).

Lee, S., Kim, J.C., Jung, H.S., Lee, M.J., Lee, S., 2017. Spatial prediction of flood susceptibility using random-forest and boosted-tree models in Seoul metropolitan city, Korea. Geomat. Nat. Hazards Risk 8 (2), 1185–1203.

Li, K., Wu, S., Dai, E., Xu, Z., 2012. Flood loss analysis and quantitative risk assessment in China. Nat. Hazards 63 (2), 737–760.

Mahmoud, S.H., 2014. Investigation of rainfall–runoff modeling for Egypt by using remote sensing and GIS integration. Catena 120, 111–121.

Mahmoud, S.H., Gan, T.Y., 2018. Multi-criteria approach to develop flood susceptibility maps in arid regions of Middle East. J. Clean. Prod. 196, 216–229.

Majidi, A., Shahedi, K., 2012. Simulation of rainfall-runoff process using green-Ampt method and HEC-HMS model (case study: Abnama watershed, Iran). Int. J. Hydraul. Eng. 1 (1), 5–9.

Meyer, V., Scheuer, S., Haase, D., 2009. A multicriteria approach for flood risk mapping exemplified at the Mulde river, Germany. Nat. Hazards 48 (1), 17–39.

Mohanta, N.R., Patel, N., Beck, K., Samantaray, S., Sahoo, A., 2021. Efficiency of river flow prediction in river using Wavelet-CANFIS: a case study. In: *Intelligent data engineering and analytics*. Springer, Singapore, pp. 435–443.

Mojaddadi, H., Pradhan, B., Nampak, H., Ahmad, N., Ghazali, A.H.B., 2017. Ensemble machine-learning-based geospatial approach for flood risk assessment using multi-sensor remote-sensing data and GIS. Geomatics Nat. Haz. Risk 8 (2), 1080–1102.

Nandi, A., Mandal, A., Wilson, M., Smith, D., 2016. Flood hazard mapping in Jamaica using principal component analysis and logistic regression. Environ. Earth Sci. 75, 465.

Ogden, F.L., Raj Pradhan, N., Downer, C.W., Zahner, J.A., 2011. Relative importance of impervious area, drainage density, width function, and subsurface storm drainage on flood runoff from an urbanized catchment. Water Resour. Res. 47 (12).

Pal, M., 2005. Random forest classifier for remote sensing classification. Int. J. Remote Sens. 26 (1), 217–222.

Pradhan, B., 2010. Flood susceptible mapping and risk area delineation using logistic regression, GIS and remote sensing. J. Spat. Hydrol. 9 (2), 4.

Pradhan, B., Hagemann, U., Tehrany, M.S., Prechtel, N., 2014. An easy to use ArcMap based texture analysis program for extraction of flooded areas from TerraSAR-X satellite image. Comput. Geosci. 63, 34–43.

Predick, K.I., Turner, M.G., 2008. Landscape configuration and flood frequency influence invasive shrubs in floodplain forests of the Wisconsin River (USA). J. Ecol. 96 (1), 91–102.

Pulvirenti, L., Pierdicca, N., Chini, M., Guerriero, L., 2011. An algorithm for operational flood mapping from synthetic aperture radar (SAR) data using fuzzy logic. Nat. Hazards Earth Syst. Sci. 11, 529–540.

Rahmati, O., Pourghasemi, H.R., 2017. Identification of critical flood prone areas in data-scarce and ungauged regions: a comparison of three data mining models. Water Resour. Manag. 31 (5), 1473–1487.

Rahmati, O., Pourghasemi, H.R., Zeinivand, H., 2016. Flood susceptibility mapping using frequency ratio and weights-of-evidence models in the Golastan Province, Iran. Geocarto Int. 31, 42–70.

Rahmati, O., Yousefi, S., Kalantari, Z., Uuemaa, E., Teimurian, T., Keesstra, S., Pham, T.D., Tien Bui, D., 2019. Multi-hazard exposure mapping using machine learning techniques: a case study from Iran. Remote Sens. 11 (16), 1943.

Reager, J.T., Thomas, B.F., Famiglietti, J.S., 2014. River basin flood potential inferred using GRACE gravity observations at several months lead time. Nat. Geosci. 7 (8), 588–592.

Rozalis, S., Morin, E., Yair, Y., Price, C., 2010. Flash flood prediction using an uncalibrated hydrological model and radar rainfall data in a Mediterranean watershed under changing hydrological conditions. J. Hydrol. 394 (1–2), 245–255.

Saaty, T., 1980. November. The analytic hierarchy process (AHP) for decision making. In: Kobe, pp. 1–69. Japan.

Samantaray, S., Sahoo, A., 2021. A Comparative study on prediction of monthly streamflow using hybrid ANFIS-PSO approaches. KSCE J. Civ. Eng. 25 (10), 4032–4043.

Samantaray, S., Sahoo, A., Ghose, D.K., 2019. Assessment of runoff via precipitation using neural networks: watershed modelling for developing environment in arid region. Pertanika J. Sci. Technol. 27 (4), 2245–2263.

Samantaray, S., Tripathy, O., Sahoo, A., Ghose, D.K., Udgata, S., 2020. Rainfall forecasting through ANN and SVM in Bolangir Watershed, India. In: Satapathy, S., Bhateja, V., Mohanty, J. (Eds.), *Smart intelligent computing and applications*. Springer, Singapore, pp. 767–774.

Skaugen, T., Onof, C., 2014. A rainfall-runoff model parameterized from GIS and runoff data. Hydrol. Process. 28 (15), 4529–4542.

Soulis, K.X., Valiantzas, J.D., 2013. Identification of the SCS-CN parameter spatial distribution using rainfall-runoff data in heterogeneous watersheds. Water Resour. Manag. 27 (6), 1737–1749.

Stefanidis, S., Stathis, D., 2013. Assessment of flood hazard based on natural and anthropogenic factors using analytic hierarchy process (AHP). Nat. Hazards 68 (2), 569–585.

Tehrany, M.S., Pradhan, B., Jebur, M.N., 2013. Spatial prediction of flood susceptible areas using rule based decision tree (DT) and a novel ensemble bivariate and multivariate statistical models in GIS. J. Hydrol. 504, 69–79.

Tehrany, M.S., Pradhan, B., Jebur, M.N., 2014a. Flood susceptibility mapping using a novel ensemble weights-of-evidence and support vector machine models in GIS. J. Hydrol. 512, 332–343.

Tehrany, M.S., Pradhan, B., Jebur, M.N., 2014b. Flood susceptibility mapping using a novel ensemble weights-of-evidence and support vector machine models in GIS. J. Hydrol. 512, 332–343.

Tehrany, M.S., Pradhan, B., Mansor, S., Ahmad, N., 2015. Flood susceptibility assessment using GIS-based support vector machine model with different kernel types. Catena 125, 91–101.

Termeh, S.V.R., Kornejady, A., Pourghasemi, H.R., Keesstra, S., 2018. Flood susceptibility mapping using novel ensembles of adaptive neuro fuzzy inference system and metaheuristic algorithms. Sci. Total Environ. 615, 438–451.

Thakur, B., Parajuli, R., Kalra, A., Ahmad, S., Gupta, R., 2017. Coupling HEC-RAS and HEC-HMS in precipitation runoff modelling and evaluating flood plain inundation map. World Environmental and Water Resources Congress 2017.

Therneau, T., B. Atkinson, and B. Ripley. 2013. *Package 'rpart'*. Accessed January 5, 2014.

Tiwari, M.K., Chatterjee, C., 2010. Uncertainty assessment and ensemble flood forecasting using bootstrap based artificial neural networks (BANNs). J. Hydrol. 382, 20–33.

Yao, L., Wei, W.E.I., Yu, Y., Xiao, J., Chen, L., 2018. Rainfall-runoff risk characteristics of urban function zones in Beijing using the SCS-CN model. J. Geogr. Sci. 28 (5), 656–668.

Zhang, Y., Smith, J.A., 2003. Space–time variability of rainfall and extreme flood response in the Menomonee River basin, Wisconsin. J. Hydrometeorol. 4 (3), 506–517.

Zhao, G., Pang, B., Xu, Z., Yue, J., Tu, T., 2018. Mapping flood susceptibility in mountainous areas on a national scale in China. Sci. Total Environ. 615, 1133–1142.

CHAPTER

14

Assessing the impact of land use and land cover changes on the water balances in an urbanized peninsular region of India

Harsh Ganapathi[a], *Mayuri Phukan*[b], *Preethi Vasudevan*[c], *and Santosh S. Palmate*[d]

[a]Wetlands International South Asia, New Delhi, India [b]The Energy and Resources Institute, New Delhi, India [c]TERI School of Advanced Studies, New Delhi, India [d]Texas A&M AgriLife Research, Texas A&M University, El Paso, TX, United States

OUTLINE

1 Introduction 226

2 Materials and methods 227
- 2.1 Study area 227
- 2.2 Data 228
- 2.3 Methodology 229

3 Results and discussion 232
- 3.1 LULC change from 1995 to 2020 232
- 3.2 Sensitivity analysis 234
- 3.3 Model calibration and validation 235
- 3.4 LULC impact on water balance 235
- 3.5 Discussion 236

4 Conclusion 239

References 239

https://doi.org/10.1016/B978-0-323-91910-4.00014-5

1 Introduction

Changes in land use and land cover (LULC) cause a significant impact on environmental, social, ecological, and hydrological systems and their processes. The factors that contribute to LULC changes in a region over a period of time are driven by anthropogenic pressures like conversion of land into urban areas, climate variability-induced changes, socioeconomic factors and nature conservation policies (Dwarakish and Ganasri, 2015; Elfert and Bormann, 2010; Mao and Cherkauer, 2009; Palmate et al., 2021). The surface runoff generated at a point in time follows a drainage path determined by the topography of the region and influenced by soil texture, surface freshwater and groundwater connectivity, infiltration rate and the contemporary LULC (Fiener et al., 2011; Wei et al., 2007). LULC changes within a watershed or a basin is likely to alter the natural drainage path. In the last few decades with the increase in population size and demand for more natural resources, LULC has been consistently modified by human-induced changes like urbanization, deforestation, agricultural expansion and construction of water storage structures like dams and reservoirs (Foley et al., 2005; Liu et al., 2008; Yin et al., 2017). These rampant changes have impacted the hydrologic processes by altering water storage and flow regimes, infiltration rate, evapotranspiration, and therefore on the overall water balance (Mao and Cherkauer, 2009; Piao et al., 2007; Scanlon et al., 2007; Sherwood and Fu, 2014; Wang et al., 2013; Yin et al., 2017).

In general, a smaller watershed (area lesser than 100 ha) responds greatly to the LULC change than a large river basin (area greater than 10,000 ha) wherein many complex processes interact with each other (Costa et al., 2003). The effects of LULC change on the hydrological response in a larger basin are difficult to distinguish; therefore, the cumulative effects of LULC change on runoff and other effects on the streams and river discharges are identified at the basin outlet (Salmoral et al., 2015). Vorosmarty (2000) had predicted that by 2025, land use change would have sterner consequences than climate change. Hence it is necessary to understand the past, present and likely future changes in LULC for suitable basin management (Ndulue et al., 2015, 2018).

A good understanding of LULC change and hydrological processes at watershed or basin scale is necessary to study the runoff patterns in the varying spatial and temporal scale (Blöschl and Sivapalan, 1995) and consequently the adverse changes on hydroecological balance. A good working water balance model provides a conducive environment to understand the changes in hydrological regimes due to LULC change as well as predict the likely effects of future LULC changes (Bewket and Sterk, 2005; Miller et al., 2002). Geographical Information System (GIS) provides a good platform to assess such rapid changes in LULC and also aid in mapping its impacts on the surface runoff, sediment and water quality (Sun et al., 2011; Li and Wang, 2009; Pandey and Palmate, 2018, 2019). To simulate the LULC impacts on a watershed, various hydrological models like Storm Water Management Model (Huber et al., 1988) developed by Environmental Protection Agency, Soil and Water Assessment Tool (SWAT) (Arnold et al., 1998) developed by United States Department of Agriculture and Agricultural Research Service etc., have been developed with an integration of GIS. These hydrological models function on continuous time-series and point data, and can be used to assess the impacts of LULC changes on hydrology and stream quality (Sun et al., 2011; Li and Wang, 2009).

Among the widely available hydrological models, the SWAT model (Gassman et al., 2007) has gained major popularity for its wide range of use and applicability in various hydrological studies and land management practices. SWAT is a GIS-based, semi-distributed, and continuous time-step hydrological model that facilitates quicker implementation of input datasets over small to large river basins and provides an effective and reasonable understanding of the impacts of LULC on water quantity and quality (Gassman et al., 2007; Schilling and Wolter, 2009; Volk et al., 2009; Palmate and Pandey, 2021). The model has been executed across numerous watersheds, environmental conditions, climatic zones, and water management systems worldwide (Samal and Gedam, 2021). Its advantage is its ability to simulate the hydrological processes using multiple GIS platforms with the help of global as well as local datasets like digital elevation model (DEM), soil types, LULC data and meteorological data. The model is also widely used and recognized as one of the leading hydrological models for addressing hydrologic and environmental problems across the globe (Akoko et al., 2021).

SWAT has been extensively used in India for runoff estimation, modeling and assessing the impact of land use and climate change on hydrological variables (Astuti et al., 2019; Sinha and Eldho, 2018; Kumar et al., 2017; Kundu et al., 2017; Patel and Nandhakumar, 2016; Singh et al., 2017).

In this study, an effort has been made to assess the impacts of past LULC changes on water balance in the Palar and other sub-basins, that form the upper basin area of the east-flowing rivers between Pennar and Cauvery basin, using the SWAT model. The basin is selected because this southern part of India has witnessed a significant LULC change in the last two decades, which must have impacted the hydrological regimes of the study area. Based on this premise, the LULC change assessment was carried out for the duration of 1995 to 2020 using classified satellite images, DEM, meteorological data, soil data, and their impacts on streamflow were assessed. An Uncertainty in Sequential Fitting (SUFI-2) algorithm was used to calibrate and validate the model simulation for streamflow.

2 Materials and methods

2.1 Study area

The Palar and other river sub-basins are a part of east flowing rivers between Pennar and Cauvery basin. It is bounded by the Eastern Ghats on the North, Tamil Nadu uplands on the West, Tamil Nadu plains on the South and the Bay of Bengal on the East. The study area (Fig. 1) encompasses an area of 35,392 sq. km; which is shared within the States of Karnataka, Andhra Pradesh and Tamil Nadu (CWC and ISRO, 2014). Major cities in the basin include Kolar, Chittoor, Chennai, and Vellore, indicating rapid urbanization in the study area over the last few decades.

The basin consists three major topographical divisions, i.e., the hilly region of eastern ghats, the plateau region and the coastal plains (CWC, 2012). Major portion of the basin area exhibits undulating topography with several local depressions. Geomorphological features present in the basin include denuding and residual hills, plateaus, valley fill, pediments, pediplains, and alluvial plains (MR, C, & Resmi et al., 2019). Major soil groups observed

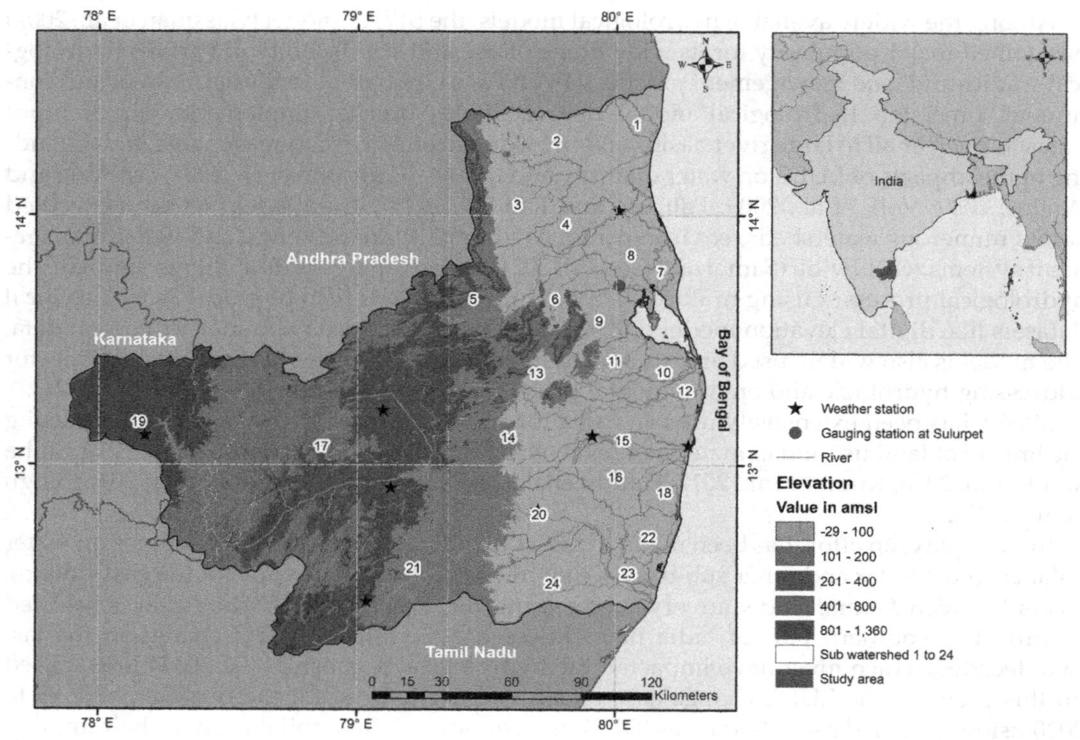

FIG. 1 Study area extent map.

in the basin are Inceptisol, Alfisol, Entisol, Ultisols, and Vertisol (National Water Mission, 2017). Land of the basin is majorly covered by agriculture and plantations.

The region experiences sub-tropical climate, with precipitation predominantly from the Northeast monsoon and occasionally from Southwest monsoon (Resmi et al., 2016). The average annual rainfall ranges between 600 mm and 1400 mm. Various independent rivers such as Swarnamukhi, Arani, Cooum, Adyar, and Palar flow through the basin, draining to the Bay of Bengal. They cover 24 sub-watersheds consisting of about 18,040 surface water bodies including major 18 dams and 16 reservoirs present in the region (CWC and ISRO, 2014). The annual surface water potential of the basin is about 1675 MCM at 75% dependability. Groundwater comprises unconsolidated and semi consolidated formations as well as in weathered, fissured and fractured crystalline rocks. Depth of groundwater table ranges between 5 to 15 m below ground level (National Water Mission, 2017).

2.2 Data

In this study, a complete data set of daily precipitation and temperature from 8 stations (Fig. 1) across the study area was obtained from NASA POWER (Power.larc.nasa.gov, 2021). Observed flow data for gauge station located in the study area (Fig. 1) was obtained from India-WRIS (India-WRIS, 2021). The topography for the study area is delineated using the SRTM DEM of 30 m resolution from EarthExplorer (EarthExplorer, 2021). The soil parameters for the study area were obtained from the Harmonized World Soil Database dataset prepared by Food and Agriculture Organization (Fischer et al., 2008). The LULC data layer for the year 1995 generated by hierarchical class merging approach (Roy et al., 2016) was derived from Landsat 5 satellite imagery, IRS 1C–LISS III of spatial resolution 30 m and 72 m (resampled to 56 m), respectively, used in this study. The 2020 LULC data was produced by Impact Observatory for Esri (Karra et al., 2021) generated using Sentinel-2 L2A/B imagery of spatial resolution 10 m and a combination of classified Landsat 8 OLI/TIRS C2 L2 imagery is used to generate the LULC 2020 raster.

2.3 Methodology

This chapter majorly covers two sections: (1) Modeling LULC change for the upper basin of the east flowing rivers between Pennar and Cauvery River, which encapsulates the LULC classification for the study area for the year 1995 and 2020, the accuracy assessment for the classified LULC of 2020 and the LULC transformation for during the study period, and (2) Hydrological modeling using SWAT model, in which the streamflow is considered as a main component representing hydrological consequences under LULC changes in the study area.

2.3.1 Modeling the LULC change from 1995 to 2020

The LULC map for the year 1995 was reclassified into five classes: cropland, built up, wetlands/waterbodies, tree cover/shrubland and barren land. The LULC for 2020 using Landsat 8 OLI/TIRS C2 L2 followed a classification scheme of Anderson et al. (1976) wherein layers pertaining to build up and wetlands/waterbodies were extracted using supervised classification technique by Abbas and Jaber (2020). This map was then used in combination with the LULC map produced by Karra et al. (2021). LULC map of 2020 is also classified into the same 5 classes. All LULC maps were resampled to a spatial resolution of 10 m.

The accuracy assessment of the LULC maps was carried out for the year 2020. The generated maps were checked from the classifiers, which were performed based on the computation of the error matrix (Congalton and Green, 1999). The statistical measures of the error matrix include overall accuracy (OA), user accuracy (UA), producer accuracy (PA) and the kappa coefficient (Kc), which were computed using Eqs. (1)–(4) (Congalton and Green, 1999; Rwanga and Ndambuki, 2017). In total, 639 random sampling points distributed throughout the study area were used to assess the classification accuracy. The Google Earth Imagery of the year 2020 was used for the ground-truthing data points.

$$OA = \frac{1}{N}\sum_{t=1}^{r} n_{ii}, \quad (1)$$

$$PA = \frac{n_{ii}}{n_{icol}}, \quad (2)$$

$$\mathrm{UA} = \frac{n_{ii}}{ni_{row}} \tag{3}$$

$$K_c = N\sum\nolimits_{t=1}^{r} n_{ii} - \sum\nolimits_{t=1}^{r} \frac{n_{icol}n_{irow}}{N^2} - \sum\nolimits_{t=1}^{r} n_{icol}n_{irow} \tag{4}$$

where n_{ii} is the number of pixels correctly classified in a LULC category; N is the total number of pixels in the confusion matrix; r is the number of rows; and n_{col} and n_{row} are the column (reference data) and row (predicted classes) total.

A LULC transformation matrix for the LULC change from 1995 to 2020 was developed by intersecting the LULC classes of the 1995 and 2020 similar to the technique used by Akbar et al. (2019) and Hu et al. (2019) by overlaying the 1995 classified LULC map over the 2020 LULC map and identifying areas of (a) no LULC change and (b) areas of LULC changed from one category to another category. This transformation matrix is a means to define the quantity of transformation from a specific LULC of 1995 to the most recent LULC of 2020.

2.3.2 *Hydrological modeling using SWAT*

The SWAT model (https://swat.tamu.edu/) is a semi-distributed hydrologic model widely used to simulate the impact of land management practices on water balance, water quality and sediment fluxes in a complex river basin (Borah and Bera, 2003). In this study, the model setup was performed in five steps (1) data preparation, (2) watershed discretization, (3) hydrologic response unit (HRU) generation, (4) parameter sensitivity analysis, also called parameterization, and (5) calibration and uncertainty analysis. The basin area was divided into sub-watersheds and discretized into Hydrological Response Units (HRUs) based on the land use type, soil type, and slope (Arnold et al., 1998). Further, the simulation at each HRU was carried out based on the water balance equation:

$$SW_t = SW_0 + \sum_{i=1}^{t}\left(R_{day} - Q_{surf} - E_a - w_{seep} - Q_{gw}\right) \tag{5}$$

where, SW_t is the final soil water content (mm), SW_0 is the initial water content (mm), t is the time-step, R is the amount of precipitation (mm), Q_{surf} is the amount of surface runoff (mm), ET is the amount of evapotranspiration (mm), W_{seep} is the amount of water entering the vadose zone (mm), and Q_{gw} is the amount of return flow (mm).

(a) *Model set-up*

ArcSWAT 2012 was used for modeling in this study. The model requires a Digital Elevation Model (DEM) for delineating the watershed. LULC maps of 1995 and 2020 prepared for the study area were used for the modeling. The soil parameters for the study area and the daily climate inputs from for eight climate stations were used. Discharge data from one gauging station at Sulurpet used for calibration and validation was obtained from the India WRIS.

The watershed and sub-basin outlets were defined based on the streamlines generated from the DEM and the monitoring gauge stations. The delineated watershed covers an area of 3,431,620.107 ha and is divided into 24 sub-watersheds. Multiple HRUs were generated with a 5% land use and soil type overlap. A total of 338 HRUs for 24 sub-watersheds were generated for the year 1995 and 431 HRUs were generated for the year 2020. The delta change was used to change the LULC inputs; therefore, different HRUs were generated.

The model was run on a monthly time-step from January 1990 to December 2020 with a three-year initial warm-up period.

(b) *Calibration and validation*

Model calibration and validation help to assess the reliability and accuracy of the model simulation. Calibration is the process of altering model parameters within their regional value range to fit the simulated output with the observed data. The validation process ensures that the calibrated parameters continue to produce output that is in accordance with the observed data (Abbaspour et al., 2004).

SWAT-CUP is a program developed for calibration, validation, and uncertainty analysis of the SWAT model (Abbaspour et al., 2007). The Sequential Uncertainty Fitting (SUFI-2) algorithm in SWAT-CUP is a semi-automated routine for a combined calibration and uncertainty analysis (Abbaspour et al., 2004).

In SUFI-2, parameter uncertainty accounts for all sources of uncertainties including uncertainty in model input, model parameters, and measured data (Singh et al., 2013; Abbaspour, 2015). The algorithm tries to bracket the measured data with all the uncertainties accounted for within the 95% prediction uncertainty (95PPU). The 95PPU is calculated at the 2.5% and 97.5% levels of the cumulative distribution of an output variable obtained through Latin hypercube sampling (Abbaspour et al., 2004; Yang et al., 2008).

The p-factor and r-factor are two parameters that help determine the performance of the model calibration and uncertainty (Abbaspour, 2015; Arnold et al., 2012). The p-factor gives the percentage of observed data that is covered by the 95 PPU. It ranges from 0 to 1 with 1 representing 100% coverage (Setegn et al., 2009). The r-factor is calculated by dividing the average thickness of the 95PPU by the standard deviation of the observed data (Yang et al., 2008). It determines the quality of the calibration with a lower value representing a better calibration. A better p-factor can only be achieved with a higher r-factor and hence there needs to be a trade-off (Abbaspour, 2015).

The model has been calibrated using SUFI-2 for the Palar basin using the observed streamflow data for one gauging station. The period 1993 to 2006 was considered for calibration and the period 2007 to 2015 was considered for validation. A 3-year warm-up period was provided for model initialization.

(c) *Performance evaluation of the model*

To assess the performance of the simulation with respect to the observed data, two goodness-of-fit indicators were considered, Nash-Sutcliffe Efficiency (NSE), and coefficient of determination (R^2) according to Moriasi et al. (2007).

NSE is a normalized dimensionless statistic that determines the relative magnitude of the residual variance compared to the measured data variance (Nash and Sutcliffe, 1970). NSE can take a value from $-\infty$ to 1, a higher value indicates a better fit between observed and simulated data.

$$NSE = 1.0 - \frac{\sum_{i=1}^{N}(o_i - p_i)^2}{\sum_{i=1}^{N}(o_i - \bar{o})^2} \tag{6}$$

where o_i represents the ith observed value, p_i represents the ith simulated value, $\bar{o}$ represents the mean of the observed data and N is the total number of observations.

R^2 gives the proportion of variances of the observed values that are replicated by the simulations and ranges from 0 to 1 (Krause et al., 2005). A higher value indicates a better agreement to the observed values.

$$R^2 = \left[\frac{\sum_{i=1}^{N} (O_i - \overline{O}) \times (P_i - \overline{P})}{\left[\sum_{i=1}^{N} (O_i - \overline{O})^2\right]^{0.5} \times \left[\sum_{i=1}^{N} (P_i - \overline{P})^2\right]^{0.5}} \right]^2 \quad (7)$$

where O_i represents the ith observed value, P_i represents the ith simulated value, $\overline{O}$ and $\overline{P}$ represents the mean values of the observed and simulated data and N is the total number of observations.

(d) *Sensitivity analysis*

Sensitivity analysis identifies the most important processes in the study region. Performing sensitivity analysis allows reducing the number of parameters in the calibration procedure by eliminating the parameters that do not influence the output. The local sensitivity analysis was carried out by changing only one parameter while other parameters are fixed to identify the effect on model output. Furthermore, the global sensitivity analysis identified sensitive parameters by varying all parameters over a large number of runs. Often the sensitivity of one parameter is dependent on the value of other parameters, and this is a limitation of the local sensitivity analysis as the value of parameters is fixed. The sensitivity of parameters is also based on the parameter ranges and the number of runs carried out which can be a limitation of the global sensitivity analysis (Abbaspour, 2015).

An initial sensitivity analysis was performed to identify the most sensitive parameters for calibration with data from one gauging station at Sulurpet.

2.3.3 *Land use land cover change impact on streamflow*

The model was further used to simulate the impact of LULC change on streamflow by replacing the 1995 LULC map with the 2020 LULC map. The model simulations under different LULC inputs were further analyzed to assess the impact on streamflow.

3 Results and discussion

3.1 LULC change from 1995 to 2020

The LULC change has been studied for the period of 1995 to 2020. Fig. 2 shows the LULC maps of the study area during 1995 and during 2020. The study area has witnessed a drastic increase of the built-up area, while water bodies/wetlands show considerable decrease. The class wise accuracy assessment of the classified LULC map for the year 2020 indicates that the study area has been accurately identified and extracted during classification. Table 1 shows the overall accuracy of 90.61% and the Kappa coefficient of 0.87 which indicates a good agreement between the classification and actual land observations.

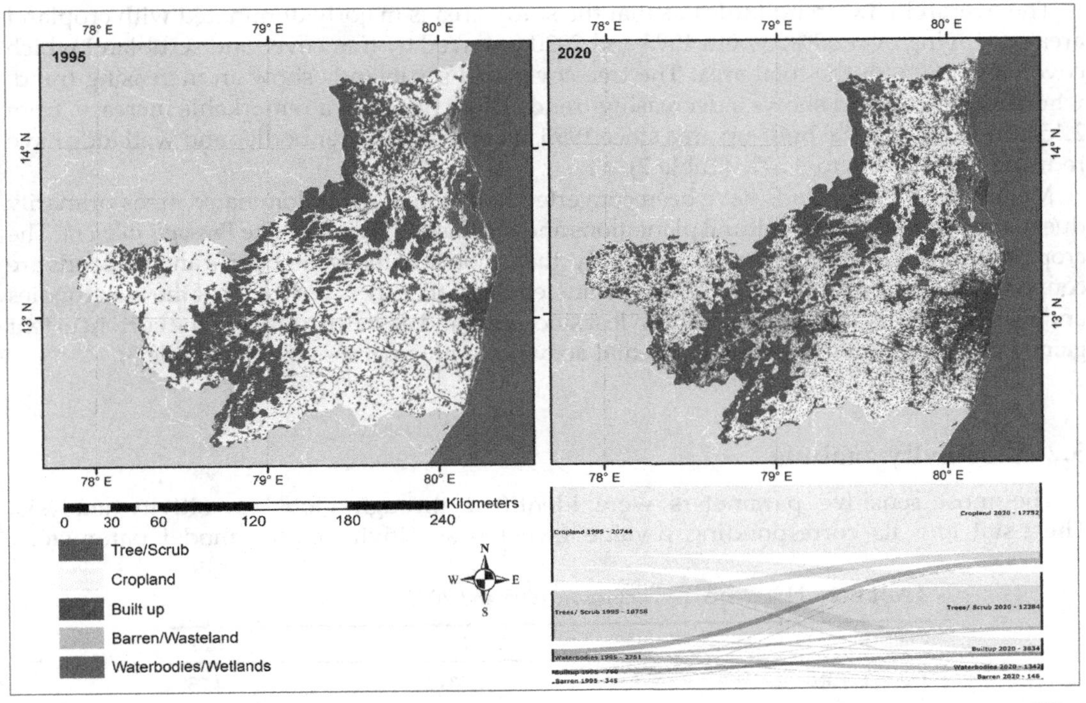

FIG. 2 LULC maps of the study area for the year 1995 and 2020 along with the Snakey diagram showing LULC change transformation. LULC class frequency is reported in terms of sq.km.

TABLE 1 Accuracy assessments of classified LULC maps for the years 2020.

	2020	
LULC classes	**UA**	**PA**
Trees/Scrubs	91.00%	95.29%
Cropland	87.74%	83.78%
Built up	94.03%	87.50%
Barren/Wasteland	87.32%	89.86%
Waterbodies/Wetlands	92.45%	88.29%
Overall accuracy	90.61%	
Kappa coefficient	87.02%	

The overall LULC trend indicates that the study area is majorly dominated with cropland area occupying over 50% in both 1995 and 2020 followed by tree-cover and scrubland which covers over 30% of the total area. The tree-cover and scrublands show an increasing trend, whereas the cropland shows a decreasing trend. There has been a remarkable increase, from 2.15% to 10.84%, in the built-up area since 1995, whereas the water bodies and wetlands have reduced from 7.78% to 3.79% (Table 2).

Majority of the croplands have been converted to tree and scrub dominated areas primarily due to the increase in horticultural plantations and increase in invasives like *Prosopis juliflora*. The croplands have also been expanded. Similarly, the majority of the water bodies and wetlands are converted to croplands and tree/scrub dominated areas (Fig. 2). Nearly 65% of the waterbodies and wetland dominated areas have been lost since 1995 and as much as 35% of the current extent gained due conversion and developmental activities with a net loss of 48% (Table 3).

3.2 Sensitivity analysis

The most sensitive parameters were identified through global sensitivity analysis. The t-stat and its corresponding p-value give the sensitivity of the model parameters.

TABLE 2 LULC area percentage in 1995 and 2020.

Classes	1995	2020
Trees/Scrubs	30.43%	34.75%
Cropland	58.68%	50.22%
Built up	2.15%	10.84%
Barren/Wasteland	0.97%	0.41%
Waterbodies/Wetlands	7.78%	3.79%

TABLE 3 LULC transformation from year 1995 to 2020 in the study area.

		2020					
Year	**LULC class**	**Trees/ Scrubs**	**Cropland**	**Built up**	**Barren/ Wasteland**	**Waterbodies/ Wetlands**	**Grand total**
1995	Trees/Scrubs	8818	1536	332	13	59	10,758
	Cropland	2697	14,977	2666	21	383	20,744
	Built up	27	34	685	1	13	760
	Barren/Wasteland	166	107	39	23	10	345
	Waterbodies/ Wetlands	576	1098	112	88	877	2751
	Grand Total	12,284	17,752	3834	146	1342	35,358

The t-stat determines the measure of the sensitivity, a larger value means a more sensitive parameter, and the p-value determines the significance of the sensitivity with a lower value meaning more significance. A parameter with absolute t-stat value ≥ 2 and P-value $\leq .05$ are considered statistically significant (Abbaspour, 2015). It can be seen that CN2 is the most sensitive parameter followed by SOL_K, RCHRG_DP, and SOL_AWC (Table 4).

3.3 Model calibration and validation

The model calibration and validation was carried out for the Sulurpet gauging station. A p-factor of 0.7 and a r-factor value of 1 are considered acceptable for streamflow simulation (Abbaspour et al., 2004, 2007). Model efficiency measures for the calibration are R^2 and NSE are 0.64 and 0.59 respectively which indicate a satisfactory fit between the monthly simulated and observed values. The 95PPU covers about 60% of the observed values with an r-factor of 0.51 which is close to the acceptable limits. The calibrated model was found to perform satisfactorily during the validation period with R^2 and NSE values of 0.79 and 0.56 respectively (Fig. 3A). The model was not calibrated for other gauging stations on the Palar river due to the extensively managed nature of the sub-watersheds resulting in high variations between simulated and observed data (Tables 5 and 6).

3.4 LULC impact on water balance

The average annual surface runoff in the watershed has increased by 12% in 2020, from 294 mm to 329 mm. The trend in evapotranspiration has decreased 4%; from 250 mm in 1995 to 240 mm in 2020. The water yield has increased 5% from 762 mm to 798 mm from 1995 to 2020 (Fig. 3B and C). The change in surface runoff is the highest and most evident among all the hydrological parameters. Overall, the annual average changes in these parameters are relatively small.

It has been observed that 23 out of the 24 sub-watersheds have experienced an increase in the annual average surface runoff with changes varying from 1% to 25%. One of the sub-watersheds exhibits a decrease in the annual average surface runoff by 4%. Chennai city covers sub-watersheds 14, 15, and 16 partially and sub-watersheds 18 completely. The 19% increase in the runoff in sub-watersheds 18 can be attributed to the increase in built-up. A relatively smaller increase in the runoff is observed in the sub-watersheds 14, 15, and 16 (Fig. 4).

TABLE 4 Most sensitive parameters of the SWAT model.

Parameter code	Description	*t*-Stat	*P*-value
R__CN2.mgt	Curve number II	−17.3082	0
R__SOL_K(..).sol	Saturated hydraulic conductivity (mm/h)	−3.88895	.000209
V__RCHRG_DP.gw	Deep aquifer percolation fraction	2.782196	.00675
R__SOL_AWC(..).sol	Available water capacity of soil layer (mm H2O/mm soil)	2.660721	.009441

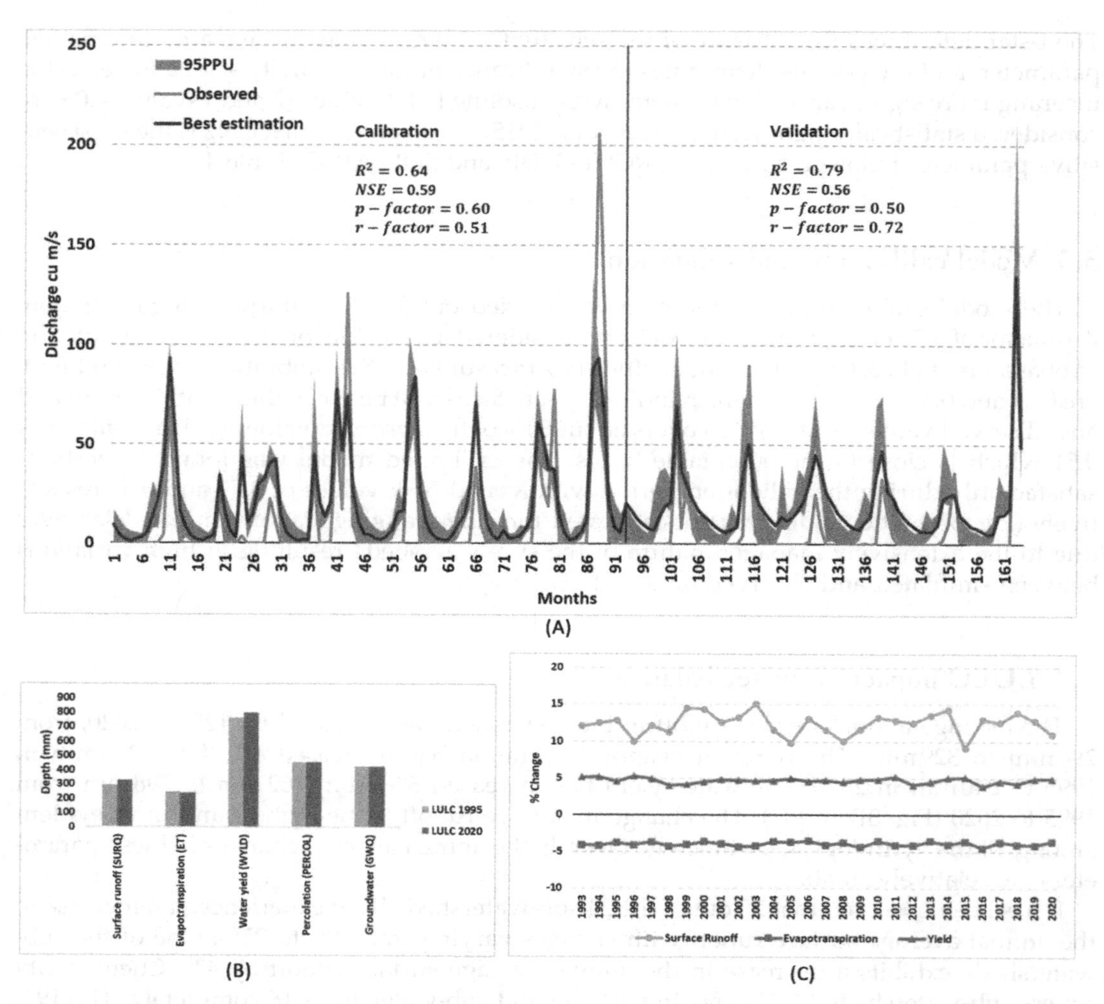

FIG. 3 (A) 95PPU plot and observed streamflow during calibration and validation, (B) LULC change impact on water balance components, (C) Annual trend changes of water balance components.

3.5 Discussion

Groundwater processes are among the important components in influencing the flow in a watershed (Kumar et al., 2017). The groundwater processes in SWAT are not simulated rigorously which impacts the results obtained (Rostamian et al., 2008). It is evident from Fig. 3A that the baseflow is not estimated accurately resulting in a less desirable value for p-factor for a good calibration and lessening the quality of the overall calibration.

To minimize the uncertainty range and obtain the right parameter values, proper calibration and uncertainty analysis must be performed (Abbaspour, 2015). The model efficiency

TABLE 5 Calibrated parameter range.

Parameter code	Description	Fitted value	Min value	Max value
R__CN2.mgt	Curve number II	−0.35125	−0.5	0.5
R__SOL_K(..).sol	Saturated hydraulic conductivity (mm/h)	19.874353	−32.964363	64.21029
R__CH_K2.rte	Effective hydraulic conductivity in main channel alluvium (mm/h)	−27.344208	−37.283699	−3.44713
R__RCHRG_DP.gw	Deep aquifer percolation fraction	−0.288321	−0.444409	0.153057
R__SOL_AWC(..).sol	Available water capacity of soil layer (mm H2O/mm soil)	0.464308	−0.167722	1
R__REVAPMN.gw	Threshold depth for deep percolation (mm H2O)	46.958263	−5.631896	79.70711
R__EPCO.hru	Plant uptake compensation factor	−0.814627	−0.9	−0.73463
R__ALPHA_BF.gw	Baseflow alpha factor (1/days)	−0.505844	−0.859694	0.10645
R__CANMX.hru	Maximum canopy storage (mm H2O)	−14.618073	−46.266869	68.29891
V__GWQMN.gw	Threshold depth for return flow (mm H2O)	−1670.091919	−1868.2513	429.2494

TABLE 6 Statistical measures for model calibration and validation.

	Station	p-factor	r-factor	R^2	NS
Calibration	Sulurpet	0.60	0.51	0.64	0.59
Validation	Sulurpet	0.50	0.72	0.79	0.56

was assessed through the R^2 and NSE between the simulated and observed values (Narsimlu et al., 2015). The calibration indicates a satisfactory agreement with satisfactory values for R^2 (0.64) and NSE (0.59) (Nash and Sutcliffe, 1970; Setegn et al., 2009).

It is observed that certain peaks both in the calibration and the validation phase are not bracketed by the 95PPU. This indicates that the model is unable to correctly predict simulations for certain extreme events. During the validation phase, smaller flows are often overestimated and the 95PPU brackets an overall of 50% of the observed data with a higher uncertainty (r-factor of 0.72).

Conceptual uncertainties in the model include natural processes that occur in the watershed but are not accounted for in the model (Yang et al., 2008). Uncertainties in the output can be caused by not including catchment processes like extraction of water, reservoir operations, soil erosion, etc. Any uncertainty in the measurement of flow values also adds to the uncertainty and furthers the inaccuracies in the estimation (Kumar et al., 2017).

The Palar River being the longest river in the study area is intensely managed with multiple check dams, and multiple diversions for irrigation and human use (Senthilkumar and Elango, 2004; National Water Mission, 2017) which are unaccounted for in the SWAT model.

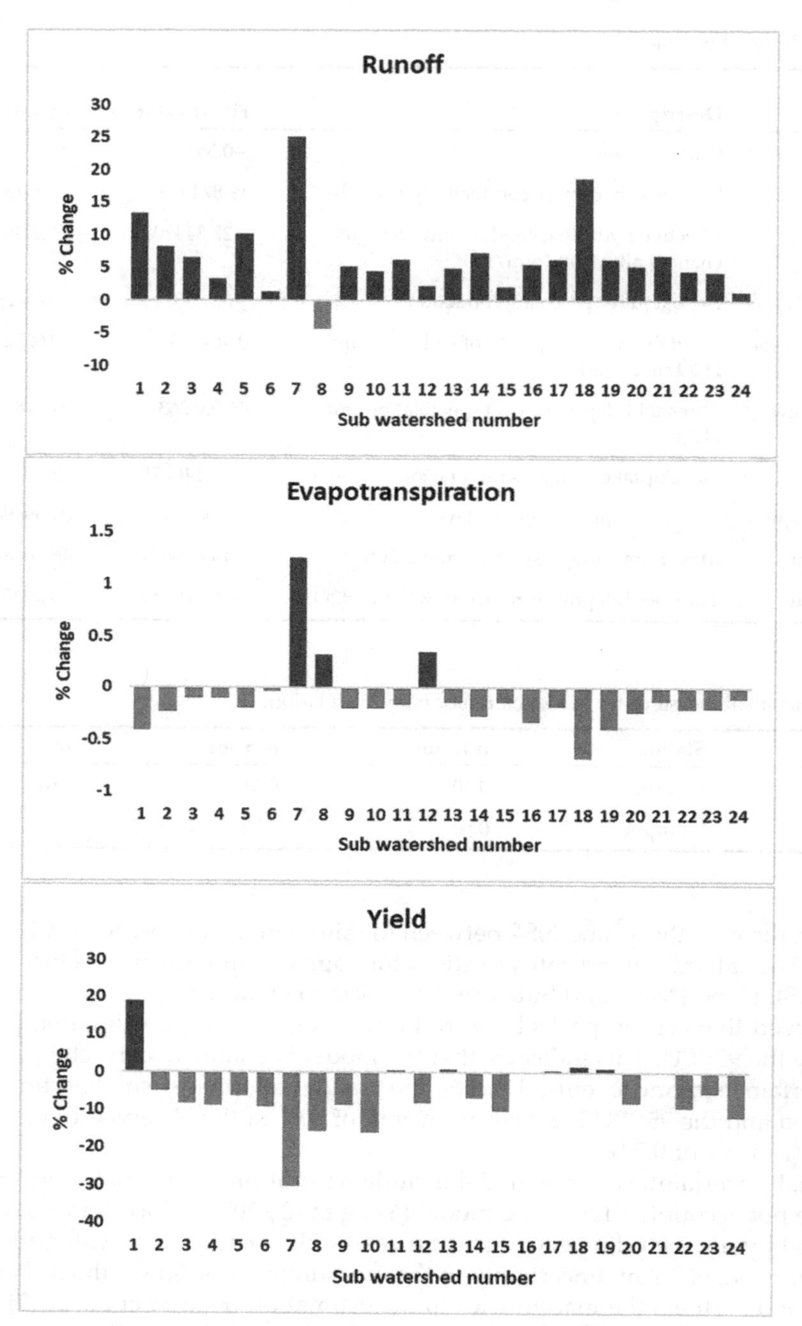

FIG. 4 Sub-watershed level changes (in %) for surface run-off, evapotranspiration and water yield.

This resulted in the simulated flow varying with the actual observed values to a large extent. This variation rendered the monitoring gauges across the Palar River unusable for calibration.

4 Conclusion

The study area of Palar and other sub-basins, which is an agriculture dominated landscape in the Southern Peninsular region of India, is experiencing a remarkable increase in built-up area from 2.15% to 10.84% during 1995 and 2020. The rise in built up area is attributed to the conversion of cultivable land into infrastructural development activities. The sensitivity analysis that was executed for the study area indicates that CN2, SOL_K, RCHRG_DP, and SOL_AWC were the most sensitive parameters. Nearly all the sub-watersheds have experienced an increase in runoff ranging from 1% to 25% increase. Sub-watershed number 18 which entirely lies in the Chennai metropolitan area experiences a 19% increase in runoff which can be attributed to the increase in built up area. The rapid change in LULC patterns observed in urban watersheds is common to various cities across the country. It has a serious impact on the urban water cycle and the overall environment of the region. These changes need to be closely monitored and modeled, to draw up environment risk management plans.

References

Abbas, Z., Jaber, H., 2020. Accuracy assessment of supervised classification methods for extraction land use maps using remote sensing and GIS techniques. IOP Conf. Ser.: Mater. Sci. Eng. 745, 012166. https://doi.org/10.1088/1757-899x/745/1/012166.

Abbaspour, K., Johnson, C., van Genuchten, M., 2004. Estimating uncertain flow and transport parameters using a sequential uncertainty fitting procedure. Vadose Zone J. 3 (4), 1340–1352. https://doi.org/10.2136/vzj2004.1340.

Abbaspour, K., Yang, J., Maximov, I., Siber, R., Bogner, K., Mieleitner, J., et al., 2007. Modelling hydrology and water quality in the pre-alpine/alpine Thur watershed using SWAT. J. Hydrol. 333 (2–4), 413–430. https://doi.org/10.1016/j.jhydrol.2006.09.014.

Abbaspour, K.C., 2015. SWAT-CUP: SWAT Calibration and Uncertainty Programs – A User Manual. SWAT.

Akbar, T.A., Hassan, Q.K., Ishaq, S., Batool, M., Butt, H.J., Jabbar, H., 2019. Investigative spatial distribution and modelling of existing and future urban land changes and its impact on urbanization and economy. Remote Sens. 11 (2), 105. https://doi.org/10.3390/rs11020105.

Akoko, G., Le, T.H., Gomi, T., Kato, T., 2021. A review of SWAT model application in Africa. Water 13 (9), 1313. https://doi.org/10.3390/w13091313.

Anderson, J., Hardy, E., Roach, J., Witmer, R., 1976. A Land Use and Land Cover Classification System for Use With Remote Sensor Data. Professional Paper. https://doi.org/10.3133/pp964.

Arnold, J., Moriasi, D., Gassman, P., Abbaspour, K., White, M., Srinivasan, R., et al., 2012. SWAT: Model use, calibration, and validation. Trans. ASABE 55 (4), 1491–1508. https://doi.org/10.13031/2013.42256.

Arnold, J.G., Srinivasan, R., Muttiah, R.S., Williams, J.R., 1998. Large-area hydrologic modeling and assessment: part I model development. J. Am. Water Res. Assoc. 34, 73–89.

Astuti, I.S., Sahoo, K., Milewski, A., Mishra, D.R., 2019. Impact of land use land cover (LULC) change on surface runoff in an increasingly urbanized tropical watershed. Water Resour. Manag. https://doi.org/10.1007/s11269-019-02320-w.

Bewket, W., Sterk, G., 2005. Dynamics in land cover and its effect on stream flow in the Chemoga watershed, Blue Nile basin, Ethiopia. Hydrol. Process. 19 (2), 445–458. https://doi.org/10.1002/hyp.5542.

Blöschl, G., Sivapalan, M., 1995. Scale issues in hydrological modeling: a review. Hydrol. Process. 9 (3–4), 251–290. https://doi.org/10.1002/hyp.3360090305.

Borah, D.K., Bera, M., 2003. Watershed-scale hydrologic and nonpoint-source pollution models: review of mathematical bases. Trans. ASAE 46 (6), 1553. https://doi.org/10.13031/2013.15644.
Congalton, R.G., Green, K., 1999. Assessing the Accuracy of Remotely Sensed Data Principles and Practices. Lewis Publishers, Boca Raton.
Costa, M.H., Botta, A., Cardille, J.A., 2003. Effects of large-scale changes in land cover on the discharge of the Tocantins River, Southeastern Amazonia. J. Hydrol. 283 (1–4), 206–217. https://doi.org/10.1016/s0022-1694(03)00267-1.
CWC, 2012. Integrated Hydrological Data Book. Central Water Commission, New Delhi.
CWC and ISRO, 2014. Watershed Atlas of India. Government of India Ministry of Water Resources, India.
Dwarakish, G., Ganasri, B., 2015. Impact of land use change on hydrological systems: A review of current modeling approaches. Cogent Geosci. 1 (1), 1115691. https://doi.org/10.1080/23312041.2015.1115691.
EarthExplorer, 2021. Earthexplorer.usgs.gov. Retrieved from https://earthexplorer.usgs.gov/. (Accessed 18 August 2021).
Elfert, S., Bormann, H., 2010. Simulated impact of past and possible future land use changes on the hydrological response of the Northern German lowland "Hunte" catchment. J. Hydrol. 383 (3–4), 245–255. https://doi.org/10.1016/j.jhydrol.2009.12.040.
Fiener, P., Auerswald, K., Van Oost, K., 2011. Spatio-temporal patterns in land use and management affecting surface runoff response of agricultural catchments—a review. Earth Sci. Rev. 106 (1–2), 92–104. https://doi.org/10.1016/j.earscirev.2011.01.0.
Fischer, G., Nachtergaele, F., Prieler, S., Velthuizen, H., Verelst, L., Wiberg, D., 2008. Global Agro-Ecological Zones Assessment for Agriculture. IIASA/FAO, Laxenburg, Austria/Rome, Italy.
Foley, J., DeFries, R., Asner, G., Barford, C., Bonan, G., Carpenter, S., et al., 2005. Global consequences of land use. Science 309 (5734), 570–574. https://doi.org/10.1126/science.1111772.
Gassman, P.W., Reyes, M.R., Green, C.H., Arnold, J.G., 2007. The soil and water assessment tool: historical development, applications, and future research directions. Trans. ASABE 50 (4), 1211–1250. https://doi.org/10.13031/2013.23637.
Hu, Y., Batunacun, Z., L., & Zhuang, D., 2019. Assessment of land-use and land-cover change in Guangxi, China. Sci. Rep. 9 (1). https://doi.org/10.1038/s41598-019-38487-w.
Huber, W., Dickenson, R., Roesner, L., Aldrich, J., 1988. Storm Water Management Model user's Manual, Version 4. US Environmental Protection Agency, Athens.
India-WRIS, 2021. India Water Resources Information System. Retrieved from https://indiawris.gov.in/wris/#/. (Accessed 30 July 2021).
Karra, K., Kontgis, C., Statman-Weil, Z., Mazzariello, J., Mathis, M., Brumby, S., 2021. Global land use/land cover with sentinel 2 and deep learning. In: International Geoscience and Remote Sensing Symposium – IGARSS. IEEE, Brussels.
Krause, P., Boyle, D., Bäse, F., 2005. Comparison of different efficiency criteria for hydrological model assessment. Adv. Geosci. 5, 89–97. https://doi.org/10.5194/adgeo-5-89-2005.
Kumar, N., Singh, S.K., Srivastava, P.K., Narsimlu, B., 2017. SWAT model calibration and uncertainty analysis for streamflow prediction of the tons River Basin, India, using sequential uncertainty fitting (SUFI-2) algorithm. Model. Earth Syst. Environ. 3 (1). https://doi.org/10.1007/s40808-017-0306-z.
Kundu, S., Khare, D., Mondal, A., 2017. Past, present and future land use changes and their impact on water balance. J. Environ. Manag. 197, 582–596. https://doi.org/10.1016/j.jenvman.2017.04.018.
Li, Y., Wang, C., 2009. Impacts of urbanization on surface runoff of the Dardenne Creek Watershed, St. Charles County, Missouri. Phys. Geogr. 30 (6), 556–573. https://doi.org/10.2747/0272-3646.30.6.556.
Liu, X., Li, X., Shi, X., Wu, S., Liu, T., 2008. Simulating complex urban development using kernel-based non-linear cellular automata. Ecol. Model. 211 (1–2), 169–181. https://doi.org/10.1016/j.ecolmodel.2007.08.0.
Mao, D., Cherkauer, K.A., 2009. Impacts of land-use change on hydrologic responses in the Great Lakes region. J. Hydrol. 374 (1–2), 71–82. https://doi.org/10.1016/j.jhydrol.2009.06.016.
Miller, S., Kepner, W., Mehaffey, M., Hernandez, M., Miller, R., Goodrich, D., et al., 2002. Integrating landscape assessment and hydrologic modeling for land cover change analysis. J. Am. Water Resour. Assoc. 38 (4), 915–929. https://doi.org/10.1111/j.1752-1688.2002.tb05534.x.
Moriasi, D.N., Arnold, J.G., Van Liew, M.W., Bingner, R.L., Harmel, R.D., Veith, T.L., 2007. Model evaluation guidelines for systematic quantification of accuracy in watershed simulations. Trans. ASABE 50 (3), 885–900. https://doi.org/10.13031/2013.23153.

Narsimlu, B., Gosain, A.K., Chahar, B.R., Singh, S.K., Srivastava, P.K., 2015. SWAT model calibration and uncertainty analysis for streamflow prediction in the Kunwari River basin, India, using sequential uncertainty fitting. Environ. Process. https://doi.org/10.1007/s40710-015-0064-8.

Nash, J.E., Sutcliffe, J., 1970. River flow forecasting through conceptual models: part I—a discussion of principles. J. Hydrol. 10, 282–290.

National Water Mission, 2017. Palar Report. National Water Mission, New Delhi.

Ndulue, E.L., Mbajiorgu, C.C., Ugwu, S.N., Ogwo, V., Ogbu, K.N., 2015. Assessment of land use/cover impacts on runoff and sediment yield using hydrologic models: a review. J. Ecol. Nat. Environ. 7 (2), 46–55. https://doi.org/10.5897/jene2014.0482.

Ndulue, E.L., Mbajiorgu, C.C., Ugwu, S.N., Ogwo, V., Ogbu, K.N., 2018. Hydrological modeling of upper Ebonyi watershed using the SWAT model. Int. J. Hydrol. Sci. Technol. 8 (2), 120–133.

Palmate, S.S., Pandey, A., 2021. Effectiveness of best management practices on dependable flows in a river basin using hydrological SWAT model. In: Water Management and Water Governance. vol. 96. Springer, Cham, pp. 335–348, https://doi.org/10.1007/978-3-030-58051-3_22.

Palmate, S.S., Pandey, A., Pandey, R.P., Mishra, S.K., 2021. Assessing the land degradation and greening response to changes in hydro-climatic variables using a conceptual framework: a case-study in Central India. Land Degrad. Dev. 32 (14), 4132–4148. https://doi.org/10.1002/ldr.4014.

Pandey, A., Palmate, S.S., 2018. Assessments of spatial land cover dynamic hotspots employing MODIS time-series datasets in the Ken River Basin of Central India. Arab. J. Geosci. 11 (17), 1–8. https://doi.org/10.1007/s12517-018-3812-z.

Pandey, A., Palmate, S.S., 2019. Assessing future water–sediment interaction and critical area prioritization at sub-watershed level for sustainable management. Paddy Water Environ. 17 (3), 373–382. https://doi.org/10.1007/s10333-019-00732-3.

Patel, D.P., Nandhakumar, N., 2016. Runoff potential estimation of Anjana khadi watershed using SWAT model in the part of lower Tapi Basin, West India. Sustain. Water Resour. Manag. 2 (1), 103–118. https://doi.org/10.1007/s40899-015-0042-3.

Piao, S., Friedlingstein, P., Ciais, P., de Noblet-Ducoudre, N., Labat, D., Zaehle, S., 2007. Changes in climate and land use have a larger direct impact than rising CO_2 on global river runoff trends. Proc. Natl. Acad. Sci. 104 (39), 15242–15247. https://doi.org/10.1073/pnas.0707213104.

Power.larc.nasa.gov, 2021. NASA POWER, Data Access Viewer. Available at: https://power.larc.nasa.gov/data-access-viewer/. (Accessed 25 July 2021).

Resmi, M., Achyuthan, H., Jaiswal, M.K., 2016. Middle to late holocene paleochannels and migration of the PalarRiver, Tamil Nadu: implications of neotectonic activity. Quat. Int. https://doi.org/10.1016/j.quaint.2016.05.002.

Resmi, M.R., Babeesh, C., Achyuthan, H., 2019. Quantitative analysis of the drainage and morphometric characteristics of the Palar River basin, Southern Peninsular India; using bAd calculator (bearing azimuth and drainage) and GIS. Geol. Ecol. Landscapes 3 (4), 295–307. https://doi.org/10.1080/24749508.2018.1563750.

Rostamian, R., Jaleh, A., Afyuni, M., Mousavi, S., Heidarpour, M., Jalalian, A., Abbaspour, K., 2008. Application of a SWAT model for estimating runoff and sediment in two mountainous basins in Central Iran. Hydrol. Sci. J. 53 (5), 977–988. https://doi.org/10.1623/hysj.53.5.977.

Roy, P.S., Meiyappan, P., Joshi, P.K., Kale, M.P., Srivastav, V.K., Srivasatava, S.K., Behera, M.D., Roy, A., Sharma, Y., Ramachandran, R.M., Bhavani, P., Jain, A.K., Krishnamurthy, Y.V.N., 2016. Decadal Land Use and Land Cover Classifications across India, 1985, 1995, 2005. ORNL DAAC, Oak Ridge, Tennessee, USA, https://doi.org/10.3334/ORNLDAAC/1336.

Rwanga, S., Ndambuki, J., 2017. Accuracy assessment of land use/land cover classification using remote sensing and GIS. Int. J. Geosci. 8, 611–622. https://doi.org/10.4236/ijg.2017.84033.

Salmoral, G., Willaarts, B.A., Troch, P.A., Garrido, A., 2015. Drivers influencing streamflow changes in the Upper Turia basin, Spain. Sci. Total Environ. 503-504, 258–268. https://doi.org/10.1016/j.scitotenv.2014.07.0.

Samal, R.D., Gedam, S., 2021. Assessing the impacts of land use and land cover change on water resources in the upper Bhima River basin, India. Environ. Challenges 5, 100251. https://doi.org/10.1016/j.envc.2021.100251.

Scanlon, B.R., Jolly, I., Sophocleous, M., Zhang, L., 2007. Global impacts of conversions from natural to agricultural ecosystems on water resources: quantity versus quality. Water Resour. Res. 43 (3). https://doi.org/10.1029/2006wr005486.

Schilling, K.E., Wolter, C.F., 2009. Modeling nitrate-nitrogen load reduction strategies for the Des Moines River, Iowa using SWAT. Environ. Manag. 44 (4), 671–682. https://doi.org/10.1007/s00267-009-9364-y.

Senthilkumar, M., Elango, L., 2004. Three-dimensional mathematical model to simulate groundwater flow in the lower Palar River basin, southern India. Hydrogeol. J. 12 (2), 197–208. https://doi.org/10.1007/S10040-003-0294-0.

Setegn, G., Srinivasan, R., Melesse, A.M., Dargahi, B., 2009. SWAT model application and prediction uncertainty analysis in the Lake Tana Basin, Ethiopia. Hydrol. Process. 24, 357–367. https://doi.org/10.1002/hyp.7457.

Sherwood, S., Fu, Q., 2014. A drier future. Science 343, 737–739.

Singh, S.K., Laari, P.B., Mustak, S., Srivastava, P.K., Szabó, S., 2017. Modeling of land use land cover change using earth observation data-sets of Tons River Basin, Madhya Pradesh, India. Geocarto Int., 1–21. https://doi.org/10.1080/10106049.2017.1343390.

Singh, V., Bankar, N., Salunkhe, S., Bera, A.K., Sharma, J.R., 2013. Hydrological stream flow modeling on Tungabhadra catchment: parameterization and uncertainty analysis using SWAT CUP. Curr. Sci. 104 (9), 1187–1199.

Sinha, R.K., Eldho, T.I., 2018. Effects of historical and projected land use/cover change on runoff and sediment yield in the Netravati river basin, Western Ghats, India. Environ. Earth Sci. 77 (3). https://doi.org/10.1007/s12665-018-7317-6.

Sun, Z., Guo, H., Li, X., Huang, Q., Zhang, D., 2011. Effect of LULC change on surface runoff in urbanization area. In: Proceedings of the ASPRS 2011 Annual Conference, Milwaukee, Wisconsin, May, pp. 1–5.

Volk, M., Liersch, S., Schmidt, G., 2009. Towards the implementation of the European water framework directive? Land Use Policy 26 (3), 580–588. https://doi.org/10.1016/j.landusepol.2008.08.

Vorosmarty, C.J., 2000. Global water resources: vulnerability from climate change and population growth. Science 289 (5477), 284–288. https://doi.org/10.1126/science.289.5477.284.

Wang, R., Kalin, L., Kuang, W., Tian, H., 2013. Individual and combined effects of land use/cover and climate change on Wolf Bay watershed streamflow in southern Alabama. Hydrol. Process. 28 (22), 5530–5546. https://doi.org/10.1002/hyp.10057.

Wei, W., Chen, L., Fu, B., Huang, Z., Wu, D., Gui, L., 2007. The effect of land uses and rainfall regimes on runoff and soil erosion in the semi-arid loess hilly area, China. J. Hydrol. 335 (3–4), 247–258. https://doi.org/10.1016/j.jhydrol.2006.11.016.

Yang, J., Reichert, P., Abbaspour, K.C., Xia, J., Yang, H., 2008. Comparing uncertainty analysis techniques for a SWAT application to the Chaohe Basin in China. J. Hydrol. 358, 1–23.

Yin, J., He, F., Xiong, Y.J., Qiu, G.Y., 2017. Effects of land use/land cover and climate changes on surface runoff in a semi-humid and semiarid transition zone in Northwest China. Hydrol. Earth Syst. Sci. 21 (1), 183.

SECTION IV

Advances in hydroinformatics mitigation

CHAPTER

15

Random vector functional link network based on variational mode decomposition for predicting river water turbidity

Salim Heddam[a], *Sungwon Kim*[b], *Ahmed Elbeltagi*[c], *and Ozgur Kisi*[d,e]

[a]Faculty of Science, Agronomy Department, Hydraulics Division, Laboratory of Research in Biodiversity Interaction Ecosystem and Biotechnology, University 20 Août 1955, Skikda, Algeria [b]Department of Railroad Construction and Safety Engineering, Dongyang University, Yeongju, Republic of Korea [c]Agricultural Engineering Department, Faculty of Agriculture, Mansoura University, Mansoura, Egypt [d]Civil Engineering Department, Ilia State University, Tbilisi, Georgia [e]Department of Civil Engineering, Technical University of Lübeck, Lübeck, Germany

OUTLINE

1 Introduction 246

2 Materials and methods 248

2.1 *Study site and data used* 248

2.2 *Performance assessment of the models* 250

2.3 *Methodology* 250

3 Results and discussion 255

4 Conclusions 259

5 Recommendations 261

Acknowledgments 262

References 262

https://doi.org/10.1016/B978-0-323-91910-4.00015-7

1 Introduction

The need of maintaining high water quality for the freshwater ecosystems has been well documented and it has become very crucial (Potes et al., 2012). Thus, control, monitoring and management of freshwater quality have received high importance during the last few years. Among the large number of water quality variables, water turbidity (TU) and suspended sediment concentration (SS) play a crucial role in the control and regulation of several aquatics operations, and they directly affect several physical, chemical and biological processes of water ecosystems (Liu et al., 2021; Campos and Pedrollo, 2021). Classified as a physical water quality variable, water TU reflect the capability of the water to absorb the light and it affect the spread of underwater light (Ding et al., 2021), and it is used as indicator of sediment resuspension (Zhao et al., 2021).

Modeling river water TU using machine learning models have been widely reported in the literature and several kind of models have been applied based on the combination of several kind of input variables. The structures of the previous studies related to river TU highlighting the input and output variables are summarized in Table 1 and all models are discussed below. Ding et al. (2021) developed a model for linking wind speed (U_2) to river water TU using data collected at Lake Taihu, China. The authors compared between three models namely, the linear regression model (LR), the decision tree regression model (DTR), and the support vector machine regression model (SVR). Using the root mean square error (RMSE) and the coefficient of determination (R^2) for models evaluation, the performances of the models varied significantly from one model to another, and the predicted TU was poorly fitted to the measured data, in addition, the LR model was slightly better than the two others models. Linking river discharge (Q) to the river TU was reported in the study of Wang et al. (2021a). Using data collected at the Esopus Creek watershed in New York, USA, the authors compared between several regression models namely, quadratic rating curves (QRC), quantile regression (QR), local regression (LR), dynamic linear models (DLMs), and Box-Jenkins (B-J) models for

TABLE 1 Variables considered by previous researchers for river water turbidity modeling.

N°	Authors	Inputs	Output
01	Ding et al. (2021)	U_2	*TU* (NTU)
02	Wang et al. (2021a)	Q	*TU* (NTU)
03	Zounemat-Kermani et al. (2021)	Q, SS, pH, DO, SC	*TU* (FNU)
04	Eze et al. (2021)	pH, Tw, DO	*TU* (NTU)
05	Villamil et al. (2021)	Turbid Image	*TU* (NTU)
06	Zhang et al. (2021a)	U_2, U_d, T_a, R_{ai}	*TU* (NTU)
07	Wang et al. (2021b)	WL, H_S, T_p	*TU* (NTU)
08	Liu and Wang (2019)	Landsat 8: band 2, 3, 4 and 5	*TU* (NTU)
09	Teixeira et al. (2020)	R_{ai}, Q and A	*TU* (NTU)
09	Alizadeh et al. (2018)	Q and *TU at previous lags time*	*TU* (NTU)

forecasting river TU up to seven day in advance. According to the obtained results, the DLM model was more accurate and provided robust forecasting accuracies compared to the others models for both low and high turbidity values. Zounemat-Kermani et al. (2021) compared between several machine learning models for predicting river TU namely, the multilayer perceptron neural network (MLPNN), classification and regression tree (CART), group method of data handling (GMDH), online sequential extreme learning machine (OS-ELM), and the response surface method (RSM) models. The models were compared according to three scenarios: (i) using only river Q, (ii) using Q and suspended sediment (SS), and (iii) using the pH, Q, SS, dissolved oxygen (DO), and specific conductance (SC), and their performances were compared base on the R^2, Nash-Sutcliffe efficiency (NSE) and RMSE values. For the first scenario using only the Q, the MLPNN was more accurate exhibiting the high R^2, NSE and lowest RMSE values of 0.636, 0.630, and 10.665, respectively. For the second scenario, the OSELM model provided the best predictive accuracies with R^2, NSE and lowest RMSE values of 0.936, 0.936, and 4.431, respectively. Similarly, for the third scenario the OSELM surpasses the others models and they provided the high R^2, NSE values of 0.916, 0.921, respectively. Eze et al. (2021) introduced a new modeling framework for better prediction of river water TU using double stage modeling strategy. First, the authors used the ensemble empirical mode decomposition (EEMD) for decomposing the river TU and water quality variables, i.e., water pH, DO and water temperature (T_w) into several subcomponent called intrinsic mode functions (IMFs). In the second stage, the obtained IMFs were combined and used by the long-short term memory (LSTM) deep learning model for better prediction of river TU. Comparison between LSTM-EEMD and LSTM showed the superiority of the hybrid model and usefulness of the EEMD in improving the models performances.

Recently, Villamil et al. (2021) used the least squares (LS) and least absolute shrinkage and selection operator (LASSO) for modeling river TU without water quality variables, and using a new modeling strategy based on the observed turbid images provided by cameras. It was highlighted that this new approach can be a good solution for better prediction of river water clarity. Another kind of machine learning model was recently used by Zhang et al. (2021a) for predicting water TU in China's South Tai Lake using only meteorological data, i.e., wind speed (U_2), wind direction (U_d), air temperature (T_a), and rainfall (R_{ai}). The authors applied the random forest regression (RFR) model and high accuracy was obtained with R^2 ranging from $\approx$0.89 to $\approx$0.93, and MAPE ranging from $\approx$14% to $\approx$23%, respectively. Wang et al. (2021b) compared between linear regression (LR), SVR, MLPNN and genetic programming (GP) models for modeling tidally-averaged sea surface turbidity using hourly water level (WL), significant wave height (H_S) and wave peak periods (T_p) as potential predictors. From the obtained results, the MLPNN ($R^2 \approx$.886, RMSE$\approx$10.83NTU) was more accurate than the GP ($R^2 \approx$.863, RMSE$\approx$11.85NTU), the SVR ($R^2 \approx$.856, RMSE$\approx$12.51 NTU), and more than the LR ($R^2 \approx$.779, RMSE$\approx$13.75NTU). Finally, a Bayesian modeling framework was proposed for modeling water TU across several rivers, lakes and ponds located in China. Obtained results were very promising and demonstrated that water TU can be predicted very well based on the available images from smartphone camera, and high NSE values ($\approx$0.73) was obtained.

Liu and Wang (2019) compared between multiple linear regression (MLR) and gene-expression programming (GEP) models in modeling river water TU using Landsat-8 satellite imagery, showing the superiority of the GEP model in modeling water TU. Teixeira et al. (2020) compared between fuzzy inference system (FIS) and the MLPNN in modeling river water TU using three input variables: R_{ai}, Q and the area of basin (A). It was found that the FIS model was more accurate showing a NSE value of 0.86. Alizadeh et al. (2018) investigated the effect of river Q on machines learning performances for river TU in coastal and estuarine waters prediction. They demonstrated that, the SVR, MLPNN and the extreme learning machine models were more accurate if the river Q was included as input variable.

In the present study, we propose a new modeling strategy for better prediction of river water TU using three machine learning models namely, the random vector functional link (RVFL), the generalized regression neural networks (GRNN) and the radial basis function neural network (RBFNN). The novelty of our study is that, the three models were developed using only river discharge (Q) and the variational mode decomposition (VMD) for decomposing the Q into several intrinsic mode functions (IMFs) which were used as input variables for the RVFL, GRNN and RBFNN models. The single and hybrid models were applied for hourly and daily times scale data.

2 Materials and methods

2.1 Study site and data used

In the present chapter, river water turbidity (TU) was modeled using only river water discharge (Q). The investigation was conducted using data collected at one station available at the *United States Geological Survey* (USGS) web site. The selected station was: (i) USGS 14182500 Little North Santiam River near Mehama, Marion County, Oregon, USA (Latitude 44°47′30″, Longitude 122°34′40″). Location of the Study area showing the USGS station is shown in Fig. 1. We used data measured at hourly and daily time scale. For hourly time scale, data were recorded during the period ranging from 01 January 2014 to 31 December 2014 with a total of 8221 patterns divided into training (70%: 5755) and validation (30%: 2466). Similarly, for the daily time scale, data were recorded during the period ranging from April 01, 2000, to December 31, 2015, with a total of 5241 patterns divided into training (70%: 3669) and validation (30%: 1572). The statistical parameters were calculated and reported in Table 2. Fig. 2 shows the graph changes of Q (cu.ft./s) and TU (FNU) at hourly and daily time scale.

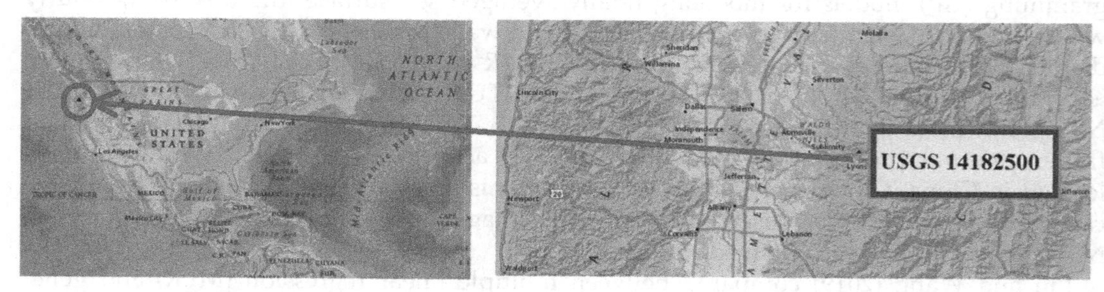

FIG. 1 Map showing the location of the USGS station North Dakota, USA.

TABLE 2 Summary statistics of river water turbidity and river discharge.

Variables	Subset	Unit	X_{mean}	X_{max}	X_{min}	S_x	C_v	R
USGS 14182500 Little North Santiam River near Mehama, Marion County, Oregon, USA. Hourly Data								
TU	Training	FNU	2.129	21.000	0.100	3.192	1.499	1.000
	Validation	FNU	2.159	21.000	0.100	3.094	1.433	1.000
	All data	FNU	2.138	21.000	0.100	3.156	1.476	1.000
Q	Training	cu.ft./s	690.531	4660.000	25.400	782.969	1.134	0.848
	Validation	cu.ft./s	711.021	4750.000	26.100	795.889	1.119	0.852
	All data	cu.ft./s	696.677	4750.000	25.400	785.775	1.128	0.849
USGS 14182500 Little North Santiam River near Mehama, Marion County, Oregon, USA. Daily Data								
TU	Training	FNU	2.020	20.800	0.100	3.072	1.521	1.000
	Validation	FNU	1.902	20.900	0.100	3.082	1.620	1.000
	All data	FNU	1.985	20.900	0.100	3.065	1.544	1.000
Q	Training	cu.ft./s	604.592	4030.000	16.400	623.488	1.031	0.759
	Validation	cu.ft./s	586.338	4330.000	16.800	609.604	1.040	0.778
	All data	cu.ft./s	599.117	4330.000	16.400	617.655	1.031	0.765

Abbreviations: X_{mean}, *mean;* X_{max}, *maximum;* X_{min}, *minimum;* S_x, *standard deviation;* C_v, *coefficient of variation;* R, *coefficient of correlation with TU; TU, water turbidity, Q, discharge; FNU, Formazin Nephelometric unit; cu.ft./s, cubic feet per second.*

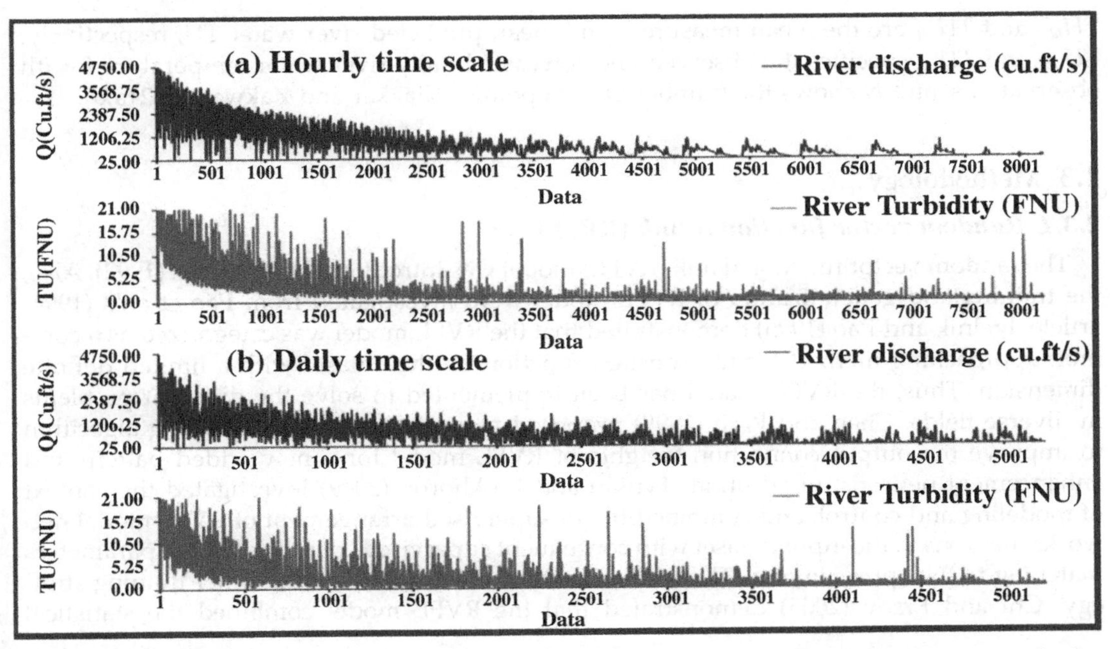

FIG. 2 Graph showing changes in river discharge (Q: cu.ft./s) and river water turbidity (TU: FNU) measured at the USGS 14182500 at little north Santiam River near Mehama, Marion County, Oregon: (A) Hourly time scale and (B) Daily time scale.

2.2 Performance assessment of the models

All models were evaluated using root mean square error (RMSE), mean absolute error (MAE), correlation coefficient (R), and Nash-Sutcliffe efficiency (NSE). Expressions are given as:

$$RMSE = \sqrt{\frac{1}{N}\sum_{i=1}^{N}\left[(TU_{obs,i}) - (TU_{est,i})_i\right]^2}, \ (0 \le RMSE < +\infty) \tag{1}$$

$$MAE = \frac{1}{N}\sum_{i=1}^{N}\left|TU_{obs,i} - TU_{est,i}\right|, \ (0 \le MAE < +\infty) \tag{2}$$

$$R = \left[\frac{\frac{1}{N}\sum_{i=1}^{N}\left(TU_{obs,i} - \overline{TU_{obs}}\right)\left(TU_{est,i} - \overline{TU_{est}}\right)}{\sqrt{\frac{1}{N}\sum_{i=1}^{n}\left(TU_{obs,i} - \overline{TU_{obs}}\right)^2}\sqrt{\frac{1}{N}\sum_{i=1}^{n}\left(TU_{est,i} - \overline{TU_{est}}\right)^2}}\right], \ (-1 < R \le +1) \tag{3}$$

$$NSE = 1 - \left[\frac{\sum_{i=1}^{N}\left[TU_{obs} - TU_{est}\right]^2}{\sum_{i=1}^{N}\left[TU_{obs,i} - \overline{TU_{obs}}\right]^2}\right], \ (-\infty < NSE \le 1) \tag{4}$$

$\overline{TU}_{obs}$ and $\overline{TU}_{est}$ are the mean measured, and mean predicted river water TU, respectively, TU_{obs} and TU_{est} specifies the observed and forecasted daily river water temperature for ith observations, and N shows the number of data points (Niazkar and Zakwan, 2021a,b).

2.3 Methodology

2.3.1 Random vector functional link (RVFL)

The random vector functional link (RVFL) model was introduced by Pao et al. (1992). Also, the training and generalization of RVFL model were investigated from Pao et al.'s (1994) article. Igelnik and Pao (1995) demonstrated that the RVFL model was categorized as a common approaching method for the constant function on the dataset with a limited definite dimension. Thus, the RVFL model has been implemented to solve the different problems in diverse fields. Chen and Wan (1999) proposed a dynamic stepwise updating algorithm to improve the output connection weights of RVFL model for a new added pattern and enhancement neurons. In addition, Tyukin and Prokhorov (2009) investigated the context of modeling and control, and combined the unsupervised arrangement of RVFL model networks' neurons to the input dataset with consequent supervised training of linear parameters matching to the approximator. The RVFL model can be combined with other training strategy. Chi and Ersoy (2005) demonstrated that the RVFL model combined the statistical hypothesis and enhancement neurons to develop the statistical self-organizing learning system. Consider the connection weights a_{ij} from the input to the enhancement neurons in the RVFL model. The activation functions $g(a_j^T x + b_j)$ are not determined perfectly when they are randomly generated. Following the approach of Alhamdoosh and Wang (2014), all connection weights are generated with a uniform distribution with [-S, +S], where S is a scale factor to be computed during the training scheme. For the RVFL model, the output connection weights β are required to solve the following formula (1).

$$t_i = d_i^T \beta, i = 1, 2, \ldots, P \tag{5}$$

where P is data number, t is target variable and d is the connection of original and random vector. Solving the formula (1) may lead to the overfitting problem. The RVFL model can be approximately classified into two categories based on the method for computing the output connection weights. First, it can be defined as the iterative RVFL model, which computes them with an iterative scheme using the function of error gradient. Second, it is the closed-pattern RVFL model, which calculates them using a single-stage (Zhang and Suganthan, 2016). A straight clarification within a single training aspect can be accomplished using the pseudo-inverse (Igelnik and Pao, 1995) among Moore-Penrose pseudo-inverse approaches, $\beta = D + T$, where D and T are the features and targets of all dataset, is frequently employed. Fig. 3 represents the typical RVFL model's strategy in this study. The input layer is converted into the enhancement layer using the corresponding neurons. The input connection weights for the enhancement neurons are developed randomly. All the input and enhancement neurons on the output layer are connected and supplied into the output neurons. The depicted description on the RVFL model's examination and employment can be tracked from the accomplished articles (Dash et al., 2018; Scardapane et al., 2016; Shi et al., 2014).

2.3.2 *Generalized regression neural networks (GRNN)*

The generalized regression neural networks (GRNN) model symbolizes a reshaped strategy of radial basis function (RBF) (Specht, 1991). The input, hidden, summation and output layers are the essential design arrangement for the GRNN model's scenario. The corresponding neurons of input, hidden, and summations are eventually connected, whereas the neuron of output layer is participated with only a few neurons matching to the summation layer with numerous summation neurons and one division neuron. The computing of each neuron matching to the output layer is evaluated based on dividing the output values from the summation neuron by the output node from division neuron in the summation layer (Tsoukalas and Uhrig, 1997; Kim and Kim, 2008).

The GRNN model can be clarified as a neural approach based on the nonlinear regression methodology. Also, the modeling procedure employing the GRNN model does not meet the regional minimal difficulty because the iteration of training process is not necessitated for the category of GRNN modeling (Sudheer et al., 2003; Kişi, 2008; Yaseen et al., 2016). The training aspect between the input and hidden layers applies the unsupervised training scheme with the particular cluster algorithms such as orthogonal least-squares and K-means, while the training aspect between the hidden and summation layers employs the supervised training

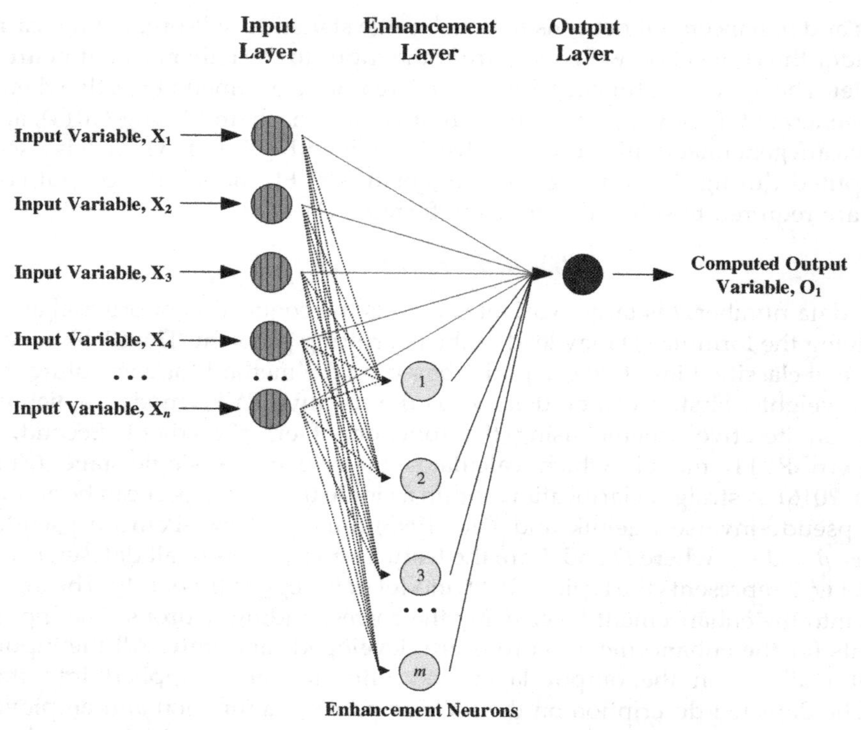

FIG. 3 The typical RVFL model's strategy.

scheme based on the mean square error's minimization. In addition, the computed optimal parameters during the training aspect can be explained as the connection weights, centers, and widths. Various researches have illustrated that the RBF is commonly appointed for the transfer function of hidden layer. Determining the parameters of RBF announces the approval of optimal centers and widths. Also, the optimal connection weights can be computed from the supervised training scheme. The detailed description on the GRNN model's development and application can be traced from the related articles (Cigizoglu and Alp, 2006; Ladlani et al., 2012; Ahmadi et al., 2019). Fig. 4 provides the typical GRNN model's strategy in this study.

2.3.3 *Radial basis function neural network (RBFNN)*

Radial basis function neural network (RBFNN) considers one of the important architectures of ANN, suggested separately by Broomhead and other researchers (Karamichailidou et al., 2021; Deng et al., 2021). RBFNs are feed-back networks having one layer for input variables, hidden and output, respectively. The input layer sends standardized input variables to the hidden layer units (Vafaei et al., 2021). Every hidden unit performs a RBF with dimension equal to the number of input variables connected with a central vector. This finding provides to a non-linear connection of the input to the hidden layer, whereas the output layer is connected linearly to the hidden layer (Karamichailidou et al., 2021). The usual construction of RBFNN is illustrated in Fig. 5. RBF networks provide major advantages in terms of speed

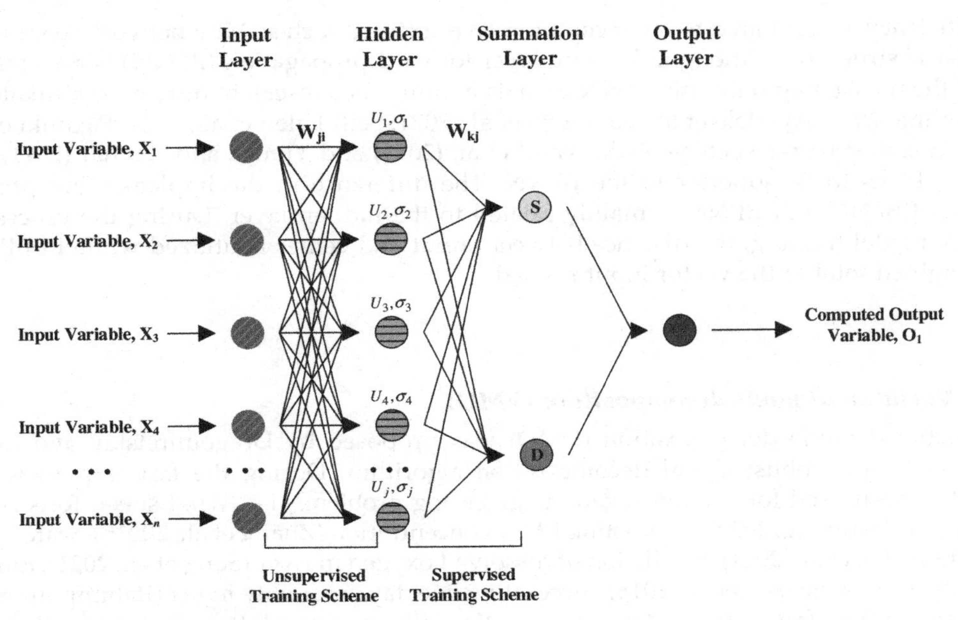

FIG. 4 The typical GRNN model's strategy.

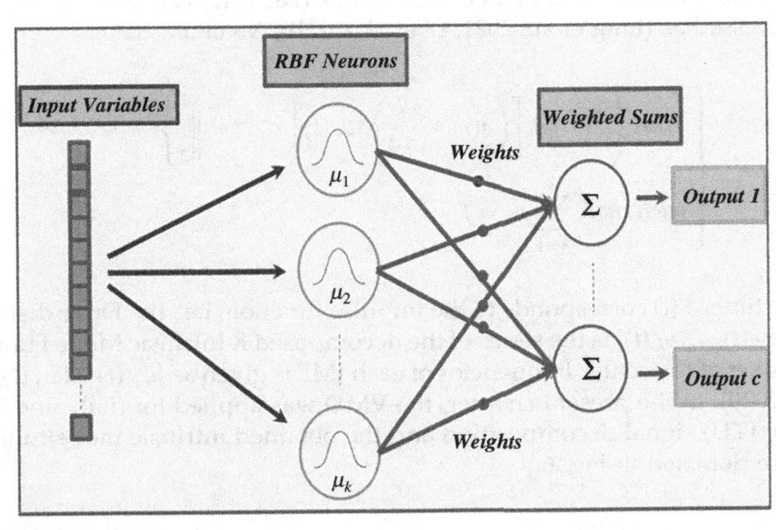

FIG. 5 The RBFNN structure.

and efficiency as compared to other common architectures such as MLP networks because of their basic structure. Although, the multilayer forward propagation (BPNN) is nonetheless one of the most commonly used ANN models in numerous research initiatives (Venkatesan and Anitha, 2006; Ayodele et al., 2021; Ojo et al., 2021; Seifi Laleh et al., 2021; Pazouki et al., 2021), in earlier works such as Shahsavand et al. (2011) and Yilmaz and Kaynar (2011), the RBFNN looks to be superior to the BPNN. The difference in the implementing process between RBFNN and BPNN is mainly related to the hidden layer. During the process of RBFNN model training, the distance between input and center is utilized while in BPNN, the weighted total of the vector input is used.

2.3.4 Variational mode decomposition (VMD)

Variational mode decomposition (VMD) was proposed by Dragomiretskiy and Zosso (2014) as a new robust signal decomposition algorithm. During the last few years, the VMD has been used for solving several engineering problems, i.e., wind speed forecasting (Yagang and Guifang, 2020), forecasting $PM_{2.5}$ concentration (Zhang et al., 2021b), water level prediction (Tiu et al., 2021), prediction of dissolved oxygen in river (Song et al., 2021), modeling daily river stage (Seo et al., 2018), forecasting one day air quality index (Rahimpour et al., 2021), streamflow forecasting (Meng et al., 2021), daily reservoir inflow forecasting (Li et al., 2021a), and for daily and monthly runoff forecasting (He et al., 2019, 2020). The VMD is used to decompose nonlinear signal into several number of band limited modes called modes (μ_k) (Ling et al., 2021) also called intrinsic mode functions (IMF). During the decomposition process, the center pulsation or frequency (ω_k) should be determined and each single mode k can be concentrated around it (Li et al., 2021b). If the river water turbidity corresponds to an approximation function $f(t)$ and it is the sum of an ensemble of k (i.e., IMF) components, the constrained problem is expressed as (Ling et al., 2021; Li et al., 2021b; Xu et al., 2020):

$$\begin{cases} \min\limits_{u_k,\,\omega_k} \left\{ \sum\limits_{k=1}^{K} \left\| \partial_t \left[\left(\delta(t) + \frac{j}{\pi t} \right) * \mu_k(t) \right] e^{-j\omega_k(t)} \right\|_2^2 \right\} \\ such.that. \sum\limits_{k=1}^{K} \mu_k = f \end{cases} \tag{6}$$

where (t) is the time, $\delta(t)$ corresponds to the impulse function, i.e., the Dirac distribution and $\{\mu_k(t) = \{\mu_1(t), \mu_2(t), \ldots, \mu_k(t)\}$ is the series of the decomposed K Intrinsic Mode Functions (IMF) modes, and the set of the center frequencies of each IMF is given as $\{\omega_k(t) = \{\omega_1(t), \omega_2(t), \ldots, \omega_k(t)\}$ (Xu et al., 2020). In the present chapter, the VMD was applied for daily and hourly river water turbidity (TU) signal decomposition and the obtained intrinsic mode functions (IMF) components are depicted in Fig. 6.

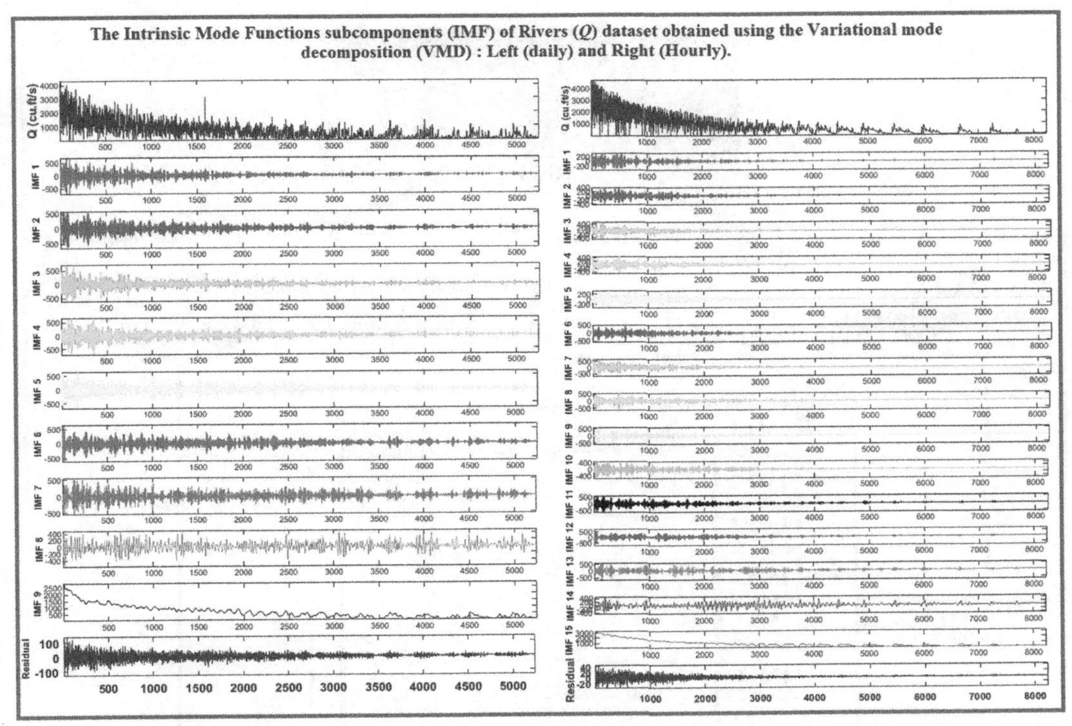

FIG. 6 the intrinsic mode functions (IMF) components of river discharge (Q) dataset decomposed by the VMD method: *left* (daily) and *right* (hourly).

3 Results and discussion

In the present chapter, we propose a modeling strategy for river water turbidity (TU) using three machines learning, i.e., radial basis function neural network (RBFNN), generalized regression neural networks (GRNN), and random vector functional link (RVFL) neural network. The proposed models were developed and compared according to two scenarios: (i) single models using only river discharge (Q), i.e., RVFL, GRNN, and RBFNN, and (ii) hybrid models using the variational mode decomposition (VMD), for which the VMD was used for decomposing the river Q into several subcomponents called the intrinsic mode functions (IMF), and used as input to the machine learning models. The flowchart of the proposed modeling approach is shown in Fig. 7. The investigation was conducted using data at hourly and daily time scale and the results are reported in Tables 3 and 4 and discussed below. Table 3 shows the models performances at hourly time scale. According to Table 3,

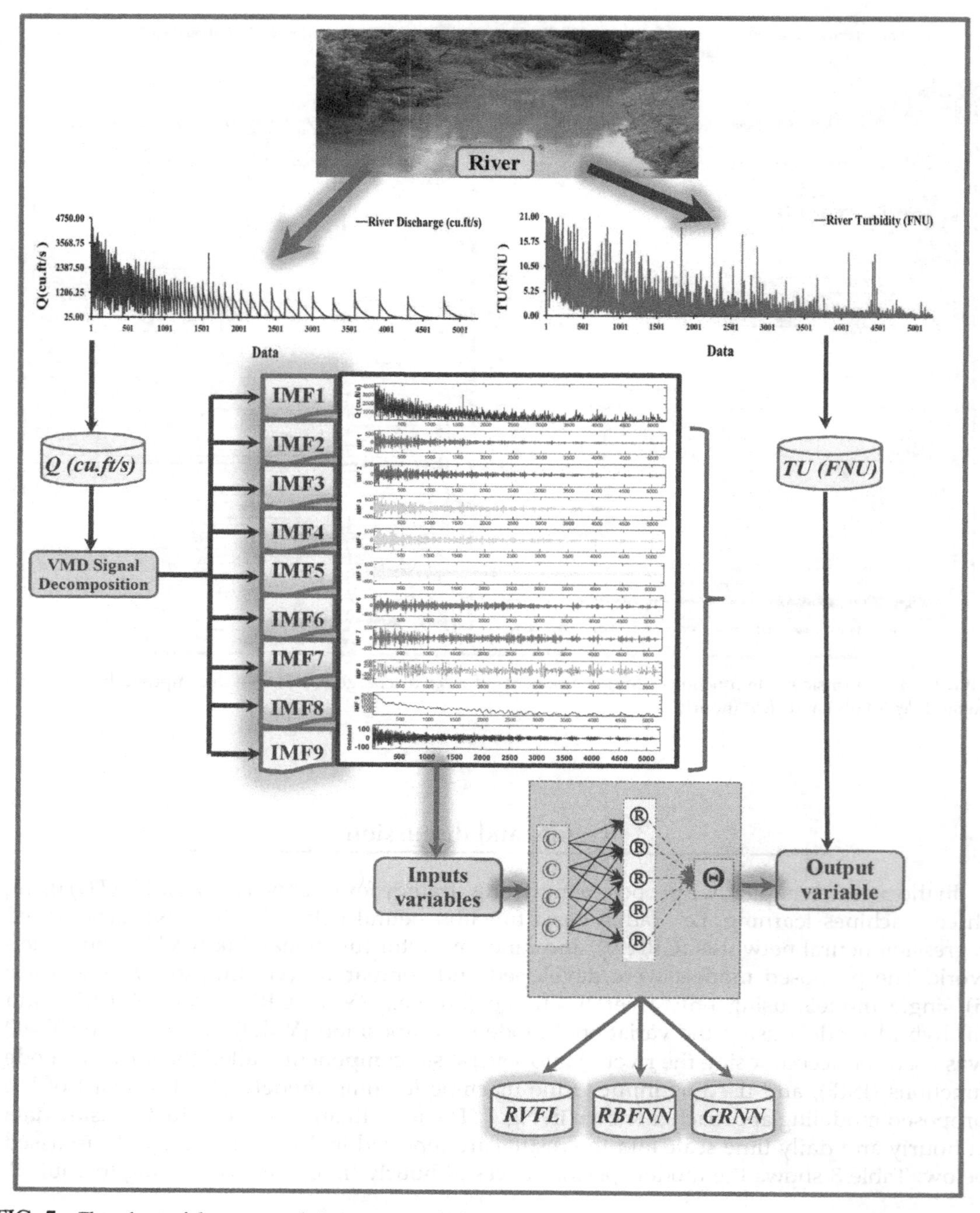

FIG. 7 Flowchart of the proposed river water turbidity modeling strategy.

TABLE 3 Performances of different river turbidity models at hourly time scale.

	Training				Validation			
Models	*R*	NSE	RMSE	MAE	*R*	NSE	RMSE	MAE
RVFL	0.873	0.761	1.554	0.747	0.870	0.756	1.516	0.742
RVFL_VMD	0.998	0.996	0.200	0.153	0.998	0.994	0.236	0.171
GRNN	0.872	0.760	1.558	0.753	0.869	0.754	1.522	0.757
GRNN_VMD	0.997	0.993	0.259	0.185	0.996	0.990	0.311	0.206
RBFNN	0.870	0.756	1.571	0.787	0.865	0.748	1.543	0.788
RBFNN_VMD	0.995	0.990	0.316	0.217	0.993	0.984	0.386	0.253

TABLE 4 Performances of different river turbidity models at daily time scale.

	Training				Validation			
Models	*R*	NSE	RMSE	MAE	*R*	NSE	RMSE	MAE
RVFL	0.779	0.607	1.917	1.055	0.747	0.532	2.083	1.007
RVFL_VMD	0.989	0.978	0.452	0.314	0.983	0.962	0.597	0.358
GRNN	0.822	0.676	1.740	0.963	0.776	0.601	1.925	1.034
GRNN_VMD	0.992	0.984	0.385	0.267	0.990	0.979	0.439	0.296
RBFNN	0.788	0.620	1.884	1.042	0.785	0.616	1.887	1.003
RBFNN_VMD	0.992	0.984	0.387	0.270	0.977	0.955	0.646	0.329

during the validation phase it is clear that, there is no clear superiority of one models compared to another for the first scenario based only on river *Q*. The three models worked equally with relatively the same performances metrics with mean R, NSE, RMSE, and MAE of $\approx$0.868, $\approx$0.753, $\approx$1.527, and $\approx$0.762, respectively, and it is clear that the RVFL model was slightly more accurate with negligible difference compared to the GRNN and RBFNN, showing an improvement ratios below $\approx$2% in terms of RMSE and MAE. In the second stage of the study, we have tried to improve the models performances using the VMD algorithm, and three models were compared, i.e., the RVFL_VMD, GRNN_VMD, and RBFNN_VMD, respectively. It is more evident that, calibrating the models using the VMD would likely lead to more predictive accuracies compared to the single models.

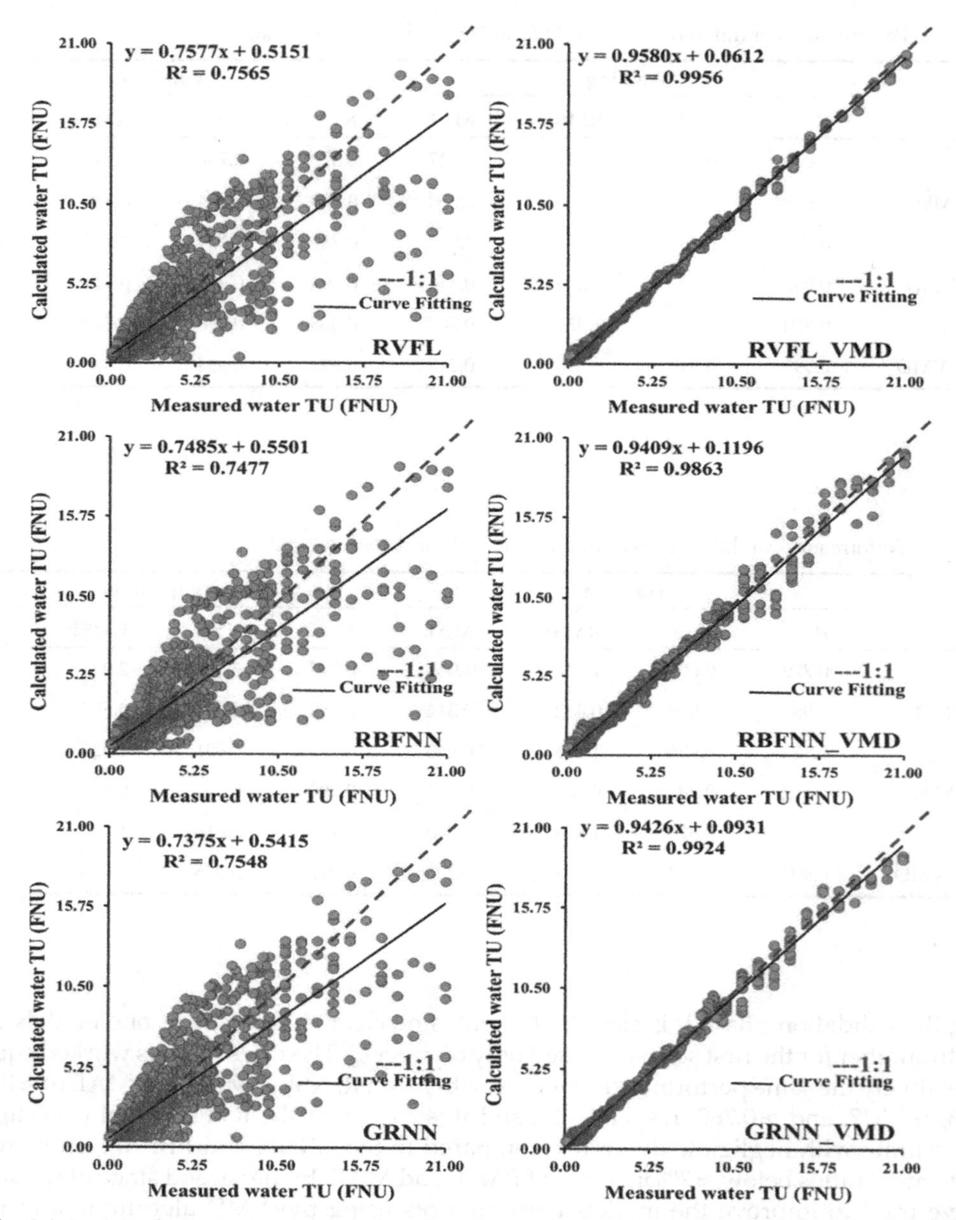

FIG. 8 Scatterplots of measured against calculated hourly river water turbidity (TU: FNU) at the USGS 14182500 for the validation stage.

The RVFL_VMD performed well in overall compared to the GRNN_VMD with RMSE=1.516, MAE=0.742, R=.998, and NSE of .994, and the predicted TU concentration was highly fitted and had high correlation with in situ measured data as depicted in Fig. 7. For comparison, the RVFL_VMD improves the performances of the RVFL by decreasing the RMSE and MAE by ≈84.43% and ≈76.95%, and increasing the R and NSE by ≈14.71% and ≈31.48%, respectively. The RVFL_VMD improves the performances of the GRNN_VMD by ≈0.20%, ≈0.40%, ≈24.11%, and ≈16.99%, in terms of R, NSE, RMSE, and MAE, respectively. In addition, the RVFL_VMD improves the performances of the RBFNN_VMD by ≈0.50%, ≈0.40%, ≈38.86%, and ≈32.41%, respectively. The scatterplots of measured against calculated river turbidity concentration at hourly time scale are depicted in Fig. 8.

Table 4 shows the models performances at daily time scale. Contrary to the results obtained at hourly time scale, the RVFL model was the poorest model and showing the lowest performances with lower R (≈0.747) and NSE (≈0.532) values. The best single model was the RBFNN having an R and NSE values of ≈0.747 and ≈0.616, respectively. Overall, using only river discharge, river turbidity was poorly to moderately predicted and the difference between the models was negligible. Using the VMD, models performances were significantly improved, showing a significant improvement in terms of R and NSE, and a remarkably decreasing in the errors metrics, i.e., the RMSE and MAE. The best accuracies were obtained using the GRNN_VMD showing the highest R (≈0.990) and NSE (≈0.979) and the lowest RMSE (≈0.439) and MAE (≈0.296). In overall, comparison between single and hybrid models it is clear that, hybrid models (i.e., RVFL_VMD, GRNN_VMD, and RBFNN_VMD) improve the performances of the single models (i.e., RVFL, GRNN, and RBFNN) by increasing the means R and NSE by ≈21.40% and 38.23%, respectively, and by achieving a remarkably decreasing rates in terms of RMSE and MAE of approximately ≈71.47% and 67.71%, respectively. The scatterplots of measured against calculated river turbidity concentration at daily time scale are depicted in Fig. 9.

4 Conclusions

The viability of random vector functional link network combined with variational mode decomposition (RVFL_VMD) was investigated in modeling hourly and daily river water turbidity. The outcomes of the RVFL_VMD was compared with the standalone RVFL, GRNN, RBFNN, GRNN_VMD and RBFNN_VMD with respect to R, NSE, RMSE and MAE assessment statistics. Hourly and daily data of rive turbidity and discharge belonging to the station USGS 14182500 situated in Marion County, Oregon were downloaded from USGS website. Only discharge variable was used as input to the models to predict 1-h and 1-day ahead turbidity. The comparison assessment provided the following conclusions:

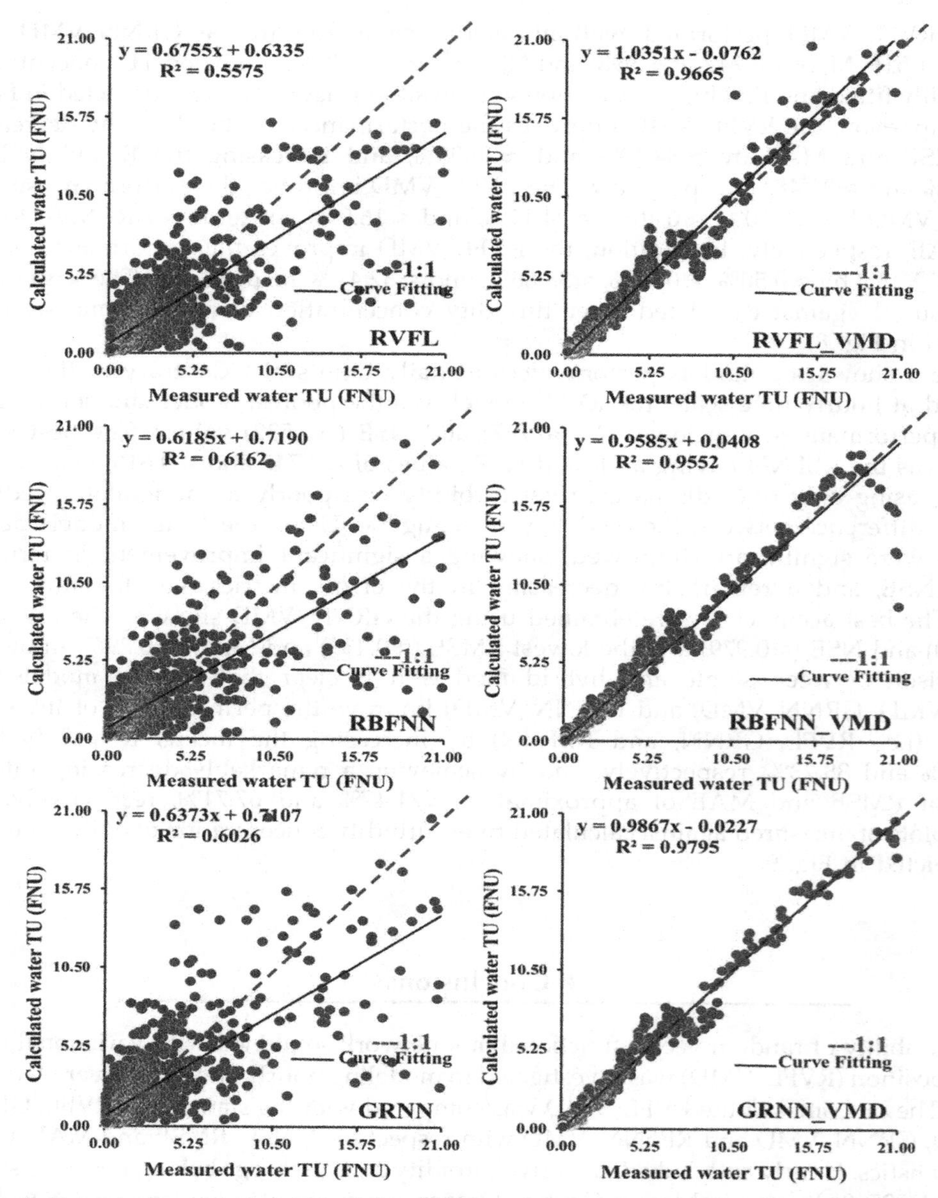

FIG. 9 Scatterplots of measured against calculated daily river water turbidity (TU: FNU) at the USGS 14182500 for the validation stage.

1. In modeling hourly river turbidity, the RVFL_VMD performed better than the RVFL, GRNN, RBFNN, GRNN_VMD and RBFNN_VMD; decrement in RMSE and MAE are 87.1% and 79.5% for the RVFL, 87.2% and 79.7% for the GRNN, 22.8% and 17.3% for the GRNN_VMD, 87.3% and 80.6% for the RBFNN, 36.7% and 29.5% for the RBFNN_VMD and increment in R and NSE are 14.3% and 30.9% for the RVFL, 14.4% and 31.1% for the GRNN, 0.1% and 0.3% for the GRNN_VMD, 14.7% and 31.7% for the RBFNN, 0.3% and 0.6% for the RBFNN_VMD in the training (simulation) stage while the corresponding improvements in R, NSE, RMSE and MAE in the testing (prediction) stage are 14.7%, 31.5%, 84.4 and 77% for the RVFL, 14.8%, 31.8%, 84.5% and 77.4% for the GRNN, 0.2%, 0.4%, 24.1%, and 17% for the GRNN_VMD, 15.4%, 32.9%, 84.7% and 78.3% for the RBFNN, 0.5%, 1%, 38.9% and 32.4% for the RBFNN_VMD, respectively. It was observed that the VMD provided considerable improvements in the models' accuracies in prediction hourly river turbidity using only discharge input in both training and testing stages. Among the standalone methods, the RVFL was ranked as the 2nd best model and followed by the GRNN.
2. The superiority of the GRNN_VMD over the other methods was observed in modeling daily river turbidity; by applying GRNN_VMD, the obtained improvements in R, NSE, RMSE and MAE are 23.3%, 62.1%, 79.9% and 74.7% for the RVFL, 0.3%, 0.6%, 14.8% and 15% for the RVFL_VMD, 20.7%, 45.6%, 77.9% and 72.3% for the GRNN, 25.9%, 58.7%, 79.6% and 74.4% for the RBFNN, 0%, 0%, 0.5% and 1.1% for the RBFNN_VMD in the training stage whereas the corresponding percentages for the testing stage are 31.6%, 80.8%, 71.3% and 64.4% for the RVFL, 26.7%, 60.1%, 69%, and 65.4% for the RVFL_VMD, 0.7%, 1.7%, 36%, and 20.9% for the GRNN, 25.2%, 56.2%, 68.4%, and 64.3% for the RBFNN, 0.6%, 0.7%, 7.6% and 8.8% for the RBFNN_VMD, respectively. In daily turbidity prediction also the VMD considerably improved the efficiency of the standalone methods in both stages of training and testing. The GRNN seems to be the best model among the standalone methods in predicting daily river turbidity and its accuracy is followed by the RVFL method.

5 Recommendations

The outcomes of the presented study recommend the use of RVFL_VMD and GRNN_VMD in modeling hourly and daily turbidity when the river discharge data are available. These methods may be beneficial for the related authority in monitoring and management of water quality. The implemented methods were only assessed by one station data and more data can be used in the future to test these methods and to derive more concrete conclusions. These methods may be compared with other machine learning methods such as deep learning or ANFIS_VMD.

Acknowledgments

This study could not have been possible without the support of the USGS data survey. The author thanks the staffs of USGS web server for providing the data that makes this research possible.

References

Ahmadi, A., Nasseri, M., Solomatine, D.P., 2019. Parametric uncertainty assessment of hydrological models: coupling UNEEC-P and a fuzzy general regression neural network. Hydrol. Sci. J. 64 (9), 1080–1094. https://doi.org/10.1080/02626667.2019.1610565.

Alhamdoosh, M., Wang, D., 2014. Fast decorrelated neural network ensembles with random weights. Inform. Sci. 264, 104–117. https://doi.org/10.1016/j.ins.2013.12.016.

Alizadeh, M.J., Kavianpour, M.R., Danesh, M., Adolf, J., Shamshirband, S., Chau, K.W., 2018. Effect of river flow on the quality of estuarine and coastal waters using machine learning models. Eng. Appl. Comput. Fluid Mech. 12 (1), 810–823. https://doi.org/10.1080/19942060.2018.1528480.

Ayodele, B.V., Mustapa, S.I., Witoon, T., Kanthasamy, R., Zwawi, M., Owabor, C.N., 2021. Radial basis function neural network model prediction of thermo-catalytic carbon dioxide oxidative coupling of methane to C2-hydrocarbon. Top. Catal. 64 (5), 328–337. https://doi.org/10.1007/s11244-020-01401-0.

Campos, J.A., Pedrollo, O.C., 2021. A regional ANN-based model to estimate suspended sediment concentrations in ungauged heterogeneous basins. Hydrol. Sci. J. 66 (7), 1222–1232. https://doi.org/10.1080/02626667.2021.1918695.

Chen, C.P., Wan, J.Z., 1999. A rapid learning and dynamic stepwise updating algorithm for flat neural networks and the application to time-series prediction. IEEE Trans. Syst. Man Cybern. B Cybern. 29 (1), 62–72. https://doi.org/10.1109/3477.740166.

Chi, H.M., Ersoy, O.K., 2005. A statistical self-organizing learning system for remote sensing classification. IEEE Trans. Geosci. Remote Sens. 43 (8), 1890–1900. https://doi.org/10.1109/TGRS.2005.851188.

Cigizoglu, H.K., Alp, M., 2006. Generalized regression neural network in modelling river sediment yield. Adv. Eng. Softw. 37 (2), 63–68. https://doi.org/10.1016/j.advengsoft.2005.05.002.

Dash, Y., Mishra, S.K., Sahany, S., Panigrahi, B.K., 2018. Indian summer monsoon rainfall prediction: a comparison of iterative and non-iterative approaches. Appl. Soft Comput. 70, 1122–1134. https://doi.org/10.1016/j.asoc.2017.08.055.

Deng, Y., Zhou, X., Shen, J., Xiao, G., Hong, H., Lin, H., et al., 2021. New methods based on back propagation (BP) and radial basis function (RBF) artificial neural networks (ANNs) for predicting the occurrence of haloketones in tap water. Sci. Total Environ. 772, 145534. https://doi.org/10.1016/j.scitotenv.2021.145534.

Ding, W., Zhao, J., Qin, B., Wu, T., Zhu, S., Li, Y., et al., 2021. Exploring and quantifying the relationship between instantaneous wind speed and turbidity in a large shallow lake: case study of Lake Taihu in China. Environ. Sci. Pollut. Res. 28 (13), 16616–16632. https://doi.org/10.1007/s11356-020-11544-y.

Dragomiretskiy, K., Zosso, D., 2014. Variational mode decomposition. IEEE Trans. Signal Process. 62 (3), 531–544. https://doi.org/10.1109/TSP.2013.2288675.

Eze, E., Halse, S., Ajmal, T., 2021. Developing a novel water quality prediction model for a south African aquaculture farm. Water 13 (13), 1782. https://doi.org/10.3390/w13131782.

He, X., Luo, J., Li, P., Zuo, G., Xie, J., 2020. A hybrid model based on variational mode decomposition and gradient boosting regression tree for monthly runoff forecasting. Water Resour. Manag. 34 (2), 865–884. https://doi.org/10.1007/s11269-020-02483-x.

He, X., Luo, J., Zuo, G., Xie, J., 2019. Daily runoff forecasting using a hybrid model based on variational mode decomposition and deep neural networks. Water Resour. Manag. 33 (4), 1571–1590. https://doi.org/10.1007/s11269-019-2183-x.

Igelnik, B., Pao, Y.H., 1995. Stochastic choice of basis functions in adaptive function approximation and the functional-link net. IEEE Trans. Neural Netw. 6 (6), 1320–1329. https://doi.org/10.1109/72.471375.

Karamichailidou, D., Kaloutsa, V., Alexandridis, A., 2021. Wind turbine power curve modeling using radial basis function neural networks and tabu search. Renew. Energy 163, 2137–2152. https://doi.org/10.1016/j.renene.2020.10.020.

Kim, S., Kim, H.S., 2008. Neural networks and genetic algorithm approach for nonlinear evaporation and evapotranspiration modeling. J. Hydrol. 351 (3–4), 299–317. https://doi.org/10.1016/j.jhydrol.2007.12.014.

Kişi, Ö., 2008. River flow forecasting and estimation using different artificial neural network techniques. Hydrol. Res. 39 (1), 27–40. https://doi.org/10.2166/nh.2008.026.

Ladlani, I., Houichi, L., Djemili, L., Heddam, S., Belouz, K., 2012. Modeling daily reference evapotranspiration (ET_0) in the north of Algeria using generalized regression neural networks (GRNN) and radial basis function neural networks (RBFNN): a comparative study. Meteorol. Atmos. Phys. 118 (3), 163–178. https://doi.org/10.1007/s00703-012-0205-9.

Li, F., Ma, G., Chen, S., Huang, W., 2021a. An ensemble modeling approach to forecast daily reservoir inflow using bidirectional long-and short-term memory (bi-LSTM), variational mode decomposition (VMD), and energy entropy method. Water Resour. Manag. 35 (9), 2941–2963. https://doi.org/10.1007/s11269-021-02879-3.

Li, R., Chen, X., Balezentis, T., Streimikiene, D., Niu, Z., 2021b. Multi-step least squares support vector machine modeling approach for forecasting short-term electricity demand with application. Neural Comput. Applic. 33, 301–320. https://doi.org/10.1007/s00521-020-04996-3.

Ling, Q., Zhang, Q., Zhang, J., Kong, L., Zhang, W., Zhu, L., 2021. Prediction of landslide displacement using multi-kernel extreme learning machine and maximum information coefficient based on variational mode decomposition: a case study in Shaanxi, China. Nat. Hazards, 1–22. https://doi.org/10.1007/s11069-021-04713-w.

Liu, C., Duan, P., Zhang, F., Jim, C.Y., Tan, M.L., Chan, N.W., 2021. Feasibility of the spatiotemporal fusion model in monitoring Ebinur Lake's suspended particulate matter under the missing-data scenario. Remote Sens. (Basel) 2021 (13), 3952. https://doi.org/10.3390/rs13193952.

Liu, L.W., Wang, Y.M., 2019. Modelling reservoir turbidity using Landsat 8 satellite imagery by gene expression programming. Water 11 (7), 1479. https://doi.org/10.3390/w11071479.

Meng, E., Huang, S., Huang, Q., Fang, W., Wang, H., Leng, G., Liang, H., 2021. A hybrid VMD-SVM model for practical streamflow prediction using an innovative input selection framework. Water Resour. Manag. 35 (4), 1321–1337. https://doi.org/10.1007/s11269-021-02786-7.

Niazkar, M., Zakwan, M., 2021a. Assessment of artificial intelligence models for developing single-value and loop rating curves. Complexity 2021a. https://doi.org/10.1155/2021/6627011.

Niazkar, M., Zakwan, M., 2021b. Application of MGGP, ANN, MHBMO, GRG, and linear regression for developing daily sediment rating curves. Math. Probl. Eng. 2021b. https://doi.org/10.1155/2021/8574063.

Ojo, S., Imoize, A., Alienyi, D., 2021. Radial basis function neural network path loss prediction model for LTE networks in multitransmitter signal propagation environments. Int. J. Commun. Syst. 34 (3), e4680. https://doi.org/10.1002/dac.4680.

Pao, Y.H., Park, G.H., Sobajic, D.J., 1994. Learning and generalization characteristics of the random vector functional-link net. Neurocomputing 6 (2), 163–180. https://doi.org/10.1016/0925-2312(94)90053-1.

Pao, Y.H., Phillips, S.M., Sobajic, D.J., 1992. Neural-net computing and the intelligent control of systems. Int. J. Control. 56 (2), 263–289. https://doi.org/10.1080/00207179208934315.

Pazouki, G., Golafshani, E.M., Behnood, A., 2021. Predicting the compressive strength of self-compacting concrete containing Class F fly ash using metaheuristic radial basis function neural network. Struct. Concr. https://doi.org/10.1002/suco.202000047.

Potes, M., Costa, M.J., Salgado, R., 2012. Satellite remote sensing of water turbidity in Alqueva reservoir and implications on lake modelling. Hydrol. Earth Syst. Sci. 16 (6), 1623–1633. https://doi.org/10.5194/hess-16-1623-2012.

Rahimpour, A., Amanollahi, J., Tzanis, C.G., 2021. Air quality data series estimation based on machine learning approaches for urban environments. Air Qual. Atmos. Health 14 (2), 191–201. https://doi.org/10.1007/s11869-020-00925-4.

Scardapane, S., Comminiello, D., Scarpiniti, M., Uncini, A., 2016. A semi-supervised random vector functional-link network based on the transductive framework. Inform. Sci. 364, 156–166. https://doi.org/10.1016/j.ins.2015.07.060.

Seifi Laleh, M., Razaghi, M., Bevrani, H., 2021. Modeling optical filters based on serially coupled microring resonators using radial basis function neural network. Soft. Comput. 25, 585–598. https://doi.org/10.1007/s00500-020-05170-6.

Seo, Y., Kim, S., Singh, V.P., 2018. Comparison of different heuristic and decomposition techniques for river stage modeling. Environ. Monit. Assess. 190 (7), 1–22. https://doi.org/10.1007/s10661-018-6768-2.

Shahsavand, A., Fard, F.D., Sotoudeh, F., 2011. Application of artificial neural networks for simulation of experimental CO_2 absorption data in a packed column. J. Nat. Gas Sci. Eng. 3 (3), 518–529. https://doi.org/10.1016/j.jngse.2011.05.001.

Shi, H., Li, B., Wang, X., Wang, Y., Wong, D., 2014. Estimation of biodiesel yield using fast decorrelated neural network ensemble. J. Eng. Sci. Technol. Rev. 7 (3), 158–163. https://doi.org/10.25103/jestr.073.25.

Song, C., Yao, L., Hua, C., Ni, Q., 2021. A water quality prediction model based on variational mode decomposition and the least squares support vector machine optimized by the sparrow search algorithm (VMD-SSA-LSSVM) of the Yangtze River, China. Environ. Monit. Assess. 193 (6), 1–17. https://doi.org/10.1007/s10661-021-09127-6.

Specht, D.F., 1991. A general regression neural network. IEEE Trans. Neural Netw. 2 (6), 568–576. https://doi.org/10.1109/72.97934.

Sudheer, K.P., Gosain, A.K., Ramasastri, K.S., 2003. Estimating actual evapotranspiration from limited climatic data using neural computing technique. J. Irrig. Drain. Eng. 129 (3), 214–218. https://doi.org/10.1061/(ASCE)0733-9437(2003)129:3(214).

Teixeira, L.C., Mariani, P.P., Pedrollo, O.C., dos Reis Castro, N.M., Sari, V., 2020. Artificial neural network and fuzzy inference system models for forecasting suspended sediment and turbidity in basins at different scales. Water Resour. Manag. 34 (11), 3709–3723. https://doi.org/10.1007/s11269-020-02647-9.

Tiu, E.S.K., Huang, Y.F., Ng, J.L., Aldahoul, N., Ahmed, A.N., Elshafie, A., 2021. An evaluation of various data pre-processing techniques with machine learning models for water level prediction. Nat. Hazards, 1–33. https://doi.org/10.1007/s11069-021-04939-8.

Tsoukalas, L.H., Uhrig, R.E., 1997. Fuzzy and Neural Approaches in Engineering. John Wiley and Sons Inc., NY, USA.

Tyukin, I.Y., Prokhorov, D.V., 2009. Feasibility of random basis function approximators for modeling and control. In: IEEE Control Applications & Intelligent Control. IEEE, pp. 1391–1396, https://doi.org/10.1109/CCA.2009.5281061.

Vafaei, A., Ghaedi, A.M., Avazzadeh, Z., Kiarostami, V., Agarwal, S., Gupta, V.K., 2021. Removal of hydrochlorothiazide from molecular liquids using carbon nanotubes: radial basis function neural network modeling and culture algorithm optimization. J. Mol. Liq. 324, 114766. https://doi.org/10.1016/j.molliq.2020.114766.

Venkatesan, P., Anitha, S., 2006. Application of a radial basis function neural network for diagnosis of diabetes mellitus. Curr. Sci. 91 (9).

Villamil, J., Victorino, J., Gómez, F., 2021. The effect of mobile camera selection on the capacity to predict water turbidity. Water Sci. Technol. https://doi.org/10.2166/wst.2021.238.

Wang, K., Gelda, R.K., Mukundan, R., Steinschneider, S., 2021a. Inter-model comparison of turbidity-discharge rating curves and the implications for reservoir operations management. J. Am. Water Resour. Assoc. 57 (3), 430–448. https://doi.org/10.1111/1752-1688.12906.

Wang, Y., Chen, J., Cai, H., Yu, Q., Zhou, Z., 2021b. Predicting water turbidity in a macro-tidal coastal bay using machine learning approaches. Estuar. Coast. Shelf Sci. 252, 107276. https://doi.org/10.1016/j.ecss.2021.107276.

Xu, Z., Li, C., Yang, Y., 2020. Fault diagnosis of rolling bearing of wind turbines based on the variational mode decomposition and deep convolutional neural networks. Appl. Soft Comput. 95, 106515. https://doi.org/10.1016/j.asoc.2020.106515.

Yagang, Z., Guifang, P., 2020. A hybrid prediction model for forecasting wind energy resources. Environ. Sci. Pollut. Res. Int. 27 (16), 19428–19446. https://doi.org/10.1007/s11356-020-08452-6.

Yaseen, Z.M., Jaafar, O., Deo, R.C., Kisi, O., Adamowski, J., Quilty, J., El-Shafie, A., 2016. Stream-flow forecasting using extreme learning machines: a case study in a semi-arid region in Iraq. J. Hydrol. 542, 603–614. https://doi.org/10.1016/j.jhydrol.2016.09.035.

Yilmaz, I., Kaynar, O., 2011. Multiple regression, ANN (RBF, MLP) and ANFIS models for prediction of swell potential of clayey soils. Expert Syst. Appl. 38 (5), 5958–5966. https://doi.org/10.1016/j.eswa.2010.11.027.

Zhang, L., Suganthan, P.N., 2016. A comprehensive evaluation of random vector functional link networks. Inform. Sci. 367, 1094–1105. https://doi.org/10.1016/j.ins.2015.09.025.

Zhang, Y., Yao, X., Wu, Q., Huang, Y., Zhou, Z., Yang, J., Liu, X., 2021a. Turbidity prediction of lake-type raw water using random forest model based on meteorological data: a case study of tai lake, China. J. Environ. Manage. 290, 112657. https://doi.org/10.1016/j.jenvman.2021.112657.

Zhang, Z., Zeng, Y., Yan, K., 2021b. A hybrid deep learning technology for PM 2.5 air quality forecasting. Environ. Sci. Pollut. Res., 1–14. https://doi.org/10.1007/s11356-021-12657-8.

Zhao, J.X., Ding, W.H., Xu, S.K., Ruan, S.P., Wang, Y., Zhu, S.L., 2021. Prediction of sediment resuspension in Lake Taihu using support vector regression considering cumulative effect of wind speed. Water Sci. Eng. 14 (3), 228–236. https://doi.org/10.1016/j.wse.2021.08.002.

Zounemat-Kermani, M., Alizamir, M., Fadaee, M., Sankaran Namboothiri, A., Shiri, J., 2021. Online sequential extreme learning machine in river water quality (turbidity) prediction: a comparative study on different data mining approaches. Water Environ. J. 35 (1), 335–348. https://doi.org/10.1111/wej.12630.

CHAPTER

16

Water quality management: Development of a fuzzy-based index in hydro informatics platform

A. Krishnakumar, S.K. Aditya, K. AnoopKrishnan, Revathy Das, and K. Anju

ESSO-National Centre for Earth Science Studies (NCESS), Ministry of Earth Sciences, Government of India, Thiruvananthapuram, Kerala, India

OUTLINE

1 Introduction 265

2 Study area 267

3 Materials and methods 269

3.1 Fuzzy logic 269

3.2 Water quality index (WQI) 270

3.3 Heavy metal pollution index (HPI) 271

3.4 Development of freshwater fuzzy model for lake systems-fuzzy lake index (FLI) 272

4 Results and discussion 274

4.1 Model validation 280

5 Conclusion 282

Acknowledgments 282

References 282

1 Introduction

The rapid increase in urbanization, industrialization, population, and higher living standards has caused an ever-increasing demand for good quality water (Bouwer, 2003) worldwide. It is estimated that by the year 2025, the total water demand in India

https://doi.org/10.1016/B978-0-323-91910-4.00016-9

(1050 BCM) will be very close to the total utilizable water resources (1122 BCM) available (Central Pollution Control Board, 2015), indicating that the country may run out of water in the due course. A time has reached to take immediate measures for conserving the quality of our existing water resources and several other unexplored lakes in India to safeguard it from further deterioration. This can be achieved by regular monitoring of quality indicators of freshwater resources by evaluating several indices which can be adopted to assess the quality of water for drinking needs. The water quality indicators generally belong to any of the three broad categories: physical, chemical, and biological. The acceptability of water for its intended use wholly depends on the magnitude of these indicators (Ocampo-Duque et al., 2006).

The literature survey reveals numerous water quality studies. Horton (1965) made a pioneering attempt to study the general water quality indices, by selecting and weighting parameters (Chang et al., 2001). One of the most widely used index called as the Water Quality Index (WQI) was developed by the National Sanitation Foundation (NSF, 2005, America) with Delphi technique (Ott, 1978). Since then considerable advancements have appeared on WQI using slightly modified concepts (Suvarna and Somashekar, 1997; Heinonen and Herve, 1994; Dojlido et al., 1994). Another index which is used for assessing drinking water is the Heavy Metal Pollution Index (HPI) which is considered as a powerful tool for ranking amalgamated influence of individual heavy metal on the overall water quality (Reza and Singh, 2010b).

Nowadays artificial intelligence, neural networks, genetic algorithms, and fuzzy logic have attracted the attention of environmental scientists and have been applied to environmental issues (Chau, 2006), since they are believed to be effective in resolving the inaccuracies and ambiguity related to the traditional indices (Gharibi et al., 2012). Fuzzy logic was introduced by Zadeh (1965) and has become one of the most appropriate tools for developing environmental indices. Fuzzy logic provides a framework to model uncertainty (Semiromi et al., 2011) and enables the user to include parameters (qualitative and quantitative) with different values as the index (Gharibi et al., 2012). Fuzzy logic includes a simple, if-then rule-based approach, e.g., "If X and Y then Z" to solving a problem rather than modeling a system mathematically (Sen, 2001; Elmas, 2003).

Numerous studies have confirmed that, traditional indices by themselves contain less information than the raw data and are at times less suited to specific questions (Semiromi et al., 2011). Moreover, another demerit of indices lies in their incapability of dealing with uncertainty and subjectivity of the environmental issues (Silvert, 2000). Therefore, it is observed from the literature that a water quality index capable of revealing the true quality of a nonflowing water body is absent. Unlike flowing water bodies like rivers, a nonflowing water bodies contain unique characteristic properties. The physiochemical parameters and heavy metal concentrations are different in nonflowing water systems like ponds and lakes because of the absence of dilution and mixing from external environments. Moreover, the qualities of a nonflowing water body have a close association with the groundwater. Hence, a comprehensive model capable of predicting the quality of lakes especially for nonflowing water resources that is representative of the overall water quality in terms of physiochemical parameters and heavy metal influence is needed for the accurate estimation of drinking water quality of lakes and ponds. Therefore, in the present study, an endeavor has been made to propose a new water quality model, to be computed with WQI and

HPI using an artificial intelligence interface based on fuzzy logic and fuzzy inference tools. This type of modeling is considered to be the first attempt in water quality analysis as per the available literature.

In this method, membership functions of the quality indices and fuzzy rule bases are defined and a fuzzy logic model for freshwater resources is generated in fuzzy logic toolbox of MATLAB package. To assess the performance of the proposed model under actual conditions, three major inland water bodies of Kerala state, located in SW side of India were chosen. The waterbodies were selected in such a way that they should have a direct interaction with ground water since quality of surface water in an inland waterbody has a profound influence on the groundwater table and groundwater quality of nearby aquifers (Ravikumar et al., 2013).

2 Study area

Kerala state, located in the extreme south region of Indian peninsula is well known for its abundant natural resources, especially surface water resources. Among these, the freshwater lakes have significant roles in ecological sustainability, flood control, groundwater recharge, pollution assimilation, etc. The Vellayani, Sasthamkotta and Pookot lakes are the three prominent freshwater lakes located in different terrains of Kerala state and are depended for various human needs.

Vellayani Lake located in Southern Kerala lies between 8°24′90″ and 8°26′30″ North latitude and 76°59′8″ and 76°59′47″ East longitudes in Thiruvananthapuram district with an altitude of 8m above sea level near to the coast. The water spread area of the lake is 5.5 km^2 and is fed by the Karamana river. The lake is situated in the low land terrain of the state with a depth of 2 to 6m. The average rainfall received in the lake is 1600mm and the temperature ranges between 25°C and 35°C. The lake is oriented parallel to the coast line and is surrounded by steep to moderately steep hills. The lake serves drinking water and irrigational purposes of Venganoor, Kalliyoor, and Vizhinjam Gram Panchayats.

Sasthamkotta lake known as the queen of lakes in Kollam district is the largest freshwater lake in Kerala. It is a RAMSAR designated wetland, located in the midland terrains of the state. The lake lies between 9°1′ and 9°4′30″N latitudes and 76°35′ and 76°40′E longitudes. The mean elevation of the lake is 33m above sea level with a water spread area of 3.75 km^2 and depth 13.9m. The mean rainfall is 2350mm with a temperature range of 25°C to 30°C. The lake is surrounded by hills on all sides except south. The lake is an important drinking water source for adjacent villages and Kollam town for around ten lakh people.

Pookot lake is an elliptical shaped freshwater basin situated at a high inter montane region of 770m above sea level between 11°32′33″ and N latitude and 76°1′38″E longitude surrounded by charnockite hills. The lake is a perennial freshwater water body in Wayanad district. The region experiences salubrious climate of 7°C to 30°C with a mean rainfall of 2786mm which is 70% of the annual rainfall. Two streams of Kabini drain into this lake. The water spread area of the lake is 0.085 km^2 with a depth of 6.5m. Fig. 1 represents the location map of the study area.

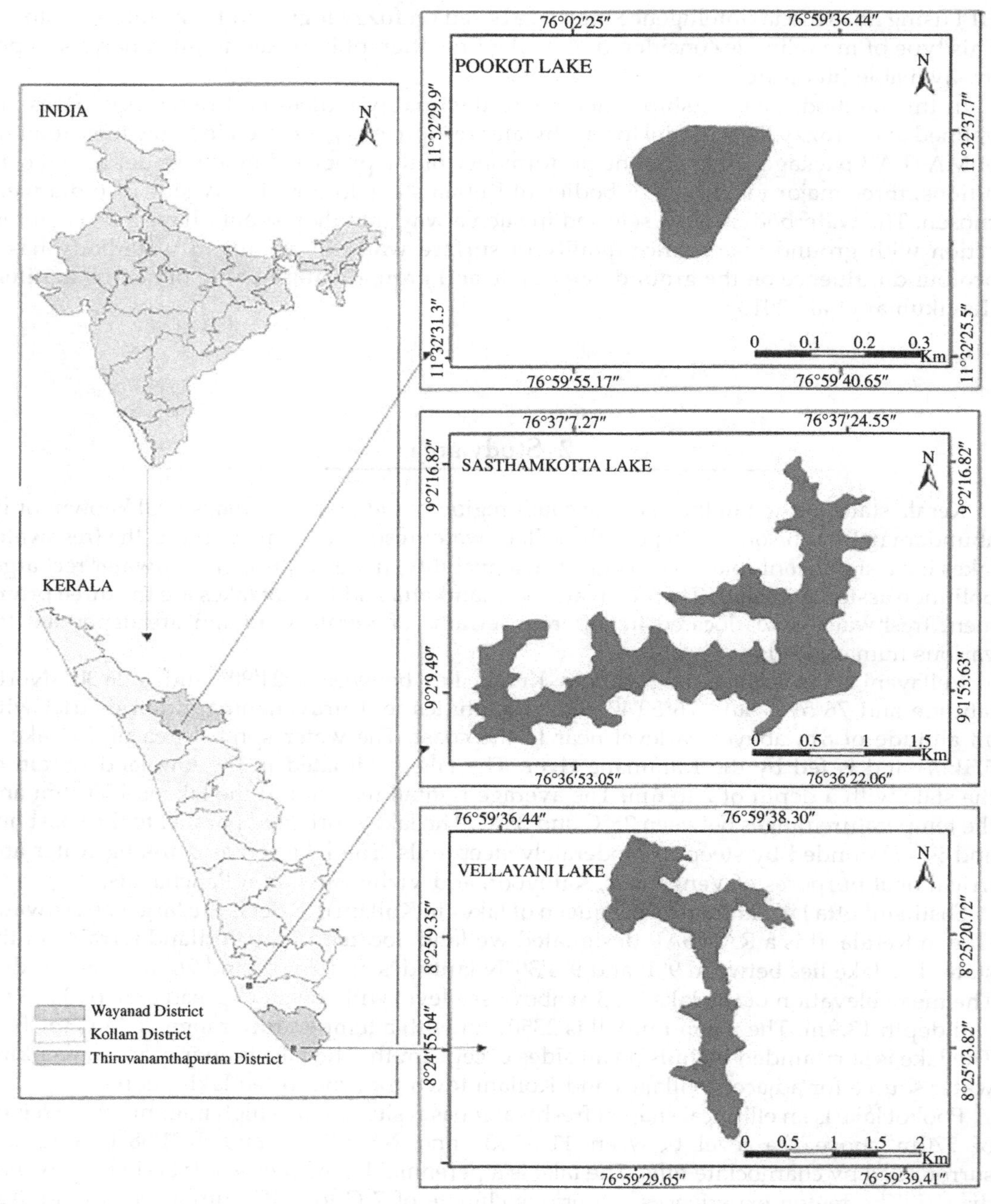

FIG. 1 Location map of the study area.

3 Materials and methods

3.1 Fuzzy logic

Fuzzy logic was introduced by Zadeh (1965) and is built around the central concept of fuzzy set. Fuzzy set theory has been developed for modeling complex systems in uncertain and imprecise environment (Ross, 2004). It was designed to supplement the interpretation of linguistic or measured uncertainties for real world random phenomena (Chang et al., 2001). Fuzzy system has the ability to provide a transparent and clear depiction of a system with the inclusion of new ideas.

Fuzzy logic comprises usually fuzzification, evaluation of inference rules, and defuzzification of fuzzy output results. Fuzzy inference is the process of formulating the mapping from a given input determinant to an output determinant using fuzzy logic reasoning (Adullah et al., 2008). Model development on fuzzy includes: membership functions, fuzzy set operations and inference rules.

3.1.1 *Fuzzification and defuzzification*

Fuzzification is the process of breaking down a system or output into fuzzy sets (Kumar et al., 2010), while defuzzification is the process of output transformation considering overall process of fuzzy set evaluation. In fuzzification, inputs and outputs are expressed in linguistic terms so as to generate input rules in a simple manner. In defuzzification the fuzzy output is transformed into a crisp domain representing the inferred fuzzy values of linguistic output variable.

3.1.2 *Membership functions*

A membership function is a curve that defines how each point in the input space is mapped into a member value between 0 and 1 (Bai et al., 2009). Each selected input or input set has a domain called the universe of discourse that is divided into subsets which are expressed by linguistic terms. The membership function links each point in fuzzy set to a membership grade between 0 and 1. A membership function value of zero implies that the corresponding element is definitely not an element of the fuzzy set, while a value of unit means that the element fully belongs to the set (Kumar et al., 2010). The input can be represented in various forms such as trapezoidal, triangular, Gaussian etc. Triangular functions are mostly applied because of their simplicity (Pedrycz, 1994). The output axis is called the membership value μ.

If X is the universe of discourse and its elements are denoted by x then a fuzzy set A can be defined as a set of ordered pair by the following nomenclature,

$$A = \{x_1, \mu_A(x)\ |x\epsilon X\}; 0 \leq \mu_A(x) = 1$$

where $\mu_A(x)$ shows the membership function of x in fuzzy set A.

Different methods such as fuzzy clustering, expert knowledge (Turksen, 1991), genetic algorithms (Karr and Gentry, 1993) and neural networks (Jang and Sun, 1995) can be used to determine membership functions.

3.1.3 *Fuzzy set operations*

Fuzzy set operators define the relationship among the fuzzy subsets. They manage the essence of fuzzy logic. The standard fuzzy set operators are:

Intersection (AND).

AND operator denotes the intersection of two or more fuzzy sets and is defined by (Ross, 2004):

$$\text{AND}: \mu_A \cap_B(x) = \mu_A(x) \cap \mu_B(x) = \min\left(\mu_A(x), \mu_B(x)\right)$$

Union (OR)

The OR operator makes a new subset from the input subsets by uniting them. Each fuzzy set operator creates a new subset from two or more subsets. Ross (2004) defined the union operator by the following equation:

$$\text{OR}: \mu_A \cup_B(x) = \mu_A(x) \cup \mu_B(x) = \max\left(\mu_A(x), \mu_B(x)\right)$$

Negation or Additive complement (NOT).

The NOT operator makes a complement set to the input sets on which it operates and is defined as (Ross, 2004):

$$\text{NOT}: \mu_{-A}(x) = 1 - \mu_A(x)$$

Inference rules.

An inference rule represents the relationship between the input variables and the output variables by means of fuzzy if-then rules. Each rule consists of two parts: "if" part called as the antecedent proposition and the "then" part called as the consequent proposition. The general form of an inference rule is:

If x is A then z is C, where

A and C are linguistic values defined by fuzzy sets in the universes of discourse X and Z respectively. In the fuzzy inference system, the degree of fulfillment for each rule is computed and applied to the conclusion part of each rule (Adriaenssens et al., 2004). Fuzzy inference system generates the set of possible rules based on the degree of fulfillment and nonfulfillment of the rule (Rani and Elangovan, 2017).

3.2 Water quality index (WQI)

WQI is the commonly used tool to express the quality of water (to Singh et al., 2013). WQI gives an understandable insight about the feasibility of water for different purposes. In this paper, Weighted Arithmetical Index method (Brown et al., 1972) was utilized to assess the quality of water. Twelve water quality parameters were considered keeping in view that more the parameters considered for evaluation greater will be the accuracy of the result. pH, Turbidity, TDS, Bicarbonates, Total hardness, Ca, Mg, Cl, NO_3, SO_4, Dissolved oxygen and fecal coliforms were assessed, in which a specific weight was assigned to the parameter according to the influence of the factor in modifying water quality. The WQI has been calculated by using the standards of drinking water quality recommended by Bureau of Indian Standards (BIS).

Calculation of WQI involves three steps:

$$Q_i = 100\{(V_{\text{actual}} - V_{\text{ideal}})/(V_{\text{standard}} - V_{\text{ideal}})\} \quad (1)$$

(Let there be *i* water parameter and quality rating (*Qi*) corresponding to i^{th} parameter is a number reflecting the relative value of this parameter in the polluted water).

Qi = Quality rating for the ith water quality parameter.
V_{actual} = Monitored value of the ith parameter at given location.
V_{standard} = Standard permissible value of the ith parameter prescribed by BIS.
V_{ideal} = Ideal value of ith parameter in pure water. (0 is taken as the ideal for all other parameters except pH and Dissolved oxygen (7.0 and 14.6 mg/L respectively).

$$Wi = K/Si \quad (2)$$

(Unit weight was calculated by a value inversely proportional to the recommended standard value *Si* of the corresponding parameter).

Wi = Unit weight for the ith parameter,
Si = Standard value for i^{th} parameter,
K = Constant for proportionality.

The overall water quality index was calculated by aggregating the quality rating with the unit weight linearly.

$$WQI = \sum QiWi/\sum Wi \quad (3)$$

3.3 Heavy metal pollution index (HPI)

The HPI is a tool which indicates the overall quality of water like WQI method. In this index, the influence of individual heavy metal ions (Reza and Singh, 2010a; Milivojevic et al., 2016) is generally considered. Generally, the critical pollution index value for drinking water should be less than 100 as given by Prasad and Bose (2001). For the study, nine key heavy metals (Fe, Mn, Al, Zn, Pb, Cr, Cu, Cd, and Ni) were taken depending on their influence on humans. The monitored concentrations of heavy metals in ppm (mg/L) were converted to μg/L prior to calculation since HPI was computed with μg/L (Reza and Singh, 2010a; Giri and Singh, 2014).

HPI was calculated using the equation

$$HPI = \sum_{i=1}^{n}(QiWi)/\sum_{i=1}^{n} Wi \quad (4)$$

(Let there be *i* water parameter and quality rating (*Qi*) corresponding to i^{th} parameter is a number reflecting the relative value of this parameter in the polluted water).

Qi = sub index of ith parameter.
Wi = unit weightage for i^{th} parameter and n is the number of parameters considered.

$$Wi = k/Si \quad (5)$$

(Unit weight was calculated by a value inversely proportional to the recommended standard value Si of the corresponding parameter).

where,

Wi = unit weightage
Si = recommended standard for i^{th} parameter
k = constant of proportionality.

$$Qi = 100\,(V/Si) \quad (6)$$

where,

Qi = sub index of ith parameter,
Vi = monitored value of the ith parameter in μg/L
Si = standard or permissible limit for the ith parameter.

3.4 Development of freshwater fuzzy model for lake systems-fuzzy lake index (FLI)

The procedure carried out within a FIS is described in this section. Twenty quality parameters were included in the model development based on their relative importance in the overall quality and health of the water. The parameters included were pH, turbidity, TDS, Total Hardness, Ca, Mg, Cl, NO_3, SO_4, Dissolved oxygen, Fe, Mn, Al, Zn, Pb, Cr, Cu, Cd and Ni. We have hypothesized that, Water Quality Index (WQI) and Heavy Metal Pollution Index (HPI) are sufficient to evaluate water quality and hence taken as inputs. Five kinds of membership functions were defined for WQI (Fig. 2) according to the ranges set by Ramakrishnaiah et al. (2009). Similarly, the membership functions of HPI were defined based on expert's knowledge (Fig. 3). The membership functions of the fuzzy sets were defined linguistically as "excellent," "good," "poor," "very poor," and "unsuitable for drinking" fuzzy sets as inputs for WQI and "less," "moderate," "slightly high," "high," and "extreme" fuzzy sets as inputs for HPI. Each parameter as an input was assigned to one of the ten fuzzy sets in terms of membership functions. The ranges for the input variables were set considering

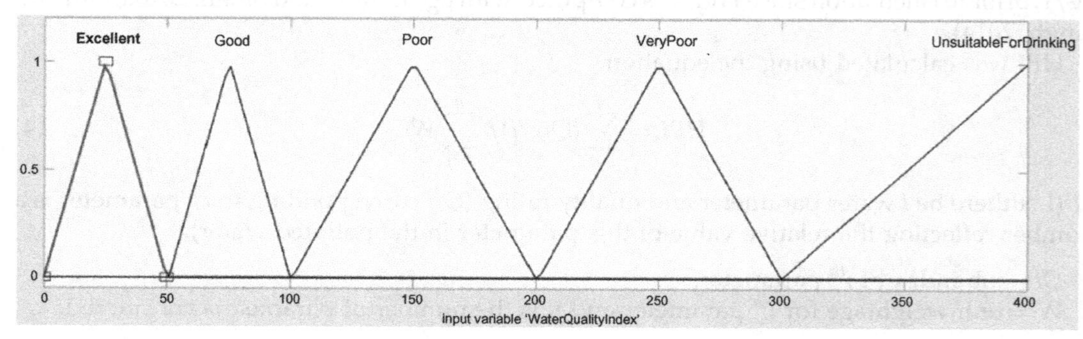

FIG. 2 Membership functions of WQI.

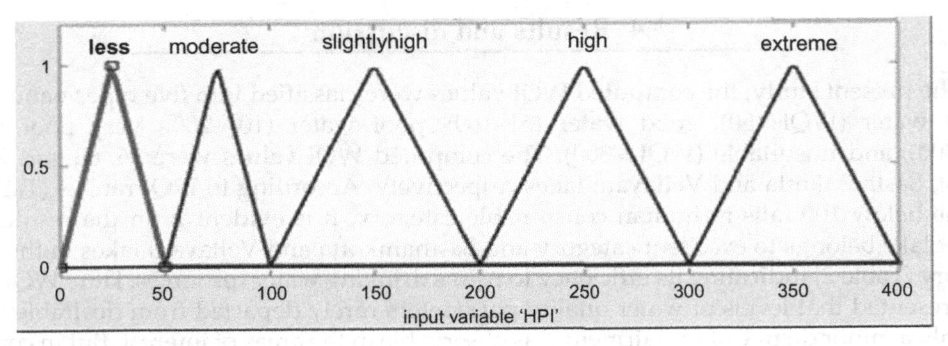

FIG. 3 Membership function of HPI.

the maximum and minimum values, an input variable can have. The output ranges of fuzzy sets were "excellent," "ideal," "slightly polluted," "moderately polluted," and "severely polluted." Regarding the ranges of the output variables, a scoring system of 1 to 10 was assigned, in which values near 10 represent excellent quality of water in drinking needs.

Fuzzy sets and rules have been constructed for implementation in water quality classification (Lee et al., 1997; Jensen et al., 2000). Expressions as the following are frequently used in water quality assessment by experts: "if the water quality is good and heavy metal pollution in the freshwater is low," then the expected quality of freshwater is "excellent." In fuzzy language, this could be enunciated as follows:

Rule 1. If WQI is excellent and HPI is less, then FWQ is Excellent.

In the same way, other rules can be enunciated like.

Rule 2. If WQI is higher and HPI is higher, then FWQ is moderately polluted.

Rule 3. If WQI is good and HPI is moderate then FWQ is ideal.

In rules with one antecedent the degree of support is the degree of membership. The degree of support for the entire rule is used to shape the output fuzzy set. We tried to cover maximum possible number of relations of WQI and HPI through inference rules. The total numbers of rules generated were 118. Centre of Gravity method (COG) was utilized for defuzzification of outputs since it is the most conventionally and physically applicable method (Gharibi et al., 2012).

Derivation of COG is based on the following equation (Ross, 2004).

$$Z = \frac{\int \mu(z) z dz}{\int \mu(z) dz} \tag{7}$$

Fuzzy inference model generation layers for the present study:

Layer 1: input layer: given inputs related to WQI and HPI are transferred to the next layer.

Layer 2: membership layer: membership functions are created and ranges for each membership is fixed.

Layer 3: inference layer: fuzzy rule set generation to integrate WQI and HPI.

Layer 4: normalization layer: combining of the overall process.

Layer 5: output layer: the output of the freshwater quality model with rules n.

4 Results and discussion

In the present study, the computed WQI values were classified into five types namely excellent water (WQI < 50), good water (51–100), poor water (101–200), very poor water (201–300), and unsuitable (WQI > 300). The computed WQI values were 46, 68 and 77 for Pookot, Sasthamkotta and Vellayani lakes respectively. According to WQI ranges (Table 1), a value below 100 falls in human consumable category. It is evident from the results that Pookot lake belongs to excellent category and Sasthamkotta and Vellayani lakes in the good category (Table 2) indicating its efficiency to use as drinking water resources. Here WQI analysis presented that levels of water quality parameters rarely departed from desirable levels and only a minor degree of impairment was observable in the areas of interest. But in contrast to the WQI, the HPI analysis unveiled the unhealthy status of lakes (Table 3). The evaluated HPI values of Vellayani, Sasthamkotta and Pookot were found exceeding the critical limit 100. This confirmed that the lakes were polluted with heavy metals. In these situations, the water quality assessment involving any one index has no relevance to explain the overall quality of water for drinking since from each index differing conclusions are drawn.

The efficacy of the developed model was checked under actual conditions taking case studies from Vellayani, Sasthamkotta and Pookot lakes. Here fuzzy logic has been used as an estimator of status of water quality generated by WQI and HPI and named as Fuzzy lake index with values ranging from 0 to 10. Accordingly, a quality category with specific ranges indicating the quality of freshwater was derived from the output functions attained (Table 4). The freshwater quality model developed in this study is entirely applicable for nonflowing water systems since this study encompasses the lacustrine systems with stagnant water conditions and hence devoid of any particular environmental issue that is being affected by the flowing of water. A nonflowing water system like lake extents over only a small area and the quality of water is unique. If the surface water quality of rivers is to be explained the parameters are different since the values differ at several places due to the human influence. Hence, separate ranges have to be fixed in order to delineate the quality of water as per the Fuzzy lake index.

A total of 118 rules were formulated in such a way that lakes with excellent WQI and less HPI will always generate an output belonging to excellent category (Fig. 4). Likewise, poor WQI and maximum HPI will result in a range belonging to severely polluted category (Fig. 5). From the generated model, the results were interpreted in 2 methods: (1) rule viewer and (2) surface viewer. A rule viewer diagram shows the whole Fuzzy Inference System (FIS)

TABLE 1 Water quality ranges.

Range	Water quality
<50	Excellent
50–100	Good water
101–200	Poor water
201–300	Very poor
>300	Water unsuitable for drinking

TABLE 2 Water Quality Index (WQI) for Vellayani, Sasthamkotta and Pookode lakes.

Parameter		pH	Turbidity	TDS	HCO_3	TH	Ca^{2+}	Mg^{2+}	Cl^-	NO_3	SO_4^{2-}	DO	
IS WQ standard value		6.5–8.5	5	500	200	200	75	30	250	45	200	6.5–8.5	**Water Quality Index =** $\frac{\sum QiWi}{\sum Wi}$
Relative weight (*Wi*) $\sum Wi = 0.556$		0.117	0.2	0.002	0.005	0.005	0.013	0.03	0.004	0.022	0.005	0.11	
Vellayani lake	Actual measured value	6.48	7.99	307.45	84.89	18.47	10.35	8.97	13.71	0.4	2.94	7.1	77
	Quality rating (*Qi*)	−34.66	158	61.49	42.44	9.23	13.8	29.9	5.48	0.88	1.47	109	
	Weighted value (*Qi.Wi*) $\sum Qi.Wi = 42$	−4.055	31.6	0.122	0.212	0.046	0.179	0.897	0.021	0.017	0.007	13.52	
Sasthamkotta lake	Actual measured value	7.28	5.51	429.33	67.48	16.84	6.83	6.13	22.27	0.25	1.88	6.86	68
	Quality rating (*Qi*)	18.66	102.2	85.86	33.74	8.42	9.10	20.43	8.90	0.55	0.94	126.88	
	Weighted value (*Qi.Wi*) $\sum Qi.Wi = 37.7$	2.18	20.44	0.17	0.16	0.04	0.11	0.61	0.03	0.01	0.0047	13.95	
Pookode lake	Actual measured value	5.86	5.90	42.66	21.49	12.19	7.32	5.21	6	0.045	0.74	6.77	46
	Quality rating (*Qi*)	−102.66	118	8.53	10.74	6.09	9.76	17.36	2.4	0.08	0.37	128.36	
	Weighted value (*Qi.Wi*) $\sum Qi.Wi = 25.44$	−12.01	22.6	0.017	0.05	0.03	0.12	0.52	0.0096	0.0017	0.0018	14.11	

TABLE 3 Heavy metal Pollution Index (HPI) for Vellayani, Sasthamkotta and Pookode lakes.

Parameter		Fe	Mn	Al	Zn	Pb	Cr	Cu	Cd	Ni	
Standard value		**300**	**100**	**30**	**5000**	**10**	**50**	**50**	**3**	**20**	**Heavy Metal Index =** $\frac{\sum QiWi}{\sum Wi}$
Relative weight (*Wi*) $\sum Wi = 0.566$		**0.003**	**0.011**	**0.033**	**0.0002**	**0.1**	**0.02**	**0.02**	**0.33**	**0.05**	
Vellayani lake	Actual measured value	261	136	138	22	8	27	12	6	8	167
	Quality rating (*Qi*)	87	136	460	0.44	80	54	24	200	40	
	Weighted value (*Qi.Wi*) $\sum Qi.Wi = 94.361$	0.261	1.36	15.18	0.000088	8	1.02	0.48	66	2	
Sasthamkotta lake	Actual measured value	353	235	209	11	15	130	150	11	11	309
	Quality rating (*Qi*)	117.6	235	696	0.22	150	260	300	366	55	
	Weighted value(*Qi.Wi*) $\sum Qi.Wi = 175.41$	0.35	2.35	22.98	0.000044	15	5.2	6	120.78	2.75	
Pookode lake	Actual measured value	235	134	137	13	16	20	14	10	10	261
	Quality rating (*Qi*)	78.33	134	456.66	0.26	160	40	28	333	50	
	Weighted value (*Qi.Wi*) $\sum Qi.Wi = 146.37$	0.234	1.34	15.04	0.000052	16	0.8	0.56	109.89	2.5	

TABLE 4 Fuzzy Lake Index derived from the model.

Ranges	Category
0–0.9	Severely polluted
1–1.9	Moderately polluted
2–3.9	Slightly polluted
4–7.9	Ideal
8–10	Excellent

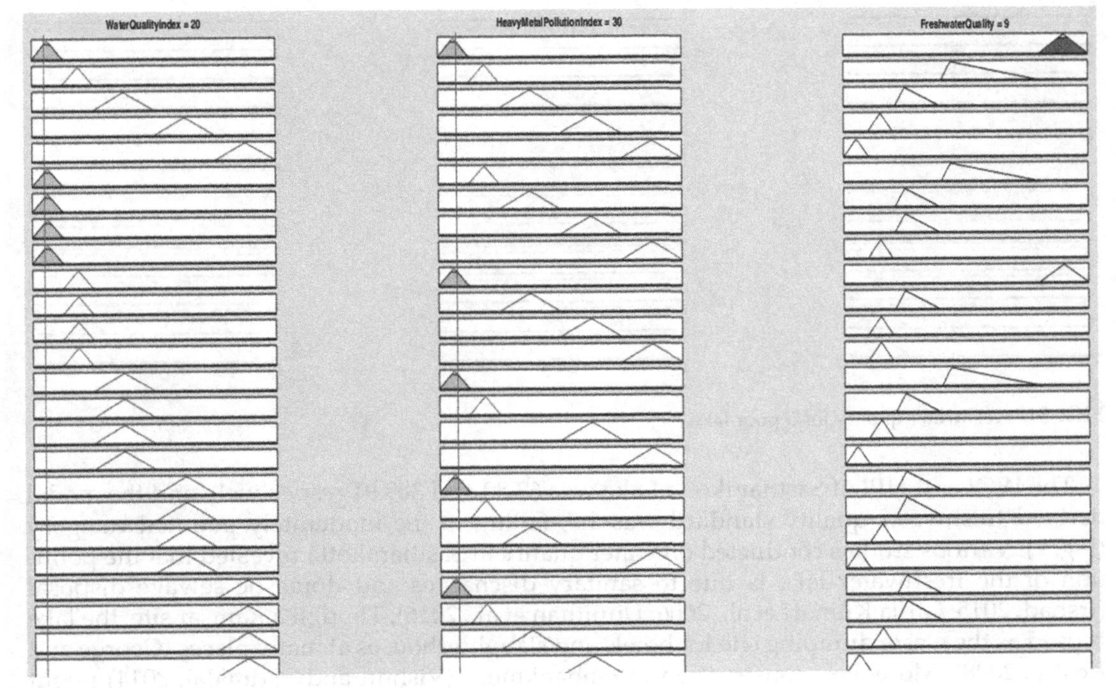

FIG. 4 Freshwater quality for an excellent lake.

processes presenting one calculation at a time in detailed manner. Each rule is a row of plots and each column is a variable. The first two column shows the membership functions, i.e., the if part of each rule (antecedent) and the last column shows the membership functions referenced by the then part of each rule (consequent).

Vellayani lake had a WQI of 77.34 and HPI of 166.71. The overall freshwater quality assessed from the proposed model was 5.52 (FLI) which falls under the ideal category (Fig. 6). Studies pointed that, if the lake is left unattended, it may get vulnerable to pollution and degradation (Dharmapalan, 2014) and hence measures have to be taken for its proper management and conservation.

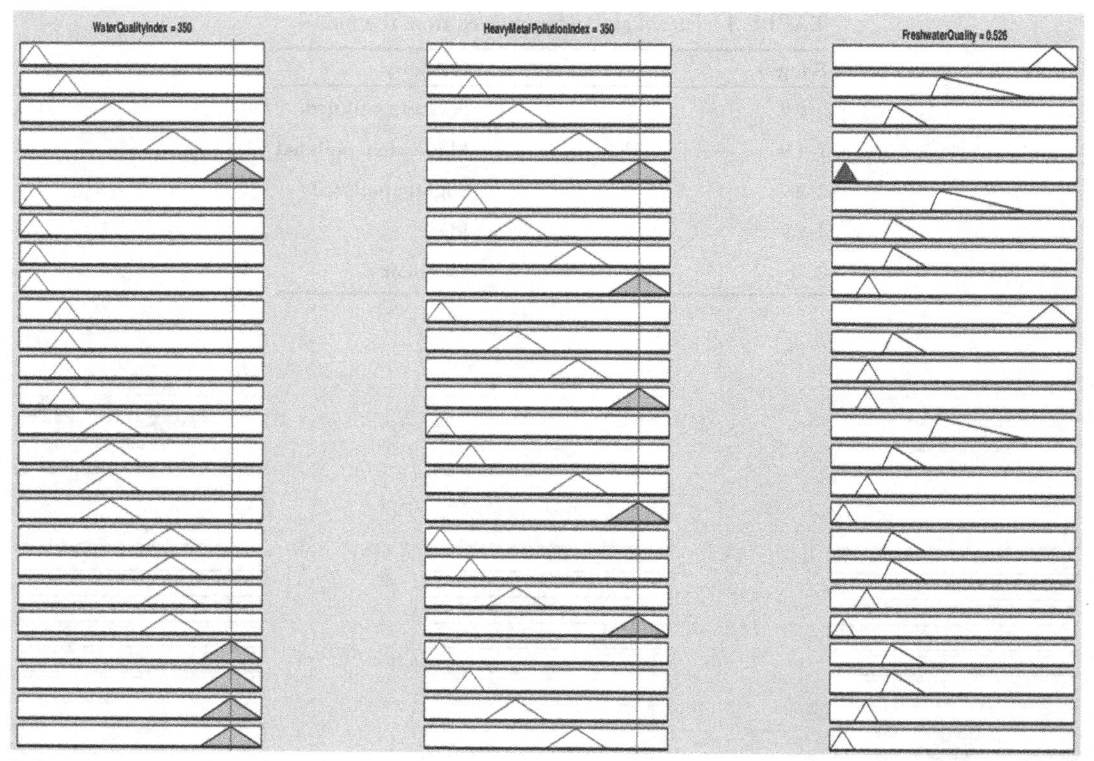

FIG. 5 Freshwater quality for a poor lake.

The WQI and HPI of Sasthamkotta Lake was 67.80 and 309.91 respectively and the model derived freshwater quality standard was 1.5, falling in the moderately polluted category (Fig. 7). Various studies conducted on water quality in Sasthamkotta revealed that the pollution of the freshwater lake is due to sanitary discharges and domestic sewage disposal (Irshad, 2015; Girija Kumari et al., 2006; Unnithan et al., 2016). Though a Ramsar site, the lake is used as the waste dumping site for hotels and slaughterhouses at many places (George and Koshy, 2008). Moreover, construction of embankment (Vishnu and Padmalal, 2014) might have isolated the lake system, accumulating heavy metals and pollution load.

According to the derived result from the model, Pookot lake (5.84) belongs to ideal category (Fig. 8). The sources of pollution in Pookot lake are minimal since a major portion of the lake is surrounded by dense forest and the rest by cash crops like coffee and tea in the hilly regions around the lake. Since other sources of heavy metal accumulation are absent, mineralization and weathering by natural processes should be taken into consideration. Also, tourism along with intense plantation activities in the catchments of the lake may have resulted in slight pollution especially by fertilizer and pesticide runoff (Veena and Achyuthan, 2012). Excessive weed growth, weak boundary conditions, deforestation, excess cultivation of plantation crops on the catchment were reported earlier by Thomas et al. (2017) which may have contributed for the degraded environmental conditions to the lake.

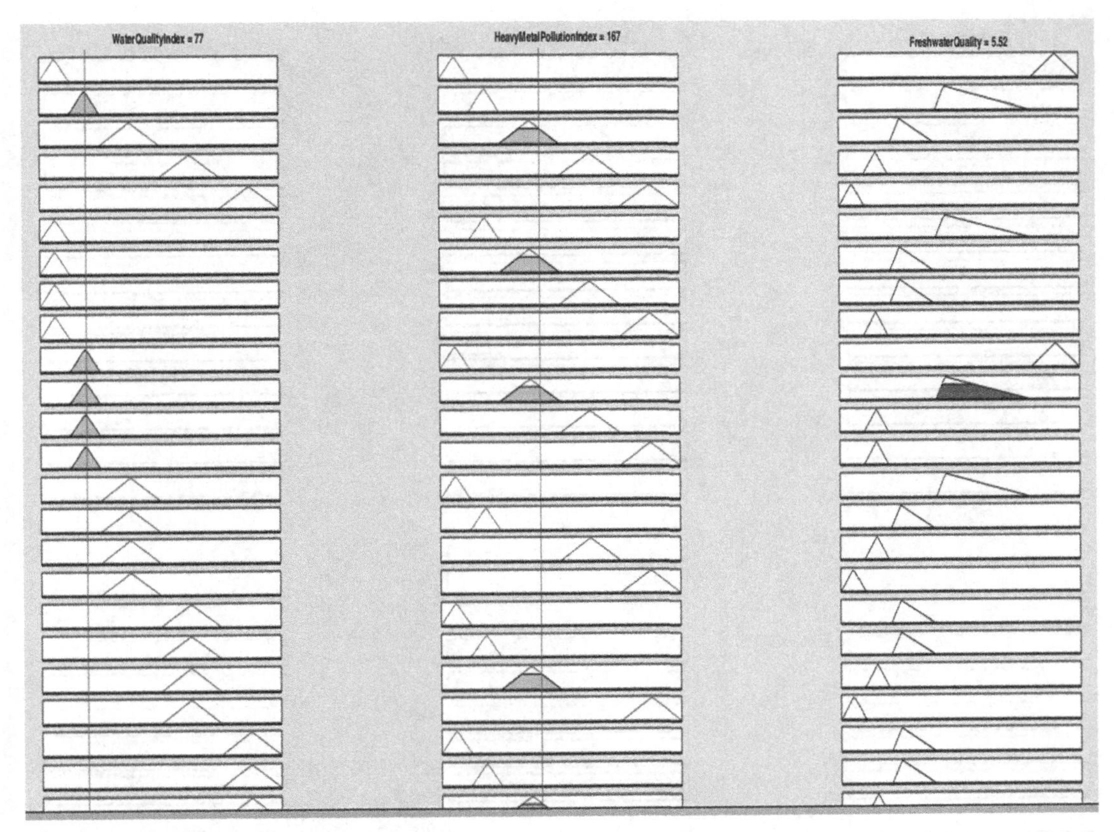

FIG. 6 Freshwater quality for Vellayani lake.

The surface viewer is a three-dimensional curve displaying the mapping from Water Quality Index and Heavy Metal Pollution Index to Freshwater Quality. Generally a surface viewer is a Graphical User Interface (GUI) that examines the output surface of a Fuzzy Inference System (FIS) from different angles. From the surface plot, it is observed that freshwater quality increases when WQI and HPI values decreases while freshwater quality decreases, when WQI and HPI value increases. Here Vellayani lake (5.58), Sasthamkotta lake (1.5) and Pookode lake (5.84) are spatially represented in the surface viewer of the freshwater quality model (Fig. 9). Likewise, the quality of others lakes can be determined by analyzing the surface viewer. Thus, an entire freshwater quality mapping of lakes based on WQI and HPI can be viewed in a single plot.

From the results obtained it is confirmed that the developed fuzzy logic model proposed in the present study produces a clear output than WQI and HPI evaluated separately. This is probably due to the fact that environmentally critical pollutants like Cd, Pb, Ni, etc. are taken into consideration in the model development since the aggregation of important parameters can lead to a relevant and confirming result. No work has hitherto been carried out by considering WQI and HPI together for analyzing the water quality. The model presented here has

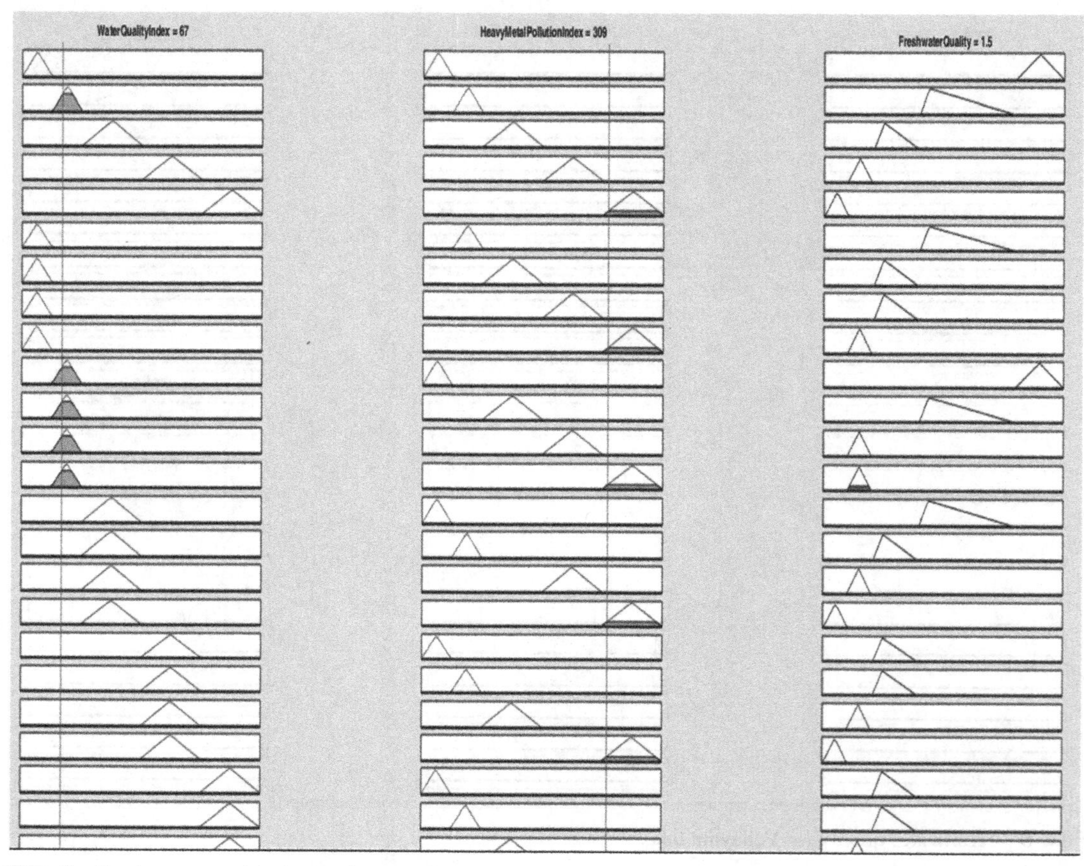

FIG. 7 Freshwater quality for Sasthamkotta lake.

given the integrated results regarding the quality of the lakes accurately. Therefore, this model can be adopted to estimate the state of other freshwater systems and the developed FLI can serve as a useful tool in categorizing the freshwater resources based on drinking needs. Moreover, the freshwater quality model can also be used to analyze stress exerted on the water body taking into account the natural and anthropogenic agents.

4.1 Model validation

Validation of the proposed model is attempted by the following approaches. Selection of fuzzy logic methodology for the development of freshwater model is one of the most relevant approach. This methodology has been proven to be appropriate since it has the ability to reflect human thoughts and expertise involving complex equations into words. The logical

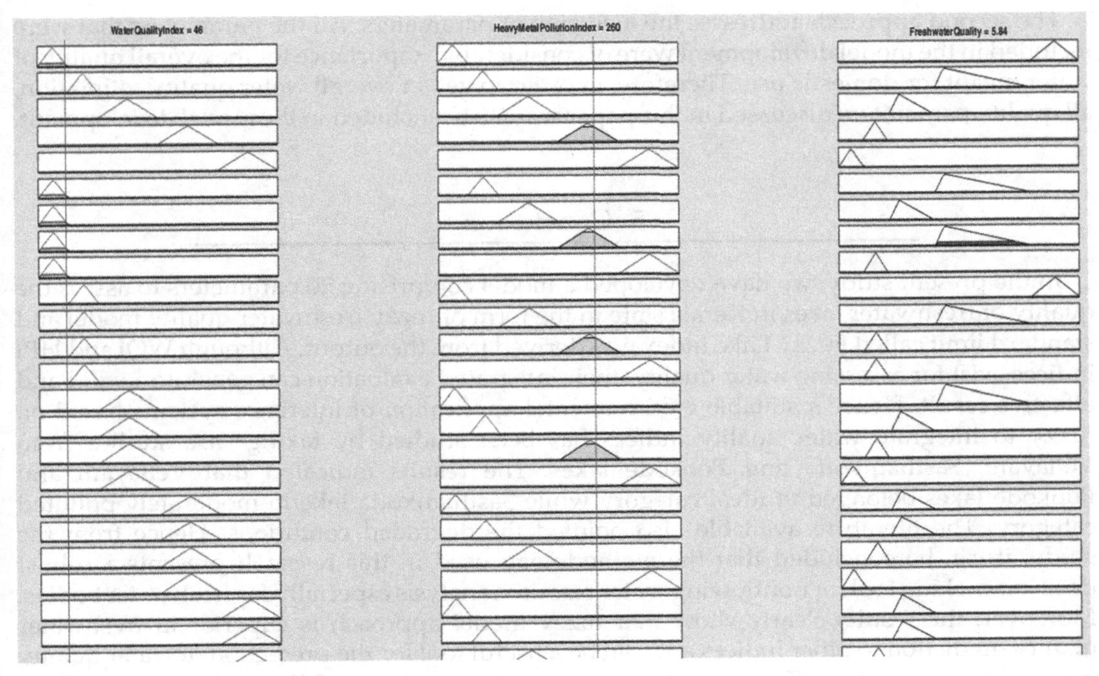

FIG. 8 Freshwater quality for Pookode lake.

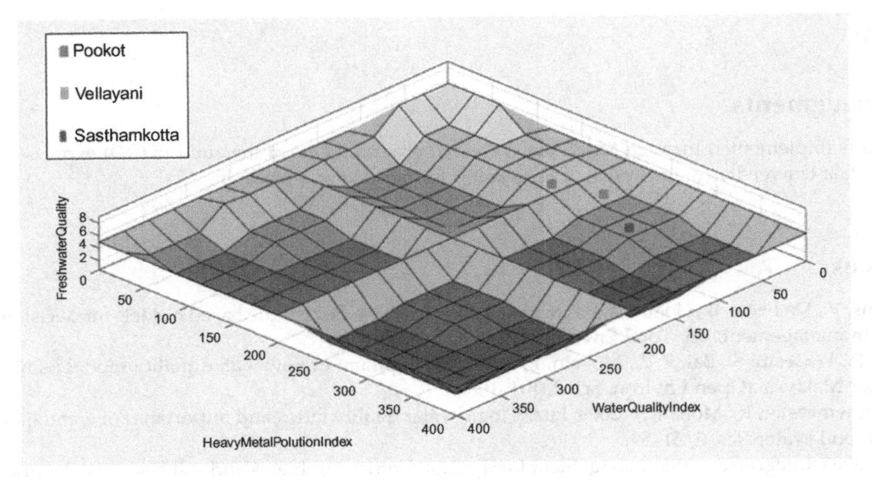

FIG. 9 Surface viewer of freshwater quality model.

structure of the rules facilitates understanding and analysis of the model in a semi qualitative manner, close to how human interprets about the real world. In addition, since the synthesized model is in linguistic terms, the results and the process are more understandable to the public, managers and nonexperts.

The second approach addresses the inclusion of parameters. All the parameters that were included in the model development were of considerable importance for the overall quality of water meant for domestic use. Therefore, in order to get an overall water quality estimation, all quality parameters discussed in this paper were to be included in the model development.

5 Conclusion

In the present study, we have developed a model comprising 20 parameters to assess the quality of freshwater lakes in Kerala state in the form of fuzzy freshwater quality model and standard limit called Fuzzy Lake Index was derived from the output. Although WQI and HPI indices exist for assessing water quality, their integrated evaluation can generate a valid and effective result. Hence a suitable environmental application of inference systems based on fuzzy to integrate water quality indices has been studied by taking case studies from Vellayani, Sasthamkotta and Pookode lakes. The results indicated that Vellayani and Pookode lakes belonged to ideal category while Sasthamkotta lake in moderately polluted category. The literature available also pointed the degraded conditions. Hence from the study, it can be concluded that the methodology used in this research presents a robust decision-making tool for nonflowing water resource analysis especially for freshwater bodies. Moreover, the results clearly show that fuzzy model approach is superior in freshwater quality prediction to other indices and can be a useful tool for the prediction of water quality with sufficient accuracy. Therefore, this model and FLI can be worthily used as a comprehensive tool for succeeding studies on inland freshwater bodies to reveal the quality status of a water body.

Acknowledgments

The study was implemented through MoES plan research program. One of the authors (RD) expresses her sincere thanks to Kerala University Junior Research Fellowship for her doctoral research.

References

Adriaenssens, V., De Baets, B., Goethals, P.L.M., De Pauw, N., 2004. Fuzzy rule based models for decision support in ecosystem management. Sci. Total Environ. 319, 1–12.

Adullah, M.P., Waseem, S., Bai, R.V., Mohsin, I., 2008. Development of new water quality model using fuzzy logic system for Malaysia. Open Environ. Sci. 2, 101–106.

Bai, R.V., Bouwmeester, R., Mohan, S., 2009. Fuzzy logic water quality index and importance of water quality parameters. Air Soil Water Res. 2, 51–59.

Bouwer, H., 2003. Integrated water management for the 21st century: problems and solutions. Food Agric. Environ. 1 (1), 118–127.

Brown, R.M., McClelland, N.I., Deininger, R.A., O'Connor, M.F., 1972. A water quality index—crashing the physiological barrier. In: Indicators of Environmental Quality. vol. 1, pp. 173–182.

Central Pollution Control Board, 2015. Water Quality of Rivers at Interstate Borders. Interstate River Boundary Monitoring Programme. Series: IRBM/01/2015.

Chang, N.B., Chen, H.W., Ning, S.K., 2001. Identification of river water quality using the fuzzy synthetic evaluation approach. J. Environ. Manage. https://doi.org/10.1006/jema.2001.0483. pp 293-305.

Chau, K.W., 2006. A review on integration of artificial intelligence into water quality modeling. Mar. Pollut. Bull. 52, 726–733.

Dharmapalan, B., 2014. Conserving the Vellayani Lake. Science Reporter.

Dojlido, J., Raniszewski, J., Woyciechowska, J., 1994. Water quality index- application for rivers in vistula river basin in Poland. Water Sci. Technol. 30 (10), 57–64.

Elmas, C., 2003. Fuzzy Logic Inspections (Theory, Application, Neural Fuzzy Logic). Seckin, Ankara, p. 230 (in Turkish).

George, A.V., Koshy, M., 2008. Water quality studies of sasthamkotta lake of Kerala. Pollut. Res. 27, 419–424.

Gharibi, H., Mahvi, A.H., Nabizadeh, R., Arabalibeik, H., Yunesian, M., Sowlat, M.H., 2012. A novel approach in water quality assessment based on fuzzy logic. J. Environ. Manage., 87–95.

Giri, S., Singh, A.K., 2014. Assessment of surface water quality using heavy metal pollution in Subarnarehka river, India. Water Qual. Expo. Health 5, 173–182.

Girija Kumari, S., Abraham, M.P., Santhosh, S., 2006. Assessment of fecal indicating bacteria of sasthamkotta lake. Indian Hydrobiol., 159–167.

Heinonen, P., Herve, S., 1994. The development of a new water quality classification system for the Finland. Water Sci. Technol. 30 (10), 21–24.

Horton, R.K., 1965. An index number system for rating water quality. J. Water Pollut. Control Fed. 37 (3), 300–305.

Irshad, M.S., 2015. Cashing in on natural resource mismanagement: a study on depleting Sasthamkotta Fresh Water Lake in Kerala. Nat. Resour. Conserv. 3 (3), 50–56.

Jang, J., Sun, C., 1995. Neuro-fuzzy modeling and control. Proc. IEEE 83, 378–406.

Jensen, M.e., Reynolds, K., Andreasen, J., Goodman, I.A., 2000. A knowledge based approach to the assessment of watershed condition. Environ. Monit. Assess. 64, 271–289.

Karr, C., Gentry, E., 1993. Fuzzy control of pH using genetic algorithms. IEEE Trans. Fuzzy Syst. 1, 46–53.

Kumar, N.V., Mathew, S., Swaminathan, G., 2010. Multifactorial fuzzy approach for the assessment of groundwater quality. J. Water Resour. Prot. 2, 597–608.

Lee, H.K., Oh, K.D., Park, D.H., Jung, J.H., Soon, S.J., 1997. Fuzzy expert system to determine stream water quality classification from ecological information. Water Sci. Technol. 36 (12), 199–206.

Milivojevic, J., Krstic, D., Smit, B., Djekic, V., 2016. Assessment of heavy metal contamination and calculation of its pollution index for Ugljesnica River, Serbia. Bull. Environ. Contam. Toxicol. 97 (5), 737–742.

NSF National Sanitation Foundation International, 2005. Available at: http://www.nsf.org. (Accessed October 2005).

Ocampo-Duque, W., Ferre-Huguet, N., Domingo, J.L., Schuhmacher, M., 2006. Assessing water quality in rivers with fuzzy inference systems: a case study. Environ. Int. 32, 733–742.

Ott, W.R., 1978. Water Quality Indices: A Survey of Indices Used in the United States, EPA-600/4–78-005. US Environmental Protection Agency, Washington, DC, p. 128.

Pedrycz, W., 1994. Why triangular membership functions. Fuzzy Set. Syst. 64, 21–30.

Prasad, B., Bose, J.M., 2001. Evaluation of heavy metal pollution index for surface and spring water near a limestone mining area of the lower Himalayas. Environ. Geol. 41, 183–188.

Ramakrishnaiah, C.R., Dadashivaiah, C., Ranganna, G., 2009. Assessment of water quality index for the ground water in Tumkur Taluk, Karnataka State, India. E-J. Chem. 6 (2), 523–530.

Rani, R., Elangovan, K., 2017. An emerging intuitionistic fuzzy based groundwater level pollution. Indian J. Geo-Mar. Sci. 46 (06), 1213–1219.

Ravikumar, P., Mehmood, M.A., Samasekhar, R.K., 2013. Water quality index to determine the surface water quality of Sankey tank and Mallathahalli lake, Bangalore urban district, Karnataka, India. Appl. Water Sci. 3, 247–261.

Reza, R., Singh, G., 2010a. Assessment of ground water quality status by using water quality index method in Orissa, India. World Appl. Sci. J. 9 (12), 1392–1397.

Reza, R., Singh, G., 2010b. Heavy metal contamination and its indexing approach for river water. Int. J. Environ. Sci. Technol. 7 (4), 785–792.

Ross, T.J., 2004. Fuzzy Logic with Engineering Applications. John Wiley and Sons, New York.

Semiromi, F.B., Hassani, A.H., Torabian, A., Karbassi, A.R., Lotfi, H.F., 2011. Water quality index development using fuzzy logic: a case study of the Karoon River of Iran. Afr. J. Biotechnol. 10 (50), 10125–10133.

Sen, Z., 2001. Fuzzy Logic and Modeling Principles. Bilge Kultur Sanat, Istanbul, p. 174 (in Turkish).

Silvert, W., 2000. Fuzzy indices of environmental conditions. Ecol. Model. 130, 111–119.

Singh, P.K., Tiwari, A.K., Panigarhy, B.P., Mahato, M.K., 2013. Water quality indices used for water resources vulnerability assessment using GIS technique: a review. Int. J. Earth Sci. Eng. 6 (6-1), 1594–1600.

Suvarna, A.C., Somashekar, R.K., 1997. Evaluation of water quality index of river Cauvery and its tributaries. Curr. Sci. 72, 640–646.

Thomas, T.T., Sony, D.C., Kuruvila, E.C., 2017. Rapid environmental impact assessment of eco-tourism in Pookode lake Wayanad. Int. Res. J. Eng. Technol. 4, 3149–3154.

Turksen, I., 1991. Measurement of membership functions and their acquisition. Fuzzy Set. Syst. 40, 5–38.

Unnithan, C.S., George, P.M., Shubhashree, N.S., Unnithan, H., 2016. Hydrological studies of selected lakes of Kerala, India. EM Int. 35 (3), 619–622.

Veena, M.P., Achyuthan, H., 2012. Distribution and assessment of heavy metals in the tropical lake sediments, Pookode, Kerala: impact of contamination and pollution. J. Appl. Geochem. 14, 173–183.

Vishnu, M.S., Padmalal, D., 2014. Holocene evolution of the Freshwater Lakes of Southern Kerala, SW India. Gondwana Geol. Mag. 15, 67–78.

Zadeh, L.A., 1965. Fuzzy Sets Inference Control. pp. 338–353.

CHAPTER

17

Appraisal of multigene genetic programming for estimating optimal properties of lined open channels with circular shapes incorporating constant and variable roughness scenarios

Majid Niazkar

Department of Civil and Environmental Engineering, School of Engineering, Shiraz University, Shiraz, Iran

OUTLINE

1 Introduction 285

2 Methods and materials 286

2.1 *Optimum design of circular channels* 286

2.2 *Available design equations of circular channels* 287

2.3 *Multigene genetic programming* 289

2.4 *Application of MGGP to develop design equations* 289

2.5 *Performance evaluation criteria* 290

3 Results and discussion 290

4 Conclusions 293

Funding 295

Conflict of interest 295

References 297

1 Introduction

One of the main problems in water resources management is how to design canals in a bid to convey sufficient amount of water. Due to the large distance between water supply and

https://doi.org/10.1016/B978-0-323-91910-4.00017-0

demand nodes, water conveyance projects usually are inevitably expensive. In this context, an optimum design of manmade channels requires minimizing the construction budget, as the first priority. Hence, many studies have been conducted to design suitable canals by treating the design problem as an optimization problem, whose objective function is to minimize excavation and lining costs (Swamee et al., 2000; Aksoy and Altan-Sakarya, 2006). In addition, hydraulic relationships, which govern flow movements throughout an open channel, have been utilized as the constraint of the optimization problem.

Solving the optimization algorithm of the channel design will provide a single value of channel geometric variables. However, the design process in practice requires considering discharge values with various return periods. Therefore, in practice, the optimization problem needs to be solved much more than one time, while the obtained channel geometries should be tested for different scenarios. This requirement brought hydraulic researchers to develop design charts or design equations based on solving the optimization problem for numerous scenarios. As a result, some studies have been used different techniques, including optimization algorithms, to propose new design equations to estimate channel properties for any possible situation in question (Swamee and Chahar, 2015; Niazkar and Afzali, 2015; Niazkar et al., 2018). Recently, a few studies employed machine learning (ML) techniques, e.g., Artificial Neural Network (ANN) and Genetic Programming (GP), for predicting optimum values of channel geometries in the literature (Niazkar, 2020; Tawfik, 2020; Niazkar, 2021b). Nevertheless, further investigations may be required in light of improving the estimation of optimum channel geometries using ML techniques as a slight improvement in a channel geometry can significantly mitigate the construction required.

Among different shapes for open canals, circular channels are one of the most common geometries that have been used in practice. Therefore, this study focuses on applying a new ML method, i.e., Multigene GP (MGGP), for designing optimum properties of circular channels. Based on the literature, MGGP has been used successfully for a few problems in water resources field of research (Niazkar and Zakwan, 2021a,b; Zakwan and Niazkar, 2021; Niazkar, 2022). Consequently, it has been used to develop new design equations for estimating optimum values of geometries of lined circular canals. The design equations were developed for both constant and variable Manning's coefficients. According to the current literature, it is the first time that MGGP has been used to predict optimum values of properties in a lined circular channel. The performance of the new explicit equations developed by MGGP was compared with that of the existing design equations.

2 Methods and materials

2.1 Optimum design of circular channels

Circular channels are one of the widely-used types of canals to convey water in different parts of the world. The problem statement of a channel design is nothing but determining suitable values of canal geometries. The optimum value of a channel property is the one achieved by solving an optimization problem, which intends to minimize construction cost by taking into account hydraulic aspects of flow conveyance. For instance, in optimum design of circular channel, the aim is to find adequate values of circular radius (r) and water depth (y). In order to consider a variety of scenarios, some of previous studies presented design equations using dimensionless geometric variables. To be more specific, a design equation of circular canals gives dimensionless circular radius (r_*) and dimensionless water depth (y_*), where $r_* = \frac{r}{\lambda}$, $y_* = \frac{y}{\lambda}$, $\lambda = \left(\frac{Qn}{\sqrt{S}}\right)^{3/8}$, Q is discharge, n is Manning's coefficient, and S is channel slope.

Based on previous studies, the problem statement of the design of a lined circular channel with variable $n$is shown in Eqs. (1), (2). The former is the objective function, while the latter is Manning's resistance equation, which plays as the hydraulic constraint of the design problem (Niazkar et al., 2018):

$$\text{minimize} \quad C_* = \beta_{L*}\theta r_* + 0.5r_*(\theta - \sin\theta) + 0.5\beta_{A*}r_*\left[(y_* - r_*)\theta + \frac{4r_*^2 + 2(y_* - r_*)^2}{3r_*^2}\sqrt{r_*^2 - (y_* - r_*)^2}\right] \tag{1}$$

$$\text{subjected to} \quad n_f \times \{1 + 0.18(2\pi - \theta)[0.1 + \exp(-0.3\theta)\sin^2(0.38\theta)]\} - [0.5r_*(\theta - \sin\theta)]^{5/3}(\theta r_*)^{-2/3} = 0 \tag{2}$$

where C_* is the dimensionless total construction cost per unit length, β_{L*} is the dimensionless lining cost per unit area, θ is the water depth angle, β_{A*} is the dimensionless additional earthwork cost associated with different earthwork costs in different depths, and n_f is Manning's coefficient when the whole channel cross section is filled with water.

Since a cross section is not always filled with water in a circular canal, $n_f$may be substituted with Manning's coefficient associated with a partially full cross section (n) in the design problem. For this purpose, an experimental-based relationship presented in Eq. (3) can be used (Yarnell and Woodward, 1920; Wilcox, 1924; Zaghloul, 1992; Akgiray, 2004):

$$n_f = \frac{n}{1 + 0.18(2\pi - \theta)[0.1 + \exp(-0.3\theta)\sin^2(0.38\theta)]} \tag{3}$$

In general, Manning's coefficient may vary with flow in natural steams or manmade canals (McKay and Fischenich, 2011; Niazkar et al., 2019; Niazkar, 2021a). The design problem defined in Eqs. (1), (2) can take into account flow-dependent Manning's coefficient as it utilized n_f in the constraint. On other hand, when Manning's coefficient is assumed to be constant with flow, Eq. (2) can be further simplified, as shown in Eq. (4) (Aksoy and Altan-Sakarya, 2006, Niazkar, 2021b):

$$\left\{r_*\left[\pi - 2\sin^{-1}\left(1 - \frac{y_*}{r_*}\right)\right]\right\}^{2/3} - \left\{0.5r_*^2\left[\pi - 2\sin^{-1}\left(1 - \frac{y_*}{r_*}\right) - 2\left(1 - \frac{y_*}{r_*}\right)\frac{\sqrt{r_*^2 - (r_* - y_*)^2}}{r_*}\right]\right\}^{5/3} = 0 \tag{4}$$

2.2 Available design equations of circular channels

The design problem defined in Eqs. (1), (2) or Eqs. (1), (4) can be solved by any suitable optimization algorithm, e.g., Modified Honey Bee Mating Optimization (MHBMO) algorithm for a set of β_{L*} and β_{A*}. Additionally, Eqs. (1), (2) or Eqs. (1), (4) gave optimum results for variable and constant roughness scenarios, respectively. According to the literature, a few design equations have been developed by solving the described design problems. The design equations are presented in the following:

1. Swamee et al.'s (2000) model: They developed one equation for estimating the optimum value of $2r$ and one equation for predicting the optimum value of y, which are shown in Eqs. (5), (6), respectively:

$$2r = 0.78065L + \frac{0.19375\beta_A L^3}{\beta_E L + 13.6232\beta_L} \tag{5}$$

$$y = 0.39032L\left(1 + \frac{0.12631\beta_A L^2}{\beta_E L + 12.9379\beta_L}\right)^{-1} \tag{6}$$

where $L = \lambda_1\left(\frac{\varepsilon}{\lambda_1} + \frac{8\upsilon\lambda_1}{Q}\right)^{0.04}$, $\lambda_1 = \left(\frac{Q}{\sqrt{gS}}\right)^{0.4} = \frac{\lambda^{32/30}}{n^{0.4}g^{0.2}}$, g is the gravitational acceleration, ε is the average roughness height of canal surface, υ is kinematic viscosity, β_E is the earthwork cost per unit volume, $\beta_A = \frac{\beta_{A*}\beta_E}{\lambda}$, and $\beta_L = \lambda\beta_E\beta_{L*}$.

2. Aksoy and Altan-Sakarya's (2006) models: They proposed two models for estimating r_* and y_* considering constant n. The design equations of the first and second models developed by Aksoy and Altan-Sakarya (2006) are given in Eqs. (7), (8) and Eqs. (9), (10), respectively:

$$r_* = 1.004 + \frac{0.113\beta_{A*}}{\beta_{L*}} \tag{7}$$

$$y_* = 1.004\left(1 + \frac{0.055\beta_{A*}}{\beta_{L*}}\right)^{-1} \tag{8}$$

$$r_* = 1.004 + \frac{0.58\beta_{A*}}{1 + 5.008\beta_{L*}} \tag{9}$$

$$y_* = 1.004\left(1 + \frac{0.277\beta_{A*}}{1 + 4.937\beta_{L*}}\right)^{-1} \tag{10}$$

3. Niazkar and Afzali's (2015) model: They exploited the MHBMO algorithm to develop a new design equation for predicting optimum values of r_* and y_* when n is constant. Their model is presented in Eqs. (11), (12):

$$r_* = 1.004 + 0.2358\beta_{A*}^{0.9978}\beta_{L*}^{-0.7749} \tag{11}$$

$$y_* = 1.004\left(1.0091 + 3.3182\beta_{A*}^{1.3175}\beta_{L*}^{-1.0878}\right)^{-0.0708} \tag{12}$$

4. Niazkar et al. (2018): They proposed two new models for estimating r_* and y_* for variable n. Eqs. (13), (14) and Eqs. (15), (16) show the first and second models developed by Niazkar et al. (2018):

$$r_* = 0.9493\beta_{L*}^{0.0202} + 0.3542\beta_{A*}^{1.0351}\beta_{L*}^{-0.702} \tag{13}$$

$$y_* = 1.1663\beta_{L*}^{-0.01} - 0.1557\beta_{A*}^{0.9322}\beta_{L*}^{-0.5489} \tag{14}$$

$$r_* = 0.9539\beta_{L*}^{0.0175} + 0.3501\beta_{A*}^{1.0352}\beta_{L*}^{-0.705} \tag{15}$$

$$y_* = 1.1598\beta_{L*}^{-0.0114} - 0.1618\beta_{A*}^{1.0695}\beta_{L*}^{-0.5676} \tag{16}$$

2.3 Multigene genetic programming

Generally, MGGP is a modified version of GP, which is grouped as a machine learning method. Like GP, MGGP assumes that each mathematical relation can be defined as a tree structure with three types of node (root node, function node, and terminal node). At the beginning, MGGP generates a generation with a predefined number of individuals. Each individual is a mathematical estimation model, which has a few nodes. The focus of the MGGP process is to build a new generation based on the individual with relatively highest fitness index. As a result, genetic operations (selection, crossover, and mutation) are applied to nodes of best-fitted individuals (parent genes) in favor of producing new individuals (offspring genes) for the next generation. The generation of new population is repeatedly done in light of developing an estimation model (individual) with a desirable accuracy. The main difference between GP and MGGP is that an individual in the former consist of only one gene (tree), whereas an individual in the latter can possess more than one gene (tree). This characteristic enables MGGP to perform better than GP when it comes to tackle complex relationships (Zakwan and Niazkar, 2021).

In this study, an open source toolbox of MGGP in the MATLAB environment was used. This version was adopted form the literature (Searson, 2009) and has been successfully used for solving a few problems in water resources and hydraulic engineering (Niazkar and Zakwan, 2021a; Zakwan and Niazkar, 2021). In addition, the controlling parameters of MGGP were set as those considered in previous studies (Niazkar and Zakwan, 2021b).

2.4 Application of MGGP to develop design equations

In this study, the described design problem of circular channels was solved by the MHBMO algorithm, and optimum values of r_* and y_* were obtained for various values of β_{A*} and β_{L*} for both constant and variable n. To be more precise, 210 pairs of β_{A*} and β_{L*} were considered for each roughness scenario. Similar to previous studies, $0 < \frac{\beta_{A*}}{\beta_{L*}} < 2$ was checked and satisfied for each pair of β_{A*} and β_{L*} (Aksoy and Altan-Sakarya, 2006; Niazkar and Afzali, 2015; Niazkar et al., 2018). Furthermore, each set of normalized β_{L*} and β_{A*} was introduced to the MGGP code as the input data. To be more specific, the train and test parts comprise of 150 and 60 data points, respectively. The former was used to train MGGP, while the latter was utilized to conduct a comparative analysis. Moreover, the root mean square of errors between the estimated and optimum values of r_* and y_* was minimized as the objective function. Also, the MGGP code was run more than 50 times to develop each design equation, while the relation with the best accuracy was reported as the MGGP-based design equation.

2.5 Performance evaluation criteria

In a bid to determine how accurate the MGGP-based models performed, a few metrics were used in this study (Niazkar and Afzali, 2018). These performance evaluation criteria are (1) Root Mean Square Error (RMSE), (2) Mean Absolute Relative Error (MARE), (3) Relative Error (RE), and (4) coefficient of determination (R^2), which are written for y_* in Eqs. (17)–(20) (Niazkar and Afzali, 2018):

$$RMSE = \sqrt{\frac{1}{N}\sum_{i=1}^{N}\left(y_{*,database} - y_{*,estimated}\right)^2} \tag{17}$$

$$MARE = \frac{1}{N}\sum_{i=1}^{N}\left|\frac{y_{*,database} - y_{*,estimated}}{y_{*,database}}\right| \times 100 \tag{18}$$

$$RE = \frac{y_{*,estimated} - y_{*,database}}{y_{*,database}} \tag{19}$$

$$R^2 = \left(\frac{\sum_{i=1}^{N}\left[\left(y_{*,database} - \frac{\sum_{i=1}^{N} y_{*,database}}{N}\right)\left(y_{*,estimated} - \frac{\sum_{i=1}^{N} y_{*,estimated}}{N}\right)\right]}{\sqrt{\sum_{i=1}^{N}\left[\left(y_{*,database} - \frac{\sum_{i=1}^{N} y_{*,database}}{N}\right)^2\left(y_{*,estimated} - \frac{\sum_{i=1}^{N} y_{*,estimated}}{N}\right)^2\right]}}\right)^2 \tag{20}$$

where $y_{*,\ database}$ and $y_{*,\ estimated}$ are the database and estimated dimensionless water depth in the circular canal, respectively.

3 Results and discussion

MGGP was applied to the database generated by solving the design problem of lined circular channels for both constant and variable roughness scenarios. As a result, four design equations were developed by MGGP for estimating r_* and y_*:

(1) For flow-independent n scenario: The MGGP-based design equations for predicting optimum values of r_* and y_* are shown in Eqs. (21), (22), respectively:

$$r_* = 1.00395 + 0.29532\Big\{0.05209\tanh\left[\exp\left(\beta_{L*}\right) - 8.328\beta_{A*}\right] + 1.754\tanh\left(8.328\beta_{L*} + \beta_{A*}\right) - 1.547\tanh\left(\beta_{A*}^2 + 8.558\beta_{L*}\right) - 0.2443\tanh\left(8.602\beta_{L*} - 2\beta_{A*}\right) + 0.9964\tanh\left[\exp\left(-\beta_{L*}\right)\sin\left(\beta_{A*}\right)\right] - 0.01279\Big\} \tag{21}$$

$$y_* = 0.88811 + 0.11584\Big\{23.73\tanh\left(2\beta_{L*} + \beta_{A*}\right) - 20.01\tanh\left[2\beta_{L*} + \tanh\left(\beta_{A*}\right)\right] + 6.795\sin\left(\beta_{A*} + 2.906\right) + 0.2263\left(\beta_{A*} - 9.774\right)\tanh\left(2\beta_{L*}\right) - 6.311\tanh\left(2\beta_{L*}\right)\sin\left(\beta_{A*} + 2.899\right) - 0.5786\Big\} \tag{22}$$

(2) For flow-dependent n scenario: When n varies with flow, MGGP resulted in Eqs. (23), (24) for estimating optimum values of r_* and y_*, respectively:

$$r_* = 0.86366 + 0.92487\left\{0.0171\beta_{A*} + 0.1252\tanh\left(3.173\beta_{L*}\right) + \frac{2.684\tanh\left[\sin\left(\beta_{A*}\right)\right]}{7.897\beta_{L*} + 0.9965} - \frac{0.2908\tanh\left[\sin\left(\beta_{A*}\right)\right]}{\beta_{L*} + 0.1717\beta_{A*} + 0.1366} - \frac{0.2376\tanh\left(\beta_{A*}^2\right)}{7.865\beta_{L*} - 0.1361} + \frac{0.2111\left(0.1631\beta_{L*} + \beta_{A*}^2\right)}{7.897\beta_{L*} - 0.1357} + 0.0198\right\} \tag{23}$$

$$y_* = 0.89532 + 0.34627\left\{\frac{102.2\beta_{A*} + 37.23}{180.7\beta_{L*} + 8.321} - \frac{2.848}{8.037\beta_{L*} + \beta_{A*} + 0.3777} + \frac{3.618}{8.037\beta_{L*} + \beta_{A*} + 0.416} - \frac{5.289\left(\beta_{A*} + 0.416\right)}{8.037\beta_{L*} + 0.416} - \frac{0.03665\beta_{L*}\beta_{A*}}{1.416\beta_{L*} + 0.416} + 0.6708\right\} \tag{24}$$

The results of RE calculated for r_* and y_* assuming constant roughness are depicted as contour plots in Figs. 1 and 2, respectively. The benefit of contour plots for RE values is that it gives a clear perspective how the proposed models perform for each range of independent variables (i.e., β_{L*} and β_{A*}). As shown in Fig. 1, the MGGP-based design equation proposed for estimating r_* under flow-independent roughness scenario achieved RE values in the range of (−0.24, 0.24) and (−0.012, 0.024) for the train and test data, respectively. Additionally, Fig. 1 indicates that the highest and lowest RE values are obtained around $\beta_{L*}=0.5$ and $\beta_{L*}=2.8$ for the train data, respectively. Similarly, the highest RE values for the test data is associated with $\beta_{L*}=0.5$. Moreover, the lowest RE values in Fig. 1B are in the area of $\beta_{L*}=1.4$. Furthermore, Fig. 2 demonstrates that the RE values calculated for y_* is with the range of (−0.011, 0.014) and (−0.011, 0.009) for the train and test data, respectively. The contour plots shown in Fig. 2 indicates that for most of the domain in which $0 < \frac{\beta_{A*}}{\beta_{L*}} < 2$ was satisfied, Eq. (22) gave satisfactory results in terms of RE. In other words, the MGGP-based design equation for predicting y_* under constant roughness coefficient yielded accurate results, particularly in the right-hand side of plots in Fig. 2.

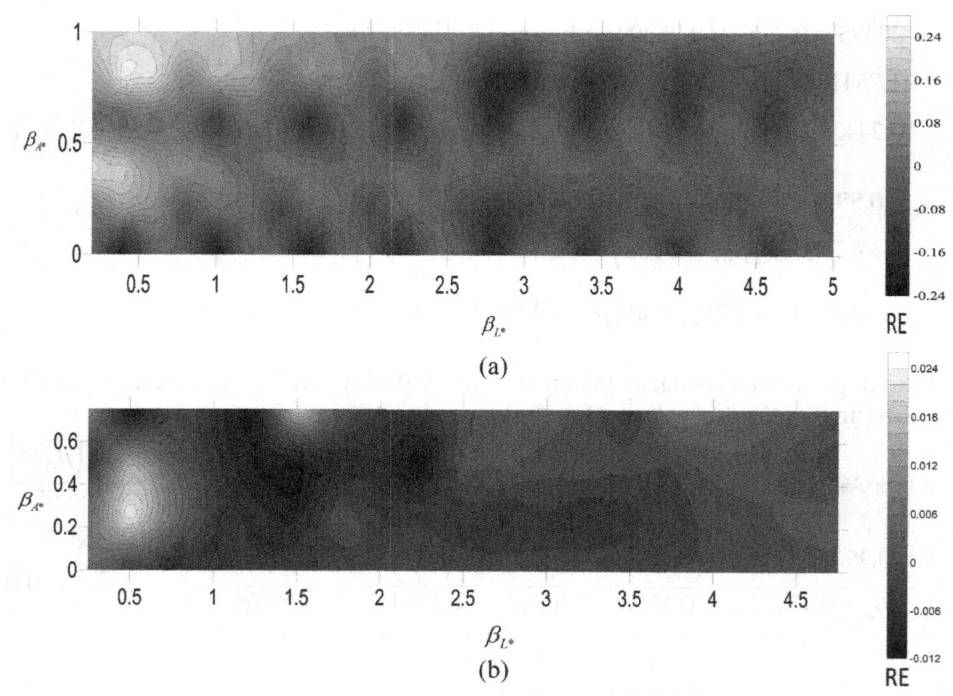

FIG. 1 Contour lines of relative errors calculated for r_* assuming constant roughness for (A) train data and (B) test data.

Figs. 3 and 4 illustrate the contour plots of RE values calculated for r_* and y_* considering variable roughness scenario, respectively. In Fig. 3, Eq. (23) resulted in a unique and close to zero value of RE for most of the points with $\beta_{L*} > 2.0$, the highest and lowest values of RE are observed for $\beta_{L*} = 0.0$. Hence, it may be deduced that the proposed MGGP-based model for estimating r_* under variable roughness gives better results for higher values of β_{L*}. Likewise, Fig. 4 depicts quite the same pattern for RE values. To be more specific, Eq. (24) yielded quite the same RE values for $\beta_{L*} > 1.0$ and $\beta_{L*} > 2.0$ for the train and test data, respectively. Similar to Fig. 3, the highest and lowest values of RE are associated with $\beta_{L*} < 0.5$ for the train and test data in Fig. 4. In addition, Figs. 3 and 4 imply that the proposed design equations estimate optimum values of r_* and y_* of circular channels in the range of (−0.05, 0.06) and particularly much better values of RE for $\beta_{L*} > 2.0$.

The performance of the new design equations developed by MGGP are compared with previous explicit equations (Eqs. 5–16), ANN (Niazkar, 2021b), and GP (Niazkar, 2021b). It should be noted that the train and test data used in this study is the same as those utilized by Niazkar (2021b), which applied ANN and GP to solve this design problem. The results of the comparative analysis in terms of RMSE, MARE and R^2 are presented in Fig. 5. As shown, the explicit design equations developed by MGGP yield better estimations than explicit equations available in the literature for both constant and variable roughness scenarios for RMSE, MARE and R^2. Moreover, Fig. 5 indicates that the MGGP-based models perform

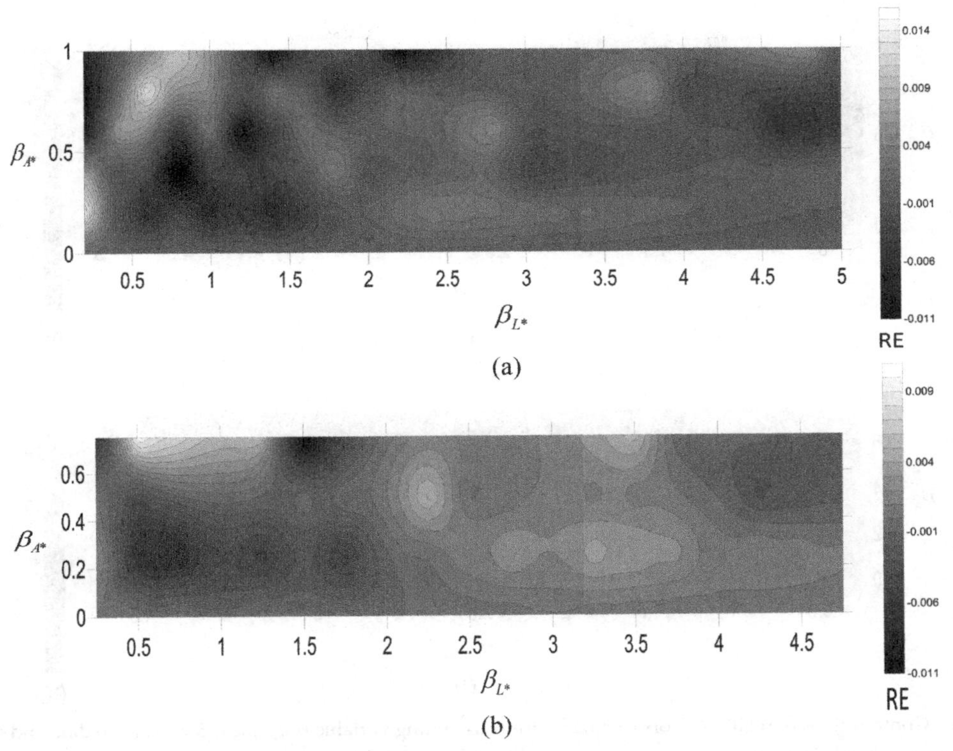

FIG. 2 Contour lines of relative errors calculated for y_* assuming constant roughness for (A) train data and (B) test data.

close to those of other ML methods (ANN and GP), while the former gave explicit equations that may be simpler to be used by hydraulic engineers. Also, an explicit equation for channel design can be implemented in MS Excel or other engineering software for further use and applications. Since a few studies combined MGGP with Generalized Reduced Gradient (GRG) (Niazkar and Zakwan, 2021a,b) and improved the performance of MGGP, it is postulated that application of hybrid MGGP-GRG technique may enhance the precision of the proposed design equation. Therefore, application of the hybrid MGGP-GRG is recommended for future studies.

4 Conclusions

Channel design is one of the problems associated with water conveyance projects in water resources and hydraulic engineering. One of the prominent key factors in such project is to minimize the construction cost. Therefore, an optimum canal is the one which has the lowest construction cost, and consequently, hydraulic engineers seek the channel properties that

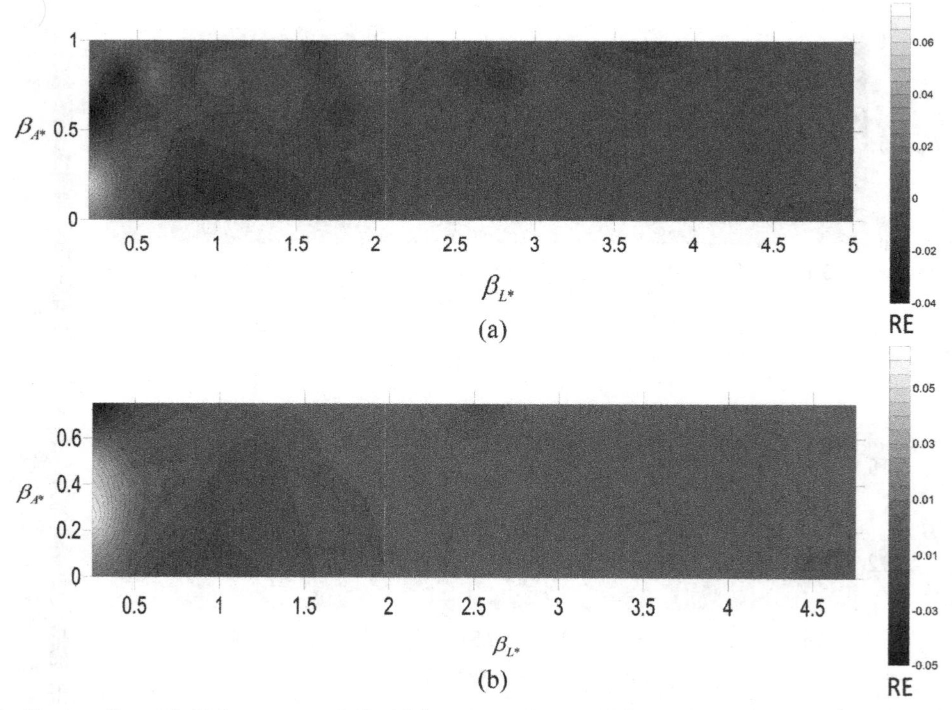

FIG. 3 Contour lines of relative errors calculated for r_* assuming variable roughness for (A) train data and (B) test data.

yield the minimum construction cost. In this context, the optimum channel design is an optimization problem, whose objective function is to minimize construction costs. The solution of such optimization problem consists of a single set of channel properties, while channel design requires testing various flow conditions and roughness scenarios. Therefore, some contributions have been made to develop design equations that provide optimum values of dimensionless channel geometries. This chapter explores the application of MGGP to develop explicit design equations for estimating optimum values of channel geometries with circular shapes. For this purpose, the design equations were developed under constant and variable roughness scenarios. The comparative analysis between the proposed MGGP-based models and the available methods in the literature demonstrates that the new design equations perform better than available explicit equations according to RMSE, MARE and R^2, while they yield quite the same estimations as ANN and GP. As the proposed MGGP-based equations are explicit, they can be implemented in MS Excel or other engineering software in favor of the optimum design of circular canals. Because of the improvement made in the

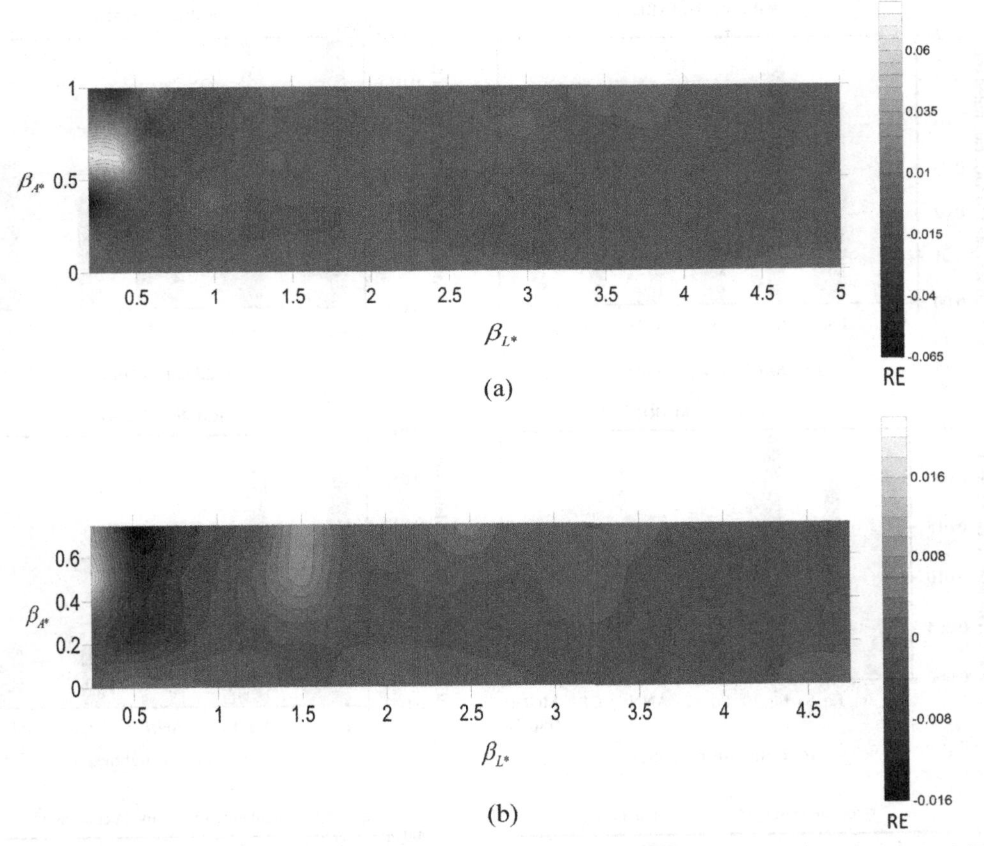

FIG. 4 Contour lines of relative errors calculated for y_* assuming variable roughness for (A) train data and (B) test data.

current study, it is stipulated that applications of hybrid techniques incorporating MGGP may further enhance the accuracy of predicting optimum values of channel properties, which is suggested for future investigation of the topic.

Funding

None.

Conflict of interest

None.

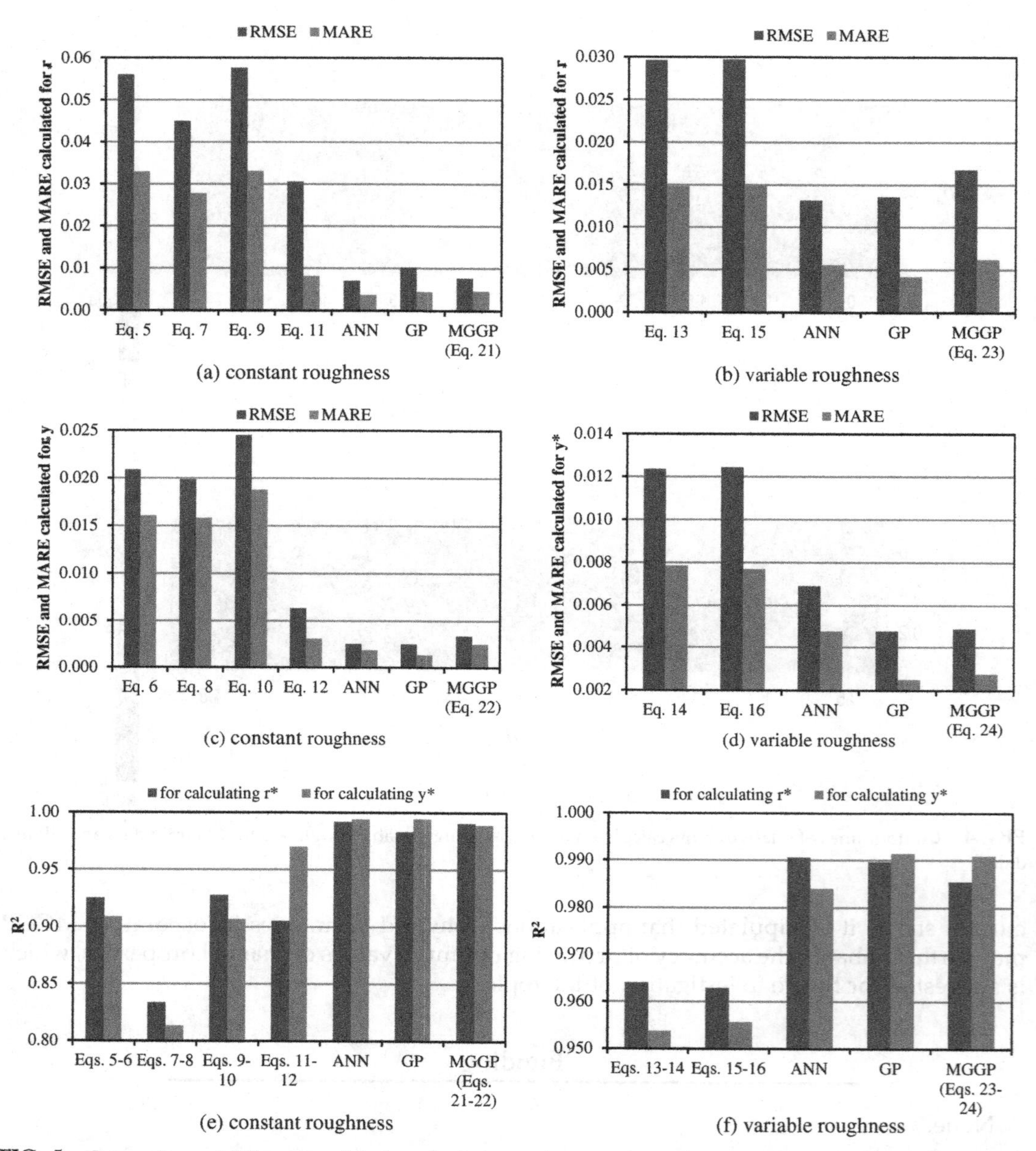

FIG. 5 Comparison of different models for calculating optimum values of (A) and (B) RMSE and MARE for r_*, (C) and (D) RMSE and MARE for y_*, and (E) and (F) R^2.

References

Akgiray, O., 2004. Simple formulae for velocity, depth of flow, and slope calculations in partially filled circular pipes. Environ. Eng. Sci. 21 (3), 371–385. https://doi.org/10.1089/109287504323067012.

Aksoy, B., Altan-Sakarya, A.B., 2006. Optimal lined channel design. Can. J. Civ. Eng. 33 (5), 535–545. https://doi.org/10.1139/l06-008.

McKay, S.K., Fischenich, J.C., 2011. Robust prediction of hydraulic roughness. In: ERDC/CHL CHETN-VII-11. U.S. Army Engineer Research and Development Center, Vicksburg, Mississippi.

Niazkar, M., 2020. Assessment of artificial intelligence models for calculating optimum properties of lined channels. J. Hydroinf. https://doi.org/10.2166/hydro.2020.050.

Niazkar, M., 2021b. Optimum design of straight circular channels incorporating constant and variable roughness scenarios: assessment of machine learning models. Math. Probl. Eng. 2021, 1–21. https://doi.org/10.1155/2021/9984934.

Niazkar, M., 2021a. An excel VBA-based educational module for bed roughness predictors. Comput. Appl. Eng. Educ. 29 (5), 1051–1060. https://doi.org/10.1002/cae.22358.

Niazkar, M., 2022. Multi-gene genetic programming and its various applications. In: 3-Volume Handbook of HydroInformatics. HandHyd, Elsevier. Edited by Professor Saeid Eslamian, HandHyd Chief Editor, Elsevier (Current status: Accepted for Publication).

Niazkar, M., Afzali, S.H., 2015. Optimum design of lined channel sections. Water Resour. Manag. 29 (6), 1921–1932. https://doi.org/10.1007/s11269-015-0919-9.

Niazkar, M., Afzali, S.H., 2018. Developing a new accuracy-improved model for estimating scour depth around piers using a hybrid method. Iran. J. Sci. Technol. Trans. Civ. Eng. 43 (2), 179–189. https://doi.org/10.1007/s40996-018-0129-9.

Niazkar, M., Rakhshandehroo, G., Afzali, S.H., 2018. Deriving explicit equations for optimum design of a circular channel incorporating a variable roughness. Iran. J. Sci. Technol. Trans. Civ. Eng. 42 (2), 133–142. https://doi.org/10.1007/s40996-017-0091-y.

Niazkar, M., Talebbeydokhti, N., Afzali, S.H., 2019. One dimensional hydraulic flow routing incorporating a variable grain roughness coefficient. Water Resour. Manag. 33 (13), 4599–4620. https://doi.org/10.1007/s11269-019-02384-8.

Niazkar, M., Zakwan, M., 2021a. Application of MGGP, ANN, MHBMO, GRG and linear regression for developing daily sediment rating curves. Math. Probl. Eng. 2021a, 1–13. https://doi.org/10.1155/2021/8574063.

Niazkar, M., Zakwan, M., 2021b. Assessment of artificial intelligence models for developing single-value and loop rating curves. Complexity 2021b, 1–21. https://doi.org/10.1155/2021/6627011.

Searson, D., 2009. GPTIPS: Genetic programming and symbolic regression for MATLAB.

Swamee, P.K., Chahar, B.R., 2015. Design of Canals. Springer India, New Delhi, ISBN: 978-81-322-2321-4.

Swamee, P.K., Mishra, G.C., Chahar, B.R., 2000. Minimum cost design of lined canal sections. Water Resour. Manag. 14 (1), 1–12. https://doi.org/10.1023/A:1008198602337.

Tawfik, A.M., 2020. Design of channel section for minimum water loss using Lagrange optimization and artificial neural networks. Ain Shams Eng. J. https://doi.org/10.1016/j.asej.2020.04.017.

Wilcox, E.R., 1924. A Comparative Test of the Flow of Water in 8-Inch Concrete and Vitrified Clay Sewer Pipe. Washington (State) University. Engineering Experiment Station, Seattle, Washington. Engineering Experiment Station Series. Bulletin, no. 27.

Yarnell, D.L., Woodward, S.M., 1920. The Flow of Water in Drain Tile. US Department of Agriculture (854).

Zaghloul, N.A., 1992. Gradually varied flow in circular channels with variable roughness. Adv. Eng. Softw. 15 (1), 33–42. https://doi.org/10.1016/0965-9978(92)90042-E.

Zakwan, M., Niazkar, M., 2021. A comparative analysis of data-driven empirical and artificial intelligence models for estimating infiltration rates. Complexity 2021, 1–13. https://doi.org/10.1155/2021/9945218.

CHAPTER

18

Geoinformatics-based assessment of gross irrigation requirement of different crops grown in the south-western region of Haryana, India

Arvind Dhaloiya[a,b], *RS Hooda*[c], *Devendra Kumar*[a], *Anurag Malik*[d], *and Ajay Kumar*[e]

[a]Haryana Space Applications Centre, Hisar, Haryana, India [b]Department of Soil and Water Engineering, Punjab Agricultural University, Ludhiana, Punjab, India [c]Advisor to Agriculture and Farmer Welfare Department, Panchkula, Haryana, India [d]Punjab Agricultural University, Regional Research Station, Bathinda, Punjab, India [e]Department of Microbiology, Chaudhary Charan Singh Haryana Agricultural University, Hisar, Haryana, India

OUTLINE

1 Introduction 300

2 Study area 301

3 Material and methods 301

3.1 Data set and characteristics 301

3.2 Procedure of GIR computation 303

4 Results and discussion 304

4.1 Spatial database creation and cropping pattern 304

4.2 Reference evapotranspiration (ET_o) 304

4.3 Crop evapotranspiration (ET_c) 306

4.4 Effective rainfall (P_{eff}) 307

4.5 Net irrigation requirement (NIR) 307

4.6 Gross irrigation requirement (GIR) 309

5 Conclusions 315

Conflict of interest 315

References 315

Current Directions in Water Scarcity Research, Volume 7
https://doi.org/10.1016/B978-0-323-91910-4.00018-2

1 Introduction

Growing population requires an increase in food grain production, which puts water, land, and other essential resources under stress (Degirmenci et al., 2005). India has the world's 2.4% geographical area but serves the world's 17% population. It is expected that the Indian population will be around 1390 million by 2025. Food production must be increased from the present 252 million tons to 350 million tons by 2025 to meet the growing population demand (MoWR, 2012). The current water use in India's agricultural sector is nearly 605×109 m^3, which represents 83% of total water use and is expected to decrease by 10% to 20% by 2025 to meet the increasing domestic and industrial requirements (Chaudhary, 1995; GOI, 2012). Agricultural irrigation management in water shortage areas needs sustainable and innovative research ideas which can help in the transfer of effective technology (Pereira et al., 1999). The management of irrigation is a priority for effective and timely distribution of water in canal command areas, taking into account crop factors, appropriate and always updated details on the irrigation system is required for irrigation management. One of the most significant irrigation management methods is proper irrigation scheduling based on timely measurements of soil moisture and crop water needs. The demand for crop water is determined by calculating the crop evapotranspiration (ET_c) by using the Food and Agriculture Organization (FAO) (2016) suggested technique. Meteorological parameters are used to calculate ET_c. The evapotranspiration data can be used to preserve irrigated water.

Recently, many studies have explored the potential of GIS in water management for irrigation. Schmugge et al. (2002) summarized the information about appropriate RS methods to estimate water resources at different places. Current studies have shown that water management systems based on GIS and RS are very appropriate (Stehman and Milliken, 2007) to improve the efficiency of water use. GIS systems have multiple purposes in irrigation and water management-related projects like mapping tools, database integration tools, planning or management tools, and modeling (Kumar et al., 2021b; Fipps and Leig, 2003). Likewise, GIS-based systems and models have an imperative role in irrigation project management in inputting data, drawing the layers, outputting reports and results, and processing and analyzing the data (Kumar et al., 2019; Al-Safi, 2013). RS is considered to offer several benefits for such tasks (Kumar et al., 2021a). In recent decades, geoinformatics-based monitoring of water resources is well recognized and received many successful applications in operational situations (FAO, 1995; Belmonte et al., 1999; Shultz and Engman, 2000; D'Urso and Menenti, 1995; Stehman and Milliken, 2007; Elbeltagi et al., 2020). Similarly, Elbeltagi et al. (2021) recommended the application of remote sensing in mapping the spatial-temporal variability of blue and green evapotranspiration of Wheat in the Egyptian Nile Delta from 1997 to 2017.

With related background, the strict water distribution schedules and the degree of water inadequacy give farmers very restricted scope for decision-making regarding the management of canal water. Farmer choices are mainly limited to the operation of the tube wells to mitigate a portion of the canal water deficit. The extremely insufficient supply of canal water and poor soil quality generate variations in farmers' choice of crops during the summer season, while wheat is cultivated as the only crop with scattered mustard areas in the winter season. Therefore, to break the productivity barriers in the south-western region of Haryana

State, irrigation water management has become critical. Gross irrigation requirement (GIR) for various crops needs to be studied using state-of-art techniques like geoinformatics to plan irrigation scheduling and alternate cropping patterns in the area. The present study has been taken up to assess GIR for different crops grown during the Rabi and Kharif seasons using RS and GIS techniques for optimal management of water resources in the study region.

2 Study area

The study was undertaken in the command regions of the Barwala sub-branch of the Sirsa Branch of Western Yamuna Canal with a command area of approximately 89,481.6ha (latitudes: 29°10′0″ to 29°33′0″ N and longitudes: 75°34′0″ to 76°13′0″ E) as shown in Fig. 1. Barwala sub-branch meanders through Barwala tehsil of Hisar district. The monsoonal tropical climate is distinguished as a semi-arid climate type due to its continental position on the outer edges of the monsoon region south-west. The period from October to next June continues almost dry, except few light showers owing to western depression/disturbances. Generally, summers are quite warm and winters are fairly cool. The primary features of the climate in the study region are its dryness, temperature extremes, and low rainfall. Temperatures are quite high during summer; the highest value sometimes occurs in May. With the onset of the summer monsoon, however, the temperature drops around 35.0°C. In January, the lowest temperature is reached. During the postmonsoon season and summer months, thunderstorms also happen. Evaporation stays above 12mm/day in the summer months. However, the average evaporation value stays around 6.8 and 4.0mm/day during the rainy and winter seasons, respectively.

3 Material and methods

3.1 Data set and characteristics

Both conventional and satellite data were used for the computation of GIR of different crops grown in the study area. The meteorological data including rainfall and temperature of 36-years (1980–2015) were collected from Agricultural Agrometeorology Department, Chaudhary Charan Singh Haryana Agricultural University, Hisar. Soil, land resources, and canal network data of the study area were received from Haryana Space Applications Centre, Hisar. The Landsat-8 image scene of 147/40 (row/path) for September 15, 2015, October 1, 2015, for Kharif, and February 6 and 22, and March 9, 2016, for Rabi season were used for crop classification.

For the classification of the images, an unsupervised classification iterative self-organizing (ISO)-Data Clustering approach was used with blocks outside and nonagriculture masks based on some defined conditions such as the number of clusters, threshold, number of iterations, standard deviation, etc. For crop mapping, a multiphase unsupervised ISO-Data classification was used. The pixels found to be correctly classified were masked and the second

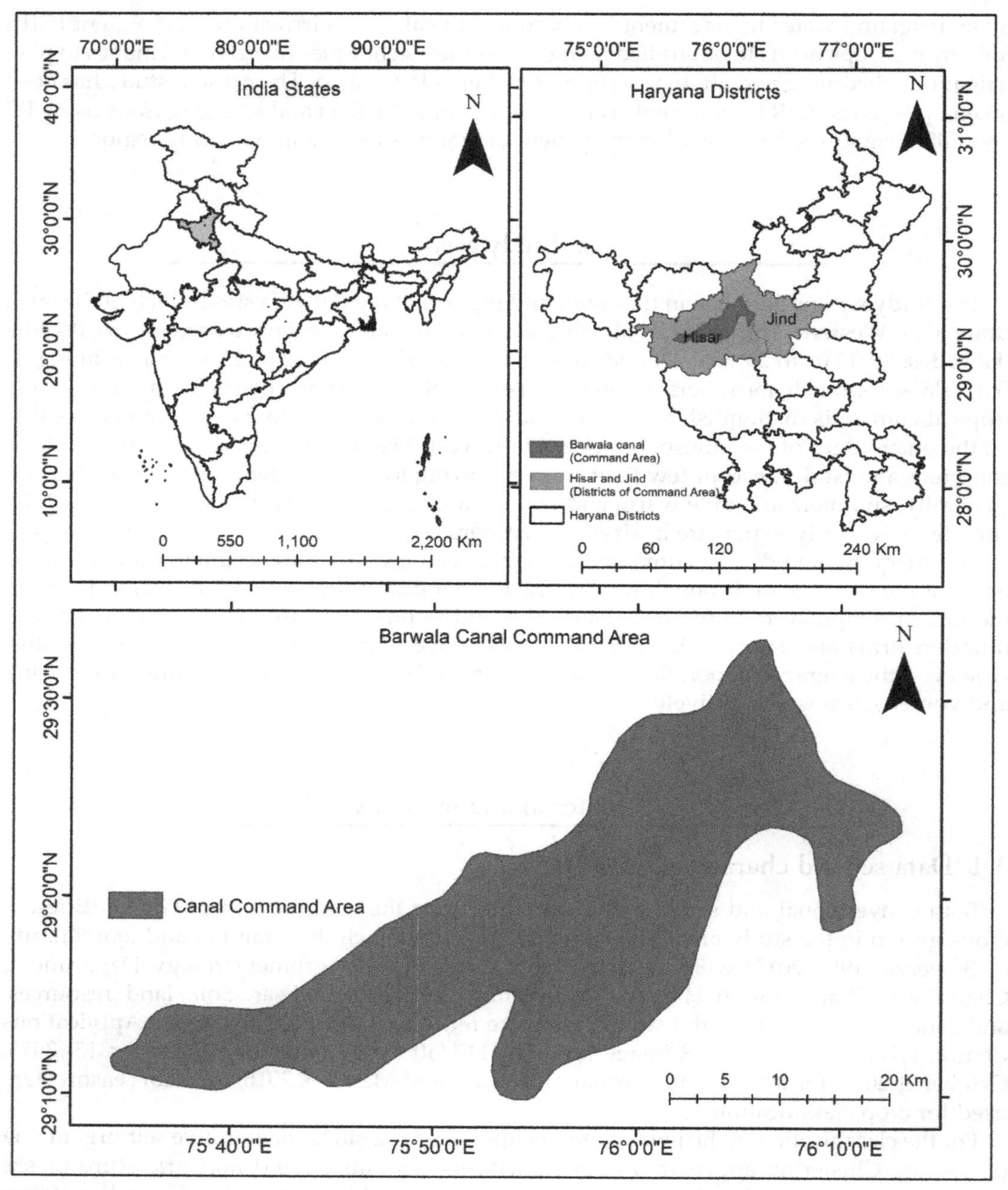

FIG. 1 Location map of the study area.

classification for the remaining region was performed until no further increase in class was achieved. Ground truth/field verification was conducted to ascertain the accuracy of the interpreted details with the help of the Global Positioning System (GPS) during field verification.

3.2 Procedure of GIR computation

Fig. 2 demonstrates the flowchart of the adopted methodology used for GIR computation. Remote sensing data, i.e., Landsat-8 images were used for crop classification of Rabi and Kharif season crops using ArcGIS and Geomatica software (Arvind et al., 2020). The rainfall and temperature data were interpolated using Inverse Distance Weighted (IDW) technique. The reference evapotranspiration (ET_o) was calculated for each position using minimum and maximum temperature and latitude of the meteorological stations using the Hargreaves-Samani formula (Hargreaves and Samani, 1985). To arrive at the crop water requirement of a particular crop, potential evapotranspiration is to be multiplied by the respective crop coefficients (K_c). It is affected by crop type, time of sowing, stage of crops, and climatic conditions including rain or irrigation frequency. K_c was considered for five main crops, i.e., Wheat and Mustard is 0.79 and 0.77 of Rabi season and Rice, Cotton and Pearl millet of Kharif season is 1.10, 0.83 and 0.73 (Michael, 2009), respectively. Every month (January to December) ET_c was determined for each crop by multiplying the respective crop coefficients with ET_o.

Effective rainfall (P_{eff}) is available in the plant root area that allows the plant to germinate or retain growth. The P_{eff} was calculated on monthly basis by multiplying the rainfall coefficient

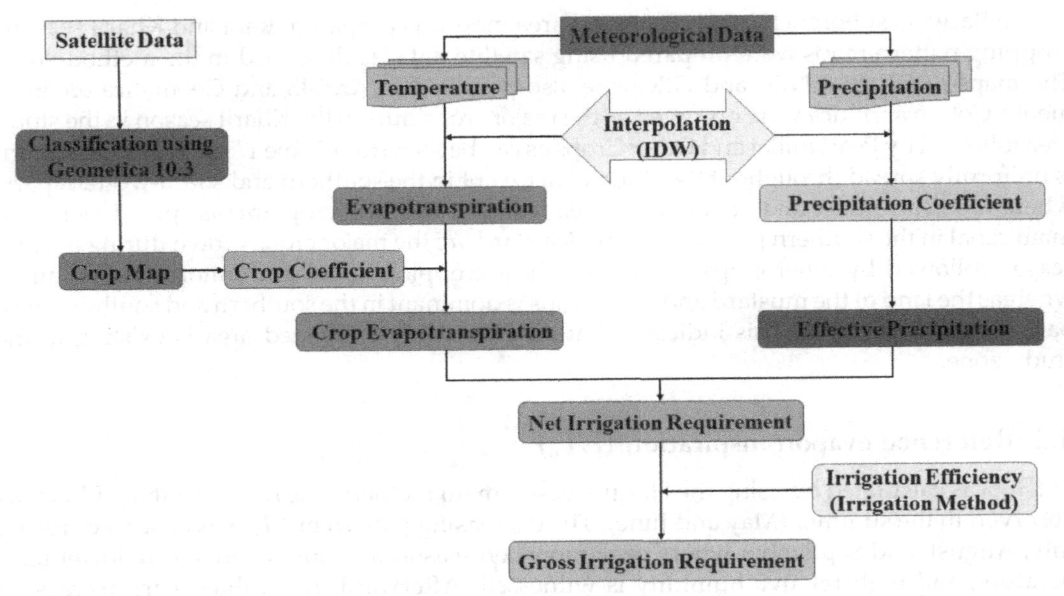

FIG. 2 Flowchart of the adopted methodology used for GIR computation.

with the monthly rainfall. The rainfall coefficient is fixed to 0.8 as a default value (Neelam and Hooda, 2016). Net irrigation requirement (NIR) is analyzed as a difference between monthly ET_c and P_{eff} on a monthly and annual basis in mm of water.

GIR depends on the effectiveness of the irrigation methods applied and calculated by the ratio of NIR to the efficiency of the irrigation method applied (Brouwer et al., 1989). In this study, the efficiency of the irrigation method (IRR_{eff}) is taken as 57% and calculated as (Brouwer et al., 1989):

$$IRR_{eff} = \frac{e_c \times e_a}{100} \tag{1}$$

In Eq. (1), e_c = conveyance efficiency (%), e_a = field application efficiency (%). The efficiency of the applied irrigation method (flood irrigation) $= \frac{95 \times 60}{100} = 57\%$. Value of e_c and e_a were taken from Brouwer et al. (1989).

4 Results and discussion

Based on conventional data collected from various sources and satellite data, a complete analysis of the water requirement of different crops grown in the Rabi and Kharif seasons was calculated in this study. Subsequently, the conventional data obtained from different sources was not indirectly usable form, it was first converted into the uniform spatial input format and then analysis was done in ArcGIS environment.

4.1 Spatial database creation and cropping pattern

The Barwala subbranch canal command area map was prepared. Rabi and Kharif seasons cropping pattern maps were prepared using satellite data, as described in the methodology. The maps of ET_c, P_{eff}, NIR, and GIR were also prepared in ArcGIS and Geomatica environments. Cotton and Rice were observed as the major crops during the Kharif season in the study area followed by Pearl millet and Other Crops as can be seen from Table 1. The cropping pattern is uniformly spread throughout the study area except in the southern and south-western parts where the Pearl millet and Fallow land area is more. The Rice crop area is spread along the main canal in the southern part. Wheat and Mustard are the major crops grown during the rabi season followed by other crops (Table 1). Wheat cropping zones are uniformly distributed whereas the land of the mustard and other crops is dominant in the southern and south-western parts of the study area. This indicates that an extensively cultivated area is existing in the study zone.

4.2 Reference evapotranspiration (ET_o)

ET_o was calculated by using the Hargreaves-Samani method. The highest value of ET_o was observed in the summer (May and June). The decreasing pattern of ET_o was observed during July, August, and September where peak monsoon season with low wind speed, lower temperature, and high relative humidity is witnessed. Afterward, ET_o values start decreasing continuously and showed the minimum during January that has the winter season with

TABLE 1 Kharif and Rabi seasons cropping pattern in the study area.

Crops/ category	Area (ha)	% of the total area
***Kharif* season**		
Cotton	42,531.12	47.53
Rice	4700.25	5.25
Pearl millet	4095.90	4.58
Other crops	16,485.21	18.42
Fallow land	15,177.69	16.70
Non-agriculture	6491.43	7.25
Total	89,481.60	100
***Rabi* season**		
Wheat	40,651.29	45.43
Mustard	23,383.08	26.13
Other crops	9866.43	11.03
Fallow land	9089.55	10.16
Non-agriculture	6491.43	7.25
Total	89,481.6	100

TABLE 2 Monthly variation in ET_o in the study area.

Month	Average ET_o (mm)	Month	Average ET_o (mm)
January	2.96	July	7.99
February	4.08	August	6.55
March	6.01	September	6.37
April	8.31	October	5.40
May	9.54	November	4.09
June	9.39	December	2.99

low temperature which is responsible for low evaporation rates. Values again start increasing in February, reaching their peak in May and June (Table 2). This comprises the peak summer season with low relative humidity, high wind speed, and higher temperature. In comparison, the lower values of ET_o have been obtained by Khandelwal and Dhiman (2015), because their study area Limbasi canal command area is located in Gujarat which experiences a lower average annual temperature. ET_o value depends on radiation and temperature parameters.

4.3 Crop evapotranspiration (ET_c)

ET_c was calculated by the multiplication of ET_o and K_c values and given in Table 3 for both Kharif and Rabi crops. It was noted from Table 3 that ET_c continuously decreases from 7.92 mm (May) to 4.48 mm (October) for the cotton crop, 6.85 mm (June) to 3.94 mm (October) for Pearl millet crop, and 10.32 mm (June) to 5.94 mm (October) for rice crop. A higher value of ET_c for rice crop has been spread by Saravanan and Saravanan (2014), because of the meteorological conditions, crop breed, and different crop cycles.

Similarly, the value of ET_c during the Rabi season (Table 3), was noted to decrease from 4.16 mm (October) to 2.30 mm (December) and increased from January (2.28) to March (4.63) for the Mustard crop. It continuously decreased due to the onset of the winter season and then increased due to a temperature rise. Raut et al. (2010) obtained the higher value for ET_c in the Mustard crop in the western Yamuna canal command area. For the wheat crop, ET_c varies between 2.33 mm (January) and 6.56 mm (April), and continuously decreased from November to January due to winter season-low temperature and then increased from February to April with an increase in temperature. The highest value of ET_c was observed in the Rice crop, i.e., 10.32 mm in June, and the lowest value in the Mustard crop, i.e., 2.28 in January. In comparison, Pakhale et al. (2010) reported higher values of ET_c for the wheat crop as the location of both the study areas is different.

TABLE 3 Monthly ET_c (mm) for Kharif and Rabi season crops in the study area.

	Kharif crop		
Month	**Cotton**	**Pearl millet**	**Rice**
May	7.92	–	–
June	7.79	6.85	10.32
July	6.63	5.83	8.79
August	5.43	4.78	7.20
September	5.29	4.65	7.00
October	4.48	3.94	5.94

	Rabi crop	
Month	**Mustard**	**Wheat**
October	4.16	–
November	3.15	3.23
December	2.30	2.36
January	2.28	2.33
February	3.14	3.22
March	4.63	4.75
April	-	6.56

4.4 Effective rainfall (P_{eff})

Precipitation data of 36 years (1980–2015) were used in the current study. The amount of monthly rainfall was found to be varying from 3.25 mm in November to 107.26 mm in July (Table 4). The highest effective rainfall of 107.26 mm was observed in July followed by 99.96 mm in August, and minimum P_{eff} was observed in November (3.25 mm), followed by October (3.51 mm) and December (4.46 mm).

4.5 Net irrigation requirement (NIR)

It appears from the result that the highest value of NIR occurred in the months where temperature (minimum, and maximum) was high and effective rainfall was low. Increased ET_c leads to water deficiency in the active root zone, which harms plant growth and development. NIR for Kharif season crops, i.e., Cotton, Pearl millet, and Rice are given in Table 5.

TABLE 4 Monthly variations of effective rainfall (P_{eff}) in the study region.

Month	Average P_{eff} (mm)	Month	Average P_{eff} (mm)
January	9.11	July	107.26
February	10.76	August	99.96
March	12.70	September	52.32
April	10.34	October	3.51
May	19.60	November	3.25
June	43.35	December	4.64

TABLE 5 NIR for Kharif season crops in the canal command area.

	NIR (mm/day)		
Month	Cotton	Pearl millet	Rice
May	7.29	–	–
June	6.39	5.45	8.93
July	3.17	2.37	5.33
August	2.21	1.55	3.98
September	3.60	2.96	5.32
October	4.37	3.83	5.82

As observed from Table 5 for the Cotton crop NIR ranges from 2.21 mm/day in August to 7.29 mm/day in May, which continuously decreased from May to August due to the onset of monsoon season and then increased from September to October, as the monsoon started receding. For Pearl millet NIR varies between 1.55 mm/day in August and 5.45 mm/day in June, which continuously decreases from May to August and then increased from September to October. NIR for Rice crops ranges from 3.98 mm/day in August to 8.93 mm/day in June. It decreased from June to August and then increased from September to October as for other Kharif crops. NIR in Kharif season is maximum for Rice crop and lowest in Pearl millet crop. Saravanan and Saravanan (2014) obtained lower values of NIR for the Rice crop in a study carried out in Tamil Nadu which has different climatic conditions.

Table 6 summaries the NIR values for the Rabi season crop and results show that in Mustard crop NIR ranges from 1.98 mm/day in January to 4.22 mm/day in March, which decreases from October to December due to the onset of the winter season and then increases from January to March due to rise the temperature in the spring season. Raut et al. (2010) obtained a higher value of NIR for the Mustard crop. For the wheat crop, NIR varies between 2.04 mm/day in January and 6.23 mm/day in April (Table 6). NIR decreased from November to January due to the onset of the winter season and then increased from February to April due to an increase in temperature in the spring season. NIR in Rabi season is maximum for Wheat crop and lowest in Mustard. NIR is dependent on ET_c and P_{eff} which fluctuate according to crop-specific K_c values as described in the methodology. In comparison, lower NIR values for Wheat crops have been obtained by Pakhale et al. (2010), as the study area was different from the present study experiencing different meteorological conditions.

TABLE 6 NIR for Rabi season crops in the canal command area.

Month	NIR (mm/day)	
	Mustard	**Wheat**
October	4.04	–
November	3.05	3.13
December	2.16	2.22
January	1.98	2.04
February	2.79	2.87
March	4.22	4.34
April	-	6.23

4.6 Gross irrigation requirement (GIR)

Table 7 recaps the GIR for Kharif season crops (Cotton, Pearl millet, and Rice), which indicates that for Cotton crop GIR varies between 3.88 mm/day in August and 12.79 mm/day in May. GIR continuously decreased from May to August due to the onset of the monsoon season and then increased from September to October due to the recession of the monsoon. The spatial variability of GIR for Cotton during the crop growing season is shown in Fig. 3. For pearl millet crop, GIR fluctuated between 2.73 mm/day in August and 9.57 mm/day in June. It continuously decreased from May to August due to the onset of the monsoon and then increased from September to October due to the recession of the monsoon season. Fig. 4 illustrates the spatial variability of GIR for Pearl millet during different months of crop growing season. GIR for rice crop varies from 6.98 mm/day in August to 15.66 mm/day in June, GIR decreased from June to August and then increased from September to October for the reason already described for other Kharif crops above. Fig. 5 shows the spatial variations in GIR for Rice during the crop growing season in the command area. A lower value of GIR for Rice crop was obtained by Saravanan and Saravanan (2014) because the applied irrigation efficiency and NIR were different from the current study.

Table 8 summarizes the GIR values for Rabi season crops including Mustard and Wheat. It was noted from Table 8 that for Mustard crop GIR extended from 3.48 mm/day in January to 7.41 mm/day in March, which decreased from October to December due to the onset of the winter season and then increased from January to March due to rise in temperature in the spring season. Spatial variability in GIR in the command area for the Mustard crop during the growing season is shown in Fig. 6. For wheat, crop GIR ranges from 3.58 mm/day in January to 10.93 mm/day in April. GIR decreased from November to January and then increased from February to April for the reason already described above for the Mustard crop. Fig. 7 shows the spatial variability of GIR during different months of the Wheat crop growing season in the command area. GIR in Rabi season was maximum for Wheat crop and lowest for the Mustard crop as the water requirement of the Wheat crop is highest and lowest for

TABLE 7 GIR for Kharif season crops in the canal command area.

	GIR (mm/day)		
Month	**Cotton**	**Pearl millet**	**Rice**
May	12.79	9.57	15.66
June	11.21	4.16	9.34
July	5.56	2.73	6.98
August	3.88	5.19	9.33
September	6.31	6.71	10.22
October	7.66	-	-

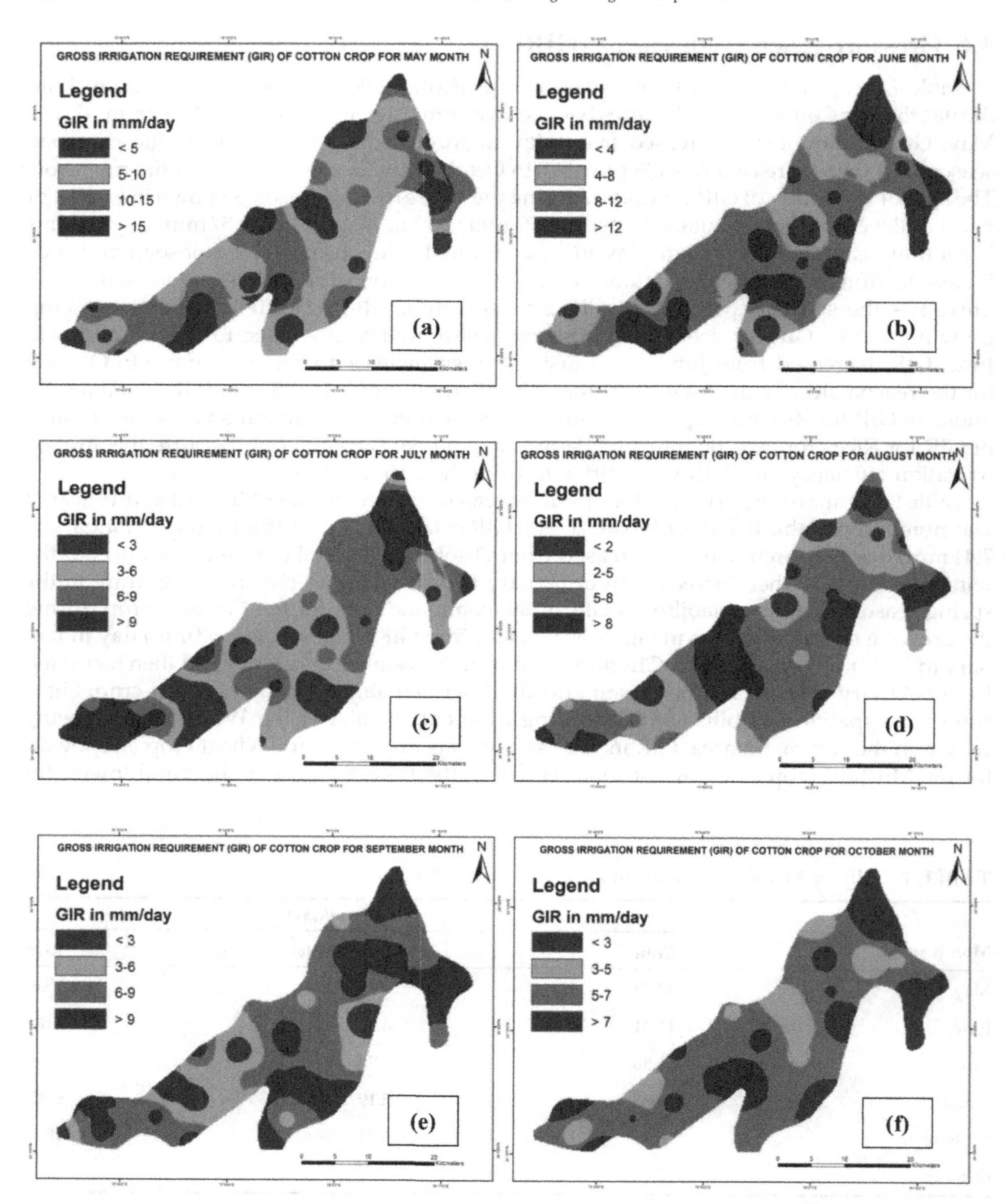

FIG. 3 GIR of the Cotton crop during (A) May, (B) June, (C) July, (D) August, (E) September, and (F) October months in the canal command area.

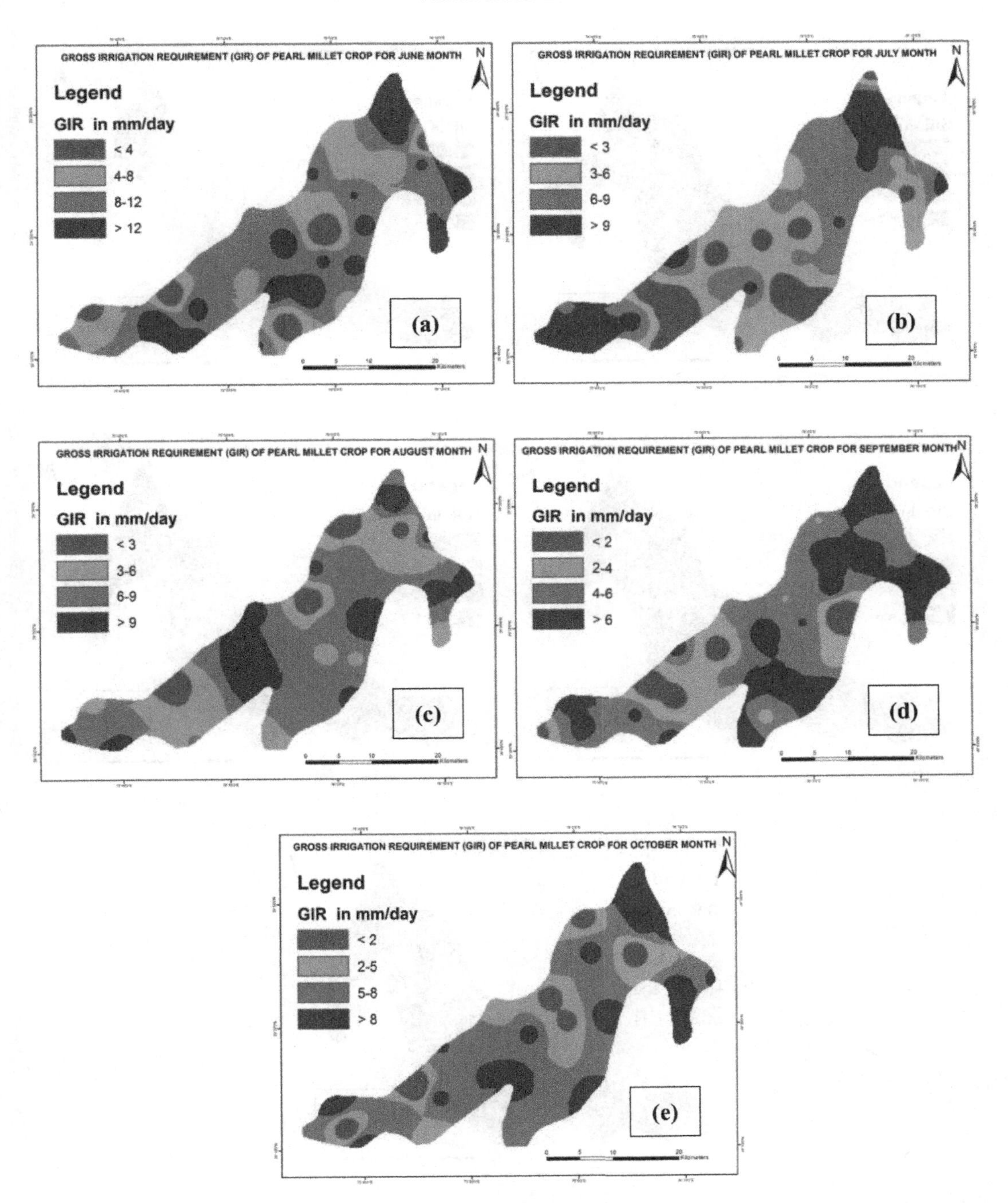

FIG. 4 GIR of Pearl Millet crop during (A) June, (B) July, (C) August, (D) September, and (E) October months in the canal command area.

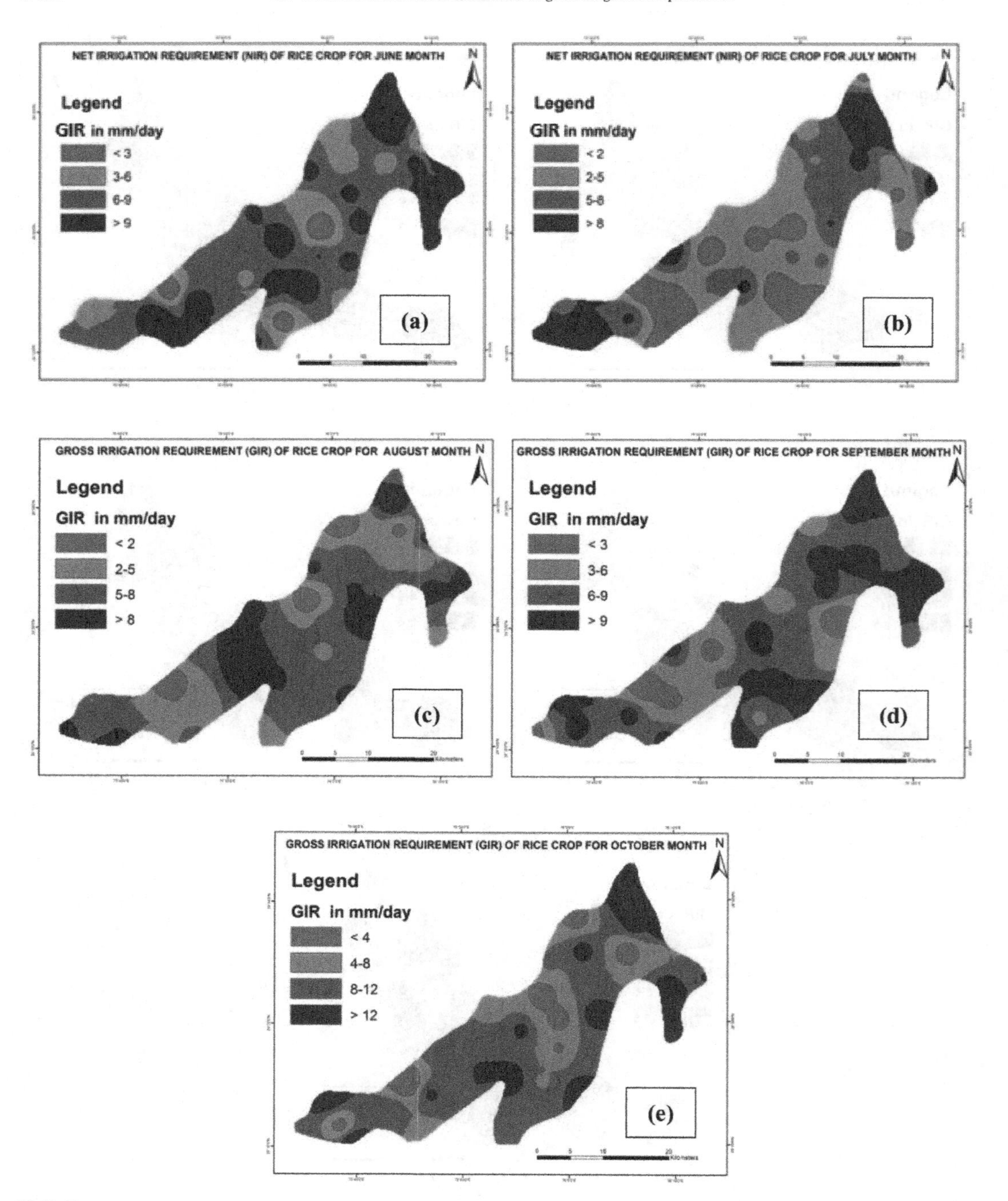

FIG. 5 GIR of Rice crop during (A) June, (B) July, (C) August, (D) September, and (E) October months in the canal command area.

TABLE 8 GIR for Rabi season crops in the canal command area.

Month	GIR (mm/day)	
	Mustard	Wheat
October	7.09	–
November	5.34	5.49
December	3.78	3.89
January	3.48	3.58
February	4.90	5.04
March	7.41	7.62
April	–	10.93

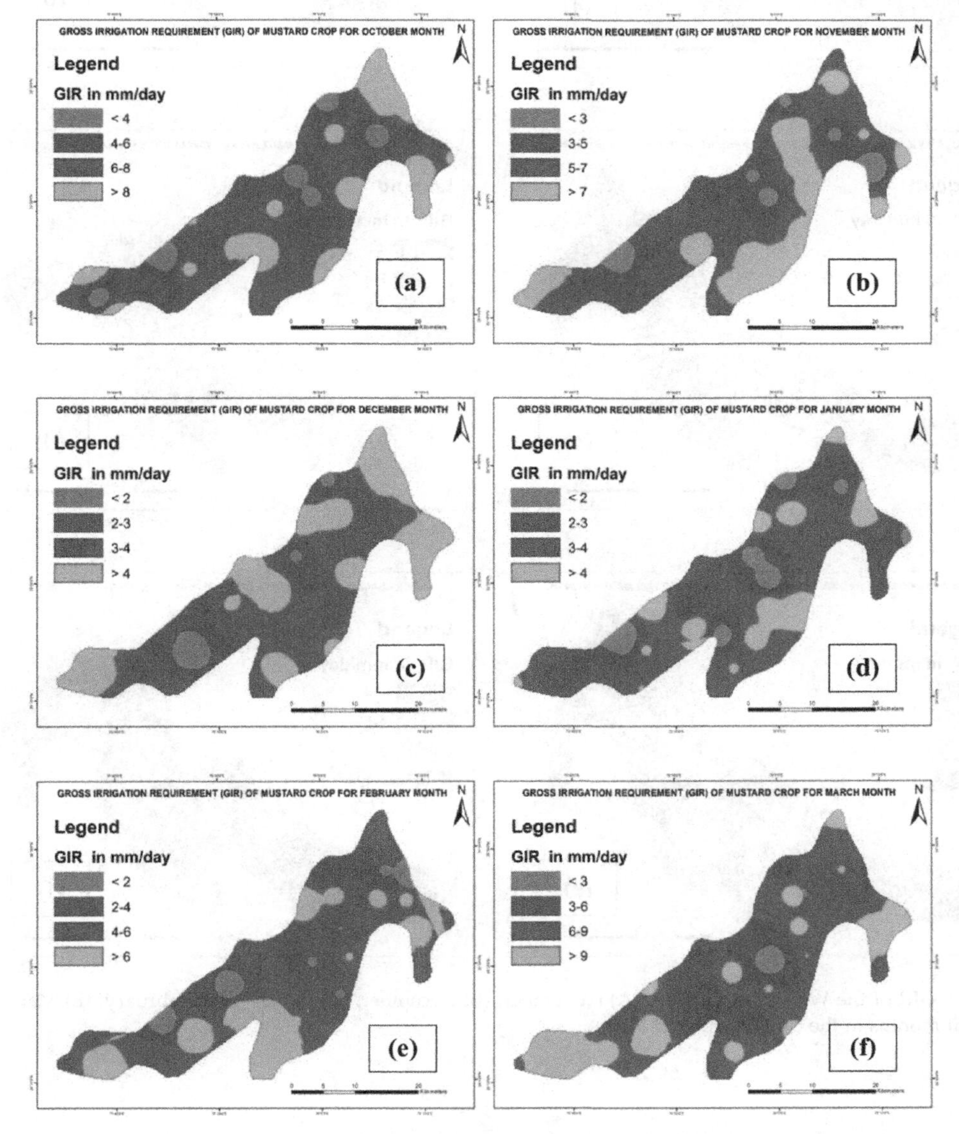

FIG. 6 GIR of Mustard crop during (A) October, (B) November, (C) December, (D) January, (E) February, and (F) March months in the canal command area.

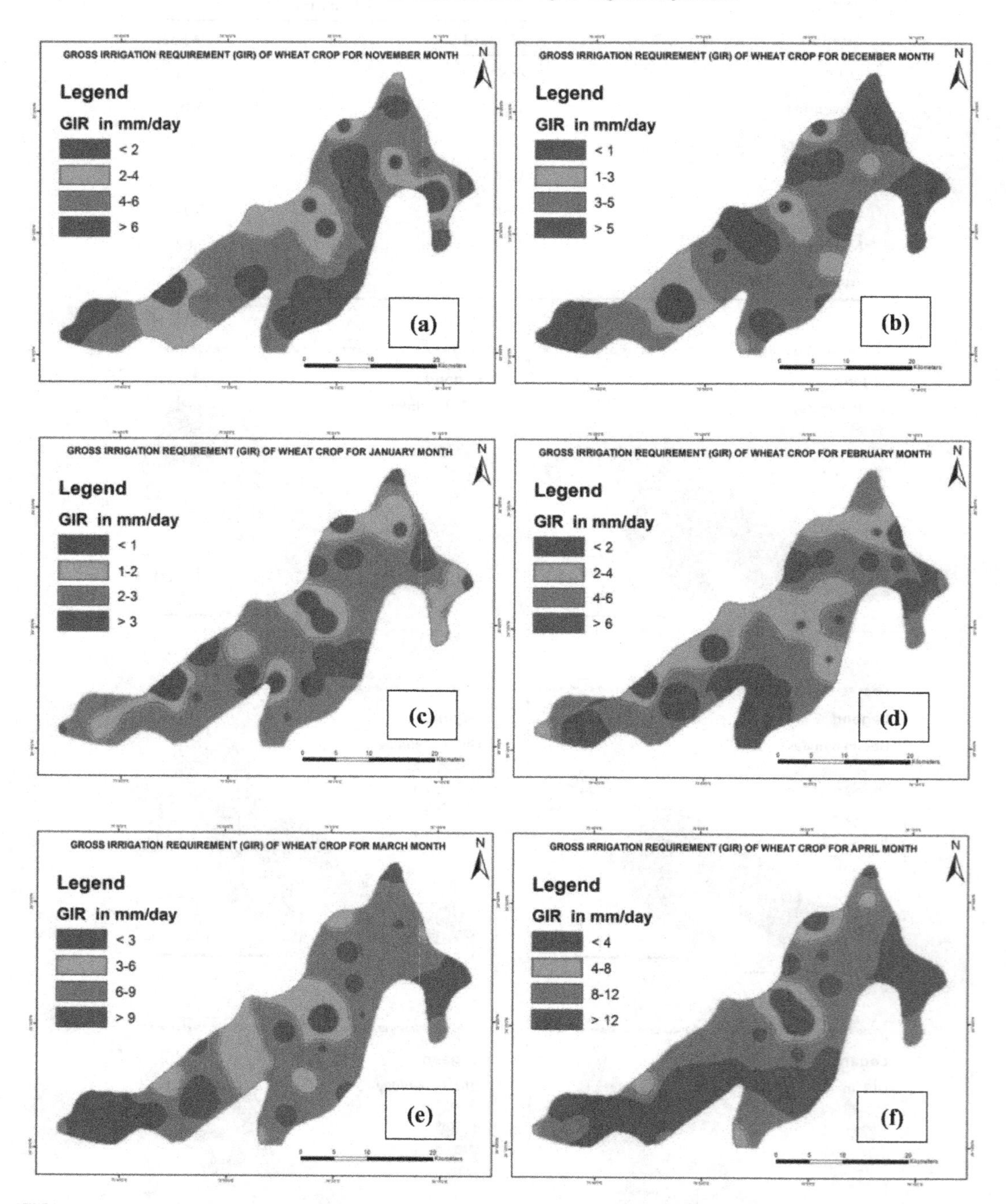

FIG. 7 GIR of the Wheat crop during (A) November, (B) December, (C) January, (D) February, (E) March, and (F) April months in the canal command area.

Mustard. GIR depends on NIR and $IRR_{eff.}$, which differ according to the crop. Lower values of GIR for the wheat crop were obtained by Pakhale et al. (2010), in the Karnal district because the applied irrigation efficiency and NIR were different from the present study area.

5 Conclusions

The current study was conducted to assess the Gross Irrigation Requirement (GIR) of Rabi and Kharif crops grown in the canal command area of Barwala sub-branch of western Yamuna Canal in Haryana State (India) by using Remote Sensing and GIS tools. Digital classification of Landsat-8 TM data of different seasons was used to generate the cropping patterns of different crops during Kharif and Rabi seasons. In the Kharif season, major crops identified in the command area are Cotton, Rice, and Pearl millet with an area of 42,531.12 ha, 4700.25 ha, and 4095.9 ha, respectively. Wheat and Mustard are observed as the main crops in the Rabi season with an area of 40,651.29 ha and 23,383.08 ha, respectively. ET_o ranges from 2.96 to 9.54 mm, which is affected by temperature and radiation values, which vary according to location. The higher ET_c was found in May and June, and lowest in August and September during Kharif season, while the highest ET_c was noted in March and April, and December and January were lowest during Rabi season. NIR of Kharif season crops was higher in May and June and lowest in August and September. Likewise, the NIR of the Rabi season crop was maximum in March and April, and lowest in December and January. GIR of Kharif season crop was maximum in May and June and minimum in August and September, whereas GIR of Rabi season crop was maximum in March and April, and minimum in December and January. According to the obtained results, it was concluded that the remote sensing and GIS techniques can be used for effective irrigation scheduling and sustainable management of available water resources in the command area. Also, the results reveal that the GIS-based irrigation management is useful for saving rain and canal water and applying the same water during a shortage in the study zone.

Conflict of interest

None.

References

Al-Safi, H.I.J., 2013. Applications of GIS software in irrigation project management. J. Univ. Babylon 21 (5), 1905–1915.

Arvind, Hooda, R.S., Sheoran, H.S., Kumar, D., Bhardwaj, S., 2020. RS-based regional crop identification and mapping: a case study of Barwala sub-branch of Western Yamuna Canal in Haryana (India). Indian J. Tradit. Knowl. 19 (1), 182–186.

Belmonte, A.C., González, J.M., Mayorga, A.V., Fernández, S.C., 1999. GIS tools applied to the sustainable management of water resources: application to the aquifer system 08-29. Agric. Water Manag. 40 (2–3), 207–220.

Brouwer, C., Prins, K., Heibloem, M., 1989. Irrigation Water Management: Irrigation Scheduling, Training Manual No. 4. Vol. 3 Food and Agriculture Organization, Rome, Italy, pp. 14–92.

Chaudhary, T.N., 1995. Water management: issues and strategies. J. Indian Soc. Soil Sci. 43 (4), 537–541.

Degirmenci, H., Buyukcangaz, H., Korukcu, A., 2005. Stakeholders and their Information Requirements in Monitoring and Evaluation (M&E) of Irrigation Projects. University of Uludag, Faculty of Agriculture, Agricultural Engineering Department Bursa, Turkey, pp. 1–10.

D'Urso, G., Menenti, M., 1995. Mapping crop coefficients in irrigated areas from Landsat TM images. In: Remote Sensing for Agriculture, Forestry, and Natural Resources. vol. 2585. International Society for Optics and Photonics, pp. 41–48.

Elbeltagi, A., Aslam, M.R., Malik, A., Mehdinejadiani, B., Srivastava, A., Bhatia, A.S., Deng, J., 2020. The impact of climate changes on the water footprint of wheat and maize production in the Nile Delta, Egypt. Sci. Total Environ. https://doi.org/10.1016/j.scitotenv.2020.140770.

Elbeltagi, A., Aslam, M.R., Mokhtar, A., Deb, P., Abubakar, G.A., Kushwaha, N.L., Venancio, L.P., Malik, A., Kumar, N., Deng, J., 2021. Spatial and temporal variability analysis of green and blue evapotranspiration of wheat in the Egyptian Nile Delta from 1997 to 2017. J. Hydrol. 594, 1–15. https://doi.org/10.1016/j.jhydrol.2020.125662.

FAO, 1995. Use of Remote Sensing Techniques in Irrigation and Drainage. FAO Water Report No.4. Cemagref-FAO, Montpellier, France.

FAO, 2016. World Agriculture: Towards 2015/2030, an FAO Study. FAO, Rome (forthcoming).

Fipps, G., Leig, E., 2003. GIS a tool in irrigation districts and projects. In: Biological and Agricultural Department, Texas A&M University, Presented in Second International Conference Conducted at the USCID on May 13, p. 200.

GOI, 2012. XII Five Year Plan. Planning Commission, Government of India, New Delhi.

Hargreaves, G.H., Samani, Z.A., 1985. Reference crop evapotranspiration from temperature. Appl. Eng. Agric. 1 (2), 96–99.

Khandelwal, S.S., Dhiman, S.D., 2015. Irrigation water requirements of different crops in Limbasi branch canal command area of Gujarat. J. Agrometeorol. 17 (1), 114.

Kumar, D., Arvind, N.A.S., Darshana, D., Arya, S., Bhardwaj, S., Abhilash, 2019. Soil loss estimation using geo-spatial technology in north western trai region of India. J. Agrometeorol. 21, 182–188.

Kumar, D., Dhaloiya, A., Nain, A.S., Sharma, M.P., Singh, A., 2021a. Prioritization of watershed using remote sensing and geographic information system. Sustainability 13 (16), 9456. https://doi.org/10.3390/su13169456.

Kumar, D., Nain, A.S., Singh, A., Mor, A., Bhardwaj, S., 2021b. Geo-spatial technology application for prioritization of land resources in Udham Singh Nagar district of Uttarakhand, India. Indian J. Tradit. Knowl. 20 (2), 595–603.

Michael, A.M. (Ed.), 2009. Irrigation: Theory and Practice. Vikas Publishing House.

MoWR, 2012. National Water Policy. Ministry of Water Resources, Government of India.

Neelam, Hooda, R.S., 2016. Determination of climatic water deficit using RS & GIS for Morni sub-watershed in Panchkula District, Haryana. Int. J. Tech. Res. Sci., 1–8.

Pakhale, G., Gupta, P., Nale, J., 2010. Crop and irrigation water requirement estimation by remote sensing and GIS: a case study of Karnal district, Haryana, India. Int. J. Eng. Technol. 2 (4), 207–211.

Pereira, L.S., Perrier, A., Allen, R.G., Alves, I., 1999. Evapotranspiration: concepts and future trends. J. Irrig. Drain. Eng. 125 (2), 45–51.

Raut, S., Sarma, K.S.S., Das, D.K., 2010. Study of irrigation and crop water requirements and growth of two Rabi crops grown in a semi arid region using agrometeorology and remote sensing. J. Indian Soc. Remote Sens. 38 (2), 321–331.

Saravanan, K., Saravanan, R., 2014. Determination of water requirements of main crops in the tank irrigation command area using CROPWAT 8.0. Int. J. Interdiscip. Multidiscip. Stud. 1 (5), 266–272.

Schmugge, T.J., Kustas, W.P., Ritchie, J.C., Jackson, T.J., Rango, A., 2002. Remote sensing in hydrology. Adv. Water Resour. 25 (8–12), 1367–1385.

Shultz, G.A., Engman, E.T. (Eds.), 2000. Remote Sensing in Hydrology and Water Management. Springer Verlag, Berlin Heidelberg, New York, pp. i–xx.

Stehman, S.V., Milliken, J.A., 2007. Estimating the effect of crop classification error on evapotranspiration derived from remote sensing in the lower Colorado River basin, USA. Remote Sens. Environ. 106 (2), 217–227.

SECTION V

Advances in watershed modelling

CHAPTER

19

Theoretical background and application of numerical modeling to surface water resources

Oscar Herrera-Granados

Faculty of Civil Engineering, Wrocław University of Science and Technology, Wrocław, Poland

OUTLINE

1 Introduction 319

1.1 What is a hydraulic numerical model? 320

1.2 Practical application of numerical models 320

2 Materials and methods 321

2.1 Model classification and theoretical background 322

2.2 Steps to be followed for building a numerical model 330

2.3 Example cases and discussions 333

2.4 Results 336

2.5 Limitations and discussion 337

3 Conclusions 338

References 339

1 Introduction

Water in or as a background in any landscape or townscape represents more than a "static" item that decorates the view of the town (Herrera-Granados, 2009). In many issues concerning hydraulic engineering and water resources management, the water bodies are assumed to be hydrostatic to facilitate the analysis of the phenomena that affect the water bodies. Nevertheless, water is practically always in motion, and the proper analysis of water resources management must consider this fact. Nowadays and due to the rapid evolution of the available computational tools, the behavior of water in motion can be easily forecasted thanks to the

https://doi.org/10.1016/B978-0-323-91910-4.00019-4

usage of numerical models (Herrera-Granados, 2009b). Since this chapter deals with water in motion, it is possible to apply the principles of hydraulics for noncompressible fluids to formulate numerical approaches for water flow and for their application in real-world cases.

1.1 What is a hydraulic numerical model?

Modeling is part of all engineering designs, and all models require knowledge of their background, good data, and careful interpretation. However, what is a hydraulic numerical model? What are its applications and how can this tool help engineers and practitioners with modeling of water resources?

As stated by Novak et al. (2010), a model is a term that is used in *hydraulics* to describe a physical or mathematical simulation of a "prototype." These simulations may be *direct* by using physical models, *semidirect* by using analogues, or *indirect* by applying computer-based analyses. The last type is what is possible to recall as a "numerical model." Another definition of a numerical model is the process of approximating the solution of equations that describe physical processes using stepwise approximations (Herrera-Granados, 2010). However, it is possible to add or redefine the term "hydraulic model" or simply "numerical model," as the application of hydraulic principles to forecast the water behavior of real-world cases in scaled prototypes under established conditions in a defined computational domain. It is necessary to highlight that the proper usage of numerical models is responsibility of the person who is utilizing this tool. That means, the engineer/scientist (the modeler) who is trying to analyze a natural phenomenon using numerical modeling shall possess the knowledge and experience concerning hydraulics to properly build, carry out, and interpret the input and output of the model.

1.2 Practical application of numerical models

Water specialists use numerical models to better understand natural phenomena that involve water flow. The usage of these computational tools is popular, and it is applied to analyze (among other phenomena):

- River basin management
- Coastal and maritime engineering
- Water flow in open channels
- River training infrastructure & sediment transport phenomena
- Design and optimization of hydraulic structures
- Effect of turbulence on water flow
- Complex water flow behavior

The number of issues that can be analyzed with the usage of numerical modeling is considerable and the dimension of the analyzed problem can be from a very detailed and complex water flow in a small and defined domain up to the water flow in a river basin or large water body. Thus, the first trade-off arises for the modeler, namely, to choose an appropriate mathematical framework according to the problem or phenomenon that is necessary to be studied. It is necessary to take into consideration the scale of the problem that needs to be analyzed. A wide number of examples can be found in the literature that demonstrates the application of different kinds of hydraulic numerical models for the management of water resources.

Large and medium scale water flow phenomena can be analyzed using hydraulic models, e.g., flooding. Among the tools used to prevent the impact of this phenomenon, flood risk assessment—commonly based on numerical modeling—is the preferred approach to obtain flood maps, which in turn permit the delineation of vulnerable areas. For example, Herrera-Granados (2008) applied the recommendations of the EU Water Framework Directive to analyze the flood extension within the city of Tuxpan in Mexico using Geographical Information Systems (GIS) and the software HEC-RAS (USACE—U.S. Army Corps of Engineers, 2021). Pinos et al. (2019) evaluated the performance of three different 1D numerical models (HEC-RAS, MIKE 11 & Flood Modeler) to estimate inundation water levels for the mountainous Santa Barbara River in Ecuador using different scenarios and data. Tavakolifar et al. (2020) applied a combined model (SWMM-2DCA) to simulate the inundation characteristics and flow interaction between urban drainage system and surface runoff. Herrera-Granados and Kostecki (2016) applied a 2D (SMS) numerical model to simulate the flow patterns within the reservoir Niedów in Poland and to establish the boundary conditions for a three-dimensional model for the experiments to calculate the weir coefficient of the Ogee weir of the hydraulic complex.

Herrera-Granados and Kostecki (2016) calculated the weir coefficient of the Ogee weir of the Niedów barrage and compared the numerical results (RANS) with the obtained ones in their physical model. The modelers compared the results of both physical and numerical models, obtaining good agreement between both modeling approaches. Herrera-Granados (2018) also tested three different CFD approaches (with Flow-3D) to analyze the turbulence behavior of the water flow behavior around a spur dike at the laboratory, finding that the RANS models are good enough to predict the time-averaged flow parameters of water but leading a misinterpretation of the turbulence behavior. This behavior is properly simulated using a more complex technique (Nezu and Nakagawa, 1993) such as Large Eddy Simulations (LES) or Direct Numerical Simulations (DNS). Nonetheless, these techniques are costly in terms of computational costs, e.g., RAM memory, disk space availability, and computational time.

As showed above, there are phenomena that cannot simply be analyzed with the usage of numerical models of water flow. Fortunately, there is also a large amount of research that demonstrates the reliability of the numerical models applied to real-world hydraulic engineering cases. Kiraga (2020) verified the sediment transport formulae that are implemented into the HEC-RAS software. Viti et al. (2019) highlighted the importance of considering air entrainment in the analysis of accurate flow motion forecasting along hydraulic jumps regardless of the complexity of the theoretical framework of modern CFD. Herrera-Granados (2021) also demonstrated that regardless of the accuracy of the modern CFD software, there are still some gaps to be filled to properly analyze complex hydraulic phenomena.

2 Materials and methods

In this section, the theory behind the most popular hydraulic numerical techniques is presented as it is a must for the modelers to understand and properly interpret before applying numerical models for real-world cases.

2.1 Model classification and theoretical background

Depending on the complexity of the governing equations of the mathematical model, the modeler relies on algorithms to perform various tasks in a quick and effective way. Thus, it is a must that the modeler understands how this complete process functions, otherwise, this can lead to what it is known as the "black box error." Fig. 1 depicts a simple framework that describes the interaction between the input of data into a numerical model (e.g., to analyze the rainfall-runoff of a storm event), and the output of the model (the hydrograph of the storm). The "black box" represents the numerical model that is used in this analysis. The modeler introduces the necessary data (input) into the black box, he/she blindly trusts in this model and assumes that the output is correct. As modelers commonly are the end-users of already developed computational software, they are not familiar with the development of the theory and algorithms that are inside this black box and how it estimates the final output of the simulation, but the modeler does possess the necessary knowledge on fluid dynamics/hydrology to properly input the necessary data and to interpret the validity of the model outcomes. Otherwise, the modeler can confront what is called the "black box dilemma" which means that the user wants to optimize the time of the analysis, but at the end, the theoretical knowledge of the modeler or his/her experience is not sufficient to properly interpret the output of the model (Fig. 1), thus the "black box dilemma" becomes a "black box error."

Models are categorized according to the aim dimensions to be resolved or to the applied numerical methods. The most common classification of numerical models is with respect to the spatial dimension. A model can be one-(1D), two-(2D), or three-dimensional (3D) depending on whether the leading variables vary in one, two, or three dimensions:

- 1D models: The velocity that is resolved in this type of models is the cross-sectional averaged velocity in the direction of the flow along the axis of a waterway
- 2D models: The computational domain is divided by a mesh of "n" elements or cells. At each cell (2D depth averaged models), the streamwise (x) and spanwise (y) velocity components are discretized.
- 3D models: In analogy to 2D models, the computational domain is divided by a mesh of "n" three-dimensional elements or cells. At each cell, the streamwise (x), spanwise (y) and vertical (z) velocity components are discretized.

The first created programs were mainly 1D because the computational tools were still a constraint to develop more powerful approximations/simulations. These types of models

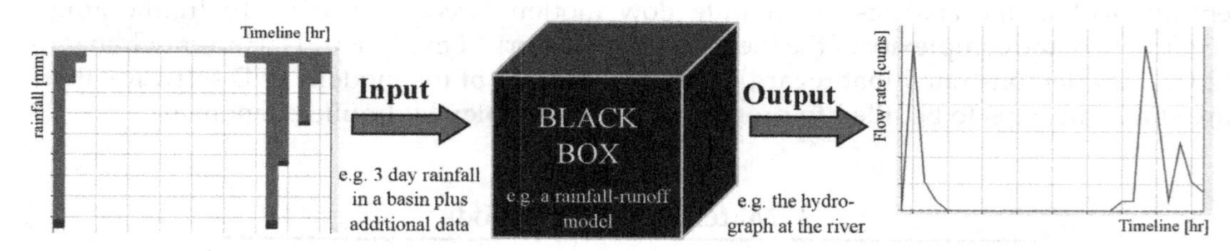

FIG. 1 The Black Box concept and the interpretation of the output of a model.

are useful tools that simplify the 3D world into a 1D framework. While 2D models are becoming the primary tool for hydraulic analysis, 1D models are still useful for the analysis of hydraulic and environmental issues and are relatively simple, providing reliable results.

2.1.1 1D numerical modeling

1D numerical modeling applied for the motion of water is limited to the calculation/estimation of the cross-sectional averaged velocity (see Fig. 2A and B). Some general assumptions are to be considered, e.g., the flow is mainly one-dimensional. This means that this kind of model is appropriate to analyze general quantities such as the Water Surface Elevation (WSE) or water depth along a river or the averaged velocity at a specific cross section of a river path. It is necessary to highlight that the calculated parameters are functions exclusively of the location and time. In this case, it is possible to consider that a system of maximal two-equations provides the modeler with the solution of the cross-sectional area (A) [L^2] and the flow rate (Q) [L^3T^{-1}] at each cross section (XS) that composes the computational domain.

For the mathematical analysis of the water flow in this small river reach, the analysis of the forces within a defined control volume are to be considered as depicted in Fig. 2C and D.

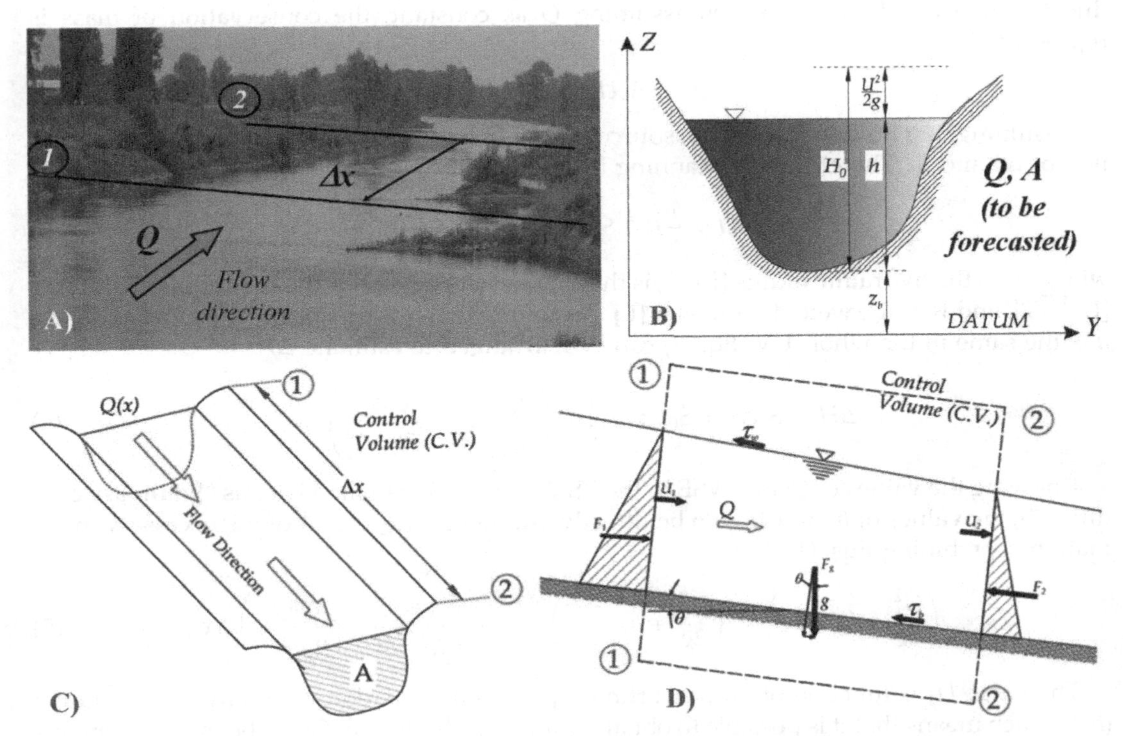

FIG. 2 (A) A quasi-straight river reach for 1D analysis. (B) The parameters to be calculated for 1D numerical model of water motion. (C) and (D) The main forces acting in a CV in a fluid in motion, delimited by XS 1 and 2.

The Control Volume (CV) is delimited by two cross sections upstream (1) and downstream (2). F_1 and F_2 are the resultant forces [MLT^{-2}] of the direction of the flow acting along the water depth of cross sections (XS) 1–2, respectively. F_g is the weight of the liquid [MLT^{-2}], in this case the weight of water. τ_b represents the total force due to the frictional resistance [MLT^{-2}] acting along the contact surface between the water and the channel boundary (the wetted perimeter) and τ_w is the resistance force due to the contact between the free surface and the wind [MLT^{-2}]. U_1 and U_2 are the area-averaged flow directions [LT^{-1}] at cross sections 1 and 2.

Considering the Volume flow rate Q as constant, the modeler can apply several assumptions for the analysis of the WSE. Moreover, neglecting the wind effects in the flow and assuming the value of the angle θ as very small ($\theta \approx 0$). Thus, the friendly Bernoulli Eq. (1) represents the simplest mathematical model of fluid motion for steady state:

$$\left(\frac{U_1^2}{2g} + h_1 + z_1\right) - \left(\frac{U_2^2}{2g} + h_2 + z_2\right) = \Delta H = S_f \Delta x \tag{1}$$

where h_1 and h_2 are the water depths [L] at XS1-2, z_1 and z_2 are the elevation of the lowest part of the XS 1 and 2 [L], ΔH represents the energy losses [L] between XS1-2; Δx is the distance [L] between the two cross sections, S_f represents the friction slope [−], and g is the acceleration due to gravity [LT^{-2}]. As one is assuming Q as constant, the conservation of mass is represented by:

$$Q_1 = A_1 U_1 = Q_2 = A_2 U_2 \tag{2}$$

Assuming that friction is the main source of energy losses, the unknown value ΔH is function of Δx and S_f. Therefore, the Manning formula can be applied

$$U = \frac{1}{n} R_h^{2/3} S_0^{1/2} \quad \rightarrow \quad R_h = \frac{A}{W} \tag{3}$$

where R_h is the hydraulic radius [L], S_0 is the channel slope [−], n is the roughness coefficient [L$^{-1/3}$T], and W is the wetted perimeter [L]. Assuming that S_f is practically the same as S_0 and n is the same in the whole CV, Eq. (3) can be rearranged to estimate ΔH:

$$\Delta H = S_f \Delta x = S_0 \Delta x = \frac{1}{2} n^2 \left(\frac{Q|Q|}{R_{h1}^{2/3} A_1^2} + \frac{Q|Q|}{R_{h2}^{2/3} A_2^2}\right) \Delta x \tag{4}$$

Knowing the values of Q and WSE in the XS-2 (what will be called later as "boundary condition"), the values of h_2 and U_2 can be directly estimated as z_2 is part of the necessary information. Combining Eqs. (1), (4):

$$\left(\frac{U_1^2}{2g} + h_1 + z_1\right) - \left(\frac{U_2^2}{2g} + h_2 + z_2\right) = \frac{1}{2} n^2 \left(\frac{Q|Q|}{R_{h1}^{2/3} A_1^2} + \frac{Q|Q|}{R_{h2}^{2/3} A_2^2}\right) \Delta x \tag{5}$$

The value U_1 is function of h_1, converting Eq. (5) as an equation with only one unknown (h_1), which means that it is possible to obtain its analytical solution. Once the modeler obtains the value of h_1, it is possible to calculate U_1 applying Eq. (2). For Eq. (5) there are three solutions of h_1 that solve this equation, two positives and one negative. The negative solution can be discarded, while the positive values are valid and correspond to the values of the water depth h_1 for two different flow regimes, one for subcritical (where the Froude number $Fr<1$) and the other for supercritical flow ($Fr>1$).

An example of the practical applications of 1D steady flow models is the estimation of the rating curve-RC (chart depicting *WSE* versus *Q*) at any XS of a river (Fig. 3B). Knowing the value of the bed slope, the values of *n* of the materials that compose the main channel and the riverbanks and applying Eq. (3), it is possible to obtain the RC analytically. However, this process is time-demanding and tedious, which can lead to committing mistakes and uncertainty while adding more information. Analyzing the same RC with a computer program is faster and more reliable. Fig. 3C depicts the computational domain of 200 m straight river reach with the same geometry of the river from Fig. 3A using HEC-RAS. The data that was also used for this analysis was the bed slope and *n*, Fig. 3D depicts the comparison of this chart using the Manning formula and the RC from HEC-RAS. As observable, the output of the numerical analysis and the analytical solution is the same, demonstrating that computer models are reliable.

However, for 1D modeling problems involving time dependency, a more complex mathematical framework is necessary. Thus, the 1D Saint-Venant system of equations (Eqs. 6, 7) can be applied:

$$\frac{\partial A}{\partial t} + \frac{\partial Q}{\partial x} = q_a \tag{6}$$

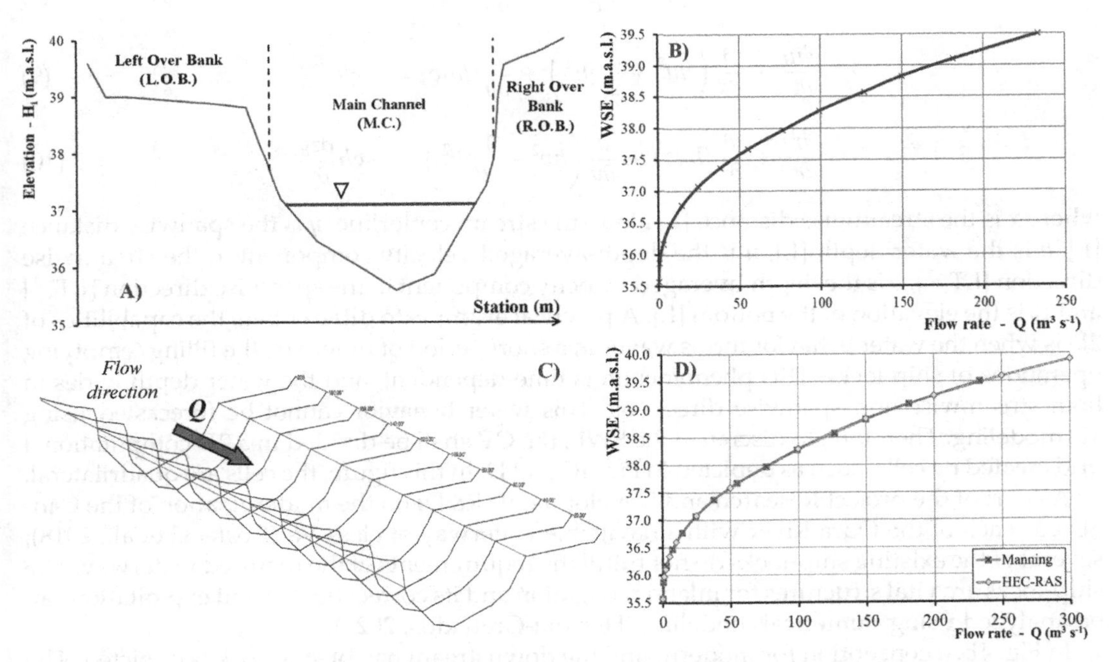

FIG. 3 (A) Shape of a river XS and its flood plains (LOB-ROB). (B) Estimated RC of this XS using the Manning formula; (C) The computational domain of a straight river reach using the geometry of (A); and (D) Comparison of the rating curves using analytical and numerical modeling.

$$\frac{1}{gA}\frac{\partial Q}{\partial t} + \frac{1}{gA}\frac{\partial}{\partial x}\left(\beta\frac{Q^2}{A}\right) + \frac{\partial y}{\partial x} + S_f - S_0 = 0 \tag{7}$$

where x is the longitudinal distance [L] along the stream centerline, t is time [T], A is the XS sectional area [L^2], Q is the water flow rate [L^3T^{-1}], q_a is the lateral flow per unit channel length [L^2T^{-1}], h is the water depth [L], and β is the Boussinesq's momentum correction coefficient [−]. Eq. (6) is known as the "Continuity equation" assuring the conservation of mass and Eq. (7) is the "Momentum equation," which means that the amount of momentum remains constant within the analyzed C.V.

2.1.2 2D numerical modeling

Two-dimensional numerical modeling is becoming the most common tool to analyze flow in water bodies where the vertical depth is small in comparison to the horizontal dimensions of the water bodies. Such phenomena as flow in open channels and rivers, floods, or tsunami modeling can be represented by nonlinear systems of partial differential equations, whose derivation involves the assumption that water is shallow. Thus, the Saint-Venant equations in two dimensions (SWE—Shallow Water Equations) can be applied. In the SWE (García-Navarro et al., 2019), the mass and momentum conservation are represented by (Eqs. 8–10):

$$\frac{\partial h}{\partial t} + u\frac{\partial hu}{\partial x} + v\frac{\partial hv}{\partial y} = 0 \tag{8}$$

$$\frac{\partial hu}{\partial t} + \frac{\partial}{\partial x}\left(hu^2 + \frac{1}{2}gh^2\right) + \frac{\partial}{\partial y}(huv) = -gh\frac{\partial z_b}{\partial x} \tag{9}$$

$$\frac{\partial hv}{\partial t} + \frac{\partial}{\partial x}(huv) + \frac{\partial}{\partial y}\left(hv^2 + \frac{1}{2}gh^2\right) = -gh\frac{\partial z_b}{\partial y} \tag{10}$$

where x is the streamwise distance [L] along the stream centerline, y is the spanwise distance [L], h is the water depth [L], u is the depth-averaged velocity component in the streamwise direction [LT^{-1}], v is the depth-averaged velocity component in the spanwise direction [LT^{-1}] and z_b is the elevation of the bottom [L]. A practical example to differentiate the capabilities of 2D is when the water behavior forms waves in a short period of time, e.g., the filling/emptying operations of ship locks. This phenomenon is time-dependent, and the water depth varies in both streamwise and spanwise directions. This water behavior cannot be forecasted using 1D modeling. Therefore, to discretize the SWE, the CV shall be divided in a 2D computational grid created by cells, such as depicted in Fig. 4C and D. In this figure, the cells are quadrilateral.

As part of the project Research and Development (R&D) on the modernization of the Canalized reach of the Odra River with a navigable waterway of class Va (Kostecki et al., 2018), several of the existing ship-locks do not fulfill the requirements of the required waterway. The ship-locks are vital structures for inland navigation and its correct design and exploitation can be analyzed using numerical modeling (Herrera-Granados, 2020).

In Fig. 4B, a conception for modernizing the downstream bay of ship-lock is depicted. The sudden opening of the sluice gates that control the movement of water from downstream to upstream generates waves that can improve the design of future hydraulic structures and

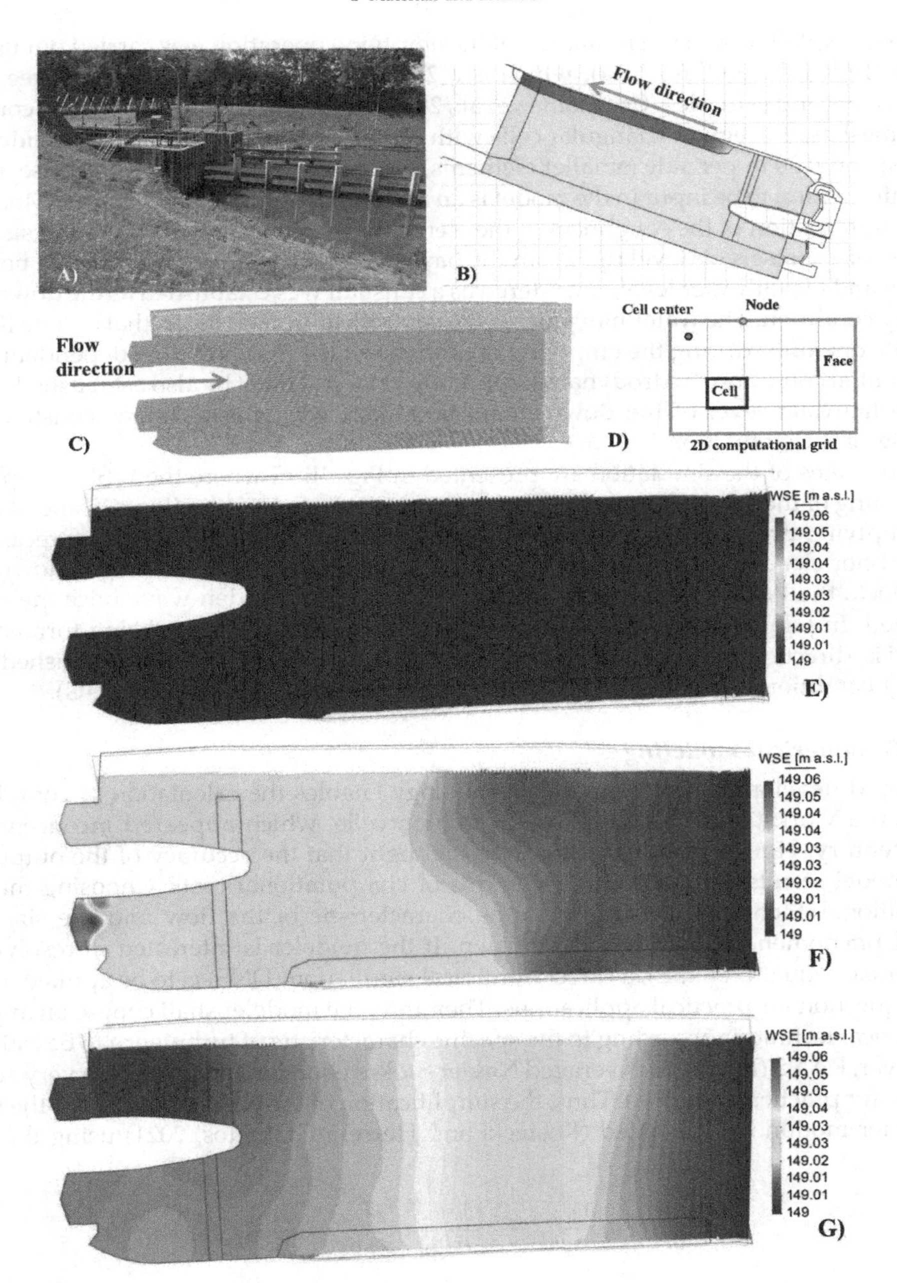

FIG. 4 (A) Outlet to the downstream bay of a shiplock of the Odra Cascade to be modernized; (B) The geometry (plan view) of the conception of a downstream visualized in IBER; (C) The triangular numerical mesh of the geometry of (B); (D) Example of a quadrilateral structured mesh. Simulation of the water flow motion of the downstream bay (B and C) at (E) The beginning; (F) During and (G) After the emptying operation of the chamber.

optimize its exploitation. The simulation of the emptying operation was carried out using the software IBER (Bladé et al., 2014), which is a 2D model for the simulation of free surface flows in rivers and estuaries (Cea Gómez et al., 2019). The computational grid was composed of three meshes of irregular triangular cells with dimensions of around 1.0 m per side (larger elements) and 0.30 m per side (smaller elements) as depicted in Fig. 4C. In this case, the necessary information to be input in the model is an accurate DEM (Digital Elevation Model) and accurate information of the roughness of the elements of the terrain. For this analysis, the information of the reservoir level upstream the bay was established as the upstream boundary condition and the tailwater level, considered as a constant was established as the downstream boundary condition. The water movement was controlled by algorithms that dictate the rules of the gate openings during the emptying operation. As the SWE are time dependent, an additional initial condition (hydrodynamic condition at $t=0$ s) must be also established. For this case, the tailwater level of the downstream boundary was considered as constant at the beginning of the simulation.

The outcomes of the simulation are presented in Fig. 4E–G where the WSE is depicted at the beginning of the simulation ($t=0$), during the emptying operation (Fig. 4F) and at the end of the emptying process. It is observable how the computer model properly forecasted the flow behavior close to the culverts at where in the middle of the chamber just downstream the shiplock, the WSE is larger in the center, depicting the sudden wave once the culverts are opened. In Fig. 4F, the WSE elevation values show how a wave is being formed in the streamwise direction and how once this wave faces the constant WSE established as the boundary condition downstream, this fact changes the wave direction (Fig. 4G).

2.1.3 3D numerical modeling

The rapid development of computer technology enables the calculation of complex 3-D models on a workstation PC with several million cells, which appeared inconceivable in the last century. However, it is necessary to highlight that the accuracy of the output from the 3D model results very expensive in terms of computational costs. Choosing the better Computational Technique according to the characteristic of the flow and the size of the analyzed phenomenon is a fundamental step. If the modeler is interested in resolving the Navier-Stokes equation (NSE), a direct numerical simulation (DNS) is to be applied, but this is out of question for practical applications. Therefore, the modeler shall choose an appropriate numerical technique according to the specific characteristic of turbulence to be estimated.

However, RANS (Reynolds-Averaged Navier-Stokes equations) represents a very accurate technique for practical purposes. Thus, the simplification of the NSE equations for the motion of 3D water motion can be stated (Kostecki and Herrera-Granados, 2021) using the RANS model as:

$$\frac{\partial \bar{u}_i}{\partial t} + \bar{u}_j \frac{\partial \bar{u}_i}{\partial x_j} = -\frac{1}{\varrho}\frac{\partial \bar{p}}{\partial x_i} + v\frac{\partial^2 \bar{u}_i}{\partial x_j \partial x_j} - \frac{\partial}{\partial x_j} R_{ij} \tag{11}$$

$$\frac{\partial \bar{u}_i}{\partial x_i} = 0 \tag{12}$$

where $\bar{u}_i$ is the mean averaged velocity component [LT^{-1}] in three directions; $\bar{p}$ is the mean pressure [$ML^{-1}T^{-2}$]; v is the molecular viscosity [L^2T^{-1}], ϱ is the water's density [ML^{-3}]; and R_{ij} is the Reynolds stress tensor [L^2T^{-2}], defined as:

$$R_{ij} = -u_i'u_j' \tag{13}$$

where u' is the fluctuating part of the velocity [LT^{-1}]. To get a closure for the system of Eqs. (11), (12), The Boussinesq approximation represents a way to compute turbulent stresses (Herrera-Granados, 2021):

$$-u_i'u_j' = v_T\left(\frac{\partial \bar{u}_i}{\partial x_j} + \frac{\partial \bar{u}_j}{\partial x_i}\right) - \frac{2}{3}k\delta_{ij} \tag{14}$$

where δ_{ij} is the Kronecker delta [−] and k is the turbulent kinetic energy per unit mass [L^{2}T^{-2}] and v_T is the Eddy viscosity [L^{2}T^{-1}], calculated with:

$$-v_T = c_\mu \frac{k^2}{\varepsilon} \tag{15}$$

Using the $k-\varepsilon$ model, the system can be closed with the equation (Wu, 2007):

$$\frac{\partial k}{\partial t} + \bar{u}_j\frac{\partial k}{\partial x_j} = \frac{\partial}{\partial x_j}\left(v + \frac{v_T}{\sigma_k}\frac{\partial k}{\partial x_j}\right) + P_k - \varepsilon \tag{16}$$

where ε is the dissipation rate of k [L^{-2}T^{-3}], and P_k is the production of turbulence by shear [L^{-2}T^{-3}]:

$$P_k = -u_i'u_j'\frac{\partial \bar{u}_i}{\partial x_j} \tag{17}$$

moreover, ε can be modeled with the expression:

$$\frac{\partial \varepsilon}{\partial t} + \bar{u}_j\frac{\partial \varepsilon}{\partial x_j} = \frac{\partial}{\partial x_j}\left(v + \frac{v_T}{\sigma_\varepsilon}\frac{\partial \varepsilon}{\partial x_j}\right) + c_{\varepsilon 1}\frac{\varepsilon}{k}P_k - c_{\varepsilon 2}\frac{\varepsilon^2}{k} \tag{18}$$

The constants of the $k-\varepsilon$ model are defined as:

$$C_\mu = 0.09,\ C_{\varepsilon 1} = 1.44,\ C_{\varepsilon 2} = 1.92,\ \sigma_k = 1.0,\ \sigma_\varepsilon = 1.3 \tag{19}$$

A simple example of the application of CFD water resources engineering is related to the design of hydraulic structures for flood alleviation schemes, such as the new labyrinth spillway proposed for the dam Witka in Poland (Fig. 5B). This new alleviation structure was designed and built along the axis of the former earth dam, destroyed during the flood of August 2010 (Kostecki et al., 2013).

Before its reconstruction, a physical model (geometric scale 1:27) was built at the laboratory (Fig. 5A) of the Wroclaw University of Science and Technology (WrUST) to test the draining capacity of the new alleviation scheme (Herrera-Granados and Kostecki, 2016). Furthermore, a numerical analysis was carried out using the CFD code Flow 3D (Isfahani and Brethour, 2009). In Fig. 5C, the 3D numerical mesh is depicted, composed by two blocks of hexahedral elements. Flow3D incorporates a special technique, known as the FAVOR method, to define the general geometric regions within a rectangular grid. This FAVORized view generated by Flow3D is depicted in Fig. 5D. The boundary conditions were established based on the 2010 flood events. The output of the computations is depicted in Fig. 5E and F. The results of the model were compared with the experiments carried out at the WrUST, presenting good agreement, and demonstrating that RANS techniques are sufficient for 3D modeling for practical or engineering purposes.

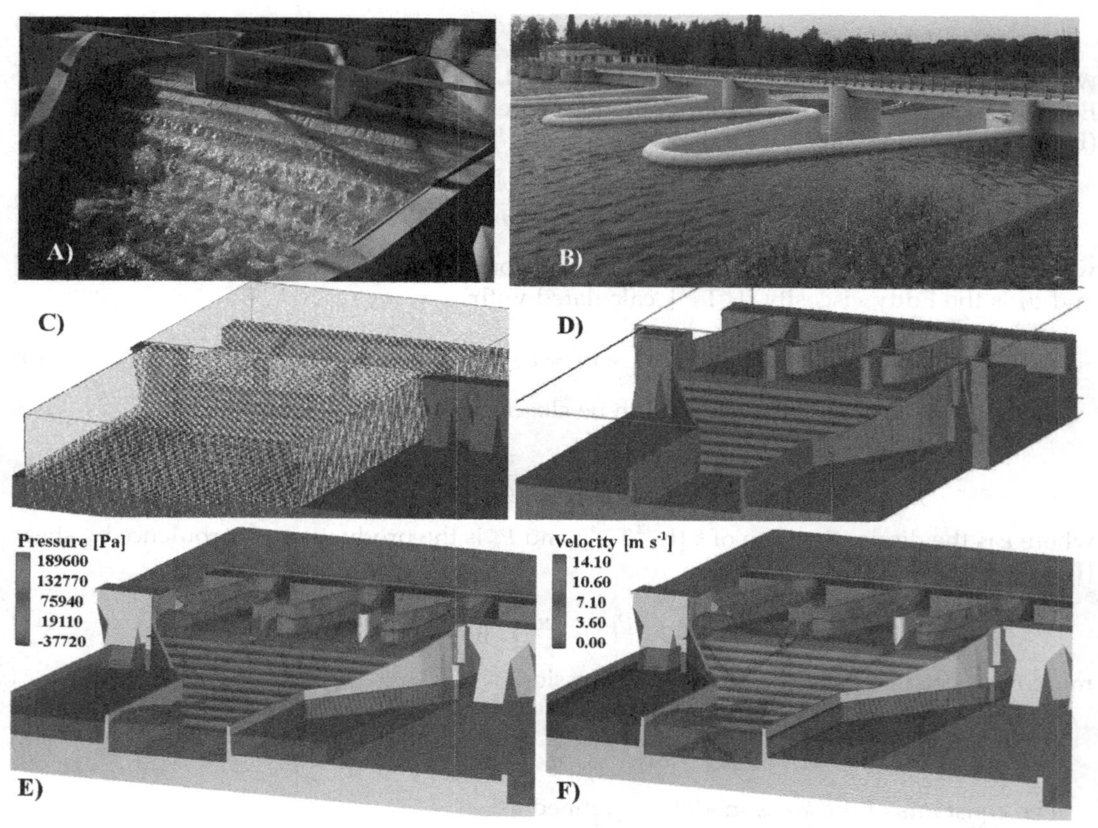

FIG. 5 (A) The physical model of the Witka labyrinth; (B) The new reconstructed labyrinth spillway of the Witka barrage; (C) The numerical mesh that was used for the analysis and its numerical domain representation using Flow3D (D). Outcomes of the CFD: (E) the pressure within the fluid domain once the computations reached almost quasi-steady state. (F) The values of the velocity magnitude.

2.2 Steps to be followed for building a numerical model

The author proposes five simple steps (based on his experience) to accurate analyze any phenomenon related to water flow:

i. Data retrieval and geometry construction of the computing domain
ii. Choosing the mathematical framework of the analyzed phenomenon
iii. Choosing an adequate numerical method to discretize the mathematical model
iv. Defining the boundary and initial conditions of the problem
v. Calibrating and validating the Model

In the next subsubsections, each of these steps are briefly described considering the difficulties that the second and third steps bring. Fortunately, for the end-users of numerical models, there are plenty computational codes (commercial, freeware, or noncommercial) that are already available for practitioners. Thus, these two steps are simplified for end-users of these available software, not forgetting to be careful with the "black-box dilemma."

2.2.1 *Data retrieval and geometry construction*

To simulate the flood extension in 1D along a river reach, it is necessary to get the information of transversal cross sections along the axis of the reach to be analyzed and its flood plains. The detailed information of the land cover is necessary as well as the hydrological conditions to be established further on as boundary and initial conditions. The retrieval of all this information is time demanding and often requires measurements in the natural environment at full scale which are expensive. Nowadays, there is plenty of information available using GIS. Furthermore, the newest versions of computer programs that simulate water flow contain GIS tools to import the information of terrain models (raster files) and other important information such as land cover, imperviousness and so on available as vector files. This GIS tools are very helpful to simplify the correct process of building the geometry within the model interface, minimizing the number of errors.

In the case of 2D modeling, the information provided by the transversal cross sections can be insufficient. Therefore, an accurate bathymetric modeling is necessary. Digital Bathymetric Modeling (DBM) integrated with LIDAR (Light Detection and Ranging) data can provide the geometry for 2-D models. However, thanks to the usage of economical echo sounders for large-scale bathymetry, the survey can be a fast, economical, and reliable technique (Uciechowska-Grakowicz and Herrera-Granados, 2021). Nonetheless, this process is time demanding. Thus, the modeler discovers that for a particular problem, there is not enough data, or in some cases none. The enterprise of acquiring sufficient data with which to solve this specific problem is a process needing careful consideration. It is always a good practice to assess oneself the quality of any primary data, using own experience, data mining techniques, and other devices for sorting out good data (Knight, 2013).

2.2.2 *Choosing the mathematical framework*

This step depends particularly on what the modeler wants to forecast and the scale of the problem to be analyzed. On the one hand, it is practically impossible to acquire the map of the flood extension along a long river reach using LES because of the demands of the computational technique. On the other hand, a 1D numerical approach is not able to estimate the secondary currents along the groyne fields of heavily modified water bodies. First, the modeler shall establish the dimension of the problem to analyze and its complexity. CFD codes can analyze the water flow along a long river reach, but in case only one parameter of interest is the WSE, the usage of CFD for calculating the WSE is unnecessary, complex, time demanding and a waste of computational resources. 1D modeling in this case represents a better solution for analyzing large-scale phenomena. However, in case the information that we have is accurate and sufficient, it is preferred to analyze water flow problems using 2D models.

2.2.3 *Choosing the appropriate numerical method*

It is not the same to analytically solve the Bernoulli equation, to approximate the solution of the Saint-Venant equations or to discretize the $k-\varepsilon$ (RANS) model. Explaining the numerical methods and schemes that are utilized for approximating the solution of the equations previously presented is extensive and out of the scope of this chapter. There are several positions in the literature that explain the numerical schemes that are used for the Saint-Venant equations (Murillo and García-Navarro, 2010) or for the discretization of the RANS equations (Wu, 2007).

There are computer codes, such as FlexPDE, where the modeler can introduce his own mathematical framework and the solver applies the Finite Element Method (FEM) to approximate the solution along a mesh that the same program generates. Fortunately, most of the computer programs already contain their own mathematical approach and the numerical scheme to discretize the system of equations.

The end-user of these computer programs (the modeler) is frequently not facing this problem. Nonetheless, it is always a healthy practice to try to solve step by step a small water flow problem to value the effort that the scientists who created these programs put to facilitate the modelers' work, not forgetting to avoid the "black box error."

2.2.4 Boundary and initial conditions

Once the geometry is defined and the numerical model has been chosen. The next step is to establish appropriate boundary and initial conditions. A numerical model is not showing the reality, but it is forecasting the water flow along a computational domain for the conditions, established by the modeler. As presented in the previous section. In the case of analyzing 1D problems for steady state in unidirectional flows, the necessary boundary conditions to be considered are: (i) a known water surface in the downstream boundary and (ii) the flow discharge coming from upstream. In case the problem is 1D and time dependent, the flow hydrograph (Q vs. time) upstream is the minimal information necessary to run the model and the rating curve downstream is convenient to be applied as the downstream boundary condition. Moreover, for time-dependent problems, the hydraulic conditions at the beginning of the simulations must be established.

In the case of 2D and 3D simulations, the definition of boundary conditions is similar considering that there can be more than only two boundaries as it is the case of 1D modeling. Moreover, there is additional information to be input into the model, such as lateral inflows or outflows, pumping stations, and so on. Additionally, there are other options to be considered as appropriate boundary conditions. Summarizing, the outcomes of the model are function of the boundary conditions that are input into the model.

2.2.5 Calibration and validation of the model

Is it possible to blindly believe in the outcomes of a numerical model? The best practice of assuring confidentially of our models is the process of validation and calibration.

The most used calibration procedure is although the optimization of the model performance, which means that the model's output is compared with observed data. This can be done using trial-error procedure or changing the parameters that are introduced as data in the model. For the case of 1D and 2D models, the roughness coefficient is the parameter that most often is changed for calibration purposes. If the simulation is not satisfactory, this value is changed, and the model is run again. The simulation is repeated until a satisfactory solution is obtained. Once the model is calibrated, it is recommended that the model is tested to check its performance, after calibration and before using it in practice (Brath et al., 2006). Such a testing procedure is called validation. A simple and intuitive way of performing validation for a model is to use the so-called "split-sample" procedure. The observed data are divided into two groups: one group is used for calibration and the other group is used to test the model by emulating a real-world application. In case the model's output is like the observed data for validation, thus the model uncertainty is practically eliminated.

Unfortunately, this practice depends much on the availability of data. For numerical models that can be compared with the data of laboratory (physical) models, this calibration and validation procedure is a must.

2.3 Example cases and discussions

Applying the previously mentioned five steps, an example of a 2D model along a quasi-straight reach is presented for the Odra River in Poland (Fig. 6A). This model was carried out as part of the project to enhance the navigability conditions of the Odra Cascade (Kostecki et al., 2018).

The canalized part of the Odra River is classified as a waterway of regional importance. Twenty-five low head barrages compose this part of the Odra River (Fig. 6B). These structures allow the control of the water surface from one structure to the other for exploitation conditions. For this case, it is of interest of the modeler to analyze the flow behavior inside a small reach of the proposed waterway (Fig. 6C and D). Therefore, two exploitation conditions analyzed, the minimal allowed flow rate for navigation (Q_{Nmin}), and the maximal allowed flow rate for navigation (Q_{Nmax}).

2.3.1 *Retrieve of information and geometry building*

The information that it is necessary to be considered for geometry is the topography of the terrain that was obtained from the ISOK project and the DBM of the river. The HEC-RAS model was applied. The greatest advantages of HEC-RAS is that the program is freeware & reliable (USACE—U.S. Army Corps of Engineers, 2021). The DEM that was used for this study has a resolution of 1x1 meters per cell. The DBM data was obtained from interpolation information from the bathymetric survey on the Odra River (Uciechowska-Grakowicz and Herrera-Granados, 2021).

The DEM and DBM were joint with GIS techniques prior their import to the computer software. In addition, Cadastral data were employed to approximate the values of the roughness coefficient of the flood plains of the river to define the main axis and banks of the waterway (Fig. 6D). The computational grid was defined by three meshes: one for the LOB (Left OverBank), one for the main channel, and one for the ROB (Right OverBank). As noticeable in Fig. 6E, the meshes of the flood plains are coarser than the mesh of the main channel. In the middle of the channel, the mesh is finer, tends to be structured, and almost orthogonal, obtaining more information of the flow structure in the center of the channel, which in this case is more important to be analyzed than the water flow in the flood plains.

2.3.2 *Mathematical framework and numerical methods*

For 2D modeling, the computer program applies the following assumptions: Incompressible flow, uniform density and that the vertical length is much smaller than the horizontal lengths. Therefore, the vertical velocity is small, and the pressure is hydrostatic, making the SWE an appropriate tool for this analysis. The momentum equation then becomes the two-dimensional form of the Diffusion Wave Approximation (DWA). Combining this

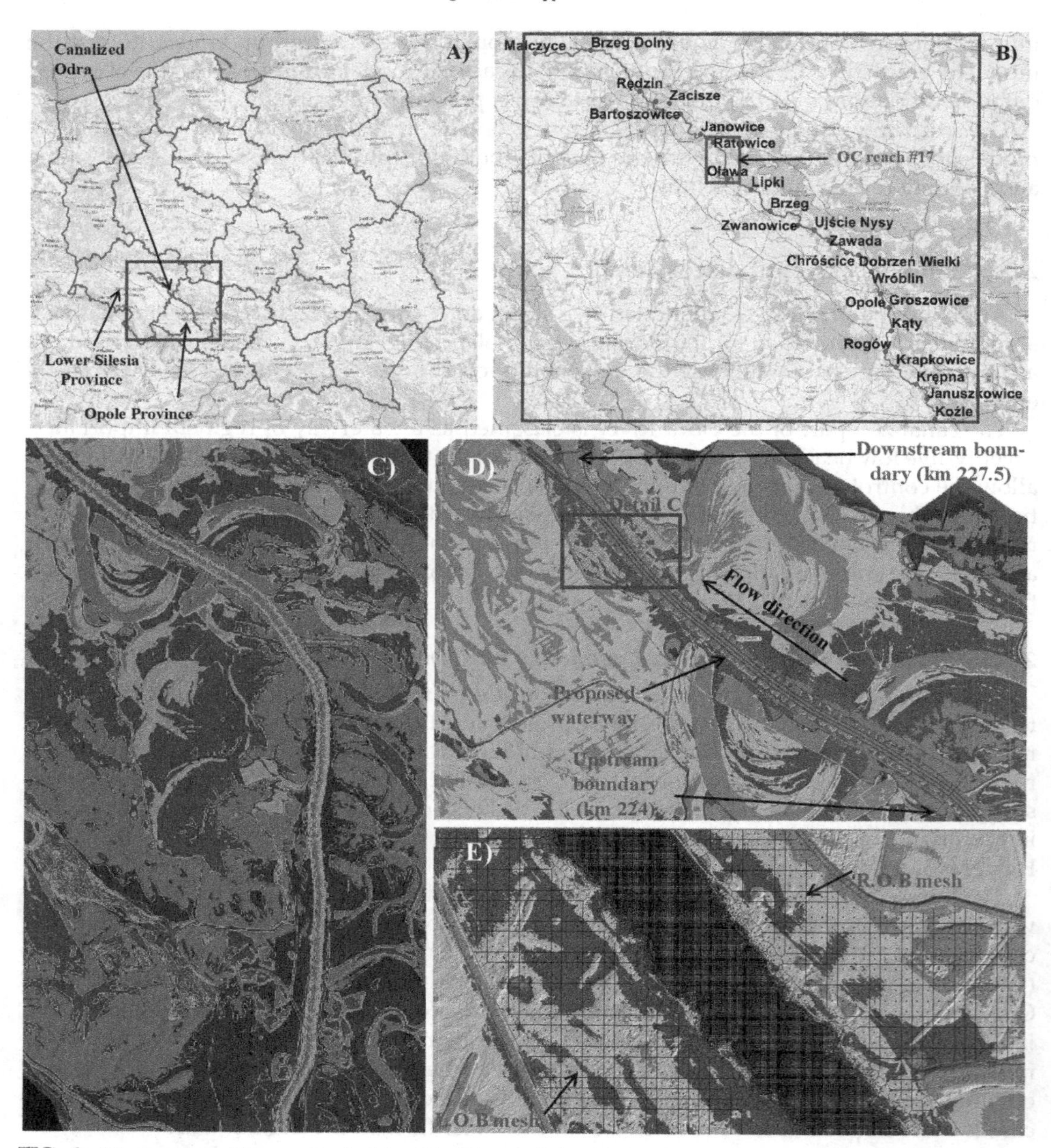

FIG. 6 The location of the canalized of the Odra River in Poland (A) and the twenty-five hydraulic structures that conform the Odra Cascade (B). Credit: Background map from the OpenStreetMap project. (C) DEM and DBM of the whole reach no. 17 of the Odra Cascade; (D) Computational domain and generated numerical meshes and (E) Detailed view of the meshes. *(B) Credit: Background map from the OpenStreetMap project.*

equation with mass conservation yields a one equation model, known as the DWA of the SWE (DSWa):

$$\frac{n^2}{R_h^{4/3}}|V|V = -\nabla z_s \tag{20}$$

where ∇z_s is the WSE gradient [−]. Dividing both sides of the equation by the square root of their norm, Eq. (20) can be rearranged into a more classical form:

$$V = -\frac{R_h^{2/3}}{n}\frac{\nabla z_s}{|\nabla z_s|^{1/2}} \tag{21}$$

Direct substitution of the DSWa of the momentum equation in the mass conservation equation yields the classical (DWE) Diffusion-Wave Equation:

$$\frac{\partial h}{\partial t} = \nabla \cdot (\beta \nabla z_s) + q \tag{22}$$

where $\beta = \frac{R_h^{2/3} h}{n|\nabla z_s|^{1/2}}$; h is the water depth [L] and q is source/sink flux term [LT^{-1}]. Once the DWE equation has been solved, the velocity can be recovered by substituting the WSE back into the DWE. This equation is very useful for its simplicity, nonetheless it can be applied in a narrower scope than the general SWE.

2.3.3 *Boundary and initial conditions*

At any given time, the boundary conditions shall give to establish the data of the problem to be analyzed at the edges of the domain. HEC-RAS allows three types of boundary conditions for 2D modeling, namely: ***Water surface elevation, normal depth*** or to provide information of the ***flow rate***. As depicted in Fig. 6D, for the analyzed reach, there are two boundaries, one upstream and one downstream, which is bounded by two hydraulic structures: the barrage "Olawa" upstream and the barrage "Ratowice" downstream. As the river-flow is controlled with these hydraulic structures for exploitation, the operational level (WSE) of the hydraulic structure "Ratowice" is to be used as downstream boundary condition. As two exploitation conditions are to be simulated. Thus, the values of the boundary conditions are (Kostecki et al., 2018):

Boundary condition downstream: WSE = 123.89 m a.s.l.
Scenario 1: $Q_{Nmin} = 49.0\,m^3/s$.
Scenario 2: $Q_{Nmax} = 389.6\,m^3/s$.
Initial condition: WSE = 123.89 m a.s.l at the beginning of the computations in the grid.

2.3.4 *Validation and calibration of the model*

The output of the model is in this case compared with the observed data of a 1D numerical model that was calibrated and validated (Fig. 7 above). Therefore, as the 1D model is already verified, the 2D model was calibrated comparing the output of the 2D model. The WSE is compared with Q_{Nmin}. The difference between the first run of the 2D model and the values of the 1D model varies from 10 up to 50 cm in five compared locations.

Therefore, varying the value of the roughness coefficient using GIS, after three runs, the values of the WSE and water depth were practically the same for both models. Using the

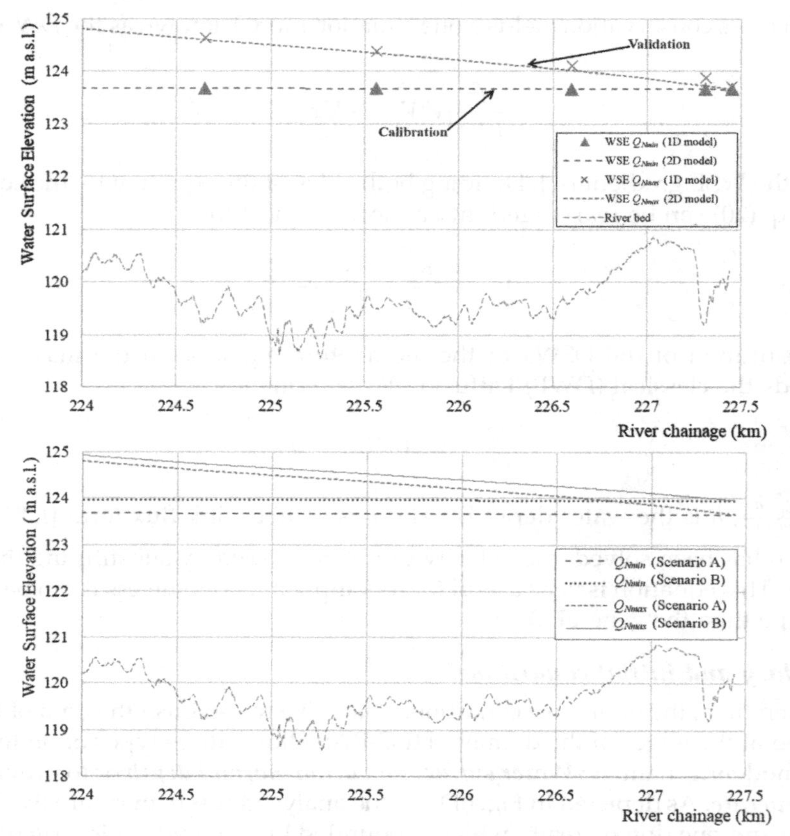

FIG. 7 *Above*: Comparison of the 2D modeling outcomes with the output of the verified 1D model for calibration and validation purposes. *Below*: Computed WSE elevation of the two analyzed scenarios for Q_{Nmin} and Q_{Nmin}.

"split-sample" procedure and considering the parameters of the already calibrated 2D model for Q_{Nmin}, the WSE values of the 2D model (blue dotted line) were compared with the values of the 1D model for Q_{Nmax} (yellow crosses) to validate the model. The differences between the values of the WSE were minimal.

2.4 Results

For navigation purposes, scenario B is more convenient (Fig. 7 below) since increasing the water level at the barrage "Ratowice" will therefore increase the water depth along the whole reach and this fact will help to fulfill the requirements of an international waterway.

It is a fact that the modeler can get the same output using 1D modeling with less computational demands. Nevertheless, as depicted in Fig. 8A, the output of the 2D model is immediately generated as a raster file. This facilitates the analyses of other related problems, such as the water depth deficit analysis to localize the places where the water depth does not fulfill the minimal depth for navigation, with the output of the 1D model, it was necessary to build

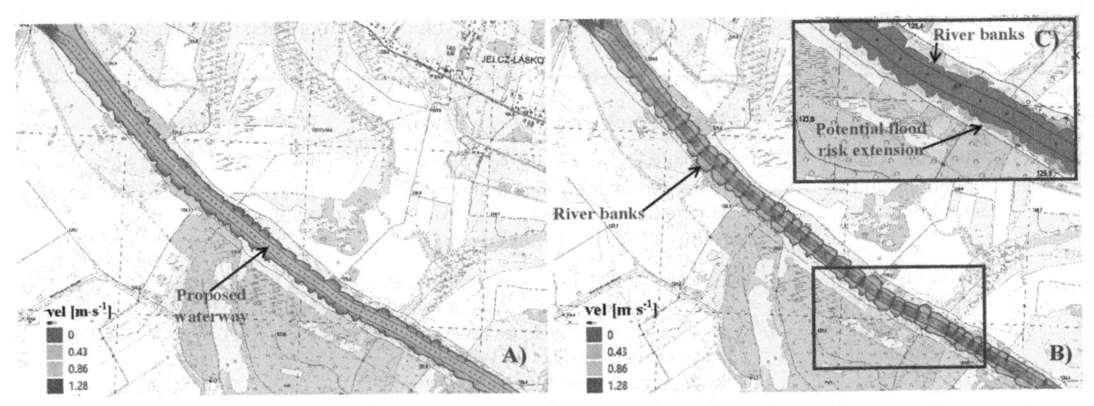

FIG. 8 (A) Calculated velocity magnitude for Q_{Nmin} and B) Q_{Nmax} along the analyzed reach (scenario B) and the potential flood risk extension for Q_{Nmax} (detail C). *Modified from Kostecki, S., Machajski, J., Herrera-Granados, O., Sawicki, E., Uciechowska-Grakowicz, A.K., Maniecki, Ł., 2018. Projekt badawczy w zakresie poprawy żeglowności rzeki Odry. Issue 46.*

another algorithm using GIS, while with the output of the 2D model, it is just necessary to classify the values of the raster generated by the model (Fig. 8B and C).

Additionally, other valuable detailed information can be obtained from the outcomes of the 2D model, such as the depth averaged velocity in two directions as well as the potential extension of the water in the flood plains for Q_{Nmax}. Thus, regardless that the computational time was longer, the output of the 2D model was more useful for further analysis of water resources management. Similar analyses were carried out along the whole canalized Odra.

2.5 Limitations and discussion

The presented case study is an example on how, thanks to the application of numerical modeling, it is possible to carry out additional analyses for real case projects. Nonetheless, the modeler is responsible also to be aware of the limitations of this computational tools. First, most of the numerical models consider important assumptions to facilitate the discretization of the mathematical framework to be approximated, e.g., neglecting the role of seepage in the flow behavior of water flow (Sharma et al., 2019). Seepage is commonly neglected in the analyses of free surface flow behavior, and regardless that the magnitude of seepage is very small, this small flow behavior is affecting the morphodynamics of river flow (Herrera-Granados, 2012). Thus, the modeler shall seriously consider whether the analyzed problem can neglect this kind of assumptions or to look for a more appropriate model.

Additionally, there are limitations regarding the numerical methods and mesh. It must be pointed out that the accuracy of the numerical solution is related to the grid refinement. Some numerical results such as used for discretizing the WSE are properly obtained with triangular grid cells (García-Navarro et al., 2019), while other methods work better with quadrilateral (HEC-RAS) or polygonal elements.

Regardless the fact that nowadays, numerical 1-D and 2-D flow simulations represent state-of-the-art solutions in hydraulic engineering. Many problems that deal with the motion of water have been solved using CFD-RANS techniques (Herrera-Granados, 2021.

TABLE 1 Type of models according to the dimensions of the velocity components, their applications, and examples of available computer software.

Model type	Governing equations	Type of problems that can be analyzed	Available software
1-D	The Saint-Venant equations and its simplifications.	Global scale problems, such long lasting flood waves. Problems that involve kinematic waves and/or hydrologic routing. Flood risk assessments	HEC-RAS, TELEMAC, DWOPER, MIKE, FLOOD MODELLER
2-D	2D Saint-Venant, Laplace equation together with the Darcy's Law.	Global scale problems of groundwater flow. Global scale problems of river flow through medium reaches. Water quality and sediment transport, Watershed management. Combined urban flow with natural runoff.	FESWMS, RIVER2D, SMS, MIKE, SOBEK SMS, IBER, GEO HEC-RAS. FLOOD MODELLER.
3-D	Vorticity Eqs. RANS, LES, DA-LES	Project scale problems with complicated geometries. Problems that involve complex physical phenomena, water quality, and transport of pollutants. Scientific research.	FLOW3D, CFX, TELEMAC, I-RIC, OPENFOAM, DELFT3D.

Modified from Herrera-Granados, O., 2010. Numerical modelling as a key tool in the decision-making process for water resources management. In: Infrastructure Planning and Development in Developing Countries—"'The Way Forward'": MIPALCON 2010: First International Alumni Conference, Stuttgart, Germany, 28th September–1st October 2010, 83–94.

Table 1 presents a summary of the computational techniques that can be applied in the analysis of problems concerning water motion. In the algorithm presented in this chapter, the second step deals with the mathematical framework of the numerical model. As depicted in Table 1, there are three kinds of models depending on the dimensionality of the velocity component. In 1D models, the cross-sectional averaged velocity is simulated or resolved in defined cross sections along the axis of the flow direction, while in 2D and 3D, two/three components of the velocity vector are resolved in each element of the computational grid. In Table 1, the mathematical frameworks that are applied for each kind of numerical model are mentioned as well as the kind of problems that the modeler can analyze using a specific type of numerical model. For example, if the modeler must prepare a flood risk assessment, there is no sense to use CFD as the mathematical model is more complex, the computational demands are higher, and the required information to be obtained can be properly simulated with a 1D or 2D modeling in a more economical form and faster.

3 Conclusions

It is necessary to remark that the available computational tools are very useful for industry cases and research. However, the modeler should properly calibrate and validate his/her models to decrease the uncertainty of the outcomes and avoid the "black box error." A useful recommendation is that modelers, above all young modelers, consult the output of their models with experienced engineers. The use of the validation of numerical models with well-instrumented and monitored physical models in the laboratory is also very helpful.

As previously mentioned, for decision makers, 1-D and 2-D models became deep-rooted in engineering companies in a cost-effective way (Herrera-Granados, 2010) for the analysis of issues related with water resources modeling. However, with increasing computer capacities, the usage of 3D models has become increasingly popular in industry since the beginning of the 21st century.

Regarding the model presented as example, it is necessary to mention that before applying and analyzing real cases using numerical models, it is necessary to pass toward all the steps of the methodology presented in this chapter. Due to the calibration and validation processes, the differences between the final models and the data available were minimal, and this fact allows the modeler to rely on the validity of the model output. In the end, for navigation purposes, scenario B is more convenient, since increasing the level of water in the "Ratowice" barrage will therefore increase the depth of the water throughout the reach and this fact will help the Odra river to fulfill the requirements of a waterway of international importance.

References

Bladé, E., Cea, L., Corestein, G., Escolano, E., Puertas, J., Vázquez-Cendón, E., Dolz, J., Coll, A., 2014. Iber: herramienta de simulación numérica del flujo en ríos. Rev. Int. Metodos Numer. para Calc. Diseno Ing. 30 (1), 1–10. https://doi.org/10.1016/J.RIMNI.2012.07.004.

Brath, A., Montanari, A., Moretti, G., 2006. Assessing the effect on flood frequency of land use change via hydrological simulation (with uncertainty). J. Hydrol. 324 (1), 141–153. https://doi.org/10.1016/j.jhydrol.2005.10.001.

Cea Gómez, L., Bladéi Castellet, E., Sanz Ramos, M., Bermúdez Pita, M., Mateos Alonso, Á., 2019. Iber applications basic guide. Two-dimensional modelling of free surface shallow water flows. In: Iber Applications Basic Guide. Two-Dimensional Modelling of Free Surface Shallow Water Flows., https://doi.org/10.17979/SPUDC.9788497497176.

García-Navarro, P., Murillo, J., Fernández-Pato, J., Echeverribar, I., Morales-Hernández, M., 2019. The shallow water equations and their application to realistic cases. Environ. Fluid Mech. 19, 1235–1252. https://doi.org/10.1007/s10652-018-09657-7.

Herrera-Granados, O., 2008. The Tuxpan river basin: a case study of the WFD as a guideline for protecting tropical river systems. In: The 7th International Conference Environmental Engineering: Selected Papers, Vilnius, Lithuania, May 22–23, 2008. Vol. 2, Water Engineering: Energy for Buildings, pp. 530–542.

Herrera-Granados, O., 2009a. Considering the importance of water dynamics in sustainable planning. In: Januchta-Szostak, A. (Ed.), Water in the Townscape. vol. 2. Publishing House of Poznan University of Technology, pp. 101–110.

Herrera-Granados, O., 2009b. Computational fluid dynamics (CFD) in river engineering: a general overview. In: II International Interdisciplinary Technical Conference of Young Scientists, Inter Tech 2009, 20–22 May 2009, Poznań, Poland: Proceedings, pp. 259–263.

Herrera-Granados, O., 2010. Numerical modelling as a key tool in the decision-making process for water resources management. In: Infrastructure Planning and Development in Developing Countries—"'The Way Forward'": MIPALCON 2010: First International Alumni Conference, Stuttgart, Germany, 28th September - 1st October 2010, pp. 83–94.

Herrera-Granados, O., 2012. Seepage Influence on River Dynamics. PhD Thesis, Tech. Reports of the Wrocław University of Science and Technology, Poland.

Herrera-Granados, O., 2018. Turbulence flow modeling of one-sharp-groyne field. In: Free Surface Flows and Transport Processes: 36th International School of Hydraulics, pp. 207–218, https://doi.org/10.1007/978-3-319-70914-7_12.

Herrera-Granados, O., 2020. Numerical analysis of filling/emptying operation proposals for ship-locks chambers used for inland navigation. In: River Flow 2020: Proceedings of the 10th Conference on Fluvial Hydraulics : Delft, Netherlands, 7–10 July, 2020, pp. 2350–2357.

Herrera-Granados, O., 2021. Numerical analysis of flow behavior in a rectangular channel with submerged weirs. Water 13 (10). https://doi.org/10.3390/w13101396.

Herrera-Granados, O., Kostecki, S., 2016. Numerical and physical modeling of water flow over the ogee weir of the new Niedów barrage. J. Hydrosci. Hydraul. Eng. 64 (1), 67–74.

Isfahani, A.H.G., Brethour, J.M., 2009. On the Implementation of Two-Equation Turbulence Models in FLOW-3D (Issue FSI-02-TN86).

Kiraga, M., 2020. Local scour modelling on the basis of flume experiments. Acta Sci. Pol. Architectura 18 (4), 15–26. https://doi.org/10.22630/aspa.2019.18.4.41.

Knight, D., 2013. Hydraulic problems in flooding: From data to theory and from theory to practice. GeoPlanet 11, 19–52. https://doi.org/10.1007/978-3-642-30209-1_2.

Kostecki, S., Herrera-Granados, O., 2021. Trójwymiarowe modelowanie przepływu przez bystrza ze zwiększoną szorstkością. In: Gorczyca, E., Radecki-Pawlik, A., Krzemień, K. (Eds.), Procesy fluwialne a utrzymanie rzek i potoków górskich. Instytut Geografii i Gospodarki Przestrzennej UJ, pp. 233–256.

Kostecki, S., Machajski, J., Herrera-Granados, O., Sawicki, E., Uciechowska-Grakowicz, A.K., Maniecki, Ł., 2018. Projekt badawczy w zakresie poprawy żeglowności rzeki Odry. (Issue 46).

Kostecki, S., Rędowicz, W., Herrera-Granados, O., Muszyński, Z., 2013. Opracowanie możliwości manewrowania zamknięciami segmentowymi na zaporze Witka (Issue 3).

Murillo, J., García-Navarro, P., 2010. Weak solutions for partial differential equations with source terms: application to the shallow water equations. J. Comput. Phys. 229 (11), 4327–4368. https://doi.org/10.1016/j.jcp.2010.02.016.

Nezu, I., Nakagawa, H., 1993. Turbulence in Open Channel Flows. Balkema.

Novak, P., Guinot, V., Jeffrey, A., Reeve, D.E., 2010. Hydraulic Modelling: An Introduction: Principles, Methods and Applications. CRC Press. https://books.google.pl/books?id=6lS6zzhKiFsC.

Pinos, J., Timbe, L., Timbe, E., 2019. Evaluation of 1D hydraulic models for the simulation of mountain fluvial floods: a case study of the Santa Bárbara River in Ecuador. Water Pract. Technol. 14 (2), 341–354. https://doi.org/10.2166/WPT.2019.018.

Sharma, A., Herrera-Granados, O., Kumar, B., 2019. Bedload transport and temporal variation of non-uniform sediment in a seepage-affected alluvial channel. Hydrol. Sci. J. 64 (8), 1001–1012. https://doi.org/10.1080/02626667.2019.1615621.

Tavakolifar, H., Abbasizadeh, H., Nazif, S., Shahghasemi, E., 2020. Development of 1D–2D urban flood simulation model based on modified cellular automata approach. J. Hydrol. Eng. 26 (2), 04020065. https://doi.org/10.1061/(ASCE)HE.1943-5584.0002036.

Uciechowska-Grakowicz, A.K., Herrera-Granados, O., 2021. Riverbed mapping with the usage of deterministic and geo-statistical interpolation methods: the Odra River case study. Remote Sens. (Basel) 13 (21), 1–16. https://doi.org/10.3390/rs13214236.

USACE—U.S. Army Corps of Engineers, 2021. HEC-RAS, River Analysis System Hydraulic Reference Manual. Hydrologic Engineering Center (HEC), May.

Viti, N., Valero, D., Gualtieri, C., 2019. Numerical simulation of hydraulic jumps. Part 2: recent results and future outlook. Water 11 (1), 28. https://doi.org/10.3390/w11010028. https://doi.org/10.3390/w11010028.

Wu, W., 2007. Computational River Dynamics. Taylor & Francis, https://doi.org/10.4324/9780203938485.

CHAPTER

20

Prophecy of groundwater fluctuation through SVM-FFA hybrid approaches in arid watershed, India

Sandeep Samantaray[a], *Abinash Sahoo*[b], *Deba Prakash Satapathy*[a], *and Shaswati S. Mishra*[c]

[a]Department of Civil Engineering, NIT Srinagar, Jammu & Kashmir, India [b]Department of Civil Engineering, NIT Silchar, Silchar, Assam, India [c]Department of Philosophy, Utkal University, Bhubaneswar, Odisha, India

OUTLINE

1 Introduction 342

2 Study area 345

3 Material and methods 347

3.1 RBFN 347

3.2 SVM 348

3.3 SVM-FFA 349

3.4 Data collection and model performance 351

3.5 Evaluating criteria 351

4 Results and discussion 352

4.1 Assessment of actual versus predicted GWL at Boden and Khariar 355

4.2 Comparison of model performance for proposed watersheds 359

5 Conclusions 362

Conflict of interest 362

Research funding 362

References 362

Further reading 365

Current Directions in Water Scarcity Research, Volume 7
https://doi.org/10.1016/B978-0-323-91910-4.00020-0

1 Introduction

Groundwater is a precious natural resource supporting human health and livelihood, economic development, and environmental assortment. Prediction of GWL fluctuation is considered a critical hydrological and ecological aspect in groundwater management. From a utilization viewpoint, groundwater is used predominantly for irrigation, accounting for about 80%. Apart from agricultural uses, it is utilized for supply of drinking water, industrial usages, domestic consumption, etc. GWL is subjected to deviations ensuing from alterations in groundwater release and recharge, variation in stream flow, effect of tides, impact of meteorological parameters, and global climate variations (Todd and Mays, 2004).

Processes involved in assessment of groundwater variations are complex and hidden in nature. For this purpose, computer-based modeling has evolved as an influential approach for protection and management of this significant natural resource. On the other hand, range of data necessary for evolving conceptual or physical-based GW prediction models is huge. These significant quantities of data are generally expensive and tough to acquire, specifically in developing and underdeveloped countries (Nikolos et al., 2008; Coulibaly et al., 2001). Hence, black-box machine learning models can serve as appropriate alternatives. With increasing computational ability of computers, there is an increase in applicability of data-driven techniques in GWL simulations and forecasts. Among these techniques, ANNs (artificial neural networks), SVM, and ANFIS (adaptive neuro-fuzzy inference system) can be considered a few approaches. A number of studies reported usage of ANNs (Wunsch et al., 2018; Coppola et al., 2003), ANFIS (Moosavi et al., 2013; Gong et al., 2016) for simulation and forecasting of GWL. Affandi and Watanabe (2007) applied ANFIS, Levenberg Marquardt (LM), and RBFN models for predicting daily GWL fluctuations at two observational wells. Results suggested that all three data-driven models could predict daily GWL with better precision by means of time lag as input networks. Yan and Ma (2016) used a combination of RBFN and ARIMA (autoregressive integrated moving average) for predicting monthly GWL fluctuations for two wells in Xi'an city, China, and compared with conventional ARIMA and RBFN models. Findings indicated that proposed robust model has a higher rate of accurateness and is more effective and feasible than conventional models. Li et al. (2017) employed GM (1, 1) and RBFN models to predict monthly GWL variations in Longyan city, Fujian Area (South China). Results demonstrated that both models forecasted GWL with high accurateness; however, RBFN model proved to be a favorable modeling approach for simulating and forecasting GWL at proposed study area. Zhang et al. (2017) assessed and compared validity of gray self-memory model (GSM), RBFN, and ANFIS models to forecast GWL at five stations in City of Jilin located in northeast China. Comparison of results indicated that ANFIS was a more efficient method compared to other two at all proposed stations. Pradhan et al. (2019) investigated relative performance of RBFN, coactive ANFIS, and fuzzy logic (FL) models utilized for estimating water level in lower portion of Ganga and Ram-Ganga basin. Results from present study indicated superior performance of FL rule-based model compared to that of RBFN and CANFIS to predict depth of water table at desired site. Even though these techniques provide a certain degree of accuracy and flexibility in groundwater modeling, several disadvantages have been stated. Yoon et al. (2011) described high uncertainty related to ANNs for GWL forecasting in a coastal aquifer, and Fallah-Mehdipour et al. (2013) reported overfitting of ANFIS in Karaj plain. Understanding of outcomes from various studies has boosted popularization of such methods involving future scenarios with climate uncertainty. Resistance of specific areas of academics for accepting different equiprobable outcomes is justified where accuracy is a preeminent criterion for adequacy. In ecological analysis, there is a necessity for models proficient in predicting and describing complicated ecological procedures. Investigation on environmental time series is essential for the higher aim of adaptation and sustainability.

In hydrology, a relatively new data-driven technique called SVM was applied, which Vapnik first proposed (1995). SVM has been employed for forecasting GWL. Hipni et al. (2013) presented a novel solution to a proficient system, utilizing SVM for water level forecasting of Klang reservoir on a daily basis situated in Peninsular Malaysia with different input scenarios and compared with that of ANFIS. Comparison results proved that SVM performed superiorly than ANFIS with the most precise outputs. Sujay Raghavendra and Deka (2015) focused on an extensive range of applicability of SVM in various hydrological fields. Gong et al. (2016) explored the potential of ANN, ANFIS, and SVM to predict GWL considering interaction amid groundwater and surface water in Florida, United States. Based on different performance measures results of SVM and ANFIS were more precise compared to ANN. Arabgol et al. (2016) developed SVM to estimate concentration of nitrate in groundwater of Arak plain, Iran. Results specified that SVM could be a quick, cost-effective, and consistent technique for groundwater quality prediction and assessment. Salem et al. (2017) applied SVM model for evaluating optimal abstraction of groundwater for irrigation and MLR for estimating decline in cost of irrigation because of GWL elevation. Different machine learning methodologies predict water table fluctuation at various gauge stations (Samantaray et al., 2019; Samantaray et al., 2020a; Sridharam et al., 2021). Rajaee et al. (2019) presented an overview of various artificial intelligence techniques for modeling and forecasting GWL at different study sites from 2001 to 2018.

Previous studies indicated that accuracy of SVM results is mostly governed by tuning of its constraints. The theory of SVM doesn't give any particular procedure for the selection of its parameters (Su et al., 2014). Previous studies are observed to perform a trial and error approach (Bhagwat and Maity, 2013; Behzad et al., 2010). Trial and error method is a time-taking procedure and significantly affects accuracy of SVM results. Because models need to be assessed at different points inside a grid for each parameter, achieving a higher accurateness using trial and error method bears a high computational cost. Metaheuristic optimization algorithms can be utilized for overcoming these limitations (Sujay Raghavendra and Deka, 2015) and have been effectively used for modeling several water resources problems (Fayaz et al., 2020; Hosseini-Moghari et al., 2017; Bozorg-Haddad et al., 2017).

Weaknesses and strengths of these algorithms have been studied (Moravej, 2018; Maier et al., 2014; Sörensen, 2015). Several optimization algorithms were applied for finding optimal parameters of SVM. A genetic algorithm (GA) was applied for choosing SVM parameters (SVM-GA) for forecasting water temperature (Quan et al., 2020), daily reference evapotranspiration (Yin et al., 2017), water quality parameters in rivers (Bozorg-Haddad et al., 2017), and monthly streamflow (Kalteh, 2015). Bat algorithm, Artificial Bee Colony (ABC), ant colony optimization (ACO), and Particle Swarm Optimization (PSO) were integrated with SVM for forecasting river-stage (Seo et al., 2016; Agnihotri et al., 2022) and forecasting streamflow (Xing et al., 2016). Results indicated that SVM-ABC and SVM-PSO provided better results than SVM-GA. Similar work is missing on GWL simulations and forecasts. Additionally, all meta-heuristic algorithms utilized to tune parameters of SVM suffer to have two or more constraints for tuning themselves. Sensitivity of optimization algorithms is stated in earlier studies (Moravej et al., 2020). Yang (2009) proposed FFA, which showed its advantage over other optimization algorithms in a broad variety of problems.

Safavi and Esmikhani (2013) proposed a hybrid model integrating SVM-GA for conjunctive utilization of groundwater and surface water of Koohpayeh plain, located in west-central Iran. Results showed that accurateness of SVM-GA model as a simulator for interaction amid groundwater and surface water is better. Jha and Sahoo (2015) intended to discover heuristic ANN models by incorporating 2 data-driven methods, such as ANN and GA, to simulate/predict time series of GWL at 17 sites of Konan basin, Japan. It was observed that hybrid ANN modeling approaches could effectively predict spatial and temporal variations of GWL at subbasin or basin levels. Kisi et al. (2015) used SVM-FFA model to develop predictive models for multistep ahead daily lake level prediction and compared the suitability and efficiency of the developed model with that of ANN and GP models. Experimental outputs revealed that SVM-FFA model improved prediction accurateness and generalization ability compared to GP and ANN. Olatomiwa et al. (2015) developed a hybrid SVM-FFA model for predicting average horizontal global solar radiation monthly, utilizing various meteorological constraints. Prediction accurateness of projected hybrid model is assessed in comparison to GP and ANN models. Obtained results revealed that developed hybrid models provided more correct prediction outputs than GP and ANN models. Moghaddam et al. (2016) employed SVM-FFA model for predicting fatigue life of PET reformed asphalt mixture and compared it with GP, fuzzy logic, and ANN models. Results demonstrated that SVM-FFA gave more precise outcomes in comparison to observed experimental data. Roy et al. (2016) applied a hybrid SVM model for predicting significant wave height using 1-year buoy data and compared it with GP and ANN models. Findings from the present study revealed that use of SVM-FFA was a potential alternative technique to estimate wave height. Gani et al. (2016) proposed a novel SVM-FFA hybrid model for estimating wind speed distribution. Accuracy of proposed hybrid model is verified based on comparisons with conventional SVM and ANN models. Results indicated that SVM-FFA is a suitable and effective method for estimating monthly wind speed distribution. Ghorbani et al. (2018) investigated applicability of SVM-FFA model for improving model prediction accuracies of river flows for Zarrineh River located in northwest Iran and compared with other artificial intelligence models. Results revealed that SVM-FFA model provided more closed outputs to observed values than other models. Natarajan and Sudheer (2019) investigated accuracy of the extreme learning method (ELM), SVM, GP, ANN, hybrid SVM-PSO, and SVM-RBF models to predict GWL at six positions in Vizianagaram district, India. Results indicated that ELM's performance was preeminent compared to all other proposed models at all study sites.

GW is an essential basis for farming, municipal and industrial practices in Khariar and Boden cities of Nuapada, India, assisting socio-economical accomplishments of over 1 million people. Hence, development of a reliable model to simulate and forecast water level is of high interest in this area. The focal aim of present study is to incorporate SVM and FFA methods for developing a hybrid model to estimate GWL fluctuation. Although in several studies, hybrid SVM-FFA is applied for developing several hydrological-based models, no study from literature was found that implemented SVM-FFA optimization algorithm to predict fluctuations in GWL. This approach, according to the authors, is novel in this study. Also, it can be observed from literature that the FFA optimization algorithm performed better than other selected algorithms in finding optimal SVM parameters which is a major motive for its usage. This research also intends to assess performance of the proposed hybrid model by comparing it with conventional SVM and RBFN models in context to different input parameters.

2 Study area

The District of Nuapada lies in western zone of Odisha, India. It lies between 21°0′ and 21°06′ N latitudes and 82°19′ and 82°60′ E longitude. Boundaries of Nuapada extend south, west, and north to District of Raipur in Chhattisgarh and east to districts of Kalahandi, Bargarh, and Balangir in Odisha. Nuapada has spread over an area of 3852 km^2. It has diverse agricultural based on climatic environments. Although precipitation here is temperately high, however, irrigation amenities are insufficient. Only 9% of irrigational region is cultivated. A significant source of precipitation is south-west monsoon. The mean annual precipitation of Nuapada is 1378.2 mm. Of total precipitation, around 75% is expected during June to September period. The climate in Nuapada is subtropical with dry and hot summer along with satisfying winter. Boden and Khariar are two subwatersheds from Nuapada district are selected for predicting GWL fluctuations. The proposed study area is represented in Fig. 1.

Hydrogeological structure of Nuapada is essentially governed by rainfall distribution, geological setup, and degree of primary and secondary porosities in geological formation for movement and storage of groundwater. As central portions of this district underlie hard rocks of varied lithological structure and composition, water bearing property of formations also fluctuates to a great extent. Hydrogeological surveys reveal lithological features and role of tectonic distortion on the occurrence and circulation of groundwater reservoirs along with their water yielding and water bearing properties. The statistical analysis of the input datasets are provided in Table 1.

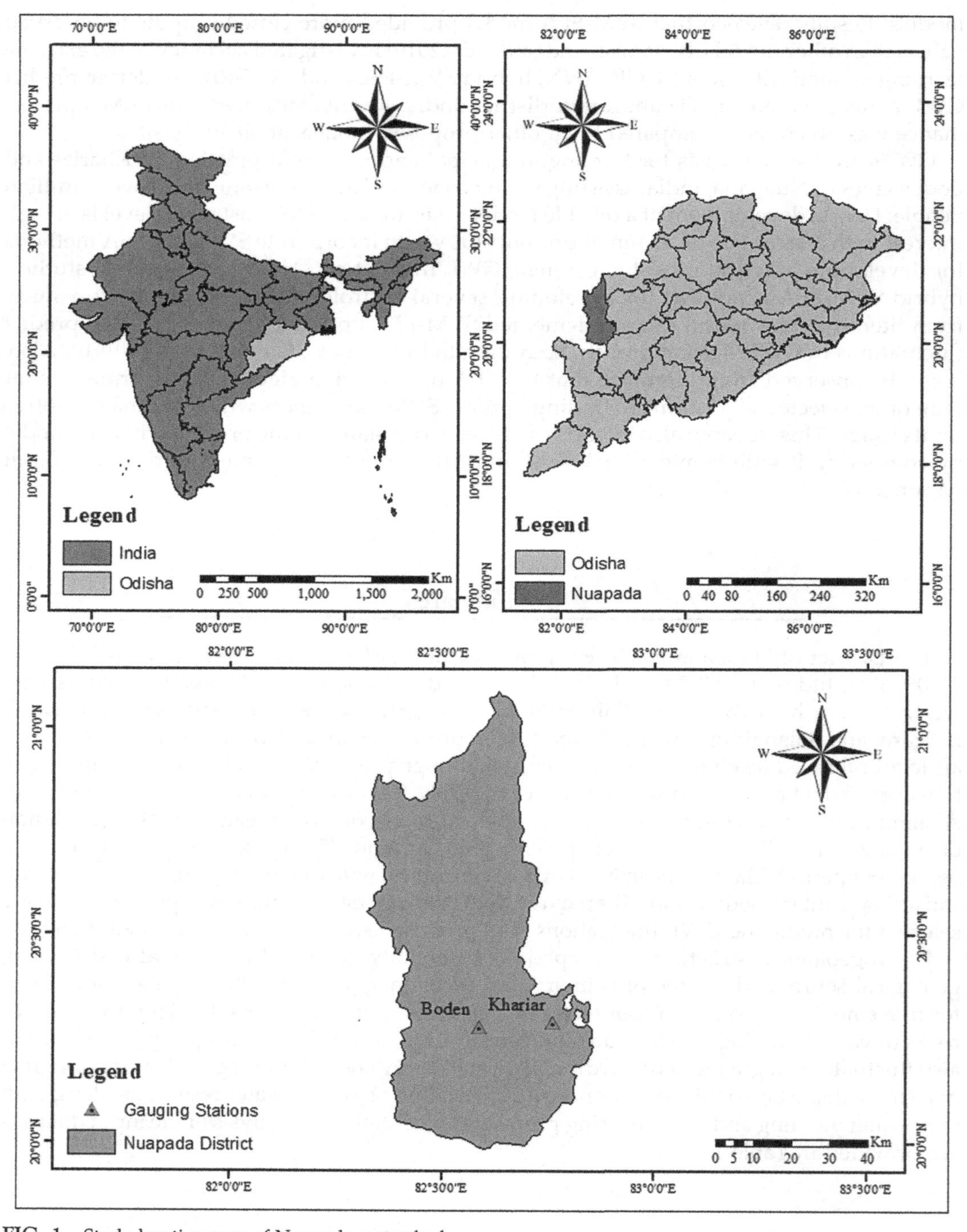

FIG. 1 Study location map of Nuapada watershed.

TABLE 1 Basic statistics of training, testing, and total available datasets.

Statistical parameters	Training set (180)	Testing set (60)	Total data set (240)
Sherkhanchawk			
Min	0.022	0.02	0.02
Max	8.081	7.801	8.081
Mean	3.474	4.142	3.641
Kurt	−0.389	−0.944	−0.673
Skew	0.566	−0.046	0.408
SD	1.782	1.962	1.847
Basantia high school			
Min	0.056	0.082	0.056
Max	8.675	8.081	8.675
Mean	3.784	3.41	3.69
Kurt	−0.589	−0.874	−0.692
Skew	0.36	0.474	0.36
SD	1.923	2.25	2.011

3 Material and methods

3.1 RBFN

NNs have gone through two key development phases: the initial 1960s and mid-1980s. So far, there are several kinds of ANN tools that have been utilized for forecasting time series. Development of different types of neural networks remained a crucial improvement in the machine learning field. ANNs were stimulated by biological verdicts concerning to behavior of brain as a grid of elements known as neurons (Rumelhart et al., 1986).

RBFN is a classic feed-forward network comprising three layers viz. input, hidden, and output (Gevindaraju and Rao, 2000). The input layer comprises of source node which connects network to its surroundings. Hidden neuron converts data from input to hidden layer utilizing nonlinear functions. The output layer being linear assists as a summary unit (Samantaray and Sahoo, 2020).

RBFN computation procedure is defined in following equations. After a network obtains a p dimensional input vector $Z=[z_1, z_2, \ldots z_p]^T$, input layer aids simply as a supplier to hidden layer. Reaction from jth hidden component for ith input data is shown by the subsequent formula:

$$\varnothing_j(z_i) = \varnothing(\|z_i - c_j\|)\ j = 1,2,\ldots,Q \tag{1}$$

where $\varnothing$ is the activation function; $\|0\|$ signifies Euclidean norm; c_j is the center of jth component of hidden layer; and Q is quantity of hidden elements. There exist a number of activation functions. Among those most regularly utilized, is the Gaussian function developed by (Schwenker et al., 2001).

$$\varnothing(Z) = \exp\left[-\frac{\|Z-c\|}{2\beta^2}\right] \tag{2}$$

where β is width constraint. Various quantities of neurons in the hidden layer are analyzed utilizing the trial and error technique (Sahoo et al., 2021; Samantaray and Ghose, 2020a,b). Dot product metrics, standard full competitive rules were utilized for building RBFN.

3.2 SVM

SVM technique is based on the concept of structural risk minimization (Vapnik, 1995, 1998). It minimizes error of upper margin extension, whereas other machine learning approaches generally minimize the error of local training. Moreover, the SVM technique has certain additional benefits compared to other systematic algorithms, such as exclusive solution and applying collection of multidimensional space kernel functions comprising astute nonlinear conversion.

In SVM arrangement, a data collection is presented by $\{x_i, d_i\}_i^n$, where x_i is a vector containing input trial data, d_i provides target values, and n denotes size of data. SVM can be expressed in mathematical terms by resulting equations.

$$f(x) = w\varphi(x) + b \tag{3}$$

$$R_{SVMs} = \frac{1}{2}\|w\|^2 + C\frac{1}{n}\sum_{i=1}^{n} L(x_i, d_i) \tag{4}$$

In Eq. (3), x signifies input vector that is plotted by $\varphi(x)$—a multidimension space requirement, w represents a normal vector, b signifies a scalar. $C\frac{1}{n}\sum_{i}^{n} L(x_i, d_i)$ is an experimental error. w and b can be evaluated by minimizing risk function. Slack variables (ξ_i, ξ_i^*) definition as can be observed in the following equation. ξ_i^* and ξ_i signify lower and upper permitted deviation boundaries, respectively.

$$\text{Minimize } R_{SVMs}(w, \xi^*) = \frac{1}{2}\|w\|^2 + C\sum_{i=1}^{n}(\xi_i + \xi_i^*)$$
$$\text{Subject to } \begin{cases} d_i - w\varphi(x_i) + b_i < \varepsilon + \xi_i \\ w\varphi(x_i) + b_i - d_i < \varepsilon + \xi_i^* \\ \xi_i, \xi_i^* \geq 0, i = 1, \dots, l \end{cases} \tag{5}$$

here, $\frac{1}{2}\|w\|^2$ denotes adjustment constraint; C signifies error forfeit constraint that is applied for controlling difference amid experimental error and adjustment parameter. In addition, ε denote loss function equaling with training data accurateness, finally training dataset number is specified by l.

3.3 SVM-FFA

FFA is a metaheuristic exploration algorithm (Shojafar et al., 2015; Shamshirband et al., 2014) based on characteristics of societal crusade of fireflies in their usual situation. The major limitations of this algorithm are (1) attraction formulation and (2) light intensity difference. With the objective to refine the considered design for maximizing objective function, it is fixed to be proportional to intensity of light generated by a firefly. The intensity of light with changing spaces in the Gaussian system can be expressed as:

$$I = I_o e^{-\gamma r^2} \tag{6}$$

where e is the exponential function, I signifies intensity light at r distance with position of firefly, I_o is the intensity of light at a distance $r=0$ from firefly reference, γ is the coefficient of light immersion.

The desirability of a firefly tracks a fraction of intensity of light while being witnessed by other fireflies. Attraction ω at distance r from firefly reference is defined as

$$\omega(r) = \omega_0 e^{-\gamma r^2} \tag{7}$$

where ω_0 is attractiveness at $r=0$ and ω is coefficient of light absorption.

Space amid any two fireflies i and j at x_i andx_j, correspondingly, can be described as Cartesian space:

$$r_{ij} = \|x_i - x_j\| = \sqrt{\sum_{k=1}^{n}(x_{i,k} - x_{j,k})} \tag{8}$$

where n signifies problem dimensionality, $x_{i,k}$ is kth spatial coordinate $x_i$constituent of ith firefly, and $x_{j,k}$ is kth spatial coordinate x_j—constituent of jth firefly. **Below** summarizes the basic FFA development steps in pseudocode.

```
start
    Define the objective function, f(x), x = (x_i, ... ... ..., x_d)T
    Generate initial population of fireflies x_i(i = 1,2,3, .... n)
    Determine light intensity l_i at x_i from f(x_i)
    Define light absorption coefficient γ
        while t< maximum generation
                Make a copy of population for movement function
            for i = 1:n for n fireflies
                for j = 1:i all n fireflies
                    If l_j > l_i
                    Move fireflies i and j in d-dimension
                    end if
                    Attractiveness varies with distance r via exp (−γr)
                    Evaluate new solution and update light intensity
                end
            end
            Rank the fireflies and find the current best
        end
        Post process result and visualization
end
```

TABLE 2 Initial parameter setting for FFA algorithm.

Algorithm	Parameter	Value
FFA	Population (β)	50, 100, 150, 200, 250, 300
	Attractiveness (β_0)	0.8
	Light absorption coefficient (γ)	0.6
	Randomization parameter (α)	0.4
	Number of iterations	1000

The performance of SVM is influenced by a suitable selection of constraints, which are calculated by utilizing FFA (Samantaray et al., 2020b). Initial parameters of proposed FFA algorithm are presented in Table 2.

3.4 Data collection and model performance

Altogether 20 years (2000 to 2019), Precipitation (P_t), Average temperature (T_{avg}), Monthly average discharge (Q_t), Evaporation (E_t), Infiltration loss (I_l), water table depth data were composed from IMD (India Meteorological Department), Groundwater survey and investigation (GWSI), Nuapada, Orissa. Data collected are utilized for the development of the model for studying the effect of ground water fluctuation. The basic equation used for predicting the GWL in two watersheds based upon the equation $GWL_t = f\ (P_t)$.

$$GWL_t = f\ (P_t, T_{avg})$$

$$GWL_t = f\ (P_t, T_{avg}, Q_t)$$

$$GWL_t = f\ (P_t, T_{avg}, Q_t, E_t)$$

$$GWL_t = f\ (P_t, T_{avg}, Q_t, E_t, I_l)$$

The data collected from 2000 to 2014 (75% of data set) are used to train, and from 2015 to 2019 (25%) are used to test the network. Input and output data are gauged so that each data fall in a quantified range prior to training. This is known as normalization with the purpose of confining normalized values in a series of 0 to 1. Normalization equation which is utilized to scale data is

$$M_t = \frac{M - M_{min}}{M_{max} - M_{min}} \tag{9}$$

where M_t = transformed data, M = actual data, M_{min} = minimum of actual data, M_{max} = maximum of actual data. Subsequent GWL arrangements are employed as input presented in Table 3.

3.5 Evaluating criteria

MSE, WI, and RMSE are quantitative statistical evaluation criteria for determining best model. For selecting best performing model, the condition is MSE (Samantaray and Ghose,

TABLE 3 Model name with different combination.

		Model name		
Scenario	**Input**	**RBFN**	**SVM**	**SVM-FFA**
1	P_t	R#1	S#1	S-FFA#1
2	P_t, T_{avg}	R#2	S#2	S-FFA#2
3	P_t, T_{avg}, Q_t	R#3	S#3	S-FFA#3
4	P_t, T_{avg}, Q_t, E_t	R#4	S#4	S-FFA#4
5	$P_t, T_{avg}, Q_t, E_t, I_l$	R#5	S#5	S-FFA#5

Where P_t*: precipitation,* T_{avg}*: average temperature,* Q_t*: monthly average discharge,* E_t*: evaporation,* I_l*: infiltration loss.*

2019), RMSE (Samantaray and Ghose, 2020a,b) must be least, and WI (Samantaray and Ghose, 2020a,b) must be maximum.

$$\mathrm{WI} = 1 - \frac{\left(\sum_{i=1}^{N} V_{comp} - \overline{V}_{comp}\right)^2}{\left(\sum_{i=1}^{N} V_{obs} - \overline{V}_{obs}\right)^2} \tag{10}$$

$$\mathrm{MSE} = \frac{1}{n}\sum_{j=1}^{n}\left(V_{comp} - V_{obs}\right)^2 \tag{11}$$

$$\mathrm{RMSE} = \frac{\sum_{i=1}^{N}\left(V_{comp} - \overline{V}_{comp}\right)\left(V_{obs} - \overline{V}_{obs}\right)}{\sqrt{\sum_{i=1}^{N}\left(V_{comp} - \overline{V}_{comp}\right)^2\left(V_{obs} - \overline{V}_{obs}\right)^2}} \tag{12}$$

where V_{comp} = estimated data, V_{obs} = observed value, $\overline{V}_{comp}$ = mean predicted value, and $\overline{V}_{obs}$ = mean observed value

4 Results and discussion

Here three different techniques (RBFN, SVM, and SVM-FFA) with five different scenarios (P_t; P_t, T_{avg}; P_t, T_{avg}, Q_t; P_t, T_{avg}, Q_t, E_t; P_t, T_{avg}, Q_t, E_t, I_l) are considered for model formulation. Three different indices MSE, RMSE, WI are applied to find model's performance during training and testing phases.

The total dataset, with regards to MSE, RMSE, WI, between target unit and output of RBFN for the observation well, is shown in Table 4. Obtained results from Table 4 interpret that RBFN with the various spared values was efficiently comparable in the entire data set. Five different scenarios are considered for the assessment of model performance. It clearly shows that for both the stations scenario 5 provides improved performance over other scenarios performed. The result shows that paramount values of MSE, RMSE, WI are 0.00786, 0.07262, 0.93136 and 0.00742, 0.03115, 0.91582 during training and testing phases for Boden gauge station. Correspondingly for Khariar watershed, the prominent values of WI are 0.93265 and 0.91846 during training and testing period, respectively. The comparison plot shown in Fig. 2 presents the analysis of the RBFN models' capability for both periods.

In case of the SVM approach, minimum RMSE was seen with scenario 5 for both the watershed (0.06415 and 0.06236). For other sets 1, 2, 3, and 4 RMSE was found to be 0.06855, 0.06843, 0.06803, 0.06677 and 0.0684, 0.06836, 0.06799, 0.06636 for Boden and Khariar watershed, respectively. The maximum WI was found to be 0.95025 for scenario 5 followed by sets 1, 2, 3, and 4 (0.93946, 0.94287, 0.94578, and 0.94761, respectively) at station Boden. Similarly, for scenario 5 at Khariar watershed, best values of WI are 0.96869 and 0.94893 in training, testing periods. The performance value for both the station during training and testing period for different scenarios was given in Table 5. Fig. 3 shows the scatter plot representation of the SVM model for training and testing phases of proposed watersheds.

The SVM-FFA model established in current research is urged to give superior GWL forecasting outputs. Table 6 signifies developed SVM-FFA model performance. For both

TABLE 4 Outcomes of RBFN.

Station	Input	Architecture	MSE Training	MSE Testing	RMSE Training	RMSE Testing	WI Training	WI Testing
Boden	R#1	1-0.8-1	0.00923	0.00879	0.07955	0.05202	0.91815	0.89992
	R#2	2-0.5-1	0.00919	0.00875	0.07922	0.05038	0.92078	0.90087
	R#3	3-0.2-1	0.00861	0.00819	0.07453	0.03607	0.92428	0.90548
	R#4	4-0.1-1	0.00822	0.00786	0.07305	0.03301	0.92735	0.91214
	R#5	**5-0.7-1**	**0.00786**	**0.00742**	**0.07262**	**0.03115**	**0.93136**	**0.91582**
Khariar	R#1	1-0.6-1	0.00904	0.00862	0.07836	0.04199	0.92053	0.90515
	R#2	2-0.9-1	0.00899	0.00859	0.07793	0.04004	0.92348	0.90946
	R#3	3-0.7-1	0.00869	0.00815	0.07426	0.03552	0.92691	0.91059
	R#4	4-0.3-1	0.00817	0.00782	0.07292	0.03272	0.92992	0.91497
	R#5	**5-0.4-1**	**0.00757**	**0.00701**	**0.07236**	**0.03102**	**0.93265**	**0.91846**

Bold signifies the values of the best porforming model.

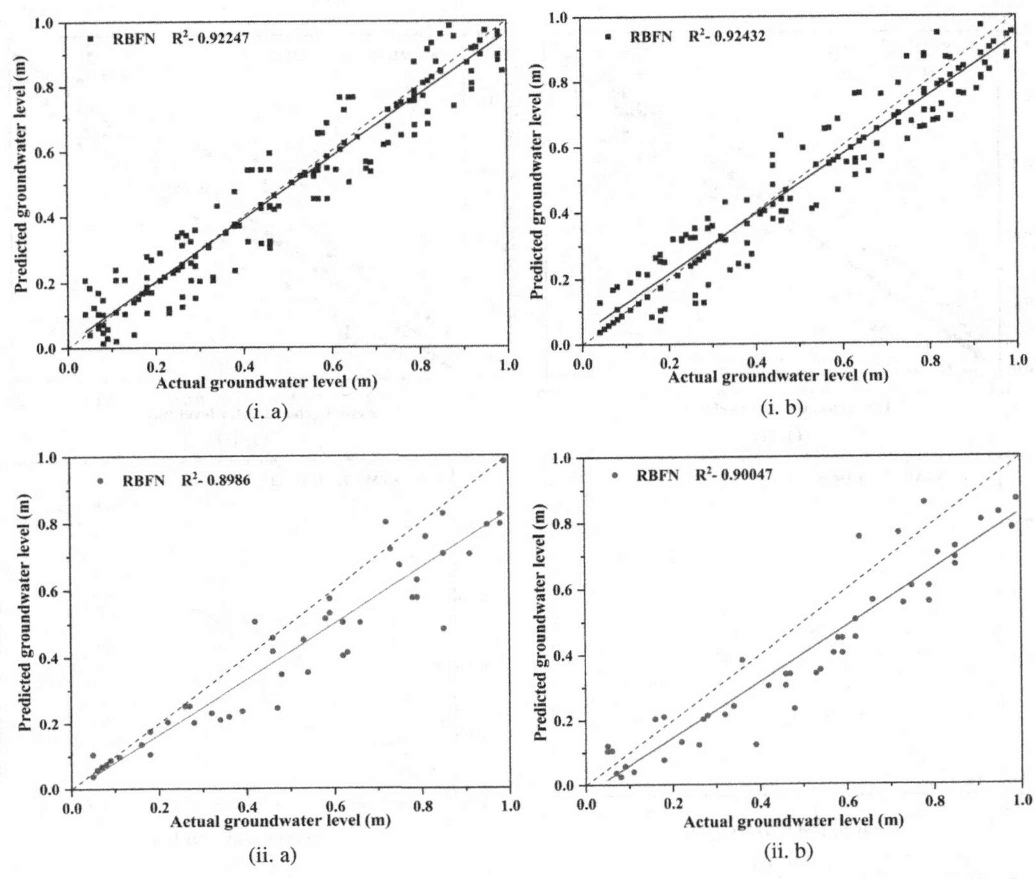

FIG. 2 Actual verses predicted GWL using RBFN algorithm at (A) Boden, (B) Khariar watersheds during (i) training, (ii) testing phase.

TABLE 5 Results of SVM.

Station	Input	MSE		RMSE		WI	
		Training	Testing	Training	Testing	Training	Testing
Boden	S#1	0.0055	0.0059	0.06855	0.02262	0.93946	0.92589
	S#2	0.00515	0.0057	0.06843	0.0224	0.94287	0.92824
	S#3	0.00542	0.00524	0.06803	0.02188	0.94578	0.93122
	S#4	0.00521	0.00499	0.06677	0.02151	0.94761	0.93546
	S#5	**0.00496**	**0.0041**	**0.06415**	**0.02131**	**0.95025**	**0.94527**
Khariar	S#1	0.00569	0.0054	0.0684	0.02233	0.95112	0.92945
	S#2	0.00565	0.00539	0.06836	0.02229	0.95787	0.93118
	S#3	0.00539	0.00519	0.06799	0.02184	0.96297	0.93565
	S#4	0.00513	0.00467	0.06636	0.02149	0.96583	0.93754
	S#5	**0.0049**	**0.00409**	**0.06236**	**0.02127**	**0.96869**	**0.94893**

Bold signifies the values of the best porforming model.

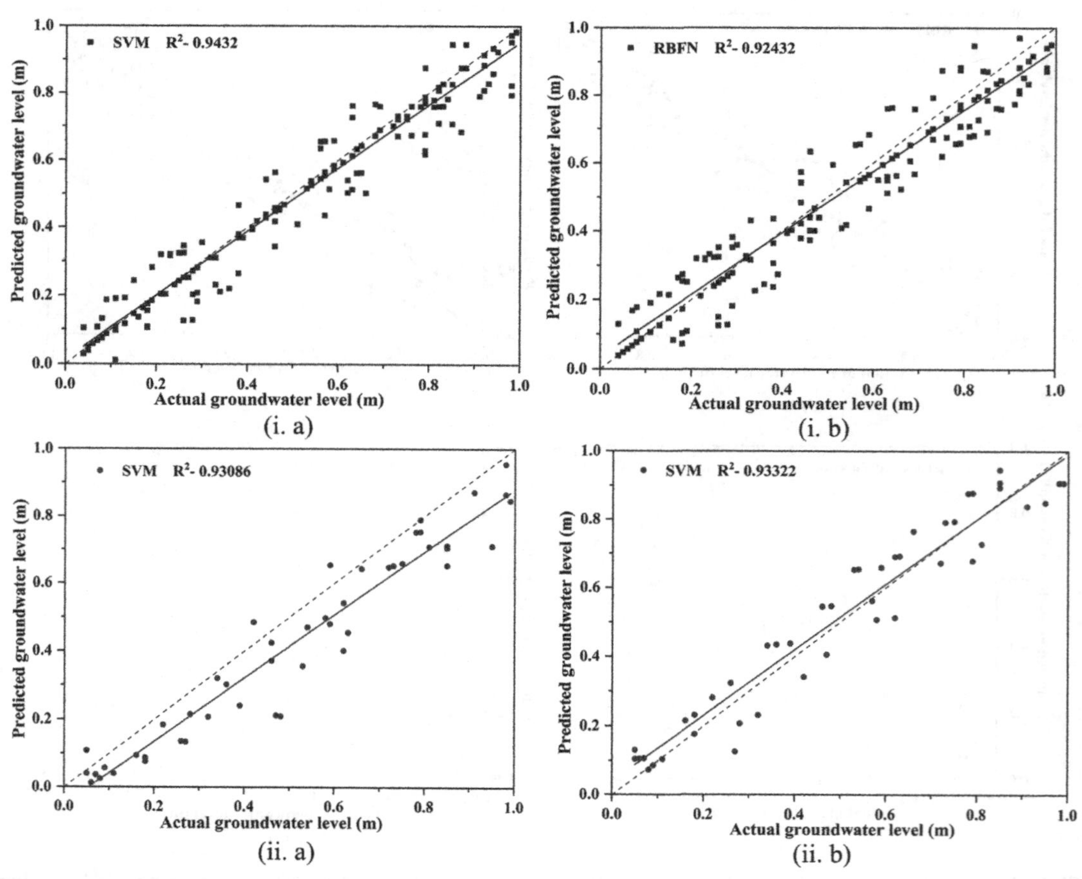

FIG. 3 Actual verses predicted GWL using SVM algorithm at (A) Boden, (B) Khariar watersheds during (i) training, (ii) testing phase.

TABLE 6 Results of SVM-FFA.

Station	Input	MSE		RMSE		WI	
		Training	Testing	Training	Testing	Training	Testing
Boden	S-FFA#1	0.0042	0.00256	0.05729	0.01794	0.97814	0.95949
	S-FFA#2	0.00399	0.00242	0.05622	0.01668	0.97975	0.96364
	S-FFA#3	0.00371	0.00225	0.05421	0.01414	0.98404	0.96646
	S-FFA#4	0.00362	0.0021	0.05284	0.01297	0.98832	0.96947
	S-FFA#5	**0.0032**	**0.00187**	**0.04996**	**0.01058**	**0.99019**	**0.97586**
Khariar	S-FFA#1	0.00387	0.00242	0.05579	0.01626	0.97921	0.96189
	S-FFA#2	0.00383	0.00241	0.05562	0.01599	0.98098	0.96597
	S-FFA#3	0.00368	0.00219	0.05367	0.01368	0.98542	0.96804
	S-FFA#4	0.00339	0.00197	0.05115	0.0115	0.98939	0.97021
	S-FFA#5	**0.00328**	**0.00183**	**0.04985**	**0.01027**	**0.99163**	**0.97807**

Bold signifies the values of the best performing model.

watersheds considered (Boden, Khariar), SVM-FFA models are expected to deliver more precise GWL forecasting than SVM and RBFN models for all scenarios. The S-FFA#5 model for Boden and Khariar watersheds have a training WI of 0.99019 and 0.99163, respectively, and are better than other model forecasts. In a similar way for testing phase, paramount values of WI are 0.97586, 0.97807 at Boden and Khariar gauge station, respectively. Fig. 4 shows actual versus predicted GWL time-series utilizing SVM-FFA models for training and testing phases. It can be witnessed from Fig. 4 that SVM-FFA can proficiently simulate actual GWL time-series superior to other models during both training and testing forecasting.

4.1 Assessment of actual versus predicted GWL at Boden and Khariar

Figs. 2, 3, and 4 represents scatter plot comparing actual and predicted GWLs utilizing RBFN, SVM, and SVM-FFA models for scenario 5 at Boden and Khariar sites. It can be witnessed that scatter plot band is very slender and near to line of perfect fitting in the SVM-FFA forecasting case. Contrariwise SVM indicates slightly reduced performance in comparison to SVM-FFA in testing period. Overall, it is determined that the SVM-FFA model gave more precise prediction outcomes at both stations followed by SVM and RBFN for all scenarios. Time series plot of actual versus predicted groundwater levels for Boden and Khariar watersheds are presented in Fig. 5. The result shows that, for Boden watershed, the projected peak GWL are 7.2405 m, 7.5217 m, and 7.7642 m for RBFN, SVM, SVM-FFA against the actual peak of 8.081 m. Correspondingly the appraised peak GWL are 7.8109 m, 8.0955 m, and 8.3479 m for RBFN, SVM, SVM-FFA in contradiction with actual peak of 8.675 m for the watershed Khariar.

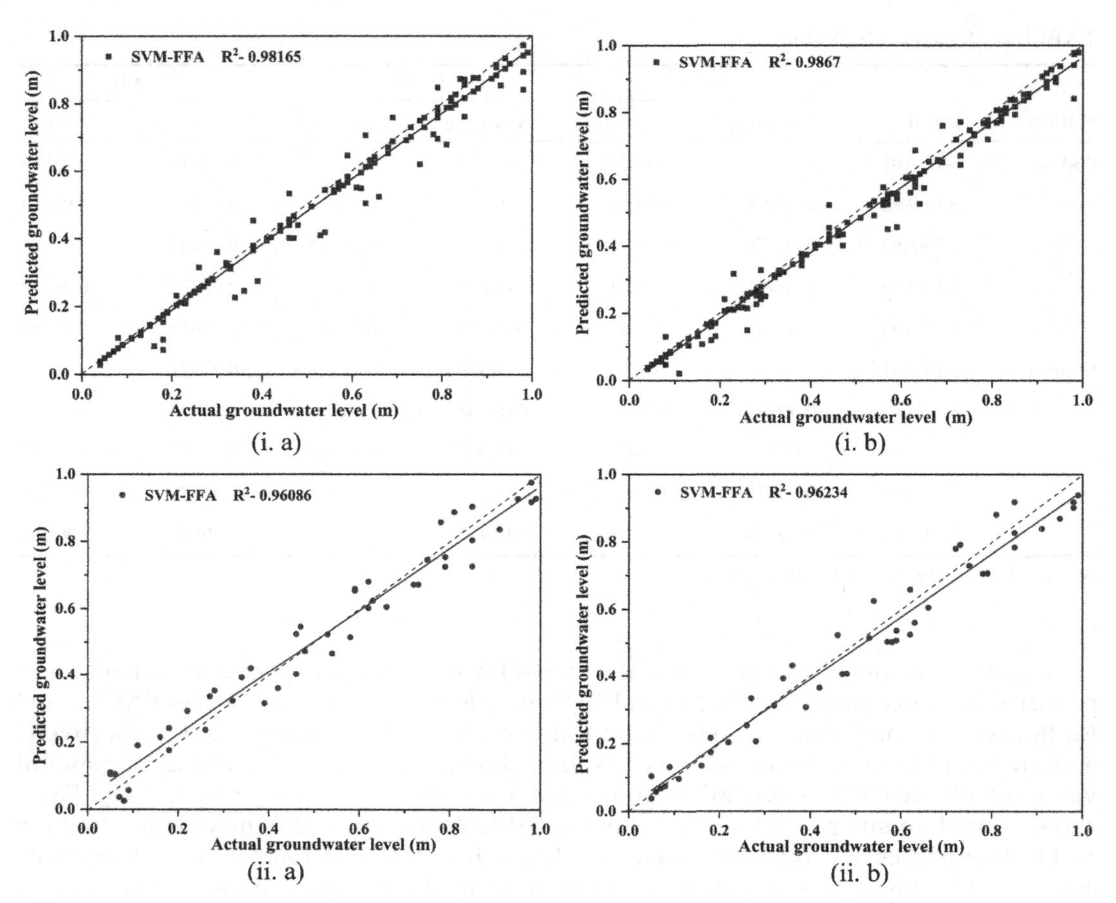

FIG. 4 Actual verses predicted GWL using SVM-FFA algorithm at (A) Boden, (B) Khariar watersheds during (i) training, (ii) testing phase.

Scattering of predicted values utilizing proposed methods by boxplots (Fig. 6) illustrates that SVM-FFA method estimates groundwater level with more accuracy than SVM and RBFN. In each box, the solid line in middle symbolizes median, lower and upper limits are lowest and highest quartiles in respective order. Near perfect covenant of the SVM-FFA model recommends its appropriateness and applicability for GWL forecasting. Nonlinear patterns of GWL time series is accurately replicated with a good generalization in forecasting of hybrid SVM-FFA model.

Fig. 7 represents histogram plots which show observed and forecasted GWL values for RBFN, SVM, and SVM-FFA models. This representation has been plotted for assessing frequency of data points in some selected error bins. An analysis is conducted on a total number of months binned in each GWL value on y-axis. It is vital for noting that forecasting by conjoint SVM-FFA model is nearer to observed values compared to those of simple SVM and conventional RBFN models.

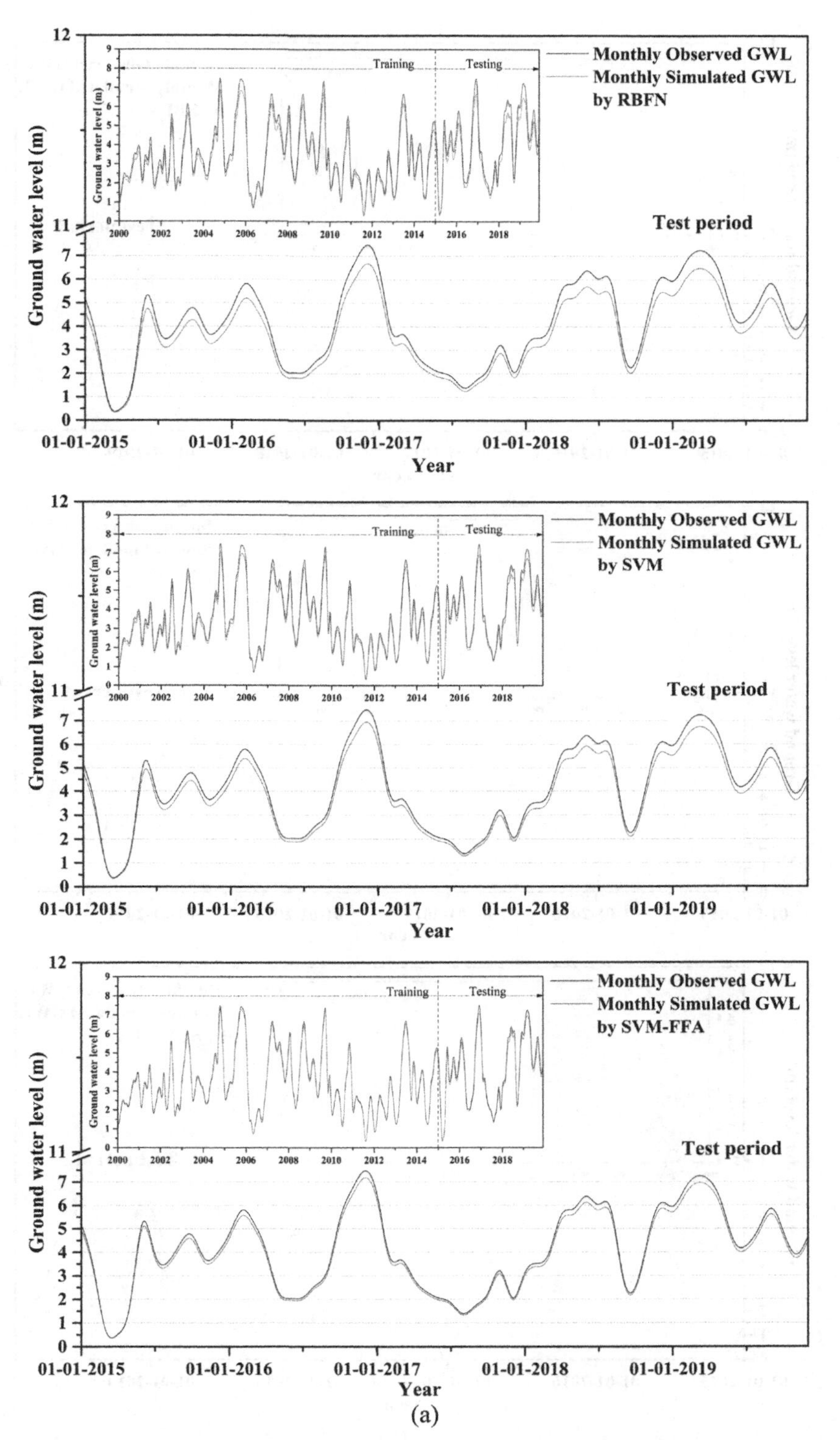

FIG. 5 Comparison of actual verses predicted GWL at (A) Boden, (B) Khariar watersheds during testing phase.

(Continued)

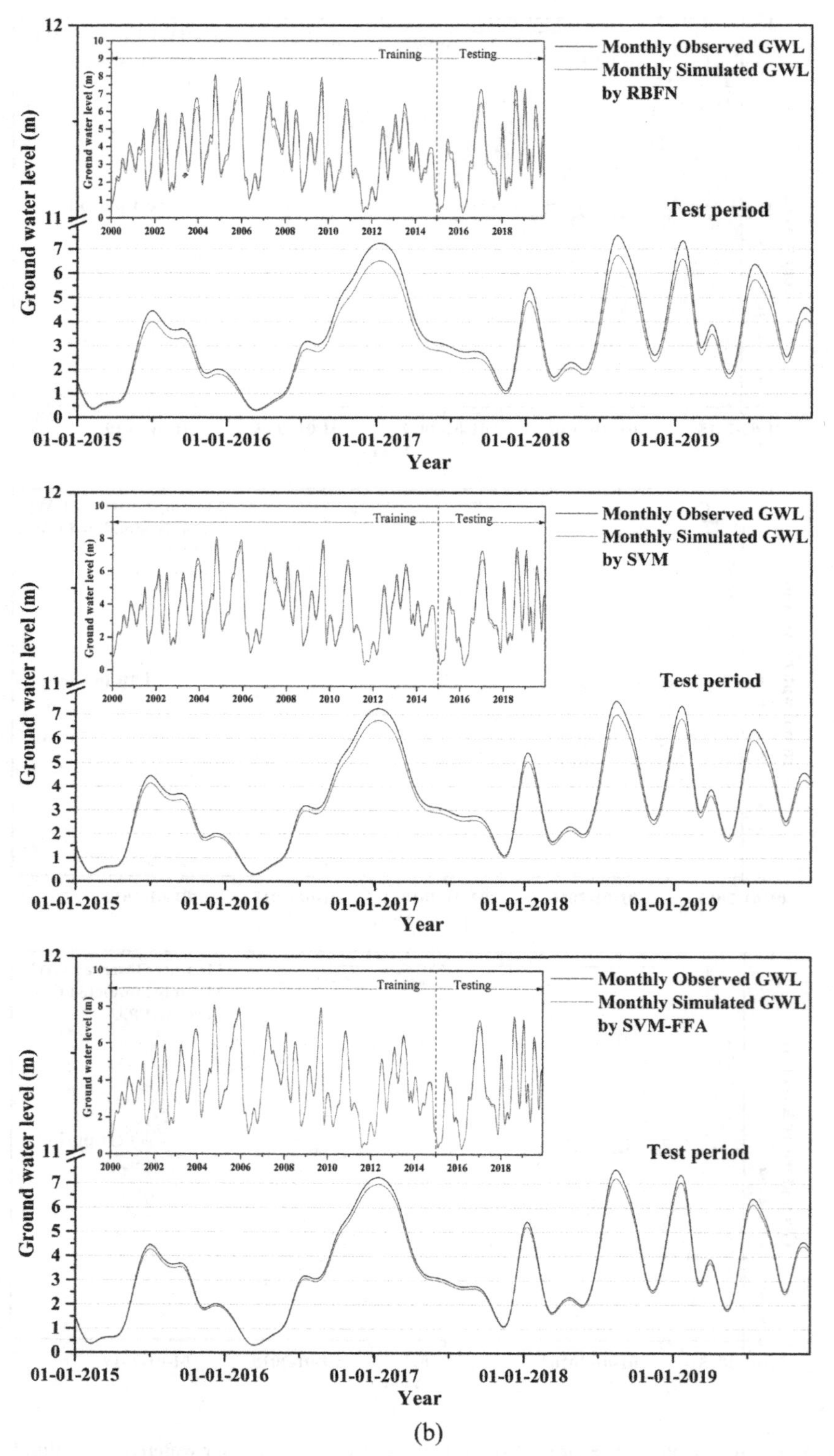

Monthly Observed GWL
Monthly Simulated GWL by RBFN
Training
Testing
Test period
Ground water level (m)
Year
01-01-2015
01-01-2016
01-01-2017
01-01-2018
01-01-2019
Monthly Observed GWL
Monthly Simulated GWL by SVM
Training
Testing
Test period
Ground water level (m)
Year
Monthly Observed GWL
Monthly Simulated GWL by SVM-FFA
Training
Testing
Test period
Ground water level (m)
Year
(b)

FIG. 5—Cont'd

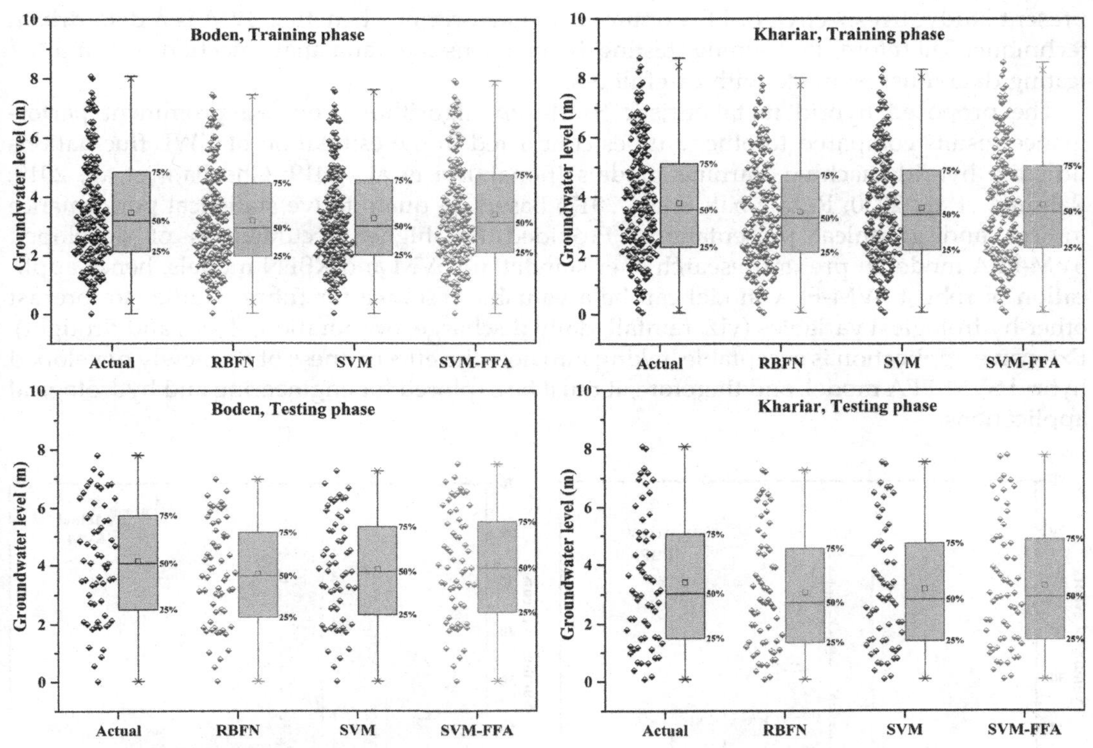

FIG. 6 Comparison of box plot.

4.2 Comparison of model performance for proposed watersheds

For checking the advantages of proposed SVM-FFA model, a comparison is made between its performance and other frequently utilized neural network models, namely SVM, and RBFN. In the above section, a brief description regarding the development of conventional techniques and their performance comparison with the hybrid model is given. Subsequently, detailed performance evaluations of SVM-FFA with SVM and RBFN are presented in this section. It can be found (from Table 7) that the WI value of SVM-FFA models is high than the SVM and RBFN. But, in the course of training period, SVM-FFA is better than SVM. It is notable that SVM-FFA shows better performance than RBFN and SVM models in the case of both watersheds.

In general, based on Table 7 and Figs. 4–7 it can be proposed that SVM-FFA is the preeminent performing model for forecasting GWL fluctuations in Boden and Khariar gauge stations, followed by SVM and RBFN. As discussed above, performance of the SVM model is mainly governed by the selection of its optimal parameters determined using FFA in present study. Irrespective of FFA optimization algorithm's benefits, SVM is a data-driven technique. Therefore, its training/testing is data sensitive, and that selection of training/testing data must be made with carefulness.

The proposed hybrid metaheuristic SVM-FFA algorithm provided prominent performance results compared to other studies conducted in the estimation of GWL fluctuations utilizing hybrid machine learning models (Bonakdari et al., 2019; Ghorbani et al., 2019; Rahman et al., 2020; Rezaie-balf et al., 2017) based on quantitative statistical performance indices and graphical presentations. Provided the higher accurateness of developed SVM-FFA model in present research over standalone SVM and RBFN models, hence application of robust SVM-FFA model can be a valuable resource for future studies to forecast other hydrological variables (viz. rainfall, daily discharge, evaporation, flood, and drought). Extensive application is acceptable, taking into note the effectiveness of the newly developed hybrid SVM-FFA model, and therefore, it must be explored for engineering and hydrological applications.

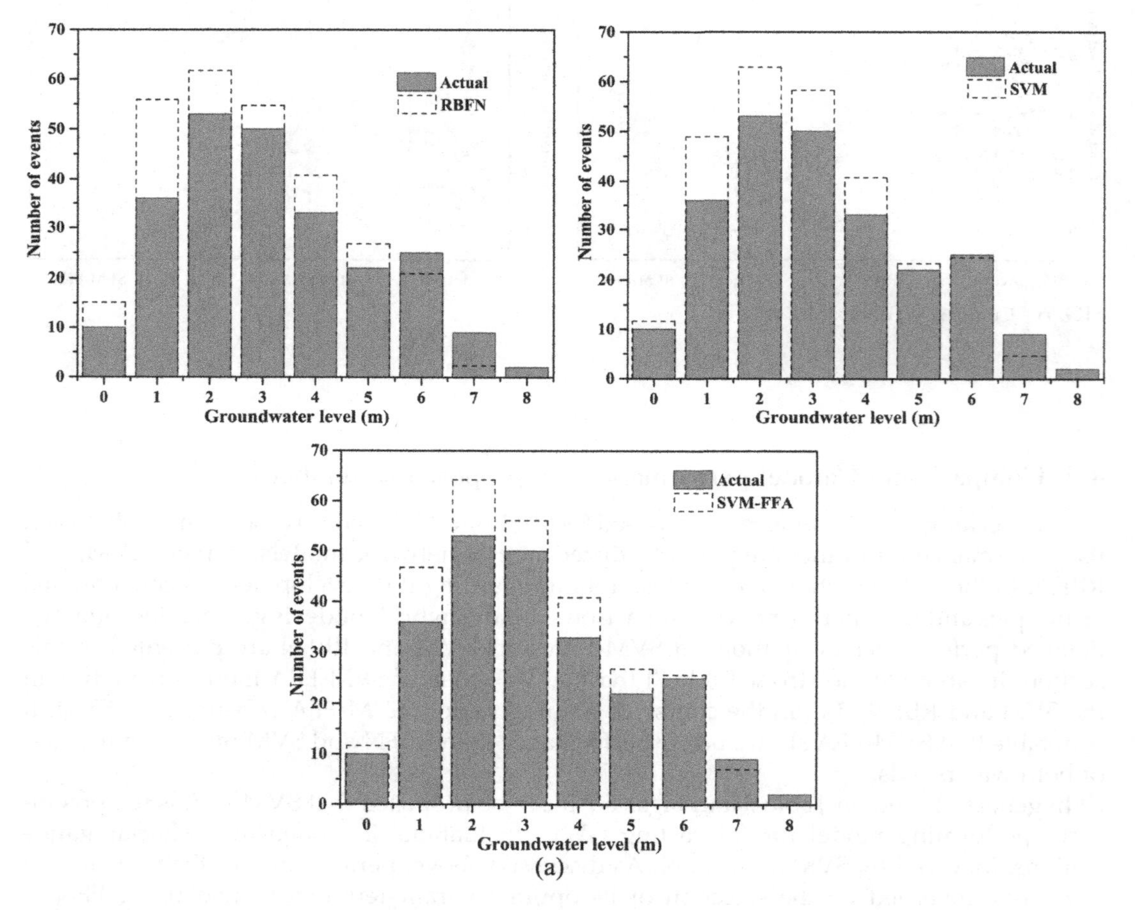

FIG. 7 Histogram plot for RBFN, SVM SVM-FFA technique at (A) Boden (B) Khariar.

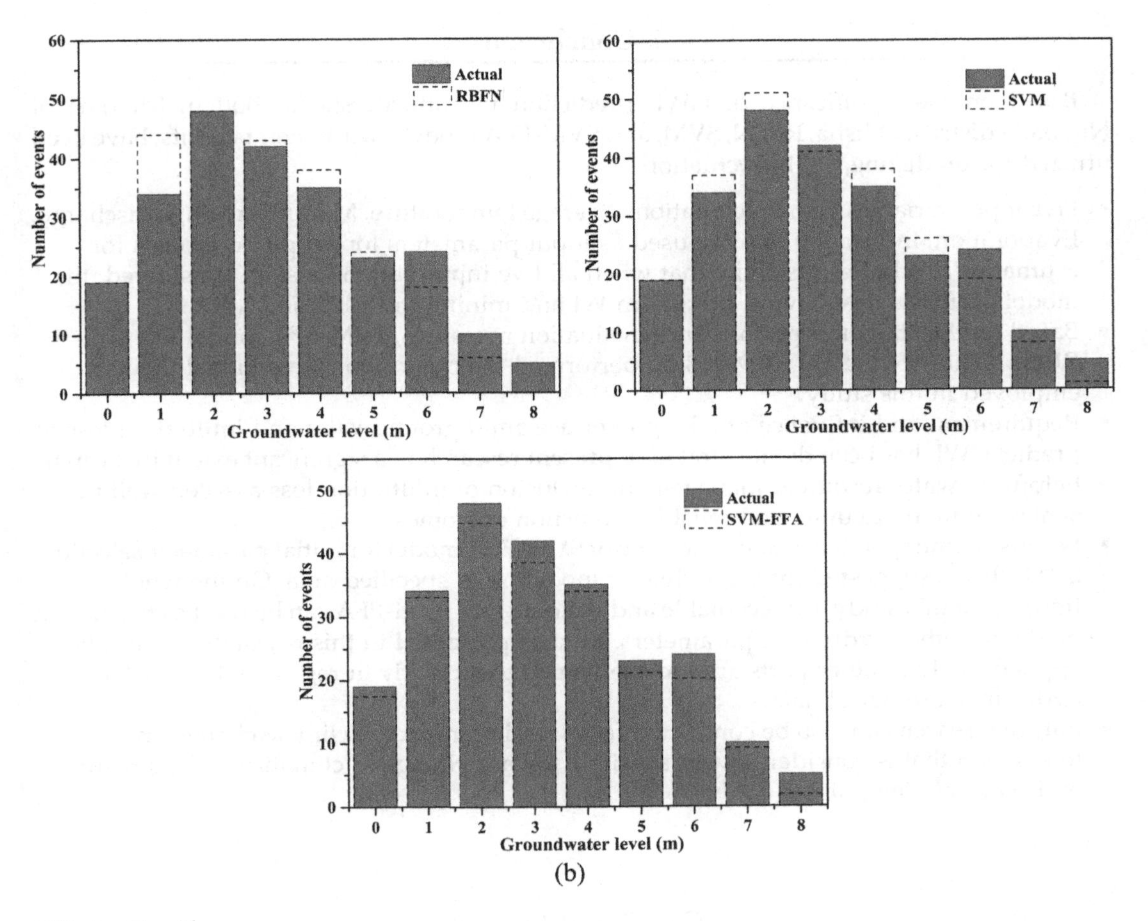

FIG. 7—Cont'd

TABLE 7 Results comparison for proposed watersheds.

Station	Techniques	MSE		RMSE		WI	
		Training	Testing	Training	Testing	Training	Testing
Boden	RBFN	0.00786	0.00742	0.07262	0.03115	0.92247	0.91582
	SVM	0.00496	0.0041	0.06415	0.02131	0.95025	0.94527
	SVM-FFA	0.0032	0.00187	0.04996	0.01058	0.99019	0.97586
Khariar	RBFN	0.00757	0.00701	0.07236	0.03102	0.93265	0.91846
	SVM	0.0049	0.00409	0.06236	0.02127	0.96869	0.94893
	SVM-FFA	0.00328	0.00183	0.04985	0.01027	0.99163	0.97807

5 Conclusions

Based on the significance of GWL prediction in subwatersheds (Boden, Khariar) of Nuapada district, Odisha, RBFN, SVM, and SVM-FFA models, w.r.t. their benefits, have been utilized for predicting GWL fluctuation.

- Five input variables viz., Precipitation, Average temperature, Monthly average discharge, Evaporation, Infiltration loss are used as input parameters for proposed models for estimating GWL. Results show that when all five input parameters are considered, the model performs best having maximum WI and minimum RMSE and MSE.
- Based on the statistical performance evaluation measures, SVM-FFA model with RMSE = 0.04985 and WI = 0.99163 outperformed the other two conventional models employed in this study.
- Requirement and influence of taking interface amid groundwater and infiltration loss to predict GWL has been demonstrated in present research to a significant extent that can be helpful to water resources management. Inclusion of infiltration loss assisted well in achieving more accurate and reliable prediction outcomes.
- Because of independency and efficiency of SVM-FFA model for initial parameter selection, SVM-FFA is suggested for groundwater modeling of specified area. On the whole, findings of this study are acceptable and validate that SVM-FFA can be used as prediction model for other hydrologic parameters. Models presented in this research can also be applied easily to other parts around the world, particularly in regions which lack of hydro-meteorological data.
- Future research can also be conducted for assessing impact of climate change on GWL in this region that is considered very sensitive toward change in climatic conditions from hydrological viewpoint.

Conflict of interest

None.

Research funding

Not applicable.

References

Affandi, A.K., Watanabe, K., 2007. Daily groundwater level fluctuation forecasting using soft computing technique. Nat. Sci. 5 (2), 1–10.

Agnihotri, A., Sahoo, A., Diwakar, M.K., 2022. Flood prediction using hybrid ANFIS-ACO model: a case study. In: Inventive Computation and Information Technologies. Springer, Singapore, pp. 169–180.

Arabgol, R., Sartaj, M., Asghari, K., 2016. Predicting nitrate concentration and its spatial distribution in groundwater resources using support vector machines (SVMs) model. Environ. Model. Assess. 21 (1), 71–82.

Behzad, M., Asghari, K., Coppola Jr., E.A., 2010. Comparative study of SVMs and ANNs in aquifer water level prediction. J. Comput. Civ. Eng. 24 (5), 408–413.

Bhagwat, P.P., Maity, R., 2013. Hydroclimatic streamflow prediction using least square-support vector regression. ISH J. Hydraul. Eng. 19 (3), 320–328.

Bonakdari, H., Ebtehaj, I., Samui, P., Gharabaghi, B., 2019. Lake water-level fluctuations forecasting using minimax probability machine regression, relevance vector machine, Gaussian process regression, and extreme learning machine. Water Resour. Manag. 33 (11), 3965–3984.

Bozorg-Haddad, O., Soleimani, S., Loáiciga, H.A., 2017. Modeling water-quality parameters using genetic algorithm–least squares support vector regression and genetic programming. J. Environ. Eng. 143 (7), 4017021.

Coppola, E., Poulton, M., Charles, E., Dustman, J., Szidarovszky, F., 2003. Application of artificial neural networks to complex groundwater management problems. Nat. Resour. Res. 12 (4), 303–320.

Coulibaly, P., Anctil, F., Aravena, R., Bobée, B., 2001. Artificial neural network modeling of water table depth fluctuations. Water Resour. Res. 37 (4), 885–896.

Fallah-Mehdipour, E., Haddad, O.B., Mariño, M.A., 2013. Prediction and simulation of monthly groundwater levels by genetic programming. J. Hydro Environ. Res. 7 (4), 253–260.

Fayaz, N., Condon, L.E., Chandler, D.G., 2020. Evaluating the sensitivity of projected reservoir reliability to the choice of climate projection: a case study of bull run watershed, Portland, Oregon. Water Resour. Manag. 34 (6), 1991–2009.

Gani, A., Mohammadi, K., Shamshirband, S., Altameem, T.A., Petković, D., 2016. A combined method to estimate wind speed distribution based on integrating the support vector machine with firefly algorithm. Environ. Prog. Sustain. Energy 35, 867–875.

Gevindaraju, R.S., Rao, A.R., 2000. Artificial Neural in Hydrology. Kluwer, The Netherlands.

Ghorbani, M.A., Khatibi, R., Karimi, V., Yaseen, Z.M., Zounemat-Kermani, M., 2018. Learning from multiple models using artificial intelligence to improve model prediction accuracies: application to river flows. Water Resour. Manag. 32, 4201–4215.

Ghorbani, M.A., Deo, R.C., Karimi, V., Kashani, M.H., Ghorbani, S., 2019. Design and implementation of a hybrid MLP-GSA model with multi-layer perceptron-gravitational search algorithm for monthly lake water level forecasting. Stoch. Env. Res. Risk A. 33 (1), 125–147.

Gong, Y., Zhang, Y., Lan, S., Wang, H., 2016. A comparative study of artificial neural networks, support vector machines and adaptive neuro fuzzy inference system for forecasting groundwater levels near Lake Okeechobee, Florida. Water Resour. Manag. 30 (1), 375–391.

Hipni, A., El-shafie, A., Najah, A., Karim, O.A., Hussain, A., Mukhlisin, M., 2013. Daily forecasting of dam water levels: comparing a support vector machine (SVM) model with adaptive neuro fuzzy inference system (ANFIS). Water Resour. Manag. 27 (10), 3803–3823.

Hosseini-Moghari, S.-M., Araghinejad, S., Azarnivand, A., 2017. Drought forecasting using data-driven methods and an evolutionary algorithm. Model. Earth Syst. Environ. 3 (4), 1675–1689.

Jha, M.K., Sahoo, S., 2015. Efficacy of neural network and genetic algorithm techniques in simulating spatio-temporal fluctuations of groundwater. Hydrol. Process. 29, 671–691.

Kalteh, A.M., 2015. Wavelet genetic algorithm-support vector regression (wavelet GA-SVR) for monthly flow forecasting. Water Resour. Manag. 29 (4), 1283–1293.

Kisi, O., Shiri, J., Karimi, S., Shamshirband, S., Motamedi, S., Petković, D., Hashim, R., 2015. A survey of water level fluctuation predicting in Urmia Lake using support vector machine with firefly algorithm. Appl. Math Comput. 270, 731–743.

Li, Z., Yang, Q., Wang, L., Martín, J.D., 2017. Application of RBFN network and GM (1, 1) for groundwater level simulation. Appl. Water Sci. 7 (6), 3345–3353.

Maier, H.R., Kapelan, Z., Kasprzyk, J., Kollat, J., Matott, L.S., Cunha, M.C., Marchi, A., 2014. Evolutionary algorithms and other metaheuristics in water resources: current status, research challenges and future directions. Environ Model Softw. 62, 271–299.

Moghaddam, T.B., Soltani, M., Shahraki, H.S., Shamshirband, S., Noor, N.B.M., Karim, M.R., 2016. The use of SVM-FFA in estimating fatigue life of polyethylene terephthalate modified asphalt mixtures. Measurement 90, 526–533.

Moosavi, V., Vafakhah, M., Shirmohammadi, B., Behnia, N., 2013. A wavelet-ANFIS hybrid model for groundwater level forecasting for different prediction periods. Water Resour. Manag. 27 (5), 1301–1321.

Moravej, M., 2018. Discussion of "application of the firefly algorithm to optimal operation of reservoirs with the purpose of irrigation supply and hydropower production" by Irene Garousi-Nejad, Omid Bozorg-Haddad, Hugo A. Loáiciga, and Miguel A. Mariño. J. Irrig. Drain. Eng. 144 (1), 07017019.

Moravej, M., Amani, P., Hosseini-Moghari, S.-M., 2020. Groundwater level simulation and forecasting using interior search algorithm-least square support vector regression (ISA-LSSVR). Groundw. Sustain. Dev. 11, 100447.

Natarajan, N., Sudheer, C., 2019. Groundwater level forecasting using soft computing techniques. Neural Comput. Applic., 1–18.

Nikolos, I.K., Stergiadi, M., Papadopoulou, M.P., Karatzas, G.P., 2008. Artificial neural networks as an alternative approach to groundwater numerical modelling and environmental design. Hydrol. Process: Int. J. 22 (17), 3337–3348.

Olatomiwa, L., Mekhilef, S., Shamshirband, S., Mohammadi, K., Petković, D., Sudheer, C., 2015. A support vector machine–firefly algorithm-based model for global solar radiation prediction. Sol. Energy 115, 632–644.

Pradhan, S., Kumar, S., Kumar, Y., Sharma, H.C., 2019. Assessment of groundwater utilization status and prediction of water table depth using different heuristic models in an Indian interbasin. Soft. Comput. 23 (20), 10261–10285.

Quan, Q., Hao, Z., Xifeng, H., Jingchun, L., 2020. Research on water temperature prediction based on improved support vector regression. Neural Comput. Applic. 34, 1–10.

Rahman, A.S., Hosono, T., Quilty, J.M., Das, J., Basak, A., 2020. Multiscale groundwater level forecasting: coupling new machine learning approaches with wavelet transforms. Adv. Water Resour. 141, 103595.

Rajaee, T., Ebrahimi, H., Nourani, V., 2019. A review of the artificial intelligence methods in groundwater level modeling. J. Hydrol. 572, 336–351.

Rezaie-balf, M., Naganna, S.R., Ghaemi, A., Deka, P.C., 2017. Wavelet coupled MARS and M5 model tree approaches for groundwater level forecasting. J. Hydrol. 553, 356–373.

Roy, C., Motamedi, S., Hashim, R., Shamshirband, S., Petković, D., 2016. A comparative study for estimation of wave height using traditional and hybrid soft-computing methods. Environ. Earth Sci. 75, 590.

Rumelhart, D.E., Hinton, G.E., Williams, R.J., 1986. Learning representations by back-propagating errors. Nature 323, 533–536.

Safavi, H.R., Esmikhani, M., 2013. Conjunctive use of surface water and groundwater: application of support vector machines (SVMs) and genetic algorithms. Water Resour. Manag. 27, 2623–2644.

Sahoo, A., Samantaray, S., Ghose, D.K., 2021. Prediction of flood in Barak River using hybrid machine learning approaches: a case study. J. Geol. Soc. India 97 (2), 186–198.

Salem, G.S.A., Kazama, S., Komori, D., Shahid, S., Dey, N.C., 2017. Optimum abstraction of groundwater for sustaining groundwater level and reducing irrigation cost. Water Resour. Manag. 31 (6), 1947–1959.

Samantaray, S., Ghose, D.K., 2019. Dynamic modelling of runoff in a watershed using artificial neural network. In: Smart Intelligent Computing and Applications. Springer, Singapore, pp. 561–568.

Samantaray, S., Ghose, D.K., 2020a. Modelling runoff in a river basin, India: An integration for developing un-gauged catchment. Int. J. Hydrol. Sci. Technol. 10 (3), 248–266.

Samantaray, S., Ghose, D.K., 2020b. Modelling runoff in an arid watershed through integrated support vector machine. h2Open J 3 (1), 256–275.

Samantaray, S., Sahoo, A., 2020. Appraisal of runoff through BPNN, RNN, and RBFN in Tentulikhunti Watershed: a case study. In: Frontiers in Intelligent Computing: Theory and Applications. Springer, Singapore, pp. 258–267.

Samantaray, S., Sahoo, A., Ghose, D.K., 2019. Assessment of groundwater potential using neural network: a case study. In: International Conference on Intelligent Computing and Communication. Springer, Singapore, pp. 655–664.

Samantaray, S., Sahoo, A., Ghose, D.K., 2020a. Infiltration loss affects toward groundwater fluctuation through CANFIS in arid watershed: a case study. In: Smart Intelligent Computing and Applications. Springer, Singapore, pp. 781–789.

Samantaray, S., Sahoo, A., Ghose, D.K., 2020b. Assessment of sediment load concentration using SVM, SVM-FFA and PSR-SVM-FFA in Arid Watershed, India: a case study. KSCE J. Civ. Eng., 1–14.

Schwenker, F., Kestler, H.A., Palm, G., 2001. Three learning phases for radial-basis-function networks. Neural Netw. 14, 439–458.

Seo, Y., Kim, S., Singh, V.P., 2016. Physical interpretation of river stage forecasting using soft computing and optimization algorithms. In: Harmony Search Algorithm. Springer, pp. 259–266.

Shamshirband, S., Shojafar, M., Hosseinabadi, A.A.R., Abraham, A., 2014. A solution for multi-objective commodity vehicle routing problem by NSGA-II. In: 2014 14th International Conference on Hybrid Intelligent Systems. IEEE, pp. 12–17.

Shojafar, M., Cordeschi, N., Amendola, D., Baccarelli, E., 2015. Energy-saving adaptive computing and traffic engineering for real-time-service data centers. In: 2015 IEEE International Conference on Communication Workshop (ICCW). IEEE, pp. 1800–1806.

Sörensen, K., 2015. Metaheuristics—the metaphor exposed. Int. Trans. Oper. Res. 22 (1), 3–18.

Sridharam, S., Sahoo, A., Samantaray, S., Ghose, D.K., 2021. Estimation of water table depth using wavelet-ANFIS: a case study. In: Lecture Notes in Networks and Systems. vol. 134. https://doi.org/10.1007/978-981-15-5397-4_76.

Su, J., Wang, X., Liang, Y., Chen, B., 2014. GA-based support vector machine model for the prediction of monthly reservoir storage. J. Hydrol. Eng. 19 (7), 1430–1437.

Sujay Raghavendra, N., Deka, P.C., 2015. Forecasting monthly groundwater level fluctuations in coastal aquifers using hybrid wavelet packet–support vector regression. Cogent Eng. 2 (1), 999414.

Todd, D.K., Mays, L.W., 2004. Groundwater Hydrology. John Wiley & Sons.

Vapnik, V.N., 1995. The Nature of Statistical Learning. Theory. Springer.

Vapnik, V.N., 1998. Statistical Learning Theory. Wiley, New York.

Wunsch, A., Liesch, T., Broda, S., 2018. Forecasting groundwater levels using nonlinear autoregressive networks with exogenous input (NARX). J. Hydrol. 567, 743–758.

Xing, B., Gan, R., Liu, G., Liu, Z., Zhang, J., Ren, Y., 2016. Monthly mean streamflow prediction based on bat algorithm-support vector machine. J. Hydrol. Eng. 21 (2), 4015057.

Yan, Q., Ma, C., 2016. Application of integrated ARIMA and RBF network for groundwater level forecasting. Environ. Earth Sci. 75 (5), 396.

Yang, X.S., 2009 October. Firefly algorithms for multimodal optimization. In: International Symposium on Stochastic Algorithms. Springer, Berlin, Heidelberg, pp. 169–178.

Yin, Z., Wen, X., Feng, Q., He, Z., Zou, S., Yang, L., 2017. Integrating genetic algorithm and support vector machine for modeling daily reference evapotranspiration in a semi-arid mountain area. Hydrol. Res. 48 (5), 1177–1191.

Yoon, H., Jun, S.-C., Hyun, Y., Bae, G.-O., Lee, K.-K., 2011. A comparative study of artificial neural networks and support vector machines for predicting groundwater levels in a coastal aquifer. J. Hydrol. 396 (1–2), 128–138.

Zhang, N., Xiao, C., Liu, B., Liang, X., 2017. Groundwater depth predictions by GSM, RBF, and ANFIS models: a comparative assessment. Arab. J. Geosci. 10 (8), 189.

Further reading

Samantaray, S., Ghose, D.K., 2022. Prediction of S12-MKII rainfall simulator experimental runoff data sets using hybrid PSR-SVM-FFA approaches. J. Water Clim. Change. https://doi.org/10.2166/wcc.2021.221.

CHAPTER

21

Basin-scale subsurface hydrology: Modeling of a stressed and data-scarce aquifer using hillslope-based approach

Soumyaranjan Sahoo[a] *and Suraj Jena*[b]

[a]School of Water Resources, Indian Institute of Technology Kharagpur, Kharagpur, West Bengal, India [b]School of Infrastructure, Indian Institute of Technology Bhubaneswar, Bhubaneswar, Odisha, India

OUTLINE

1 Introduction 367

2 Materials and methods 370

2.1 Overview of available groundwater models 370

2.2 Hillslope-storage Boussinesq model 370

3 Study area and data collection 373

4 Field application 377

4.1 Model setup 377

4.2 Model calibration and validation 379

5 Results and discussion 379

5.1 Geomorphic analysis of the study area 379

5.2 Derived hillslope width function 380

5.3 Prediction of groundwater level 381

5.4 Estimated subsurface discharge time series 382

6 Summary and conclusions 383

References 384

1 Introduction

With the rapid urbanization, increased population growth, and improved socio-economic status globally, the consumption of freshwater resources is ever increasing. Further, due to erratic patterns in the spatiotemporal distribution of hydro-meteorological variables, the

https://doi.org/10.1016/B978-0-323-91910-4.00021-2

domestic, agricultural, industrial, and ecological water supplies have been significantly affected. To meet these water demands under such a highly uncertain situation, emphasis is given worldwide for groundwater development and use as the surface water sources alone are not being sufficient (Jena and Panda, 2018). However, due to its location and slow movement, it is difficult to know and predict how the groundwater system will respond to the current use or future management plans (Sahoo and Sahoo, 2020b). Here comes the role of a model that represents nature, nature-identical system, or phenomenon. Modeling involves replicating the natural processes physically, conceptually, or numerically, and hence can aid as a predictive tool to foresee how the subsurface processes will respond to different management scenarios.

Based on the local geological formations of the subsurface flow domain, surface topography and geomorphology, groundwater movement takes place along different paths (Jena et al., 2021), known as the flow systems. According to the spatial extension, the groundwater flow systems can be classified as: local, intermediate and regional (Tóth, 1963). In the local flow system, water flows to the discharge area in the vicinity, such as ponds or intermittent streams. In the regional flow system, the involved flow domain is the largest, extending up to river basins, lakes or even oceans. However, the intermediate system is characterized by topographic highs and lows at the intermediate scale, between the local and regional scales. The local flow systems dominate in the terrain with high topographic relief, such as the headwater catchments; whereas, the areas with relatively flat relief have dominant intermediate or regional flow systems, such as the alluvial basins. According to Brunner et al. (2017), the saturated subsurface discharge process can be classified as (i) hyporheic-scale, where hyporheic flux exchange takes place (1–100 m); (ii) reach-scale, where groundwater release is predominately determined by the water level in the aquifers (100–1000 m); and (iii) river basin-scale that corresponds to the entire basin with the regional flow ($\geq$1000 m).

As emphasized by Barthel (2014), the interaction between groundwater and surface water is preferred at the regional (river basin) scale for the following reasons. Firstly, interaction study at the river basin or regional scale is essential for integrated water resources management through integrative assessment (Bouwer, 2002; Refsgaard et al., 2010). At the regional scale, hydro-meteorological, ecological, socio-economical, and political impacts can be considered due to the larger spatial domains among the recharge-withdrawal-discharge processes, between the pollutant source and affected area, and between the water source and stakeholders (McGonigle et al., 2012; Højberg et al., 2013). Secondly, the climate change impact study on the groundwater resources necessitates regional hydrogeological study (Green et al., 2011). The global climate models predict the hydro-meteorological flux at a much larger spatial scale. To incorporate future scenarios of the global climate in solving the local issues, regional- or river basin-scale modeling is the necessary intermediate step (Varis et al., 2004; Holman et al., 2012). Thirdly, the availability of high-performance hardware-software interface, robust solution schemes and codes (Zhou and Li, 2011) in the field of hydro-climatology along with the organized and improved regional test-datasets, though only at fewer places in the developed countries, augments regional-scale study. Moreover, as discussed by Barthel (2014), although a good number of publications use a regional hydrological dataset on groundwater or surface water, very few combine these observations in a truly integrated manner. Conversely, the large volume of literature on field studies, looking at the exchange processes and feedback, are at the local scales dealing with the floodplains, hyporheic zones or riparian systems. Hence, river basin-scale subsurface modeling involving the hydrological process components is deemed necessary for futuristic integrated water resources management.

However, data scarcity at the required spatiotemporal scales for the desired hydrogeological parameter is the principal challenge in basin-scale modeling (Carter et al., 1994; Candela et al., 2014). A tremendous scientific challenge is encountered while using point-scale groundwater observations for catchment- or regional-scale modeling through upscaling the point-scale parameter into a larger scale and through regionalization of the parameter to obtain the characteristic representative value from several point observations. Although theories dealing with these limitations are well established, their practical implementation at the regional scale is still a challenging task (de Marsily et al., 2005). From the above discussion, it is evident that a trade-off is necessary between the model complexity and data availability. Moreover, for modeling the subsurface processes in a data-scarce or ungauged catchment, there is the need for a simplified approach that includes all the hydrological processes and parameters without compromising the modeling accuracy.

Most of the models available in the literature to study the groundwater storage-discharge dynamics estimate the discharge by the Darcy's approach considering the transient hydraulic gradient. Further, the Darcy-Richards equation-based groundwater models simulate the flow through homogeneous porous media at point-scale; however, they cannot simulate the processes at the hillslope- or catchment-scales. While applying such models, the underlying hydrological processes are compromised by the effective parameter values (Tromp-van Meerveld et al., 2007). To address this limitation, there are several efforts to model the hydrological processes at the hillslope-scale (e.g., Troch et al., 2003; Weiler and McDonnell, 2007).

Hillslopes, as the building block of catchments, partitions the input precipitation flux to surface runoff, soil moisture storage, groundwater recharge, and subsurface runoff. However, despite the growing enthusiasm among the hydrologic community toward the hillslope-scale process-based modeling, their application in large areas is limited because of several reasons: (i) a complex model usually is associated with an increase in the number of parameters making it difficult to parameterize, (ii) data (specifically subsurface) is hardly available at the required spatiotemporal scales, (iii) lack of theoretical advancement to link and transform the hydrological processes at the hillslope-scale to that of the catchment or river basin scale, and (iv) high computational demand of the process-based models.

With the rapid increase in the computational power during the last decade followed by the use of advanced techniques in hydrology, such as cloud computing (e.g., Hunt et al., 2010) and parallelization of simulation of numerical models through multiple processors (e.g., Kollet et al., 2010), computational demand is not the issue anymore. Hence, data requirement is the major constraint as most of the groundwater basins worldwide are ungauged (Jena et al., 2020). This necessitates a simplified model that needs fewer input data and yet is physically-based.

Incidentally, the hillslope-based hillslope storage Boussinesq (HSB) model (Troch et al., 2003) is one such process-based and semidistributed groundwater model that estimates the unconfined aquifer storage discharge dynamics. It accounts for the catchment geomorphology by considering the planform geometry of the hillslopes. Further, the HSB model needs the aquifer hydraulic characteristic values in one direction, unlike the traditional approaches. This highlights the potential of the HSB model to be used in data-scarce catchments that are often encountered in developing countries. Therefore, this chapter presents a framework for groundwater modeling of catchments with limited geohydrologic data by using a hillslope-based approach.

2 Materials and methods

2.1 Overview of available groundwater models

Subsurface hydrologic models are simplified representation of the movement of water in the subsurface domain that includes the unsaturated (vadose) zone and saturated (groundwater) zone. In the case of saturated flow modeling, the modular three-dimensional groundwater flow (MODFLOW) model by McDonald and Harbaugh (1988) is one such model that is widely acclaimed by the hydrologic community. It gives an estimate of groundwater level and flows in response to recharge forcing and pumping scenarios. However, the aquifer hydraulic characteristics are seldom available at the required spatiotemporal scale, thus necessitating simplification. The simplifications could be regarding the flow direction, aquifer extent and geometry, subsurface formations (heterogeneity and anisotropy), identification of boundary and flux through the boundary. Hence, based on the data and resources available, and the desired accuracy of investigation, the complexity of the subsurface models is decided (Bredehoeft, 2005). One such measure is to reduce the model dimensions from three to two- or one-dimensional (2D/1D) forms. The absence of the third dimension reduces the number of parameters and uncertainty in parameter estimation (Lancia et al., 2018). For example, Sena and Molinero (2009) developed a 2D groundwater model and coupled it with a lumped hydrologic water balance model.

Consequently, the Boussinesq equation-based models are there to simulate the storage-discharge of sloping aquifers (Sahoo and Sahoo, 2018). Hence, for the river basin-scale modeling, considering the groundwater basin identical and underlying the surface water basin, the groundwater basin can be decomposed to several hillslope aquifers (Sahoo and Sahoo, 2020a), where the Boussinesq model variants (e.g., Troch et al., 2003) can be used. Incidentally, the strict conditions regarding hillslope geometry for applying the Boussinesq model to natural hillslopes were relaxed first by Fan and Bras (1998). As the hillslope plan shape and profile curvature have significant control over the storm flow and saturation, they developed a method to collapse the 3D soil mantle into a 2D cross-section made of pore space to preserve the essence of the physics of flow in simplified cross-section. The following sections detail the development and application of a low-dimensional subsurface flow model to a data-scarce catchment in eastern India.

2.2 Hillslope-storage Boussinesq model

The hillslope-storage Boussinesq (HSB) model operates on hillslopes as elements. This model is semidistributed as the groundwater storage in the hillslopes is modeled as a function of the distance, x from the river network of the river basin. The assumptions/simplifications used for developing the HSB model are: (i) The rate of rainfall recharge, R_e is uniformly distributed over the hillslope; (ii) the Dupuit-Forchheimer assumptions for groundwater flow

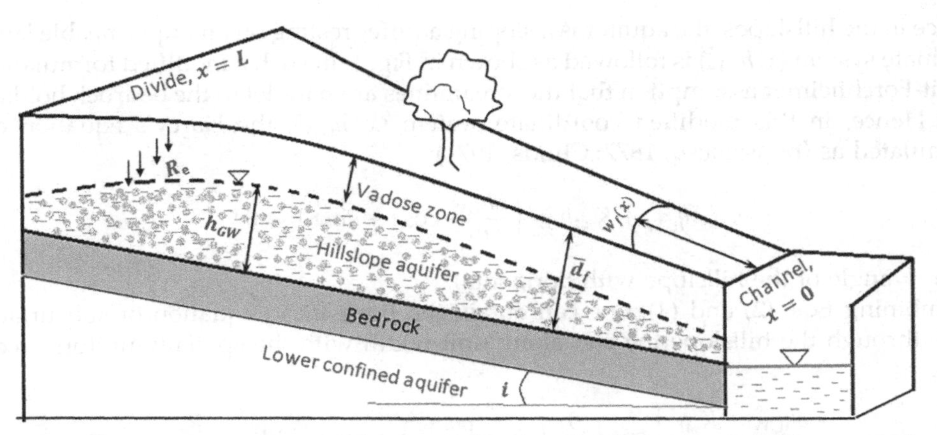

FIG. 1 Conceptual sketch of a hillslope detailing the cross-section of the flow domain resting on a bedrock layer inclined at an angle i with horizontal, in which, the hillslope length $L \gg D$, and $\bar{d}_f$ = total depth of the hillslope unconfined aquifer.

holds that consider parallel streamlines to the bedrock; and (iii) the saturated subsurface storage in the hillslopes varies with respect to the longitudinal distance x from the hillslope toe/river network; however, this variability is ignored in the transverse direction of x. Fig. 1 shows the schematic of a hillslope aquifer draining to the nearby stream.

2.2.1 *Model derivation*

According to Darcy's law, the subsurface saturated discharge per unit width, q_{GW} can be expressed as:

$$q_{GW} + K_{sat} h_{GW} \frac{\partial h_{GW}}{\partial x} = 0 \tag{1}$$

where K_{sat} = saturated hydraulic conductivity of the aquifer forming material [LT^{-1}]; h_{GW} = elevation of groundwater table [L]; x = flow distance of a location from the river network [L]; and $\frac{\partial h_{GW}}{\partial x}$ = hydraulic gradient [−].

The continuity form of the equation for the groundwater discharge from hillslopes can be expressed as (Troch et al., 2013):

$$p \frac{\partial h_{GW}}{\partial t} + \frac{\partial h_{GW}}{\partial x} dx - R_e dx = 0 \tag{2}$$

where p = drainable porosity [−]; t = time [T]; and R_e = effective rainfall recharge rate [LT^{-1}] assumed to be constant.

Combining Eqs. (1) and (2), the Boussinesq equation for horizontal aquifer can be written as:

$$\frac{\partial h_{GW}}{\partial t} + \frac{K_{sat}}{p} \frac{\partial}{\partial x}\left(h_{GW} \frac{\partial h_{GW}}{\partial x}\right) - \frac{R_e}{p} = 0 \tag{3}$$

Since in the hillslopes, the aquifer is a sloping aquifer resting on an impermeable layer, the coordinate system (x, h_{GW}) is followed as shown in Fig. 1. In such a modified formulation, the Dupuit-Forchheimer assumption that the streamlines are parallel to the bedrock holds (Bear, 1972). Hence, in this modified coordinate system (x, h_{GW}), the Darcy's equation can be reformulated as (Boussinesq, 1877; Childs, 1971):

$$q_{GW} + K_{sat} h_{GW} \left(\frac{\partial h_{GW}}{\partial x} \cos i + \sin i \right) = 0 \tag{4}$$

where i = angle of the hillslope with horizontal.

Combining Eqs. (2) and (4), one can obtain the Boussinesq equation of subsurface discharge through the hillslope toe-end along unit-width with the spatially uniform recharge rate, R_e as:

$$\frac{\partial h_{GW}}{\partial t} - \frac{K_{sat}}{p} \left[\cos i \frac{\partial}{\partial x} \left(h_{GW} \frac{\partial h_{GW}}{\partial x} \right) + \sin i \frac{\partial h_{GW}}{\partial x} \right] - \frac{R_e}{p} = 0 \tag{5}$$

where h_{GW} is measured perpendicular to the underlying impermeable layer in the coordinate system (x, h_{GW}).

Eq. (5) describes the toe-end discharge along a hillslope of unit width with sloping bedrock. Application of this equation to describe the subsurface discharge in complex hillslopes has the limitation that it does not consider the 3D soil mantle, through which the real flow occurs. Moreover, to fully capture the subsurface storage-discharge dynamics along complex hillslopes, the hillslope geometry has to be taken into account.

Incidentally, Fan and Bras (1998) developed a methodology to account for the topographic control on the hillslope hydrologic processes. This approach reduces a natural 3D regolith on the top of a bedrock layer into a 1D soil profile, made up of pore space. Considering a hillslope with 3D regolith on top of an impermeable bedrock with a slope angle of i (Fig. 1), the soil moisture storage capacity function, $S_{mc}(x)$ can be defined as (Fan and Bras, 1998):

$$S_{mc}(x) = w_f(x) \bar{d}_f(x) p \tag{6}$$

where $w_f(x)$ = width of hillslope [L] at a flow distance of x (the hillslope width function); and $\bar{d}_f(x)$ = average hillslope unconfined aquifer depth [L] at any flow distance x (the hillslope soil depth function).

Eq. (6) describes the soil strata as the thickness of the pore space along the hillslope accounting for the plan shape in terms of the hillslope width function and the profile curvature using the soil depth function. Analogous to the concept of S_{mc}, the saturated aquifer storage, S_{GW} can be presented as: $S_{GW} = w_f \times h_{GW} \times p$, where $h_{GW}(x)$ is the average elevation of the water table above the confining bedrock at a flow distance x.

Generally, the flow processes along the hillslopes are predominately governed by the plan shape and profile curvature. According to the plan shape, hillslopes can be broadly divided into three types, viz., convergent, characterized by $(\partial w_f/\partial x) > 0$; divergent, characterized by $(\partial w_f/\partial x) < 0$; and uniform $((\partial w_f/\partial x) \approx 0)$, where the width of the hillslope remains constant with distance from the channel. Similarly, in nature, there are three kinds of profile curvature encountered, viz., convergent, divergent and planar.

Incorporating the HWF and S_{mc} in continuity form of Eq. (5) and combining it with the Darcy's equation for sloping aquifers, i.e., Eq. (4), the 1D hillslope-storage Boussinesq model of Troch et al. (2003) can be expressed as:

$$p\frac{\partial S_{GW}}{\partial t} - \frac{K_{sat}\cdot\cos i}{p}\frac{\partial}{\partial x}\left[\frac{S_{GW}}{w}\left(\frac{\partial S_{GW}}{\partial x} - \frac{S_{GW}}{w}\cdot\frac{\partial w}{\partial x}\right)\right] + K_{sat}\sin i\frac{\partial S_{GW}}{\partial x} = pR_e w_f \quad (7)$$

2.2.2 *Model parameterization*

The HSB model considers four types of spatiotemporally varying variables and parameters, viz., (i) atmospheric, (ii) topographic, (iii) geomorphologic, and (iv) pedo-hydrogeologic that are detailed as follows:

The effective rainfall recharge $R_e(x, t)$ is the atmospheric forcing term which is the principal input to the saturated subsurface storage and causing the subsurface discharge. This is obtained by solving the water balance equation for the overlying unsaturated (vadose) zone, considering the different fluxes of infiltration from the land surface, I; evaporation from bare land and evapotranspiration by vegetation, ET_a; and capillary contribution to the vadose zone from the water table for shallow aquifers, C. The water balance equation can be written as (Sahoo et al., 2018):

$$R_e = I + C - ET_a - \frac{\partial}{\partial t}\left(\int_0^{\overline{d}_f - h_{GW}(t-1)/\cos i}[\theta_{VZ}(x,t) - \theta_r(x)dz]\right) \quad (8)$$

where θ_{VZ} = root zone volumetric soil moisture $[L^3L^{-3}]$ that can be measured in situ using tensiometers.

The slope angle of the lower confining bedrock, i and plan form geometry of the constituting hillslope are the topographic parameters. The bedrock slope can be estimated from well-log information. The HWF and hillslope profile curvature are the geomorphologic parameters. The HWF accounts for plan shape which can be determined from the digital elevation model (DEM) data. A detailed algorithm for the determination of the HWF is described in the later sections. The profile curvature can be determined by knowing the soil depth at different places across a hillslope using the well-log data which may be available at the local geological department.

The soil depth function, $\overline{d}_f(x)$; effective porosity of the aquifer forming material, p; and saturated hydraulic conductivity of the aquifer forming material, K_{sat} are the pedo-hydrogeologic parameters. The soil depth function can be estimated from the available drilling logs, and the parameters p and K_{sat} can be obtained from the aquifer test data.

3 Study area and data collection

The field application of the HSB model was carried out at the Kanjhari watershed, a tributary to the Baitarani River in Odisha State, India. This study area is confined between the latitudes of 21°33′N to 21°41′N and longitudes of 85°38′E to 85°48′E. The index map of the

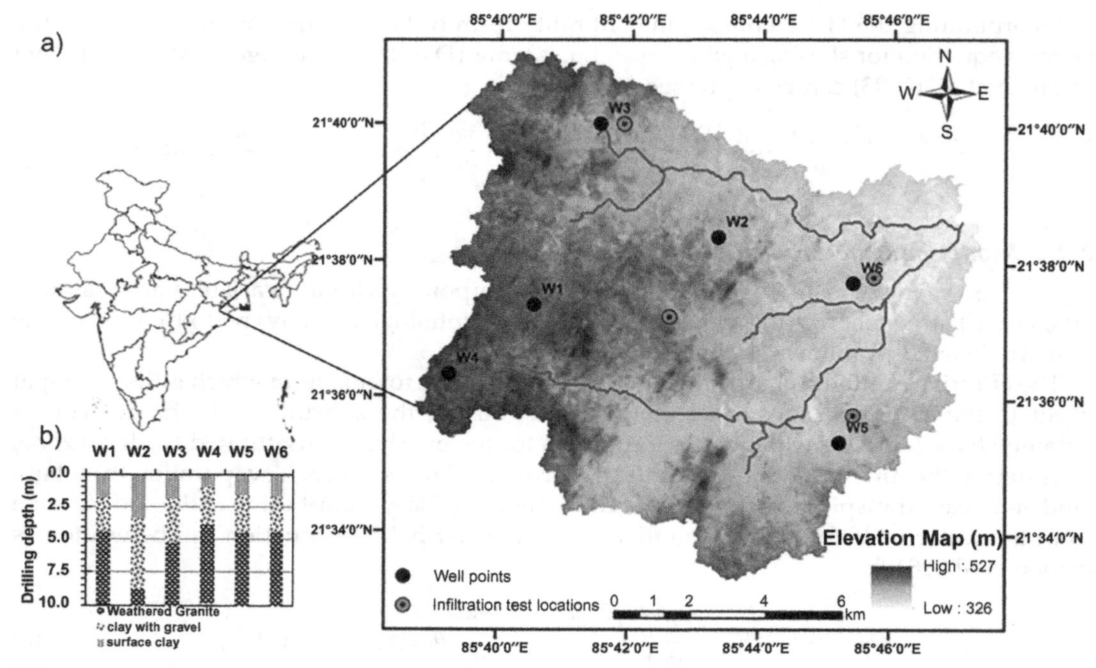

FIG. 2 The elevation map of the study area as extracted from DEM along with the locations of wells for groundwater data collection and double-ring infiltration test.

study area is illustrated in Fig. 2. The average annual rainfall in the area is about 1480 mm that is received within 75 number of rainy days per year on an average. Physiographic variations constituting moderately high hills and hillocks, undulating plains with rolling topography, and extensive tablelands on their summits along with alluvial tract are dominant in the Kanjhari watershed.

The predominant type of surface soil is the lateritic clay loam which may be categorized under Alfisol. The other soil classes distributed in the study area are coarse-loamy, fine-loamy, and loamy. The detailed soil classes encountered in the study area are illustrated in Fig. 3. The subsurface zone has developed secondary porosity that aid subsurface water storage in the different weathered formations. The subsurface weathered formations have resulted in unconfined aquifers, the groundwater source. Similarly, from the land use map of the study area as depicted in Fig. 4, it is observed that croplands are the major land use encountered followed by the forest area. Shrubs are also a major land-use class.

The daily meteorological data of precipitation and temperature were obtained for the nearest meteorological station from Meteorological Centre, Bhubaneswar. Similarly, for estimation of the evapotranspiration flux, other meteorological variables were downloaded following the link (http://power.larc.nasa.gov) by the National Aeronautics and

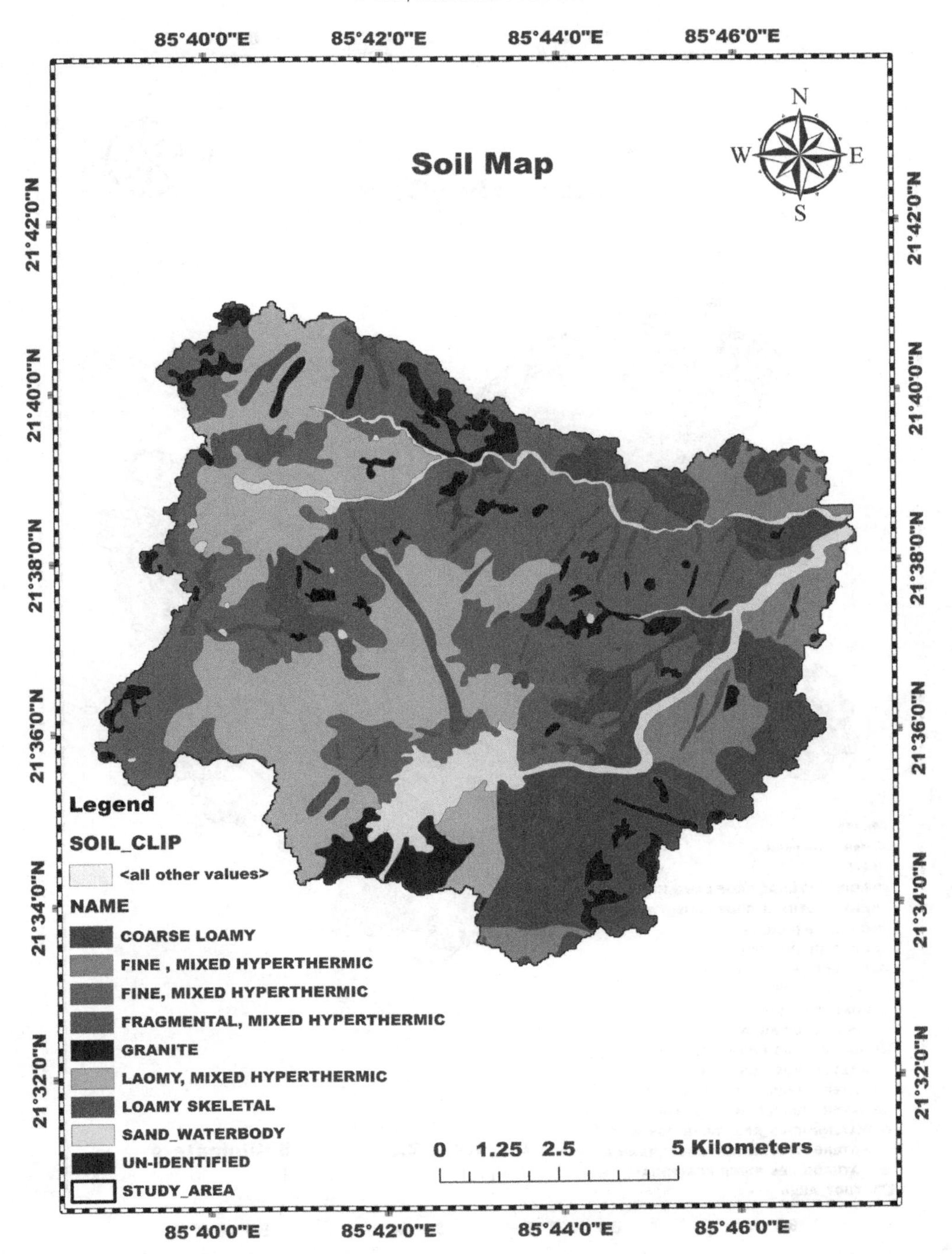

FIG. 3 Soil map of the Kanjhari watershed.

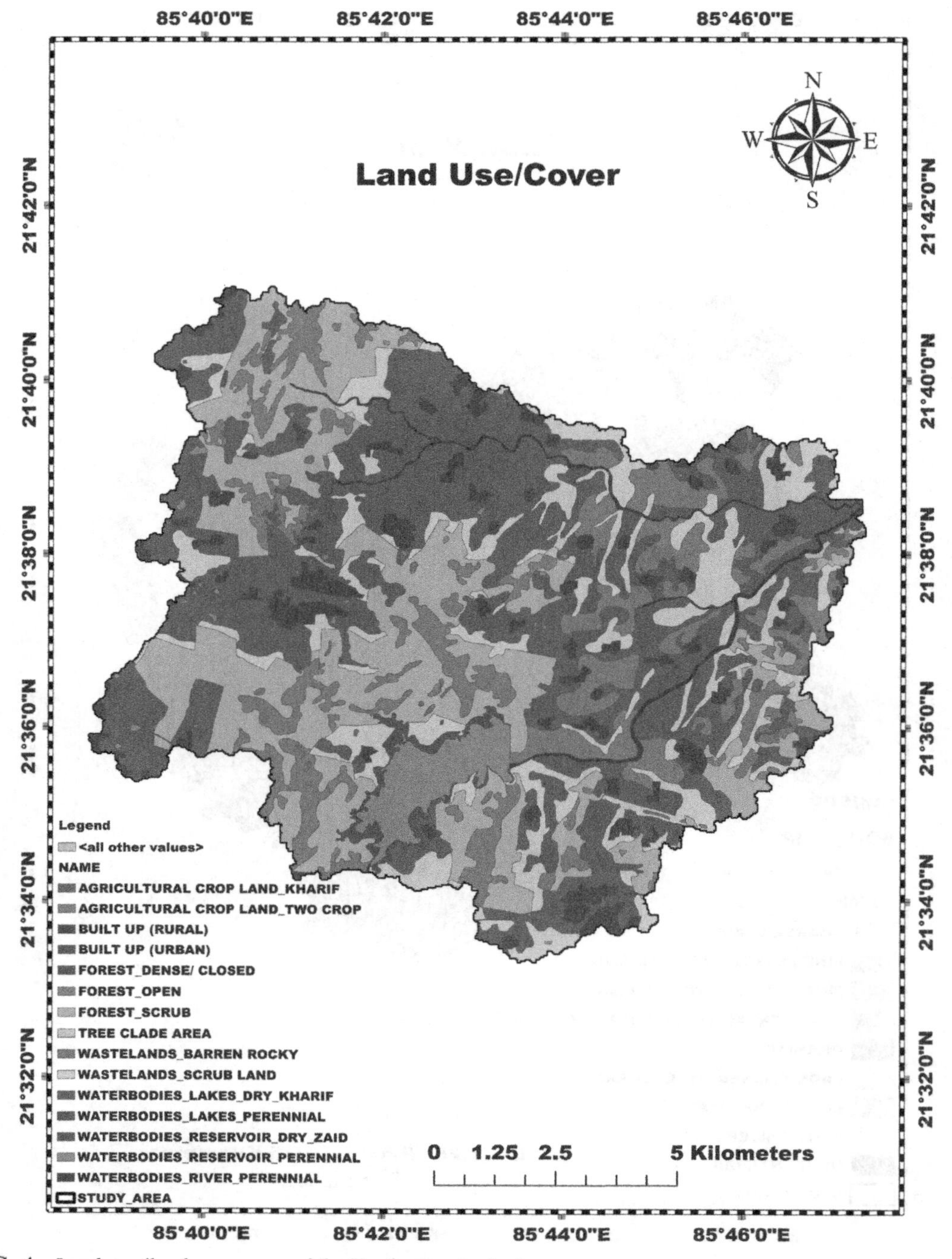

FIG. 4 Land use/land cover map of the Kanjhari watershed.

FIG. 5 Filed instrumentation, experiments and primary data collection in the study site for subsurface characterization and groundwater modeling.

Space Administration. The land use and soil maps for the study area were obtained from the Odisha Remote Sensing Application Centre. The authors have carried out multiple field experiments (such as double-ring infiltrometer and pumping tests, Fig. 5) at different locations to determine the infiltration characteristics of the land surface and aquifer hydraulic properties of the underlying aquifer. The values obtained from these experiments guided in setting the limit for model calibration and validation. Two observation wells were also installed (Fig. 5) and four additional open wells were identified from which the water table data was recorded for the study period at 15 days interval. Similarly, multiple electronic tensiometers were installed for keeping the record of soil moisture variation spatially and across the vertical profile.

4 Field application

4.1 Model setup

For field application of the hillslope-based HSB model, firstly, the catchment is to be identified and decomposed to the constituent hillslopes as the HSB model considers hillslope as

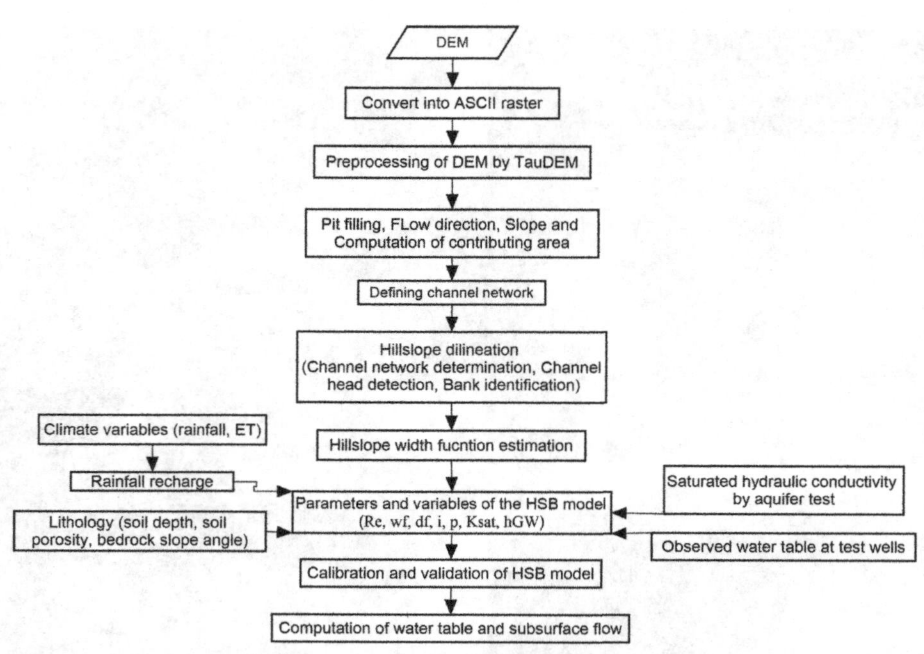

FIG. 6 Step-by-step procedure for field application of the HSB model.

the elements. The detailed step-by-step procedure to generate the hillslope, parameterize the model, and apply the model to a real-world watershed is illustrated in Fig. 6. For hillslope delineation, the different processing of the available DEM was carried out in the geographic information system (GIS) using the TauDEM toolbox (Tarboton, 2014) available for ArcGIS. From the identified streamlines and resulting hillslopes, the flow distance maps were generated that were subsequently used for determining the probability density function (PDF) of flow distance. The detailed procedure for estimation of HWF is available in Sahoo and Sahoo (2019).

The subsequent step after identifying the domain was the selection of the different initial and boundary conditions. For this study, the entire study watershed is considered as the flow domain assuming the watershed overlies on the groundwater basin. The catchment divide is considered as a no-flow boundary (i.e., $Q_{GW} = 0$, at $x = L$). As per the assumptions in the derivation of the HSB model, any flow between the constituting hillslopes was neglected. The model can handle a variety of downstream conditions possible. As these hillslopes drain to a river network, the downhill boundary condition was assigned as a time-varying specific head that varies according to the river stage. Further, the top surface was assigned as a recharge boundary. Conversely, the assignment of appropriate initial conditions is challenging in data-scarce conditions. To combat this issue, model spin-up is the alternate option. During spin-up, the model is run in a steady state until equilibrium of the state variables (water table depth in this case) is achieved (Ajami et al., 2014). In this case, the model was simulated for 12-years to describe the initial water table.

4.2 Model calibration and validation

Model calibration is performed to estimate the parameter values for which the numerical solutions could replicate the state variables of the natural system under study. Herein, calibration was performed to match the simulated water table depth with the corresponding observed values with the objective function to minimize the root mean squared error (RMSE). The calibration parameters were K_{sat} and p. The range within which the values of K_{sat} and p were varied was selected from their field-observed values collected during different experiments and field explorations (as described in Section 3). Postcalibration, the model was validated to check the efficiency of the numerical model in reproducing the water table using the calibrated parameters. The efficiency of the model during calibration and validation was reported using the well-known error statistics of root mean square error (RMSE) and coefficient of determination, R^2.

5 Results and discussion

5.1 Geomorphic analysis of the study area

The entire catchment was subdivided into 29 individual hillslopes, in which each hillslope behaves as a hydrological response unit (HRU) (see Fig. 7). Based on the plan shape, each

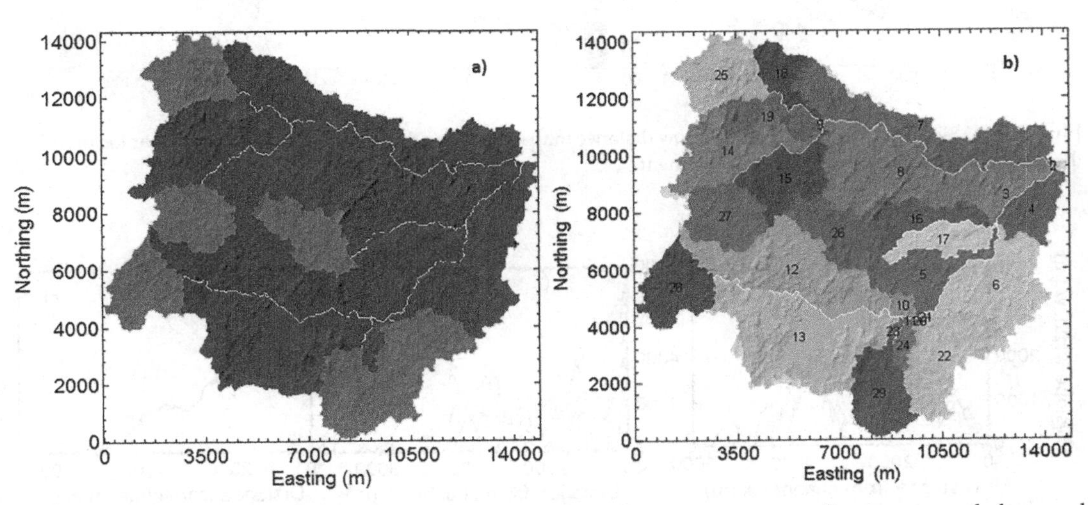

FIG. 7 (A) Classification of hillslopes into 6 convergent and 23 divergent types according to geomorphology; and (B) individual hillslope map of the Kanjhari catchment.

hillslope is classified into: (i) cells draining to the channel head comprising of six headwater hillslopes (shaded red (gray in print version) in Fig. 7A), and (ii) cells draining to the channel directly comprising of 23 divergent hillslopes (shaded blue (dark gray in print version) in Fig. 7A). The slope of the study area, as extracted from the DEM, varies significantly from 5% to 45% while considering it as a single hillslope (Fig. 8). However, while considering that the study area consists of 29 individual hillslopes, the average slope at hillslope-scale varies in the range of 6.55% (hillslope#21) to 14.86% (hillslope#2).

5.2 Derived hillslope width function

The hillslope plan form geometry as expressed in terms of the HWF, $w_f(x)$ was estimated from the PDF of flow distance along the flow path toward the channel as illustrated in Fig. 9. The results reveal that the flow distance from the river network, computed for headwater

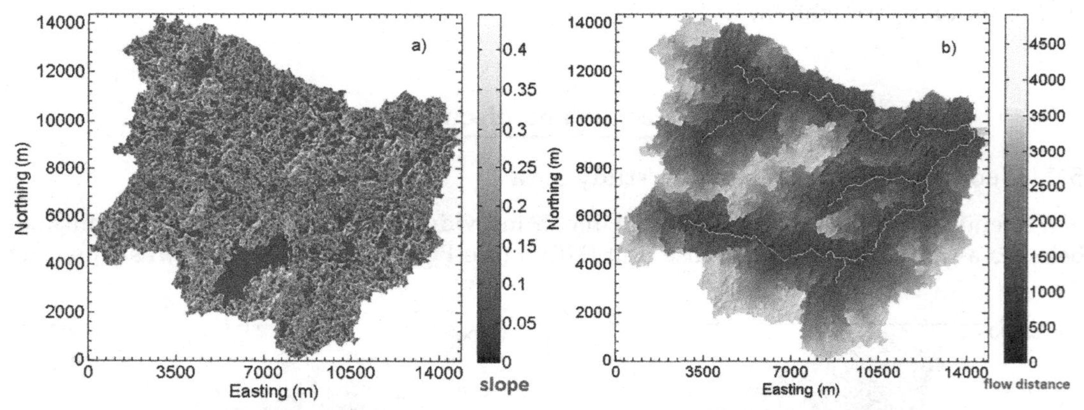

FIG. 8 (A) Slope gradient map; and (B) flow distance map illustrating the flow distance of a particular location from the channel network of the Kanjhari catchment.

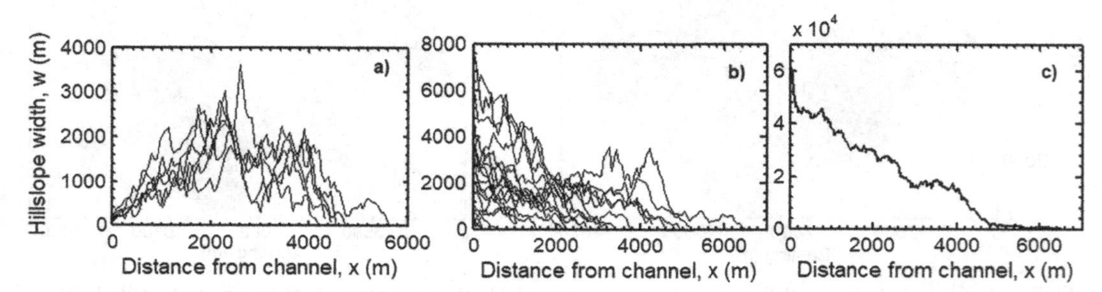

FIG. 9 Hillslope width functions for (A) convergent and (B) divergent hillslopes of the Kanjhari catchment; and for the (C) the whole catchment considering a single hillslope around the river network.

hillslopes, is more than that of the side slopes. The HWF is determined for each hillslope at the hillslope-scale and the entire catchment (treating the whole study area as one compound hillslope), surrounding the stream network. Fig. 9 illustrates the hillslope width as a function of the distance from the stream network for both the convergent and divergent hillslopes. For the convergent type headwater hillslopes, the hillslope width gradually increases with the distance from the stream network up to a certain maximum value in between and then decreases to zero at the catchment boundary. This suggests the existence of the "Fan shape" geometry of the headwater hillslopes in the study area, which can also be verified from Fig. 7. However, for divergent hillslopes of the study area, the hillslope width decreases with the distance from the channel network and approaches zero near the catchment boundary.

5.3 Prediction of groundwater level

The HSB model was calibrated using the collected water table data at six locations in the study area. The 15-daily data from June 01, 2012 to May 31, 2014 was used for calibration. Similarly, 1 year of data from June 01, 2014 to May 31, 2015 at a 15-daily temporal scale was used for model validation. The calibrated values of hydraulic conductivity and effective porosity for the different constituting hillslopes were found to be in the range 1.8 to 3.4 m/day and 0.25 to 0.32, respectively. From Fig. 10, it could be seen that the HSB model was able to reproduce the seasonality closely in all the observation wells. Some deviation in model simulated values as compared to the observed result of wells W1 and W2 during monsoon can be

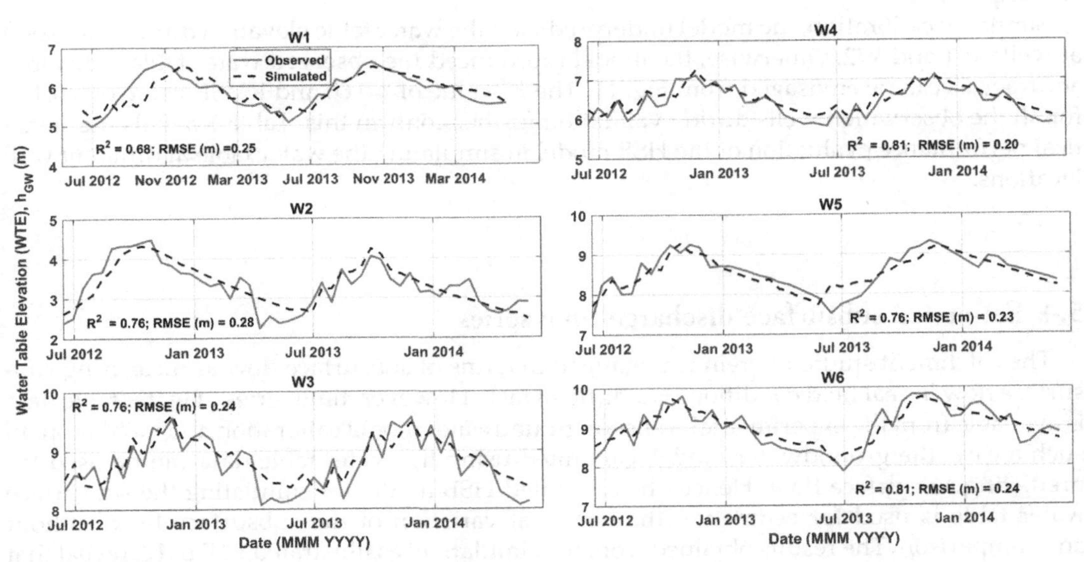

FIG. 10 Calibration of the HSB model reproducing the water table elevation above the bedrock.

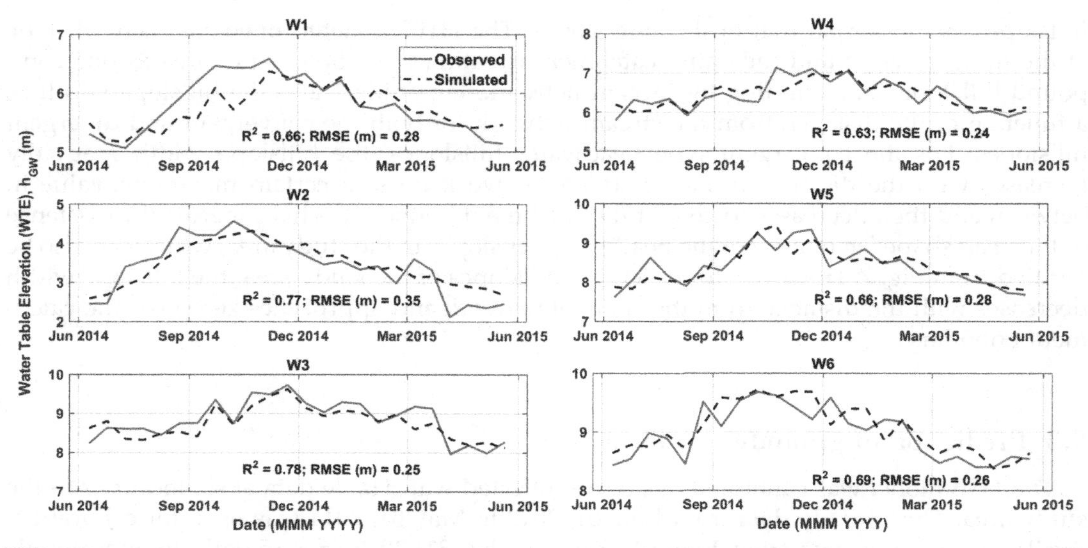

FIG. 11 Validation of the HSB model reproducing the water table elevation above the bedrock.

attributed to the local groundwater recharge from the agricultural land which was not accounted for in modeling. Note that the ponded water in paddy fields contributes to a higher recharge rate.

Similar to calibration, the model underpredicted the water table elevation during monsoon at wells W1 and W2. Otherwise, the model reproduced the observed water table with close accuracy as can be envisaged from Fig. 11. The R^2 value of >0.60 and RMSE value of <0.4 m for all the observation wells during validation further confirm this. Table 1 details the statistical performance evaluation of the HSB model in simulating the water table at different well locations.

5.4 Estimated subsurface discharge time series

The catchment studied herein is ungauged in terms of subsurface flow as measuring subsurface flow in real field conditions is a difficult task. However, time series of water table data is available from the experimental wells for groundwater model calibration and validation. In such a case, the groundwater model, calibrated using the water table data can be used for predicting subsurface flow. Hence, the calibrated HSB model for simulating the subsurface water table is used for estimating the temporal variation of the subsurface flow without any comparison. The results obtained from the simulation, as illustrated in Fig. 12, reveal that the model has a very good ability in replicating the temporal variability in recharge.

TABLE 1 Performance evaluation statistics of the HSB model in predicting the water table at the experimental open wells of the Kanjhari watershed.

	Calibration		Validation	
	R^2	RMSE (m)	R^2	RMSE (m)
W1	0.68	0.25	0.66	0.28
W2	0.76	0.28	0.77	0.35
W3	0.76	0.24	0.78	0.25
W4	0.81	0.20	0.63	0.24
W5	0.7	0.23	0.66	0.28
W6	0.81	0.24	0.69	0.26

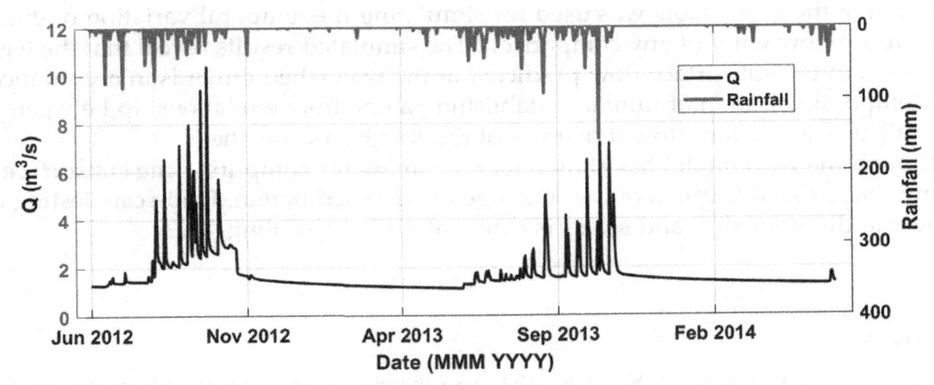

FIG. 12 Simulated subsurface discharge (Q) time series by the HSB model.

The sudden increase in discharge corresponds to the rainfall event in the catchment. Moreover, the subsurface discharge pattern is varying in both the calibration and validation periods according to the rainfall pattern. This proves the ability of the HSB model to simulate the temporal variability in subsurface discharge, which also depends on the recharge from the variable source areas in the catchment (Sahoo et al., 2021).

6 Summary and conclusions

Since most of the watersheds worldwide are ungauged in terms of subsurface flow, quantification of this subsurface flow and storage is not an easy task in these watersheds. The earlier models dealing with subsurface flow and storage have got limitations associated with the underlying assumption in the model framework since these models do not consider the 3D hillslopes through which the real flow occurs. Consequently, in this study, an attempt has been made to model the subsurface flow and storage dynamics of the ungauged Kanjhari watershed by using a linearized form of the Boussinesq equation, namely, the hillslope storage Boussinesq (HSB) model advocated by Troch et al. (2003) and further enhanced by Sahoo et al. (2018) for complex hillslopes. The governing equation of this model uses a 3-D flow domain by considering the hillslope width function accounting for the plan shape (x-y plane), the soil depth function describing the profile curvature (y-z plane), and bedrock slope angle (x-z plane). In this way, the model considers a more realistic 3D flow domain than the idealized traditional ones. The model was applied to a data-scarce watershed in eastern India to model the subsurface storage-discharge dynamics. The following conclusions can be drawn based on the modeling outcome:

i. The results reveal that the HSB model could simulate the temporal variation of the groundwater head in the shallow dug wells at the test location very well with the mean square error of <40 cm.
ii. Since the watershed studied herein is fully ungauged, the calibrated HSB model for simulating the water table was used for simulating the temporal variation of the subsurface flow without any comparison. The simulated results reveal that the temporal variation of the subsurface flow predicted at the watershed outlet is in accordance with the temporal variation of rainfall, establishing a one-to-one relationship between the rainfall and subsurface flow dynamics of the Kanjhari watershed.
iii. Although the HSB model has shown its robustness for computing the subsurface flow using the surficial features of the ungauged watershed herein, field-scale testing of the model at different sites and scales is essential for its upscaling.

References

Ajami, H., McCabe, M.F., Evans, J.P., Stisen, S., 2014. Assessing the impact of model spin-up on surface water-groundwater interactions using an integrated hydrologic model. Water Resour. Res. 50, 2636–2656. https://doi.org/10.1002/2013WR014258.

Barthel, R., 2014. Integration of groundwater and surface water research: an interdisciplinary problem? Hydrol. Earth Syst. Sci. 18 (7), 2615–2628. https://doi.org/10.5194/hess-18-2615-2014.

Bear, J., 1972. Dynamics of Fluids in Porous Media. Elsevier, New York.

Boussinesq, J., 1877. Essai sur la théorie des eaux courantes. Mem. Acad. Sci. Inst. Fr. 23 (1), 252–260.

Bouwer, H., 2002. Integrated water management for the 21st century: problems and solutions. J. Irrig. Drain. Eng. 128 (4), 193–202. https://doi.org/10.1061/(ASCE)0733-9437(2002)128:4(193).

Bredehoeft, J., 2005. The conceptualization model problem—surprise. Hydrogeol. J. 13, 37–46. https://doi.org/10.1007/s10040-004-0430-5.

Brunner, P., Therrien, R., Renard, P., Simmons, C.T., Franssen, H.-J.H., 2017. Advances in understanding river-groundwater interactions. Rev. Geophys. 55, 818–854.

Candela, L., Elorza, F.J., Tamoh, K., Jim, J., Aureli, A., 2014. Groundwater modelling with limited data sets: the Chari-Logone area (Lake Chad Basin, Chad). Hydrol. Process. 28 (11), 3714–3727. https://doi.org/10.1002/hyp.9901.

Carter, R.C., Morgulis, E.D., Dottridge, J., Agbo, J.U., 1994. Groundwater modelling with limited data: a case study in a semi-arid dunefield of northeast Nigeria. Q. J. Eng. Geol. Hydrogeol. 27 (Suppl), S85–S94. https://doi.org/10.1144/GSL.QJEGH.1994.027.0S.09.

Childs, E.C., 1971. Drainage of groundwater resting on a sloping aquifer. Water Resour. Res. 7 (5), 1256–1263. https://doi.org/10.1029/WR007i005p01256.

de Marsily, G., Delay, F., Gonçalvès, J., Renard, P., Teles, V., Violette, S., 2005. Dealing with spatial heterogeneity. Hydrogeol. J. 13 (1), 161–183. https://doi.org/10.1007/s10040-004-0432-3.

Fan, Y., Bras, R.L., 1998. Analytical solutions to hillslope subsurface storm flow and saturation overland flow. Water Resour. Res. 34 (4), 921. https://doi.org/10.1029/97WR03516.

Green, T.R., Taniguchi, M., Kooi, H., Gurdak, J.J., Allen, D.M., Hiscock, K.M., Treidel, H., Aureli, A., 2011. Beneath the surface of global change: impacts of climate change on groundwater. J. Hydrol. 405 (3–4), 532–560. https://doi.org/10.1016/j.jhydrol.2011.05.002.

Højberg, A.L., Troldborg, L., Stisen, S., Christensen, B.B.S., Henriksen, H.J., 2013. Stakeholder driven update and improvement of a national water resources model. Environ. Model. Softw. 40, 202–213. https://doi.org/10.1016/j.envsoft.2012.09.010.

Holman, I.P., Allen, D.M., Cuthbert, M.O., Goderniaux, P., 2012. Towards best practice for assessing the impacts of climate change on groundwater. Hydrogeol. J. 20 (1), 1–4. https://doi.org/10.1007/s10040-011-0805-3.

Hunt, R.J., Luchette, J., Schreuder, W.A., Rumbaugh, J.O., Doherty, J., Tonkin, M.J., Rumbaugh, D.B., 2010. Using a cloud to replenish parched groundwater modeling efforts. Ground Water 48, 360–365. https://doi.org/10.1111/j.1745-6584.2010.00699.x.

Jena, S., Panda, R.K., 2018. Prediction of groundwater level anomaly under varying demand rate due to climate and land use change scenarios. In: AGU Fall Meeting 2018. AGU Paper #H33N-2279.

Jena, S., Panda, R.K., Ramadas, M., Mohanty, B.P., Pattanaik, S.K., 2020. Delineation of groundwater storage and recharge potential zones using RS-GIS-AHP: application in arable land expansion. Remote Sens. Appl. Soc. Environ. 19, 100354. https://doi.org/10.1016/j.rsase.2020.100354.

Jena, S., Mohanty, B.P., Panda, R.K., Ramadas, M., 2021. Toward developing a generalizable pedotransfer function for saturated hydraulic conductivity using transfer learning and predictor selector algorithm. Water Resour. Res. 57, e2020WR028862. https://doi.org/10.1029/2020WR028862.

Kollet, S.J., Maxwell, R.M., Woodward, C.S., Smith, S., Vanderborght, J., Vereecken, H., Simmer, C., 2010. Proof of concept of regional scale hydrologic simulations at hydrologic resolution utilizing massively parallel computer resources. Water Resour. Res. 46, W04201. https://doi.org/10.1029/2009WR008730.

Lancia, M., Saroli, M., Petitta, M., 2018. A double scale methodology to investigate flow in karst fractured media via numerical analysis: the Cassino plain case study (central Apennine, Italy). Geofluids 2018 (2937105), 1–12. https://doi.org/10.1155/2018/2937105.

McDonald, M.G., Harbaugh, A.W., 1988. A Modular Three-Dimensional Finite Difference Ground-Water Flow Model, Techniques of Water-Resources Investigations, 6-A1., https://doi.org/10.1016/0022-1694(70)90079-X.

McGonigle, D.F., Harris, R.C., McCamphill, C., Kirk, S., Dils, R., Macdonald, J., Bailey, S., 2012. Towards a more strategic approach to research to support catchment-based policy approaches to mitigate agricultural water pollution: a UK case-study. Environ. Sci. Pol. 24, 4–14. https://doi.org/10.1016/j.envsci.2012.07.016.

Refsgaard, J.C., Højberg, A.L., Møller, I., Hansen, M., Søndergaard, V., 2010. Groundwater modeling in integrated water resources management-visions for 2020. Ground Water 48 (5), 633–648. https://doi.org/10.1111/j.1745-6584.2009.00634.x.

Sahoo, S., Sahoo, B., 2018. Modeling the hillslope storage-discharge dynamics under changing climate. In: AGU Fall Meeting 2018. AGU Paper #H11T-1721.

Sahoo, S., Sahoo, B., 2019. Modelling the variability of hillslope drainage using grid-based hillslope width function estimation algorithm. ISH J. Hydraul. Eng. 25 (1), 71–78. https://doi.org/10.1080/09715010.2018.

Sahoo, S., Sahoo, B., 2020a. Is hillslope-based catchment decomposition approach superior to hydrologic response unit (HRU) for stream-aquifer interaction modelling: inference from two process-based coupled models. J. Hydrol. 591, 125588.

Sahoo, S., Sahoo, B., 2020b. Travel time distribution-based river basin management plan for sustained dry season flow. In: AGU Fall Meeting 2020. AGU Paper #H090-0010.

Sahoo, S., Sahoo, B., Panda, S.N., 2018. Hillslope-storage Boussinesq model for simulating subsurface water storage dynamics in scantily-gauged catchments. Adv. Water Resour. 121, 219–234. https://doi.org/10.1016/j.advwatres.2018.08.016.

Sahoo, S., Aryan, M.K., Sahoo, B., 2021. Identification of hydrogeological hotspots for mitigating the future water scarcity due to land use and climate change in a river basin prone to mining activities. In: AGU Fall Meeting 2021. AGU Paper #EP41C-08.

Sena, C., Molinero, J., 2009. Water resources assessment and hydrogeological modelling as a tool for the feasibility study of a closure plan for an open pit mine (La Respina mine, Spain). Mine Water Environ. 28, 94–101. https://doi.org/10.1007/s10230-009-0067-7.

Tarboton, D., 2014. TauDEM Documentation. http://hydrology.usu.edu/taudem/taudem5/. (Accessed 8 January 2015).

Tóth, J., 1963. A theoretical analysis of groundwater flow in small drainage basins. J. Geophys. Res. 68 (16), 4795–4812. https://doi.org/10.1029/JZ068i016p04795.

Troch, P.A., Paniconi, C., Emiel van Loon, E., 2003. Hillslope-storage Boussinesq model for subsurface flow and variable source areas along complex hillslopes: 1. Formulation and characteristic response. Water Resour. Res. 39, 1316. https://doi.org/10.1029/2002wr001728.

Troch, P.A., Berne, A., Bogaart, P., Harman, C., Hilberts, A.G.J., Lyon, S.W., Paniconi, C., Pauwels, V.R.N., Rupp, D.E., Selker, J.S., Teuling, A.J., Uijlenhoet, R., Verhoest, N.E.C., 2013. The importance of hydraulic groundwater theory in catchment hydrology: the legacy of Wilfried Brutsaert and Jean-Yves Parlange. Water Resour. Res. 49, 5099–5116. https://doi.org/10.1002/wrcr.20407.

Tromp-van Meerveld, H.J., Peters, N.E., McDonnell, J.J., 2007. Effect of bedrock permeability on subsurface stormflow and the water balance of a trenched hillslope at 192 the Panola Mountain research watershed, Georgia, USA. Hydrol. Process. 21 (6), 750–769. https://doi.org/10.1002/hyp.6265.

Varis, O., Kajander, T., Lemmelä, R., 2004. Climate and water: from climate models to water resources management and vice versa. Clim. Chang. 66 (3), 321–344. https://doi.org/10.1023/B:CLIM.0000044622.42657.d4.

Weiler, M., McDonnell, J.J., 2007. Conceptualizing lateral preferential flow and flow networks and simulating the effects on gauged and ungauged hillslopes. Water Resour. Res. 43, W03403. https://doi.org/10.1029/2006WR004867.

Zhou, Y., Li, W., 2011. A review of regional groundwater flow modeling. Geosci. Front. 2 (2), 205–214. https://doi.org/10.1016/j.gsf.2011.03.003.

SECTION VI

Advances in numerical modelling in water resources

CHAPTER

22

Multiphysics modeling of groundwater flow on the example of a coupled thermo-hydro-mechanical model of infiltration of water warmer or cooler than the surroundings

Anna Uciechowska-Grakowicz

Faculty of Civil Engineering, Wrocław University of Science and Technology, Wrocław, Poland

OUTLINE

1 Introduction	389
2 Mathematical models of fluid flow in porous media	391
2.1 Basic assumptions and equations for the groundwater flow modeling	391
2.2 Poroelasticity model	392
2.3 Fluid flow coupled with heat transfer	394
2.4 Coupled thermo-hydro-mechanical model	395
2.5 Numerical model	397
3 Results and discussion	400
4 Conclusions	403
Acknowledgments	404
References	404

1 Introduction

Soil, which consists of a solid skeleton and pores, filled with water and/or gas (such as air of water vapor), is usually treated as one medium described by its effective parameters;

https://doi.org/10.1016/B978-0-323-91910-4.00022-4

in fact, it consists of two interacting phases, and the effective parameters result from the parameters of the substances of which it consists and their relations. All phenomena that occur in the soil are affected by its composition and structure, and the physical fields in the soil components interact with each other. Groundwater flow is no exception. In classical modeling of groundwater flow, assumptions such as isothermal and nonreactive flow, not affected by the electric field nor the magnetic earth field, are usually employed. Additional assumptions are laminar flow and incompressibility of the fluid or both fluid and pore space. These assumptions are valid only to a certain extent. A great influence on the seepage process has the compressibility of the water and skeleton, which requires the inclusion of mechanical properties and constitutive laws. Some analyses also demand including other phenomena, such as temperature gradient, electric field, phase transition, particle transport, etc. Typical cases where groundwater flow coupled with heat transfer has to be taken into account are ground heat exchangers (GHE), such as ground source heat pumps or thermoactive foundations; in the case of the latter, it also must be taken into account that the stress/strain field is affected by the temperature changes and the groundwater flow is affected by the mechanical load of the new construction. Such problems need a multiphysics approach, which can be defined as coupled processes or systems involving more than one simultaneously occurring physical field and studies and knowledge about these processes and systems (Liu, 2018). As nature-friendly solutions, such as heat exchangers or infiltration devices become increasingly popular, one must also assess their effect on the surroundings and/or find the best location concerning their interaction, which can affect the performance and have impact on the environment. The mutual impact of groundwater flow and energy structures is in the scope of interest of many researches dealing with heat extraction. Xu et al. (2022) found that groundwater flow around a mine tunnel larger than $1.15 \cdot 10^{-7}$ m/s can have a significant effect on temperature variations and heat transfer around and inside the tunnel, Zhang et al. (2016) report a large increase in the efficiency of GHE with a velocity change from $1 \cdot 10^{-4}$ to $5 \cdot 10^{-4}$ m/s. On the other hand, in the results of Stylianou et al. (2019), there is an upper limit of $3 \cdot 10^{-5}$ m/s, above which the increase of velocity does not affect the heat exchange efficiency. Lou et al. (2021), Ma et al. (2021), Zhang et al. (2016), and Bidarmaghz and Narsilio (2018) reported a significant impact of groundwater flow on the temperature distribution near thermoactive constructions, stating that to be able to utilize the ground as a source of energy in an efficient and sustainable manner, areas of significant temperature gradient in the ground must be identified, where they are severely dependent on the velocity and direction of groundwater flow. The articles mentioned refer to heating/cooling and used groundwater flow–temperature coupling (usually by a convective term in the thermal conduction equation), but sometimes mechanical effects are also needed in the model. The second group of coupled models concerns thermal-mechanical coupling. Saaly et al. (2020) investigated the effects of building heat loss on thermoactive piles, where they emphasized the mechanical effects of thermal loads. Temperature changes affect soil parameters, which can affect the pile bearing capacity (Xie and Qin, 2021). Thermo-hydro-mechanical (THM) coupling is also essential for modeling geothermal reservoirs (Buchwald et al., 2021; Burnell et al., 2015; Fan et al., 2022; Liu et al., 2022), underground coal gasification (UCG) (Gao et al., 2021; Uciechowska-Grakowicz and Strzelecki, 2021) or nuclear engineering (Groth-Jensen et al., 2021; Liu et al., 2021a; Radman et al., 2021).

The development of computational methods and the increase in computing resources allow engineers to include multiphysics in modeling underground flows, avoiding high time costs (Liu, 2018). The most popular commercial software for modeling coupled phenomena, not only during groundwater flow, is Comsol Multiphysics (Ma et al., 2021; Pryor, 2012; Saaly et al., 2020; Zhelnin et al., 2022). A bit less popular are FEFLOW (Abesser et al., 2021; Dahash et al., 2021) or FlexPDE (Maghoul, 2017; Stylianou et al., 2019; Sweidan et al., 2021), which models are reported to perform similarly (Dahash et al., 2021; Rühaak et al., 2014). Multiphysics modeling of groundwater flow is also possible with the use of open-source software, such as PorousFlow (Wilkins et al., 2020), OpenFOAM (Fiorina et al., 2015; German et al., 2021), or OpenGeoSys (Buchwald et al., 2021; Fan et al., 2022).

In this chapter, the procedure for building the NLTE fully coupled THM model based on the poroelasticity framework and its application in the modeling of coupled phenomena that occur in the soil near infiltration devices will be presented. To organize the information regarding groundwater flow modeling, starting from the basic underground flow equations, additional physical phenomena taking place in the soil will be presented, resulting in the coupled thermo-hydro-mechanical model for the two-phase medium which performance is presented on the example of injection of water of different temperature in the presence of groundwater flow.

2 Mathematical models of fluid flow in porous media

2.1 Basic assumptions and equations for the groundwater flow modeling

In the simplest case, the flow of water through the pores of the soil can be compared to the flow through the capillary channels, where the amount of flow of the Newtonian fluid could be described by the Poiseuille (1842):

$$Q = A \bullet \frac{r^2}{8\eta} \bullet \frac{\Delta p}{L}, \tag{1}$$

where A is the channel cross-section, r is the channel radius, η is the dynamic viscosity, Δp is the pressure difference, and L is the length of the channel. As the channels in the soil are irregular, a description of the flow should include the assumption of the pore size and shape, which are represented by the material constant. Law describing the laminar flow in soil, coherent with the Hagen-Poiseuille law (Fiorillo et al., 2022), has been described by Darcy (1856):

$$Q = A \bullet k \bullet \frac{\Delta p}{L}, \tag{2}$$

where k is the hydraulic conductivity, which depends on the pore network properties and viscosity, A is the cross-section of the medium. For the modeling of three-dimensional isothermal flows, a more general formulation of Darcy's Law is employed (Hubbert, 1957):

$$v_D = -K_{ij}\, grad(H), \tag{3}$$

where v_D is the seepage rate (Darcean velocity, defined as $v_D = \phi v$, where ϕ is the porosity, and v is the actual fluid velocity), K_{ij} is the conductivity matrix and H is the hydraulic head. For coarse soils, the inertia effect can become significant, and the fluid flow is described by the Forchheimer (1901):

$$\frac{\partial p}{\partial x} = -kv_D - k_1 v_D^2. \tag{4}$$

The classical seepage model is based on the continuity equation with the use of the Darcy Eq. (3) and the incompressibility assumption. If we imagine a Representative Elementary Volume (REV) Ω filled with porous medium, enclosed within the surface S, the change in the fluid content in the element Ω is described by the continuity equation (Euler, 1757):

$$\int_S \rho v_i dS + \int_\Omega \frac{\partial \rho}{\partial t} d\Omega = 0, \tag{5}$$

where ρ is the density of the fluid. Thus, after applying the divergence theorem, the above equation can be written in local form:

$$\frac{D\rho}{Dt} + \rho\, div(v_D) = \vartheta, \tag{6}$$

where $\frac{D}{Dt}$ is the material derivative and ϑ is the change in fluid content. Assuming a steady, non-compressible fluid, one can get:

$$div(v_D) = 0, \tag{7}$$

which equation after employing Darcy's law takes the form:

$$div(k\, grad(H)) = 0, \tag{8}$$

which is the basic equation of stationary fluid flow in porous media. For nonstationary flows, the compressibility of fluid and solid soil constituents is somehow taken into account as constituents of the storage coefficient (Jacob, 1940).

$$S = \rho g(\beta_s + \phi \beta_l), \tag{9}$$

where β_s is the compressibility of the solid constituent, β_l is the fluid compressibility, so the equation of non-stationary fluid flow takes the form:

$$div(k\, grad(H)) = S\frac{\partial H}{\partial t}, \tag{10}$$

which, despite doubts regarding mathematical consistency (De Wiest, 1966; Ingebritsen and Sanford, 1999; Strzelecki and Strzelecki, 2015), is widely accepted and used in many groundwater modeling software.

2.2 Poroelasticity model

In the theory of poroelasticity, first introduced by Maurice A. Biot (1941), porous medium is assumed to consist of two separate phases that interact with each other, so this theory enables modeling deformation of the medium together with the fluid flow. In this model, fluid flow is not a separate phenomenon, but a result of continuity and momentum conservation equations for both phases. In the poroelastic model, stresses and other values of a given phase are described as "partial" which means that they are recalculated with respect to the given phase fraction, for example, the partial stress in fluid $\sigma = -p\phi$, where p is the pressure.

In the poroelasticity model, the fields of stress/strain in both phases are coupled in both directions (a change in stress in the fluid causes a strain in the solid and a change in pore pressure causes a change in strain in the solid), thanks to the constitutive equations for both constituents, resulting straight from the first law of thermodynamics (Coussy, 2010; Derski et al., 1982):

$$\sigma_{ij} = 2N\varepsilon_{ij} + (A\varepsilon + Q\vartheta)\delta_{ij},$$
$$\sigma\delta_{ij} = Q\varepsilon\delta_{ij} + R\vartheta\delta_{ij}, \tag{11}$$

where σ_{ij} and σ are partial stresses in the solid and fluid phases of the medium, ε_{ij} is the strain tensor, A and N are Lamé's coefficients for porous medium, Q is the coefficient defining the amount of fluid that may enter the body without change of its volume, R is change of fluid content as a result of change in the pressure (Biot and Willis, 1957).

During the derivation of the continuity equation for the medium, both fluid and solid constituents are taken into account (Strzelecki et al., 2018):

$$\frac{D^s\rho}{Dt} + \rho\dot{\varepsilon} = -\left(\rho^f v_i^r\right)_{,i} \tag{12}$$

where $\frac{D^s}{Dt} = \frac{\partial}{\partial t} + v_i^s\frac{\partial}{\partial x_i}$ is material derivative, v^s is velocity of solid constituent, $\dot{\varepsilon}$ is the rate of volumetric strain ($\dot{\varepsilon} = v_{,i}^s$), ρ^f is fluid density and v^r is a velocity of the fluid related to velocity of skeleton ($v_{,i}^r = \dot{\vartheta}$), which in case of an incompressible fluid skeleton simplifies to the basic equation of fluid flow through porous media (8). When only the fluid component is taken into account, we get the following continuity equation:

$$\frac{D^r\rho_2}{Dt} + \rho_2\left(\dot{\theta} - \dot{\varepsilon}\right) = -\left(\rho_2 v_i^s\right)_{,i} \tag{13}$$

where $\frac{D^r}{Dt}$ is a material derivative with respect to the relative velocity between fluid and skeleton, ρ_2 is partial fluid density ($\rho_2 = \rho^f\phi$), and $\dot{\theta}$ is the rate of fluid volume change ($\dot{\vartheta} = \dot{\theta} - \dot{\varepsilon}$). Displacements can be calculated using the momentum conservation principle:

$$\sigma_{ij,j} + \sigma_{,i} + X_i\rho = 0 \tag{14}$$

for the skeleton, and

$$\sigma_{,j} + X_i\rho_2 = bv_i^r \tag{15}$$

for the fluid, where X_i is body force and b is a coefficient responsible for the viscous drag. For a porous medium subjected only to gravitational acceleration ($X_3 = -g$) the last equation can be rewritten to calculate the fluid velocity:

$$v_i^r = \frac{1}{b}(\sigma_{,i} + X_i\rho_2) = \frac{1}{b}(\sigma_{,j} - g\rho_2) = -\frac{\rho_2 g}{b}\left(-\frac{\sigma_{,i}}{\rho_2 g} + 1\right). \tag{16}$$

Using the definition of the hydraulic head $H = \frac{p}{\rho^f g} + z$ and assuming constant porosity:

$$\phi v_i^r = \frac{\phi^2\rho^f g}{b}H_{,i}. \tag{17}$$

When comparing the equation obtained with Darcy's law (3), it can be seen that Darcy's law is a consequence of the principle of momentum conservation in a porous medium filled with fluid, and $b = \frac{\phi^2\rho^f g}{k}$.

The basic model for the calculation of the displacement and fluid velocity field needs a continuity equation with momentum conservation principle (1–3 equations: one equation for every direction). This basic model can be of course extended to include models of additional phenomena taking part in one or both phases, such as thermoelasticity (Coussy, 2004), plasticity (Li et al., 2019), viscoelasticity (Bartlewska and Strzelecki, 2009; Makhnenko and Podladchikov, 2018), thermoviscoelasticity (Bartlewska-Urban et al., 2015), reactive flow (Evans et al., 2018), electrokinetic flow (Strzelecki and Bartlewska-Urban, 2006) and other. A rigorous mathematical description of poroelastic flows (with additional phenomena) is presented in (Meirmanov, 2014).

2.3 Fluid flow coupled with heat transfer

2.3.1 Thermal flow in local thermal non-equilibrium

There are 3 mechanisms of heat transfer: convection, conduction, and radiation; in soils, the two former mechanisms play the main role. If the rate of heat transfer between the pore fluid and the skeleton is rapid, it can be assumed that both phases are in local thermal equilibrium (LTE), which assumption is used in most models. When the permeability is low and/or seepage rates are slow, or when the soil is filled with gas, heat is being transferred by the conduction and that process can be described by only one unsteady heat transfer equation (Taler and Duda, 2006):

$$\rho c\dot{T} = \nabla(\lambda \nabla T) + q, \tag{18}$$

where λ is the heat transfer coefficient, c is the specific heat of the medium and q is a heat source. In such an approach, soil is treated as one medium. When fluid flow is present, the heat transfer equation should be coupled with the flow governing equation by the convective terms (the time derivative on the left side should become the material derivative).

2.3.2 Thermal flow in local thermal non-equilibrium (LTNE)

Another approach to modeling unsteady heat transfer in soils is to model the temperature simultaneously in both phases with two coupled equations, which allows LTNE conditions (different temperatures of both phases at the given moment) (Rees et al., 2008):

$$\begin{aligned} \nabla(\lambda_1 \nabla T_s) &= \rho_1\, c_f \frac{\partial T_s}{\partial t} + h(T_f - T_s), \\ \nabla(\lambda_2 \nabla T_f) &= \rho_2\, c_s \frac{DT_f}{Dt} + h(T_s - T_f), \end{aligned} \tag{19}$$

where the subscripts s and f stand for, respectively, solid and fluid, subscripts 1 and 2 mean partial effective values (including volume fraction of given phase), $\frac{D}{Dt}$ is material derivative with respect to fluid velocity, calculated with the use of the fluid flow governing equation. LTNE models are more precise, particularly in the case of rapid transient heating (Faizurrahmany et al., 2019; Minkowycz et al., 1999). Factors that increase the temperature difference between phases are the large difference between the thermal conductivity of the solid and liquid phases and the high porosity (Lin et al., 2016). The LTNE model is closer to reality since the averaged porous medium is only conception used to simplify the description of physical phenomena. In fact, the LTE model is a special case where one equation can be used due to the material parameters and the time/length scale used, and the LTE/LTNE criteria refer to this special case, where LTE can be assumed (Carbonell and Whitaker, 1984; Kaviany, 1995; Kim and Jang, 2002; Minkowycz et al., 1999). Conditions that require a two-equation LTNE model of heat transfer are, in particular (Rees et al., 2008):

- natural geothermal conditions where convection effects cause heat transfer from the fluid to the solid phase,
- very high solid phase conductivity (e.g., metal in a heat exchanger), where convection and conduction effects are overtaken by conduction in the solid phase,
- high flow rates.

2.4 Coupled thermo-hydro-mechanical model

The LTNE model was first adapted to thermoporoelasticity by Pecker and Deresiewicz (Pecker and Deresiewicz, 1973), who studied the impact of temperature on wave propagation in a porous medium and developed the fully coupled model, where the thermal coupling between the phases is realized by parameters describing specific heats at constant volume and interstitial heat transfer coefficient, while most LTNE models use only interstitial heat transfer coefficient. Youssef (2007) derived the model with one relaxation time based on Lord-Shulman thermoelasticity, then Ezzat (Ezzat and Ezzat, 2016), introduced the model with a heat conduction equation with fractional order of time derivative for asphaltic materials. He et al. (2012), Wu and Jin (2010) proposed a simplified model with 2 heat transfer equations for both phases and averaged temperature in the thermoporoelasticity constitutive equations. The most known criterion regarding LTE/LTNE assumption is based on Sparrow number was proposed by Minkowycz et al. (1999). More criteria for LTE/LTNE heat transfer were collected and analyzed by Al-Sumaily et al. (2021).

The continuity equation for the fluid and the momentum conservation principle in thermoporoelasticity are the same as for the isothermal poroelasticity model. They are complemented by heat transfer equations and constitutive relations that can be derived from the first law of thermodynamics for the element Ω bounded by the surface S (Strzelecki et al., 2018):

$$\dot{L} + \dot{Q} = \frac{\partial}{\partial t}(W + K), \tag{20}$$

where: $\dot{L}$is the work of internal forces, gravitational force and the viscous resistance of fluid, $\dot{Q}$is generated heat, K is kinetic energy and W is the internal energy of the medium. For the skeleton and fluid components of the medium that fill the volume Ω, the first law of thermodynamics for quasistatic media can be expressed as follows (Uciechowska-Grakowicz and Strzelecki, 2017):

$$\begin{aligned}
&\int_{\Omega} \nu^S(\sigma_{ij,j} + X_i\rho_1 + bv^r)d\Omega + \int_{\Omega}\left(\sigma_{ij}\dot{\varepsilon}_{ij} - q^s_{i,i} - \dot{q}_{fs}\right)d\Omega = \int_{\Omega} \dot{w}_s d\Omega,\\
&\int_{\Omega} \nu^f\left(\sigma_{,j} + X_i\rho_2 + bv^r\right) d\Omega + \int_{\Omega}\left(\sigma\dot{\theta} - q^l_{i,i} - \dot{q}_{sf}\right)d\Omega = \int_{\Omega} \dot{w}_s d\Omega,
\end{aligned} \tag{21}$$

where q_{fs} is the heat entering the solid phase from the fluid phase and q_{sf} is the heat entering the fluid phase from the solid phase. After applying the momentum conservation principle, one can get the change of internal energy in the porous medium in the local form:

$$\dot{w} = \sigma_{ij}\dot{\varepsilon}_{ij} + \sigma\dot{\theta} - q^s_{i,i} - q^f_{i,i}. \tag{22}$$

The local rate of free energy change is then described by the formula:

$$\dot{\chi} = \dot{w} - s_1\dot{T}^s - \dot{s}_1 T^s - s_2\dot{T}^f - \dot{s}_2 T^f, \tag{23}$$

where χ is the Helmhotz free energy, s_1 is the entropy of solid phase, $\dot{s}_1 = -\frac{q^s_{i,i}}{T^s}$, s_2 is the entropy of liquid phase, $\dot{s}_2 = -\frac{q^l_{i,i}}{T^l}$, so including Eq. (22) in Eq. (23):

$$\sigma_{ij}\dot{\varepsilon}_{ij} + \sigma\dot{\theta} - s_1\dot{T}^s - s_2\dot{T}^f - \dot{\chi} = 0. \tag{24}$$

Assuming that the Helmholtz free energy is a function of strain and temperatures in both phases:

$$\dot{\chi} = \frac{\partial\chi}{\partial\varepsilon_{ij}}\dot{\varepsilon}_{ij} + \frac{\partial\chi}{\partial\theta}\dot{\theta} + \frac{\partial\chi}{\partial T^s}\dot{T}^s + \frac{\partial\chi}{\partial T^l}\dot{T}^f, \tag{25}$$

Eq. (24) can be rewritten as follows:

$$\left(\sigma_{ij} - \frac{\partial\chi}{\partial\varepsilon_{ij}}\right)\dot{\varepsilon}_{ij} + \left(\sigma - \frac{\partial\chi}{\partial\theta}\right)\dot{\theta} + \left(-s_1 - \frac{\partial\chi}{\partial T^s}\right)T^s + \left(-s_2 - \frac{\partial\chi}{\partial T^f}\right)T^f = 0, \tag{26}$$

and after expanding Eq. (25) to the MacLaurin series (assuming that the temperature is a sum of the initial temperature T_0 and the temperature difference ϑ) around the initial state where no strain is present:

$$\begin{aligned} 2\chi\left(\varepsilon_{ij}, \theta, \vartheta^s, \vartheta^l\right) &= 2\sigma_a\theta + c_{ijkl}\varepsilon_{ij}\varepsilon_{kl} + \gamma\theta^2 + \xi^s(\vartheta^s)^2 + \xi^l\left(\vartheta^f\right)^2 + \\ &+ 2\beta_{ij}\varepsilon_{ij}\theta + 2\eta^s_{ij}\varepsilon_{ij}\vartheta^s + 2\eta^f_{ij}\varepsilon_{ij}\vartheta^f + 2\omega^s\theta\vartheta^s + 2\omega^f\theta^f + 2\xi^{fs}\vartheta^s\vartheta^f. \end{aligned} \tag{27}$$

Constitutive equations can be obtained after identification of material constants $c_{ijkl} = \frac{\partial^2\chi(0,0)}{\partial\varepsilon_{ij}\partial\varepsilon_{kl}}$, $\gamma = \frac{\partial^2\chi\left(0,0,T^s_0,T^f_0\right)}{\partial\theta\,\partial\theta}$, $\beta_{ij} = \frac{\partial^2\chi\left(0,0,T^s_0,T^f_0\right)}{\partial\varepsilon_{ij}\,\partial\theta}$, $\eta^s_{ij} = \frac{\partial^2\chi\left(0,0,T^s_0,T^f_0\right)}{\partial\varepsilon_{ij}\,\partial T^s}$, $\eta^f_{ij} = \frac{\partial^2\chi\left(0,0,T^s_0,T^f_0\right)}{\partial\varepsilon_{ij}\,\partial T^f}$, $\omega^s = \frac{\partial^2\chi\left(0,0,T^s_0,T^f_0\right)}{\partial\theta\,\partial T^s}$ and $\omega^f = \frac{\partial^2\chi\left(0,0,T^s_0,T^f_0\right)}{\partial\theta\,\partial T^f}$ as the poroelastic coefficients (the whole procedure is described in detail in (Uciechowska-Grakowicz and Strzelecki, 2017)):

$$\begin{aligned} \sigma_{ij} &= \frac{\partial\chi}{\partial\varepsilon_{ij}} = 2N\varepsilon_{ij} + \left(A\varepsilon + Q\theta - 3K\alpha^s\vartheta^s - Q\alpha^f\vartheta^f\right)\delta_{ij}, \\ \sigma &= \frac{\partial\chi}{\partial\theta} = \sigma_a + Q\varepsilon + R\theta - 3Q\alpha^s\vartheta^5 - R\alpha^f\vartheta^f, \end{aligned} \tag{28}$$

which corresponds to the Duhamel-Neumann equations for the solid body. Analogically, the following entropy equations can be obtained:

$$\begin{aligned} -s_1 &= -3K\alpha^s\varepsilon - 3Q\alpha^s\theta + \xi^s\vartheta^s + \xi^{fs}\vartheta^f, \\ -s_2 &= -Q\alpha^f\varepsilon - R\alpha^f\theta + \xi^{fs}\vartheta^f + \xi^f\vartheta^f. \end{aligned} \tag{29}$$

The entropy for both phases in REV can be described as a sum of the external entropy (resulting from the heat transfer), the internal entropy, and the end entropy resulting from the internal heat transfer between the phases:

$$\dot{s_1} = -\left(\frac{q_i^s}{T}\right)_{,i} + \frac{q_i T_{,i}}{T^2} - \frac{\dot{q_{12}}}{T},$$

$$\dot{s_2} = -\left(\frac{q_i^l}{T}\right)_{,i} + \frac{q_i T_{,i}}{T^2} + \frac{\dot{q_{12}}}{T}, \tag{30}$$

where q_{12} is the amount of heat transferred between the phases. According to the Newton's cooling law, $q=h(T_1-T_2)dA$, it can be described with the formula $q_{12}=h(T_1-T_2)a_{spec}$, where h is the interstitial heat transfer coefficient and a_{spec} is specific surface area. Using the definition of specific heat at a constant volume and Fourier's law, the coupled heat conduction equations are obtained:

$$\begin{aligned}
\nabla\lambda_1\nabla\vartheta^s &= -3TK\alpha^s\dot{\varepsilon} - 3TQ\alpha^s\dot{\theta} + \frac{T^s}{T_0}\left(\rho_1 c_V^s\dot{\vartheta}^s + \rho c_V^{fs}\dot{\vartheta}^f\right) + h(T_f - T_s)a_{spec}, \\
\nabla\lambda_2\nabla\vartheta^f &= -TQ\alpha^f\dot{\varepsilon} - TR\alpha^f\dot{\theta} + \frac{T^f}{T_0}\left(\rho c_V^{sf}\dot{\vartheta}^f + \rho_2 c_v^f\dot{\vartheta}^f\right) + h(T_s - T_f)a_{spec}.
\end{aligned} \tag{31}$$

Assuming local thermal equilibrium ($T_s=T_f=T$), those equations can be summed up to obtain one heat transfer equation:

$$\nabla\lambda^{eff}\nabla\vartheta = -(3TK\alpha^s + TQ\alpha^f)\dot{\varepsilon} - (3TQ\alpha^s + TR\alpha^f)\dot{\theta} + +\frac{T}{T_0}c^{eff}\dot{\vartheta}, \tag{32}$$

Where c^{eff} and λ^{eff} are the effective coefficients for the medium. Using the simplest assumption of two superposed phases, λ^{eff} is a sum of partial values: $\lambda^{eff}=\lambda_1+\lambda_2=(1-\phi)\lambda^s+\phi\lambda^f$—which in this form is being used in modeling (Bidarmaghz and Narsilio, 2018; Ma et al., 2021; Saaly et al., 2020), but it has to be noted, that regarding porous media structure, such calculated average value can be treated as an upper bound, corresponding to a layered medium, with layers parallel to the heat transfer direction (Farouki, 1981). The calculation of the actual effective value depends on many factors and is a subject of intensive research (Drzyzga, 2021; Liu et al., 2021b; Różański and Kaczmarek, 2020; Tong et al., 2009). It can be noticed that in this way the LTNE model simplifies to a coupled LTE thermo-hydro-mechanical model.

2.5 Numerical model

The numerical model presented in the following section concerns the thermal fluid flow in a porous medium using the thermoporoelasticity LTNE model. The couplings of the model are presented in Fig. 1. It is assumed that water of higher or lower temperature is being pressed into the medium (soil), where horizontal flow is present. Dimensions of the model with marked cross-section and points, for which the displacement graphs will be generated, are depicted in Fig. 2A.

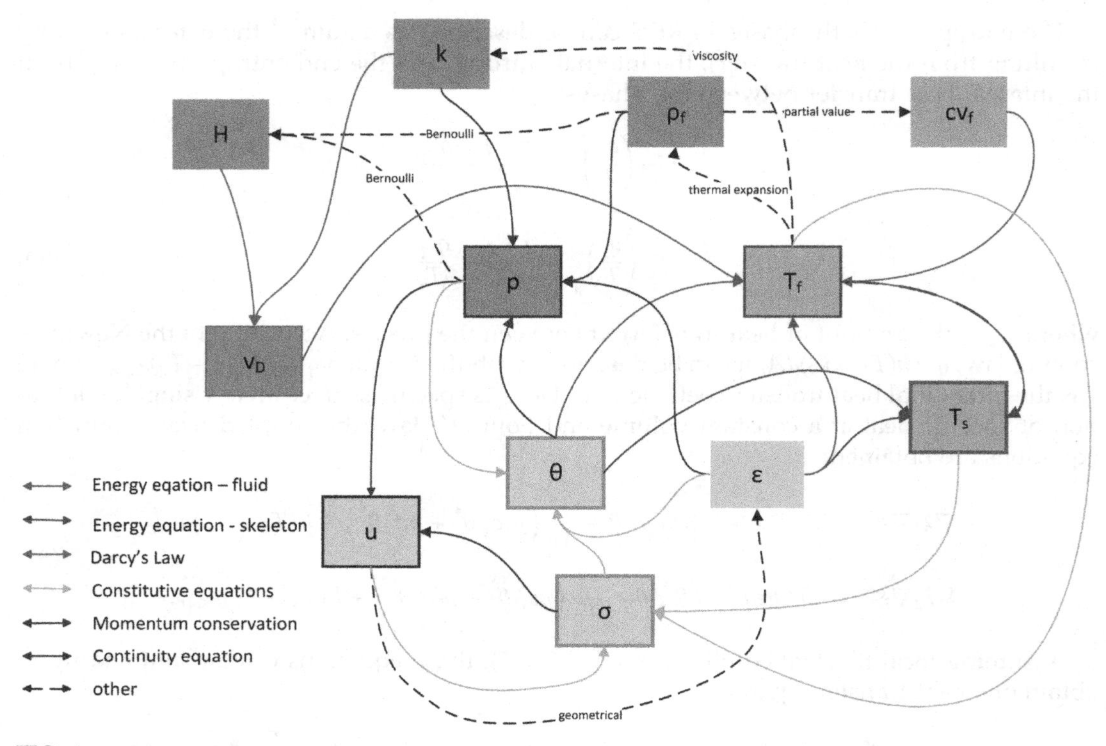

FIG. 1 Couplings of the model.

In the model, there are 15 unknown values of which 5 are independent. They can be solved with the system of equations:

- displacements (u_x, u_y) can be solved with the use of internal equilibrium Eq. (14)
- partial fluid stress (σ) can be solved with the use of momentum conservation for fluid (15):

$$\sigma_{,ii} + (X_i\rho_2)_{,i} = \frac{\rho_2\phi g}{k}(\dot{\theta} - \dot{\varepsilon}),$$

- temperature in both phases T^f and T^s (Eq. 31)
- seepage rates (v_x, v_y): Darcy's law for the thermally expandable fluid (Ene and Polisevski, 1987):

$$v^r = -k\left(H_{,i} + \alpha^l\vartheta^l\rho X_i\right), \tag{33}$$

- strain in skeleton (ε_{xx}, ε_{xy}, ε_{yy},ε): geometric relations:

$$\varepsilon_{ij} = \frac{1}{2}\left(\frac{\partial u_i}{\partial x_j} + \frac{\partial u_j}{\partial x_i}\right),$$
$$\varepsilon = \varepsilon_{11} + \varepsilon_{22} + \varepsilon_{33}, \tag{37}$$

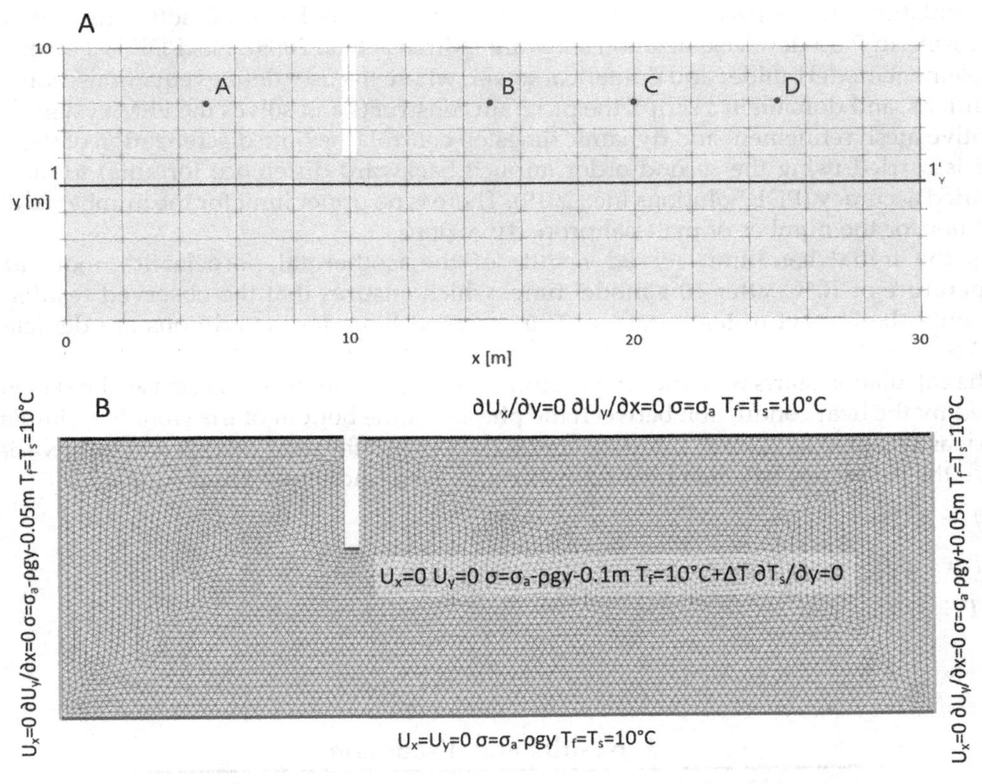

FIG. 2 Geometry of the model with the location of cross-section and points used for displacement analysis (A) with boundary conditions and initial mesh (B).

- strain in fluid (θ) and stress in the skeleton ($\sigma_{xx}, \sigma_{xy}, \sigma_{yy}$): constitutive relations (28):

The material properties used for the model are as follows:

- mechanical parameters: $N = 2.5 \cdot 10^8$Pa, $R = 1 \cdot 10^7$Pa, $A = 5 \cdot 10^8$Pa, $Q = 3 \cdot 10^7$Pa,
- thermal expansion coefficients: $\alpha_f = 69 \cdot 10^{-6}\ m/^{\circ}C$, $\alpha_s = 5 \cdot 10^{-6}\,\mathrm{m}/^{\circ}\mathrm{C}$,
- skeleton density $\rho_s = 2400 \frac{\mathrm{kg}}{\mathrm{m}^3}$, partial skeleton density $\rho_1 = (1-\phi)\rho_s$,
- fluid density: $\rho_f = \frac{\rho_{f0}}{1+\vartheta_f}$, where $\rho_{f0} = 1000\ \mathrm{kg/m^3}$, partial fluid density $\rho_2 = \phi \cdot \rho_f$,
- porosity $\phi = 0.25$,
- conductivity $k_{10} = 1.36 \cdot 10^{-4} \frac{\mathrm{m}}{\mathrm{s}}$, viscosity $\mu = \frac{179 \cdot 10^{-3}}{1 + 3.37 \cdot 10^{-2} \cdot T^f + 2.2 \cdot 10^{-4} T^{f^2}}$.
- thermal parameters: $\lambda_s = 3 \frac{\mathrm{W}}{\mathrm{mK}}$, $\lambda_f = 0.6 \frac{\mathrm{W}}{\mathrm{mK}}$, $ha_{spec} = 400 \frac{\mathrm{W}}{\mathrm{m^3K}}$, $c_{vs} = 700$, $c_{sf} = 0.1 c_{vs}$, $\frac{J}{\mathrm{kg \cdot K}}$, $c_{vf} = 4150 \frac{J}{\mathrm{kg \cdot K}}$.

Calculations were performed with the use of FEM in FlexPDE software with post-processing in Paraview visualization software (Ahrens et al., 2005). FlexPDE is a scripted finite element model builder and numerical solver, where the user defines equations, boundary conditions, and domain in a script, the program constructs and solves the FEM system, using adaptive mesh refinement and dynamic timestep control (the time discretization of the equations is carried using the second-order implicit backward difference formula) to maintain specified accuracy (PDE Solutions Inc., 2019). There is no upper limit for the number of PDE's used nor for the number of material property regions.

As the initial condition served results of the isothermal poroelastic model at the temperature of 10°C, after 10^8s model time, which ensures that the observed results concern only the impact of temperature. The accepted boundary conditions are depicted in Fig. 2B.

The calculation starts with the temperature change (only in the water phase, the skeleton is heated by the heat conduction between the phases) at the bottom of the well. The simulation time was accepted as 200 days, and the thermal fluid injection was modeled in three variants, where the difference between pressed water and surrounding medium equaled:

- $\Delta T = 5^{\circ}C$
- $\Delta T = -5^{\circ}C$
- $\Delta T = 15^{\circ}C$

3 Results and discussion

As the simulation time goes on, a wider area is affected by the higher temperature of the pressed fluid, and the temperature distribution follows the direction of flow. In Fig. 3 the vector fields of seepage rate are depicted on the background of the temperature distribution at given instances of time for variant 1. First, in the area where seepage rates are relatively high in every direction, the area of higher temperature expands in every direction (Fig. 3A and B).

When the area of increased temperature reaches the left point, where the rightward flow becomes dominant, the cooling effect of flowing water impedes the process of heat transfer, and the area of increased temperature develops only in the direction of flow (Fig. 3C and D).

In the case of variant 2, the final shape of the temperature distribution is similar (Fig. 4A), but more downward seepage rate vectors can be observed as a result of the change in density. Seepage rates are lower as a result of the change in viscosity with temperature, which also impedes the process of heat transfer.

The effect of velocity change due to viscosity and density change with temperature is even more visible for case 3 as the higher flow rates feedback heat transfer, and as a result heat is transferred faster in the vertical direction (Fig. 4B).

These three cases are compared in Fig. 4C. For cases 1 and 2, the isotherms for every 1°C temperature change are drawn, respectively, in orange and blue. For case 3, the 3°C isotherms are drawn in red. It becomes evident that the seepage rate-temperature distribution feedback

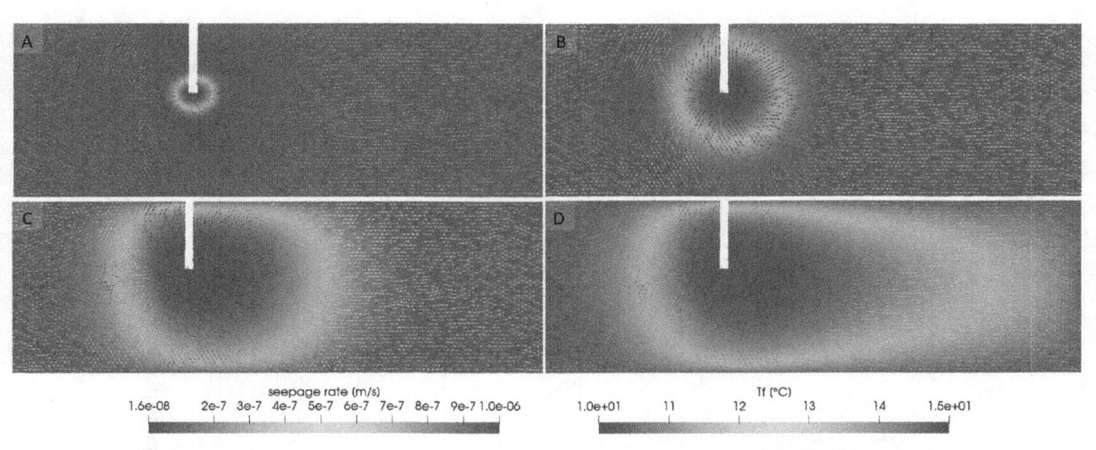

FIG. 3 Temperature distribution and seepage rates after (A) 24 h, (B) 10 days, (C) 50 days, (D) 200 days.

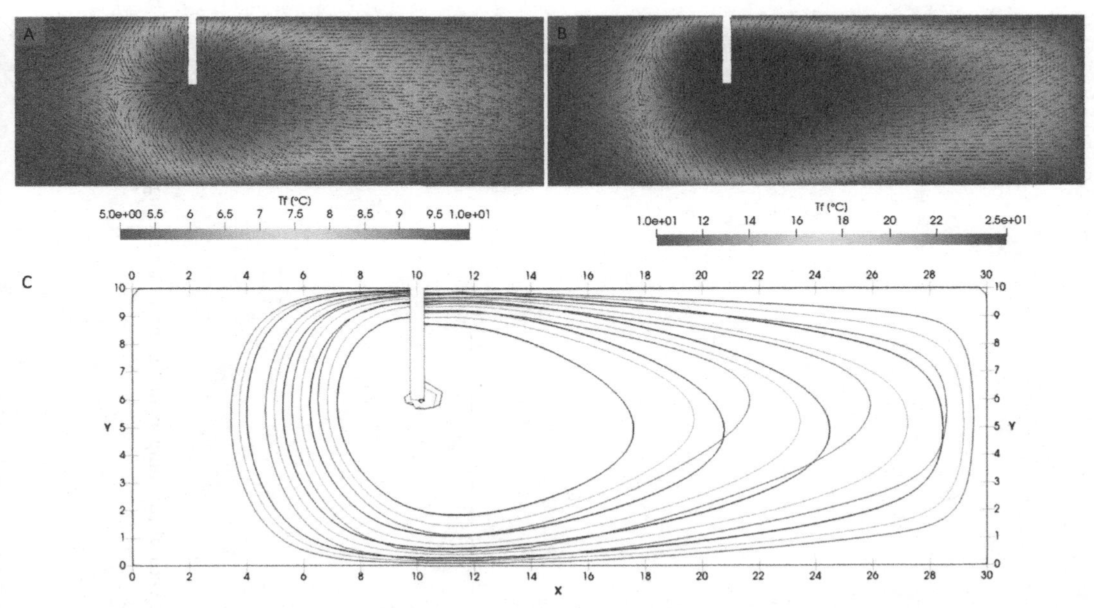

FIG. 4 Distribution of temperature and seepage rates after 200 days in the case of (A) water 5°C colder than the surrounding medium, (B) water 15°C warmer than the surrounding medium, (C) Isotherms after 200 days for cases 1, 2, and 3.

causes higher heat transfer rates for higher temperatures, and the upward direction becomes more significant for case 3.

For a better understanding of the impact of temperature on the seepage rate, their values in the cross section 1–1′ (Fig. 2) are shown in Fig. 5A (horizontal component) and Fig. 5B (vertical component). In the case of the horizontal component, for $\Delta T = +5°C$ the effect of temperature

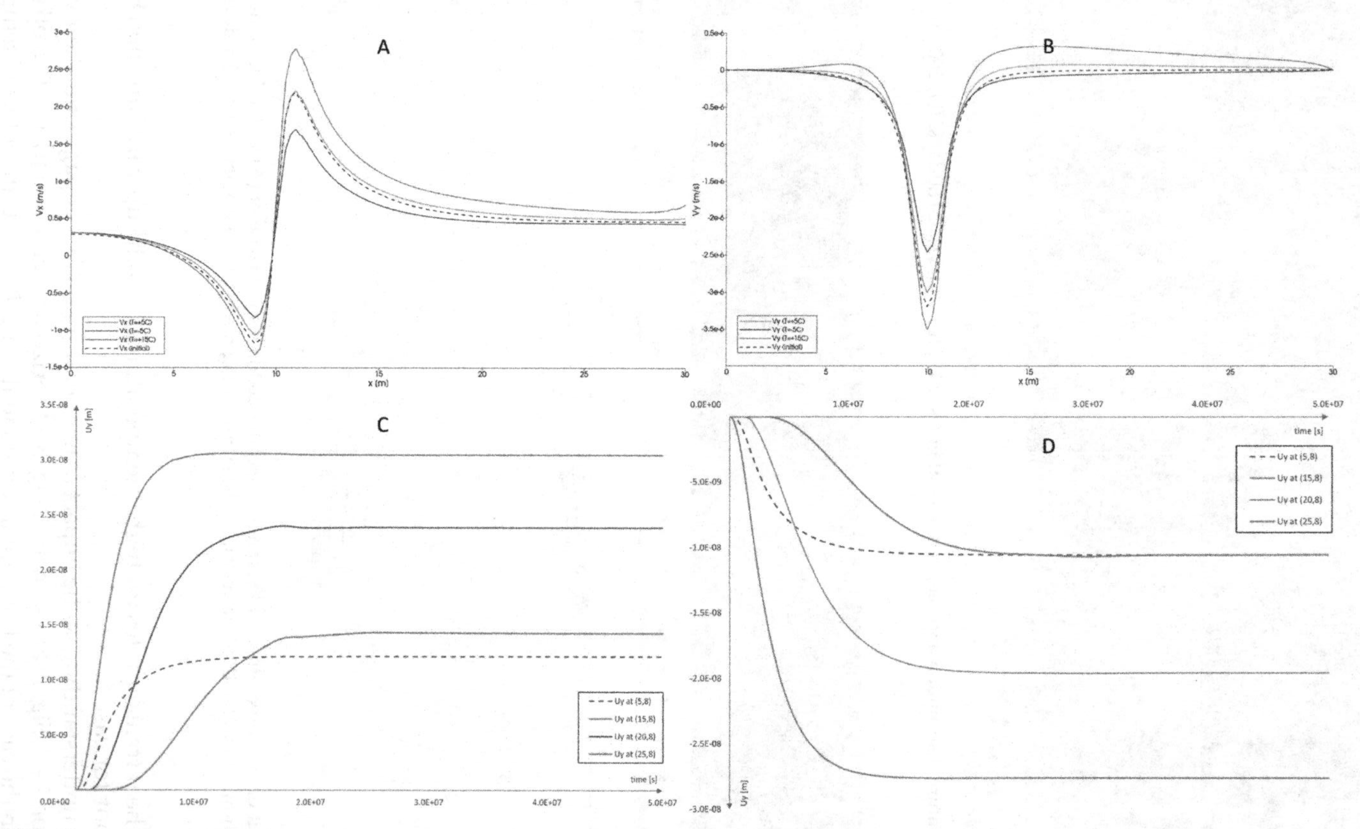

FIG. 5 Horizontal (A) and vertical (B) components of the water seepage rate, time course of vertical displacements at four points for the case of (C) $\Delta t = +5°C$, (D) $\Delta t = -5°C$.

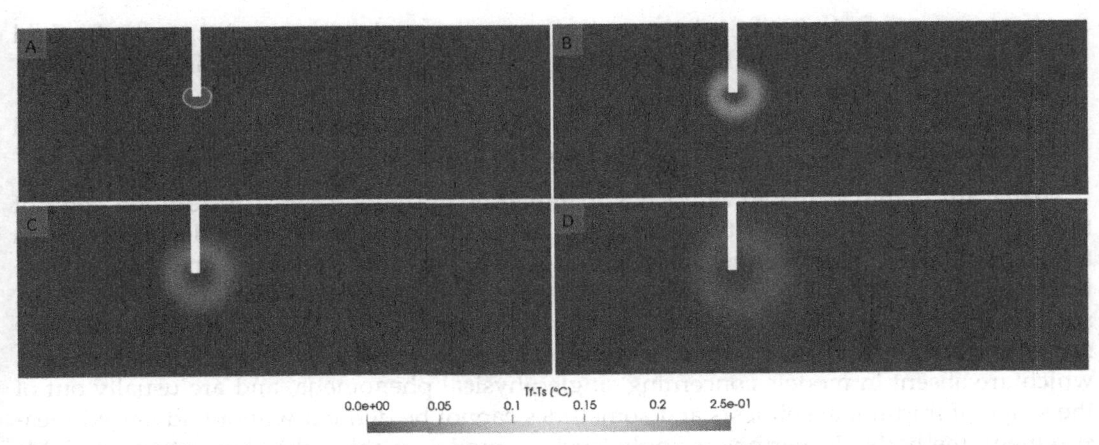

FIG. 6 Temperature difference between the phases for (A) $t=6$h, (B) $t=2$days, (C) $t=4$days, (D) $t=8$days.

is rather small, in contrast to the cases of $\Delta T = -5°C$ and $\Delta T = +15°C$. The highest seepage rates for every case occur at the same distance from the well, so the impact of temperature on to localization of v_{max} is negligible.

In Fig. 5B, the effect of temperature on the vertical flow can be observed—as before, most noticeable in the case of $\Delta T = +15°C$.

For displacement calculations, the initial state was accepted as reference—in Fig. 5C and D, there are time courses of displacements (referred to the initial state), in four points (depicted in Fig. 2A).

For such a small temperature change (+/−5°C) the changes are not large and in the simulation scale they do not affect the surface in a noticeable manner, but the way the temperature change (coupled with the flow direction) affects the displacements can be observed: for example, vertical displacements on the left side of the well develop in a different manner than at the point of similar distance, but on the right side of the well.

Because the temperature change is set as the boundary condition only in the fluid, the heat transfer in the solid constituent is the result of the transfer from the liquid phase. The fluid flowing through the medium transmits heat to further areas. The difference in temperature between the phases is most visible at the beginning of the simulation (Fig. 6A, for $t=6$h), then the temperatures of both phases equalize, and the LTNE area takes the shape of a ring (Fig. 6B) and expands. As the area of LTNE increases, the temperature difference between the phases decreases (Fig. 6C and D) and finally disappears.

4 Conclusions

The mathematical THM model enabled the numerical simulation of fluid flow in a porous medium, coupled with displacement and temperature fields. This model is coupled in both directions, which means that:

- the temperatures in the fluid in the skeleton affect each other (coupling in the heat flow PDEs),

- temperature (both in fluid or skeleton) has an impact on the stress and strain in the fluid and skeleton, stress/ strain has an impact on the temperature (coupling in the heat flow PDEs and the constitutive relations).
- stress in each phase has an impact on strain in both phases and strain in each phase has an impact on the stress in both phases (coupling in the constitutive relations),

so that any of the calculated values has an impact on the others (Fig. 1). In addition, some of the material parameters (density and viscosity of the fluid) are temperature dependent. Such a full coupling allows modeling simultaneously all phenomena of interest and their interaction on each other, which is not possible with the use of single phenomena models. Of course, such an approach also has downsides. This model consists of five coupled PDE's and clearly needs more computation time and material constants, particularly coupling parameters, which are absent in models concerning single physical phenomena, and are usually out of the scope of traditional soil tests and sometimes cannot be defined without advanced measurement methods. Nevertheless, only such a model enables the simulation of fields interacting with each other—the choice of the model should depend on the medium parameters, the length/time scale used, and the required accuracy—in this example, as shown in Fig. 6, the temperature differences between the phases of the medium are noticeable only in short length and time scale, so in models concerning long-time simulations the LTE model is sufficient, while in short-time and length scale coupled groundwater flow-heat conduction nonstationary models the LTNE approach should be taken into consideration. Coupled models need mode material constants, which, especially coupling parameters can be difficult to measure directly or may involve additional equipment. To avoid high costs, coupling constants could be obtained with the use of optimization methods, whose development will enable a fast and cheap assessment of coupling constants.

Acknowledgments

Calculations have been carried out using resources provided by the Wroclaw Centre for Networking and Supercomputing (http://wcss.pl), Grant No. 427.

References

Abesser, C., Schincariol, R.A., Raymond, J., García-Gil, A., Drysdale, R., Piatek, A., Giordano, N., Jaziri, N., Molson, J., 2021. Case studies of geothermal system response to perturbations in groundwater flow and thermal regimes. Groundwater, 1–19.

Ahrens, J., Geveci, B., Law, C., 2005. ParaView: An End-User Tool for Large Data Visualization. vol. 836 Elsevier.

Al-Sumaily, G.F., Al Ezzi, A., Dhahad, H.A., Thompson, M.C., Yusaf, T., 2021. Legitimacy of the local thermal equilibrium hypothesis in porous media: A comprehensive review. Energies 14 (23), 1–47.

Bartlewska, M., Strzelecki, T., 2009. Equations of Biot's consolidation with kelvin-Voight rheological frame. Stud. Geotech. Mech. 31 (2), 3–15.

Bartlewska-Urban, M., Strzelecki, T., Urban, R., 2015. Determination of effective parameters of Biot model with the Kelvin-Voight rheological skeleton. Arch. Civil Mech. Eng. 15 (4), 1173–1179.

Bidarmaghz, A., Narsilio, G.A., 2018. Heat exchange mechanisms in energy tunnel systems. Geomech. Energy Environ. 16, 83–95.

Biot, M.A., Willis, D.G., 1957. The elastic coefficients of the theory of consolidation. J. Appl. Mech. 24, 594–601.

Biot, M.A., 1941. Reprinted series general theory of three-dimensional consolidation. J. Appl. Phys. 12 (2), 155–164.

Buchwald, J., Kaiser, S., Kolditz, O., Nagel, T., 2021. Improved predictions of thermal fluid pressurization in hydrothermal models based on consistent incorporation of thermo-mechanical effects in anisotropic porous media. Int. J. Heat Mass Transf. 172 (March).

Burnell, J., O'sullivan, M., O'sullivan, J., Kissling, W., Croucher, A., Pogacnik, J., Pearson, S., Caldwell, G., Ellis, S., Zarrouk, S., Climo, M., 2015. Geothermal Supermodels: the Next Generation of Integrated Geophysical, Chemical and Flow Simulation Modelling Tools. Proceedings World Geothermal Congress, April, 19–25.

Carbonell, R., Whitaker, S., 1984. Heat and mass transfer in porous media. In: Bear, J., Corapcioglu, M.Y. (Eds.), Fundamentals of Transport Phenomena in Porous Media. Martinus Nijhoff Publishers.

Coussy, O., 2004. Poromechanics. John Wiley & Sons, Ltd.

Coussy, O., 2010. Mechanics and physics of porous solids. In: Mechanics and Physics of Porous Solids.

Dahash, A., Ochs, F., Giuliani, G., Tosatto, A., 2021. Understanding the interaction between groundwater and large-scale underground hot-water tanks and pits. Sustain. Cities Soc. 71 (February), 102928.

Darcy, H., 1856. (1803-1858). A. du texte. Les fontaines publiques de la ville de Dijon: exposition et application des principes à suivre et des formules à employer dans les questions de distribution d'eau… / par Henry Darcy,… Victor Dalmont.

De Wiest, R.J.M., 1966. On the storage coefficient and the equations of groundwater flow. J. Geophys. Res. 71 (4).

Derski, W., Izbicki, R., Kisiel, I., Mróz, Z., 1982. Mechanika techniczna. Tom VII. Mechanika skał i gruntów. PWN.

Drzyzga, A., 2021. The latest mathematical and empirical models to calculate the thermal conductivity of the soils. In: E3S Web of Conferences. vol. 323. 00007.

Ene, H., Polisevski, D., 1987. Thermal Flow in Porous Media. D. Reidel Publishing Company.

Euler, L., 1757. Principes généraux du mouvement des fluides. 11 Mémoires de l'Acad. Des Sciences de Berlin, pp. 274–315.

Evans, O., Spiegelman, M., Kelemen, P.B., 2018. A Poroelastic model of Serpentinization: Exploring the interplay between rheology, surface energy, reaction, and fluid flow. J. Geophys. Res. Solid Earth 123 (10), 8653–8675.

Ezzat, M., Ezzat, S., 2016. Fractional thermoelasticity applications for porous asphaltic materials. Pet. Sci. 13 (3), 550–560.

Faizurrahmany, Z., Faridy, E., Trabelsi, A., Kuznik, F., Claude, U., Lyon, B., Umr, C., 2019. Brief overview on local thermal non-equilibrium modeling of heat and mass transfer in open porous materials. Acad. J. Civil Eng. 37 (1), 374–377.

Fan, Y., Zhang, S., Huang, Y., Pang, Z., Li, H., 2022. Determining the recoverable geothermal resources using a numerical thermo-hydraulic coupled modeling in geothermal reservoirs. Front. Earth Sci. 9.

Farouki, O.T., 1981. Thermal Properties of Soils, CRREL Mono-Graph 81–1. US Army Corps of Engineers, Cold Regions Research and Engineering Laboratory.

Fiorillo, F., Esposito, L., Leone, G., Pagnozzi, M., 2022. The relationship between the Darcy and Poiseuille Laws. Water (Switzerland) 14 (2).

Fiorina, C., Clifford, I., Aufiero, M., Mikityuk, K., 2015. GeN-foam: a novel OpenFOAM® based multi-physics solver for 2D/3D transient analysis of nuclear reactors. Nucl. Eng. Des. 294 (February 2019), 24–37.

Forchheimer, P., 1901. Wasserbewegung durch boden. Z. Ver. Dtsch. Ing. 45 (50).

Gao, W., Zagorščak, R., Thomas, H.R., 2021. Numerical study of ground deformation during underground coal gasification through coupled flow-geomechanical modelling. Fuel 122833.

German, P., Tano, M.E., Fiorina, C., Ragusa, J.C., 2021. Data-driven reduced-order modeling of convective heat transfer in porous media. Fluids 6 (8).

Groth-Jensen, J., Nalbandyan, A., Klinkby, E.B., Lauritzen, B., Sabbagh, P., Pedersen, A.V., 2021. Verification of multiphysics coupling techniques for modeling of molten salt reactors. Ann. Nucl. Energy 164, 108578.

He, L.W., Jin, Z.H., Zhang, Y., 2012. Convective cooling/heating induced thermal stresses in a fluid saturated porous medium undergoing local thermal non-equilibrium. Int. J. Solids Struct. 49 (5), 748–758.

Hubbert, M.K., 1957. Darcy's law and the field equations of the flow of underground fluids. In: International Association of Scientific Hydrology. Bulletin. vol. 2. issue 1.

Ingebritsen, S.E., Sanford, W.E., 1999. Groundwater in Geologic Processes. Cambridge University Press.

Jacob, C.E., 1940. On the flow of water in an elastic artesian aquifer. Trans. Am. Geophys. Union 21 (2).

Kaviany, M., 1995. Convection heat transfer. In: Principles of Heat Transfer in Porous Media. Mechanical Engineering Series. Springer.

Kim, S.J., Jang, S.P., 2002. Effects of the Darcy number, the Prandtl number, and the Reynolds number on local thermal non-equilibrium. Int. J. Heat Mass Transf. 45 (19), 3885–3896.

Li, X., El Mohtar, C.S., Gray, K.E., 2019. 3D poro-elasto-plastic modeling of breakouts in deviated wells. J. Petrol. Sci. Eng. 174, 913–920.

Lin, W., Xie, G., Yuan, J., Sundén, B., 2016. Comparison and analysis of heat transfer in aluminum foam using local thermal equilibrium or nonequilibrium model. Heat Transfer Eng. 37 (3–4), 314–322.

Liu, B., He, S., Moulinec, C., Uribe, J., 2021a. Coupled porous media approaches in sub-channel CFD. Nucl. Eng. Des. 377.

Liu, J., Xue, Y., Zhang, Q., Wang, H., Wang, S., 2022. Coupled thermo-hydro-mechanical modelling for geothermal doublet system with 3D fractal fracture. Appl. Therm. Eng. 200.

Liu, L., He, H., Dyck, M., Lv, J., 2021b. Modeling thermal conductivity of clays: A review and evaluation of 28 predictive models. Eng. Geol. 288.

Liu, Z., 2018. Multiphysics in porous materials. In: Multiphysics in Porous Materials.

Lou, Y., Fang, P.F., Xie, X.Y., Chong, C.S.A., Li, F.Y., Liu, C.Y., Wang, Z.J., Zhu, D.Y., 2021. Numerical research on thermal response for geothermal energy pile groups under groundwater flow. Geomech. Energy Environ. 28, 100257.

Ma, C., Di Donna, A., Dias, D., Zhang, T., 2021. Thermo-hydraulic and sensitivity analyses on the thermal performance of energy tunnels. Energ. Buildings 249, 111206.

Maghoul, P., 2017. Numerical simulation for foundations energy efficiency in cold region. In: Poromechanics 2017—proceedings of the 6th Biot Conference on Poromechanics, July 2017.

Makhnenko, R.Y., Podladchikov, Y.Y., 2018. Experimental Poroviscoelasticity of common sedimentary rocks. J. Geophys. Res. Solid Earth 123 (9), 7586–7603.

Meirmanov, A., 2014. Mathematical Models for Poroelastic Flows. vol. 1 Atlantis Press.

Minkowycz, W.J., Haji-Sheikh, A., Vafai, K., 1999. On departure from local thermal equilibrium in porous media due to a rapidly changing heat source: The sparrow number. Int. J. Heat Mass Transf. 42 (18), 3373–3385.

PDE Solutions Inc, 2019. FlexPDE 7. pp. 1–336.

Pecker, C., Deresiewicz, H., 1973. Thermal effects on wave propagation in liquid-filled porous media. Acta Mech. 16 (1–2), 45–64.

Poiseuille, J.L.M., 1842. Recherches expérimentales sur le mouvement des liquides dans les tubes de trés-petits diamétres [Experimental research into the movement of liquids in pipes of very small diameter]. Imprimerie Royale.

Pryor, R.W., 2012. Multiphysics Modeling Using COMSOL 4. Mercury Learning and Information.

Radman, S., Fiorina, C., Pautz, A., 2021. Development of a novel two-phase flow solver for nuclear reactor analysis: Validation against sodium boiling experiments. Nucl. Eng. Des. 384 (January 2022), 111422.

Rees, D.A.S., Bassom, A.P., Siddheshwar, P.G., 2008. Local thermal non-equilibrium effects arising from the injection of a hot fluid into a porous medium. J. Fluid Mech. 594 (October 2016), 379–398.

Rózański, A., Kaczmarek, N., 2020. Empirical and theoretical models for prediction of soil thermal conductivity: a review and critical assessment. Stud. Geotech. Mech. 42 (4), 330–340.

Rühaak, W., Bense, V.F., Sass, I., 2014. 3D hydro-mechanically coupled groundwater flow modelling of Pleistocene glaciation effects. Comput. Geosci. 67, 89–99.

Saaly, M., Maghoul, P., Holländer, H., 2020. Investigation of the effects of heat loss through below-grade envelope of buildings in urban areas on thermo-mechanical behaviour of geothermal piles. In: E3S Web of Conferences, 205.

Strzelecki, T., Bartlewska-Urban, M., 2006. Zagadnienie jednowymiarowe elektrohydrokonsolidacji. In: Kostecki, S. (Ed.), Problemy hydrotechniki. Modelowanie i hydro-informatyka oraz wybrane zagadnienia ochrony przeciwpowodziowej. DWE, pp. 307–319.

Strzelecki, T., Bartlewska-Urban, M., Kaźmierczak, A., Overchenko, L., Strzelecki, M., Uciechowska-Grakowicz, A., 2018. Mechanika ośrodków porowatych. Dolnośląskie Wydawnictwo Edukacyjne.

Strzelecki, T., Strzelecki, M., 2015. Relation between filtration and soil consolidation theories. Stud. Geotech. Mech. 37 (1), 105–114.

Stylianou, I.I., Tassou, S., Christodoulides, P., Aresti, L., Florides, G., 2019. Modeling of vertical ground heat exchangers in the presence of groundwater flow and underground temperature gradient. Energ. Buildings 192, 15–30.

Sweidan, A.H., Niggemann, K., Heider, Y., Ziegler, M., Markert, B., 2021. Experimental study and numerical modeling of the thermo-hydro-mechanical processes in soil freezing with different frost penetration directions. Acta Geotech. 17 (1), 231–255.

Taler, J., Duda, P., 2006. Solving Direct and Inverse Heat Conduction Problems. Springer.

Tong, F.G., Jing, L., Zimmerman, R.W., 2009. An effective thermal conductivity model of geological porous media for coupled thermo-hydro-mechanical systems with multiphase flow. Int. J. Rock Mech. Min. Sci. 46 (8), 1358–1369.

Uciechowska-Grakowicz, A., Strzelecki, T., 2017. Non-isothermal constitutive relations and heat transfer equations of a two-phase medium. Stud. Geotech. Mech. 39 (3), 67–78.

Uciechowska-Grakowicz, A., Strzelecki, T., 2021. Application of the thermoporoelasticity model in numerical modelling of underground coal gasification influence on the surrounding medium. Stud. Geotech. Mech. 43 (2), 116–134.

Wilkins, A., Green, C., Ennis-King, J., 2020. PorousFlow: a multiphysics simulation code for coupled problems in porous media. J. Open Source Softw. 5 (55), 2176.

Wu, H.-L., Jin, Z.H., 2010. A local thermal nonequilibrium poroelastic theory for fluid saturated porous media. J. Therm. Stresses 33 (8), 799–813.

Xie, J., Qin, Y., 2021. Heat transfer and bearing characteristics of energy piles: review. Energies 14 (20).

Xu, Y., Li, Z., Wang, J., Chen, Y., Li, R., Wang, Q., Jia, M., 2022. Ventilation and heat exchange characteristics in high geotemperature tunnels considering buoyancy-driven flow and groundwater flow. Int. J. Therm. Sci. 173 (September 2021).

Youssef, H.M., 2007. Theory of generalized porothermoelasticity. Int. J. Rock Mech. Min. Sci. 44 (2), 222–227.

Zhang, G., Guo, Y., Zhou, Y., Ye, M., Chen, R., Zhang, H., Yang, J., Chen, J., Zhang, M., Lian, Y., Liu, C., 2016. Experimental study on the thermal performance of tunnel lining GHE under groundwater flow. Appl. Therm. Eng. 106, 784–795.

Zhelnin, M., Kostina, A., Prokhorov, A., Plekhov, O., Semin, M., Levin, L., 2022. Coupled thermo-hydro-mechanical modeling of frost heave and water migration during artificial freezing of soils for mineshaft sinking. J. Rock Mech. Geotech. Eng. 14 (2), 537–559. https://doi.org/10.1016/j.jrmge.2021.07.015.

CHAPTER

23

Hydrosalinity modeling of water and salt dynamics in irrigated soil groundwater systems

Samanpreet Kaur[a], *Mehraj U. Din Dar*[a], *and S.K. Kamra*[b]

[a]Department of Soil and Water Engineering, Punjab Agricultural University, Ludhiana, India

[b]Emeritus Scientist (Hon.), ICAR-Central Soil Salinity Research Institute, Karnal, Haryana, India

OUTLINE

1 Introduction 410

2 Materials and methods 411

2.1 *Processes occurring in soil water system in irrigated regions* 411

2.2 *Components of hydrosalinity model for an irrigation project* 412

2.3 *Hydrologic submodel* 412

2.4 *Salinity submodel* 414

2.5 *Criteria for evaluating hydrosalinity models* 414

2.6 *Field investigations for different hydrosalinity models* 415

3 Hydrosalinity models 420

4 Results and discussions 421

4.1 *Case studies on applications of different hydrosalinity models in northern India* 421

4.2 *Regional scale soil salinity modeling* 422

5 Subsurface drainage modeling 423

6 Conclusions 427

References 427

Current Directions in Water Scarcity Research, Volume 7
https://doi.org/10.1016/B978-0-323-91910-4.00023-6

1 Introduction

The transport of solute in groundwater such as fertilizers or pesticides, radioactive wastes, nitrogenous species, and many contaminants in soils has degraded the quality of groundwater. One of the important aspects of groundwater management is the control of contamination by identifying pollution sources, field measurements, and mathematical modeling in different geological formations (Katsifarakis, 2000). Hence, subsurface solute transport has been increasingly capturing the attention of hydrologists and engineers. The sources of groundwater pollution can be classified into point sources (e.g., leaking underground tanks, septic tank, waste water lagoons) and nonpoint sources (e.g., agricultural activities, precipitation).

In order to better study water and salt transport at the regional scale, the use of regional water and salt models, model coupling, and a combination of modeling and GIS has been proposed based on the principle of regional water and salt balances (Mirlas, 2012). Akram et al. (2008) simulated the groundwater levels and soil salinity changes using the SAHYSMOD model and evaluated the performance of different biodrainage system designs. It is basic to regulate the movement of soil water and salt to correctly understand the governing mechanisms and principles to devise water saving technologies, especially in the arid and semiarid irrigation areas where water resources are scarce and salinization is an eventual outcome. Previous studies on water and salt dynamics only scarcely considered the continuity of groundwater flow in the aquifer and the relationship between water and salt movement in different regions. Due to the differences in topography, irrigation and drainage, climate conditions, etc., the soil salinity dynamics in the cultivated and salt afflicted wasteland also vary in the different subirrigation regions.

Hydrological and economic analyses are needed to shape and guide society's vision for sustainable land and water management (Khan et al., 2008). Accordingly, in the recent decade, there have been many efforts to develop hydroeconomic modeling (Bekchanov et al., 2017). From the salinity management point of view, hydroeconomic studies can be classified into two important categories of (1) farm or field scale and (2) catchment or regional scale studies. Farm- or field-scale hydroeconomic studies adopted different approaches of field works and data analysis (e.g., Yorobe et al., 2016), simulation models, optimization models (e.g., Reca et al., 2018), and simulation-optimization models (e.g., Zekri et al., 2017).

Water fluxes are of particular concern as water is the weathering reactive agent as well as the transporting force. Changes in water fluxes affect chemical weathering and transport of solutes and particles through soils (Chadwick and Chorover, 2001) and lead to waterlogging and secondary soil salinity in canal irrigated systems. Irrigated agriculture increases the amount of water flowing through soils. Artificial subsurface drainage, designed to remove excess water from soils, reduces the residence time of water in soils, increases soil aeration and increases the amount of infiltrating water and induces changes in the water pathways by either reducing the runoff (Grazhdani et al., 1996) or intercepting water fluxes at regular intervals. Artificial drainage seems to affect water fluxes in soils to a greater extent than irrigation.

The survival of irrigated agriculture as an economically profitable and environmentally sustainable activity requires optimal management of irrigation and drainage on a regional scale. It is important to foresee the long-term effects of current and projected irrigation and agricultural practices to plan remedial and mitigation measures. There are various solutions that could be considered to address the problems of waterlogging and salinization associated with irrigated agriculture; however, the effectiveness of all the solutions and their combinations cannot be verified with field experiments (Singh and Ramana Rao, 2014). Simulation models are widely used for the management of waterlogging and salinization problems (Zhao et al., 2004; Morway et al., 2013) because these models usually help to find an answer to "What if?" due to their predictive capability. The most important step toward problem solving is the delineation of water-and salt flow regimes using hydro salinity models. A hydrosalinity model links the hydrological and salinity interactions of different subsystems with the help of suitable equations or empirical relationships to provide an acceptable representation for water and salt transport in a large irrigation project. These models provide increased insight and better conceptual understanding of interacting physical, chemical and biological processes and are likely to play in future an increasing role for the management of land and water resources. The literature also revealed that researchers are mainly concerned about pollutant storage in soil and plant uptake, but there is little effort for pollutant transport and ground water contamination estimates by comparing the various developed solute transport models which is a subject matter of interest for this investigation.

This chapter presents an overview of the conceptual framework, relative complexity of water-and salt-balance analysis and limitations of different approaches of hydrosalinity modeling. It also briefly summarizes the essential feature and procedures of calibration, field validation and sensitivity analysis at the basin level and a field scale model that have been tested in northern India. The guidelines and methodology for collection of important relevant data, required in most modeling exercises (though with varying degree of intensity and periodicity), are also discussed.

2 Materials and methods

2.1 Processes occurring in soil water system in irrigated regions

Irrigation is a major user of water, which affects the water balance of the irrigated area and its related basin (Fig. 1). An irrigation system can be broadly divided into three subsystems of water delivery, farm and water removal.

The **water delivery subsystem** can be further divided into two components—one starting from the headworks to the section of river where water is diverted to irrigate croplands, and second from river diversion works to the individual farm.

The **farm subsystem** begins at the point where water is delivered to the farm and continues to a point where surface water is removed from the farm. It consists vertically as beginning at the ground surface and terminating at the bottom of the root zone.

The principal function of **water removal subsystem**, consisting of surface run-off and water moving below the root zone, is to facilitate adequate drainage for aeration and or

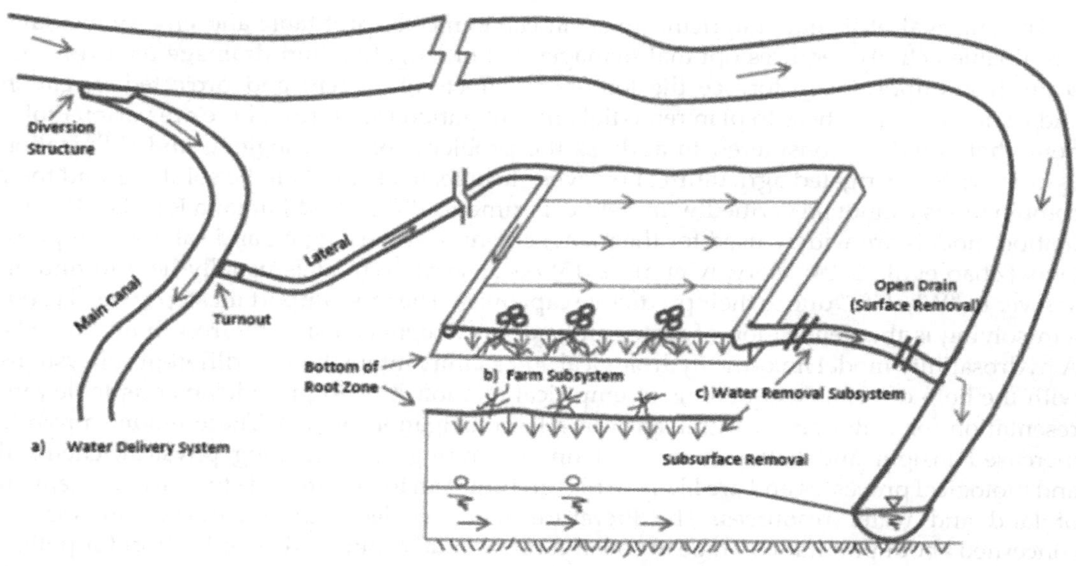

FIG. 1 Sketch of an irrigation system.

salinity control in the root zone. The most satisfactory mechanism for minimizing drainage needs is through proper operation of water delivery subsystem and on-farm water management.

Levels of soil water salinity vary by location and over time in irrigated soils with differences in the quality, quantity, and application patterns of irrigation and natural waters and in the chemical, geohydrologic, and biological properties of the soil.

2.2 Components of hydrosalinity model for an irrigation project

The hydrosalinity model explicitly implies that hydrologic and salinity submodels are combined. As the flow of salts is directly dependent upon the flow of water, salinity simulation requires output from the former. The modeling approaches for simulation of water and salt flows, however, may be quite diverse.

2.3 Hydrologic submodel

The hydrologic model (Fig. 2) from Wood (1976) is an idealized hydrologic unit of an integrated system, consisting of three basic subsystems or elements: surface water (stream), soil water (land) and ground water (subsurface). Each hydrologic unit in a river basin contains two or three of these subsystems, between which water (as well as salts) are transferred according to certain physical relationships or some operational plan. A proper accounting of different quantities, which also get translated or routed in space and time, is made in the

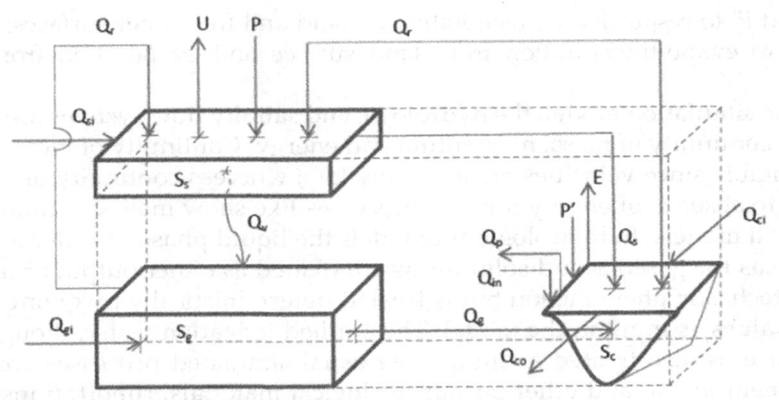

FIG. 2 A hydrologic model consisting of soil water, ground water and stream subsystems. *Adapted from Wood, E. F., 1976. An analysis of the effects of parameter uncertainty in deterministic hydrologic models. Water Resour. Res. 12 (5), 925–932.*

model. Each of these subsystems may be further subdivided into pools (storage compartments) and fluxes. The pools contain the state variable water (and salts) whereas the fluxes describe the pathways and rate of transfer from one pool to another. For each hydrologic element, Wood (1976) invoked the continuity equation such that:

$$\frac{dS}{dt} = \sum\nolimits_{k=0}^{n} Q \tag{1}$$

where S is the total volume of stored water; Q is the rate of flow into or out of the element; and t is time. This hydrologic unit has water inputs of precipitation (P) and land surface (Q_{ls}), stream (Q_{ci}) and groundwater (Q_{gi}) rim inflows and outputs of evapo-transpiration (U,E), exported water (Q_p) and stream outflows (Q_{cv}) from each subsystem of the unit. One can add land surface (Q_{so}) and ground water (Q_{go}) imported water (Q_i), and deep percolation beyond the reach of groundwater pumpage (Q_{dp}). By including above additions to idealized unit of Wood (1976), the changes in the internal water balance in the stream, soil water and ground water elements may be respectively given by:

$$\frac{dS_c}{dt} = Q_{ci} - Q_{co} + Q_s + Q_g - Q_d + Q_{in} - Q_p + P' - E \tag{2}$$

$$\frac{dS_s}{dt} = Q_{si} - Q_s + Q_d + Q_t - U + P - Q_v - Q_{so} + Q_i \tag{3}$$

$$\frac{dS_g}{dt} = Q_{gi} - Q_g + Q_v - Q_r - Q_{go} - Q_{dp} \tag{4}$$

where the subscripts c, s, and g refer respectively to the stream-soil water and groundwater storage elements; "i" and "o" to inflow and outflow from the element; d to diversion for irrigation use; p and i respectively to export from stream and import to land surface, v and dp to vertical seepage from soil to ground water and deep percolation out of groundwater domain; r to recirculation or pumping and may include rising ground waters to the land surface; P and P' to respectively precipitation on land and free water surfaces; and U and E respectively to evapo-transpiration from land surface and evaporation from free water surface.

In dynamic simulation model, the hydrologic and salinity flow systems are based on the principles of continuity of mass, momentum and energy. Continuity of momentum is considered negligible since velocities are generally low; whereas continuity of energy is also ignored due to absence of energy transfer processes like snow melt. Continuity of mass is essential for all models. In hydrologic submodels the liquid phase of water is usually simulated, whereas the gaseous and solid forms are treated as either output or input. Precipitation is a stochastic phenomenon but is treated deterministically by giving its temporal and spatial values as input in the model. The applied irrigation water, though varying in time and space, is also treated as input. The usual simulated processes are infiltration, water movement in soil and other porous geological materials, runoff, transpiration and evaporation. Most mathematical models for water flow in hydrologic simulation are based upon Darcy's law:

$$q = -K\frac{d\varnothing}{dZ} \tag{5}$$

where q is the flux of water, K is the hydraulic conductivity, $\varnothing$ is the potential, Z is the distance, and $d\varnothing/dZ$ is the potential gradient. The transient soil water flow, a major process in unsaturated soils, was described by Nimah and Hanks (1973).

2.4 Salinity submodel

The sources and sinks of salinity in soil are manifold. Soil water plays a major role in salinity considerations, since it affects both salt concentration as well as salt movement. With the exception of precipitation and evapo-transpiration, all water quantities considered in hydrologic model have quality parameters associated with them. All salinity models for irrigated soils represent the dissolved salts by the lumped parameter of either total dissolved solids (TDS) or electrical conductivity (EC). TDS is used to calculate the mass of salts as a product of salt concentration and water volume; EC, being related to osmotic potential, provides a better estimate of the impact of salinity on plants. If all quantity components (defined in Eqs. 2–4) and the associated quality parameters can be identified and grouped into m surface water and n ground water sources and the storage elements.

2.5 Criteria for evaluating hydrosalinity models

In different hydrosalinity models, the mechanisms of water and salt flow are studied in different degrees of complexity and emphasis, depending upon specific objectives and relative significance perceived for different subsystems in the study. The complexity of solute transport models ranges from simple applications of plate theory assuming piston-flow movement of solute and water (Tanji, 2002), to detailed models which attempt to represent the complex chemical reactions within the soil profile by the use of both hydrodynamic dispersion and diffusion principles. For example, if the primary objective of a study is to characterize mechanisms of moisture and salt flow in a soil profile considering chemical reactions like cation-exchange, mineral precipitation-dissolution, oxidation-reduction, or ion-association, the emphasis on surface water distribution and or other subsystems will be proportionately diluted. Under such conditions, salinity is treated as a reactive parameter and movement of solute species of Na^{2+}, Ca^{2+}, Mg^{2+}, Cl^-, HCO_3^-, CO_3^{2-}, So_4^{2-} in liquid phase, and of gypsum ($CaSO_4$ $2H_2O$) and lime ($CaCO_3$) in solid (adsorbed) phase are modeled. For such conditions, chemical equilibrium approach is adopted. The transient one-dimensional transport of an adsorbing solute species in the porous media may be described as, by (Bresler, 1973)

$$\frac{\delta(\theta C)}{\delta t} = \frac{\delta}{\delta Z}\left\{D(v,\ \theta)\frac{\delta C}{\delta Z}\right\} - \frac{\delta(qC)}{\delta Z} - \frac{\delta(\rho Z)}{\delta t} \tag{6}$$

where C is the dissolved solute concentration; D is the apparent diffusion coefficient of solute species; v is the average pore water velocity; q is the volumetric water flux; and p is the soil bulk density. The first term on RHS of Eq. (6) is for diffusive transport, the second for convective transport, and the third for adsorption. Other sink or source terms may be added to Eq. (6) depending upon the area of emphasis of the study. Similarly, if an irrigated agricultural region of the size of a few hundred or thousands of hectares underlain by a shallow perched aquifer is considered, detailed chemical modeling of soil-water in farm subsystem becomes of secondary importance. Though a thorough treatment of the unsaturated water and salt-flow dynamics in the root zone will be necessary at a local scale and over relatively short time periods, a simple mass-balance approach for describing water and salt balance in the zone above water-table is generally used in regional-scale hydrosalinity modeling (Gates and Grismer, 1989). Further guidelines for model formulation, validation and application, and model selection and scope will be discussed during the course of discussion of individual basin-level (e.g., SALTMOD, RSSM) and field-level hydrosalinity (DRAINSAL) models. Table 1 provides a generalized criterion for evaluation of hydrosalinity models that could be applied for specific purpose.

2.6 Field investigations for different hydrosalinity models

Field investigations for collecting information to be used as input into hydrosalinity models aim primarily to establish the boundaries of study area and to determine the source of problem. This involves establishing a monitoring network and conducting field studies to assess or estimate parameters of the delivery, farm and water-removal subsystem. The intensity and frequency of the investigations depends on the cost, required degree of accuracy and the time frame. Parameters like irrigated area, canal and drainage network, size of fields, irrigation supply and methods, soil texture and area under different crops in the command are available from existing records and maps. Boundaries of the study area are governed by natural characteristics such as hydrologic divide of watersheds and canal or drainage system network. Initially the whole area is considered; the boundaries are subsequently reduced to

TABLE 1 Criteria for evaluating hydrosalinity models.

1	Model capabilities (1) Applicable situations (2) Constituents modeled
2	Model assumptions (1) Within root zone (2) Within the unsaturated zone below root zone (3) Within the saturated groundwater zone
3	Salt pick up methodology
4	Representation of groundwater salt component
5	Data requirements (1) For model inputs (2) Additional for model verification
6	Model costs (1) Initiation costs (2) Utilization costs
7	Model accuracy (1) Representation of physical system (2) Numerical accuracy (3) Sensitivity to input errors (4) Sensitivity of management options
8	Ease of application (1) Adequacy of available documentation (2) Output form and content updateability of data
9	Model credibility

select representative areas for detailed studies. Integration of results of representative areas with additional variables estimated for the remaining area facilitates identifying the source of water logging or soil salinity problem in the command.

2.6.1 Investigations on delivery subsystem

In a salinity-control program of irrigated areas, field investigations related to delivery subsystem include primarily the measurement of flow rate and seepage in different stations of the network. Flow measurement methods can be grouped into three categories involving either the measurement of velocity (pitot tubes, current meters, venturi and propeller meters) or of hydraulic head (parshall, cut throat, trapezoidal and H flumes, weirs and orifices) or miscellaneous techniques like chemical salt and dye dilution, total count radio-isotopes, magnetic and some other methods. Monitoring flow rate of supply and drain water at gauging stations requires a calibration curve of flow versus depth of water. Surface and subsurface inflows to the study area, not generally measured, can be estimated from changes in base line of the discharge hydrographs. Chemical Analysis for major cations and anions and electrical conductivity (EC) of irrigation and drain water samples need to be conducted regularly, whereas those of surface run-off water occasionally.

Seepage from canals, laterals or field channels is influenced by the characteristics of channel bed and its operational life, water head and its salinity and sediment load, depth to groundwater, time of conducting tests during the year and several biological factors. Three common methods for measuring seepage rates include the inflow-outflow method, ponding method, and seepage meter method. The inflow-outflow method consists of measuring the inflow to an outflow from a specific section and its length and average wetted perimeter. The ponding method involves measuring the rate of fall of water surface in a *pool* created in the canal section. It is probably the best method when canal discharges are large in relation to seepage rates. Seepage meters determine seepage rates over a small area under normal operating conditions; a realistic average is made from observations taken at several points along the canal section. The method is difficult to apply under high flow depth or velocity conditions. Computation of seepage from conveyance network requires data on designed discharge, daily discharge, number of running days, length of conveyance system (lined vs unlined) for each major and minor channel. Empirical relations linking seepage rates, upstream discharge and length of the reach are developed for individual sites from observed measurements.

On the basis of a number of field studies in India it can be generalized that seepage through unlined canals may vary from 1.8 to 2.5 cumecs km^{-2} wetted area in normal soils with some clay contents and 3 to 3.5 cumecs km^{-2} of wetted area in sandy soils. For lined canals the seepage losses may be taken as 20% of these values. Studies in Punjab have indicated that canal seepage is of the order of 115 ha m $year^{-1}$ km^{-1} length of main canal and 38.24 ha m $year^{-1}$ km^{-1} of distributary canals. The seepage from tanks may vary from 9% to 20% of their live storage capacity. As a thumb rule, seepage from the field channels can be taken as 10% to 15% of water delivered at the channel head.

2.6.2 *Investigations on farm subsystem*

The amount of water diverted to a farm or field must be measured and accounted for in terms of evapo-transpiration, infiltration and tail-water run-off. In shallow water-table areas, the contribution of ground water through capillary rise also needs to be estimated. Four types of basic data are required for on-farm water-use investigations: crop parameters, soil- and water quality parameters and climatic data such as evapo-transpiration and precipitation. Crop parameters help select the irrigation method and evaluate the sensitivity of plants to salinity, specific ion toxicity and evapo-transpiration demands. Soil-classification maps along with depth wise information are required each for field capacity, permanent wilting point, water-holding capacity, saturation percentage. Average rooting depth, EC of soil-saturation extract, percent gypsum and bulk density. Representative moisture-characteristic curves of the soil profile or of individual layers in a layered soil are almost indispensable for most transient distributed models for the unsaturated zone. Basic soil chemistry reactions, electrical conductivity and ionic contents of soil solution as well as of irrigation water should also be determined.

Precipitation can be easily measured from meteorological observatories in the vicinity of study area. Recording and nonrecording gauges are used to compute rainfall parameters and average depth over a given area. Reference evapo-transpiration (ET) can be calculated by Class A evaporation method, whereas actual evapo-transpiration (ET) of grown crops grown can be determined using crop coefficients (K) adapted to local conditions. For more elaborate studies, ET can be computed from meteorological data using approaches varying from temperature-dependent analysis (Blaney-Criddle method) to an analysis of energy balance and convective transport (Penman-combination method). The meteorological data accordingly may include daily solar radiation, air temperature, dew point temperature, relative humidity, wind speed and precipitation. Some formulae may also require information on monthly percentage of daylight hours, latitude, altitude, crop height, root-zone depth, crop, and phenotype growth-stage coefficients. The yields and areal percentage of different crops should be measured each year in the study area, and at least once for the entire irrigated area.

2.6.3 Deep percolation

Deep percolation is probably the most difficult and elusive segment of water-salt budgeting. Methods using concentration distribution of conservative ions in the soil profile as tracers are fairly inexpensive and usually provide fair estimate of the quantity of deep percolation. Porous ceramic cups are effective tools to collect the samples of soil-water quality at different soil depths; chloride (Cl) and bromide (Br) being the commonly used ions in these investigations. Deep percolation may take place on account of recharge due to rainfall or excess or nonuniform irrigation water application.

2.6.4 Rainfall recharge

The rainfall recharge is influenced by the type of strata and formulations through which water infiltrates to join the ground water-table The Soil Conservation Service, USDA, has classified soils into four hydrological groups based on their basic infiltration rates. Group A consists of deep sand and aggregate silts with basic infiltration rate of 7.6 to 1.4 $mm h^{-1}$. Group B constitutes sandy loam soils having basic rate of 3 8 to 7.6 $mm h^{-1}$. Group C includes clay loams, shallow sandy loam soils with low organic content and high clay content, with basic infiltration rate of 1.3 to 3.8 $mm h^{-1}$ and Group D covers swelling soils, heavy plastic clay and alkali soils, which have a basic intake rate of as low as 0.0 to 1.3 $mm h^{-1}$. In temperate climates where rainfall storms of high intensity and short duration occur, rainfall recharge may be estimated by evaluating the infiltration and soil-moisture deficit during individual storm events. In humid regions having shallow and fluctuating ground water-table, estimation of recharge from rainfall-water-table depth relationship is recommended. In the absence of adequate data, norms given in Table 2 may be adopted to estimate the rainfall recharge under different types of geological formations.

2.6.5 Recharge from irrigation

The maximum deep percolation losses from applied irrigation need to be determined from available information and field tests. Where such information is inadequate, percolation losses corresponding to different soils and irrigation methods may be estimated using Table 3. These figures are applicable for crops other than rice, for which the rates may be increased by 5%. If ground water is used for irrigation, the rates of deep percolation losses may be reduced by 5%.

TABLE 2 Rainfall infiltration (%) to ground water body in different formations.

	Rock type or formation	Rainfall infiltration (%)
(a)	**Alluvial area**	
	Indo-Gangetic and inland areas	22
	East coast	16
	West coast	10
(b)	**Hard rock areas**	
	Weathered granite, gneiss and schist with low clay content	11
	Weathered granite, gneiss and schist with significant clay content	8
	Granulite facies like charnockite, etc.	5
	Vesicular and jointed basalt	13
	Weathered basalt	7
	Laterite	7
	Semiconsolidated sandstone	12
	Consolidated sandstone, quartzite, limestone (except cavernous limestone)	6
	Phyllites, shales	4
	Massive poorly fractured rock	1

An additional 2% of rainfall recharge factor may be used in such areas or part of the areas where watershed development with associated soil conservation measures are implemented. This additional factor is subjective and is separate from the contribution due to the water conservation structures such as check dams, nalla bunds, percolation tanks, etc.

TABLE 3 Estimated deep percolation losses as related to irrigation method and soil type.

		Deep percolation (% of water delivered to the field)	
Method	**Application process**	**Heavy**	**Light**
Basin	Poorly leveled and shaped	30	40
	Well leveled and shaped	20	30
Furrow, border	Poorly graded and sized	30	40
	Well graded and sized	25	35

2.6.6 Investigations on water removal subsystem

The surface contributions to groundwater due to head ditch seepage and deep percolation losses have been discussed. The data on uniform distribution of on-farm water use is very critical in the analysis of water removal subsystem. Measurement of discharge and operational days of drains as well as collection of water samples for quality analysis follows the same pattern as for water delivery system. A water level recording station is required at main outlet to measure the total volume of return flow and to take samples for chemical analysis.

2.6.7 Drainage investigations

Three types of agricultural drainage systems are used, i.e., the interceptor drains, relief drains and pump drains. These are installed to lower the water-table and to provide effective leaching to control soil salinity. Data required for most drainage evaluations include drain spacing, depth, size, envelope thickness, soil-hydraulic characteristics, time-distributed quantity and quality of applied irrigation and drainage water. Discharge, horse power and pumping hours of tube wells need to be collected to estimate the amount of ground-water withdrawal.

3 Hydrosalinity models

Over the past four decades significant advances have been made in various aspects of basin level hydrosalinity models. The models range from simple mass-balance based models to complex salinity-management models, which consider in much detail the individual processes contributing to the salinity of irrigation return flow. These models synthesize both hydrologic (water flow) and chemical (solute flow) processes. The representation of the hydrology component varies substantially depending upon the watershed studied and the desired application.

The different simulation computer models used for salinity management can also be classified as basin scale model (Tanji, 1977) and field scale model (Droogers et al., 2001). The water flow and solute transport equation or mass balance equation, are based on field-scale models such as LEACHC model (Wagenet and Hutson, 1987), SWAP model (Feddes et al., 1978; Van Dam et al., 1997), SOWACH model (Dudley and Shani, 2003), HYDRUS model (Šimůnek et al., 1998), UNSATCHEM model (Šimůnek and Van Genuchten, 1996) DRAINSAL (Kamra et al., 1991a), DRAINMOD, SALTMOD. Spatial variations owing to variations in topography can be simulated and predicted using salinity cum groundwater models like SahysMod and RSSM.

The majority of the computer models for water and solute transport in the soil (e.g., SWAP, DrainMod-S, UnSatChem, and Hydrus) are based on Richard's differential equation for the movement of water in unsaturated soil in combination with Fick's differential convection-diffusion equation for advection and dispersion of salts. Simpler models, like SaltMod, based on monthly or seasonal water and soil balances and an empirical capillary rise function, are also available. They are useful for long-term salinity predictions in relation to irrigation and drainage practices. LeachMod, using the SaltMod principles, helps in analyzing leaching experiments in which the soil salinity is monitored in various root zone layers while the model will optimize the value of the leaching efficiency of each layer so that a fit is obtained of observed with simulated soil salinity values. Spatial variations owing to variations in topography can be simulated and predicted using salinity cum groundwater models, like SahysMod.

4 Results and discussions

4.1 Case studies on applications of different hydrosalinity models in northern India

Most of the existing models require inputs which are not easy to measure and also, they use short time steps and need at least a daily database of hydrologic phenomena. Keeping aforementioned reservations in view, Rao et al. (1992) applied SALTMOD model to predict fluctuations of water table and soil salinity in the root zone under existing irrigation and cropping practices and for delineating the areas prone to waterlogging and soil salinity in a 1580 ha area in Tungabhadra irrigation command in India. Construction of an interceptor drain was found to help mitigate the twin problems. Singh (2012a,b) also used an agrohydrosalinity model, SaltMod (Oosterbaan and de Lima, 1989; Jain and Srinivasulu, 2004), to analyze water and salt balances of an irrigated semiarid area located in the Haryana State of India where the groundwater level has risen in the last few decades (Singh et al. 2010). Improved and efficient irrigation methods along with better cropping pattern, canal lining, and reduced canal water use and increased groundwater use were suggested to manage the waterlogging and salinization problems of the area.

The DRANSAL model was validated using the experimental field data of 2 years (1984 and 1985) from a highly saline tile-drained site at Sampla in Haryana State. Thereafter, 10-year predictions on salt distribution in soil, ground water and drain effluent were made. The model results corroborated the field observations that subsurface drains of about 75 m spacing and 1.5 to 1.8 m depth are reasonably effective in ameliorating the root zone of sandy loam saline soils of Haryana State. The salt concentration in soil and of drainage and ground water were found significantly influenced in time and space by aquifer stratification, depth of impervious layer and adsorption (Kamra et al., 1991b, 1994). Kamra and Gupta (1993) applied DRAINSAL to estimate the salt loads of drainage effluent from a 2000 ha initially highly saline area under alternate drainage designs. Assuming soil, climate, irrigation and geohydrological conditions in whole area to be similar to those at Sampla, five drainage designs of different drain spacing (25)—drain depth (*d*) combinations, viz., $2S=48$ m, $d=1.0$ m; $2S=67$ m, $d=1.5$ m; $2S=75$ m, $d=1.8$ m; $2S=77$ m, $d=2.0$ m; $2S=85$ m, $d=2.5$ m, were considered. It was observed that, irrespective of the salinity of irrigation water up to 5 dS/m, the annual amount of salts discharged during initial 2 years of reclamation from drains wider than 75 m and deeper than 1.8 m is likely to be significantly higher than the closer and shallower drains. In the fifth year the salt discharged from all design options was almost identical.

The water management simulation model DRAINMOD-S was calibrated and validated using 3 years (1995–97) experimental field data from the installed subsurface drainage system at 1.8 m drain depth with 40, 60, and 80 m drain spacing at Golewala watershed, Faridkot (Kiran, 2010). The parameters that were determined to be most effective in changing the model output were lateral saturated hydraulic conductivity, drain depth and drain spacing. The reliability of model has been evaluated by comparing observed and simulated values. The calibrated and validated model was used to predict the salt concentration for 5 consecutive years (1998–2003). The average soil salinity of root zone decreased from 1417 to 1083, 1523 to 1097 and 1167 to 789 ppm, at the end of 5 years (2003) for 40, 60 and 80 m drain spacing respectively. The Root mean square error, model efficiency and correlation coefficient between observed and simulated salt concentration ranged from 6.49 to 38.43 ppm, 0.647 to 0.834 and 0.957 to 0.999 for three mentioned drain spacing respectively. These values are quite comparable to values reported by others in similar model validation studies. The overall performance of DRAINMOD-S was satisfactory and can be applied for design and evaluation of subsurface drainage system at Punjab and other parts of country in similar conditions.

Singh (2013) applied the model SGMP for evaluating the impacts of potential policy changes on the management of waterlogging problems in an irrigated area of Haryana State of India. The agro-hydrological model SWASALT, which is an extended version of SWATRE (Feddes et al., 1978; Belmans et al., 1983), was utilized by Singh (2010) for managing the problems of waterlogging and salinization in Hisar district of Haryana State, India. The simulation revealed that in most agrohydroclimatic conditions, saline water of up to 7.5 dS/m can be used safely on a long-term basis for mustard-based cropping systems.

4.2 Regional scale soil salinity modeling

The quasi three-dimensional regional soil salinity model (RSSM) has been applied in a 6000-ha salinity affected irrigated tract in south-west Punjab (India), using the available data on hydrological systems combined with information on soil properties, salinity levels and groundwater quality from a standard grid survey. Data on the operation of the irrigation channels, groundwater levels and cropping patterns were monitored during the period 1983–86 and a survey on soil salinity and quality of canal and drain water was undertaken during 1983. A model application indicated drainage to be effective, especially in saline ground water areas. Makin and Goldsmith (1988) advocated the use of such hydrosalinity models at all stages of the development of a project and for identification of priority areas for phased development of drainage over time. Linking the outputs of soil water balance/numerical models can help to estimate spatially varying recharge.

The field scale eco-hydrological model SWAP including detailed crop growth simulations was applied for identification of appropriate strategies to improve water management and productivity in an irrigated area of 4270 km^2 in Sirsa district (Singh et al., 2006). Field experiments, satellite images and existing geographical data were used to aggregate the representative input parameters of all so-called homogeneous *"simulation units"* and their boundary conditions. The simulated water and salt limited crop yields showed a good correspondence with the independent crop yields data obtained from remote sensing, field measurements and statistical records. Simulation results for the period 1991–2001 show that the water and salt limited crop production is 1.2 to 2.0 times higher than the actual recorded crop production. Improved crop husbandry in terms of improved crop varieties, timely sowing, better nutrient supply and more effective weed, pest and disease control, will increase crop yields and water productivity in Sirsa district. The scenario results further showed that reduction of seepage losses to 25% to 30% of the total canal inflow and reallocation of 15% canal water inflow from the northern to the central canal commands will improve significantly the long term water productivity, halt the rising and declining groundwater levels, and decrease the salinization in Sirsa district.

5 Subsurface drainage modeling

The original subsurface drainage model was the Hooghoudt equation (Hooghoudt, 1940), which is a one-dimensional steady-state simplification of the two-dimensional transient flow to parallel drains. It calculates the midpoint water table elevation between the drains. Bouwer and van Schilfgaarde (1963) modified the Hooghoudt equation for transient analysis. The US Bureau of Reclamation developed drainage equations for transient analysis of midpoint water table elevation. Kirkham (1958) developed a Laplace analytic solution for transient two-dimensional Sub Surface Drainage (SSD) to simulate water table height as a function of distance from the drain rather than just the midpoint water table elevation.

Lumped hydrosalinity models based on conservation of water and salt have been developed and applied to predict irrigation return flow and associated salt loads (Hornsby and Davidson, 1973; Walker, 1978; Tanji, 1977; Aragüés et al., 1985). Of these models, SALTMOD has been widely applied to predict soil and water salinities and water table depth in cropped land under different hydrological conditions and water management practices in Nile Delta of Egypt and Tungabhadra Irrigation Project in Karnataka state of India (Rao et al., 1992). Singh et al. (2002) used an extended version of SALTMOD, in a subsurface drained saline clay textured rice field in coastal Andhra Pradesh (India) and observed that water table depth and root zone salinity were predicted better than salinity of drainage water.

The lumped models and analytical solutions of subsurface drainage systems have been replaced by a variety of numerical models developed for simulating the saturated-unsaturated water flow, in particular DRAINMOD (Skaggs, 1978) and solute transport (Pickens et al., 1979; El-Din et al., 1987; Šimůnek and Suarez, 1994) in SSD systems. The DRAINMOD simulation model is based on a simplified water balance approach and simulates water flow in irrigated land with shallow water tables (Skaggs, 1978; Dar et al., 2020). Feddes et al. (1988) developed SWACROP by combining a drainage simulation model SWATRE (Belmans et al., 1983) with a crop production model. SWACROP numerically solves Richard's equation for unsaturated flow using a finite difference scheme.

An extended version, DRAINMOD-S, uses soil water fluxes predicted by DRAINMOD and a numerical solution of the convection-dispersion equation to simulate salt transport in subsurface drained soils (Kandil et al., 1992). DRAINMOD and other models have been successfully applied for simulating water management in irrigated subsurface drained lands in Eastern Delta of Egypt (Kandil et al., 1992), simulation of controlled drainage in Western Delta of Egypt (Wahba et al., 2002) and for irrigated areas of South-eastern Australia (Wahba and Christen, 2006). In above solute transport models, standard finite difference and finite element techniques were employed to transform the governing partial differential equations of water and solute transport into a finite set of approximate algebraic equations. The time derivative was also discretized by finite differences based on iterations to march through the intermediate time steps to develop solution at the desired time. As expected, data requirement and methodology for applying numerical models based on solution of partial differential equations is much more difficult and complex than lumped models.

Recently DRAINMOD, SWAP (Soil-Water-Atmosphere-Plant), and HYDRUS software have been widely used to evaluate performance of SSD in controlling water table. DRAINMOD and SWAP are 1-D models based on hydrological processes (Van Dam et al., 1997; Kroes et al., 2000, Kelleners et al., 2000), while HYDRUS is a window-based model that can simulate two- and three-dimensional water, heat and solute transport in variably saturated media. HYDRUS can handle flow domain delineated by irregular boundaries and nonuniform soils with an arbitrary degree of anisotropy (Šimůnek et al., 2006). HYDRUS 2D has been applied for simulating water flow into subsurface drains for a layered soil profile (Öztekin, 2002), in paddy fields for various drain spacing-depth combinations, soil textures, and crack conditions (Ebrahimian and Noory, 2014) and in comparative performance evaluation of pipe drains, pipe drains with a gravel trenches and mole drains for water table control (Filipović et al., 2014).

In India, a semidiscrete model, DRAINSAL (Kamra et al., 1991a) has been comprehensively applied to predict the transient movement and distribution of dissolved chemicals in a tile-drained soil-aquifer system (Fig. 3) in Haryana. DRAINSAL is a field scale 2-D finite element model that provides long-term predictions on desalinization of a tile-drained soil and changes in quality of groundwater and drain effluent. The model considers steady state water movement through the partially saturated soil and to the drains in the saturated zones, and includes the effect of convective transport, dispersion and linear adsorption. The numerical solution is exact in time and explicitly calculates the concentration field at any future time without needing to iterate through the intermediate times. In this chapter, features of DRAINSAL are summarized and an example of application of this field scale model to estimate regional salt load of subsurface drained fields in Haryana is presented.

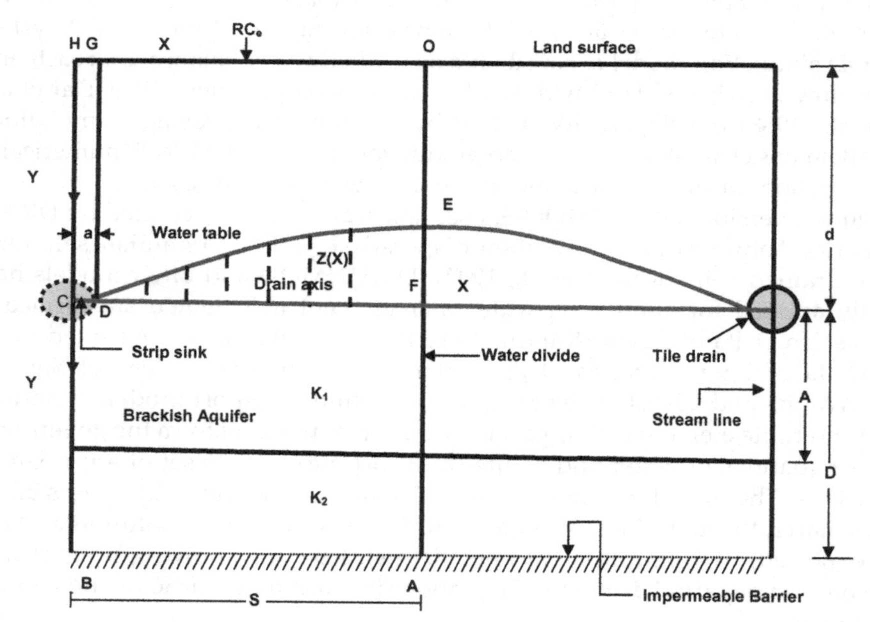

FIG. 3 Flow domain for solute transport in a tile-drained soil aquifer system.

The model inputs consist of information on finite element discretization of flow domain, boundary conditions and of drainage system (depth, spacing and radius of drains), aquifer (porosity, hydraulic conductivity, depth to impervious layer and groundwater salinity), soil (water retention and unsaturated hydraulic conductivity functions and initial salinity), inflow parameters (rainfall, quantity and quality of irrigation water, evapo-transpiration) and adsorption parameters (if to be considered). This data (except moisture distribution in soil) can be derived from field observations or historical records. Longitudinal and transverse dispersivities, decided from comparison of observed and predicted results during calibration and validation phases, are also part of input parameters. Representative values of selected model parameters, including steady annual water fluxes for Sampla SSD site in Haryana, are listed in Table 4.

The model was calibrated and validated against two independent yearly data sets of CSSRI's drainage experimental fields and 10 years predictions on salt distribution in soil, groundwater and drainage effluent were made. The model results for alternate drainage designs corroborated the field observations that subsurface drains of 75 m spacing and 1.5 to 1.8 m depth are quite effective in ameliorating the saline sandy loam soils of Haryana State. The model results indicated drainage effluent to be much saline during the initial years of reclamation in drains wider and deeper than these limits, especially if installed at depths $\geq$2.5 m. The observations remained similar even while irrigating with waters of 5 dS/m salinity. The salinity of soil, drainage water and groundwater were found to be significantly influenced by aquifer stratification, initial groundwater salinity, depth of impervious layer and adsorption.

TABLE 4 Values of selected soil hydraulic and drainage system parameters.

Parameter	Value(s)
Drain spacing, S	**25, 50, 75 m**
Drain depth, d	**1.8 m**
Depth of impervious layer below drain axis, D	1.2, 2.0, 5.0 m
Saturated hydraulic conductivity of aquifer, K_s	3.0 m/day
Soil water retention parameters (Van Genuchten, 1978)	
K	1.0 m/day
θ_s	0.4486
θ_r	0.1004
a	0.0088 1/cm
n	1.6715
m (=1-1/n)	0.4017
Soil bulk density, ε	1.5 g/cm^3
Distribution coefficient, K_d (cm^3/g)	0.0
Longitudinal dispersivity, $\propto_L$	0.8 m
Transverse dispersivity, $\propto_T$	0.08 m
Annual steady water flux for 25, 50, 75 m drain spacing	**1.0, 0.7, 0.4 mm/day**

A notable outcome of the application of DRAINSAL (Kamra et al., 1991b) has been the abandoning of critical water table concept in favor of the net downward water flux for deciding drain depth and consequent acceptance of shallower subsurface drains in arid and semiarid regions across world. The results on higher salt load in wider and deeper SSD systems under steady (Kamra et al., 1991b) and transient flow (Grismer, 1993) conditions also found resonance with results based on water flow pathways in subsurface drains (Fio and Deverel, 1991) especially under conditions where salinity increases with depth in soil profile.

DRAINSAL was applied to estimate the salt load of drainage effluent under alternate drainage designs from a 2000 ha highly saline area. Assuming that soil, climatic, irrigation and geohydrological conditions in whole area were similar to those of Sampla experiment farm in Haryana (Kamra et al., 1991), five designs of drain spacing (2S)-depth (d) combinations, i.e., $2S=48$ m, $d=1.0$ m; $2S=67$ m, $d=1.5$ m; $2S=75$ m, $d=1.8$ m; $2S=77$ m, $d=2.0$ m; $2S=85$ m, $d=2.5$ m were considered. Table 5 presents predicted annual drainage salt load of these combinations for two salinities, C_{in} ($=0.5$ dS/m and 5.0 dS/m), of irrigation water. Irrespective of irrigation water salinity, the annual amount of salts discharged during initial 2 years of reclamation from drains wider than 75 m and deeper than 1.8 m depth, were found to be significantly higher than closer and shallower drain combinations. In the fifth year, the salt load discharged from all design options was almost identical.

This example demonstrates that a field level salinity simulation model can be effectively utilized to obtain estimates of required parameters at regional scale. Large area can be divided into a number of representative blocks and the model results of individual blocks could be combined to obtain regional estimates of required parameters. These estimates may not truly represent actual regional results, but are better than no results and definitely useful for comparative qualitative evaluation of alternative management options. Comprehensive window based models like HYDRUS have totally replaced DOS based models like DRAINSAL due to better graphics and user interface for data input.

TABLE 5 Annual amount of salt load (*tons*) of drainage effluent under alternate SSD designs and using different quality of irrigation water (C_{in}).

Drainage design		$C_{in}=0.5$ dS/m			$C_{in}=5.0$ dS/m		
		Time (years)			Time (years)		
2S	d	1	2	5	1	2	5
48	1.0	4.6	4.1	2.7	4.6	4.3	3.1
67	1.5	4.6	4.4	2.9	4.6	4.4	3.3
75	1.8	4.6	4.3	2.9	4.6	4.4	3.2
77	2.0	4.9	4.6	2.8	5.0	4.7	3.3
85	2.5	5.8	5.0	2.9	5.9	5.1	3.0

6 Conclusions

The limitation of such holistic models is extensive data requirements. The potential numerical models, however, need to be locally calibrated and validated for reliable application of model outputs. Modeling soil-water salt-plant relationships is important for the use of scaling and extension of technologies and decision support system. The hydrosalinity models are categorized by the scale of application. Complex salinity- management models of irrigation system have advantages over simple mass balance-based models, their routine use, however, is restricted mainly due to nonavailability of their input data. Detailed, sophisticated models of a subsystem can sometimes be usefully utilized to compare management options for the main irrigation system. Physical budget analysis is never straightforward; dynamic mechanisms in response to hydrologic and economic activity could be at work to increase the salinity in an area. Concurrent flows and water quality records at desired locations and times are seldom available. Data intensive, interdisciplinary efforts need to be made to identify and quantify the salinity-producing mechanisms in an irrigated basin. Many diverse pieces of data must be integrated to understand the movement of water and salt through the system. Sufficient evidence should be obtained from water and salt budgets to conclude whether salinity will be reduced by implementing the proposed management strategies. Irrespective of the scale, each model should be seen as a mean, and not an end in itself, to get better insight into the flow mechanisms of a system and to get better estimates of required variables than otherwise possible. The modeler has a moral and professional responsibility to verify his basic data; quantitative model results and intuitive solutions must be viewed with caution.

Based on a comparative analysis of the different water-salt interactions, we conclude that the amount of salt introduced, discharged, and accumulated in the irrigation groundwater regimes decreases with a decrease in total water diversion, but the salt concentration of the drainage water will increase, which will have an impact on the salinity of the region's water body. Although a certain amount of water can be saved through various measures, the minimum ecological water demand required to maintain the existing water surface area

and salinity in the area, and the optimal water management scheme should be determined based on the analysis of the actual ecological water supplement conditions in the study region.

References

Akram, S., Kashkouli, H.A., Pazira, E., 2008. Sensitive variables controlling salinity and water table in a bio-drainage system. Irrig. Drain. Syst. 22 (3), 271–285.

Aragüés, R., Tanji, K.K., Quilez, D., Alberto, F., Faci, J., Machin, J., Arrue, J.L., 1985. Calibration and verification of an irrigation return flow hydrosalinity model. Irrig. Sci. 6 (2), 85–94.

Bekchanov, M., Sood, A., Pinto, A., Jeuland, M., 2017. Systematic review of water-economy modeling applications. J. Water Resour. Plan. Manag. 143 (8), 04017037.

Belmans, C., Wesseling, J.G., Feddes, R.A., 1983. Simulation model of the water balance for the cropped soil: SWATRE. J. Hydrol. 63, 271–286.

Bouwer, H., van Schilfgaarde, J., 1963. Simplified method of predicting fall of water table in drained land. Trans. ASAE 6 (4), 288–296.

Bresler, E., 1973. Anion exclusion and coupling effects in nonsteady transport through unsaturated soils: I. Theory. Soil Sci. Soc. Am. J. 37 (5), 663–669.

Chadwick, O.A., Chorover, J., 2001. The chemistry of pedogenic thresholds. Geoderma 100 (3–4), 321–353.

Dar, M.U.D., Singh, J.P., Ali, S.R., 2020. Watertable behaviour under subsurface drainage system in Thehri Muktsar District of Punjab. J. Soil Salinity Water Qual. 12 (2), 241–249.

Droogers, P., Seckler, D., Makin, I., 2001. Estimating the potential of rain-fed agriculture (Vol. 20). IWMI.

Dudley, L.M., Shani, U., 2003. Modeling plant response to drought and salt stress: reformulation of the root-sink term. Vadose Zone J. 2 (4), 751–758.

Ebrahimian, H., Noory, H., 2014. Modeling paddy field subsurface drainage using HYDRUS-2D. Paddy Water Environ. 13 (4), 477–485.

El-Din, M.N., King, I.P., Tanji, K.K., 1987. Salinity management model: I. Development. J. Irrig. Drain. Eng. 113 (4), 440–453.

Feddes, R.A., Kowalik, P.J., Zaradny, H., 1978. Simulation of Field Water Use and Drop Yield. Wageningen CAD, Wageningen, The Netherlands.

Feddes, R.A., Kabat, P., Van Bakel, P., Bronswijk, J.J.B., Halbertsma, J., 1988. Modelling soil water dynamics in the unsaturated zone—state of the art. J. Hydrol. 100 (1–3), 69–111.

Filipović, V., Mallmann, F.J.K., Coquet, Y., Šimůnek, J., 2014. Numerical simulation of water flow in tile and mole drainage systems. Agric. Water Manag. 146, 105–114.

Fio, J.L., Deverel, S.J., 1991. Groundwater flow and solute movement to drain laterals, Western San Joaquin Valley, California: 2. Quantitative hydrologic assessment. Water Resour. Res. 27 (9), 2247–2257.

Gates, T.K., Grismer, M.E., 1989. Irrigation and drainage strategies in salinity-affected regions. J. Irrig. Drain. Eng. 115 (2), 255–284.

Grazhdani, S., Jacquin, F., Sulçe, S., 1996. Effect of subsurface drainage on nutrient pollution of surface waters in south eastern Albania. Sci. Total Environ. 191 (1–2), 15–21.

Grismer, M.E., 1993. Subsurface drainage system design and drain water quality. J. Irrig. Drain. Eng. 119 (3), 537–543.

Hooghoudt, S.B., 1940. General consideration of the problem of field drainage by parallel drains, ditches, watercourses, and channels. In: Contribution to the Knowledge of Some Physical Parameters of the Soil, p. 7.

Hornsby, A.G., Davidson, J.M., 1973. Solution and adsorbed fluometuron concentration distribution in a water-saturated soil: experimental and predicted evaluation. Soil Sci. Soc. Am. J. 37 (6), 823–828.

Jain, A., Srinivasulu, S., 2004. Development of effective and efficient rainfall-runoff models using integration of deterministic, real-coded genetic algorithms and artificial neural network techniques. Water Resour. Res. 40 (4).

Kamra, S.K., Gupta, R.K., 1993. Hydro-salinity modeling in irrigated command areas. In: Tyagi, N.K., Kamra, S.K., Minhas, P.S., Singh), N.T. (Eds.), Sustainable Irrigation in Saline Environment. Central Soil Salinity Research Institute, Karnal, India, pp. 37–58.

Kamra, S.K., Singh, S.R., Rao, K.V.G.K., van Genuchten, 1991a. A semi- discrete model for water and solute movement in tile- drained soils: I. Governing equations and solution. Water Resour. Res. 27 (9), 2439–2447.

Kamra, S.K., Singh, S.R., Rao, K.V.G.K., van Genuchten, M.Th., 1991b. A semi-discrete model for water and solute movement in tile-drained soils: II Field validation and applications. Water Resour. Res. 27 (9), 2448–2456.

Kamra, S.K., Singh, S.R., Rao, K.V.G.K., 1994. Effect of depth of impervious layer and adsorption on solute transport in tile-drained irrigated lands. J. Hydrol. 155 (1–2), 251–264.

Kandil, H., Miller, C.T., Skaggs, R.W., 1992. Modeling long-term solute transport in drained unsaturated zones. Water Resour. Res. 28 (10), 2799–2809.

Katsifarakis, K.L. (Ed.), 2000. Groundwater pollution control. Wit Press.

Kelleners, T.J., Kamra, S.K., Jhorar, R.K., 2000. Prediction of long term drainage water salinity of pipe drains. J. Hydrol. 234 (3–4), 249–263.

Khan, S., Mushtaq, S., Hanjra, M.A., Schaeffer, J., 2008. Estimating potential costs and gains from an aquifer storage and recovery program in Australia. Agric. Water Manag. 95 (4), 477–488.

Kiran, P, 2010. *Simulation of salt load through sub-surface drainage system using DRAINMOD-S model*. In: Tech. Dissertation. Punjab Agricultural University, Ludhiana, India.

Kirkham, D., 1958. Seepage of steady rainfall through soil into drains. Eos, Trans. Am. Geophys. Union 39 (5), 892–908.

Makin, I.W., Goldsmith, H., 1988. Selection of drainage and its phased implementation for salinity control. Irrig. Drain. Syst. 2 (1), 109–121.

Mirlas, V., 2012. Assessing soil salinity hazard in cultivated areas using MODFLOW model and GIS tools: A case study from the Jezre'el Valley, Israel. Agric. Water Manag. 109, 144–154.

Morway, E.D., Gates, T.K., Niswonger, R.G., 2013. Appraising options to reduce shallow groundwater tables and enhance flow conditions over regional scales in an irrigated alluvial aquifer system. J. Hydrol. 495, 216–237.

Nimah, M.N., Hanks, R.J., 1973. Model for estimating soil water, plant, and atmospheric interrelations: I. Description and sensitivity. Soil Sci. Soc. Am. J. 37 (4), 522–527.

Oosterbaan, R.J., de Lima, P., 1989. SALTMOD manual. International Institute for Land Reclamation and Improvement, Wageningen, The Netherlands.

Öztekin, T., 2002. Simulating water flow to a subsurface drain in a layered soil. Turk. J. Agric. For. 26 (4), 179–185.

Pickens, J.F., Gillham, R.W., Cameron, D.R., 1979. Finite-element analysis of the transport of water and solutes in title-drained soils. J. Hydrol. 40 (3–4), 243–264.

Rao, K.V.G.K., Ramesh, G., Chauhan, H.S., and Oosterbaan, R.J. 1992. Salt and water balance studies to evaluate remedial measures for waterlogged saline irrigated soils. Proc. 5th International Drainage Workshop, Lahore, Pakistan, Vol. II: 267- 277.

Reca, J., Trillo, C., Sánchez, J.A., Martínez, J., Valera, D., 2018. Optimization model for on-farm irrigation management of Mediterranean greenhouse crops using desalinated and saline water from different sources. Agric. Syst. 166, 173–183.

Šimůnek, J., Suarez, D.L., 1994. Two-dimensional transport model for variably saturated porous media with major ion chemistry. Water Resour. Res. 30 (4), 1115–1133.

Šimůnek, J., Van Genuchten, M.T., 1996. Estimating unsaturated soil hydraulic properties from tension disc infiltrometer data by numerical inversion. Water Resour. Res. 32 (9), 2683–2696.

Šimůnek, J., van Genuchten, Wendroth, O., 1998. Parameter estimation analysis of the evaporation method for determining soil hydraulic properties. Soil Sci. Soc. Am. J. 62 (4), 894–905.

Šimůnek, J.I.R.K.A., Van Genuchten, M.T., Sejna, M., 2006. The HYDRUS software package for simulating the two- and three-dimensional movement of water, heat, and multiple solutes in variably-saturated media. Tech-man 1.

Singh, A., 2010. Decision support for on-farm water management and long-term agricultural sustainability in a semi-arid region of India. J. Hydrol. 391 (1–2), 63–76.

Singh, A., 2012a. Development and application of a water-table model for the assessment of waterlogging in irrigated semi-arid regions. Water Resour. Manag. 26 (15), 4435–4448.

Singh, A., 2012b. Validation of SaltMod for a semi-arid part of Northwest India and some options for control of waterlogging. Agric. Water Manag. 115, 194–202.

Singh, A., 2013. Groundwater modelling for the assessment of water management alternatives. J. Hydrol. 481, 220–229.

Singh, R., Ramana Rao, K.V., 2014. Agricultural drainage technologies for temporary waterlogged *Vertisols*. In: Gupta, S.K. (Ed.), Agricultural Land Drainage in India. Agro-tech Publishing Academy, Udaipur, India, pp. 207–225.

Singh, M., Bhattacharya, A.K., Singh, A.K., Singh, A., 2002. Application of SALTMOD in coastal clay soil in India. Irrig. Drain. Syst. 16 (3), 213–231.

Singh, R., Jhorar, R.K., van Dam, J.C., Feddes, R.A., 2006. Distributed ecohydrological modelling to evaluate irrigation system performance in Sirsa district, India II: impact of viable water management scenarios. J. Hydrol. 329 (3–4), 714–723.

Singh, A., Krause, P., Panda, S.N., Flugel, W.A., 2010. Rising water table: A threat to sustainable agriculture in an irrigated semi-arid region of Haryana, India. Agric. Water Manag. 97 (10), 1443–1451.

Skaggs, R.W., 1978. A Water Management Model for Shallow Water Table Soils. Water Resources Research Institute of the University of North Carolina.

Tanji, K.K., 1977. A conceptual hydrosalinity model for predicting salt load in irrigation return flow, Managing saline water for irrigation. Texus Tech University, Lubbock, Texas.

Tanji, K.K., 2002. Salinity in the soil environment. In: *Salinity: Environment-plants-molecules*. Springer, Dordrecht, pp. 21–51.

Tanji, K.K., Fried, M., Van De Pol, R.M., 1977. A steady-state conceptual nitrogen model for estimating nitrogen emissions from cropped lands, (Vol. 6, No. 2, American Society of Agronomy, Crop Science Society of America, and Soil Science Society of America, pp. 155–159.

Van Genuchten, M. Th., 1978. Calculating the unsaturated hydraulic conductivity with a new closed form analytic model. In: Research Rep. 78- WR-08. Water Resource Program, Dept. of Civil Engg., Princeton University, Princeton, N.J., USA.

Van Dam, J.C., Huygen, J., Wesseling, J.G., Feddes, R.A., Kabat, P., Van Walsum, P.E.V., Van Diepen, C.A., 1997. Theory of SWAP version 2.0. Technical Document, 45, p. 167.

Wagenet, R. J., & Hutson, J. L. (1987). LEACHM: Leaching estimation and chemistry model. *A process-based model of water and solute movement, transformations, plant uptake and chemical reactions in the unsaturated zone, Continuum*, 2.

Wahba, M.A.S., Christen, E.W., 2006. Modeling subsurface drainage for salt load management in South-Eastern Australia. Irrig. Drain. Syst. 20 (2–3), 267–282.

Wahba, M.A.S., El-Ganainy, M., Abdel-Dayem, M.S., Kandil, H., Gobran, A.T.E.F., 2002. Evaluation of DRAINMOD-S for simulating water table management under semi-arid conditions. Irrig. Drain. 51 (3), 213–226.

Walker, R.G., 1978. Deep-water sandstone facies and ancient submarine fans: models for exploration for stratigraphic traps. AAPG Bull. 62 (6), 932–966.

Wood, E.F., 1976. An analysis of the effects of parameter uncertainty in deterministic hydrologic models. Water Resour. Res. 12 (5), 925–932.

Yorobe, Ali, J., Pede, V.O., Rejesus, R.M., Velarde, O.P., Wang, H., 2016. Yield and income effects of rice varieties with tolerance of multiple abiotic stresses: the case of green super rice (GSR) and flooding in the Philippines. Agric. Econ. 47 (3), 261–271.

Zekri, M., El Kafhali, S., Aboutabit, N., Saadi, Y., 2017. DDoS attack detection using machine learning techniques in cloud computing environments. In: 2017 3rd International Conference of Cloud Computing Technologies and Applications (CloudTech). IEEE, pp. 1–7.

Zhao, Q., Armfield, S., Tanimoto, K., 2004. Numerical simulation of breaking waves by a multi-scale turbulence model. Coast. Eng. 51 (1), 53–80.

SECTION VII

Optimization techniques and analytical formulations in water resource

CHAPTER

24

Multiobjective optimization techniques for integrated urban water management: A case study of Varanasi city

Satya Prakash Maurya and Ramesh Singh

Department of Civil Engineering, Indian Institute of Technology (BHU), Varanasi, India

OUTLINE

1 Introduction 433
1.1 Existing frameworks for total water cycle management 434
1.2 Water sensitive urban design planning support system (WSUD-PSS) 436
2 Material and methods 436
2.1 Integrated assessment modeling and decision-making context 436
2.2 Models supporting urban water development planning (UWDP) 436
2.3 Mathematical optimization approaches in water resource management 438
2.4 Water resources optimization 439
3 Case study of Varanasi City 440
4 Conclusion 444
References 445

We live in the best of all possible worlds **G.W. Leibniz**

1 Introduction

Water is fundamental to our quality of life, to economic growth and to the environment across the world. Industrial and commercial activities along with increasing urbanization

Current Directions in Water Scarcity Research, Volume 7
https://doi.org/10.1016/B978-0-323-91910-4.00024-8

created serious water problem in different region. In year 2000, the Millennium Development Goals addressed concerns of water availability and suggested integration of consumption practices and availability of water for efficient management under Agenda 21. However, sustainability of systems remains under question, and still, most of the urban centers are facing water stress to an extent of modest to serious. Booming economy and growing population, further, increased the pressure on water resources. These pressures are compounded by the impact of climate variability and accelerating climate change.

Scientific community realized the complex nature of water problem that covers multiple dimensions such as water availability, consumption in various uses, quality and human health, surface water condition and guidelines for humans consumption (direct or indirect). The complexity of sustainable water management system draws the attention of academician and various integrated approaches like water sensitive urban design (WSUD) (Wong, 2006), sustainable urban drainage system (SUDS) (Woods-Ballard et al. 2007), best management practices (BMP) US EPA, 2011, etc. are developed to resolve the issues of water problems along with suppressing the impact of hydrological domains of surrounding environment.

Further, WSUD is associated with the integration of long-term sustainability, equity, water security, public health, flood protection, and economic vitality (Dam et al. 2012; Wong et al., 2013; Martin and Monroy, 2019). Thus, it may be seen that integrated urban water management (IUWM) gained enough momentum, but scope and context of various frameworks are limited to region's urban water issues. Water problems are identified as nonhomogeneous and inconsistent in spatiotemporal domain. Therefore, the best practices for a sustainability of urban water system require an optimal multiscale holistic planning process through a decision-support system (DSS). DSS addressing sustainable urban water development (SUWDP) must be a part of (i) urban water cycle, (ii) urban form (3) water governance, as dimensions of spatial planning time, space and human decision-making. In this chapter, an attempt is made to illustrate material and methods for multiobjective optimization technique addressing SUWDP.

1.1 Existing frameworks for total water cycle management

1.1.1 Three steps multibenefit framework

The basic philosophy of TWCM considers that all elements of water cycle are independent (Weber et al., 2018). TWCM approach has been applied to decrease water demand, reduce stormwater runoff, and improve pollutant wash-off from urban catchment and has been incorporated into numerous water planning and management practices. It is a systematic and comprehensive approach-based framework presented for multibenefit which empowers water management decisions. It is a three-step process to improve trade-offs among the considered benefits with robust solutions. The detailed framework is shown in Fig. 1.

In the initial stage defining the problem, scope, and costs and benefits identification have been done whereas the characterization of costs and benefits done at step 2. The final step of the framework deals to incorporate the costs and benefits into decision making and policy preparation. Here, to assess each benefit among different potential benefits is a real challenge considering context relevance and availability of adequate quality data. This multibenefit framework is useful to empower the decision makers of different corporations, organizations,

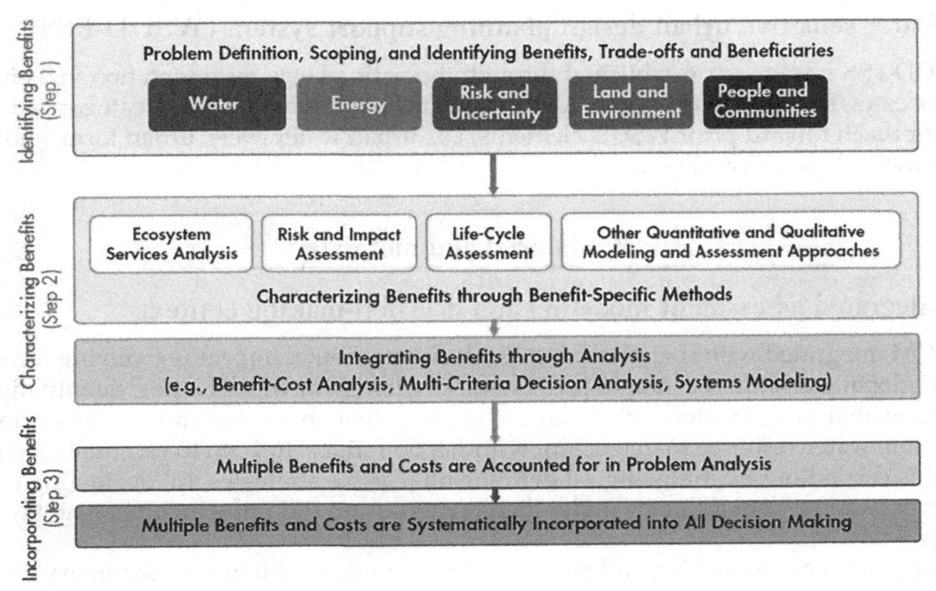

FIG. 1 A three steps multibenefit framework for urban water management and decision making (Diringer et al., 2019).

and government agencies to consider the broader perspectives of costs and benefits during decision making of water management.

1.1.2 Multipattern approach

A theoretical framework "Multi-Pattern Approach" has been applied from transitional theory for analytical modeling of an urban water management system (de Haan et al., 2011). In a multidisciplined problem like urban water where different disciples, viz., environment, technology, ecology, and sociology are handled, the adaptive nature of the system must describe the behavior of systems, especially inter-relating the transitions from one stable state to another. In such a multidisciplined problem, a complex system theory with principles can be applied over a different disciplines relating urban development (Scheffer, 2009).

1.1.3 Agent based modeling (ABM)

An agent-based model (ABM) simulates the behavior of complex social systems and has been identified in various disciplines accurately imitating the behavior of complex societal systems. Generally, ABMs in urban water management are used for simulating demand-supply patterns but a few can model more substantial parts too. In some of agent-based models households are considered as its agents to simulate the spatial uptake of rain gardens and green roofs (Montalto et al., 2013); however, a model simulating WSUD under different scenarios considers households as well as local government and developers as their agents (Lu et al., 2013).

1.2 Water sensitive urban design planning support system (WSUD-PSS)

WSUD-PSS framework established through thematic planes representation viz. biophysical processes, spatial strategies and adaptive strategies which are further categorized PSS in their approach toward prior WSUD elements, i.e., urban water cycle, urban form, and water governance, respectively (Fig. 2).

2 Material and methods

2.1 Integrated assessment modeling and decision-making context

TWCM integrated with socio-economic and environmental impact is receiving increasing interest globally as socio-economic uses of water, ecology of water bodies, quantitative and qualitative analyses of water cycle in an urban setup that envelopes various dimensions for sustainable water resource management within a boundary. In DSS, to formulate a strategy TWCM is carried out by following: (i) defining all feasible strategies, (ii) evaluating the best strategy which can achieve goals under given criteria, and (iii) implementation ability of the strategy.

To support decision making and policy analysis and deal with inter-disciplinary nature of environmental science, integrated assessment modeling (IAM) is a good tool. However, IAM can be very challenging in terms of validation, and therefore important to consider data as key components in the modeling activity (Voinov and Shugart, 2013).

Typically, no optimal solution exists to solve the decision problem where several conflicting points of view must be taken into considerations. However, multicriteria decision analysis (MCDA) is an effective and preferred tool to identify the most preferred solution out of a set of "efficient" solutions, rather than any feasible set of solutions.

Multicriteria assessment techniques mainly based on (i) *Multiattribute Utility Theory (MAUT)* methods and, (ii) *Outranking methods*, are used in MCDA, MCA with subjective logic (SL), multiobjective optimization (MOO), etc. However, *a real concern is availability and quality of data*.

Usually, a problem facing under multicriteria analysis along with multiobjective evaluations, may have different objectives with no relation which can ensure the fairness of every solution. MOO process presents many optimal solutions to decision-makers, rather than producing a single score for every decision set each being a different trade-off among the objectives. However, evaluating the choices either manually or exhaustively, is difficult to execute due to the sheer number of possible solutions. Thus, an algorithm may be more useful which can perform search through the many possible decision sets and find the most optimal ones.

Presently, intervention of computer applications such as machine learning (ML), artificial intelligence (AI), and MCDM techniques within spatiotemporal domain in urban water management has been considered as a subclass of tools to implement an appropriate DSS.

2.2 Models supporting urban water development planning (UWDP)

Recently, water problems framed under international policy, including social, economic, environmental and political has been reinforced the water security with safety and qualifying

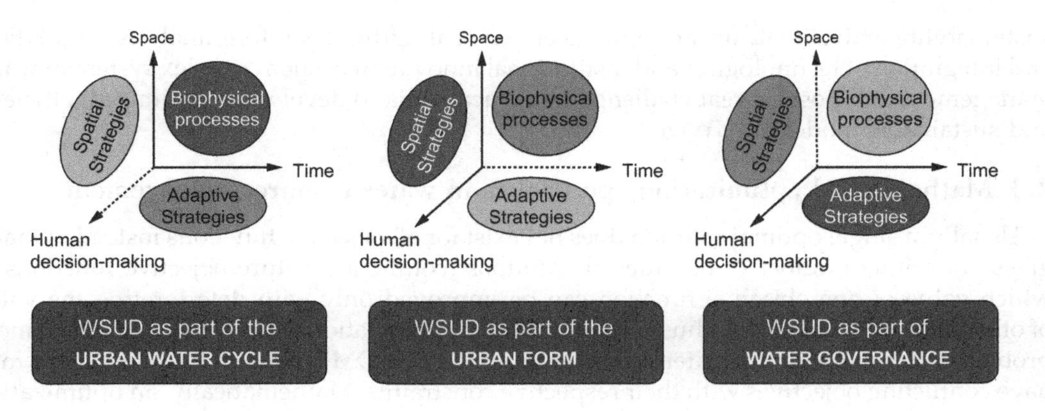

<u>Water Balance Model</u> UWOT (Makropoulos et al., 2008) Aquacycle (Mitchell et al., 2001) UVQ (Mitchell et al., 2003) Urban Developer (eWater, 2011b) City Water Balance (Last, 2011) <u>Hydrological and Hydraulic Models</u> MUSIC (eWater, 2011a) PURRS (Coombes, 2002) SWMM (Rossman, 2010) Stormwater BMP interactive model (WERF)	<u>Planning Simulators</u> UrbanBEATS (Bach, 2014) Sustain-EPA (Lee et al., 2012) Adaptation Support Tool (Voskamp & Ven, 2015) <u>Technology Selection</u> Scholz (2006) SUDASLOC (Ells & Viavattene, 2008) SuDS Retrofit (Stovin and Swarn, 2007) Climate app (Bosch Slabbers et al., 2016) Green-blue grids (Atelier GrounBlauw, 2016) <u>Technology evaluation</u> BeST (Digman et al., 2015) E2STORMED (Morales-Torres et al., 2016) Green Value Calculator (CNT, 2009) NYC GI Cost-Benefit calculator (NYC-EP, 2014) GI valuation toolkit (NEN, 2010) Chow et al. (2015)	<u>Complex System Model</u> Societal Transition Workbench (De Haan et al., 2011) Montalto et al. (2013) Lu et al. (2013) <u>Transition Framework</u> Water Sensitive Cities Continuum and Index (Bech et al., 2016; Brown et al., 2009) Diagnostic Procedure for Transformative Change (Ferguson et al., 2013) <u>Scenario Analysis</u> DAnCE4Water (Rauch et al., 2012) RDM Scenario Approach (Groves & Lempert, 2006) VIBe (Sitzenfrei et al., 2013) ReVISIONS (Ward et al., 2012)

Low Level — High Level

Implementation — Design — Planning — Conceptualization — Strategy — Vision

FIG. 2 Integrated urban water management using models with WSUD (Kuller et al., 2017).

sustainability with climate uncertainties (Zeitoun et al., 2016). Therefore, analysis, simulation, and integration of technological and institutional innovation of such complex system of water management still pose a great challenge to the academia to develop an optimized, efficient, and sustainable model for UWDP.

2.3 Mathematical optimization approaches in water resource management

Usually, a single optimal solution does not exist for all objective functions instead of that it gives an infinite number of noninferior solutions (conflicting nature objective functions in which value of one objective function can be improved only with deteriorating the value of other objective functions). Thus, a multiobjective optimization water resource management problem is ultimate a multicriteria decision-making (MCDM) problem in which there may have conflicting objectives with their respective constraints. Mathematically, an optimization may have elements such as Objective Function: Decision Variables, Constraints which are very well addressed in various optimization programming and so arranged that it established relationship among all objectives leads to an optimal solution which is crucial for making the decision.

Some commonly used optimization approaches like Linear Programming, No Preference Method (decision-maker has no choice to accept or reject the solution), global criteria method (GCM) (minimize the distance among the objective feasible regions) and Posteriori Method (facilitating the noninferior solutions and decision-maker's choice is the most preferred alternative) to optimize the problem considering data as a vector points. *The Posteriori Method*, the most popular as in pragmatic solution of a complex problem, each objective function cannot be considered as equally important (conflicting objective functions may have different priorities based on expert's opinion) whereas other methods evaluates equal weightage biased solutions.

The weighing scheme in the Posteriori Method is based on importance of the objectives; different weights may be assigned to each objective function according to priority. Finally, multiple objective functions can be translated into a single objective to minimize it and convert the vector optimization problem into weighing problem for finding the optimal solution. The formulation of the MOO after assigning weights to each objective function can be written as (Emmerich, Deutz, 2018):

$$\begin{aligned}&\text{Min}\{O_1(d), O_2(d), O_3(d), \ldots, O_n(d)\}\\&\text{Subject to}\\&d\varepsilon S; \lambda_i \geq 0; \sum\nolimits_{j=1}^{n} \lambda_i = 1; d \geq 0\end{aligned} \tag{1}$$

where d refers to a vector of decision variables that belong to the feasible region set S which is a subset of decision variable space R^n.

Here, $O_i(d)$ is the individual objective function expression; λ_i represents the weight vector. Introducing weights to the objective function the formulation becomes:

$$\begin{aligned}&\text{Min} \sum\nolimits_{j=1}^{n} \lambda_i O_i(d)\\&\text{Subject to}\\&d\varepsilon S; \lambda_i \geq 0; \sum\nolimits_{j=1}^{n} \lambda_i = 1; d \geq 0\end{aligned} \tag{2}$$

2.4 Water resources optimization

Initially, two sets of indicators (water quality and water resources) are identified to manage water resources, in which water quality concerns runoff of untapped wastewater and other pollutants which pollute the existing surface water and water resources considered as quantitative measure of water exploitation intensity at spatial and temporal scale within the boundary. The major objective was to establish an implementable approach which may evaluate a long-term sustainability of water resources among three aspects, i.e., social, economic, and technical.

A number of attempts have been made to establish such evaluator which covers every aspect of water resources and fulfills the objectives to achieve desired sustainability level. Urban water development planning index (WDPI) was defined to assess the sustainability of an urban area and ensuring the future sustainable water development (Maurya et al., 2020). The index is based on PSR framework consisting of seven indicators viz. Water Security, Investment Scope, Water Quality, Water Quantity, Infrastructure, Reuse, Recycle & Recharge, and Governance; under which 22 measures have been defined. Originally this framework has been formulated taking identical weight for each objective (test case of Varanasi City), but to make it a dynamic and accurate predictive model, there is need to introduce a technique which can inter-relates objectives more tightly coupled to ensure the impacts of pressure-state-response (PSR) on the final objective function, i.e., WDPI.

Sustainability Index can be calculated by taking product of weight value and indicator value (through its subindicators values and its corresponding weight value) for each indicator.

$$SDI = \sum_{j=1}^{m} W_j \times I_j$$

$$I_j = \sum\nolimits_{i=1}^{n} SI_i \times w_i \tag{3}$$

$$SDI = \sum\nolimits_{j=1}^{m} W_j \times \left(\sum\nolimits_{i=1}^{n} SI_i \times w_i\right) \tag{4}$$

where W = weight of indicator; I = indicator value; SI = subindicator value; w = weight of subindicator.

Water for development planning may be calculated as WDPI (Maurya et al., 2020).

$$\text{WDPI} = W_P \times \text{PR} + W_S \times \text{ST} + W_R \times \text{RE}$$

where PR = pressure; W_P = weight factor of pressure; ST = state; W_S = weight factor of state; RE = response; W_R = weight factor of response.

$$\text{Subjected to } W_P + W_S + W_R = 1$$

WDPI evaluated on a scale of 10 in which index value indicates <3: poor, >3 but <5: critical, >5 but <8: Good and >8: Excellent water condition.

The above equation may be translated to a posteriori formulation as below:

Optimum (O_P W_P, O_S W_S, O_R W_R).

Subject to $W_P + W_S + W_R = 1$; and W_P, W_S, $W_R \geq 0$.

Since pressure is a negative factor in term of sustainability, so its measures have been taken such as with increase in its value will improve the sustainability. Other two factors state and response are already positive terms as it improves WDPI enhances. Objective function with sustainability goals (i) maximize surface runoff storage, (ii) maximize reuse potential, (iii) improve economic efficiency, (iv) adding resource recovery, (v) minimize extra water consumption, and (vi) reduction in fresh water demand through rainwater harvesting.

3 Case study of Varanasi City

Varanasi is the fourth largest city of Uttar Pradesh, the most populous state in India. The location map along with demographic data given below in Fig. 3.

The water condition of Varanasi is observed through various surveys (primary data via field survey and secondary data from government documents). With the context to above, PSR is identified as three different objective functions and to get a single WDPI, 22 measures (listed in Table 1) are considered. These measures are associated with either single or more than one objective, which make the evaluation process complicated. The weight values of each measure have been decided in consultation with water experts and field survey. Individual objective functions are calculated with weight factors of pressure, state and response (Table 1) using Eq. (3).

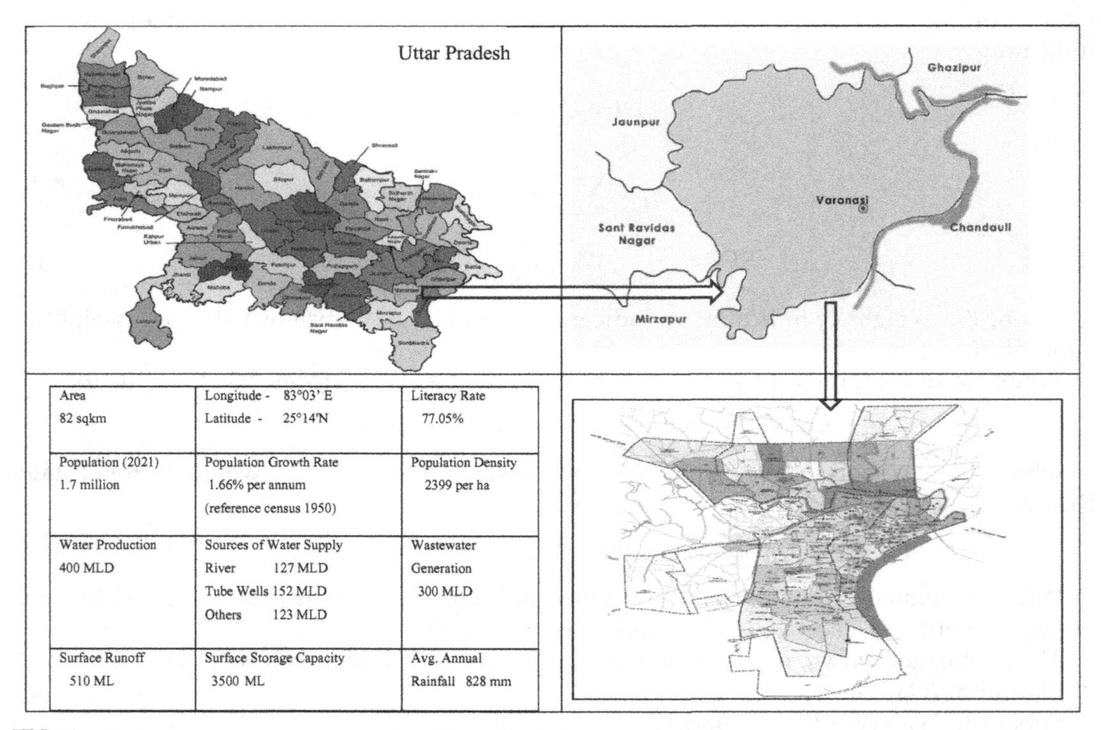

Area 82 sqkm	Longitude - 83°03' E Latitude - 25°14'N	Literacy Rate 77.05%
Population (2021) 1.7 million	Population Growth Rate 1.66% per annum (reference census 1950)	Population Density 2399 per ha
Water Production 400 MLD	Sources of Water Supply River 127 MLD Tube Wells 152 MLD Others 123 MLD	Wastewater Generation 300 MLD
Surface Runoff 510 ML	Surface Storage Capacity 3500 ML	Avg. Annual Rainfall 828 mm

FIG. 3 Study area on map representation (Varanasi city).

TABLE 1 Objectives and measures along with weight values in WDPI evaluation scheme.

Objective (PR/ST/RE)	Indicator (I)	Subindicator (SI)	Weight within each category (w)	Weight factor	Weight within objective (W)	Weight of objective ($W_P/W_S/W_R$)
Pressure	Water security	Urbanization rate	0.20	0.16	0.80	0.23
		Water withdrawal	0.40	0.32		
		Fresh water scarcity	0.30	0.24		
		Pollution risk vulnerability	0.10	0.08		
	Investment scope	Economic pressure	1	0.20	0.20	
State	Water quality	Surface water quality	0.50	0.125	0.25	0.37
		Ground water quality	0.50	0.125		
	Water quantity	Adequacy	0.40	0.15	0.375	
		Reliability	0.40	0.15		
		Consumption	0.20	0.075		
	Infrastructure	Water supply coverage area	0.35	0.131	0.375	
		Wastewater collection coverage area	0.35	0.131		
		Separation of wastewater and storm water	0.30	0.115		
Response	Reuse, recycle and recharge	% Availability of treated wastewater for reuse	0.10	0.067	0.67	0.40
		Surface runoff storing capacity	0.20	0.134		
		Reuse potential of city, region	0.20	0.134		
		Economic efficiency	0.05	0.034		
		Resource recovery	0.15	0.100		
		Groundwater Recharge potential	0.20	0.134		
	Governance	Management and action plan	0.40	0.132	0.33	
		Public participation	0.40	0.132		
		People's acceptability	0.20	0.066		

In a test case study, equal weightage is assigned to each measure and WDPI is calculated by applying Eq. (4) for sustainability goal by (i) maximize surface runoff storage, (ii) maximize reuse potential, (iii) improve economic efficiency, (iv) adding resource recovery (v) minimize extra water consumption, and (vi) reduction in fresh water demand through rainwater harvesting. The outcome of the study is shown in Table 2. It indicates that Varanasi is moving toward water stress zone as WDPI is <5. However, applying the above course of action which will improve the overall sustainability of the city, a comparison for the year 2015 and 2030 has been presented in Fig. 4.

For the further sensitivity analysis of weight, an attempt is made to calculate WDPI by applying posterior method using Eqs. (1), (2). The posterior method shows an improvement over identical method and a value obtained for WDPI is closer to 5 (Table 3).

Wastewater discharge observed is 380 MLD whereas the reported value is 300 MLD which means there is some uncontrolled water supply exists. Wastewater treatment capacity is 100

TABLE 2 Calculation of WDPI for Varanasi city for the base year 2015.

Subindicator	*N_SI*	*w*	*I*	*W*	OF	$W_P/W_S/W_R$	WDPI
Urbanization rate	3.5	0.20	2.77	0.80	3.8	0.23	4.09
Water withdrawal	3.3	0.40					
Fresh water scarcity	1.5	0.30					
Pollution risk vulnerability	3	0.10					
Economic pressure	8	1	8.00	0.20			
Surface water quality	5.5	0.50	6.50	0.25	6.45	0.37	
Ground water quality	7.5	0.50					
Adequacy	10	0.40	8.66	0.375			
Reliability	10	0.40					
Extra consumption	3.3	0.20					
Water supply coverage area	6.5	0.35	4.22	0.375			
Wastewater collection	3	0.35					
% Availability of treated wastewater for reuse	9.4	0.10					
Surface runoff storing capacity	4.6	0.20					
Reuse potential of city	0	0.20	2.89	0.67	2.08	0.40	
Economic efficiency	4	0.05					
Resource recovery	1.5	0.15					
Groundwater recharge potential	1	0.20					
Management and action plan	1	0.40	0.43	0.33			
Public participation	0.5	0.40					
People's acceptability	0.5	0.20					

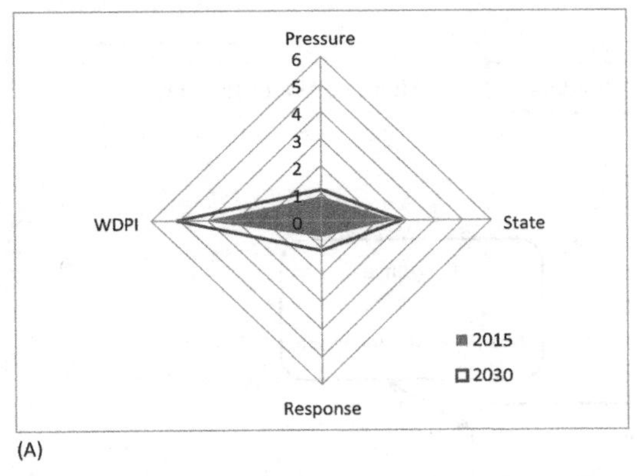

(A)

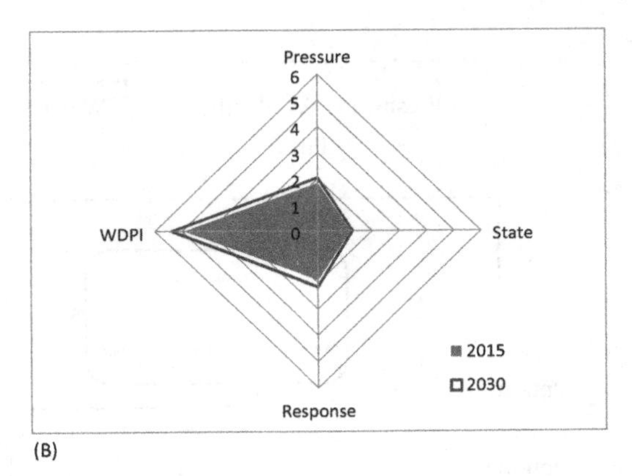

(B)

FIG. 4 Scenario comparison for the year 2015 and 2030 after applying improvements on WDP indicator to achieve sustainability goals.

TABLE 3 Analysis showing objective function values in identical vs. posterior weight scheme (for the base year 2015).

	Objective function		Objective weight		Weighted function value		WDPI	
Objective	**Identical**	**Posterior**	**Identical**	**Posterior**	**Identical**	**Posterior**	**Identical**	**Posterior**
Pressure	3.8	5.2	0.23	0.35	0.874	1.82	4.09	4.91
State	7.0	6.0	0.37	0.20	2.59	1.20		
Response	1.5	4.2	0.40	0.45	0.6	1.89		

MLD which is being expanded to 400 MLD. Total annual runoff through storm water has been estimated as 51 MCM. These resources can be utilized to supply for other than drinking and household usage. The consideration of reclaimed and reuse of alternate water resources (storm water and treated wastewater) can change the evaluations of sustainability. However, to show its effectiveness within the model, its weight must be exercised with the help of different field experts. However, incorporation of any improvement must be done carefully according to its complexity. In order to achieve such sensitive analysis, the agent-based modeling may be adopted.

An agent modeling based conceptual framework for urban water sustainability evaluation has been attempted by Maurya and Singh (2021). The result shows that there is further scope for ABM sensitivity analysis and evaluation of various management scenarios which deals with complex formulas. A framework for multiobjective optimization of IUWM based on PSR model has been proposed in which posterior method can be utilized to obtain a better result (Fig. 5).

Conceptualization of ABM with a PSR framework for IUWM may change the paradigms of sustainable urban water simulation model that may help to optimize the evaluation quality and sensitivity analysis, by reducing conflict levels. Such models can be a powerful tool that helps to set up rules based scenario evaluations, especially focusing on water demands, fresh water availability and environmental concerns.

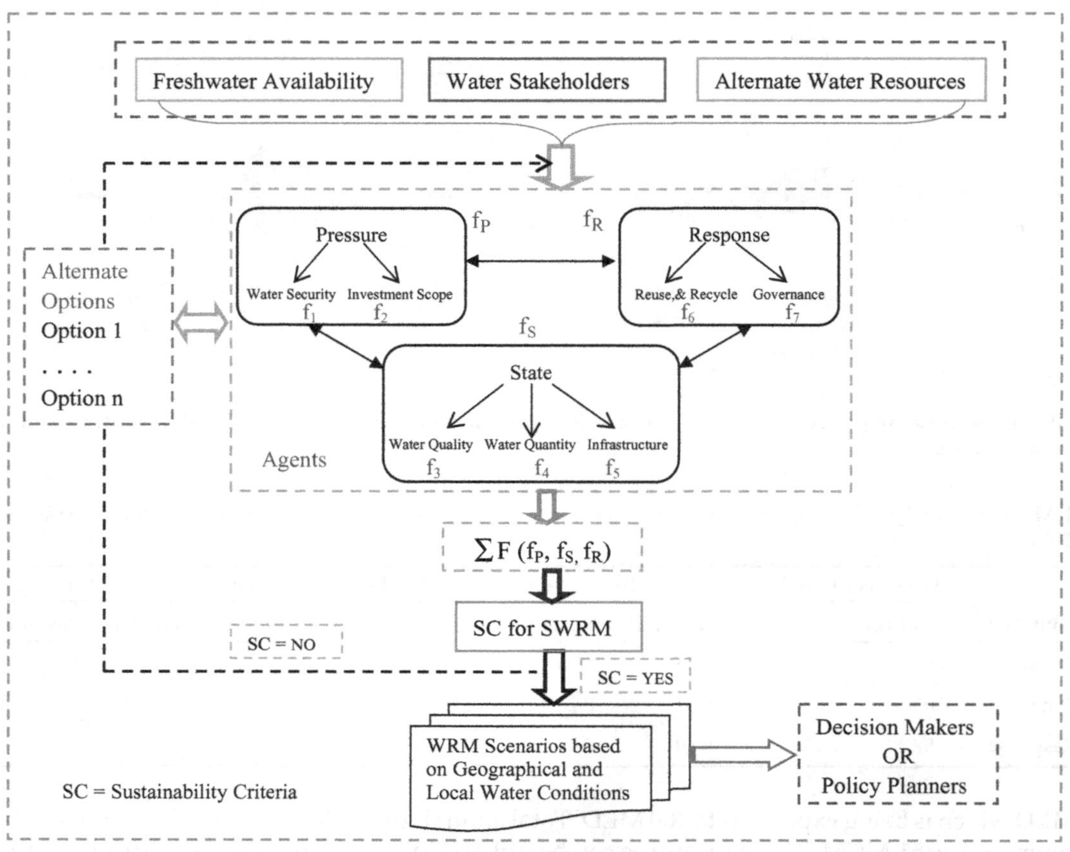

FIG. 5 Conceptual framework of agent based sustainable water management (Maurya and Singh, 2021).

4 Conclusion

Water problems are nonhomogeneous and inconsistent in spatiotemporal domain. Therefore, the sustainability of urban water system requires an optimal multiscale holistic planning process through a DSS focusing on water demands, fresh water availability and environmental concerns. The upcoming computer applications such as ML, AI, and MCDM techniques within spatiotemporal domain in urban water management has been considered as a subclass of tools to implement an appropriate DSS. The adaptive nature of the system must describe the behavior of systems, especially inter-relating the transitions from one stable state to another. In this chapter, an attempt has been made to review the various frame works and modeling approaches already practiced in water management. Further, an agent base modeling approach has been introduced along with the concept of single index measure, i.e., WDPI for sustainability assessment with nonuniform weighting distribution scheme and posterior method. This method applied and tested for Varanasi city to evaluate a case study. It is evident from the test case study of Varanasi city that multiobjective optimization may be much effective tool if its sensitivity could be tested and tuned with different weighting scheme for measurement of environmental adaptive decision making. Thus, an ABM approach for multiobjective optimization in urban water resource management and future sustainability evaluation can play a vital role and make the existing model robust for accurate decision making for future water developments.

References

Dam, T., Fryd, O., Backhaus, A., Jensen, M.B., 2012. Stormwater agents in a political perspective. In: Barton, A.C.T. (Ed.), WSUD 2012: Water Sensitive Urban Design, Building the Water Sensitive Community, 7th International Conference on Water Sensitive Urban Design. Engineers Australia, pp. 694–702.

De Haan, J., Ferguson, B., Brown, R., Deletic, A., 2011. A workbench for societal transitions in water sensitive cities. In: Proceedings from the 12th International Conference on Urban Drainage, Porto Alegre, Brazil.

Diringer, S., Anne, T., Heather, C., Robert, W., Morgan, S., McKenzie, B., 2019. Moving Toward a Multi-Benefit Approach to Water Management. Pacific Institute, Oakland, CA.

Emmerich, M.T.M., Deutz, A.H., 2018. A tutorial on multiobjective optimization: fundamentals and evolutionary methods. Nat. Comput. 17, 585–609. https://doi.org/10.1007/s11047-018-9685-y.

Kuller, M., Bach, P.M., Ramirez-Lovering, D., Deletic, A., 2017. Framing water sensitive urban design as part of the urban form: a critical review of tools for best planning practice. Environ Model Softw. 96, 265–282.

Lu, Z., Noonan, D., Crittenden, J., Jeong, H., Wang, D., 2013. Use of impact fees to incentivize low-impact development and promote compact growth. Environ. Sci. Technol. 47 (19), 10744–10752. https://doi.org/10.1021/es304924w.

Martín, G.B., Monroy, C.R., 2019. Sustainability and water sensitive cities: analysis for intermediary cities in Andalusia. Sustainability 11 (17), 4677. https://doi.org/10.3390/su11174677.

Maurya, S.P., Singh, P.K., Ohri, A., Singh, R., 2020. Identification of indicators for sustainable urban water development planning. Ecol. Indic. 108.

Maurya, S.P., Singh, R., 2021. Sustainable water resources. In: Sustainable Resource Management. Elsevier, pp. 147–162.

Montalto, F.A., Bartrand, T.A., Waldman, A.M., Travaline, K.A., Loomis, C.H., McAfee, C., Geldi, J.M., Riggall, G.J., Boles, L.M., 2013. Decentralised green infrastructure: the importance of stakeholder behaviour in determining spatial and temporal outcomes. Struct. Infrastruct. Eng. 9 (12), 1187–1205. https://doi.org/10.1080/15732479.2012.671834.

Scheffer, M., Bascompte, J., Brock, W.A., Brovkin, V., Carpenter, S.R., Dakos, V., Held, H., van Nes, E.H., Rietkerk, M., Sugihara, G., 2009. Early-warning signals for critical transitions. Nature 461, 53–59. https://doi.org/10.1038/nature08227.

United States Environmental Protection Agency (US EPA), 2011. Environmental Cleanup Best Management Practices: Effective Use of the Project Life Cycle Conceptual Site Model. https://www.epa.gov/sites/default/files/2015-04/documents/csm-life-cycle-fact-sheet-final.pdf. (Accessed 28 January 2022).

Voinov, A.A., Shugart, H.H., 2013. Integronsters, integral and integrated modeling. Environ. Model. Softw. 39, 149–158.

Weber, T., Leinster, S., McKeough, E., Dubowski, P., Hoban, A., 2018. Healthy Land and Water (2018) MUSIC Modelling Guidelines. Healthy Land and Water Limited, Brisbane, Queensland.

Woods-Ballard, B., Kellagher, R., Martin, P., Jefferies, C., Bray, R., Shaffer, P., 2007. The SUDS Manual CIRCA, London, UK.

Wong, T.H.F., Allen, R.A., Brown, R.R., Deletic, A., Gangadharan, L., Gernjak, W., Jakob, C., Reeder, M.J., Tapper, N.J., Walsh, C.J., 2013. Stormwater Management in a Water Sensitive City: Blueprint 2013. The Centre for Water Sensitive Cities.

Wong, T.H.F., 2006. Water sensitive urban design—the journey thus far. Aust. J. Water Resour. 10 (3), 213–222.

Zeitoun, M., Lankford, B., Krueger, T., Forsyth, T., Carter, R., Hoekstra, A.Y., Taylor, R., Varish, O., Cleaver, F., Boelens, R., 2016. Reductionist and integrative research approaches to complex water security policy challenges. Glob. Environ. Chang. 39, 143–154.

CHAPTER

25

Hybrid extreme learning machine optimized bat algorithm based on ensemble empirical mode decomposition for modeling dissolved oxygen in river

Salim Heddam[a], *Sungwon Kim*[b], *Ahmed Elbeltagi*[c], and *Ozgur Kisi*[d,e]

[a]Faculty of Science, Agronomy Department, Hydraulics Division, Laboratory of Research in Biodiversity Interaction Ecosystem and Biotechnology, University 20 Août 1955, Skikda, Algeria [b]Department of Railroad Construction and Safety Engineering, Dongyang University, Yeongju, Republic of Korea [c]Agricultural Engineering Department, Faculty of Agriculture, Mansoura University, Mansoura, Egypt [d]Civil Engineering Department, Ilia State University, Tbilisi, Georgia [e]Department of Civil Engineering, Technical University of Lübeck, Lübeck, Germany

OUTLINE

1 Introduction 448

2 Materials and methods 449

2.1 *Study site and data used* 449

2.2 *Performance assessment of the models* 450

2.3 *Methodology* 450

3 Results and discussion 456

4 Conclusions 462

5 Recommendations 463

References 463

Further reading 466

Current Directions in Water Scarcity Research, Volume 7
https://doi.org/10.1016/B978-0-323-91910-4.00025-X

1 Introduction

During the last few years, the use of metaheuristics optimization algorithms (MOA) and preprocessing signal decomposition (PSD) for improving the performances of machines learning (ML) models applied for modeling and forecasting water quality variables has gained much popularities. While the usefulness of the MOA has been proven best in a large numbers of applications all over the world (Chia et al., 2021), the combination of the ML models with PSD have attracted the interest of researchers and many researchers have drawn attention to the benefits that may be gained from the use of them for water quality prediction (Jamei et al., 2020; Ahmadianfar et al., 2020; Shi et al., 2018; Ebrahimi-Khusfi et al., 2021; Song et al., 2021a, b; Xiao et al., 2019; Saber et al., 2019). Given the known influence of river dissolved oxygen (DO) concentration on the overall water quality variation, an important and consistent works have been done for developing robust models for accurately predicting DO concentration, and numerous studies have assessed the PSD algorithms for further improving the accuracies of DO prediction (Song et al., 2021a, b; Huang et al., 2021; Cao et al., 2019; Huan et al., 2018; Li et al., 2018).

Song et al. (2021a) applied a hybrid model (LSSVM-SSA-VMD) based on a combination of least squares support vector machine (LSSVM), the variational mode decomposition (VMD) and the sparrow search algorithm (SSA), for modeling weekly DO at the Yangtze River, China. As a modeling strategy, they used DO measured at the previous 12 weeks as input variables for predicting DO at the next week. The SSA was used for optimizing the model parameters and the VMD was used as a preprocessing algorithm for decomposing the 12 input variables into several intrinsic mode functions (IMF). The hybrid LSSVM-SSA-VMD was compared to the LSSVM, LSSVM-SSA, LSSVM-VMD, the support vector regression (SVR), and the multilayer perceptron neural network (MLPNN). It was found that, the best accuracies were obtained using the LSSVM-SSA-VMD with a correlation coefficient (R), Nash-Sutcliffe efficiency (NSE), mean absolute error (MAE), and root mean square error (RMSE) of $\approx$0.954, $\approx$0.910, $\approx$0.217, and $\approx$0.293 mg/L, respectively, thus, showing the high contribution of the VMD in improving the performances of the standalone machine learning models. Song et al. (2021b) used the hybrid model (LSTM-ISSA-SWT) based on the combination of long short-term memory (LSTM) deep learning model, the synchrosqueezed wavelet transform (SWT) and the improved sparrow search algorithm (ISSA), for predicting weekly DO concentration measured in the Yongding and Haihe Rivers, China. Numerical comparison between LSTM-ISSA-SWT and the LSTM, LSTM-SWT, LSTM-ISSA, LSTM-SSA-SWT, SVR, and the MLPNN revealed the superiority of the hybrid LSTM-ISSA-SWT who exhibited MAE, RMSE, R, and NSE of $\approx$0.555, $\approx$0.825, $\approx$0.879, and $\approx$0.774, respectively, and the gain in models performances improvement was highly demonstrated. Another kind of SPD algorithms, i.e., the complete ensemble empirical mode decomposition with adaptive noise (CEEMDAN) which is an improved version of the original empirical mode decomposition (EMD), has been exclusively and successfully used for DO concentration prediction. For example, Huang et al. (2021) proposed a new hybrid model, i.e., the GRU-CEEMDAN-GOBLPSO for predicting DO measured at 10 min interval of time. The hybrid model combines three different paradigm, i.e., the gated recurrent unit (GRU), the generalized opposition-based learning particle swarm optimization algorithm (GOBLPSO), and the CEEMDAN. According to the obtained results, the GRU-CEEMDAN-GOBLPSO model improves the performances of the LSTM-CEEMDAN-GOBLPSO and performed much better showing and improvements rates of approximately, ≈0.930%, ≈3.32%, ≈0.51%, in terms of mean absolute percentage error (MAPE), RMSE, and R^2 values, respectively. In another study, Cao et al. (2019) used the ensemble empirical mode decomposition (EEMD) for improving the performances of the regularized extreme learning machine (RELM) used for predicting DO concentration measured at half hour interval of time. They used the water pH and water temperature as predictors which are then decomposed using the EEMD and the IMFs were then calculated and used as new inputs variables. Comparison between, RELM-EEMD, LSSVM-EEMD, ELM-EEMD and the radial basis function neural network (RBFNN-EEMD) revealed that the RELM-EEMD performed much better with R^2 equal to 0.977.

Several other examples of the application of SPD algorithms for improving the accuracies of DO prediction can be found in the literature and we can cite a whole number of example: (*i*) LSSVM-EEMD (Huan et al., 2018), ELM-EEMD (Li et al., 2018), outlier robust extreme learning machine (ORELM) combined with the empirical wavelet transform (EWT) and optimized by particle swarm optimization and gravitational search algorithm (ORELM-EWT-PSOGSA) (Liu et al., 2021), hybrid wavelet analysis (WA) coupled with Cauchy particle swarm optimization (CPSO) algorithm and LSSVM (LSSVM-WA-CPSO) (Liu et al., 2014), wavelet transform (WT) coupled with multiple linear regression (MLR), SVM, MLPNN, and random forest regression (RFR), i.e., MLR-WT, SVM-WT MLPNN-WT, RFR-WT (Xu et al., 2021a) and discrete wavelet transform (DWT) coupled MLPNN (MLPNN-DWT) (Alizadeh and Kavianpour, 2015), multiple nonlinear regression (MNLR) coupled discrete wavelet transforms (MNLR-DWT) (Evrendilek and Karakaya, 2014), CEEMDAN coupled extreme gradient boosting (XGBoost-CEEMDAN) (Lu and Ma, 2020), ELM and LSSVM coupled CEEMDAN and VMD, i.e., LSSVM-CEEMDAN and ELM-CEEMDAN (Fijani et al., 2019). Indeed, after in-depth analysis of the above-reported literature, it is clear that the use of signal decomposition for modeling DO concentration, especially the EMD and its two variants, i.e., the EEMD and the CEEMDAN has been broadly reported in the literature; however, using only river water temperature as a single predictor for modeling DO in river has not been reported in the literature, which constitutes the major motivation of our study. Hence, in the present chapter, we introduce a new ELM optimized using Bat algorithm (Bat-ELM) for modeling hourly DO concentration in river, and we compare its performances with those of the standalone ELM and relevance vector machine (RVM). The three models were compared according to two scenarios: (*i*) single models using river water temperature, i.e., Bat-ELM, ELM and RVM, and (*ii*) hybrid models using the EMD, EEMD and CEEMDAN for decomposing the water temperature into several subcomponents. Specifically, this study aims to demonstrate the significant contribution of the signal decomposition in improving the accuracies of ML models applied for modeling DO in river.

2 Materials and methods

2.1 Study site and data used

The present investigation was conducted using data collected at two stations available at the USGS web site. The selected stations were: (*i*) USGS 11509370 at Klamath River above

Keno dam, keno, Oregon, USA (Latitude 42°07′41″, Longitude 121°55′40″), and (*ii*) USGS 420853121505500 Klamath River at Miller Island boat ramp, Oregon, USA (Latitude 42°08′53″, Longitude 121°50′55″). Location of the Study area showing the USGS stations is shown in Fig. 1. For modeling DO concentration (DO: mg/L), we use only river water temperature (T_w: °C) as a single predictor. We used data measured at hourly time scale. For the USGS 11509370 station, data were recorded during the period ranging from 01 September 2020 to 17 September 2021 with a total of 9147 patterns divided into training (70%: 6403) and validation (30%: 2744). Similarly, For the USGS 420853121505500 station, data were recorded during the period ranging from 01 August 2020 to 17 August 2021 with a total of 9147 patterns divided into training (70%: 6403) and validation (30%: 2744). The statistical parameters were calculated and reported in Table 1. Fig. 2 shows the graph changes of DO (mg/L) and T_w (°C) at hourly time scale.

2.2 Performance assessment of the models

In the present chapter, the RMSE, MAE, *R*, and NSE are adopted for models evaluation.

$$\mathrm{RMSE} = \sqrt{\frac{1}{N}\sum_{i=1}^{N}\left[(\mathrm{DO}_{obs,i}) - (\mathrm{DO}_{est,i})_i\right]^2}, \quad (0 \leq \mathrm{RMSE} < +\infty) \tag{1}$$

$$\mathrm{MAE} = \frac{1}{N}\sum_{i=1}^{N}\left|\mathrm{DO}_{obs,i} - \mathrm{DO}_{est,i}\right|, \quad (0 \leq \mathrm{MAE} < +\infty) \tag{2}$$

$$R = \left[\frac{\frac{1}{N}\sum_{i=1}^{N}\left(\mathrm{DO}_{obs,i} - \overline{\mathrm{DO}_{obs}}\right)\left(\mathrm{DO}_{est,i} - \overline{\mathrm{DO}_{est}}\right)}{\sqrt{\frac{1}{N}\sum_{i=1}^{n}\left(\mathrm{DO}_{obs,i} - \overline{\mathrm{DO}_{obs}}\right)^2}\sqrt{\frac{1}{N}\sum_{i=1}^{n}\left(\mathrm{DO}_{est,i} - \overline{\mathrm{DO}_{est}}\right)^2}}\right], \quad (-1 < R \leq +1) \tag{3}$$

$$\mathrm{NSE} = 1 - \left[\frac{\sum_{i=1}^{N}\left[\mathrm{DO}_{obs} - \mathrm{DO}_{est}\right]^2}{\sum_{i=1}^{N}\left[\mathrm{DO}_{obs,i} - \overline{\mathrm{DO}_{obs}}\right]^2}\right], \quad (-\infty < \mathrm{NSE} \leq 1) \tag{4}$$

$\overline{\mathrm{DO}_{obs}}$ and $\overline{\mathrm{DO}_{est}}$ are the mean measured, and mean predicted DO concentration, respectively, DO_{obs} and DO_{est} specifies the observed and forecasted daily river DO for *i*th observations, and *N* shows the number of data points (Niazkar and Zakwan, 2021a, b).

2.3 Methodology

2.3.1 *Extreme learning machine optimized Bat algorithm (Bat-ELM)*

The ELM model was initially recommended and utilized by Huang et al. (2006) as a rapid and effective category of feedforward neural networks (FFNN) model. It involves a sole

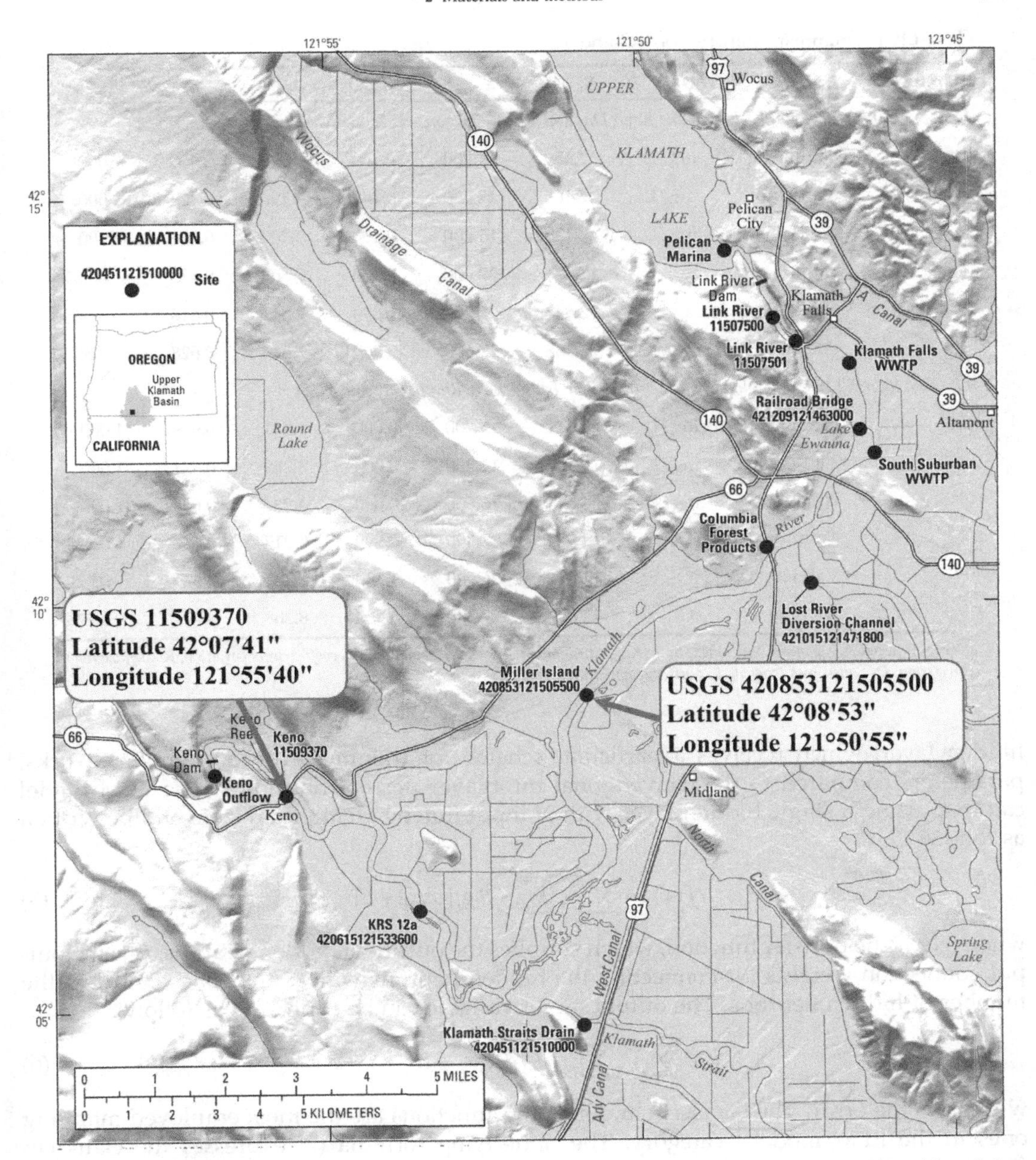

FIG. 1 Map showing the location of the two USGS stations at Klamath River, Oregon, USA. *Adopted from Sullivan, A.B., Rounds, S.A., 2020. Modeling a 2- and 4-Foot Drawdown in the Link River to Keno Dam Reach of the Upper Klamath River, South-Central Oregon (No. 2020-5001). US Geological Survey. https://doi.org/10.3133/sir20205001.*

TABLE 1 Summary statistics of DO and T_w.

Variables	Subset	Unit	X_{mean}	X_{max}	X_{min}	S_x	C_v	R
USGS 11509370 Klamath River above Keno Dam, at Keno, Oregon, USA								
DO	Training	mg/L	7.359	14.200	0.100	3.229	0.439	1.000
	Validation	mg/L	7.460	15.200	0.100	3.190	0.428	1.000
	All data	mg/L	7.429	15.200	0.100	3.200	0.431	1.000
T_w	Training	°C	12.421	27.600	0.900	7.862	0.633	−0.554
	Validation	°C	12.643	27.200	1.000	7.811	0.618	−0.546
	All data	°C	12.488	27.600	0.900	7.846	0.628	−0.552
USGS 420853121505500 Klamath River at Miller Island Boat Ramp, Oregon, USA								
DO	Training	mg/L	6.641	15.300	0.000	4.241	0.639	1.000
	Validation	mg/L	6.560	14.200	0.000	4.269	0.651	1.000
	All data	mg/L	6.648	16.500	0.000	4.235	0.637	1.000
T_w	Training	°C	12.924	27.300	0.900	8.217	0.636	−0.722
	Validation	°C	12.841	26.900	1.100	8.201	0.639	−0.729
	All data	°C	12.812	27.300	0.900	8.208	0.641	−0.725

Abbreviations: X_{mean}, *mean;* X_{max}, *maximum;* X_{min}, *minimum;* S_x, *standard deviation;* C_v, *coefficient of variation;* R: *coefficient de correlation; DO: dissolved oxygen;* T_w: *water temperature.*

hidden layer, which receives a particular scheme for training artificial neural networks' parameters compared to the conventional multilayer perceptron model. The ELM model can map using a single hidden layer with N input independent parameters and be written as follow:

$$f(x) = \sum_{j=1}^{N}\sum_{i=1}^{L} \beta_i g_i(w_i x_j + b) \tag{5}$$

where $g(.)$ is the transfer function, which supplies the output of hidden neurons, β_i is the output connection weights for connecting the hidden neurons with output ones, and L is the number of hidden neurons. The output target variable can be expressed as follow:

$$y = \sum_{j=1}^{N}\sum_{i=1}^{L} \beta_i g_i(w_i x_j + b) = t + \varepsilon \tag{6}$$

where ε is the error. The Gaussian and sigmoid functions are the most employed mapping ones in the ELM model's category. The underlying formula (7) expresses the Gaussian function.

$$g(x_i) = h(a, c, x_i) = \exp\left(-a\|x_i - c\|^2\right) \tag{7}$$

where a and c are the activation functions. During the training aspect, the connection weights are fixed in the ELM model's category. That is, random values are allowed directly to

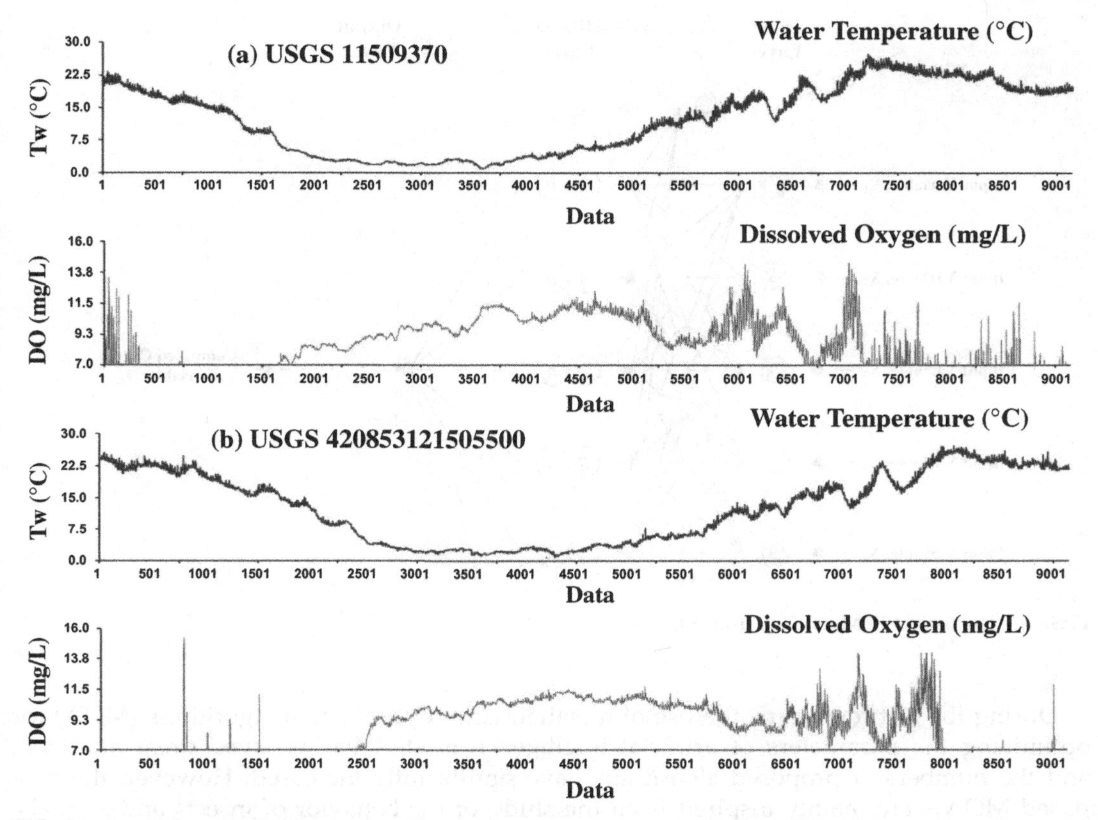

FIG. 2 Graph showing changes in river dissolved oxygen concentration (DO: mg/L) and river water temperature (T_w: °C) at the USGS 11509370 and USGS 420853121505500 stations.

neurons' activation functions instead of requesting an iterative procedure for updating them. The connection weights for output neuron can be achieved continuously utilizing the least square method. In other words, the approximation error can be reduced by computing $||\mathbf{H\beta} - \mathbf{T}||^2$ for the connection weight ($\boldsymbol{\beta}$), where **H** is the randomized matrix corresponding to the hidden layer and **T** is the target matrix.

$$\mathbf{H} = \begin{bmatrix} g(x_1) \\ \cdot \\ \cdot \\ g(x_M) \end{bmatrix}, \quad \mathbf{T} = \begin{bmatrix} t_1^T \\ \cdot \\ \cdot \\ t_M^T \end{bmatrix} \tag{8}$$

The output connection weights can be resolved based on linear equation system such as$\boldsymbol{\beta} = \mathbf{H}^+\mathbf{T}$, where $\mathbf{H}^+$ is the Moore-Penrose generalized inverse function (Yaseen et al., 2018; Kim et al., 2020). The specific description on the ELM model's development and implication can be found from the previous articles (Adnan et al., 2021; Gharib and Davies, 2021; Zounemat-Kermani et al., 2021). Fig. 3 illustrates the typical ELM model's strategy in this study.

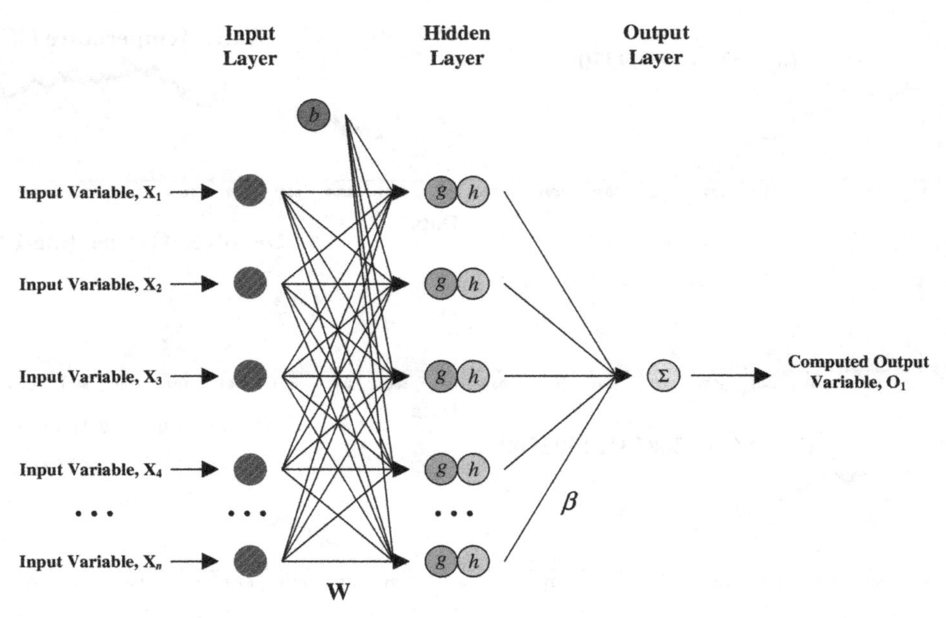

FIG. 3 The typical ELM model's strategy.

During the last few years, the use of metaheuristics optimization algorithms (MOA) for optimizing the parameters of artificial intelligence models has received great attention, and the numbers of proposed algorithms have significantly increased. However, the proposed MOA were mainly inspired from the study of the behavior of insects and animals. Among the proposed MOA, the Bat algorithm proposed by Yang (2010), and is was becoming increasingly renowned for solving optimizations problems, and several applications can be found in the literature, i.e., predicting chlorophyll-a concentration (Alizamir et al., 2021), modeling dew point temperature (Dong et al., 2020), rainfall forecasting (Ali et al., 2018), product quality monitoring (Farias Jr et al., 2014), forecasting of power transmission lines (Sun and Wang, 2019), and classification of credit card fraud detection (Zhu et al., 2020). The idea behind the Bat algorithm is the bat echolocation, and an optimization algorithm based on Bat algorithm requires three steps (Wang et al., 2021): (*i*) to make a distinction between food and obstacle was achieved by all Bats using the bat echolocation; (*ii*) in order to look and search the food, the Bat are subject to random motion and they should flying randomly at position (X_i) at speed (V_i), with a frequency (f_i) and a pulse loudness (A_0); (*iii*) the loudness A_0 is typically ranged between two values, i.e., the minimal (A_{min}) and maximal (A_{max}). The overall Bat algorithm can be summarized as follow (Wang et al., 2021; Ali et al., 2018; Dong et al., 2020; Farias Jr et al., 2014; Sun and Wang, 2019; Zhu et al., 2020):

1) The algorithms parameters are randomly initialized, i.e., objective function, population size, the A_0, the pulse emission rate (r_i), the frequency range of the pulse (f_{min}, f_{max}), and the maximum number of iteration.

2) The position and speed of the Bats are randomly initialized, i.e., Z_i and U_i
3) The pulse frequency should be updated in time, accompanied by a simultaneously updating of the Z^t and U^t.

Form a mathematical point of view, the Bats fly randomly using a velocity U_t at position Z_t, using a fixed frequency $f_{\min}$ a with a multitude of wavelengths λ. A new position and velocity are expressed as follow (Naik et al., 2020; Wang et al., 2021; Singh and Lather, 2020):

$$f_t = f_{\min} + (f_{\max} - f_{\min})\beta \tag{9}$$

$$U_t^{s+1} = U_t^s + (Z_t^s + Z^*)f_t \tag{10}$$

$$Z_t^{s+1} = Z_t^s + U_t^{s+1} \tag{11}$$

where β is a random variable ranging within the interval of [0, 1]; Z^* corresponds to the global best location, and t represents the current number of iterations. In the present study, for applying the Bat algorithm we use the MatLab code available at: https://fr.mathworks.com/matlabcentral/fileexchange?q=Bat+algorithm, and the following parameters were used: Population size (25); loudness (A_0=0.1); Pulse rate (r_0=0.1); $f_{\min}=-1$ and $f_{\max}=+1$.

2.3.2 *Relevance vector machine (RVM)*

Tipping (2001) has suggested the RVM. It is a controlled learning method based on the sparse theory (Kong et al., 2019). This procedure combines Bayesian theory, the property of Markov, the maximum probability estimate and other hypotheses. It guarantees the sparing characteristics of the model with the introduction of the Gaussian previous distribution with a zero mean value by hyper parameters for the weight vector (Lu et al., 2021). RVM considers one of the machine learning approaches used to produce parsimonious regression and probabilistic classification responses (Pan et al., 2021). The RVM performs the same tasks as the support vector machine (SVM) but it is probabilistic. The Bayesian RVM formulation eliminates the set of free factors of the SVM in comparison to SVM method (Jia et al., 2021). However, RVM is at danger of local minimum levels using expectation maximization (EM)-like learning technique (Li et al., 2021). This differs from typical sequential minimal optimization SMO-based methods used by SVM, which ensure that the world optimum is found. For instance, RVM's kernel function is not limited to Mercer and the kernel function range is broader; RVM does not require cross-validation to acquire hyper parameters and is able to generalize them more effectively (Fattahi and Hasanipanah, 2021; Lin et al., 2021).

2.3.3 *Ensemble empirical mode decomposition (EEMD)*

EMD was proposed by Huang et al. (1998) as signal decomposition algorithm, and later, two improved versions were successfully introduced namely, the EEMD (Wu and Huang, 2009), and the CEEMDAN (Torres et al., 2011). In the present chapter, the EMD, EEMD, and CEEMDAN were applied for hourly river water temperature (T_w) signal decomposition. An example of the obtained IMF using the CEEMDAN components are depicted in Fig. 4. For the EMD and the EEMD, we have the same decomposition principle and different level of decomposition. From Fig. 4, we can see that, the original input variable, i.e., the water

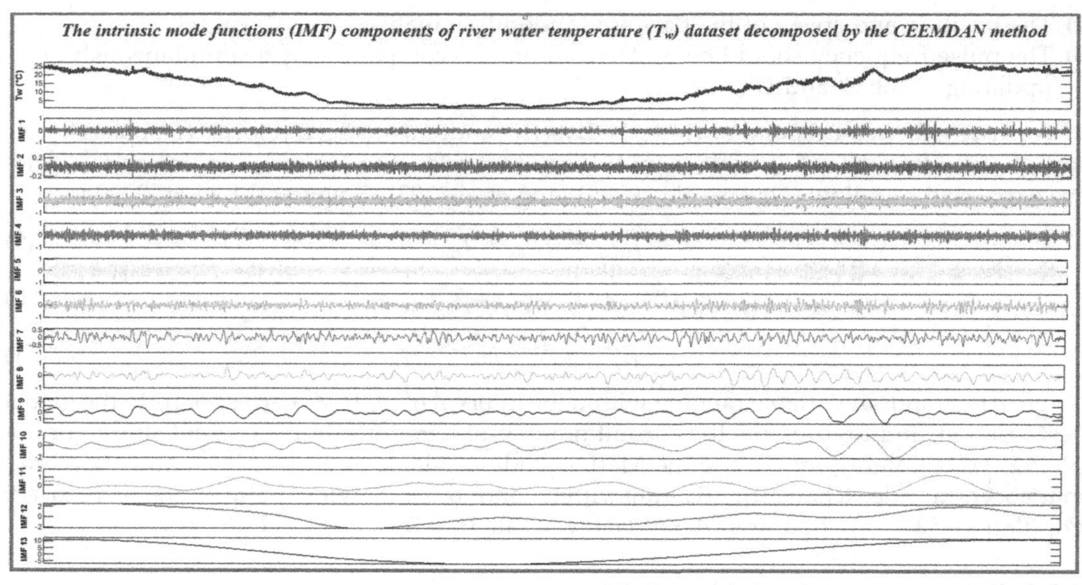

FIG. 4 The intrinsic mode functions (IMF) components of river water temperature (T_w) dataset decomposed by the CEEMDAN method.

temperature (T_w) was decomposed into several IMF, and the trend of the signal has become very condensed for the first five IMF, and thereafter start to be less condensed until the last IMF. The numerical values of the IMF were stored and used as input variable for the proposed models.

3 Results and discussion

In the present chapter, we propose a new method for modeling river DO concentration. We applied three ML models namely, ELM; the ELM optimized using Bat algorithm (Bat-ELM) and the RVM. The novelty of our investigation is that, the models were developed using only river water temperature (T_w). The models were calibrated and validation according to two scenarios: (*i*) using only river water T_w, i.e., ELM, Bat-ELM and RVM and (*ii*) the river water T_w was decomposed using the EMD, EEMD and CEEMDAN and the obtained IMF were used as input variables to the models. The flowchart of the proposed modeling approach is shown in Fig. 5. The models were applied using data from two stations and the results were reported in Table 2 and depicted in Figs. 6 and 7. The models were evaluated using four performances metrics, i.e., *R*, NSE, RMSE, and MAE.

Results obtained at the USGS 11509370 station are first discussed. Using only water T_w, all models, i.e., ELM, Bat-ELM, and RVM performed significantly worse with high RMSE and

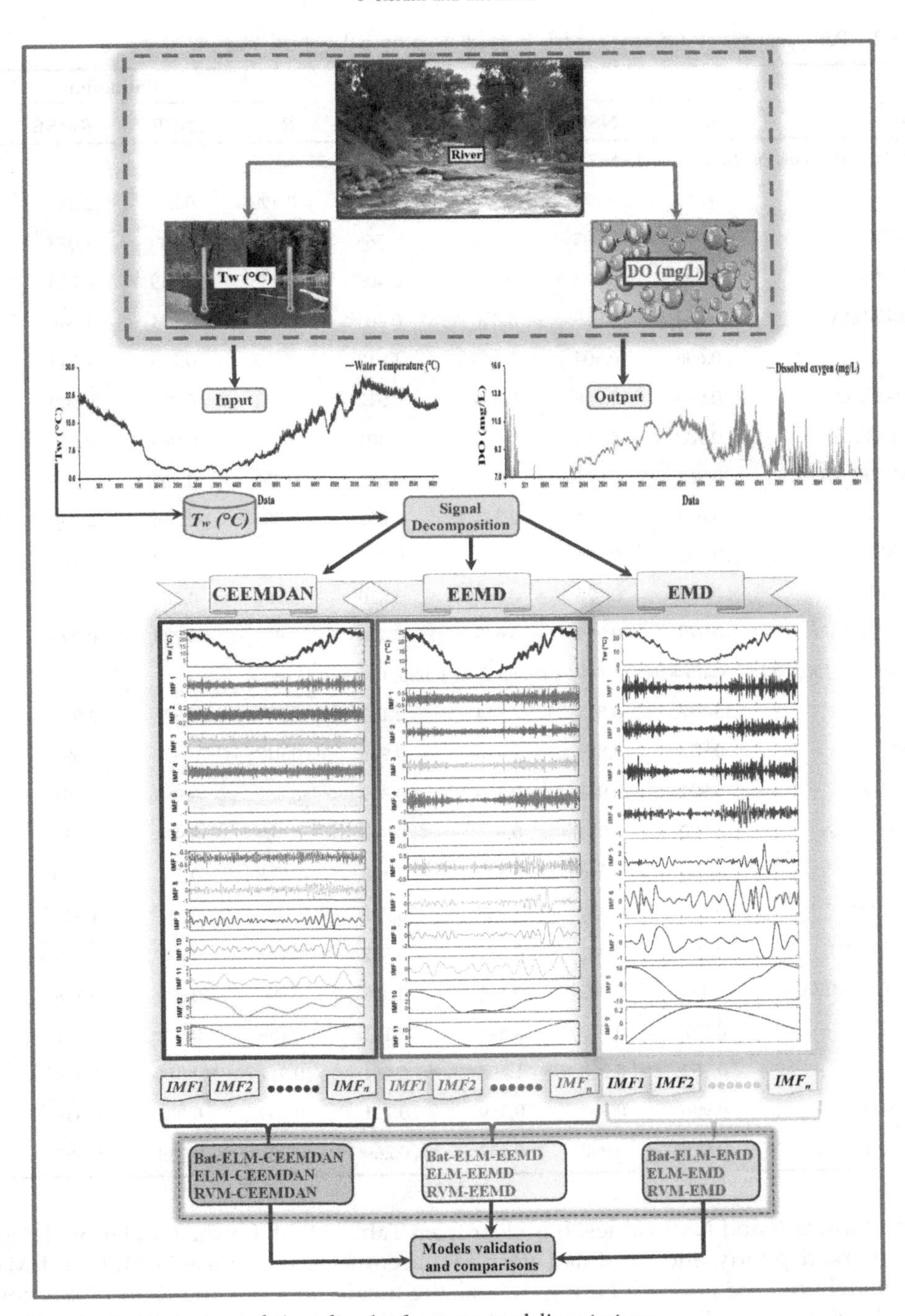

FIG. 5 Flowchart of the proposed river dissolved oxygen modeling strategy.

TABLE 2 Performances of different models in predicting river dissolved oxygen.

	Training				Validation			
Models	***R***	**NSE**	**RMSE**	**MAE**	***R***	**NSE**	**RMSE**	**MAE**
USGS 11509370 Klamath River above Keno Dam, at Keno, Oregon, USA								
ELM	0.616	0.379	2.513	2.001	0.499	0.231	2.823	2.283
ELM-EMD	0.967	0.935	0.814	0.566	0.945	0.889	1.073	0.712
ELM-EEMD	0.982	0.963	0.610	0.406	0.975	0.949	0.723	0.497
ELM-CEEMDAN	0.964	0.929	0.850	0.605	0.947	0.894	1.046	0.752
Bat-ELM	0.636	0.404	2.463	1.913	0.533	0.273	2.744	2.176
Bat-ELM-EMD	0.973	0.948	0.729	0.512	0.961	0.921	0.909	0.624
Bat-ELM-EEMD	0.982	0.964	0.605	0.404	0.980	0.959	0.650	0.450
Bat-ELM-CEEMDAN	0.962	0.926	0.871	0.621	0.952	0.905	0.985	0.715
RVM	0.604	0.364	2.544	2.033	0.501	0.246	2.796	2.236
RVM-EMD	0.992	0.985	0.392	0.265	0.976	0.952	0.703	0.443
RVM-EEMD	0.988	0.975	0.501	0.320	0.981	0.962	0.626	0.398
RVM-CEEMDAN	0.979	0.959	0.645	0.426	0.969	0.938	0.796	0.548
USGS 420853121505500 Klamath River at Miller Island Boat Ramp, Oregon, USA								
ELM	0.737	0.543	2.844	2.151	0.756	0.572	2.811	2.153
ELM-EMD	0.937	0.879	1.476	0.965	0.917	0.839	1.693	1.159
ELM-EEMD	0.991	0.982	0.559	0.379	0.985	0.969	0.751	0.516
ELM-CEEMDAN	0.992	0.984	0.534	0.345	0.989	0.978	0.636	0.431
Bat-ELM	0.738	0.545	2.839	2.128	0.758	0.575	2.800	2.123
Bat-ELM-EMD	0.967	0.934	1.085	0.732	0.957	0.915	1.231	0.857
Bat-ELM-EEMD	0.990	0.981	0.581	0.382	0.988	0.977	0.655	0.461
Bat-ELM-CEEMDAN	0.991	0.982	0.561	0.369	0.990	0.979	0.618	0.419
RVM	0.737	0.543	2.844	2.151	0.756	0.572	2.811	2.153
RVM-EMD	0.952	0.906	1.298	0.861	0.917	0.841	1.684	1.108
RVM-EEMD	0.996	0.993	0.359	0.231	0.989	0.977	0.649	0.397
RVM-CEEMDAN	0.995	0.990	0.432	0.269	0.992	0.983	0.555	0.354

MAE, and lower *R* and NSE values. It is clear from Table 2 that, for the validation dataset, all models worked poorly and exhibited very low accuracies with mean RMSE and MAE of approximately ≈2.788 and ≈2.232 mg/L, and the mean values of the R and NSE does not exceeded 0.511 and 0.250, respectively, showing the limitation of the machine learning models in better predicting DO using only T_w. For improving the models performances,

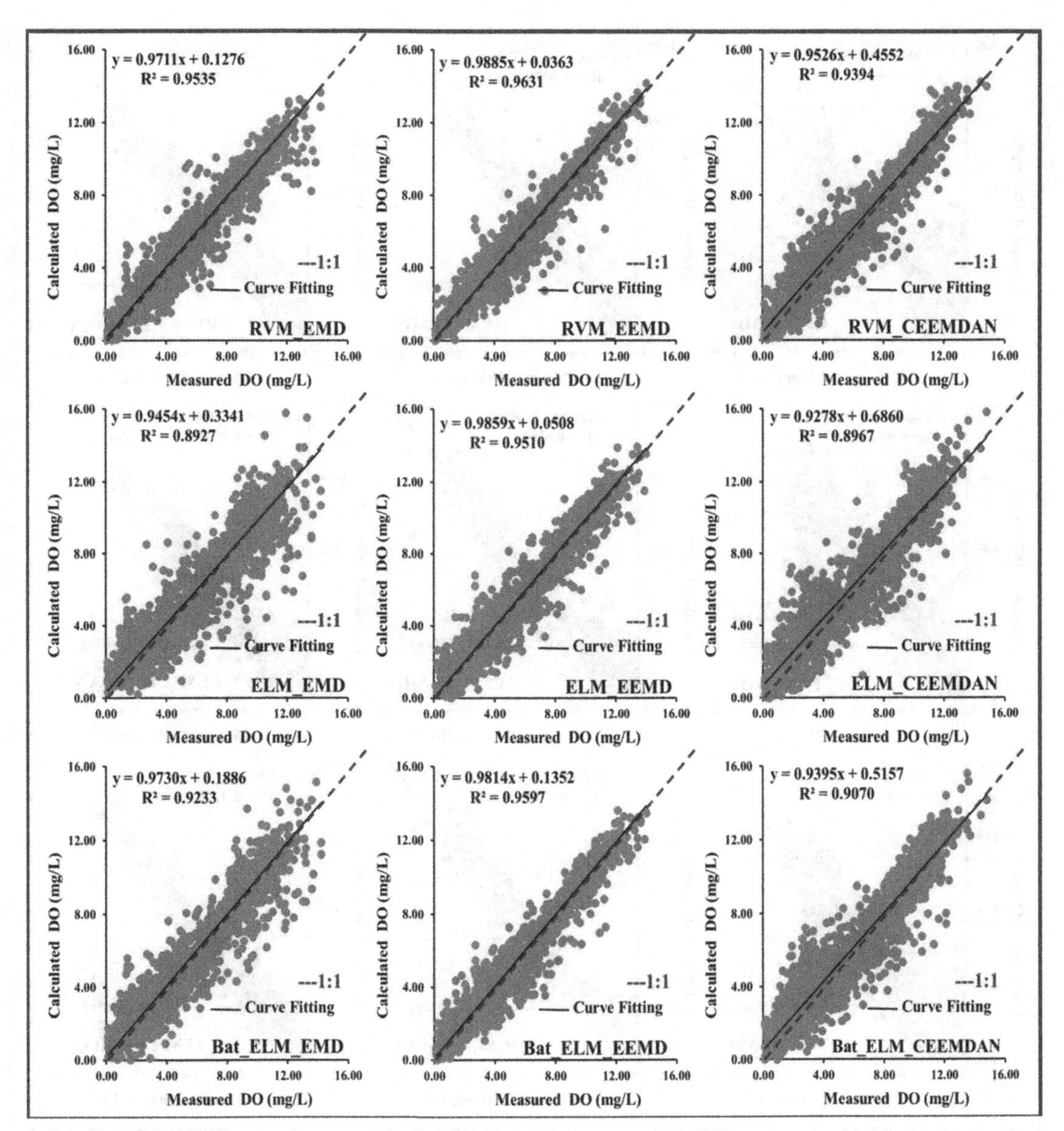

FIG. 6 Scatterplots of measured against calculated hourly river dissolved oxygen concentration (DO: mg/L) at the USGS 11509370 for the validation stage.

the PSD approaches were chosen in the present study, and it is clear from Table 2 that models performances were increased and the predictive accuracies were significantly improved, for which the *R* and NSE were remarkably increased and the RMSE and MAE were dramatically decreased.

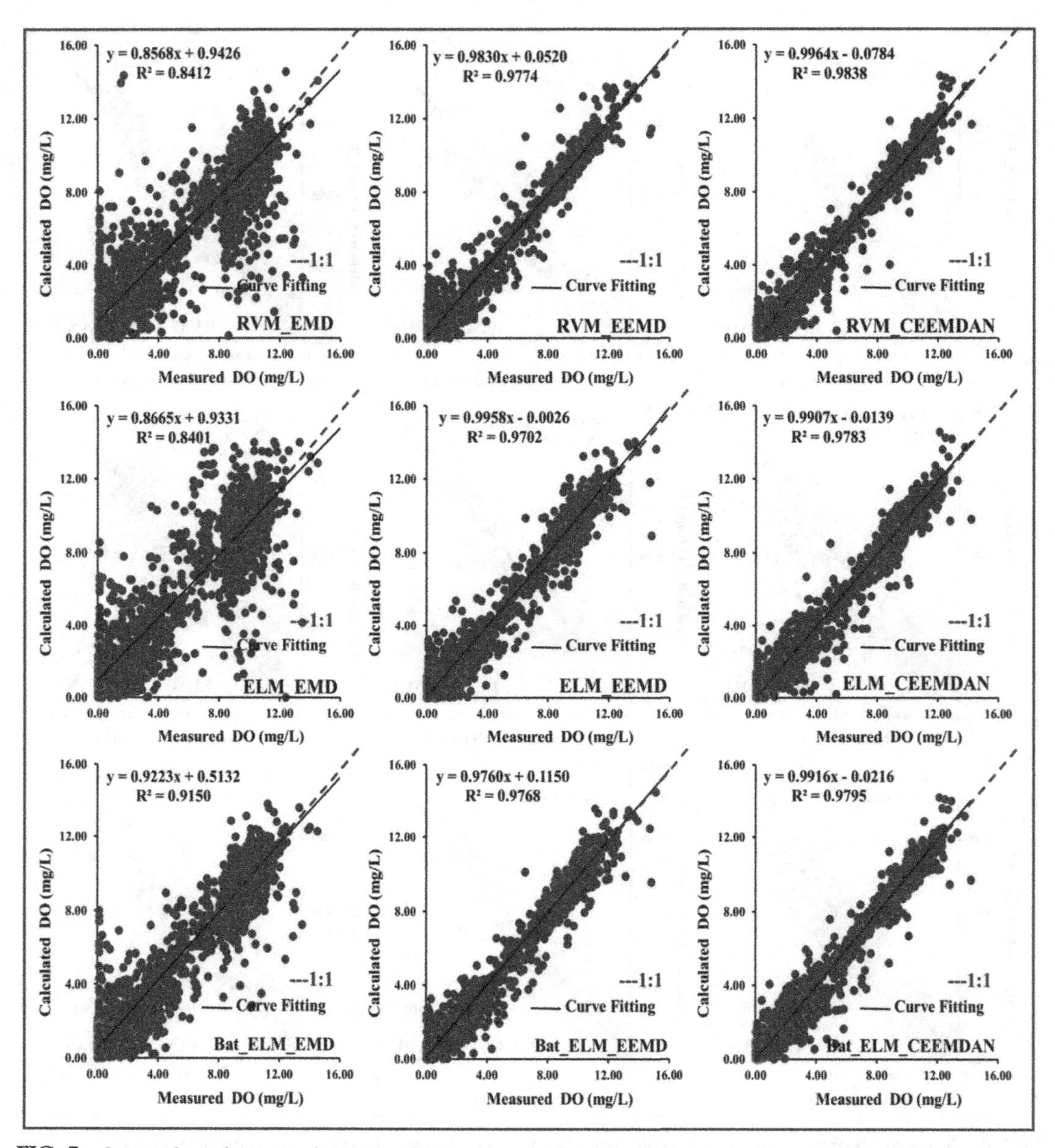

FIG. 7 Scatterplots of measured against calculated hourly river dissolved oxygen concentration (DO: mg/L) at the USGS 420853121505500 for the validation stage.

First, using the EMD algorithm, it is clear that, the best accuracies were obtained by the RVM-EMD compared to the ELM-EMD and Bat-ELM-EMD models. The RVM-EMD has the high R and NSE values of ≈0.976 and ≈0.952, and the lowest RMSE (≈0.703 mg/L) and MAE (≈0.443 mg/L), and more precisely, the RVM-EMD improves the accuracies of the ELM-EMD and Bat-ELM-EMD by decreasing the RMSE and MAE by approximately ≈22.66%, ≈34.48%, and ≈34.48%, ≈37.78%, respectively, and by increasing the R and NSE by ≈1.56%, ≈3.36%, and ≈3.28%, ≈7.087%, respectively. Also, it is clear that, the Bat-ELM-EMD was more accurate than the ELM-EMD. More precisely, the Bat-ELM-EMD surpassed the ELM-EMD by approximately ≈1.69%, ≈3.60%, ≈15.28%, and ≈12.369%, in terms of R, NSE, RMSE, and MAE, respectively. Second, according to Table 2, it is clear that the gain in models performances is too high using the EEMD compared to the EMD algorithm. Indeed, even as it is clear that, the RVM-EEMD was slightly more accurate compared to the Bat-ELM-EEMD, the differences between the three models were only significant tacking into account the RMSE and MAE improvement. More precisely, the RVM-EEMD improves the performances of the Bat-ELM-EEMD and ELM-EEMD by decreasing the RMSE and MAE by ≈3.69%, ≈11.55%, and ≈13.41%, and ≈19.92%, respectively, in addition, it is clear that the Bat-ELM-EEMD was more accurate than the ELM-EEMD showing an improvement of about 10.09%, ≈9.45%, respectively. Finally, using the CEEMDAN algorithm, models performances were significantly decreased compared to the EEMD and EMD algorithms, showing an increase in the errors metrics and a decrease in the R and NSE values. According to Table 2, the RVM-CEEMDAN was the most accurate with the high R (≈0.969) and NSE (≈0.938) values, and the lowest RMSE (≈0.796 mg/L) and MAE (≈0.548 mg/L) values. In addition, the RVM-CEEMDAN improves the performances of the Bat-ELM-EEMD by ≈1.78%, ≈3.64%, ≈19.18%, and ≈23.35%, and it improves the performances of the ELM-EEMD by ≈2.323%, ≈4.922%, ≈23.90%, and ≈27.12%, in terms of R, NSE, RMSE, an MAE respectively. More in depth analysis of the obtained results we can conclude that, models based on the EEMD were more accurate, followed by the models using the EMD, the CEEMDAN in the third place, and finally the models without signal decomposition were the poorest. According to the ascending order, the means RMSE and MAE of the single models, i.e., RVM, ELM and Bat-ELM were improved by ≈76.09% and ≈79.91% using the EEMD algorithm, and by ≈67.89% and ≈73.42% using the EMD algorithm, and by ≈66.19% and ≈73.42% using the CEEMDAN algorithm, respectively. On the other hand, taking into account the R and NSE indices, the gain in the improvement of predictive accuracies was more evident for which, an improvement of about ≈91.52% and ≈282.66%; ≈87.99% and ≈268.26%; and ≈87.08% and ≈264.93% were achieved by the EEMD, EMD, and CEEMDAN, respectively. The scatterplots of measured against calculated DO concentration at the USGS 11509370 station are depicted in Fig. 6.

Results obtained at the USGS 420853121505500 station were different to what are obtained in the USGS 11509370 station for which, the superiority of the CEEMDAN was more evident. First, using only water T_w as single input variable, the three models, i.e., RVM, ELM and Bat-ELM were equal in terms of models performances showing the same R (≈0.75) and NSE (≈0.57), RMSE (≈0.2.81 mg/L) and MAE (≈0.2.12 mg/L), respectively. Using the EMD algorithm, the Bat-ELM-EMD was more accurate and it improves the accuracies of the ELM-EMD by approximately ≈4.362%, ≈9.058%, ≈27.289%, and ≈26.057%, in terms of R, NSE, RMSE, and MAE, respectively, and it improves the accuracies of the RVM-EMD by approximately ≈4.180%, ≈8.087%, ≈26.900%, and ≈22.653%, in terms of R, NSE, RMSE, and MAE, respectively. Using the EEMD, it is clear from Table 2 that, the three models, i.e., Bat-ELM-EEMD, ELM-EEMD and RVM-EEMD, achieved relatively the same performances metrics with very negligible difference showing a means R, NSE, RMSE, and MAE of approximately ≈0.987, ≈0.974%, ≈0.685 mg/L, and ≈0.458 mg/L, respectively, showing an improvement of approximately ≈6.127%, ≈12.640%, ≈55.40%, and ≈56.01%, compared to the means values obtained using the EMD algorithms, leading to the conclusion that high improvements ratios were gained showing the *R* and NSE values for all models remarkably increased and the RMSE and MAE dramatically decreased. Finally, using the CEEMDAN, the performances metrics of all models were reaching their upper levels with maximum *R* and NSE, and minimum RMSE and MAE values, also, it is clear that the best accuracies were achieved using the RVM-CEEMDAN model, which improve the accuracies of the Bat-ELM-CEEMDAN and ELM-CEEMDAN by ≈10.19%, ≈15.51%, and ≈12.73%, ≈17.86%, in terms of RMSE and MAE, respectively. Finally, for concluding, obtained results at the USGS 420853121505500 station have clearly highlighted the utilities and robustness of the signal decomposition in improving the models performances, and we can report that the means *R*, NSE, RMSE and MAE were improved by ≈22.95%, ≈50.96%, ≈45.28%, and ≈51.40% using the EMD algorithm, and by ≈30.48%, ≈70.04%, ≈75.60%, and ≈78.62% using the EEMD algorithm, and by ≈30.88%, ≈71.03%, ≈78.52%, and ≈81.27% using the CEEMDAN algorithm, respectively. The scatterplots of measured against calculated DO concentration at the USGS 420853121505500 station are depicted in Fig. 7.

Indeed, it is clear that the use of the PSD algorithms have led to an excellent improvement of the models performances; however, these algorithms are not adopted without criticism. The important remark is that these algorithms have naturally resulted in models that are themselves very complex as the number of input variables, i.e., the IMF has increased to be more than 10 input, which make the training process hard and time consuming. It will be very interesting if an input selection variable would be proposed and used for better selection of the relevant IMF rather than the inclusion of the all obtained IMF.

4 Conclusions

This study investigated the applicability of three machine learning methods, ELM, Bat-ELM and RVM combined with EMD, EEMD, and CEEMDAN preprocessing approaches in modeling DO in river using only water temperature data as input. Data from two stations (USGS 11509370 and USGS 420853121505500) operated by United States Geological Survey were utilized for the models' development and the outcomes were assessed based on *R*, NSE, RMSE, and MAE criteria. After observing comparison outcomes, the following conclusions were reached:

1. In both stations, the preprocessing approaches considerably improved the methods' accuracy in modeling DO concentration; by employing EMD, improvements in *R*, NSE, RMSE and MAE respectively are 89%, 285%, 62%, and 69% for the ELM, 80%, 237%, 67%, and 71% for the Bat-ELM and 95%, 287%, 75%, and 80% for the RVM in the validation stage of the 1st station (USGS 11509370) while the corresponding percentages in the 2nd station (USGS 420853121505500) are 21%, 47%, 40%, and 46% for the ELM, 26%, 59%, 56%, and 60% for the Bat-ELM and 21%, 47%, 40%, and 49% for the RVM.
2. Among the preprocessing approaches, the most improvements were generally provided by the EEMD in both stations. In the 1st station, the EEMD improved the accuracy of the three methods by about 84%–96%, 251%–311%, 74%–78%, and 78%–82% with respect to R,

NSE, RMSE and MAE in the validation stage, whereas the corresponding improvements are 30%–31%, 69%–71%, 73%–77%, and 76%–82% for the 2nd station, respectively.

3. Comparison of the methods' accuracy revealed that the RVM-EEMD acted better than the other methods in modeling DO concentration in the 1st station, whereas the RVM-CEEMDAN offered the best accuracy in the 2nd station. By implementing RVM-EEMD in the 1st station, the improvements in the prediction performance in respect of *R*, NSE, RMSE, and MAE are 97%, 317%, 78%, and 83% for standalone ELM, 3.8%, 8.2%, 42%, and 44% for ELM-EMD, 0.6%, 1.4%, 13%, and 20% for ELM-EEMD, 3.6%, 7.6%, 40%, and 47% for ELM-CEEMDAN, 84%, 252%, 77%, and 82% for standalone Bat-ELM, 2.1%, 4.5%, 31% and 36% for Bat-ELM-EMD, 0.1%, 0.3%, 3.7% and 12% for Bat-ELM-EEMD, 3%, 6.3%, 36%, and 44% for Bat-ELM-CEEMDAN, 96%, 291%, 78%, and 82% for standalone RVM, 0.5%, 1.1%, 11%, and 10% for RVM-EMD, 1.2%, 2.26, 21%, and 27% for RVM-CEEMDAN, respectively. In the 2nd station, the RVM-CEEMDAN improved the performance of R, NSE, RMSE and MAE by about 31%, 72%, 80%, and 84 for the standalone ELM, Bat-ELM and RVM, by about 3.7%–8.2%, 7.4%–17%, 55%–67%, and 59%–70% for the EMD-based methods, by about 0.3%–0.7%, 0.6%–1.4%, 15%–26%, and 11%–31% for the EEMD-based methods and by about 0.2%–0.3%, 0.4%–0.5%, 10%–13%, and 16%–18% for the CEEMDAN-based ELM and Bat-ELM methods, respectively.

5 Recommendations

General outcomes showed that the combining preprocessing methods, EMD, EEMD and CEEMDAN, with ELM, Bat-ELM, and RVM methods are very useful in modeling DO using only water temperature data as input. The RVM-EEMD and RVM-CEEMDAN methods are recommended in prediction DO concentration which is very important parameter indicating the water quality and these methods can be used as tools in monitoring water quality. This study can be extended by including more effective input parameters in future and other water quality parameters can also be predicted using EMD, EEMD, and CEEMDAN methods.

References

Ali, M., Deo, R.C., Downs, N.J., Maraseni, T., 2018. Multi-stage hybridized online sequential extreme learning machine integrated with Markov Chain Monte Carlo copula-Bat algorithm for rainfall forecasting. Atmos. Res. 213, 450–464. https://doi.org/10.1016/j.atmosres.2018.07.005.

Alizamir, M., Heddam, S., Kim, S., Danandeh Mehr, A., 2021. On the implementation of a novel data-intelligence model based on extreme learning machine optimized by bat algorithm for estimating daily chlorophyll-a concentration: case studies of river and lake in USA. J. Clean. Prod. 285, 124868. https://doi.org/10.1016/j.jclepro.2020.124868.

Adnan, R.M., Mostafa, R.R., Kisi, O., Yaseen, Z.M., Shahid, S., Zounemat-Kermani, M., 2021. Improving streamflow prediction using a new hybrid ELM model combined with hybrid particle swarm optimization and grey wolf optimization. Knowl.-Based Syst. 230, 107379. https://doi.org/10.1016/j.knosys.2021.107379.

Alizadeh, M.J., Kavianpour, M.R., 2015. Development of wavelet-ANN models to predict water quality parameters in Hilo Bay, Pacific Ocean. Mar. Pollut. Bull. 98 (1–2), 171–178. https://doi.org/10.1016/j.marpolbul.2015.06.052.

Ahmadianfar, I., Jamei, M., Chu, X., 2020. A novel hybrid wavelet-locally weighted linear regression (W-LWLR) model for electrical conductivity (EC) prediction in surface water. J. Contam. Hydrol. 232, 103641. https://doi.org/10.1016/j.jconhyd.2020.103641.

Cao, W., Huan, J., Liu, C., Qin, Y., Wu, F., 2019. A combined model of dissolved oxygen prediction in the pond based on multiple-factor analysis and multi-scale feature extraction. Aquac. Eng. 84, 50–59. https://doi.org/10.1016/j.aquaeng.2018.12.003.

Chia, S.L., Chia, M.Y., Koo, C.H., Huang, Y.F., 2021. Integration of advanced optimization algorithms into least-square support vector machine (LSSVM) for water quality index prediction. Water Supply. https://doi.org/10.2166/ws.2021.303.

Dong, J., Wu, L., Liu, X., Li, Z., Gao, Y., Zhang, Y., Yang, Q., 2020. Estimation of daily dew point temperature by using bat algorithm optimization based extreme learning machine. Appl. Therm. Eng. 165, 114569. https://doi.org/10.1016/j.applthermaleng.2019.114569.

Ebrahimi-Khusfi, Z., Taghizadeh-Mehrjardi, R., Kazemi, M., Nafarzadegan, A.R., 2021. Predicting the ground-level pollutants concentrations and identifying the influencing factors using machine learning, wavelet transformation, and remote sensing techniques. Atmos. Pollut. Res. 12 (5), 101064. https://doi.org/10.1016/j.apr.2021.101064.

Evrendilek, F., Karakaya, N., 2014. Regression model-based predictions of diel, diurnal and nocturnal dissolved oxygen dynamics after wavelet denoising of noisy time series. Physica A 404, 8–15. https://doi.org/10.1016/j.physa.2014.02.062.

Fijani, E., Barzegar, R., Deo, R., Tziritis, E., Skordas, K., 2019. Design and implementation of a hybrid model based on two-layer decomposition method coupled with extreme learning machines to support real-time environmental monitoring of water quality parameters. Sci. Total Environ. 648, 839–853. https://doi.org/10.1016/j.scitotenv.2018.08.221.

Fattahi, H., Hasanipanah, M., 2021. Prediction of blast-induced ground vibration in a mine using relevance vector regression optimized by metaheuristic algorithms. Nat. Resour. Res. 30 (2), 1849–1863. https://doi.org/10.1007/s11053-020-09764-7.

Farias Jr., F.S., Azevedo, R.A., Rivera, E.C., Herrera, W.E., Rubens Filho, M., Lima Jr., L.P., 2014. Product quality monitoring using extreme learning machines and bat algorithms: a case study in second-generation ethanol production. In: Computer Aided Chemical Engineering. vol. 33. Elsevier, pp. 955–960, https://doi.org/10.1016/B978-0-444-63456-6.50160-5.

Gharib, A., Davies, E.G., 2021. A workflow to address pitfalls and challenges in applying machine learning models to hydrology. Adv. Water Resour. 152, 103920. https://doi.org/10.1016/j.advwatres.2021.103920.

Huang, J., Liu, S., Hassan, S.G., Xu, L., Huang, C., 2021. A hybrid model for short-term dissolved oxygen content prediction. Comput. Electron. Agric. 186, 106216. https://doi.org/10.1016/j.compag.2021.106216.

Huan, J., Cao, W., Qin, Y., 2018. Prediction of dissolved oxygen in aquaculture based on EEMD and LSSVM optimized by the Bayesian evidence framework. Comput. Electron. Agric. 150, 257–265. https://doi.org/10.1016/j.compag.2018.04.022.

Huang, G.B., Zhu, Q.Y., Siew, C.K., 2006. Extreme learning machine: theory and applications. Neurocomputing 70 (1–3), 489–501. https://doi.org/10.1016/j.neucom.2005.12.126.

Huang, N.E., Shen, Z., Long, S.R., Wu, M.C., Shih, H.H., Zheng, Q., Yen, N., Liu, H.H., 1998. The empirical mode decomposition and the Hilbert spectrum for nonlinear and non-stationary time series analysis. Proc. R. Soc. London, Ser. A 454 (1971), 903–995. https://doi.org/10.1098/rspa.1998.0193.

Jamei, M., Ahmadianfar, I., Chu, X., Yaseen, Z.M., 2020. Prediction of surface water total dissolved solids using hybridized wavelet-multigene genetic programming: New approach. J. Hydrol. 589, 125335. https://doi.org/10.1016/j.jhydrol.2020.125335.

Jia, S., Ma, B., Guo, W., Li, Z.S., 2021. A sample entropy based prognostics method for lithium-ion batteries using relevance vector machine. J. Manuf. Syst. https://doi.org/10.1016/j.jmsy.2021.03.019.

Kong, D., Chen, Y., Li, N., Duan, C., Lu, L., Chen, D., 2019. Relevance vector machine for tool wear prediction. Mech. Syst. Signal Process. 127, 573–594. https://doi.org/10.1016/j.ymssp.2019.03.023.

Kim, S., Alizamir, M., Zounemat-Kermani, M., Kisi, O., Singh, V.P., 2020. Assessing the biochemical oxygen demand using neural networks and ensemble tree approaches in South Korea. J. Environ. Manag. 270, 110834. https://doi.org/10.1016/j.jenvman.2020.110834.

Li, C., Li, Z., Wu, J., Zhu, L., Yue, J., 2018. A hybrid model for dissolved oxygen prediction in aquaculture based on multi-scale features. Inf. Proc. Agric. 5 (1), 11–20. https://doi.org/10.1016/j.inpa.2017.11.002.

Liu, S., Xu, L., Jiang, Y., Li, D., Chen, Y., Li, Z., 2014. A hybrid WA-CPSO-LSSVR model for dissolved oxygen content prediction in crab culture. Eng. Appl. Artif. Intell. 29, 114–124. https://doi.org/10.1016/j.engappai.2013.09.019.

Liu, H., Yang, R., Duan, Z., Wu, H., 2021. A hybrid neural network model for marine dissolved oxygen concentrations time-series forecasting based on multi-factor analysis and a multi-model ensemble. Engineering. https://doi.org/10.1016/j.eng.2020.10.023.

Li, W., Chen, L., Zhao, J., Wang, W., 2021. Embedded feature selection based on relevance vector machines with an approximated marginal likelihood and its industrial application. In: IEEE Transactions on Systems, Man, and Cybernetics: Systems, pp. 1–14, https://doi.org/10.1109/TSMC.2021.3049597.

Lin, L., Zhang, G., Wang, J., Tian, M., Wu, S., 2021. Utilizing transfer learning of pre-trained AlexNet and relevance vector machine for regression for predicting healthy older adult's brain age from structural MRI. Multimed. Tools Appl. 1-17. https://doi.org/10.1007/s11042-020-10377-8.

Lu, H., Ma, X., 2020. Hybrid decision tree-based machine learning models for short-term water quality prediction. Chemosphere 249, 126169. https://doi.org/10.1016/j.chemosphere.2020.126169.

Lu, H., Iseley, T., Matthews, J., Liao, W., Azimi, M., 2021. An ensemble model based on relevance vector machine and multi-objective salp swarm algorithm for predicting burst pressure of corroded pipelines. J. Pet. Sci. Eng. 203, 108585. https://doi.org/10.1016/j.petrol.2021.108585.

Naik, S.M., Jagannath, R.P.K., Kuppili, V., 2020. Bat algorithm-based weighted Laplacian probabilistic neural network. Neural Comput. Applic. 32 (4), 1157–1171. https://doi.org/10.1007/s00521-019-04475-4.

Niazkar, M., Zakwan, M., 2021a. Assessment of artificial intelligence models for developing single-value and loop rating curves. Complexity. https://doi.org/10.1155/2021/6627011.

Niazkar, M., Zakwan, M., 2021b. Application of MGGP, ANN, MHBMO, GRG, and linear regression for developing daily sediment rating curves. Math. Probl. Eng. https://doi.org/10.1155/2021/8574063.

Pan, Q.J., Leung, Y.F., Hsu, S.C., 2021. Stochastic seismic slope stability assessment using polynomial chaos expansions combined with relevance vector machine. Geosci. Front. 12 (1), 405–414. https://doi.org/10.1016/j.gsf.2020.03.016.

Saber, A., James, D.E., Hayes, D.F., 2019. Estimation of water quality profiles in deep lakes based on easily measurable constituents at the water surface using artificial neural networks coupled with stationary wavelet transform. Sci. Total Environ. 694, 133690. https://doi.org/10.1016/j.scitotenv.2019.133690.

Song, C., Yao, L., Hua, C., Ni, Q., 2021a. A water quality prediction model based on variational mode decomposition and the least squares support vector machine optimized by the sparrow search algorithm (VMD-SSA-LSSVM) of the Yangtze River, China. Environ. Monit. Assess. 193 (6), 1–17. https://doi.org/10.1007/s10661-021-09127-6.

Song, C., Yao, L., Hua, C., Ni, Q., 2021b. A novel hybrid model for water quality prediction based on synchrosqueezed wavelet transform technique and improved long short-term memory. J. Hydrol. 603, 126879. https://doi.org/10.1016/j.jhydrol.2021.126879.

Shi, B., Wang, P., Jiang, J., Liu, R., 2018. Applying high-frequency surrogate measurements and a wavelet-ANN model to provide early warnings of rapid surface water quality anomalies. Sci. Total Environ. 610, 1390–1399. https://doi.org/10.1016/j.scitotenv.2017.08.232.

Singh, P., Lather, J.S., 2020. Dynamic power management and control for low voltage DC microgrid with hybrid energy storage system using hybrid bat search algorithm and artificial neural network. J. Energy Storage 32, 101974. https://doi.org/10.1016/j.est.2020.101974.

Sun, W., Wang, C., 2019. Staged icing forecasting of power transmission lines based on icing cycle and improved extreme learning machine. J. Clean. Prod. 208, 1384–1392. https://doi.org/10.1016/j.jclepro.2018.10.197.

Torres, M.E., Colominas, M.A., Schlotthauer, G., Flandrin, P., 2011. A complete ensemble empirical mode decomposition with adaptive noise. In: 2011 IEEE International Conference on Acoustics, Speech and Signal Processing (ICASSP). IEEE, pp. 4144–4147, https://doi.org/10.1109/ICASSP.2011.5947265.

Tipping, M.E., 2001. Sparse Bayesian learning and the relevance vector machine. J. Mach. Learn. Res. 1, 211–244. https://doi.org/10.1162/15324430152748236.

Wang, H., Jing, W., Li, Y., Yang, H., 2021. Fault diagnosis of fuel system based on improved extreme learning machine. Neural. Process. Lett. 53, 2553–2565. https://doi.org/10.1007/s11063-019-10186-7.

Wu, Z., Huang, N.E., 2009. Ensemble empirical mode decomposition: a noise-assisted data analysis method. Adv. Adapt. Data Anal. 1 (01), 1–41. https://doi.org/10.1142/S1793536909000047.

Xiao, X., He, J., Yu, Y., Cazelles, B., Li, M., Jiang, Q., Xu, C., 2019. Teleconnection between phytoplankton dynamics in north temperate lakes and global climatic oscillation by time-frequency analysis. Water Res. 154, 267–276. https://doi.org/10.1016/j.watres.2019.01.056.

Xu, C., Chen, X., Zhang, L., 2021a. Predicting river dissolved oxygen time series based on stand-alone models and hybrid wavelet-based models. J. Environ. Manag. 295, 113085. https://doi.org/10.1016/j.jenvman.2021.113085.

Yaseen, Z.M., Deo, R.C., Hilal, A., Abd, A.M., Bueno, L.C., Salcedo-Sanz, S., Nehdi, M.L., 2018. Predicting compressive strength of lightweight-foamed concrete using extreme learning machine model. Adv. Eng. Softw. 115, 112–125. https://doi.org/10.1016/j.advengsoft.2017.09.004.

Yang, X.S., 2010. A new metaheuristic bat-inspired algorithm. In: González, J.R., Pelta, D.A., Cruz, C., Terrazas, G., Krasnogor, N. (Eds.), Nature Inspired Cooperative Strategies for Optimization (NICSO 2010). Studies in Computational Intelligence. vol. 284. Springer, Berlin, Heidelberg, https://doi.org/10.1007/978-3-642-12538-6_6.

Zounemat-Kermani, M., Batelaan, O., Fadaee, M., Hinkelmann, R., 2021. Ensemble machine learning paradigms in hydrology: a review. J. Hydrol. 598, 126266. https://doi.org/10.1016/j.jhydrol.2021.126266.

Zhu, H., Liu, G., Zhou, M., Xie, Y., Abusorrah, A., Kang, Q., 2020. Optimizing Weighted Extreme Learning Machines for imbalanced classification and application to credit card fraud detection. Neurocomputing 407, 50–62. https://doi.org/10.1016/j.neucom.2020.04.078.

Further reading

Lu, Y., Zhang, X., Jing, L., Li, X., Fu, X., 2020. Estimation of the foetal heart rate baseline based on singular spectrum analysis and empirical mode decomposition. Futur. Gener. Comput. Syst. 112, 126–135. https://doi.org/10.1016/j.future.2020.05.008.

Peng, T., Zhang, C., Zhou, J., Nazir, M.S., 2021. An integrated framework of Bi-directional Long-Short Term Memory (BiLSTM) based on sine cosine algorithm for hourly solar radiation forecasting. Energy 221, 119887. https://doi.org/10.1016/j.energy.2021.119887.

Qiao, W., Liu, W., Liu, E., 2021. A combination model based on wavelet transform for predicting the difference between monthly natural gas production and consumption of US. Energy, 121216. https://doi.org/10.1016/j.energy.2021.121216.

Xu, X., Huo, X., Qian, X., Lu, X., Yu, Q., Ni, K., Wang, X., 2021b. Data-driven and coarse-to-fine baseline correction for signals of analytical instruments. Anal. Chim. Acta 1157, 338386. https://doi.org/10.1016/j.aca.2021.338386.

CHAPTER

26

Application of machine learning models to side-weir discharge coefficient estimations in trapezoidal and rectangular open channels

Majid Niazkar[a] and *Mohammad Zakwan*[b]

[a]Department of Civil and Environmental Engineering, School of Engineering, Shiraz University, Shiraz, Iran [b]School of Technology, MANUU, Hyderabad, India

OUTLINE

1 Introduction 467

2 Material and methods 469

2.1 Problem statement 469

2.2 Data 470

2.3 Machine learning methods 471

2.4 Performance evaluation metrics 473

3 Results and discussion 474

4 Conclusion 477

References 478

1 Introduction

In the parlance of open channels engineering, weirs are hydraulic structures widely used for various applications in water resource management, water supply, and irrigation (Niazkar and Afzali, 2018). Among different types of weirs, a side weir is commonly installed on a main channel sidewall not only to divert but also to regulate flow. To be more specific, the former use is considered in case of the shortage of enough space required for conveying the downstream water. Therefore, the diversion attempts to protect hydropower turbines, agricultural lands, or other hydraulic structures placed at the downstream of a main channel.

https://doi.org/10.1016/B978-0-323-91910-4.00026-1

The key advantage of a weir is that the overflow height over a typical weir can be utilized as an index for flow measurement. In other words, the discharge flowing in the channel can be estimated using the water depth that overflows the weir. This particular feature makes it possible to measure flows in canals by installing a suitable weir. The exclusive challenge in using weirs as flow measurement structures is that each weir needs to be calibrated in advance of any operation. The calibration is nothing but predicting a discharge coefficient that relates flow to overflow depth.

According to the literature, some attempts has been made to estimate the discharge coefficient of side weirs in favor of improving the accuracy of calculating discharge through side weirs (Mirzaei and Sheibani, 2020; Zakwan and Khan, 2020). Singh et al. (1994) conducted an analytical-experimental analysis to investigate changes of discharge coefficients of rectangular side weirs under the subcritical flow condition. They reported that the corresponding discharge coefficient is inversely proportional to the Froude number. Furthermore, Keshavarzi and Ball (2014) carried out an experimental investigation to predict discharge coefficients of side weirs installed in rectangular and trapezoidal canals. They observed that the discharge coefficient is a function of the Froude number, the ratio of the crest height of the side weir to the flow depth at the upstream of the weir, and the wall slope of the main channel. In addition, Bagheri et al. (2014) studied discharge coefficients of rectangular side weirs experimentally. Their experiments aimed to determine impacts of hydraulic and geometric parameters on the corresponding discharge coefficient. Moreover, Ebtehaj et al. (2015a) explored the application of Gene Expression Programming (GEP) model to estimate discharge coefficients of rectangular side weirs. They developed an equation that incorporates hydraulic and geometric parameters in the calculation of the discharge coefficient, while their comparative analysis indicated that the GEP model performs better than those of other available methods. Similarly, Ebtehaj et al. (2015b) used Group Method of Data Handling (GMDH) to forecast discharge coefficients of side weirs. Their findings demonstrated that the GMDH yielded more accurate results than that of Artificial Neural Network (ANN). Also, Khoshbin et al. (2016) suggested a hybrid model to estimate the discharge coefficient of side weirs. The hybrid method combines three soft computing models: Adaptive Neuro-Fuzzy Inference System (ANFIS), Genetic Algorithm (GA) and the Singular Value Decomposition (SVD). In addition, Azimi et al. (2017a) employed Extreme Learning Machine (ELM) not only to predict discharge coefficients of side weirs on placed trapezoidal channels but also to delineate how each involving parameters may affect the results through a sensitivity analysis. Moreover, Azimi et al. (2017b) applied GEP to propose a relation for estimating discharge coefficients of side weirs located on trapezoidal canals under subcritical conditions. Furthermore, Azimi et al. (2019) compared six models based on Support Vector Machine for predicting discharge coefficients of weirs designed on a trapezoidal canal. Also, Zakwan and Khan (2020) applied Generalized Reduced Gradient (GRG) technique to estimate the discharge coefficient of a side weir. Additionally, Haghshenas and Vatankhah (2021) proposed a formula for semicircular side weirs using the height, approach Froude number and radius of side weirs. In addition, Maranzoni and Tomirotti (2021) employed three-dimensional computational fluid dynamics to understand flow characteristics of oblique side weirs. Despite of previous efforts for estimating discharge coefficients of side weirs, there is still a quest for further investigations on this endeavor, particularly with the emergence of new machine learning (ML) methods.

This chapter tends to investigate the application of a powerful ML named multigene genetic programming (MGGP) to estimate discharge coefficients of side weirs installed on rectangular and trapezoidal canals. For this application, a few sets of experimental results were collected from the literature. The results of the MGGP-based model were compared with that of ANN, while the observed data was considered as the benchmark solution for this comparative analysis.

2 Material and methods

2.1 Problem statement

A side weir can be used for flow measurement if it is calibrated. The calibration process of a typical weir is to determine a relationship between discharge and the water depth that flows over the weir. The relation commonly entails a coefficient, invariantly called the discharge coefficient, which cannot be measured directly in an experiment. The discharge coefficient can have a fixed value for a specific set of flow and channel geometry condition, while it may vary under different circumstances. In this regard, previous experimental studies carried out on side weirs aimed to address which parameters play a role in the magnitude of the discharge coefficient of a side weir.

Based on the results of previous experimental investigations, the discharge coefficient of a side weir located on a trapezoidal or a rectangular channel is a function of six dimensionless parameters under the subcritical flow condition (Azimi et al., 2017b):

$$C_d = f\left(Fr, \frac{L}{b}, \frac{L}{y_1}, m, \frac{y_1}{b}, \frac{W}{y_1}\right) \tag{1}$$

where C_d is the discharge coefficient of a side weir, f denotes a function, Fr is the Froude number, L is the side weir length, b is the width of the main channel, y_1 is the flow depth at the upstream of the side weir, m is the slope of channel side walls, and W is the crest height of the side weir. The geometric parameters appeared in Eq. (1) are introduced in Fig. 1 for a better clarification.

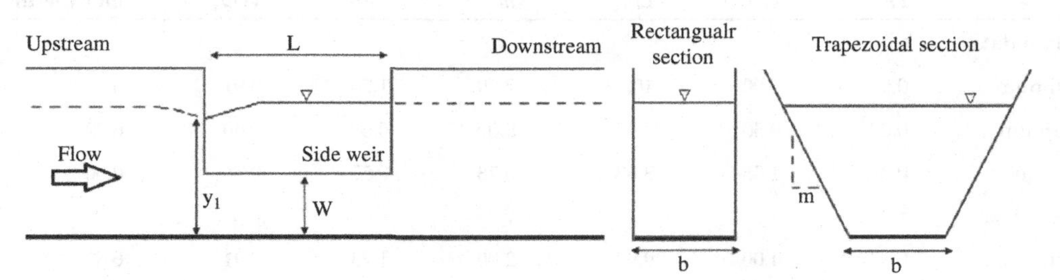

FIG. 1 A schematic view of a side weir.

According to Eq. (1), when the side weir is installed in a rectangular canal, $m=0$, whereas $m \neq 0$ if the main channel has a trapezoidal cross section. Furthermore, Eq. (1) implies that under the subcritical flow condition, the impact of the bottom channel slope on C_d can be neglected, as observed in the previous experimental studies (Borghei et al., 1999; Azimi et al., 2017b).

2.2 Data

In this study, experimental data of four previous studies conducted on side weirs installed on rectangular (Emiroglu et al., 2011; Keshavarzi and Ball, 2014) and trapezoidal channels (Cheong, 1991; Bagheri et al., 2014) were collected. To be more specific, a straight rectangular channel with 12 m length, 0.5 m width, and 0.5 m height were used by Emiroglu et al. (2011) to estimate discharge coefficient of a side weir. Moreover, Bagheri et al. (2014) conducted an experimental analysis on the side weir placed at the distance of 4.5 m from the inlet of a rectangular open channel with a horizontal slope. The length, width and length of the main channel were 8 m, 0.4 m, and 0.6 m, respectively. In contrast, the channel considered in Cheong's (1991) study was a straight one with the trapezoidal cross section. The length and bed width of the main canal were 10 m and 0.67 m, respectively. Furthermore, the side weir was placed at the two-thirds of the channel length from the inlet of the canal. Moreover, different slopes were implemented for the channel sidewalls. Additionally, Keshavarzi and Ball (2014) studied the estimation of discharge coefficients of a side weir installed at the middle of a trapezoidal channel. The channel length, height and bottom width were 36 m, 0.5 m and 0.4 m, respectively, while the length and height of the weir were 0.4 m and 0.13 m, respectively.

The total number of the collected data was 302. The total data were randomly divided into two parts named train and test data. The former was mainly used to train the ML methods, whereas the latter was utilized to compare the performance of different estimation models. The number of data points in the train and test data was 226 (about 75% of the total data) and 76 (about 25% of the total data). Additionally, Table 1 presents the minimum, maximum and average values of different parameters involved in both train and test data. As shown, both

TABLE 1 Range of different parameters in the train and test data.

	Fr	L/b	L/y_1	m	y_1/b	W/y_1	C_d (based on experiments)
a. Train data							
Maximum	0.99	4.00	10.71	2.00	1.76	0.91	0.93
Minimum	0.08	0.30	0.35	0.00	0.08	0.00	0.22
Average	0.39	1.58	3.63	0.28	0.55	0.56	0.50
b. Test data							
Maximum	0.93	4.00	9.97	2.00	1.73	0.91	0.83
Minimum	0.09	0.30	0.42	0.00	0.16	0.00	0.31
Average	0.40	1.52	3.80	0.39	0.51	0.49	0.48

train and test parts contain data associated with both rectangular and trapezoidal cross sections. Finally, the detail of the random division is clarified in the literature (Niazkar and Afzali, 2018).

2.3 Machine learning methods

In the parlance of soft computational techniques, artificial intelligence and ML techniques provide searching for new solutions in any problem without the need to know the physical background of the process (Niazkar, 2019). In essence, this fundamental characteristic becomes tangible particularly when there is no analytical solution available or the numerical modeling of the phenomenon requires considerable data measurements and computational efforts (Niazkar, 2020). Therefore, applications of ML methods, which can be served as versatile estimation tools, need to be paid more attention in the absence of an analytical solution for calculating discharge coefficients of side weirs. In this context, in this chapter, two ML methods have been used to estimate discharge coefficients for side weirs installed in open channels with rectangular and trapezoidal cross sections. In the following, these two ML methods are briefly introduced.

2.3.1 *Artificial neural network*

Artificial Neural Network (ANN) is a well-documented ML technique that has been widely used for numerous purposes in the field of water resources. Generally, ANN comprises a network of interconnected layers, which are so called input, hidden and output layers. According to the ANN architecture, each layer contains a group of neurons, whose functionality enables them not only to store information but also transfer information exclusively to other neurons in adjacent layers. In other words, in each layer, neurons do not interact with one another because they carry information of independent variables. On the other hand, the back-and-forth data flow between neurons of adjacent layers makes it possible to develop an estimation model (Niazkar et al., 2019). Hence, ANN continues to search for an estimation model until either a desirable accuracy is obtained or the number of iterations is exceeded.

In this study, a feed-forward back-propagation three-layer ANN, whose controlling parameters were selected in accordance with previous studies (Niazkar et al., 2021), was utilized to estimate discharge coefficients. To be more precise, the input layer of the ANN model, which was used in this study, consisted of six neurons, while the hidden and output layers had 10 and 1 neuron, respectively. In a bid to estimate more accurate discharge coefficients, the gathered data were normalized by the maximum and minimum values of each parameter prior to the ANN training. For instance, the normalized Froude number for the ith data point ($\overline{Fr_i}$) is computed by $\overline{Fr_i} = \frac{Fr_i - Fr_{\min}}{Fr_{\max} - Fr_{\min}}$ where Fr_i is the observed Froude number for the ith data point, and $Fr_{\min}$ and $Fr_{\max}$ are the minimum and maximum values of the observed Froude number of the total data. Finally, the ANN results were demoralized for the comparison purpose.

2.3.2 *Multigene genetic programming*

MGGP is an improved version of Genetic Programming (GP), which is classified as a ML method. Similar to GP, MGGP not only has a tree-like structure but also exploits genetic algorithm as a search engine. In other words, both ML methods assume a tree configuration for a mathematical equation, while they employ the genetic optimization algorithm to optimize different components (i.e., nodes) of an estimation model (Niazkar and Zakwan, 2021a). To be more specific, a typical estimation model in GP is a gene (or a tree), which may consist of three kinds of nodes, named as root node, function node and terminal node. In contract with GP, an estimation model in MGGP can be comprised of more than one gene (or tree), which counts as the main difference between GP and MGGP (Niazkar, 2022).

The detail of developing estimation models using MGGP is depicted in Fig. 2. As shown, the only task that that an MGGP user is required to do is nothing by selecting appropriate values for the MGGP controlling parameters, which are listed in Table 2. According to the previous studies, the maximum number of genes allowed in an individual (G_{max}) and the maximum depth of trees (d_{max}) are two crucial parameters affecting the accuracy and complexity of an MGGP-based estimation model (Niazkar and Zakwan, 2021b).

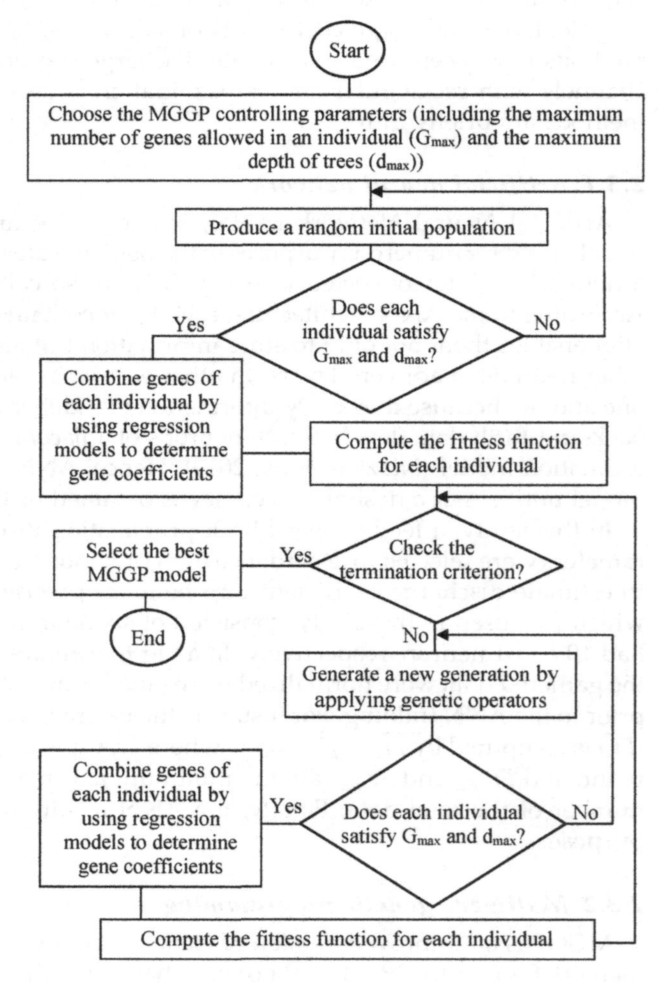

FIG. 2 Flowchart of the MGGP model for estimating discharge coefficients of side weirs.

TABLE 2 MGGP controlling parameters considered in this study.

Parameter	Values
Number of generations	120
G_{max}	5
d_{max}	4
Tournament size	4
Subtree mutation	0.9
Direct reproduction	0.05
Ephemeral random constants	[−10, 10]

Therefore, their values should be selected based on the trade-off existing in the accuracy and complexity of the estimation model in question. In this study, the values of the MGGP controlling parameters selected based on previous study, as shown in Table 2, while default values specified in the manual of the MGGP MATLAB code were used for other controlling parameters not mentioned in Table 2. The aforementioned MATLAB code was adopted from the literature and has been used in other applications in water resources (Niazkar, 2021; Zakwan and Niazkar, 2021). Moreover, the trigonometric functions, square function, exponential function and arithmetic operations are the functions allowed to be used in the function nodes of an MGGP-based model in this study. Additionally, the MGGP code was run for more than 50 times for developing the best estimation model for predicting discharge coefficients of side weirs in this study.

2.4 Performance evaluation metrics

In the comparative analysis, the performance of the ML-based models for predicting discharge coefficients of side weirs were evaluated using five metrics. These metrics are (1) Root Mean Square Error (RMSE), (2) Mean Absolute Relative Error (MARE), (3) Maximum Absolute Relative Error (MXARE), (4) Relative error (RE), and (5) determination coefficient (R^2), which are introduced in Eq. (2) to Eq. (6), respectively. Based on these equations, a lower value of RMSE, MARE, MXARE, and RE and a higher value of R^2 indicate a better performance in the comparative analysis (Niazkar and Zakwan, 2021a).

$$\mathrm{RMSE} = \sqrt{\frac{1}{N}\sum_{i=1}^{N}\left(C_{d,i}^{\mathrm{database}} - C_{d,i}^{\mathrm{estimated}}\right)^2} \tag{2}$$

$$\mathrm{MARE} = \frac{1}{N}\sum_{i=1}^{N}\left|\frac{C_{d,i}^{\mathrm{estimated}} - C_{d,i}^{\mathrm{database}}}{C_{d,i}^{\mathrm{database}}}\right| \tag{3}$$

$$\text{MXARE} = \max\left(\left|\frac{C_{d,i}^{\text{estimated}} - C_{d,i}^{\text{database}}}{C_{d,i}^{\text{database}}}\right| \quad \text{for} \quad i = 1, \ldots, N\right) \tag{4}$$

$$\text{RE} = \frac{C_{d,i}^{\text{estimated}} - C_{d,i}^{\text{database}}}{C_{d,i}^{\text{database}}} \tag{5}$$

$$R^2 = \left(\frac{\sum_{i=1}^{N}\left[\left(C_{d,i}^{\text{database}} - \frac{\sum_{i=1}^{N} C_{d,i}^{\text{database}}}{N}\right)\left(C_{d,i}^{\text{estimated}} - \frac{\sum_{i=1}^{N} C_{d,i}^{\text{estimated}}}{N}\right)\right]}{\sqrt{\sum_{i=1}^{N}\left[\left(C_{d,i}^{\text{database}} - \frac{\sum_{i=1}^{N} C_{d,i}^{\text{database}}}{N}\right)^2\left(C_{d,i}^{\text{estimated}} - \frac{\sum_{i=1}^{N} C_{d,i}^{\text{estimated}}}{N}\right)^2\right]}}\right)^2 \tag{6}$$

where $C_{d,i}^{\text{database}}$ is the ith discharge coefficient calculated based on the observed data, $C_{d,i}^{\text{estimated}}$ is the ith discharge coefficient estimated by the ML-based models, and N is the number of data.

3 Results and discussion

This study employs two ML methods, named ANN and MGGP, to estimate C_d of side weirs installed on rectangular and trapezoidal open channels. ANN, as a black box ML method, yielded a calibrated network that can be used for predictingC_d, whereas MGGP achieved an explicit equation, which can be further implemented in any numerical model if required. The MGGP-based model for estimating C_d of side weirs is given in Eq. (7):

$$\begin{aligned}\overline{C_d} = {} & \frac{12.87\overline{\frac{W}{y_1}}}{\exp\left(9.273\overline{\frac{W}{y_1}}\right)} - \frac{0.07527}{\exp\left(9.215\overline{\frac{W}{y_1}}\right)} - 0.684\overline{m}\overline{\frac{W}{y_1}} - 0.2241\overline{Fr} \\ & + \frac{5.142\overline{\frac{L}{b}}}{\exp\left(9.215\overline{\frac{L}{b}}\right)} + 0.4482\left(\overline{\frac{L}{b}}\right)^3 + 1.421\overline{Fr}\overline{\frac{L}{b}}\left(\overline{\frac{W}{y_1}}\right)^2 + 0.266\end{aligned} \tag{7}$$

where $\overline{C_d}$ is the normalized C_d, $\overline{\frac{W}{y_1}}$ is the normalized $\frac{W}{y_1}$, $\overline{m}$ is the normalized m, $\overline{Fr}$ is the normalized Fr, and $\overline{\frac{L}{b}}$ is the normalized $\frac{L}{b}$.

As shown in Eq. (7), the best MGGP-based model exploits all the involving parameters to estimate C_d of side weirs with the aid of the exponential and square functions. Furthermore, it can be directly utilized for side weirs located in both rectangular and trapezoidal channels, while the term including the side channel slope (i.e., $-0.684\overline{m}\overline{\frac{W}{y_1}}$) becomes zero for the former cross section as $m=0$ for canals with rectangular shapes. Unlike the ANN model, the explicit form of the MGGP-based model provides an opportunity to use it in numerical modeling in river engineering and channel design calculations.

Fig. 3 compares the performances of the two ML-based techniques in terms of RMSE, MARE and MXARE. As shown, ANN performs better than MGGP in respect with RMSE and MARE for both train and data. Additionally, based on the MXARE values shown in Fig. 3, ANN yield better estimations of discharge coefficients than MGGP for the train data, whereas MGGP resulted in better MXARE value than ANN for the test data.

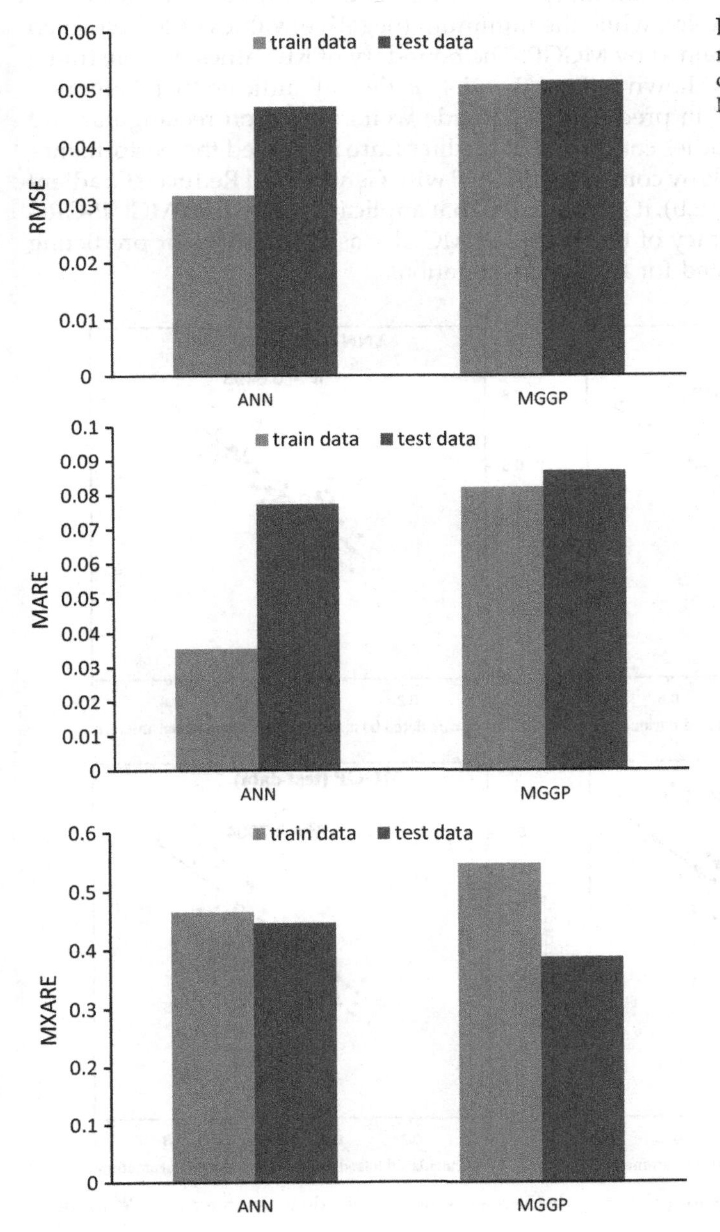

FIG. 3 Comparison of ML-based models for predicting discharge coefficients of side weirs in terms of RMSE, MARE and MXARE.

Fig. 4 depicts $C_{d,i}^{\text{estimated}}$ versus $C_{d,i}^{\text{database}}$ for both ANN and MGGP. As shown, the former obtained better R^2 values than the latter for both train and test data. To be more specific, ANN improves the values of R^2 obtained by MGGP about 11.6% and 7.4% for the train and test data, respectively.

The values of RE calculated by the ML-based prediction models for each data point are presented in Fig. 5. As shown, the maximum (positive) RE values computed by MGGP are higher than those calculated by ANN, while the minimum (negative) values of RE achieved by ANN are larger than those obtained by MGGP. The boundary of RE values demonstrated in Fig. 5 is in line with the results shown in Figs. 3 and 4, as they all indicate that ANN performs slightly better than MGGP in predicting C_d of side weirs placed on rectangular and trapezoidal canals. Since a few studies conducted in the literature improved the performance of MGGP-based estimation models by combining MGGP with Generalized Reduced Gradient (GRG) (Niazkar and Zakwan, 2021a,b), it is postulated that application of hybrid MGGP-GRG technique may improve the accuracy of the proposed MGGP-based equation for predicting C_d of side weirs, which is suggested for future investigations.

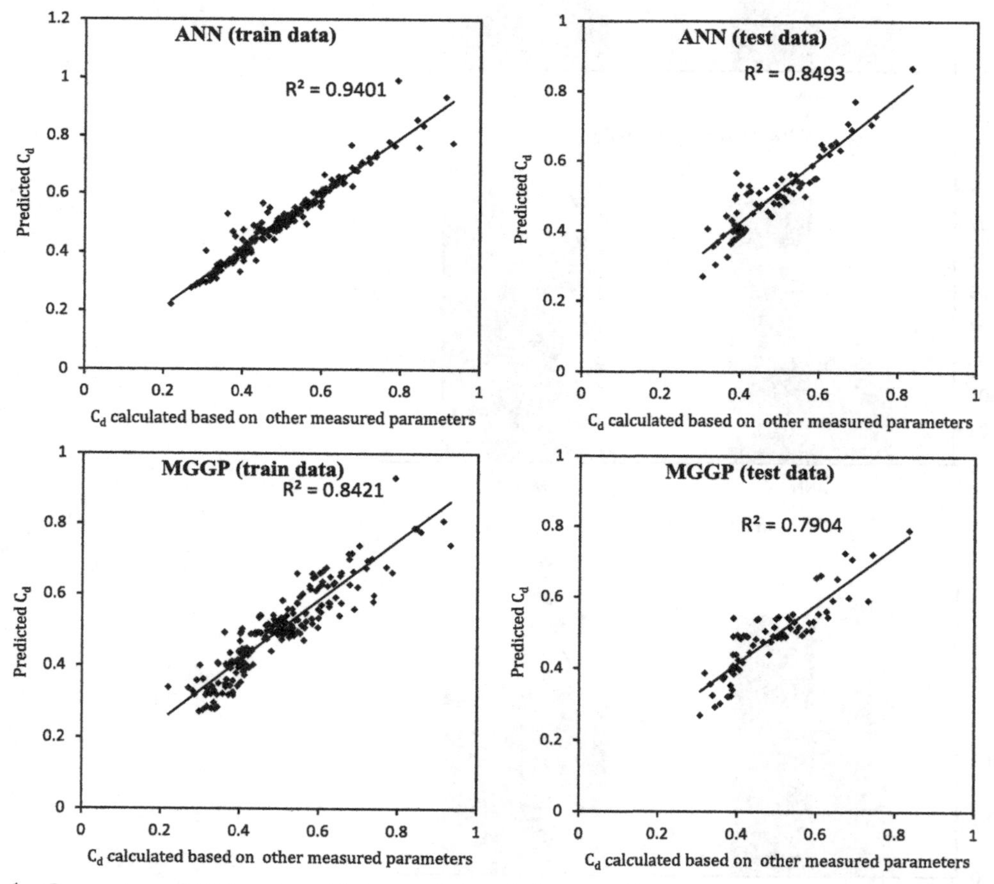

FIG. 4 Comparison of ML-based models for predicting discharge coefficients of side weirs in terms of R^2 for both train and test data.

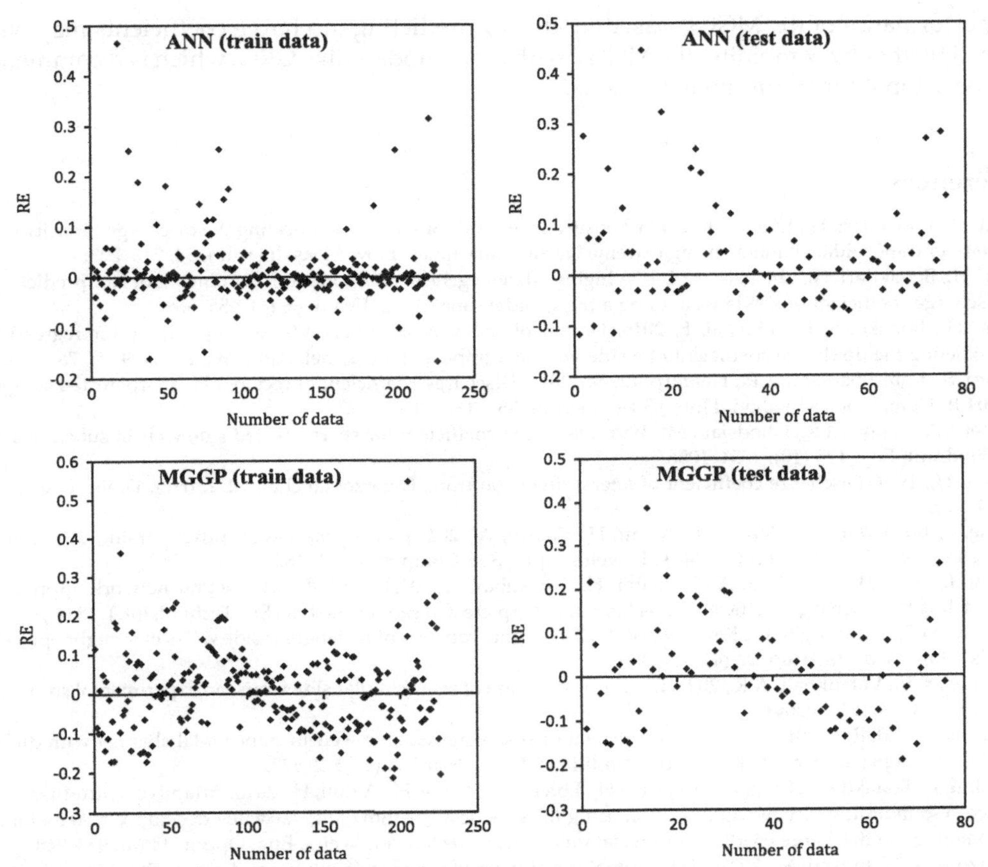

FIG. 5 Comparison of ML-based models for predicting discharge coefficients of side weirs in terms of RE for both train and test data.

4 Conclusion

This chapter explores the application of ANN and MGGP to develop new models for estimating discharge coefficients of side weirs placed on trapezoidal and rectangular canals under subcritical flow condition. Based on the current literature, it is the first time that MGGP has been used for this purpose. The ML-based prediction models are suggested as an alternative for the cumbersome process of calibration of side weirs. Although ANN develops a calibrated network that can be used for estimating discharge coefficients, MGGP provides an explicit equation that incorporates all flow and geometric parameters that play a role in changing the value of the discharge coefficient. Based on the five metrics used in this study, it was observed that ANN performs slightly better than MGGP in estimating discharge coefficients of side weirs. Nevertheless, the MGGP-based model can be utilized in numerical modeling in river engineering and the design of open channels. Finally, it is postulated that the performance of the MGGP-based model for predicting discharge coefficients may be improved further by combining the MGGP with other models, like GRG, which is recommended as a new topic for future investigations.

References

Azimi, H., Bonakdari, H., Ebtehaj, I., 2017a. Sensitivity analysis of the factors affecting the discharge capacity of side weirs in trapezoidal channels using extreme learning machines. Flow Meas. Instrum. 54, 216–223.

Azimi, H., Bonakdari, H., Ebtehaj, I., 2017b. A highly efficient gene expression programming model for predicting the discharge coefficient in a side weir along a trapezoidal canal. Irrig. Drain. 66 (4), 655–666.

Azimi, H., Bonakdari, H., Ebtehaj, I., 2019. Design of radial basis function-based support vector regression in predicting the discharge coefficient of a side weir in a trapezoidal channel. Appl. Water Sci. 9 (4), 78.

Bagheri, S., Kabiri-Samani, A.R., Heidarpour, M., 2014. Discharge coefficient of rectangular sharpcrested side weirs part II: Domínguez's method. Flow Meas. Instrum. 35, 116–121.

Borghei, S.M., Jalili, M.R., Ghodsian, M., 1999. Discharge coefficient for sharp-crested side weir in subcritical flow. J. Hydraul. Eng. 125 (10), 1051–1056.

Cheong, H., 1991. Discharge coefficient of lateral diversion from Trapezoidal channel. J. Irrig. Drain. Eng. 117 (4), 461–475.

Ebtehaj, I., Bonakdari, H., Zaji, A.H., Azimi, H., Sharifi, A., 2015a. Gene expression programming to predict the discharge coefficient in rectangular side weirs. Appl. Soft Comput. 35, 618–628.

Ebtehaj, I., Bonakdari, H., Zaji, A.H., Azimi, H., Khoshbin, F., 2015b. GMDH-type neural network approach for modeling the discharge coefficient of rectangular sharp-crested side weirs. Eng. Sci. Technol. Int. J. 18 (4), 746–757.

Emiroglu, M.E., Agaccioglu, H., Kaya, N., 2011. Discharging capacity of rectangular side weirs in straight open channels. Flow Meas. Instrum. 22 (4), 319–330.

Haghshenas, V., Vatankhah, A.R., 2021. Discharge equation of semi-circular side weirs: an experimental study. Flow Meas. Instrum. 81, 102041.

Keshavarzi, A., Ball, J., 2014. Discharge coefficient of sharp-crested side weir in trapezoidal channel with different side-wall slopes under subcritical flow conditions. Irrig. Drain. 63 (4), 512–522.

Khoshbin, F., Bonakdari, H., Ashraf Talesh, S.H., Ebtehaj, I., Zaji, A.H., Azimi, H., 2016. Adaptive neuro-fuzzy inference system multi-objective optimization using the genetic algorithm/singular value decomposition method for modelling the discharge coefficient in rectangular sharp crested side weirs. Eng. Optim. 48 (6), 933–948.

Maranzoni, A., Tomirotti, M., 2021. 3D CFD analysis of the performance of oblique and composite side weirs in converging channels. J. Hydraul. Res. 59 (4), 586–604.

Mirzaei, K., Sheibani, H.R., 2020. Experimental investigation of arched sharp-crested weir flow and comparing it with rectangular weir. Iran J. Sci. Technol. Trans. Civil Eng., 1–10. https://doi.org/10.1007/s40996-020-00425-6.

Niazkar, M., 2019. Revisiting the estimation of colebrook friction factor: a comparison between artificial intelligence models and C-W based explicit equations. KSCE J. Civ. Eng. 23 (10), 4311–4326. https://doi.org/10.1007/s12205-019-2217-1.

Niazkar, M., 2020. Assessment of artificial intelligence models for calculating optimum properties of lined channels. J. Hydroinf. 22 (5), 1410–1423. https://doi.org/10.2166/hydro.2020.050.

Niazkar, M., 2021. Optimum design of straight circular channels incorporating constant and variable roughness scenarios: assessment of machine learning models. Math. Probl. Eng. 2021, 9984934, 1–21. https://doi.org/10.1155/2021/9984934.

Niazkar, M., 2022. Multi-gene genetic programming and its various applications. In: Eslamian, S. (Ed.), 3-Volume Handbook of HydroInformatics (HandHyd, Elsevier). Elsevier. HandHyd Chief Editor. (Current status: accepted for publication).

Niazkar, M., Afzali, S.H., 2018. Application of new hybrid method in developing a new semicircular-weir discharge model. Alex. Eng. J. 57, 1741–1747. https://doi.org/10.1016/j.aej.2017.05.004.

Niazkar, M., Zakwan, M., 2021a. Application of MGGP, ANN, MHBMO, GRG and linear regression for developing daily sediment rating curves. Math. Probl. Eng. 2021a, 8574063, 1–13. https://doi.org/10.1155/2021/8574063.

Niazkar, M., Zakwan, M., 2021b. Assessment of artificial intelligence models for developing single-value and loop rating curves. Complexity 2021b, 6627011, 1–21. https://doi.org/10.1155/2021/6627011.

Niazkar, M., Talebbeydokhti, N., Afzali, S.H., 2019. Novel grain and form roughness estimator scheme incorporating artificial intelligence models. Water Resour. Manag. 33 (2), 757–773. https://doi.org/10.1007/s11269-018-2141-z.

Niazkar, M., Hajizadeh Mishi, F., Eryılmaz Türkkan, G., 2021. Assessment of artificial intelligence models for estimating lengths of gradually-varied flow profiles. Complexity 2021, 5547889, 1–11. https://doi.org/10.1155/2021/5547889.

Singh, R., Manivannan, D., Satyanarayana, T., 1994. Discharge coefficient of rectangular side weirs. J. Irrig. Drain. Eng. 120 (4), 814–819.

Zakwan, M., Khan, I., 2020. Estimation of discharge coefficient for side weirs. Water Energy Int. 62 (11), 71–74.

Zakwan, M., Niazkar, M., 2021. A comparative analysis of data-driven empirical and artificial intelligence models for estimating infiltration rates. Complexity 2021, 9945218, 1–13. https://doi.org/10.1155/2021/9945218.

SECTION VIII

Advances in sediment transport modelling and river engineering

CHAPTER

27

The hole size analysis of bursting events around mid-channel bar using the conditional method approach

Mohammad Amir Khan[a], *Nayan Sharma*[b], *Mohammad Aamir*[c], *Manish Pandey*[d], *Rishav Garg*[a], *and Hanif Pourshahbaz*[e]

[a]Department of Civil Engineering, Galgotias College of Engineering and Technology, Greater Noida, India [b]Center for Environmental Sciences & Engineering (CESE), Shiv Nadar University, Institution of Eminence, Greater Noida, UP, India [c]Department of Civil Engineering, Chaitanya Bharathi Institute of Technology, Hyderabad, India [d]National Institute of Technology Warangal, Warangal, India [e]Department of Civil and Water Engineering, Laval University, Pavillon Adrien-Pouliot, Québec, QC, Canada

OUTLINE

1 Introduction 484

2 Experimental program 485

3 Bursting events 487

3.1 *Variation of bursting events with the hole size* 488

3.2 *Temporal distribution of bursting events* 490

3.3 *Depth-wise distribution of Reynolds stress* 492

4 Conclusions 493

5 Limitation and scope for future research work 494

References 494

Current Directions in Water Scarcity Research, Volume 7
https://doi.org/10.1016/B978-0-323-91910-4.00027-3

1 Introduction

The Braiding process is found in alluvial rivers endowed with high fluvial energy environments. The river Brahmaputra is well-known on the Indian subcontinent for its extreme braided properties. Many academics have done extensive research on braided rivers and their channel pattern forms in the past (Callander, 1969; Sharma, 2004), etc. However, the braided rivers have been less studied as compared to the meandering river system.

Because sediment transport and water flow are inextricably linked, morphological changes in braided streams are related to bank erosion and deformation (Duan et al., 2020; Grass, 2020; Aamir and Sharma, 2015). Due to the widening of an initially straight channel in a braided river, the channel development process begins with the bar emergence under particular flow circumstances (Ashworth, 1996).

Nakagawa and Nezu (1978, 1981), Nezu and Nakagawa (2017), Huai et al. (2019), and Khan and Sharma (2021a) had performed experiments in a fully developed turbulent open channel flow. They studied the space time correlation structure of the bursting events using the conditional sampling technique. They observed that the temporal and spatial scale of sweep events extend more toward downstream than upstream.

According to Ashworth (1996), Schuurman et al. (2013), Khan and Sharma (2020, 2021a,b, 2022), the production and deformation of midchannel bars in an alluvial channel are temporary (2013). Taking 3-D measurements of essential microscale variables in a large alluvial river to get insight into the internal fluvial mechanism and precisely characterize turbulent bursts during channel generating/deforming flood flows is neither possible nor feasible. Significantly, the above-mentioned difficulty of faithfully and safely obtaining necessary hydraulic data for analysis is circumvented by using a commonly used alternative in hydraulic research for flow simulation in a scaled down miniature in lab hydraulic model. As a result, the idealized version of the bar model can be used for flow simulation in order to do quantitative study of turbulent burst at the bar, as reported in this chapter. Getting insight into the flow dynamics around the submerged structure requires a thorough understanding of turbulence (Raushan et al., 2018; Singh et al., 2018). Researchers Khan et al. (2016), Khan and Sharma (2019), and Jennifer et al. (2011) investigated 2-D bursting events and discovered that sediment entrainment is predominantly caused by the sweep event.

Several research workers, namely Ashworth (1996), and Khan et al. (2016), found that the creation and deformation of midchannel bars in an alluvial channel are observed to be transient. Significantly, as observed byAshworth (1996), and Ashworth (1996), the braiding behavior in alluvial streams originates from the initiation of the midchannel bar and that is the motivation for selecting its study in the present research. The midchannel bar is formed due to the deposition occurring at the center of the channel when the flow is shallow. In the natural

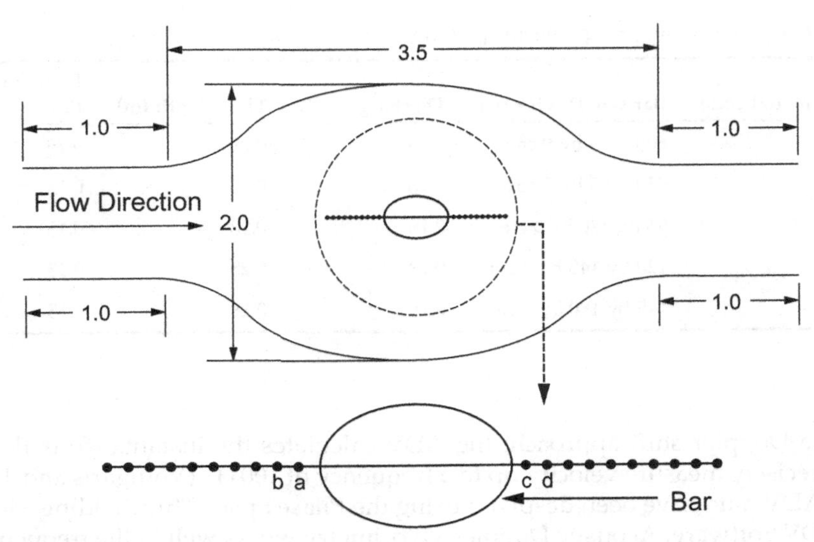

FIG. 1 The midchannel bar model.

channel, the shape of the bar is nonuniform. For the sake of simplifying the experimental setup, the elliptical-shaped bar is configured in the model tray (Fig. 1).

In a midchannel bar, the turbulent burst is critical for understanding the flow behavior at locations in the upstream flow divergence zone as well as at the downstream zone of flow convergence (Khan and Sharma, 2020). As a result, the fluctuating components of turbulent burst in the zones upstream and downstream of the midchannel bar are investigated in this chapter. The hole size (*H*) concept is utilized for separating the high intensity events from the low intensity events. The prime aim of the present study is to analyze the induced fluxes that get manifest in the turbulent structure due to the presence of a bar. Such analysis is focused on unraveling the inherent fluvial processes, which are primarily related to deposition and scouring in the proximity of the bar.

2 Experimental program

Experiments were conducted in a 10 m long, 2.6 m wide, and 1 m deep flume for this investigation. For all experimental runs, bed slope of the flume was kept constant at 0.005. For velocity measurements, Acoustic Doppler Velocimetry (ADV) is used.

TABLE 1 The details of experiments performed.

Experimental code	Bar size ($L \times B \times H_b$)	Discharge m^3/s	Flow depth (m)	Froude number Fr
1R	60 by 75 by 9 cm	0.18	0.28	0.15
2R	75 by 95 by 9 cm	0.18	0.28	0.15
3R	95 by 120 by 9 cm	0.18	0.28	0.15
4R	120 by 145 by 9 cm	0.18	0.28	0.15
5R	145 by 180 by 9 cm	0.18	0.28	0.15

Using the Doppler shift approach, the ADV calculates the instantaneous flow velocity. ADV can precisely measure velocity up to a frequency of 100 Hz (Voulgaris and Trowbridge, 1998). The ADV data have been despiked using the Phase Space Thresholding Method using the WIN ADV Software. Acoustic Doppler Velocimetry works well in the frequency range of 20–25 Hz, according to McLelland and Nicholas (2000). As a result, for velocity measurement, a frequency of 25 Hz was used. The noise is removed by applying Hurther and Lemmim's correction approach (2001).

The Acoustic Doppler Velocimetry setup is attached to the sliding rail for measuring the velocity at different location and knob is provided in experiment setup for moving the Acoustic Doppler Velocimetry up and down. The fully developed flow is maintained in the flume.

The midchannel bar model is shown in Fig. 1. The width of model is varying from 0.8 to 2 m. The width of midchannel model is maximum at the center of the bar (Fig. 1). The main reason of taking maximum width of channel at the center of bar is to minimize the effect of channel walls on the flow structure in the proximity of bar. The flow depth is controlled by tail gate. The experiments are performed in clear water condition.

For all experimental runs, the flow is kept constant at 28 cm with the help of tailgate. The flow rate of 0.18 m^3/s is used for all experimental runs (Table 1). In Table 1, B indicates the minor axis and L indicates the major axis of elliptical bar and H_b indicates the bar height.

Table 1 shows the details of experiments performed. The size of bar is varying and the depth of bar is kept constant. The velocity is measured at 20 sections in the proximity of the bar shown by dotted circles. Ten sections are taken upstream of midchannel bar and 10 sections are taken downstream of midchannel bar. Out of 20 measuring sections, only 4 sections are discussed in this chapter. These sections are named as "a," "b," "c," and "d" (Fig. 1).

The velocity measurements are taken at 20 different relative depths for each part. The ratio of the distance between the bed and the depth of flow is known as relative depth. Table 2 shows the relative depths at which velocity is observed.

TABLE 2 The relative depths at which velocity are measured for each point.

S. no.	Depth of flow h cm	Relative depth (z/h)
1	28	0.005
2	28	0.01
3	28	0.015
4	28	0.02
5	28	0.025
6	28	0.03
7	28	0.035
8	28	0.04
9	28	0.045
10	28	0.05
11	28	0.055
12	28	0.06
13	28	0.065
14	28	0.07
15	28	0.075
16	28	0.08
17	28	0.085
18	28	0.09
19	28	0.095
20	28	0.1

3 Bursting events

Turbulent bursts play an important role in sediment entrainment (Naot et al., 1993). The two-dimensional bursting process involves analyzing the relationship between the velocity fluctuations. The velocity fluctuation distribution u′ and w' in four quadrants is shown below. Here, u' and w' are the velocity fluctuations in longitudinal and vertical direction respectively.

The four quadrants depending on the sign of fluctuating velocity u' and w' are shown below.

Outward interaction in which $u'>0$, $w'>0$—Quadrant I
Ejection in which $u'<0$, $w'>0$—Quadrant II
Inward interaction in which $u'<0$, $w'<0$—Quadrant III
Sweep in which $u'>0$, $w'<0$—Quadrant IV

Nakagawa and Nezu (1981) observed that only the high intensity turbulent burst is responsible for the turbulent bursting process. Thus, hole size (H) concept is used in order to separate the low intensity events from the high intensity events. After the application of hole concept, the u'-w' plane is divided into five regions. The hole region is bounded by the curves $|u'\ w'| = H\overline{u'}\overline{w'}$ in the plane u'-w'. Here $\overline{u'}$ and $\overline{w'}$ are the local root-mean-square values of the turbulence velocity fluctuations in the longitudinal and vertical direction respectively. The hole size (H) is the threshold parameter in the Reynolds Stress signal; which is utilized to select the Reynolds Stress having magnitude more than the H times of its average value.

Contributions to the Reynolds Stress from each quadrant for Zero-hole size are calculated using Eq. (1) (Nakagawa and Nezu, 1981).

$$RS_i = -\frac{1}{N_t}\sum_{j=1}^{N_t}\left[u'w_j'(t)\right]_i \tag{1}$$

Here $[u'\ w_j'(t)]_i$ represents product of velocity fluctuations belonging to ith quadrant. Here RS_i is fractional Reynolds stress contribution from ith coordinate; N_t is the total number of products $u'(t)\ w'(t)$.

For hole size H, fractional Reynolds Stress contribution for each quadrant is given by Eq. (2) (Nakagawa and Nezu, 1981).

$$RS_{i,H} = -\frac{1}{N_t}\sum_{j=1}^{N_t}X_Q\left[u'w_j'(t)\right]_i \tag{2}$$

The indicator function X_Q, for performing conditional averaging procedure, is defined in Eq. (3) (Nakagawa and Nezu, 1981).

$$X_Q = \left\{\frac{1 \text{ for } \left|u'w_j'\right|_i \geq H\sqrt{\overline{u'^2}}\sqrt{\overline{w'^2}}}{0 \text{ for } \left|u'w_j'\right|_i < H\sqrt{\overline{u'^2}}\sqrt{\overline{w'^2}}}\right\} \tag{3}$$

$$RS_{total} = \frac{\tau}{\rho} = \sum_{i=1}^{5} RS_i \tag{4}$$

In Eq. (4), $RS_{total}$is the total stress for the whole observation time related to the chosen pairs of turbulence velocities. $|u'w_j'|_i$ represents the products of jth fluctuating components of longitudinal and vertical velocities which belongs to quadrant i, here $j=1, 2...N$.

3.1 Variation of bursting events with the hole size

Researchers Jennifer et al. (2011), Nezu et al. (1994) found that the high intensity events contribute to the flow turbulence. They used the concept of hole size for separating the high intensity events from the weak intensity quadrant events. The effect of hole size on the bursting events is studied in detail. The Reynolds Stress contributions from each quadrant events are computed using the Eq. (2). Low magnitude of hole size causes selection of both low and high intensity events and High magnitude of hole size separates the high intensity events from the low intensity events. The hole size effect on the bursting events is studied in this section. Seven different hole sizes are arbitrarily selected for analyzing the quadrant events variation with the hole sizes. These hole sizes are taken as per as Jennifer et al. (2011). The variations of quadrant events with the hole sizes are computed for all 20 measuring sections. As it is not possible to display all the graphs, so only the four sections "a," "b," "c," and "d" are selected for displaying the graph.

The variation of quadrant events with the hole size plotted at "a," "b," "c," and "d" sections for 2R experimental run (Fig. 2). For points upstream of bar, i.e., "a" and "b" sections, the value of occurrence probability of even events above 0.3 and value of occurrence probability of odd events lies below 0.2 for points upstream of bar. This indicates that the even events are dominant at the zone upstream of bar. and the value of these events increasing with value of hole size increases.

For points downstream of midchannel bar, i.e., "c" and "d" sections, the occurrence probability of odd events lies above 0.28 and occurrence probability of even event lies below 0.24.

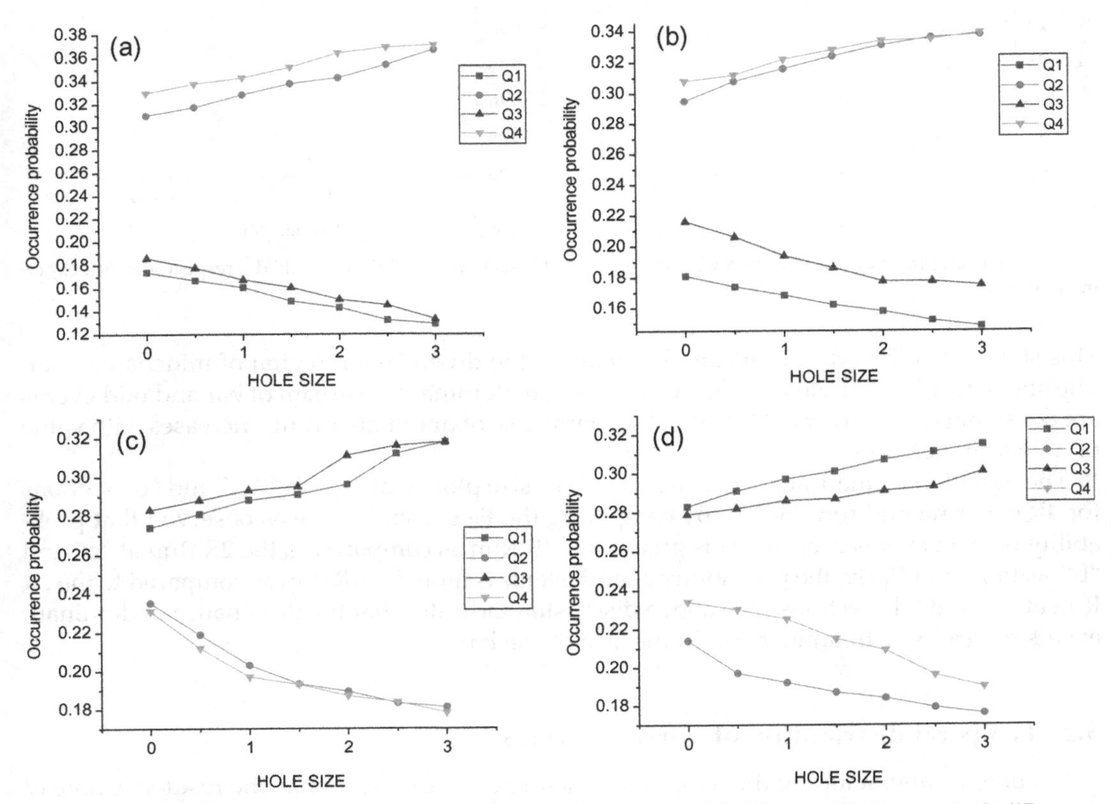

FIG. 2 The quadrant events variation with the hole size for section "a," "b," "c," and "d," respectively (2R experimental run).

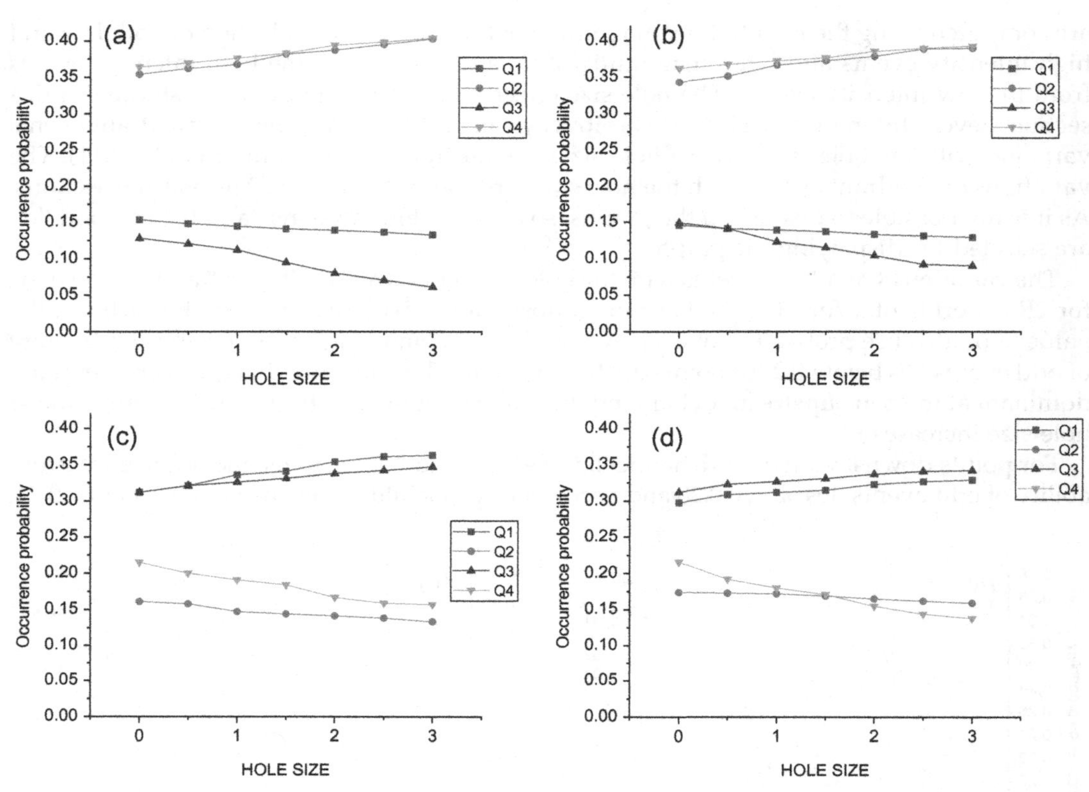

FIG. 3 The quadrant events variation with the hole size for section "a," "b," "c," and "d," respectively (4R experimental run).

This shows that the odd events are dominant at the downstream region of midchannel bar. The above results indicate that the even events are dominant upstream of bar and odd events are dominant downstream of bar and the dominance of dominant events increases with value of hole size increases.

The variation of quadrant events with the hole size plotted at "a," "b," "c," and "d" sections for 4R experimental run (Fig. 3). By comparing the Figs. 2 and 3, it was observed that probability of even events occurrence is greater for 4R Run as compared to the 2R Run at "a" and "b" sections. Similarly, the probability of odd events is more for 4R Run as compared to the 2R Run at "c" and "d" sections. The above discussion indicates that the dominance of dominant events increases with an increase in the size of the bar.

3.2 Temporal distribution of bursting events

For better understanding the temporal structure of quadrant events, the residence time of quadrant events are plotted against the frequency of occurrence of these events in this study.

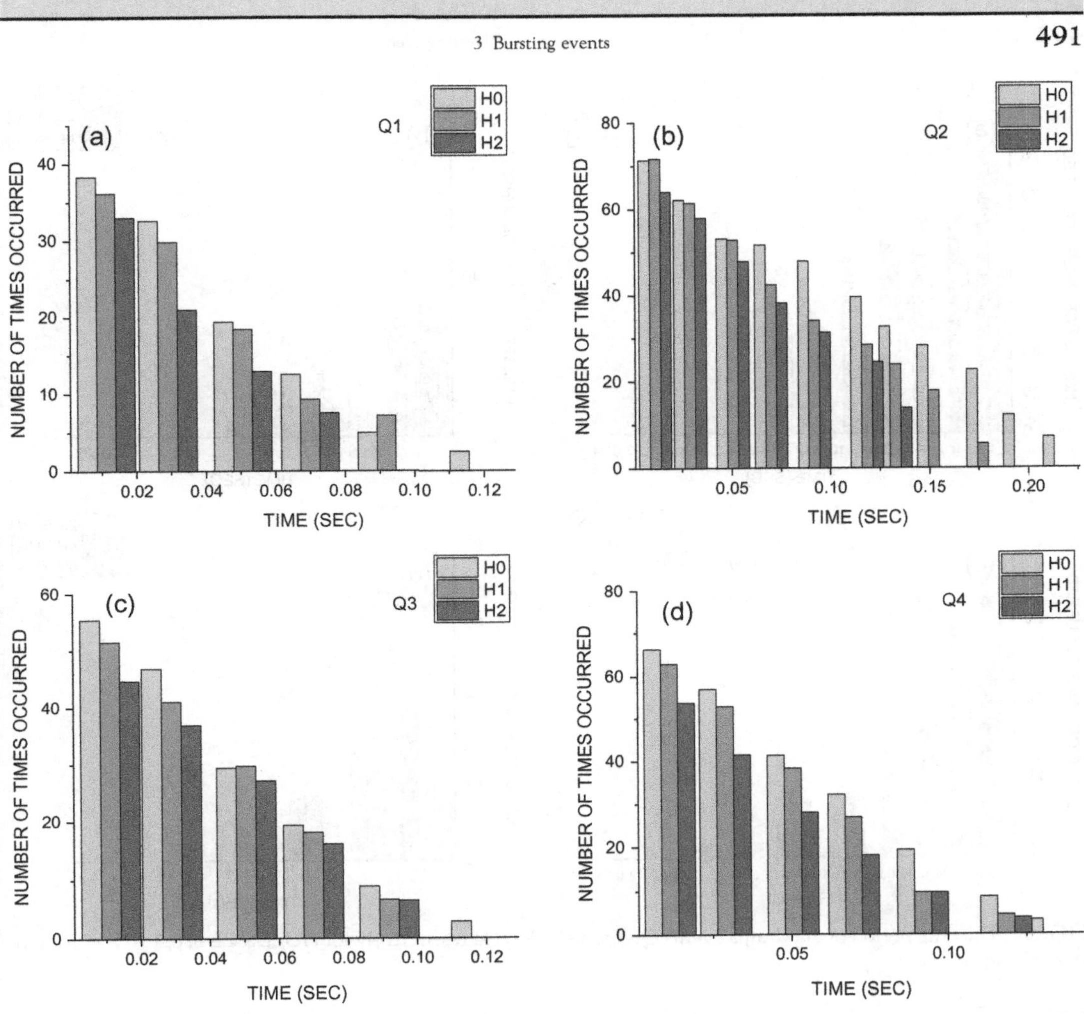

FIG. 4 The time period of quadrants events Q1, Q2, Q3 AND Q4 respectively for HOLE SIZE of 0, 1 AND 2 (upstream zone of bar).

The averaged time period of quadrant events plotted at hole size of 0, 1 and 1.5 for zone upstream of midchannel bar (Fig. 4). The Fig. 4 indicates that the Q2 events persist up to time period of 0.2 s. The other quadrant events do not last more than 0.12 s (Fig. 4). This indicates that the Q2 events are temporally more stable in zone upstream of the bar. This shows that the events of second quadrant having high occurrence time are observed at the upstream zone of midchannel bar.

The averaged time period of quadrant events plotted at hole size of 0, 1 and 1.5 for zones downstream of midchannel bar (Fig. 5). Fig. 5 indicates that the Q1 events persist up to time period of 0.2 s at region downstream of midchannel bar. This indicates that the Q1 events are temporally more stable in zone downstream of the midchannel bar.

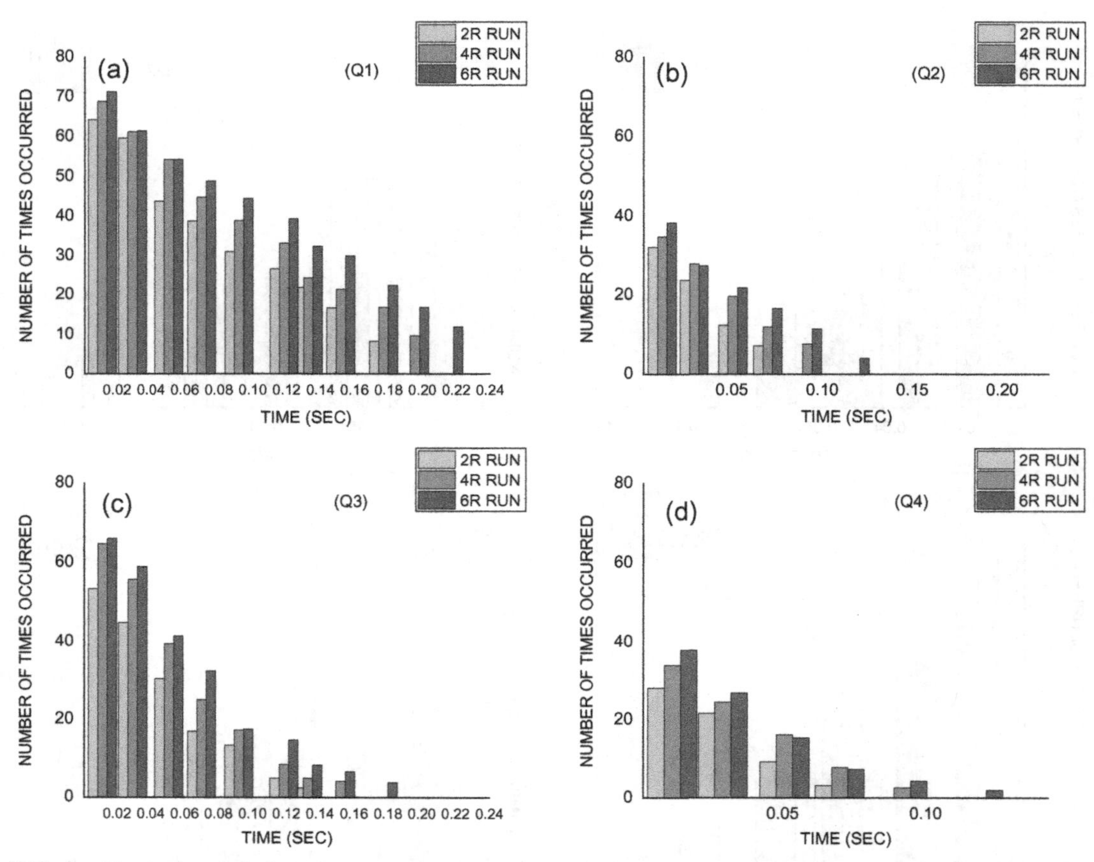

FIG. 5 The time period of quadrants events Q1, Q2, Q3 AND Q4 respectively for HOLE SIZE of 0, 1 AND 2 (downstream zone of bar).

Figs. 4 and 5 indicate that the frequency of quadrant events occurrence decreases with value of hole size increases. The above point indicates that the events of high magnitude have lower value of occurrence frequency.

3.3 Depth-wise distribution of Reynolds stress

Researchers such as Nakagawa and Nezu (1978), Wu and Jiang (2007) found that the bursting events are correlated with the sediment entrainment rather than the total shear stress. Thus, for thoroughly understanding the flow behavior, the conditional method is utilized for decomposing the Reynolds Stress into the bursting events. The Quadrant Reynolds Stress is computed by using the Eq. (1). The Depth-wise distribution of fractional Reynolds Stress is plotted at "a," "b," "c," and "d" sections for 2R experimental run (Fig. 6). The Fig. 6 shows that the Quadrant Reynolds Stress is greater in the near bed zone, i.e., relative depth < 0.02 (Fig. 6). This indicates that the turbulent burst is dominant only in the near bed zone.

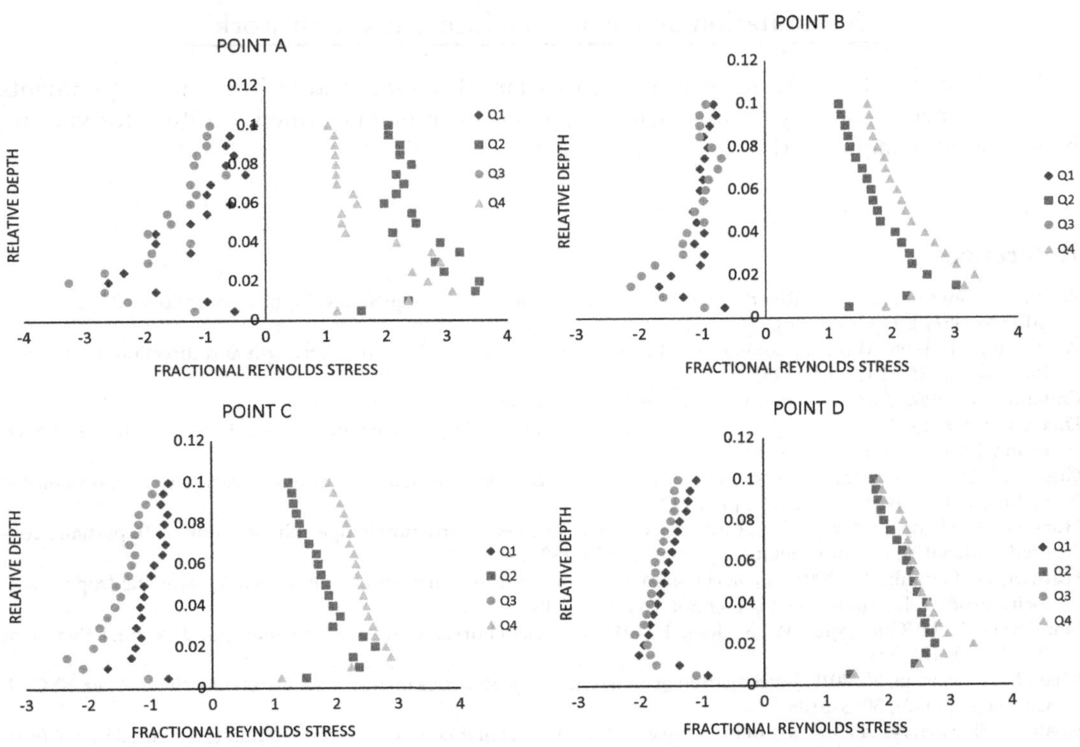

FIG. 6 The Fractional Reynold's stress of quadrant events for points A, B, C and D respectively (2R experimental run).

4 Conclusions

The value of occurrence probability of even events increases at scouring zone. The value of occurrence probability of odd events increases at depositional zone. The results indicate that the II and IV quadrants are dominant at zone upstream of bar and I and III quadrant are dominant at zone downstream of bar and the dominance of dominant events increases with increase in the hole size. The magnitude of dominant events also increases with bar size increase.

The Q2 events persist up to time period of 0.2 s at region upstream of bar. The other quadrant events do not last more than 0.12 s The results indicate that the Q2 events are temporally more stable in zone upstream of the bar. The Q1 events persist up to time period of 0.2 s at zone downstream of midchannel bar. This indicates that the Q1 events are temporally more stable in zone downstream of the bar. The occurrence frequency of quadrant events decreases with value of hole size increase. This clearly state that the high intensity quadrant events have lesser frequency of occurrence.

The Quadrant Reynolds Stress is greater in the near bed zone. This indicates that the turbulent burst is dominant only in the near bed zone of bar.

5 Limitation and scope for future research work

The experiments are performed for the constant discharge and in future the experiments can be performed at varying rate. The experiments can be performed in future for varying height of midchannel bar.

References

Aamir, M., Sharma, N., 2015. Riverbank protection with porcupine systems: development of rational design methodology. ISH J. Hydraul. Eng. 21 (3), 317–332.

Ashworth, P.J., 1996. Mid-channel bar growth and its relationship to local flow strength and direction. Earth Surf. Process. Landf. 21 (2), 103–123.

Callander, R., 1969. Instability and river channels. J. Fluid Mech. 36 (3), 465–480.

Duan, Y., Chen, Q., Li, D., Zhong, Q., 2020. Contributions of very large-scale motions to turbulence statistics in open channel flows. J. Fluid Mech. 892.

Grass, A.J., 2020. The influence of boundary layer turbulence on the mechanics of sediment transport. In: Mechanics of Sediment Transport. CRC Press, pp. 3–17.

Huai, W.-X., Zhang, J., Wang, W.-J., Katul, G.G., 2019. Turbulence structure in open channel flow with partially covered artificial emergent vegetation. J. Hydrol. 573, 180–193.

Hurther, D., Lemmin, U., 2001. A correction method for turbulence measurements with a 3D acoustic Doppler velocity profiler. J. Atmos. Ocean. Technol. 18 (3), 446–458.

Jennifer, D., Li, H., Guangqian, W., Xudong, F., 2011. Turbulent burst around experimental spur dike. Int. J. Sediment Res. 26 (4), 471–523.

Khan, M.A., Sharma, N., 2019. Investigation of coherent flow turbulence in the proximity of mid-channel bar. KSCE J. Civ. Eng. 23 (12), 5098–5108.

Khan, A., Sharma, N., 2020. Study of bursting events and effect of hole-size on turbulent bursts triggered by the fluid and mid-channel bar interaction. Water Sci. Technol. Water Supply 20 (6), 2428–2439.

Khan, M.A., Sharma, N., 2021a. Experimental Observation of Turbulent Structure at Region Surrounding the Mid-Channel Braid Bar. pp. 1–14.

Khan, M.A., Sharma, N., 2021b. Two-Dimensional Turbulent Burst Examination and Angle Ratio Utilization to Detect Scouring/Sedimentation Around Mid-Channel Bar. pp. 1–14.

Khan, M., Sharma, N., 2022. Analysis of turbulent flow structure with its fluvial processes around mid-channel bar. Sustainability 14 (1), 392.

Khan, M.A., Sharma, N., Singhal, G.D., 2016. Experimental study on bursting events around a bar in physical model of a braided channel. ISH J. Hydraul. Eng. 23 (12), 1–8.

McLelland, S.J., Nicholas, A.P., 2000. A new method for evaluating errors in high-frequency ADV measurements. Hydrol. Process. 14 (2), 351–366.

Nakagawa, H., Nezu, I., 1978. Bursting Phenomenon Near the Wall in Open-Channel Flows and Its Simple Mathematical Model. vol. 40 Kyoto University Faculty Engineering Memoirs, pp. 213–240.

Nakagawa, H., Nezu, I., 1981. Structure of space-time correlations of bursting phenomena in an open-channel flow. J. Fluid Mech. 104, 1–43.

Naot, D., Nezu, I., Nakagawa, H., 1993. Hydrodynamic behavior of compound rectangular open channels. J. Hydraul. Eng. 119 (3), 390–408.

Nezu, I., Nakagawa, H., 2017. Turbulence in Open-Channel Flows. Routledge.

Nezu, I., Nakagawa, H., Jirka, G.H., 1994. Turbulence in open-channel flows. J. Hydraul. Eng. 120 (10), 1235–1237.

Raushan, P.K., Singh, S.K., Debnath, K., Aerodynamics, I., 2018. Grid generated turbulence under the rigid boundary influence. J. Wind Eng. Ind. Aerodyn. 182, 252–261.

Schuurman, F., Marra, W.A., Kleinhans, M., 2013. Physics-based modeling of large braided sand-bed rivers: bar pattern formation, dynamics, and sensitivity. J. Geophys. Res. 118 (4), 2509–2527.

Sharma, N., 2004. Mathematical modelling and braid indicators. In: The Brahmaputra Basin Water Resources. Springer, pp. 229–260.

Singh, S.K., Raushan, P., Debnath, K., Mukherjee, M., Mazumder, B., 2018. Turbulence characteristics in boundary layers over a regular array of cubical roughness. ISH J. Hydraul. Eng., 1–14.

Voulgaris, G., Trowbridge, J.H., 1998. Evaluation of the acoustic Doppler velocimeter (ADV) for turbulence measurements*. J. Atmos. Ocean. Technol. 15 (1), 272–289.

Wu, F., Jiang, M., 2007. Numerical investigation of the role of turbulent bursting in sediment entrainment. J. Hydraul. Eng. 133 (3), 329–334.

CHAPTER

28

Magnitude frequency analysis of sediment transport: Concept, review, and application

Mohammad Zakwan[a], *Qamar Sultana*[b], *and Gulfam Ahamad*[c]

[a]School of Technology, MANUU, Hyderabad, India [b]Department of Civil Engineering, MJCET, Hyderabad, India [c]Department of Computer Sciences, BGSBU, Rajouri, India

OUTLINE

1 Introduction 497
1.1 Dominant discharge 498
1.2 Bankfull discharge 498
1.3 Discharge for a specific recurrence interval 499
1.4 Effective discharge 499
2 Literature review 503
3 Materials and method 508
3.1 Effective discharge computation 508
4 Results and discussion 508
5 Applications of effective discharge in river design and monitoring 509
6 Conclusion 510
References 510

1 Introduction

As water flows through rivers and channels, it carries with it some sediment which encompass sediment settled at the bottom of the channel and also which is in suspension form. Suspended load forms an important variable in sediment problems as it travels with approximately the same velocity as that of water. The extremely large variation in sediment concentration indicates that large numbers of factors affect the movement and quantity of sediment

https://doi.org/10.1016/B978-0-323-91910-4.00028-5

in alluvial streams. It is affected by hydrological as well as hydraulic characteristics. Various experimental relationships have been deduced to explain temporal sediment transport variability for different periods that is days to years. Primarily, hydrologist and geomorphologist were concerned with the sediment load transported by floods to shape the channels.

1.1 Dominant discharge

The alluvial streams carry highly varying discharges and sediment loads. The ratio of maximum to minimum discharge can attain values as high as 1000 in many streams and is likely to have even higher values for flashy streams. Similarly the variation in sedimentation load can also be very large. During low flows there can be very little transport, while during high flows the stream may carry high sediment loads with wide range of sediment size. It becomes tougher to select a particular discharge to study the various features of the stream as there is wide range in stream flow. Inglis (1947) introduced the concept of dominant discharge and stated that it is required for the design of a steady and well-formed stable waterway. According to Inglis (1947) controlling flow rate is the one having 60% of the safe maximum flow adopted for design which depends on the ways, the land alongside the river has been utilized. Hence the controlling flow estimated depending upon the maximum flow will not reflect the river forming features.

Till the 1960s, it was presumed that the flows of large magnitude with small reoccurrence frequency the river formation. So, it was assumed that the large magnitude flows have more potential to carry and transfer sediment and hence change the channel form. But however, this idea was contradicted by Wolman and Miller (1960), who displayed that for a period of time, the stream flow, transporting sediment of a certain magnitude and contributing to the forming of the channel is influenced by both the sediment carrying capacity and its recurrence interval. Very large flood events produce extremely large quantities of sediment, but occur very rarely and for so short duration that over a period, they contribute very less for the movement of the sediment. Even small flood events contribute very less to the sediment movement due to low sediment transport capacity despite their frequent occurrence. From this observation it follows that moderate magnitude and frequency flows contribute the largest amount of sediment (Biedenharn and Thorne, 1994).

1.2 Bankfull discharge

The bankfull discharge is defined as that largest flow a channel can carry in it without causing overflow on to the floodplain. It is of utmost morphological importance as it symbolizes the turning point from the processes of channel formation to floodplain formation. This maximum flow is obtained by initially locating the largest flow stage and then calculating the flow rate related to that stage. To identify the required attributes of the area which helps to determine the level of stream related to maximum flow will be tougher. Several ground measures to obtain the river stage are put forth, but all these seem to be not relevant or free from questionability (Williams, 1978). One common definition for bankfull stage which is widely accepted is the altitude of the flat land near the river which gets flooded during heavy flows. Another way of representing the maximum flow level is the elevation where the width to depth ratio is a minimum, which seems to be a reliable and structured idea depending on exact ground studies. The maximum altitude of waterway bars or where there are significant changes in vegetal land cover can be utilized as measures of bankfull stage in many situations. Hence it can be concluded that there can be large variability in ground finding of bankfull stage as there are innumerable ways and techniques available.

The ideal method to obtain the flow rate related to bankfull stage is by using a stage-discharge rating curve which relies on information collected on the field. Several unpredictabilities involve those related with hydraulic roughness coefficients and the channel cross-sections influencing the correctness of this rating curve. The calculation of bankfull stage for a stream reach of nearly one wavelength or ten channel widths will reduce these uncertainties.

1.3 Discharge for a specific recurrence interval

Several researchers have related the flows that alter the sections of streams to that which reoccur at a particular interval as they faced many difficulties in identifying bankfull stage and discharge. In this context, stable streams for which bankfull elevation were easily obtainable and stream measuring meters available nearby were selected for study by the researchers. In general, it is assumed that the maximum flows in stable channels without causing bank overflows corresponds to the annual flows that occur after every 1 to 2.5 years and discharge associated with 1.5 years return period is found to be a reliable one for various rivers (Leopold, 1994). Wolman and Leopold (1957), observed the frequency of the occurrence of the maximum flows as one to two years. However, there were numerous cases where it was seen that the 1-to-2.5-year recurring flows do not control the geometrical changes of a river. For example, Williams (1978) in his study of 35 floodplains of United States observed that there was a wide variation in the frequency of recurring of maximum flows and found, one third of the total flood plains to have a maximum flow with 1 to 5 years as the range of recurring frequency. Similarly, Pickup and Warner (1976) in his studies obtained the frequency of recurrence of maximum flows to be between 4 and 10 years. Considering the above discrepancies, the researchers have come to a point that the determination of bankfull flows do not have a good result by using the concept of recurring frequency and it should be supported by confirmation from field data so as to ensure that the particular chosen flows displays important physical attributes.

1.4 Effective discharge

Effective discharge is defined as those flows which carries maximum quantities of sediment over a period when compared to other flow periods (Wolman and Miller, 1960). After observing stabilized rivers, an attempt has been made by various researchers, to display a resemblance between bankfull and effective discharges (Andrews, 1980; Carling, 1988). But they are not always identical as concluded by Benson and Thomas (1966), Pickup and Warner (1976), Webb and Walling (1982) and Lyons et al. (1992). Hence, it can be concluded that effective discharge may in many cases act as substitute for the channel-forming flows.

In the beginning researchers followed theoretical approach estimating the effective discharge, later on Nash (1994) proposed analytical approach for estimating the effective discharge. At present both theoretical and analytical approach are prevalent for calculation of effective discharge. Theoretical approach involves determination of frequency of each discharge class interval and average discharge of that class interval, product of these two quantities will provide the sediment load transported by a particular discharge class over a period of time. Analytical approach involves fitting a frequency distribution to the discharge data and fitting a sediment rating curve to sediment discharge data. Most widely used sediment rating equation is presented in Eq. (1).

$$Q_s = aQ^b \tag{1}$$

where Q_s is the sediment load, Q is the corresponding discharge, and a and b are regression coefficients.

If F is discharge frequency distribution, then its effectiveness (+) in sediment transport can be expressed as

$$E = SF \tag{2}$$

For maximum effectiveness or effective discharge identification

$$\frac{\partial E}{\partial Q} = 0 \tag{3}$$

Sediment ratings are also used in theoretical approach when sufficient sediment load data is unavailable. However, Zakwan et al. (2018) suggested that use of sediment rating curve to estimate sediment load may lead to under/over estimation of effective discharge.

Over the years the popularity of effective discharge computation or magnitude frequency analysis of sediment transport has increased. Several researchers have employed this concept to address the issues with the rivers all over the world (Wolman and Miller, 1960; Pickup and Warner, 1976; Andrews, 1980; Carling, 1988; Biedenharn and Thorne, 1994; Hudson and Mossa, 1997; Orndorff and Whiting, 1999; Emmett and Wolman, 2001; Goodwin, 2004; Crowder and Knapp, 2004; Lenzi et al., 2006; Quader and Guo, 2009; Saeidi et al., 2011; Bunte et al., 2014; Downs et al., 2016; Zakwan et al., 2021). The finding of these researchers is presented in Table 1, and detailed discussion on their finding is presented in subsequent section.

TABLE 1 Contributions in the field of magnitude frequency analysis of sediment transport.

Researcher	River	Location of river	Sediment load data	Approach	Type of river	Magnitude/ frequency of effective discharge	Class size
Wolman and Miller (1960)	Colorado, Puerco, Cheyenne, Niobrara	USA	SSL and Total Load	Theoretical	Alluvial	RI ≈ 1 year	
Benson and Thomas (1966)	55 streams	USA	SSL	Theoretical		$Q_{bf} > Q_{eff}$	Equal discharge classes
Pickup and Warner (1976)	Spring Creek, Okay Creek, Thompsons Creek, Kemps Creek	England	Bed Load	Analytical	Coarse sand bed	1.15–1.45	5 equal classes

TABLE 1 Contributions in the field of magnitude frequency analysis of sediment transport.—Cont'd

Researcher	River	Location of river	Sediment load data	Approach	Type of river	Magnitude/frequency of effective discharge	Class size
Andrews (1980)	15 gauging stations of Yampa river basin - Colorado and Wyoming	USA	Total sediment	Theoretical	Alluvial	1.18–3.26 ($Q_{bf} \approx Q_{eff}$)	$50\,m^3/s$
Ashmore and Day (1988)	21 sites in Saskatchewan river	Canada	SSL	Theoretical	Sand bed	Moderate Discharge	20 equal classes
Carling (1988)	2 streams	England	Bed Load	Theoretical	Alluvial	($Q_{bf} \approx Q_{eff}$)	
Biedenharn and Thorne (1994)	Lower Mississippi river	USA	SSL	Theoretical	Alluvial	RI = 1 year	$5000\,m^3/s$
Nash (1994)	55 streams	USA	SSL and Bed Load	Analytical			
Hudson and Mossa (1997)	Rio Grande, Brazos, Mexico	USA	SSL	Theoretical	Sand bed		Unequal class size
Orndorff and Whiting (1999)	Red river	USA	Bed Load	Analytical	Alluvial	1.46	
Sichingabula (1999)	Fraser river basin	Canada	SSL	Theoretical	Sand bed	Q_{eff}/Q_{bf} = 0.55–1.90	Class number-20
Whiting et al. (1999)	23 river basins, Central and north Idaho	USA	Bed load	Theoretical	Alluvial	1.4–4.8 ($Q_{bf} \approx Q_{eff}$)	Equal discharge classes
Biedenharn and Copeland (2000)	Various rivers	USA	Bed load	Theoretical	Alluvial	1.03–3	Standard Iterative Procedure
Emmett and Wolman (2001)	5 rivers	USA	Bed Load	Theoretical	Gravel	Q_e/Q_b = 0.98–1.31	Equal discharge classes
Phillips (2002)	Hungry Mother Lake (Holston river basin)	Southwestern Virginia	SSL	Analytical	Gravel	R.I > 200 yrs	
Vogel et al. (2003)	Susquehanna, Harrisburg, Pennsylvania	USA	Total Load	Empirical and Analytical	Alluvial	R.I < 1 yr	
Goodwin (2004)	Red river and Russian river	USA	SSL and Bed Load	Analytical	Alluvial		
Simon et al. (2004)	Various rivers	USA	SSL	Analytical		R.I ≈ 1.5 year	33 equal classes
Torizzo and Pitlick (2004)	Various streams in Colorado	USA	Bed Load		Gravel bed	Q_{eff}/Q_{bf} = 0.70–1.55	
Copeland et al. (2005)	57 streams	USA	Bed Load	Analytical	Sand bed	R.I: 1–3 years	

Continued

TABLE 1 Contributions in the field of magnitude frequency analysis of sediment transport.—Cont'd

Researcher	River	Location of river	Sediment load data	Approach	Type of river	Magnitude/frequency of effective discharge	Class size
Crowder and Knapp (2004)	23 streams, Illinois	USA	SSL	Theoretical	Alluvial	R.I: 1–1.23 years	Standard Iterative Procedure
Lenzi et al. (2006)	Rio Cordon Basin	Italy	SSL and Bed Load	Analytical and Theoretical	Gravel bed	Relatively frequent floods (RI: 1.5–3 years)	4 Class intervals
Gomez et al. (2007)	Several stations on Waipoa River	New Zealand	SSL	Analytical	Gravel bed	$Q_{eff}/Q_{bf} \approx 0.25$	
Doyle et al. (2007)	4 sites Snowmelt-2 Flashy-2	USA	SSL and Bed Load	Analytical	Gravel bed	$Q_{bf} >>> Q_{eff}$	
Doyle and Shields (2008)	23 sites (Illinois)	USA	SSL	Analytical	Alluvial		
Quader and Guo (2009)	Urban Streams in Southern Ontario	Canada	SSL	Analytical		R.I: 1.5–2.5 years	
Henck et al. (2010)	44 streams	South west China and Tibet	SSL	Theoretical		R.I: 1 year	Standard Iterative Procedure
Ma et al. (2010)	Wuding river basin	China	SSL	Theoretical	Loess sand.	Variable RI	S,0.75S,0.5S and 0.25S
Klonsky and Vogel (2011)	15 rivers	USA	SSL	Analytical			
Saeidi et al. (2011)	Ghoroud watershed	Iran	SSL	Theoretical			20 Class Intervals
Ferro and Porto (2012)	27 rivers	Southern Italy	SSL	Analytical		R.I: 1.5 years	
Bunte et al. (2014)	41 streams	USA	Bed load	Analytical	Gravel bed		
Crowder and Knapp (2004)	88 sites in Illinois	USA	SSL	Analytical and Theoretical	Alluvial		25 to 75 class intervals
López-Tarazón and Batalla (2014)	Isabena river basin	Spain	SSL	Theoretical			Equal discharge classes
Roy and Sinha (2014)	Ganga	India	SSL	Theoretical	Alluvial River	RI < 1.5 year, moderate discharges	20 equal class size

TABLE 1 Contributions in the field of magnitude frequency analysis of sediment transport.—Cont'd

Researcher	River	Location of river	Sediment load data	Approach	Type of river	Magnitude/ frequency of effective discharge	Class size
Higgins et al. (2015)	Magdalena river	South America	SSL	Analytical	Alluvial River	R.I: 4.8 (1990–99) R.I: 3.5 (2000–2010)	
Downs et al. (2016)	Avon River, Devon	UK	Bed load	Theoretical	Gravel bed	$Q_{bf} \approx Q_{eff}$	$1\,m^3/s$
Sholtes and Bledsoe (2016)	95 sites on different rivers	USA	SSL and Bed Load	Analytical		$Q_b >>> Q_e$	
Yu et al. (2017)	Tarim River	China	SSL	Theoretical	Alluvial River	Moderate discharges	≈0.25S
Zakwan et al. (2018)	Ganga	India	SSL	Theoretical	Alluvial River	RI < 1.5 year, moderate discharges	Standard Iterative Procedure
Maheshwari and Chavan (2019)	Mahanadi, Godavari, Krishna and Cauvery River	India	SSL	Analytical	Alluvial River	RI represented large variation	
Chen et al. (2020)	Yangtze River	China	SSL	Theoretical	Alluvial River	$Q_b > Q_e$	15 equal class size
Zakwan et al. (2021)	Drava	Croatia	SSL	Analytical and Theoretical	Alluvial River	RI < 1 year, moderate discharges	Standard Iterative Procedure

2 Literature review

Wolman and Miller (1960) studied the four streams Colorado River, Rio Puerco River, Cheyenne River and Niobrara River from USA using theoretical approach and observed that the landforms are controlled by many processes which in turn depend upon the frequency and magnitudes of the flood events. It will not be right to state that because of their magnitude, the infrequent flood events will be highly influential. It was seen that major work was done by medium magnitude flows that are frequently recurring when compared to rare events of high magnitude and their recurring interval is found to be approximately one year.

Pickup and Warner (1976) has taken up his study in Cumberland Basin streams and determined dominant discharge, effective discharge for carrying bedload, 1.58-year flood and the bankfull discharge using analytical equations. The recurring interval for the effective flows was found as 1.15–1.40 years considering the annual flows. The maximum flow return interval usually ranges from 4 to 10 years. They stated that the dominant discharges have two groups. One set of discharges those which are larger in number, and that control the elementary dimensions and geometry of the stream. The other groups of flows are those which are fewer in number and that estimates the bed load carrying capacity. Andrews (1980) estimated effective and bankfull flows using the theoretical approach by selecting 15 flood plains in Yampa River basin of Colorado and Wyoming, and found the effective flows at these places to be same or a bit larger than the mean of 1.5 days per year (0.4% of the time) and 11 days per year (3.0% of the time). The variation of the recurring frequency ranges from 1.18 to 3.26 year for the effective flows on a yearly data. The effective (Q_{ef}) and the bankfull (Q_{bf}) flows were found to be same at all the places.

Ashmore and Day (1988) determined the effective flows by theoretical approach considering 20 equal classes for sediment for twenty one sites in Saskatchewan River basin of Canada having variety of sediment discharge histograms apart from the conventional single peak histogram, which occurs at flows with a duration of 1% to 3%. The range of variation of the period of effective flows is 0.1% to over 15% which is very high and the effective flows carrying maximum sediment were determined to be of moderate magnitude.

Biedenharn and Thorne (1994) performed the magnitude frequency analysis using the theoretical approach on lower Mississippi river of USA and found that the effective discharge is similar to or little greater than 13% of the time as evident from flow duration curves for each station. The recurring interval for the maximum flows is equal to or just less than one year.

Nash (1994) evaluated the presumptions and interpretations of Wolman and Miller (1960) by analyzing the daily flows and silt carrying capacity records of 55 US streams and found that there is high variability in the mean occurring frequency of the effective flows. The return period may vary from days to decades and normally is returns more than once a year. This changeability might be due to vast diversity of flows, silt, waterways, and flood plain attributes. He concludes that it is not correct to define a generally or largely used occurring frequency for effective flows due to this variability. Nash (1994) also developed the analytical solution for effective discharge based on normal distribution.

Hudson and Mossa (1997) performed the study on the frequency and quantity of flows transporting suspended load considering unequal class size for 3 vast flood plains of rivers flowing into the northern Gulf of Mexico of USA with different climatic regimes. The accuracy of flows and durations in transferring silt were considered. Even though there were variations of rainfall, basin size, and detentions for three rivers, most of sediment were carried by flows which occurred very often.

Orndorff and Whiting (1999) performed the magnitude frequency analysis using flow and sediment data of about 31 years for Idaho's Red River. The analysis was performed through S-PLUS, an application for calculating the effective flows from ground information, and found that the frequency of recurring of effective flows is 1.46 years. They stated that on examination of the record of the total period, these flows were either same as or greater than 2.85% of the duration or mean of 10.4 days per year.

Sichingabula (1999) applied theory of effective flows to the flood plains of Fraser River in Canada and found that the variation of durations of the class based effective discharge (Q_{ef}) in this basin range from 0.03% to 16.1% with an average of 8.8%. The durations of the classes of the class based Q_{ef} varied for about 0.02% to 19.6% and average carry of 50% of suspended load was 14% of the time with a range of 3% to 22% indicting that, in the Fraser River basin, larger flow events are responsible for moving most of the load. And it is stated that this is further supported by movement of 50% of suspended sediment load in the range of 12% to 22% of the total discharge with a mean value 19%.

Whiting et al. (1999) from the study of flood plains of rivers in central and northern Idaho, USA, spanning a vast scale of waterway slope, sediment dimensions, and basin expanse, found that average rains with recurring frequency of 2.0 year as the bankfull flows which extended to a mean of 9 days annually. They obtained the effective discharge as an average 80% of bankfull flows recurring for every 1.4 years.

Biedenharn and Copeland (2000) conducted magnitude and frequency analysis using the standard iterative method for various alluvial rivers of the United States using SAM-hydraulic design package and stated that the recurring frequency of effective flows ranged between 1.01 and 3 years.

Emmett and Wolman (2001) performed the analysis using equal discharge classes on 5 gravel bedded rivers of the United States and found effective flows to bankfull flows ratio, as varying between 0.98 and 1.31, while the average recurrence frequency for bankfull flows was near 1.6 years.

Phillips (2002) studied the association of stream features and the recurrence interval of flows altering channel by observing the effects on landforms from a flood event that occurred during July 2001 in the Hungry Mother flood plain near Marion, Virginia. Observing the study area since 1985, creeks were seen with no significant channel changes. The stream modification, banks material weathering, and carrying of large sized sediment by stream observed in the last 10 years were caused only due to the flood event of 2001 that recurred after a gap of 200 years. The presence of less quantity of minute sized sediment in streams prior to and after the flood, and the nonpresence of medium sized sand material in the accumulated sediments shows that the more often occurring flow events having low magnitudes are adequate only to carry the low quantities of fine material.

Vogel et al. (2003) has determined the 50% load and full load flow indices by applying the empirical approach and observed that these indices appear to be more meaningful than effective discharge, as they clearly display the flows that are accountable for transferring major part of the long-period accumulated material. From the analysis of data, it was seen that the 50% load flows clearly recurring intervals of a few decades to centuries. It was concluded that infrequent flows are accountable for transferring large quantities of material over a period which was is in agreement before inception of the theory of effective flows in 1960.

Crowder and Knapp (2004) applied the exponential curve and averages concept to compute effective flows for streams throughout Illinois. The effective discharge values computed by both the methods displayed higher values than a channel's average flow but lower than its 1.1-year flows. Also, these effective discharge values were typically less than its 1.5-year flood and considered to be identical to a channel's effective/dominant flows. It was observed that sediment data, technique used to establish sediment-rating association and the count of class intervals affects the values of effective discharge.

Goodwin (2004) estimated effective discharge for Red river in Idaho and Russian river in North California by using analytical approach and observed that the outcome is independent of rating parameter "a" which makes this concept more robust. He stated that suspended sediment rating curves were commonly applied to determine effective discharge due to insufficient bedload measurements. If there is an availability of a few bedload measurements also then the correlation of bedload to suspended load can be determined and converted to a curve associating suspended load to a total bed load. It was also observed that the effective flows were influenced by the composition of bed surface material also.

Lenzi (2004) selected the ground information for a duration of seventeen years for rivers having large slopes (Rio Cordon catchment) and carried out the study of recurring intervals of maximum flows and sediment load accumulation for 50% and 100% quantities. It was stated that there is a huge changeability in total quantities of bed sediment and sediment transferred per hour indicating that, presence of silt in streams and adjustment between streams and gradient varying processes of flood plains such as detritus and silt flows, that happened two times through the analysis period plays an important role.

Copeland et al. (2005) determined the three deterministic discharges for the curvy rivers of US. Using data of 57 such rivers, the stream dimensions for maximum flows width were then developed using this estimated channel-forming discharge. The statistical analysis of bankfull width and flows were performed and significant differences were obtained following an elementary categorization of flood plain greenery. It was observed that the 75% discharge with consideration of silt transfer Q_{75}, provided a good association with maximum flows with a coefficient of determination of 0.82 in comparison to the median discharge Q_{50}, which underestimated the maximum flows in many instances. The occurring frequency of channel forming flows varied from 1 to 3 years.

Lenzi et al. (2006) used the analytical concept to determine effective flows. The mean technique makes use of the recurring intervals and the mean sediment rates for each flow class (4 Class intervals—0.1 m^3/s, 0.2 m^3/s, 0.25 m^3/s, 0.5 m^3/s). They observed that two types of flows were responsible for landforms variations on terrain rivers. One, that is more often occurring flow with frequency of 1.5 to 3 years controlling the stream attributes like flow depth, and channel gradient and the other one which are rare, higher flows having recurrence interval ranging from 30 to 50 years and which causes stream dimension changes on a large scale. Doyle et al. (2007) determined the three quantities, viz. effective flows, bankfull flows, and a flow of a particular occurring frequency. These were applied on a few study areas and was carried further by combining the earlier research work which suggests that these three quantities were good measures of channel forming discharge for channels that carry snow melted waters, nonincised channels with coarse substrate. There was greater discrepancy between the three measures when there is deviation from the above conditions. They stated that Q_{bf} is far greater than Q_{eff} when there is channel incision, and flashy hydrology results in generally larger, briefer, and more frequent Q_{eff}.

Doyle and Shields (2008) used a typical measure called functional-equivalent discharge (Q_{fed}) that is defined as, the flow producing the quantity of silt that is produced by the full stream flows. According to them, Q_{fed} seems to be a better symbolic measure of streamflow changes when compared to the effective flows. Effective discharge serves as a check for the accuracy of isolated discharges, whereas Q_{fed} provides a scale of the average accuracy of the total flows.

Quader and Guo (2009) studied rivulets in southern region of Ontario, Canada to determine the effective discharge and identify the most critical discharges and silt transfer attributes controlling it. The analysis results using Global sensitivity showed the effective flows as susceptible to the power of silt loading curve, the basin storage parameter and the time of concentration (t_c) of a basin.

Ma et al. (2010) formed class intervals for flows, in fraction of standard deviation (S) like S, 0.75S, 0.5S, and 0.25S. It was observed that the class interval and properties of discharge and silt region had a great impact on mean discharge periods of effective flows. The mean discharge periods duration were nearly 0.026% in two areas, and 24.50% and 52.66% at Yulin and Hanjiamao, respectively, for a class interval of 0.25S.

Klonsky and Vogel (2011) applied kernel density estimation of the daily flows and sediment quantities to determine the effective discharge Q_e. This theoretical model predicted half load discharge $Q_{1/2}$ that exceeded average daily discharge which is supported by the values obtained using empirical equations at all the points of study area considered.

Saeidi et al. (2011) found the effective discharge as frequently occurring for SSL in Ghohroud flood plain, Iran. Ferro and Porto (2012) observed the discharges for 27 rivers in three different geographic regions. It was seen that the 50% of transport was due to dominant of low recurring interval and suggested that discharge of 1.5 years can be used to obtain the dominant discharge at first attempt. Bunte et al. (2014) studied coarse-bedded, steep Rocky Mountain streams which are snowmelt driven, having plane-bed morphology and found that for majority streams, effective discharge attained the value equal to or nearly equal to maximum discharge. Most of the long period accumulated gravel quantity is found to be moved by the highest discharges and not bankfull flow. López-Tarazón and Batalla (2014) calculated effective and dominant flows in different ways for the flood plain of Isabena River during 2005 and 2010 considering SSL. Major variations have been identified for effective flows with regard to discharge class size and data chosen.

Sholtes and Bledsoe (2016) determined process-based bank full discharge (Q_{bf}) predictors such as Q_e or Q_h for numerous river flood plains with gravel bed and fine silt bed in United States. It was seen that Q_{bf} were predicted very well by Q_h, the flows related to 50% of total sediment yield. The comparative error for prediction of Q_{bf} corresponding to coarse and fine bed streams is very less for Q_h on comparison with Q_e and flows of 1.5- and 2-years return interval. The Q_{bf} in coarse bedded plains were determined very well by Q_e, and Q_h followed by $Q_{1.5}$, whereas Q_e underestimates Q_{bf} values for flood plains with fine material.

Zakwan et al. (2018) analyzed the SSL transferred in Ganga River to compute effective discharge. The results indicated that the major sediment transfer occurred due to flows of moderate magnitude and moderate occurrences. The range of variation for bankfull and effective flow ratio is 1.13–3.02, while the return period of effective flows varied between 1.11 and 1.44 years.

Yu et al. (2017) analyzed the SSL transport in Tarim river, China, and found that the maximum sediment load is transported by moderate magnitude discharge and effective discharge in the basin is subject to temporal and spatial variation. Maheshwari and Chavan (2019) analyzed the sediment dynamics of several south Indian rivers and found variable magnitude and frequency of effective discharge. Chen et al. (2020) found that effective discharge for Yangtze River, China is approximately equal to the average discharge during monsoon season. Zakwan et al. (2021) reported that effective discharge in lower Drava river, is around the annual average discharge.

Above literature survey suggest that unlike common notion of peak discharges shaping the channel, both magnitude and frequency of discharge play an important role in shaping the channels and therefore, the concept of effective discharge becomes all more important.

3 Materials and method

In the present chapter sediment discharge data for the period 1995–2014 of Mancherial gauging site (latitude 18°50′09″ and longitude 79°26′41″) on Godavari river, India was used to estimate the effective discharge. The climate in the region remains dry except for the monsoon season in which nearly 80% of annual rainfall occurs (Zakwan and Ara, 2019).

3.1 Effective discharge computation

Effective discharge for each site was calculated using standard iterative procedure. The discharge range was divided into 30 intervals of equal size. Frequency of occurrence for specific class interval was obtained as the number of discharge values occurring within that class interval. Then iteratively number of classes were determined to assure that no class interval has zero frequency. Corresponding to each class interval, average suspended load was determined. The product of frequency and class interval's average suspended load was designated as the total suspended load carried by specific class. Finally, the midpoint of class interval which transports the maximum amount of total sediment load was recognized as the effective discharge (Zakwan et al., 2018). The computation of effective discharge was performed in excel sheet using inbuilt commands such as FREQUENCY, IF and AVERAGEIF etc.

4 Results and discussion

The concept of effective discharge appears to be quite robust as it not only identifies the discharge range transporting maximum sediment load but also the load histogram obtained through this concept provides the complete information about the proportion/percentage of sediment load transported by different range of discharges which is very useful for design of water resource projects. Scores of studies pertaining from small rivers (Guo et al., 2016) to large rivers (Roy and Sinha, 2014; Zakwan, 2018) all over the world have been subjected to this concept. Meanwhile, Nash (1994) and Goodwin (2004) developed the analytical estimates for the effective discharge computation. At present both theoretical and analytical procedure for effective discharge computation are prevalent. Theoretical approach appears to be more reliable to some researchers due to the under/over estimation of sediment load obtained through sediment rating (Crowder and Knapp, 2004; Zakwan et al., 2018), while the simplicity and independence from selecting discharge class size lure other researchers. Further, in certain cases, assumption that bed load is in certain proportion to suspended load is true in that case only the sediment rating curve coefficient "a" changes and the coefficient "b" remains unchanged and since the analytical estimate of effective discharge depends on coefficient "b," effective discharge for suspended and bed load shall remain the same which appears to be quite impractical.

In the present chapter effective discharge has been computed for Mancherial site on Godavari River. The standard iterative procedure proposed by Biedenharn and Copeland (2000) yielded 19 discharge class for the present data set. The load histogram obtained for

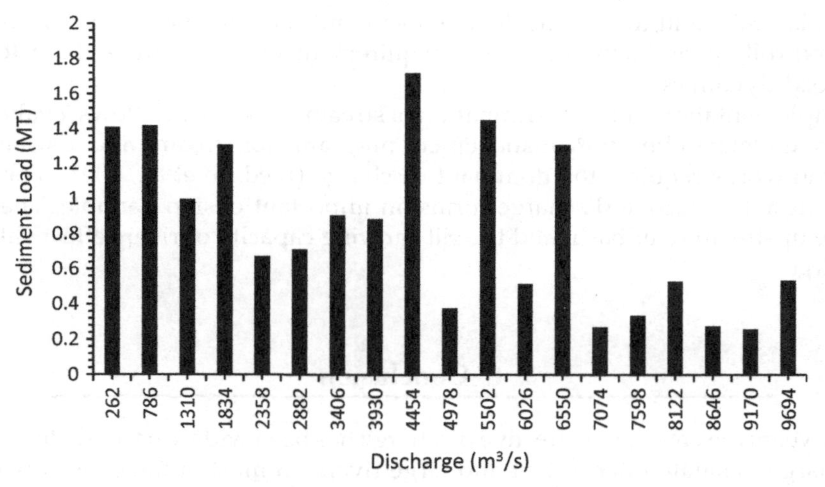

FIG. 1 Sediment load histogram at Mancherial site on Godavari river.

Mancherial is as shown in Fig. 1. It may be observed from Fig. 1 that maximum sediment load is not transported by very high discharges, instead moderate magnitude discharge transports the maximum sediment load.

5 Applications of effective discharge in river design and monitoring

The channel dimensions and flood plain attributes of a stream are changed over certain period due to flow of water and the transportation of sediment which in turn leads to a stability. The discharge associated with the changes in attributes of flow and silt carrying capacity of a stream is categorized as stream-forming or dominant flows, which is a hydraulic process, having an effective impact on the development of streams. The present condition, hidden capacity, and risk of survival to creatures living in water is affected by the equilibrium of silt quantity and its rate of accumulation. Enormous destruction to river basin and its progress is done by weathering and accumulation phenomenon related to the stream dimensions.

The relationships among river flows, shape and size of streams and their silt carrying capacity are clearly understood by study of dominant discharge and hence the concept of dominant discharge gains prominence. The layout of streams for ecological development and improvement and rehabilitation of streams needs an interpretation of streams equilibrium for large periods and hence here dominant discharge plays an important role. They can also be appraised for estimating the balanced gradients, upper side of small structures like bed sills and check dams that are commonly utilized in watershed catching more silts to stop or restrain the silt and for planning and construction of medium to massive hydraulic structures. For a knowledge of network of streams which are at risk from the consideration of varying environment, significances related to water resources and flood plains management

(Palmer et al., 2009), and to streams development and rehabilitation programs focusing the local silt controlling problems there is a requirement of an accurate characterization of sediment load dynamics.

Many implementations, like determination of stream maintenance flows, evaluation of basin changes, understanding flow management programs for streams and assistance to river rehabilitation works requires the dominant discharge (Bledsoe et al., 2007). For the stream restoration design, effective discharge forms an important design variable. The flow attributes of the upstream river basin and the silt carrying capacity of rivers effect value of effective discharge.

6 Conclusion

Over the years the concept of effective discharge has been widely used to identify the dominant discharge in small, intermittent and large rivers. In most of the cases it was observed that maximum sediment load is not transported by rarely occurring high magnitude floods, rather, the maximum sediment load is transported by much lower frequently occurring discharges, signifying the role of frequency of occurrence of discharge in channel shaping process. In the present chapter effective discharge was computed at one of the gauging site of Godavari river in India and it was observed that frequently occurring discharges played a major role in transporting the sediment load. Both theoretical and analytical approach for determining the effective discharge have received attention from the researchers. The subjectivity in the theoretical approach can be removed by following standard iterative approach while improving the accuracy of sediment rating and selecting the best fit frequency distribution can increase the reliability of analytical approach.

References

Andrews, E.D., 1980. Effective and bankfull discharge of streams in the Yampa River basin, Colorado and Wyoming. J. Hydrol. 46, 311–330.

Ashmore, P.E., Day, T.J., 1988. Effective discharge for suspended sediment transport in streams of the Saskatchewan River basin. Water Resour. Res. 24, 864–870.

Benson, M.A., Thomas, D.M., 1966. A definition of dominant discharge. Bull. Int. Assoc. Sci. Hydrol. 11, 76–80.

Biedenharn, D.S., Copeland, R.R., 2000. Effective Discharge Calculation: A Practical Guide. US Army Corps of Engineer.

Biedenharn, D.S., Thorne, C.R., 1994. Magnitude-frequency analysis of sediment transport in the lower Mississippi river. Regul. Rivers Res. Manag. 9 (4), 237–251.

Bledsoe, B.P., Brown, M.C., Raff, D.A., 2007. Geotools: a toolkit for fluvial system analysis. J. Am. Water Resour. Assoc. 43, 757–772.

Bunte, K., Abt, S.R., Swingle, K.W., Cenderelli, D.A., 2014. Effective discharge in Rocky Mountain headwater streams. J. Hydrol. 519, 2136–2147.

Carling, P., 1988. The concept of dominant discharge applied to two gravel-bed streams in relation to channel stability thresholds. Earth Surf. Process. Landf. 13 (4), 355–367.

Chen, D., Yu, M., Li, L., Liu, Y., Deng, C., 2020. Effective discharge variability for suspended sediment transport in the middle Yangtze River. Arab. J. Geosci. 13 (14), 1–15.

Copeland, R., Soar, P., Thorne, C., 2005. Channel-forming discharge and hydraulic geometry width predictors in meandering sand-bed rivers. In: Impacts of Global Climate Change, pp. 1–12.

Crowder, D.W., Knapp, H.V., 2004. Effective discharge recurrence intervals of Illinois streams. Geomorphology 64, 167–184.
Downs, P.W., Soar, P.J., Taylor, A., 2016. The anatomy of effective discharge: the dynamics of coarse sediment transport revealed using continuous bedload monitoring in a gravel-bed river during a very wet year. Earth Surf. Process. Landf. 41 (2), 147–161.
Doyle, W.M., Doug, S., Karin, F.B., Skidmore, P.B., 2007. Selection in river restoration design. J. Hydraul. Eng. 133, 831–837.
Doyle, W.M., Shields, C.A., 2008. An alternative measure of discharge effectiveness. Earth Surf. Process. Landf. 33, 308–316.
Emmett, W.W., Wolman, M.G., 2001. Effective discharge and gravel-bed rivers. Earth Surf. Process. Landf. 26, 1369–1380.
Ferro, V., Porto, P., 2012. Identifying a dominant discharge for natural rivers in southern Italy. Geomorphology 139-140, 313–321.
Gomez, B., Coleman, S.E., Sy, V.W.K., Peacock, D.H., Kent, M., 2007. Channel change, bankfull and effective discharges on a vertically accreting, meandering, gravel bed river. Earth Surf. Process. Landf. 32, 1–5. https://doi.org/10.1002/esp.1424.
Goodwin, P., 2004. Analytical solutions for estimating effective discharge. J. Hydraul. Eng. 130, 729–738.
Guo, Y., Quader, A., Stedinger, J.R., 2016. Analytical estimation of geomorphic discharge indices for small intermittent streams. J. Hydrol. Eng. 21 (7), 04016015.
Henck, A.C., Montgomery, D.R., Katharine, H.W., Liang, C., 2010. Monsoon control of effective discharge, Yunnan and Tibet. Geology 38, 975–978.
Higgins, A., Restrepo, J.C., Ortiz, J.C., Pierini, J., Otero, L., 2015. Suspended sediment transport in the Magdalena River (Colombia, South America): hydrologic regime, rating parameters and effective discharge variability. Int. J. Sediment Res. 31 (1), 25–35.
Hudson, P.F., Mossa, J., 1997. Suspended sediment transport effectiveness of three, large impounded rivers, U.S. gulf coastal plain. Environ. Geol. 32, 263–273.
Inglis, C.C., 1947. Meanders and their Bearing on River Training, seventh ed. Institution of Civil Engineering, Maritime and Waterways Engineering Division, London.
Klonsky, L., Vogel, R.M., 2011. Effective measures of "effective" discharge. J. Geol. 119, 1–14.
Lenzi, M.A., 2004. Magnitude-frequency analysis of bed load data in an alpine boulder bed stream. Water Resour. Res. 40. https://doi.org/10.1029/2003WR002961.
Lenzi, M.A., Mao, L., Comiti, F., 2006. Effective discharge for sediment transport in a mountain river: computational approaches and geomorphic effectiveness. J. Hydrol. 326, 257–276.
Leopold, L.B., 1994. A View of the River. Harvard University Press, Cambridge.
López-Tarazón, J.A., Batalla, R.J., 2014. Dominant discharges for suspended sediment transport in a highly active Pyrenean river. J. Soils Sediments 14 (12), 2019–2030.
Lyons, J.K., Pucherelli, M.J., Clark, R.C., 1992. Sediment transport and channel characteristics of a sand-bed portion of the Green River below Flaming Gorge Dam, Utah, USA. Regul. River. 7, 219–232.
Ma, Y., Huang, H.Q., Xu, J., Brierley, G.J., Yao, Z., 2010. Variability of effective discharge for suspended sediment transport in a large semi-arid river basin. J. Hydrol. 388, 357–369.
Maheshwari, S., Chavan, S.R., 2019. Magnitude–frequency analysis to assess effective discharge at daily and monthly timescales in South Indian River basins. ISH J. Hydraul. Eng. 27, 1–10.
Nash, D.B., 1994. Effective sediment-transporting discharge from magnitude–frequency analysis. J. Geol. 102, 79–95.
Orndorff, R.L., Whiting, P.J., 1999. Computing effective discharge with S-Plus. Comput. Geosci. 25, 559–565.
Palmer, M.A., Lettenmaier, D.P., Poff, L., Postel, S.L., Richter, B., Warner, R., 2009. Climate change and river ecosystems: protection and adaptation options. Environ. Manag. 44, 1053–1068.
Phillips, J.D., 2002. Geomorphic impacts of flash flooding in a forested headwater basin. J. Hydrol. 269 (3), 236–250.
Pickup, G., Warner, R.F., 1976. Effects of hydrologic regime on magnitude and frequency of dominant discharge. J. Hydrol. 29, 51–75.
Quader, A., Guo, Y., 2009. Relative importance of hydrological and sediment-transport characteristics affecting effective discharge of small urban streams in southern Ontario. J. Hydrol. Eng. 14 (7), 698–710.
Roy, N.G., Sinha, R., 2014. Effective discharge for suspended sediment transport of the Ganga River and its geomorphic implication. Geomorphology 227, 18–30.

Saeidi, P., Tabatabaei, J., Saedi, T., 2011. Effective discharge for suspended sediment transportation in Ghohruod watershed. Afr. J. Agric. Res. 6 (23), 5360–5366.

Sholtes, J.S., Bledsoe, B.P., 2016. Half-Yield discharge: process-based predictor of bankfull discharge. J. Hydraul. Eng. 142 (8), 04016017.

Sichingabula, H.M., 1999. Magnitude-frequency characteristics of effective discharge for suspended sediment transport, Fraser River, British Columbia, Canada. Hydrol. Process. 13, 1361–1380.

Simon, A., Dickerson, W., Heins, A., 2004. Suspended-sediment transport rates at the 1.5-year recurrence interval for ecoregions of the United States: transport conditions at the bankfull and effective discharge? Geomorphology 58 (1), 243–262.

Torizzo, M., Pitlick, J., 2004. Magnitude-frequency of bed load transport in mountain streams in Colorado. J. Hydrol. 290 (1–2), 137–151.

Vogel, R.M., Stedinger, J.R., Hooper, R.P., 2003. Discharge indices for water quality loads. Water Resour. Res. 39, 1273.

Webb, B.W., Walling, D.E., 1982. The magnitude and frequency characteristics of fluvial transport in a Devon drainage basin and some geomorphological implications. Catena 9, 9–23.

Whiting, P.J., Stamm, J.F., Moog, D.B., Orndorff, R.L., 1999. Sediment-transporting flows in headwaters streams. Geol. Soc. Am. Bull. 111, 450–466.

Williams, G.P., 1978. Bank-full discharge of rivers. Water Resour. Res. 14 (6), 1141–1154.

Wolman, M.G., Leopold, L.B., 1957. River Flood Plains: Some Observations on their Formation. Professional Paper No. 282-C. US Geological Survey, Washington, DC, USA, pp. 86–109.

Wolman, M.G., Miller, J.P., 1960. Magnitude and frequency of forces in geomorphic processes. J. Geol. 68, 58–74.

Yu, G.A., Li, Z., Disse, M., Huang, H.Q., 2017. Sediment dynamics of an allogenic river channel in a very arid environment. Hydrol. Process. 31 (11), 2050–2061.

Zakwan, M., 2018. Spreadsheet-based modelling of hysteresis affected curves. Appl. Water Sci. 8 (4), 101.

Zakwan, M., Ara, Z., 2019. Statistical analysis of rainfall in Bihar. Sustain. Water Resour. Manag. 5 (4), 1781–1789.

Zakwan, M., Ahmad, Z., Sharief, S.M.V., 2018. Magnitude-frequency analysis for suspended sediment transport in Ganga River. J. Hydrol. Eng. 23 (7). https://doi.org/10.1061/(ASCE)HE.1943-5584.0001671.

Zakwan, M., Pham, Q.B., Zhu, S., 2021. Effective discharge computation in the lower Drava River. Hydrol. Sci. J. 66 (5), 826–837.

CHAPTER

29

Last century evolution in local scour measuring techniques

Geeta Devi and Munendra Kumar
Delhi Technological University, New Delhi, India

OUTLINE

1 Introduction 513

2 Scour monitoring 516
- 2.1 *Fixed scour monitoring instrumentation* 516
- 2.2 *Single-use devices* 519
- 2.3 *Numbered bricks* 520
- 2.4 *Pulse devices* 521
- 2.5 *Fiber-Bragg grating* 522
- 2.6 *Sound wave devices* 522
- 2.7 *Electrical conductivity devices* 523

3 Scour monitoring measurement method in the laboratory 523
- 3.1 *Calibrated pile* 524
- 3.2 *Echo sounder and laser distance sensors* 525
- 3.3 *PIV (particle image velocimetry)* 525

4 Scour monitoring using structural dynamic 525
- 4.1 *Tiltmeters* 525
- 4.2 *Accelerometers* 526

5 Photogrammetry 526

6 Conclusion 526

References 527

Further reading 529

1 Introduction

"Man who overlooks the water under the bridge will find the bridge underwater." This citation emphasizes the adverse influence of stream-flow on the stability of piers and abutments that holds the river-based bridge. Engineers and contractors have always struggled with bridge stability, both during the designing phase and in the maintenance phase (Brandimarte and Woldeyes, 2013). Every structure in a river channel forces a relationship

https://doi.org/10.1016/B978-0-323-91910-4.00029-7

between the structure and the natural flow of the river. In general, a crossover bridge with piers and abutments in the river bed and banks represents a change in the natural geometry of the river section. This modified flow conditions at the bridge crossing acquires a strong erosive effect. As a result, flowing water behaves like a water jet, and starts eroding material from riverbeds, banks as well as adjacent bridge piers and abutments (Richardson and Davis, 2001).

Water will flow around the pier or abutment, excavating the sediment material in the form of holes is known as scour holes, as shown in Fig. 1. The formation of scour holes has a predominate effect on horseshoe vortices. Horseshoe vortices arise when the flow at the scour hole splits upstream of the pier, causing the down-flow within the scour hole to be forced close to the pier, as shown in Fig. 2. Even today, pier and abutment sinking due to scouring

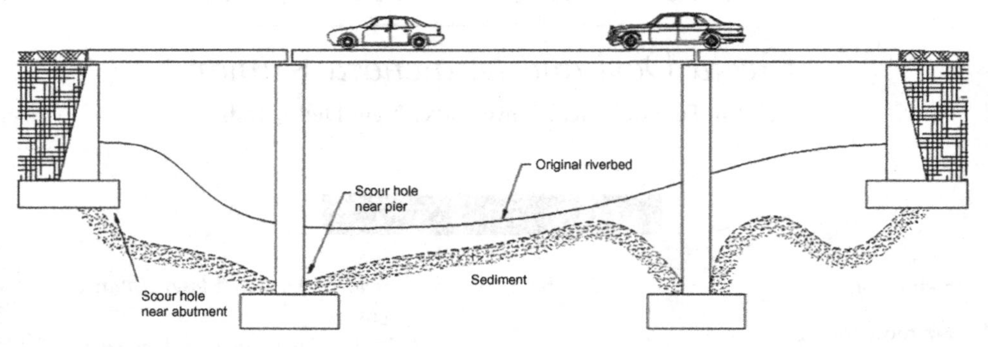

FIG. 1 Bridge Scour. *Adapted from Brandimarte, L., Paron, P., Di Baldassarre, G., 2012. Bridge pier scour: a review of processes, measurements and estimates. Environ. Eng. Manag. J. 11(5), 975–989.*

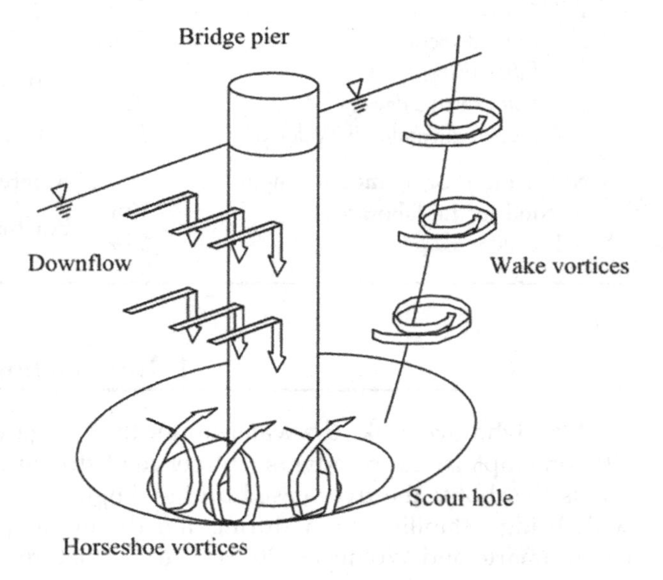

FIG. 2 Schematic of the scour process. *Adapted from Prendergast, L.J., Gavin, K., 2014. A review of bridge scour monitoring techniques. J. Rock Mech. Geotech. Eng. 6, 138–149 (Elsevier Ltd.).*

and riverbed degradation is widely acknowledged as the primary cause of bridge progressive destruction a list of bridges failure is presented in Table 1. Two bridge failures due to scour are presented in Fig. 3.

From an engineering standpoint, a precise assessment and efficient monitoring of scour depth is necessary because it affects the footing levels of bridge structures like abutments, guide banks, groyene, piers, spurs, etc. Numerous research has been carried out for the prediction of scour depth using various shapes of pier models some are listed here (Devi and Kumar, 2022a; Dey et al., 2008; Kothyari, 2007; Kothyari et al., 1992a, b, 2007; Kumar and Kothyari, 2012; Melville and Chiew, 1999; Wei et al., 2018b). Additionally, some studies have been conducted using Genetic algorithms such as (Azamathulla and Jarrett, 2013; Bateni et al., 2019; Guven and Gunal, 2008; Najafzadeh and Azamathulla, 2013; Pandey et al., 2020a, b;

TABLE 1 Bridge failures over the time.

Bridge failures	Before 1951–1990	1991–2000	2001–2010	2011–2021
Total failures	62	18	57	70
Hydraulic failure	21	03	12	21
Structural damage	07	05	09	01
Earthquake/Tornado/Tsunami	03	01	01	–
Design error	10	03	08	13
Accident	13	04	12	09
Overloading	01	02	09	11
Low maintenance	05	–	02	07
Miscellaneous	03	–	04	08

FIG. 3 Bridge failure due to scour at (A) Zagreb's Sava and (B) Dublin's Malahide viaduct. *Adapted from Prendergast, L.J., Gavin, K., 2014. A review of bridge scour monitoring techniques. J. Rock Mech. Geotech. Eng. 6, 138–149 (Elsevier Ltd.).*

Sharafati et al., 2020; Teodorescu and Sherwood, 2008). Scour can be dealt with in a variety of ways. The best way is to provide scour mitigation at the bridge design stage by including hydraulic and structural remedies (Devi and Kumar, 2022b; Schuring et al., 2017). A variety of devices for measuring the scour depth around bridge piers and abutments have been designed and several of these sensing tools are discussed in Section 2.

2 Scour monitoring

Given the seriousness of the scour phenomenon, a variety of scour hole monitoring equipment have been developed. These equipment are categorized as fixed, portable, and visual equipment. Fixed equipment can be installed on a structure, streambed, and nearby the bridge structure. Few examples of fixed monitoring devices are: sonars, magnetic sliding collars (MSCs), float-out devices, and sounding rods. According to the results of the NCHRP survey, tilt sensors and time domain reflectometers (TDRs) have also been installed at various bridges. Portable equipment is far less inexpensive than fixed equipment for monitoring a single bridge or multiple bridges but usually will not continuously monitor the components. Taking measurements from a bridge during a rainstorm might be challenging. The frequency of data gathering with portable equipment is regulated by the acceptable type of uncertainty. These equipment may be carried by hand, utilized along a bridge, and moved from one bridge to the next. Portable equipment includes sounding rods, sonars on floating boards, scour boats, and scour trucks. Visual inspection monitoring rises during high discharge periods and surface monitoring and subsurface monitoring can all be conducted on a regular schedule. Similar, to portable monitoring, during storms, inspectors are limited in their ability to visit the bridges. Positioning of the various monitoring devices is explained in Fig. 4.

2.1 Fixed scour monitoring instrumentation

2.1.1 Sonar

Sonar devices, also known as acoustic transducers, utilize sound waves to "ping" the base of a channel. A sonar device with its components is shown in Fig. 5. The majority of sonar devices are fixed to the bridge pier or substructure. They can have one sonar transducer to detect a particular location or even several transducers to detect elevation changes for larger region around the pier. This method is indirect measurement, and it decreases interference from material in water and enables more efficiency of instrument system. While, it can consistently detect and return data, consequently, it is a preferred scour monitoring method. They may be simply integrated to a monitoring station for data transmission in timely manner. These devices are very easy to accomplish because they do not require any assembly or buried in the river bottom. In addition to its value as an active monitoring device, sonar-based scour monitoring equipment are suggested in deep channels, rivers with fine particle layers, especially macro and micro areas because they are less vulnerable to biological pollutants, debris, or moving silt. However, certain above-surface detritus such as ice flows might interfere with the sonar based on the position and weather circumstances. This device consists of a low-cost recreational-type transducer that is coupled to a data acquisition device which

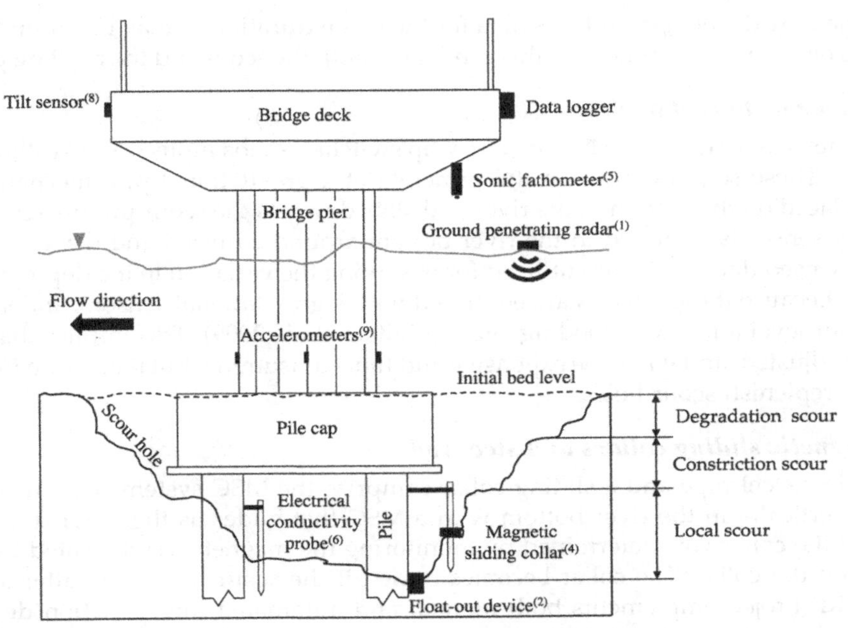

FIG. 4 A schematic representation of types of scour and bridge scour monitoring equipment. *Adapted from Maroni, A., Tubaldi, E., Ferguson, N., Tarantino, A., McDonald, H., Zonta, D., 2020. Electromagnetic sensors for underwater scour monitoring. Sensors 20(15), 1–23.*

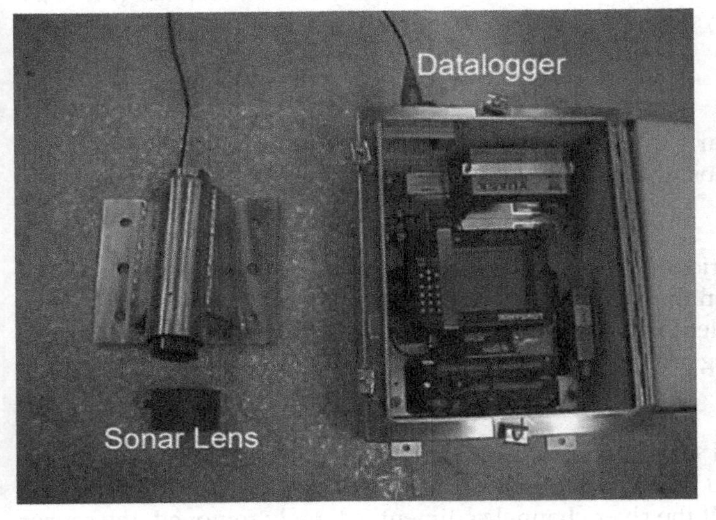

FIG. 5 Sonar with various components. *Adapted from Nassif, H., Ertekin, A.O., Davis, J., 2002. Evaluation of Bridge Scour Monitoring Methods. United States Department of Transportation, Federal Highway Administration, Trenton, pp. 1–89.*

integral the transducer and captures data for the given duration. It may be set up to take observations on a consistent time, and this can follow both the scour and the refilling procedure.

2.1.2 Driven or buried devices

Buried devices include the MSC, mercury tip switches, scuba mouse, and Wallingford tell-tail device. These sensors work on the principle of a gravitational pull mechanical probe which is placed on the bottom of the river and slips down as the scour progresses. Typically, the gravity sensor is mounted in the river bottom around a buried and driven rod device. Remotely sensed data device are utilized for observing the variation in the depth of the gravity sensor. Because these devices are equipped with a gravitational sensors, and stays at the lowest scour level after each flooding event (Mallela et al., 2009). This implies that, it might have to be adjusted, and it may be expensive and time-consuming, but there is no instructions on how to replenish scour holes.

2.1.3 Magnetic sliding collars and steel rod

A stainless steel pipe and a sliding collar comprise the MSC system. The steel pipe was mounted vertically in the river bottom with a MSC that glides as the scour develops. The location of the collar was determined by monitoring the magnetic field created by magnets mounted on the collar. The collar becomes buried if the scour hole filled after a flooding. The NCHRP Project implements both manual and automated interpretation devices. The MSC is ideal for constructing bridges across short waterways. This device can be used to assess scour at piers and vertical wall abutments, but not at spill-through or slanted abutments. The MSC device can measure scour depths up to 3 m deep. It is suggested that the pipe be pierced for at least 1 m below the predicted scour. The overall height of the metal pipe in the river bottom is estimated for being larger than 5 m. The height and flow rate of the water at the overpass must also be considered during the setup procedure. The maximum practicable depth of water without assistance of diver is roughly 3 m. Researcher (Lu et al., 2008), utilized MSC measuring system for Cho-Shui River and position of the MSC is shown in Fig. 6.

2.1.4 Scuba mouse

The scuba mouse works similarly to an MSC, other than that the collar position is detected through placing a radioactive sensing device into the supporting steel tube collar.

2.1.5 Wallingford tell- tail

This device is made up of a series of Omni-directional action sensors attached to a rod that can be placed at various depths in the ground. The motion sensors detect bed changes when the scour approaches the sensor depth of anchoring. It interacts with an information collection system, making scour monitoring a breeze (De Falco and Mele, 2002).

2.1.6 Mercury tip sensors

Mercury tip sensors mounted on a driven or buried rod might be used to check scour. Whenever a steel pipe is augured through into ground, sensors along the shaft fold up even against pipe, closing the circuit. If the river channel sediment is largely removed, the sensor opens, interrupting the circuit. This device offers a secure approach for detecting scour local

FIG. 6 Systems for monitoring scour: (A) SMC device at P16; (B) steel-rod device at P15 and MSC measurement system at Pier P16. *Adapted from Lu, J.-Y., Hong, J.-H., Su, C.-C., Wang, C.-Y., Jihn-Sung, 2008. Field measurements and simulation of bridge scour depth variations during floods. J. Hydraul. Eng. 134(6), 810–821.*

to the sensor installation region because to its simplicity. The accuracy can be improved by minimizing the sensor switch array spacing. Disadvantages of instrument is the use of mercury in the sensor, that might be environmentally destructive if spilled (Mallela et al., 2009). The sensor also cannot perceive scour hole refilling because the switches cannot be closed once they have been opened. However, this could give a decent prediction of the maximum scour depth which can be produced.

2.2 Single-use devices

Single-use devices, such as tethered buried switches and float-out devices (Briaud et al., 2011; Mallela et al., 2009), can monitor scour at their deployment points. Such sensors are placed upright in the floodplain near a scour-interesting structure, and they function on the basis that when the scour depth meets the deployment depth of device, it will gently floating out of the soil. When the device changes from upright to downward direction (after floating out), an electrical switch is tripped, signifying to a data collection system that the device is no more inside the subsurface as well as the scour depth has attained its maximum limit.

2.2.1 Float-out devices

To communicate scour process, numerous devices might be buried at various depths beneath the bed. When using float-outs devices, this is a good idea to have some consistency because the functioning of the instruments cannot be examined after it has been buried, these devices are shown in Fig. 7. When scour develops, the silt covering the instrument is removed, and the instrument floats toward the top of river. To notify its release, a signal is sent to a neighboring data acquisition system. If they are not activated in the first year, these devices have a long life expectancy and can work for several seasons. They are absolutely safe from debris since they are completely buried the riverbed. They are also simple to place in dried beds and riprap; however, expert divers can also install them in wet channels. These devices

FIG. 7 Float-out devices that are numbered and color labeled for identity. *Adapted from Hunt, B.E., 2009. NCHRP Synthesis 396: Monitoring Scour Critical Bridges—A Synthesis of Highway Practice. Transportation Research Board.*

have a number of disadvantages, while being generally affordable and low maintenance, because these devices provide a limited quantity of stored power, they must be changed on a routine basis. One disadvantage of such float-out device is that it still provides data once the scour has gone past a particular depth and the instrument has been visible. These instruments only transmit a signal once they have been exposed, and their status cannot be inspected in the meantime.

2.3 Numbered bricks

Numbered bricks were utilized by author (Lu et al., 2008), To assess the overall scour depth in his study, a column of numbered bricks was placed 100 m upstream of Piers P15 and P16. Before flooding, the face of the brick column was raised to the same level as the riverbed. A total-station transport was used to identify the precise positioning of the center of brick column. Utilizing total station method, the middle of the brickwork column was thoroughly examined again after flooding. An excavator carefully removed sediment accumulated on the brick column during the flood collapse. The uppermost brick numbering then was read and recorded on the remaining brick column. The maximum value of the total scour depth in each storm was determined by the amount of brickwork swept out by the flooding. Before and after the floods, channel cross-section inspected at the Si-Lo Bridge was also completed. The overall and pier scour measuring systems of the Si-Lo Bridge is shown in Fig. 8.

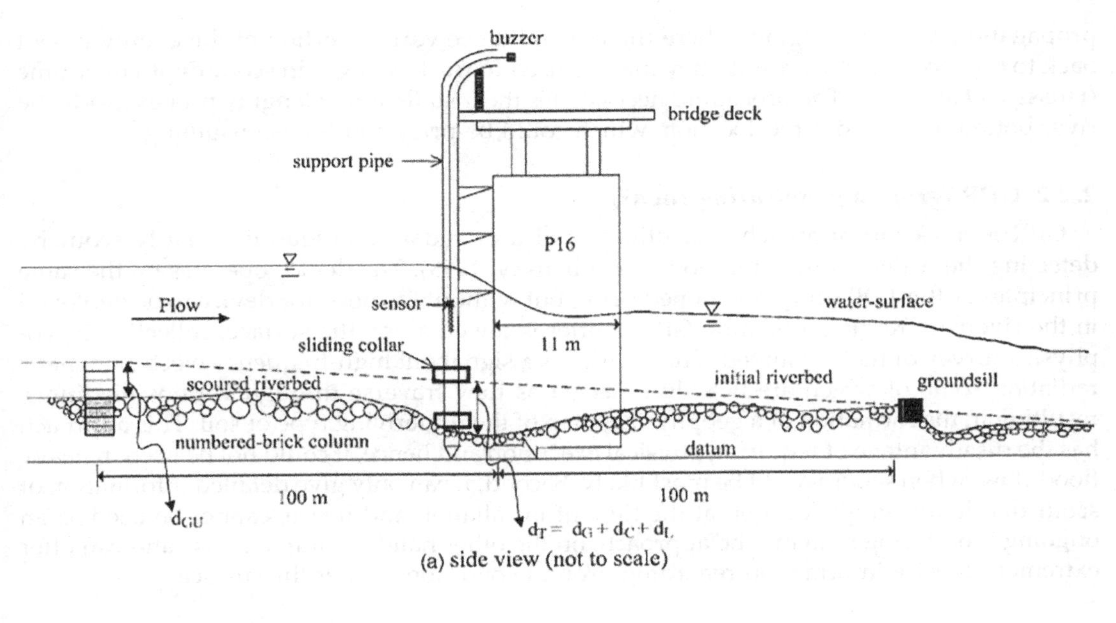

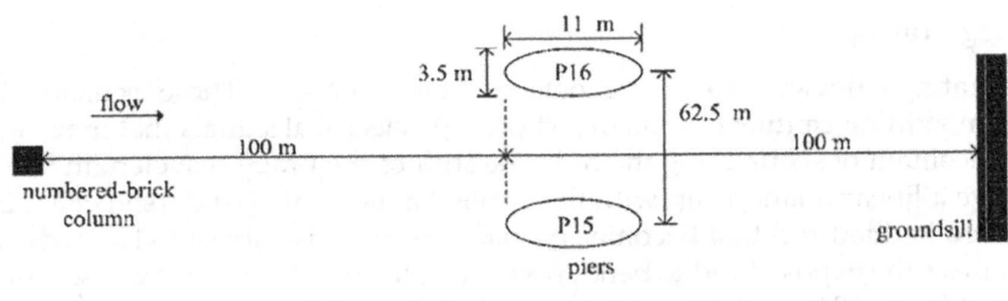

FIG. 8 Schematic diagrams of scour evaluation systems including numbered Bricks at Si-Lo Bridge. *Adapted from Lu, J.-Y., Hong, J.-H., Su, C.-C., Wang, C.-Y., Jihn-Sung, 2008. Field measurements and simulation of bridge scour depth variations during floods. J. Hydraul. Eng. 134(6), 810–821.*

2.4 Pulse devices

Pulse devices uses electromagnetic pulses or remote monitoring for recognize changes in the surface properties that occur whenever a pulse is transmitted across a turbulent environment (Zhang et al., 1999). Because this often happens at the fluid interface, this type of device can measure scour depth at a specific location.

2.4.1 TDR (time-domain reflectometers)

TDR (time-domain reflectometers) was the first way to use radar sensors. This method determines the scour depth at a certain place by comparing permittivity coefficients of materials (Zarafshan et al., 2012). In this process, measurement probes are driven into the soil at a preferred scour position, and a rapidly stepping impulses are transmitted down a tube, measuring the water-soil interface and scour depth. They operate on the premise that when the wave propagation reaches a region, where the transmittance varies, portion of the energy is sent back to the receiver. As a result, they may be used to track changes in scour depth over time (Hussein et al., 2012). The procedure necessitates the installation of lengthy probes inside the river bottom only at desired location, which could be pricy and time-consuming.

2.4.2 GPR (ground penetrating radar)

GPR is a second approach that utilizes pulse or radar technique to identify scour by detecting the water-sediment interface (Zabilansky, 1999). The device operates on the same principles as the TDR setup mentioned prior, but without the need for device to be deployed in the riverbed. Rather, a floating GPR emitter is towed along the surface, collecting a geophysics survey of the streambed. This transmits a signal out high-frequency electromagnetic radiation some of which are heavily reflected as they traverse through various mediums, resulting in the formation of a geophysical map of the underlying type of soil. This approach has the disadvantage of requiring physical execution and hence, it could not be used in heavy flood flow, when scour would be most likely. Second, it can only give detailed information of scour depth at a single location at the time of installation and hence, cannot be used as an ongoing monitoring system. The approach, on the other hand, is simple to use and can offer extremely precise information regarding ground conditions under the surface.

2.5 Fiber-Bragg grating

Fiber-Bragg gratings devices are a piezoelectric type sensors. These sensors detect straining along embedding cantilever rods to just provide electrical signals that may be used to trace the advancement of scour along the rod. The shift of the Bragg wavelength has been discovered to have a linear relationship with the applied axial strain (Anderson et al., 2007). Due to scour, an embedded rod that becomes partially exposed is subject to hydrodynamic forces, usually causes the exposed rod to bend. Because of the bending, the strain sensors can detect that the rod is free. These devices are particularly effective at identifying changes in scour depth throughout duration at its installation location and are relatively inexpensive to manufacture. The precision, on the other hand, is defined by the grid of sensing devices spacing all along rod, and it can be extremely sensitive to vibrations in the support structure generated by running water or traffic from the bridge.

2.6 Sound wave devices

They work similarly as devices that use electromagnetic waves, in which waves can reflect off different densities of materials for identifying the location of the water-sediment contact.

2.6.1 Sonic fathometers

Sonic fathometers can be installed close underneath the water level on bridge piers. They use a pulse generator to create continuous profiles of the streambed by transmitting acoustic pulse across the water-sediment contact. The device can detect and track the scour depth over time. However, the device should still be used under certain detailed limitations and is subjected to distortion from entrained air in turbulent flow. Furthermore, such type of device mainly monitors the relative shallow depth in area with varying bed topography. As a result, the width of the column in relationship with the scour hole, may have a considerable impact on the validity of the scour depth estimations (Fisher et al., 2013; Nassif et al., 2002).

2.6.2 *Reflection seismic profilers*

Reflective seismic profilers utilize sound waves to assess and identify scour. In this device, a coupled acoustical emitter and receiver are often positioned close underneath the water surface. While the device is dragged manually over the water surface, the emitter transducer generates short-period pulsed acoustic signals at scheduled interval. Seismic pulse travels down the water column and into the underground strata below. When the water-sediment interaction is contacted, acoustic radiation reflects back to the receptor, allowing it to create profile of the streambed. By combining data from various places and using expected seismic interval velocities, the temporal depth profiles may be converted into a depth profiles. The system has several drawbacks, including: (1) signal crossing-over occurs with variable streambeds; (2) source and receiver must be underwater, due to which data cannot be acquired consistently across sand bars; and (3) device needs careful manual intervention, which reduces its availability as a feasible monitoring device in many applications. Even when used correctly, it can provide a highly precise map of the channel bed.

2.6.3 *Echo sounders*

Echo sounders are similar to reflection seismic profilers in that they may be utilized to measure the depth of scour holes. The only significance is that these devices use higher frequency acoustic pulse, since high frequency pulsed acoustic energy is rapidly attenuated, only a little proportion of signals are transmitted inside the sediment bed. Plotting traces from nearby source and receiver sites yields a time-depth profile. These maps may be transformed to depth charts using estimated seismic interval velocities. As frequency pulses cannot enter the subsurface layers. This device has just one drawback against reflection seismic analysis tools: no relevant data regarding formerly loaded scour pits could be collected.

2.7 Electrical conductivity devices

This device determines the position of the water sediment interface by comparing the electrical conductivity of various material. They function by detecting the electrical current flowing between two probes. The capacity to draw a current will alter, if the substance between the probes changes. This phenomena may be utilized to detect the existence of scour, and its depth. An electrical conductivity probe is an example of a gadget that makes use of this technology. Fig. 4 represents placement of parts of this instruments. The majority of the mentioned equipment has the disadvantage of requiring underwater installation, that may be expensive, and of being utilized only seldom as part of regular check-ups; these really are important limitations. The advantages of all scour monitoring devices have been outlined in Table 2.

3 Scour monitoring measurement method in the laboratory

To determine the suitability in the measuring methods for scour monitor in the laboratory, first analyses the scour measurement system, specifications required are: (a) Experimental conditions should not be affected by the equipment utilized. As a consequence, the flow

TABLE 2 Comparison of advantages and limitations of scour monitoring devices.

Device	Advantages	Limitations
Sonar	Continuous monitor and precise data collection	Sloped rivers
MSC	Easy to operate	For installation excavation of river bed is requires; high maintenance
Steel rod	Easy to operate	For installation excavation of river bed is requires; high maintenance
ADV current profiler	Easy to carry; monitor both velocity profile and depth of water	Not suitable for higher sediment concentration
GPR	Continuous monitoring and data collection	Skilled team required for installation; time taking process during installation
Numbered bricks	Easily available and applicable to all flow conditions	For installation excavation of river bed is requires
Electrical conductivity	Easy to operate; detect the existence of scour and its depth	Expensive installation undersea installation
Sound wave	Provides accurate results	Cannot penetrate into deep layer of river

should not be interrupted or exhausted, and device positioned in it should be compact, (b) 3D profile of scour hole is expected from the system utilized, (c) scour hole profile obtained should be acquired in real time, or this same points assessed above a short period of time roughly comparable to the scour rate, (d) horizontal and vertical axes, measurements should be precise to within 1 mm, and (e) in the early phases of scour, the system should be able to capture data at second intervals.

3.1 Calibrated pile

Scour depth reading from a scale mounted onto the pile model is one of the most basic approaches utilized in the literature. This can be done by eye, using a camera mounted outside the flume, or within a transparent pile (Debnath and Chaudhuri, 2010; Roulund et al., 2005). A lathe machine can be used to mark vertical scale in the form of concentric rings on the pile at small intervals. The vertical resolution should be used, to make it easier to perceive the scale inside the flume section. A point gauge can also be relocated in a sequential way to build a 3D profile (Lança et al., 2013). Despite the development of less invasive models of point gauge, obtaining a scour hole profile with any of these devices takes a long time; therefore, dynamic profile is not practical, particularly in the early stages of the process (Ballio and Radice, 2003; Porter et al., 2015). Author (Chang et al., 2014), created a more advanced version of the vertical scale approach, Using image processing technology, the process is programmed to generate real-time scour assessment. Such nonintrusive, low-cost devices can provide high frequency estimations with good accuracy and resolution. They offer depths estimation near to the pile model; therefore, they may not reflect the deepest point of the scour hole.

3.2 Echo sounder and laser distance sensors

In the laboratory, scour depth was measured using an echo sounder and a laser distance sensor. These devices can be installed within close vicinity to the streambed, thus they can be much less obtrusive to the experiment. However, because the sensors are frequently enormous compared to the pile, in certain studies, the inflow is interrupted or even the water is drained before observations are taken such as (Hartvig et al., 2011; Jensen et al., 2006; Margheritini et al., 2006). It is also important to note that laser and echo sounder devices are significantly more expensive than the commonly used methods. While laser devices may offer relatively exact readings, echo sounder measurements are less precise due to the beam spreading as it moves away from the sensor. This implies that at each measurement position, a section of the bed is averaged as a reasonable approach to a point measurement. The frequency and diameter of the sensor have an inverse relationship with the size of the beam pattern. A high frequency device must be utilized because in small scale laboratory testing, a tiny sensor diameter is required to avoid disturbance with flow, and the beam pattern must be narrow enough to enable sufficient temporal resolution. The second approach is a General Acoustics Ultralab UWS type 1 MHz echo sounder has its tiny sensor diameter and narrow beam angle and set-up of these devices is shown in Fig. 2.

3.3 PIV (particle image velocimetry)

The PIV method offered contemporaneous and averaged velocity, vorticity, and Reynolds stress contours, as well as streamlined topology and scour depth. The observations should be carried in the symmetry zone upstream side of the pier model. The data can be collected using stereoscopic PIV, a two-dimensional three-component PIV approach (S-PIV). In recent decades, as more innovative flow measurement technology has been developed, flow image processing methods and semi evaluation instrumentation, such as the laser doppler velocimeter (LDV) and PIV, have also been used in the investigation of flow structures at cylindrical shape structures (Guan et al., 2019). Many researchers have recently used PIV and numerical modeling to examine flow structures (Apsilidis et al., 2015; Elhimer et al., 2016; Guan et al., 2019; Schanderl et al., 2017; Wei et al., 2017; Zhang et al., 2017).

4 Scour monitoring using structural dynamic

There has been a lot of research in following the dynamic reaction of structure to the development of scour in recent years. Few of them are described in this section:

4.1 Tiltmeters

Tiltmeters detect differential settlement caused by the scour process by measuring the relative rotation of a structural part. The serious shortcoming of the device is that it does not indicate scour depth clearly. Engineers will be much more successful in implementing the essential repair plans on critical structures provided devices capable of accurately evaluating structural distress are available.

4.2 Accelerometers

Accelerometers can be used to monitor the structural sensitivity to changes in boundary conditions. Observing variations of vibrational modes is a promising tool for condition monitoring since the vibration of the structure is reliant on the stiffness of the system. For the accelerometers installed on the structure, spectrum analysis methods like as FFTs or spatial frequency analysis can be utilized to evaluate the natural period of the bridge pier-foundation structure. A number of researchers have looked into the possibility of employing dynamic measures to identify scour. In these approaches, the use of accelerometer to identify fundamental properties such as frequency is commonly used throughout combination with creation of reference numerical models.

5 Photogrammetry

Photogrammetry is another technology that has been used in the laboratory in a number of setups. Author (Liang et al., 2017) used a photogrammetry system similar to that used by (Foti and Sabia, 2011). He placed the, two cameras outside the flume. A dot grid was displayed on to bed to mark out the measurement. A trigger box was used to synchronize the two cameras. The light, camera and projection position, and camera setting (aperture, focal length, and so on) have all been selected to achieve the best scour hole cover and picture quality possible and A2D data in the form of targets a specific was obtained by photography. Many of the techniques, however, necessitate pausing or draining the flow (Riezebos et al., 2016; Schleiss, 2004; Umeda et al., 2010). Authors (Baglio et al., 2001) in their study, obtained dynamic data with cameras put up outside the flume. Nevertheless, since this technique somehow does not compensate for distortions or light diffraction at the air-glass-water interface, the findings are less accurate. A similar technique was used in a recent work (Musumeci et al., 2013), which included lens distortion adjustments but did not describe refraction through the flume wall correctly. In such techniques, measuring locations inside the excavated area are detected to use a grids of dots created by a diffraction grating lasers (Baglio et al., 2001; Petersen et al., 2012) or feature matching software (Riezebos et al., 2016). In addition, VMS, a software suite, is used to do image processing and photogrammetry computations (James and Robson, 2012). The bundle adjustment approach is used in this program to offer an explicit nonlinearity solution.

6 Conclusion

Conventional scour monitoring devices are usually expensive to install and maintain, and it is also vulnerable to debris damage during flooding. There were significant advancements in scour monitoring systems, so this paper presents some more of the insights garnered from installation in a number of situations. The devices can provide 24-h surveillance during storms for bridge scour study, velocities, and river phase records, and the integration of cutting-edge scour prediction techniques with practical data gathering. The information processing from these devices can be time-consuming and difficult to handle. Use of structure dynamic behavior to identify and assess the scour depth around structures is currently a major research issue. The research paves the path for low-maintenance, semi structural analysis to recognize and monitor scour development. Dynamic measures have two benefits over typical scour depth measurement hardware: the easy installation above the water surface and the low maintenance required. The development of monitoring devices that monitor water velocity and level will produce data for improving present scour forecast techniques. The most common problems encountered while employing fixed scour instruments are system servicing and restorations, as well as funds to keep the operation and scour monitoring program functioning. Proper installation of the scour monitoring system, a well-thought-out and planned strategy can lead to successful system to secure the safety of both the structure and human life. The worldwide effect of scour may go unnoticed, except if a high density of sensors is utilized near critical scour areas, which is commonly overlooked by instrumentation deployed in the streambed. This paper provides a comprehensive evaluation of the current bridge scour studies. Different approaches of monitoring bridge scour are discussed. Every present scour monitoring tool has its own set of benefits and drawbacks. The nature of the problem, the site condition, relative efficacy and cost, and other considerations should all be considered when choosing a bridge scour monitoring equipment.

References

Anderson, N.L., Ismael, A.M., Thitimakorn, T., 2007. Ground-penetrating radar: a tool for monitoring bridge scour. Environ. Eng. Geosci. 13 (1), 1–10.

Apsilidis, N., Diplas, P., Dancey, C.L., Bouratsis, P., 2015. Time-resolved flow dynamics and Reynolds number effects at a wall-cylinder junction. J. Fluid Mech. 776, 475–511.

Azamathulla, H.M., Jarrett, R.D., 2013. Use of gene-expression programming to estimate Manning's roughness coefficient for high gradient streams. Water Resour. Manag. 27 (3), 715–729.

Baglio, S., Faraci, C., Foti, E., Musumeci, R., 2001. Measurements of the 3-D scour process around a pile in an oscillating flow through a stereo vision approach. Measurement 30 (2), 145–160.

Ballio, F., Radice, A., 2003. A non-touch sensor for local scour measurements. J. Hydraul. Res. 41 (1), 105–108.

Bateni, S.M., Vosoughifar, H.R., Truce, B., Jeng, D.S., 2019. Estimation of clear-water local scour at pile groups using genetic expression programming and multivariate adaptive regression splines. J. Waterw. Port Coast. Ocean Eng. 145 (1), 1–11.

Brandimarte, L., Woldeyes, M.K., 2013. Uncertainty in the estimation of backwater effects at bridge crossings. Hydrol. Process. 27 (9), 1292–1300.

Briaud, J.L., Hurlebaus, S., Chang, K., Yao, C., Sharma, H., Yu, O., Darby, C., Hunt, B.E., Price, G.R., 2011. Realtime monitoring of bridge scour using remote monitoring technology. Security 7 (2), 1–440.

Chang, W.Y., Lai, J.S., Yu, T.Y., Lin, F., Lee, L.C., Tsai, W.F., Loh, C.H., 2014. Pier scour monitoring system by bed-level image tracking. Int. J. Sediment Res. 29 (2), 269–277.

Debnath, K., Chaudhuri, S., 2010. Bridge pier scour in clay-sand mixed sediments at near-threshold velocity for sand. J. Hydraul. Eng. 136 (September), 597–609.

Devi, G., Kumar, M., 2022a. Characteristics assessment of local scour encircling twin bridge piers positioned side by side (SbS). Sadhana 47 (3), 109.

Devi, G., Kumar, M., 2022b. Countermeasures against local scouring at circular bridge piers using collar and combination of slot and collar. In: Jha, R., Singh, V.P., Singh, V., Roy, L.B., Thendiyath, R. (Eds.), River Hydraulics: Hydraulics, Water Resources and Coastal Engineering. vol. 2. Springer International Publishing, Cham, pp. 289–296.

Dey, S., Chiew, Y.-M., Kadam, M.S., 2008. Local scour and riprap stability at an abutment in a degrading bed. J. Hydraul. Eng. 134 (10), 1496–1502.

Elhimer, M., Harran, G., Hoarau, Y., Cazin, S., Marchal, M., Braza, M., 2016. Coherent and turbulent processes in the bistable regime around a tandem of cylinders including reattached fl ow dynamics by means of high-speed PIV. J. Fluids Struct. 60, 62–79 (Elsevier).

De Falco, F., Mele, R., 2002. The monitoring of bridges for scour by sonar and sedimetri. NDT & E Int. 35 (2), 117–123.

Fisher, M., Chowdhury, M.N., Khan, A.A., Atamturktur, S., 2013. An evaluation of scour measurement devices. Flow Meas. Instrum. 33, 55–67 (Elsevier).

Foti, S., Sabia, D., 2011. Influence of foundation scour on the dynamic response of an existing bridge. J. Bridg. Eng. 16 (2), 295–304.

Guan, D., Chiew, Y.M., Wei, M., Hsieh, S.C., 2019. Characterization of horseshoe vortex in a developing scour hole at a cylindrical bridge pier. Int. J. Sediment Res. 34 (2), 118–124 (Elsevier B.V.).

Guven, A., Gunal, M., 2008. Genetic programming approach for prediction of local scour downstream of hydraulic structures. J. Irrig. Drain. Eng. 134 (2), 241–249.

Hartvig, P.A., Thomsen, J.M.C., Frigaard, P., Andersen, T.L., 2011. Erratum: Experimental study of the development of scour and backfilling (Coastal Engineering Journal (2010) 52:2 (157–194)). Coast. Eng. J. 53 (2), 177–180.

Hussein, H.H., Juma, I.A.K., Hammo, N.I., 2012. Evaluation of the local scour downstream untraditional bridge piers. J. Eng. Sustain. Dev. 16 (3), 36–49.

James, M.R., Robson, S., 2012. Straightforward reconstruction of 3D surfaces and topography with a camera: accuracy and geoscience application. J. Geophys. Res. Earth Surf. 117 (3), 1–17.

Jensen, M.S., Juul Larsen, B., Frigaard, P., DeVos, L., Christensen, E.D., Asp Hansen, E., Solberg, T., Hjertager, B.H., Bove, S., 2006. Offshore Wind Turbines situated in Areas with Strong Currents., p. 156.

Kothyari, U.C., 2007. Indian practice on estimation of scour around bridge piers—a comment. Sadhana 32 (3), 187–197.

Kothyari, U.C., Garde, R.C.J., Ranga Raju, K.G., 1992a. Temporal variation of scour around circular bridge piers. J. Hydraul. Eng. 118 (8), 1091–1106.

Kothyari, U.C., Garde, R.C.J., Ranga Raju, K.G., 1992b. Temporal variation of scour around circular bridge piers. J. Hydraul. Eng. ASCE 118 (8), 1091–1106.

Kothyari, U.C., Hager, W.H., Oliveto, G., 2007. Generalized approach for clear-water scour at bridge foundation elements. J. Hydraul. Eng. 133 (11), 1229–1240.

Kumar, A., Kothyari, U.C., 2012. Three-dimensional flow characteristics within the scour hole around circular uniform and compound piers. J. Hydraul. Eng. 138 (5), 420–429.

Lança, R.M., Fael, C.S., Maia, R.J., Pêgo, J.P., Cardoso, A.H., 2013. Clear-water scour at comparatively large cylindrical piers. J. Hydraul. Eng. 139 (11), 1117–1125.

Liang, F., Wang, C., Huang, M., Wang, Y., 2017. Experimental observations and evaluations of formulae for local scour at pile groups in steady currents. Mar. Georesour. Geotechnol. 35 (2), 245–255 (Taylor & Francis).

Lu, J.-Y., Hong, J.-H., Su, C.-C., Wang, C.-Y., Jihn-Sung, 2008. Field measurements and simulation of bridge scour depth variations during floods. J. Hydraul. Eng. 134 (6), 810–821.

Mallela, J., Glover, L.T., Darter, M.I., Von Quintus, H., Gotlif, A., Stanley, M., Sadasivam, S., 2009. Guidelines for Implementing NCHRP 1-37A M-E Design Procedures in Ohio: Volume 1—Summary of Findings, Implementation Plan, and Next Steps. 1(134300).

Margheritini, L., Frigaard, P., Martinelli, L., Lamberti, A., 2006. Scour Around Monopile Foundations for Off-shore Wind Turbines. CoastLab06.

Melville, B.W., Chiew, Y., 1999. Time scale for local scour at bridge piers. J. Hydraul. Eng. 125 (1), 59–65.

Musumeci, R.E., Farinella, G.M., Foti, E., Battiato, S., Petersen, T.U., Sumer, B.M., 2013. Measuring sandy bottom dynamics by exploiting depth from stereo video sequences. In: Lecture Notes in Computer Science (Including Subseries Lecture Notes in Artificial Intelligence and Lecture Notes in Bioinformatics). vol. 8156. Springer, pp. 420–430 (LNCS(PART 1)).

Najafzadeh, M., Azamathulla, H.M., 2013. Group method of data handling to predict scour depth around bridge piers. Neural Comput. Applic. 23, 2107–2121. 2107–2112.

Nassif, H., Ertekin, A.O., Davis, J., 2002. Evaluation of Bridge Scour Monitoring Methods. United States Department of Transportation, Federal Highway Administration, Trenton, pp. 1–89.

Pandey, M., Zakwan, M., Khan, M.A., Bhave, S., 2020a. Development of scour around a circular pier and its modelling using genetic algorithm. Water Sci. Technol. Water Supply 20 (8), 3358–3367.

Pandey, M., Zakwan, M., Sharma, P.K., Ahmad, Z., 2020b. Multiple linear regression and genetic algorithm approaches to predict temporal scour depth near circular pier in non-cohesive sediment. ISH J. Hydraul. Eng. 26 (1), 96–103 (Taylor & Francis).

Petersen, T.U., Sumer, B.M., Meyer, K.E., Fredsøe, J., Christensen, E.D., 2012. Edge scour in current adjacent to stone covers. In: ICSE-6—6th International Conference on Scour and Erosion (2005), pp. 739–746.

Porter, K.E., Simons, R.R., Harris, J.M., 2015. Laboratory investigation of scour development through a spring-neap tidal cycle. In: Scour and Erosion—Proceedings of the 7th International Conference on Scour and Erosion, ICSE 2014, December 2014, pp. 667–677.

Richardson, E.V., Davis, S.R., 2001. Evaluating Scour at Bridges. National Highway Institute, US.

Riezebos, H.J., Raaijmakers, T.C., Tönnies-Lohmann, A., Waßmuth, S., van Steijn, P.W., 2016. Scour protection design in highly morphodynamic environments. In: Scour and Erosion—Proceedings of the 8th International Conference on Scour and Erosion, ICSE 2016, pp. 301–311.

Roulund, A., Sumer, B.M., Fredsøe, J., Michelsen, J., 2005. Numerical and experimental investigation of flow and scour around a circular pile. J. Fluid Mech. 534 (January 2016), 351–401.

Schanderl, W., Jenssen, U., Manhart, M., 2017. Near-wall stress balance in front of a wall-mounted cylinder. Flow Turbul. Combust. 99 (3–4), 665–684.

Schleiss, A., 2004. Mapping of Bed Morphology for Lateral Overflow Using Digital Photogrammetry. pp. 1–8.

Schuring, J.R., Dresnack, R., Golub, E., 2017. Design and Evaluation of Scour for Bridges Using HEC-18., p. 1.

Sharafati, A., Tafarojnoruz, A., Yaseen, Z.M., 2020. New stochastic modeling strategy on the prediction enhancement of pier scour depth in cohesive bed materials. J. Hydroinf. 22 (3), 457–472.

Teodorescu, L., Sherwood, D., 2008. High energy physics event selection with gene expression programming. Comput. Phys. Commun. 178 (6), 409–419.

Umeda, S., Yamazaki, T., Yuhi, M., 2010. An Experimental Study of Scour Process and Sediment Transport around a Bridge Pier with Foundation. pp. 66–75.

Wei, M., Chiew, Y., Asce, M., Guan, D., 2018b. Temporal development of propeller scour around a sloping Bank. J. Waterw. Port Coast. Ocean Eng. 144 (5), 1–9.

Wei, M., Chiew, Y.M., Hsieh, S.C., 2017. Plane boundary effects on characteristics of propeller jets. Exp. Fluids 58 (10), 1–15. Springer Berlin Heidelberg.

Zabilansky, N.E.Y.L., 1999. Laboratory investigation of time-domain reflectometry system for monitoring bridge scour. J. Hydraul. Eng. 125 (12), 1279–1284.

Zarafshan, A., Iranmanesh, A., Ansari, F., 2012. Vibration-based method and sensor for monitoring of bridge scour. J. Bridg. Eng. 17 (6), 829–838.

Zhang, H., Chen, S., Liang, F., 2017. Effects of scour-hole dimensions and soil stress history on the behavior of laterally loaded piles in soft clay under scour conditions. Comput. Geotech. 84, 198–209 (Elsevier Ltd.).

Zhang, H., Jennings, A., Barlow, P.W., Forde, B.G., 1999. Dual pathways for regulation of root branching by nitrate. Proc. Natl. Acad. Sci. U. S. A. 96 (11), 6529–6534.

Further reading

Further investigation is needed to analyze the various prediction approaches for the accurate prediction of the scour depth around the bridge piers. Even though some of the past techniques of mitigating the disasters were effective in some circumstances, they may no longer be appropriate in light of changing factors such as weather conditions. To develop a successful countermeasure plan, take into account factors namely the condition of the site, the relative effectiveness and expense of countermeasures, as well as the interactions and effects of multiple disasters. To facilitate scour design and countermeasure, a robust decision support system is also required.

SECTION IX

Computational intelligence in extreme hydrology: flood and droughts

CHAPTER

30

Understanding trend and its variability of rainfall and temperature over Patna (Bihar)

Nitesh Gupta[a], *Pradeep K. Mahato*[a], *Jitendra Patel*[b], *Padam Jee Omar*[c], *and Ravi P. Tripathi*[d]

[a]Department of Civil Engineering, Indian Institute of Technology (BHU), Varanasi, India
[b]Department of Civil Engineering, Samrat Ashok Technological Institute, Vidisha, India
[c]Department of Civil Engineering, Motihari College of Engineering, Motihari, India [d]Department of Civil Engineering, Rajkiya Engineering College, Sonbhadra, India

OUTLINE

1 Introduction 533
2 Area of research and information sources 535
3 Material and methods 536
3.1 Mann-Kendall test 536
3.2 Sen's slope method 537
4 Results and discussion 538
4.1 Precipitation 538
4.2 Temperature 540
5 Conclusion 541
References 542

1 Introduction

The severe weather and the climatic impacts on people and ecosystems may be detrimental in many ways, including extreme heat and cold, floods, drought, high wind damage, etc. These causes' massive losses of human life, livestock life, agriculture damage, engineering structures such as dams, bridges, and exponentially increasing costs associated with them

https://doi.org/10.1016/B978-0-323-91910-4.00030-3

(Gupta et al., 2021a, b). The rainfall distribution is a highly erratic term, and it varies over the smaller areas due to climate variation over the decades. Climate changes affect rainfall distribution all over the world (Panda and Sahu, 2019). The change in rainfall distribution would change the surface runoff, soil moisture, recharge, and storage (Omar et al., 2017). Also, the rainfall increases the discharge and water availability to the river (Shekhar et al., 2021). Agriculture is the main source of income for tribal people, and it depends on water availability. The availability of water decreases due to change in climate, and population growth. Both together would adversely affect the climate and natural resources of India up to the year 2050s (IPCC, 2007). Most of the rivers are dried up in the non-monsoon period due to climate change (Kumar et al., 2021). The dependability of water resources is increasing for fulfilling the water requirement for irrigation land and industrial growth for a growing population. The researchers have indicated that global warming is one factor that highly influences rainfall patterns at a regional scale and worldwide (Umakanth et al., 2021). The rainfall variability analysis was quite helpful in deciding the cropping pattern according to the water availability. The rise of extreme precipitation and rising temperatures leads to plant damage, excessive soil erosion, soil water reclamation, adverse surface water quality, and urban settlement disruption (Omar et al., 2021). On the other hand, the precipitation and temperature characteristics alter with the anthropogenic changes on the earth (Tripathi and Pandey, 2021). Thus, a dynamic interrelation exists between the features of precipitation and temperature with earth surface characteristics. To evolve an effective plan and implement a water policy, spatial and temporal distributions of rainfall and temperature are also essential to be estimated (Omar et al., 2020). Water supply and demand may be incompatible due to the unequal distribution of precipitation. The trend analysis technique helps in the analysis of the changing pattern of rainfall and temperature. This analysis provides a means for estimating the quantum of stream flows, soil moisture, and groundwater reserves and a framework for decision-makers who strive for the all-time availability of sufficient water to various users (Gupta et al., 2021a,b).

In the recent past, various studies have been done regarding rainfall variability and climate change. In most Indian regions, the high frequency of rainfall in the monsoon season shows an increasing trend in most central and northwest India. Warwade et al. (2016) studied the variability of climate in a part of the Brahmaputra river basin in India and a significant negative trend in annual/monsoon time series in most of the states was found whereas trend analysis of hydrological variables (discharge and sediment transport) was analyzed over Ganges river and the results stated that observed annual maximum discharge showed a negative trend at almost all the sites, annual minimum discharge showed positive trend at gauging sites upstream of confluence of Ganga River (Zakwan and Ahmad, 2021). Trend analysis of rainfall over Madhya Pradesh have shown a declining trend of rainfall from 1901 to 2011 (Duhan and Pandey, 2013). The trend variability and detection of change point of the rainfall in northern India was examined by Pranuthi et al. (2014). For trend and cumulative deviation and Worsley likelihood statistics, the author used Mann-Kendall (MK) and linear regression in detecting changes. The descending trends for the state of Jharkhand are shown in annual, monsoon, and winter precipitation between 1901 and 2011 (Chandniha et al., 2016). Omar et al. (2019) studied Varanasi and adjoining areas to identify rainfall and temperature regime trends for 30 years

(1988–2017). Kundu et al. (2015) did a similar type of study for the semiarid region of India. Pingale et al. (2016) studied the trend analysis of climatic variables (rainfall and temperature) in an arid and semiarid region of the Ajmer District Rajasthan, India. Jain et al. (2017) relate their work to seven major river basins in India. They examine the temporal trends in magnitude and intensity of rainfall and crest floods of different magnitudes. Zakwan and Ara (2019) demonstrate the statistical analysis of rainfall, whereas spatiotemporal variation of rainfall has been analyzed over Bihar state by Warwade et al. (2018). Most of the districts of Bihar state resulted as significant declining rainfall trends. As per the review process, it was seen that the climate changes mostly pronounced with the natural and anthropogenic change in space as well as time.

Several studies relating to trend analysis of precipitation in various states of India involved grid-based averaged values of precipitation. Trend analysis of precipitation and temperature relying on the station-scale data is yet scanty, particularly for Patna, the district of precipitation Bihar. Thus, the present study adds significant novelty in the current state. Therefore, the present work would focus on the Patna, capital of Bihar state. The city sprawls along the south bank of the Ganga River in Bihar, northeast India. The total geographical area of Patna city is 136 square km, and its economy primarily relies on agriculture. Hence, rainfall and temperature play a vital role in Bihar's economy, banking primarily on the agriculture sector. The varying climate condition, i.e., deficient moisture availability in the soil due to variable precipitation and temperature patterns, may offer a better explanation of the situation. The concept of the study is to carry out the trend analysis and variability of precipitation and temperature over the Patna district. In India, most of the water resources and agricultural planning are associated with the administrative boundary. The study highlights the rainfall and temperature variability and trend analysis for prospects. The adopted analysis plays an essential role in identifying climatic behavior in the past, which is also responsible for future strategic planning and mitigation over the investigation area.

2 Area of research and information sources

The current study investigated on precipitation, particularly rainfall and temperature events, over Patna, India. The latitude of *Patna*, Bihar is 25.6126, and the longitude is 85.1588 illustrated in Fig. 1. The total geographical area of study is 136 square km. Climate wise, according to the Koppen climate classification, Patna has a humid climate. Summers are hot and humid from late March to early July, while the rainy season runs from mid-July to late September. The winter season varies from November to early March. The recorded highest temperature for the study area was 46.7°C in 1966 and, the lowest recorded temperature was 0.1°C, on 9 January 2013, and maximum rainfall was 204.5 mm (8.05 in.) in 1997. Indian Meteorological Department (IMD), Pune, India, provided meteorological data such as daily rainfall, and temperature data are collected from India Water Portal (2019) (http://www.indiawaterportal.org) from 1901 to 2002. However, some data were missing in the collected data which were not considered in the calculations and analysis.

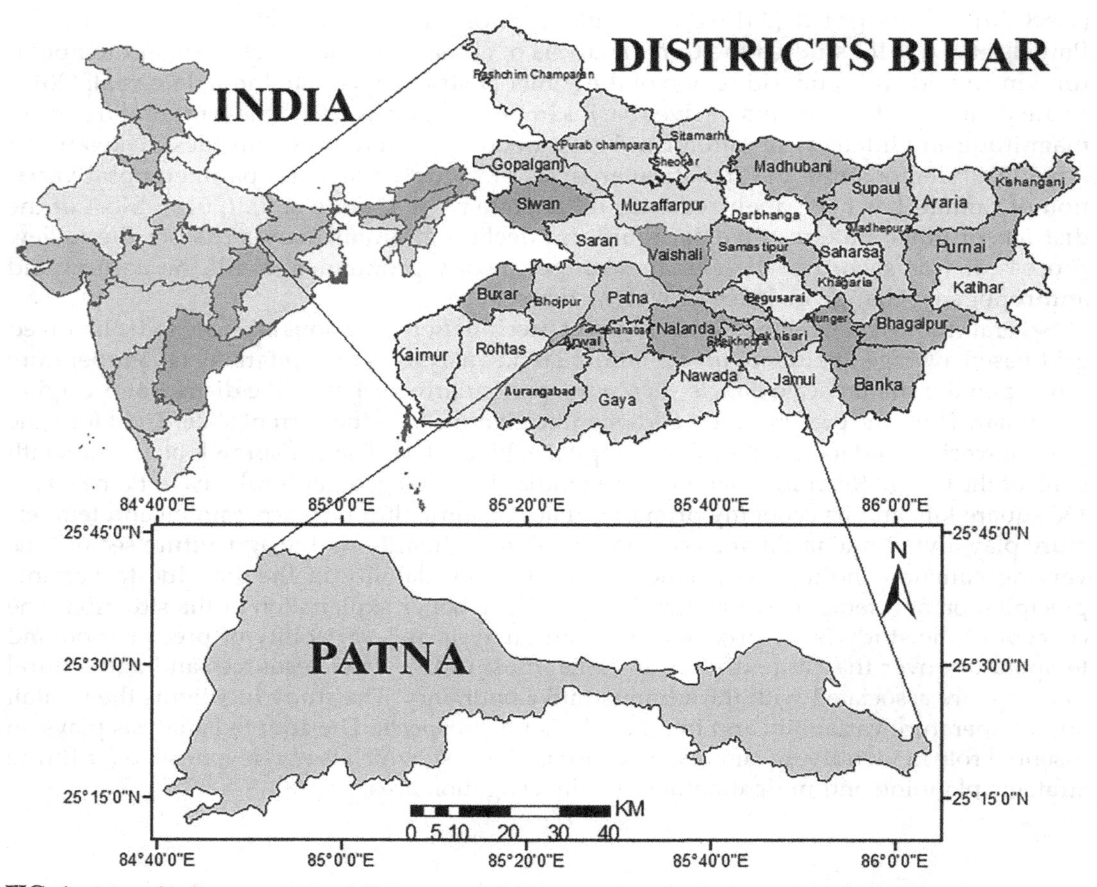

FIG. 1 Map of India representing different states with various districts of Bihar and geographical regions of Patna.

3 Material and methods

The nonparametric MK test and Sen's slope estimation were used to determine the trend of climatic variables. The interface has been developed in MATLAB® environment to analyze the trends using non-parametric statistical methods, MK test and Sen's slope estimator.

3.1 Mann-Kendall test

All of the climatic variables' time series were analyzed using the nonparametric MK trend test. Mann (1945) was the first to use this test, while Kendall (1975) was the first to calculate the test statistic distribution. The importance of the discovered trends can be determined (generally adopted as 5%). The methodology is very resistant to the effects of extremes and works well with skewed variables. It has the potential to handle missing data values effectively.

For a time series with data labeled $R_1, R_2, \ldots, R_n$, the MK test statistic (S) is determined as:

$$S = \sum_{k=1}^{n-1} \sum_{j=k+1}^{n} sgn\,(R_j - R_k) \tag{1}$$

where

$$sgn\,(R_j - R_k) = \begin{cases} +1 & \text{if } R_k < R_j \\ -1 & \text{if } R_k > R_j \end{cases} \tag{2}$$

n, R, x, k, and j reflect the total amount of time-series data, their rank, magnitude, any given data, and the remaining data in the series, respectively. Positive S values indicate an inclining trend, whereas negative S values indicate a declining trend in the dataset, respectively, with a substantially greater S indicating a significant trend in its direction.

The MK S statistic's mean is:

$$E(S) = 0 \tag{3}$$

And, variance is:

$$var(S) = \frac{1}{18}\left[n(n-1)(2+5) - \sum\nolimits_{p=1}^{q} t_p \cdot (t_p - 1) \cdot (2t_p + 5)\right] \tag{4}$$

The cluster size of p_{th} tied groups is t_p, and the number of tied groups is q.
The MK Z statistic (Z_{MK}) is:

$$Z_{MK} = \left\{ \begin{array}{c} \frac{S-1}{\sqrt{var(S)}} ifS > 0 \\ \frac{S+1}{\sqrt{var(S)}} ifS < 0 \end{array} \right\} \tag{5}$$

If $|Z| > Z_{1-}\alpha_{/2}$, where Z is the significance level and is taken from the standard normal distribution table, the hypothesis of no trend (H_0) is rejected. This shows that the trend is likely to be misinterpreted. They were generated by the expected variation of S for Z determination, the ability of the MK test to reject what seems to be a trend over a short time when the fractional data series is designed to show a minor trend. The meaning level is a criterion for rejecting the null hypothesis (H_0). Although the choice of the level remains subjective mainly, the analysts chose the level 0.05 (also known as the 5% level of meaning) or the level 0.01 (1% significance level). More evidence must differ from the null hypothesis at the lower level in order to be substantially conclusive. In this analysis of rainfall variability analysis, the 0.05 significance threshold was more cautious than the 0.01 level.

3.2 Sen's slope method

To determine the nature of the trend in the rainfall time series data and the magnitude of the trend is calculated by Sen's slope test. The slope (T_i) for $i = 1, 2, 3, \ldots, n$ is defined as the change in obtained value per change in time is determined from the equation as:

$$T_i = \frac{x_j - x_k}{j - k} \tag{6}$$

x_j and x_k are considered data measurements at times j and k ($j > k$), respectively. The median of these n data series of T_i is the magnitude of a trend. Depending upon the odd or even numbers in data series, an estimator M_i is computed as:

$$M_i = \begin{cases} T_{\frac{n+1}{2}} & n \text{ is odd} \\ \frac{1}{2}\left(T_{\frac{n}{2}} + T_{\frac{n+2}{2}}\right) & n \text{ is even} \end{cases} \tag{7}$$

Having determined M_i as above, Sen's slope estimator M_{med} is confirmed by a two-sided test at $100 \times (1 - \alpha)$ % at 0.05 significance level, and then an actual slope can be obtained. The progressive value of M_i signifies an ascending trend, whereas a negative value of M_i refers to descending trend in the datasets.

In addition, a linear trend analysis has been performed for the data set, and its statistical significance has also been tested using 95% significant ranking statistics from MK (World Meteorological Organization, WMO, 1966).

4 Results and discussion

Data were compared to years of events for the annual, premonsoon, monsoon, postmonsoon and winter seasons for the analysis of the precipitation trends. Maximum, minimum, and average time series have been arranged for temperature data. The results has been obtained from the developed tool/interface in MATLAB® environment to analyze the trends using non-parametric statistical methods, MK test and Sen's slope estimator.

4.1 Precipitation

Annual, premonsoon, monsoon, postmonsoon, and winter season precipitations time series have been analyzed and the statistical analysis has been provided in Table 1. In addition to the graphical representation of the data, the equation obtained by the linear regression has also been mentioned on these graphs, and the rate of change (slope) can be obtained from the represented equation in Fig. 2. The annual, premonsoon, monsoon,

TABLE 1 Statistical analysis of precipitation for Patna district.

Season	Mean	SD	MK statistics (S)	Kendall's statistics (Z_{MK})	Kendall's tau (τ)	P-value	Sen's slope (M_i)
Annual	1091.51	195.96	−1067	−3.0822	−0.2071	0.0021	−2.1485
Premonsoon	41.19	25.91	−93	−0.2660	−0.0181	0.7902	−0.0167
Monsoon	935.95	180.71	−1156	−3.3395	−0.2244	0.0008	−2.0250
Postmonsoon	75.53	54.88	173	0.4973	0.0336	0.6190	0.1390
Winter	38.82	23.30	−67	−0.1908	−0.0130	0.8487	−0.0110

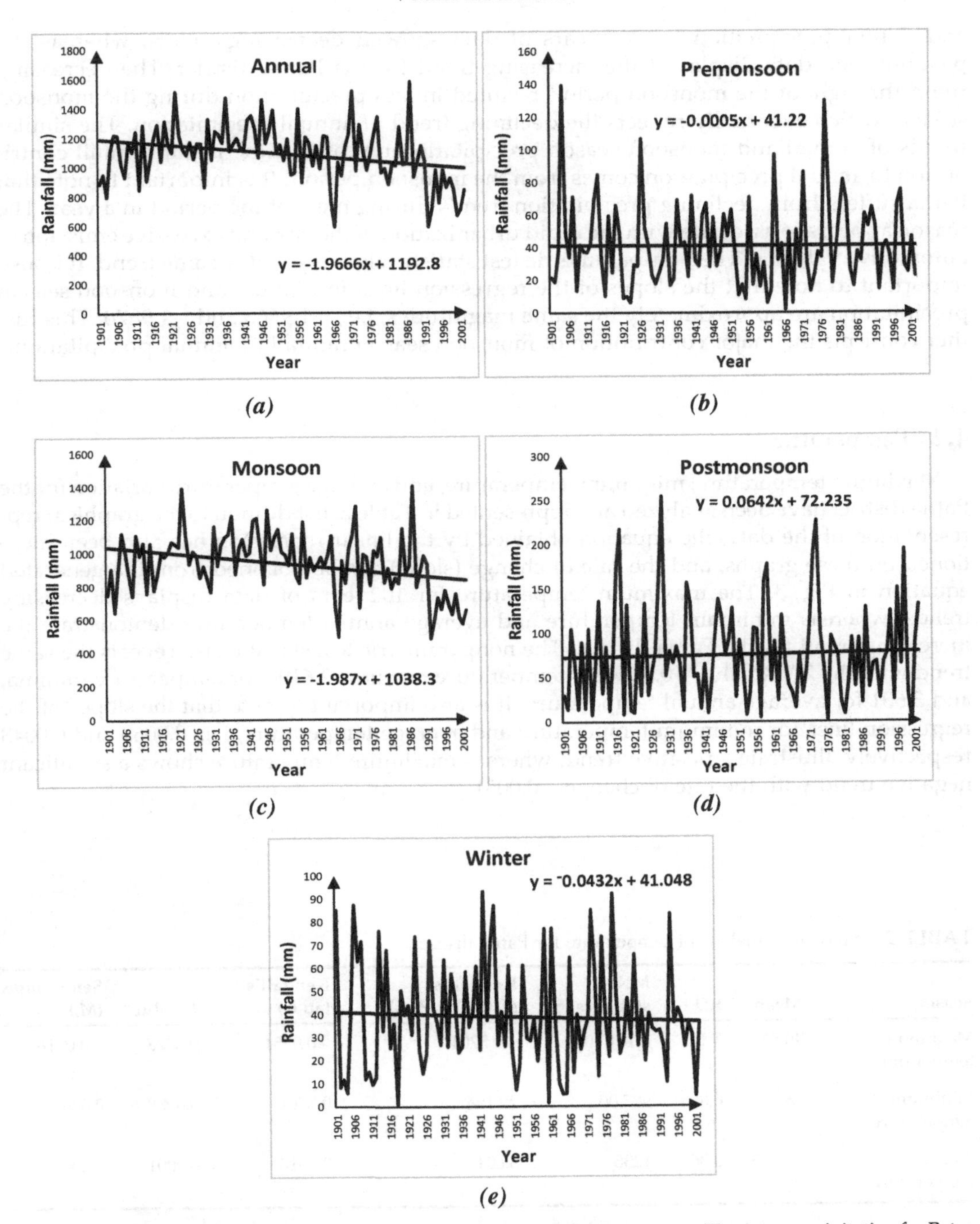

FIG. 2 Variation of (A) annual, (B) premonsoon, (C) monsoon, (D) postmonsoon, (E) winter precipitation for Patna district and corresponding linear regression equation for the trend.

and winter precipitation for 102 years of data showed decreasing trends, whereas the postmonsoon data displayed the increasing trend for the Patna district. The decreasing trend throughout the monsoon period resulted in less precipitation during the monsoon season, which eventually reflects the declining trend of annual precipitation. The similar trends of annual and monsoon season precipitation reveal that the major rainfall contribution to annual precipitation comes from the monsoon period. It is important to note that Patna suffers from declining precipitation trends during most of the period in a year. The reason appears to associate with the rapid urbanization of the city and excessive emission of automotive vehicles. The non-parametric test statistic also reveals the same trend. It is also important to note that the slopes of the regression lines for annual and monsoon season precipitation are approximately the same magnitudes, i.e., −1.9666 and −1.9870. This further confirms the major contribution of monsoon season rainfall to annual precipitation.

4.2 Temperature

Maximum temperature, minimum temperature, and average temperature variation for the Patna district have been analyzed and represented in Table 2. In addition to the graphical representation of the data, the equation obtained by the linear regression has also been mentioned on these graphs, and the rate of change (slope) can be obtained from the generated equation in Fig. 3. The maximum temperature for 102 years of data displays decreasing trends, whereas minimum temperature and average annual temperature demonstrate the increasing trend for the Patna district. The nonparametric test statistic also reveals the same trend with the *ZMK* value −0.5204 for temperature maxima, 2.6196 for temperature minima, and 3.601 for average annual temperature. It is also important to note that the slopes of the regression lines for minimum temperature and average temperature, i.e., 0.0063 and 0.0043 respectively, illustrate a positive trend, whereas maximum temperature shows a significant negative trend with the rate of change −0.0031.

TABLE 2 Statistical analysis of temperature for Patna district.

Season	Mean	SD	MK statistics (S)	Kendall's statistics (Z_{MK})	Kendall's tau (τ)	*P*-value	Sen's slope (M_i)
Maximum temperature	40.572	1.115	−181.00	−0.5204	−0.0351	0.6028	−0.0018
Minimum temperature	8.970	0.855	907.00	2.6196	0.1761	0.0088	0.0068
Average temperature	25.815	0.367	1238	3.601	0.2010	0.0001	0.0043

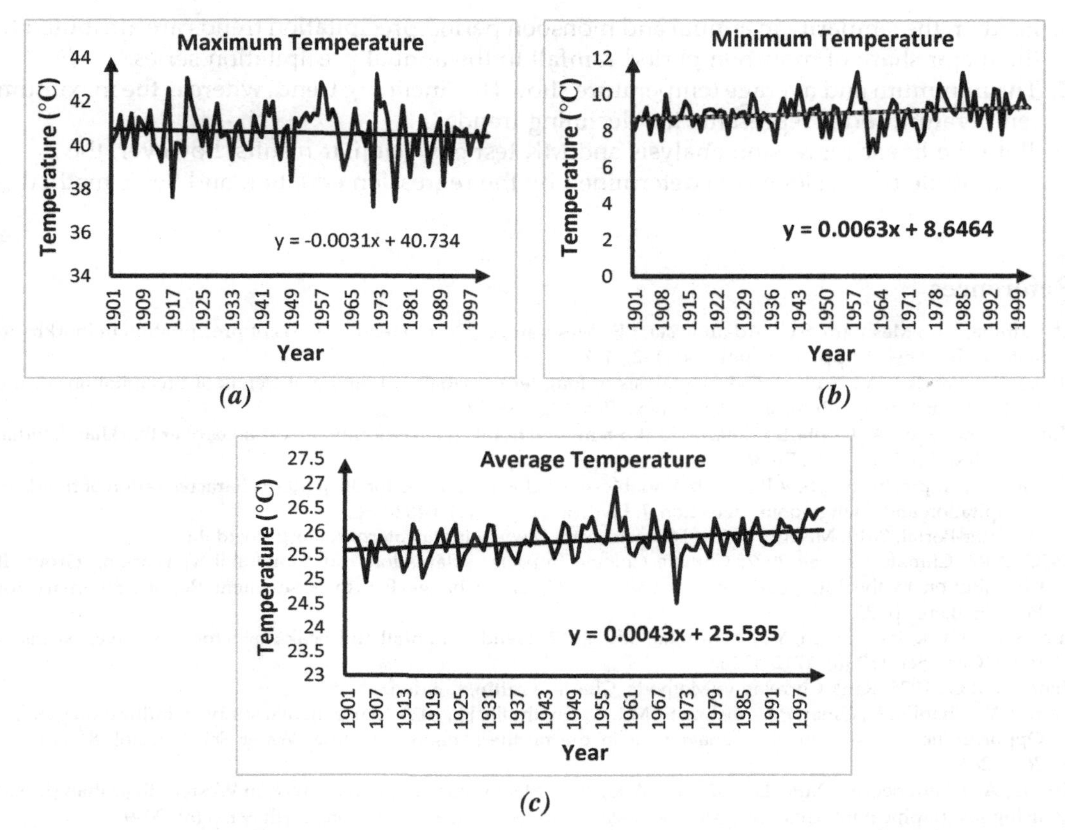

FIG. 3 Variation of (A) Maximum temperature (T_{max}). (B) Minimum temperature (T_{min}). (C) Average temperature (T_{avg}) for Patna district and corresponding linear regression equation for the trend.

5 Conclusion

The trend analysis of precipitation and temperature data for the Patna districts of Bihar was performed. Instead of using the spatially averaged values of rainfall data for a greater area, station size data were used to conduct the research. Precipitation data were analyzed using annual, premonsoon, monsoon, postmonsoon, and winter season time scales, while temperature data were analyzed using maximum, minimum, and average values. In order to achieve a natural precipitation and temperature trend in the study station, the MK test, the linear regression method and Sen Methodology were used. The conclusions drawn based on methodologies applied to the time series data and scope for the future work are as follows:

A. The annual and monsoon period precipitation shows the declining trend for the Patna district, and the reason for this decrease can be attributed to the anthropogenic activities taking place at study locations.

B. Further, the similarity in annual and monsoon period precipitation trends are attributed to the major share of monsoon period rainfall to the annual precipitation series.
C. The minimum and average temperature shows the inclining trend, whereas the maximum temperature series represents the declining trend.
D. Both the linear regression analysis and MK test gave similar results; however, the magnitude of the slope was determined by the regression equation and Sen's method.

References

Chandniha, S.K., Meshram, S.G., Adamowski, J.F., Meshram, C., 2016. Trend analysis of precipitation in Jharkhand state, India. Theor. Appl. Climatol. 130 (1–2), 1–14.

Duhan, D., Pandey, A., 2013. Statistical analysis of long term spatial and temporal trends of precipitation during 1901–2002 at Madhya Pradesh India. Atmos. Res. 122, 136–149.

Gupta, N., Banerjee, A., Gupta, S.K., 2021a. Spatio-temporal trend analysis of climatic variables over Jharkhand, India. Earth Syst. Environ. 5 (1), 71–86.

Gupta, S.K., Gupta, N., Singh, V.P., 2021b. Variable-sized cluster analysis for 3D pattern characterization of trends in precipitation and change-point detection. J. Hydrol. Eng. 26 (1), 04020056.

India Water Portal, 2019. Meteorological Datasets. http://www.indiawaterportal.org/metdata.

IPCC, 2007. Climate Change 2007: Climate Change Impacts, Adaptation and Vulnerability. Working Group II Contribution to the Intergovernmental Panel on Climate Change Fourth Assessment Report. Summary for Policymakers., p. 23.

Jain, S.K., Nayak, P.C., Singh, Y., Chandniha, S.K., 2017. Trend in rainfall and peak flows for some river basins in India. Curr. Sci. 112 (8), 1712–1726.

Kendall, M.G., 1975. Rank Correlation Methods. Charles Griffin, London.

Kumar, V., Chaplot, B., Omar, P.J., Mishra, S., Md. Azamathulla, H., 2021. Experimental study on infiltration pattern: Opportunities for sustainable management in the northern region of India. Water Sci. Technol. 84 (10–11), 2675–2685.

Kundu, A., Chatterjee, S., Dutta, D., Siddiqui, A.R., 2015. Meteorological trend analysis in Western Rajasthan (India) using geographical information system and statistical techniques. J. Environ. Earth Sci. 5 (5), 90–99.

Mann, H.B., 1945. Nonparametric tests against trend. Econometrica 13 (3), 245–259.

Omar, P.J., Bihari, D.S., Kumar, D.P., 2019. Temporal variability study in rainfall and temperature over Varanasi and adjoining areas. Disaster Adv. 12 (1), 1–7.

Omar, P.J., Dwivedi, S.B., Dikshit, P.K.S., 2020. Sustainable development and management of groundwater in Varanasi, India. In: Advances in Water Resources Engineering and Management. Springer, Singapore, pp. 201–209.

Omar, P.J., Gaur, S., Dikshit, P.K.S., 2021. Conceptualization and development of multi-layered groundwater model in transient condition. Appl. Water Sci. 11 (10), 1–10.

Omar, P.J., Gupta, N., Tripathi, R.P., Shekhar, S., 2017. A study of change in agricultural and forest land in Gwalior city using satellite imagery. Samriddhi—J. Phys. Sci. Eng. Technol. 9 (02), 109–112.

Panda, A., Sahu, N., 2019. Trend analysis of seasonal rainfall and temperature pattern in Kalahandi, Bolangir and Koraput districts of Odisha, India. Atmos. Sci. Lett. 20 (10), e932.

Pingale, S.M., Khare, D., Jat, M.K., Adamowski, J., 2016. Trend analysis of climatic variables in an arid and semi-arid region of the Ajmer District Rajasthan India. J. Water Land Dev. 28, 3–18.

Pranuthi, G., Dubey, S.K., Tripathi, S.K., Chandniha, S.K., 2014. Trend and change point detection of precipitation in urbanizing districts of Uttarakhand in India. Indian J. Sci. Technol. 7 (10), 1573–1582.

Shekhar, S., Chauhan, M.S., Omar, P.J., Jha, M., 2021. River discharge study in river ganga, Varanasi using conventional and modern techniques. In: The Ganga River Basin: A Hydrometeorological Approach. Springer, Cham, pp. 101–113.

Tripathi, R.P., Pandey, K.K., 2021. Experimental study of local scour around T-shaped spur dike in a meandering channel. Water Supply 21 (2), 542–552.

Umakanth, N., Satyanarayana, T., Madhav, B.T.P., Rao, M.C., 2021. Study of rainfall over Patna Region, India during 2013. AIP Conf. Proc. 2352 (1), 030007 (AIP Publishing LLC).

Warwade, P., Sharma, N., Pandey, A., Ahrens, B., 2016. Analysis of climate variability in a part of Brahmaputra river basin in India. In: Sharma, N. (Ed.), River System Analysis and Management. Springer, Singapore, pp. 113–142.
Warwade, P., Tiwari, S., Ranjan, S., Chandniha, S.K., Adamowski, J., 2018. Spatio-temporal variation of rainfall over Bihar state, India. J. Water Land Dev. 36, 183–197.
WMO, 1966. Measurement and Estimation of Evaporation and Evapotranspiration. WMO No. 201, Technical Note 83.
Zakwan, M., Ahmad, Z., 2021. Trend analysis of hydrological parameters of Ganga River. Arab. J. Geosci. 14, 163.
Zakwan, M., Ara, Z., 2019. Statistical analysis of rainfall in Bihar. Sustain. Water Resour. Manag. 5, 1781–1789.

CHAPTER

31

A review of climate change trends and scenarios (2011–2021)

Deborah Ayodele-Olajire[a] *and Adeyemi Olusola*[a,b]

[a]Department of Geography, University of Ibadan, Ibadan, Nigeria [b]Department of Geography, University of the Free State, Bloemfontein, South Africa

OUTLINE

1 Introduction 545
2 Methodology 547
2.1 *Data source* 547
2.2 *Data retrieval* 547
2.3 *Data processing* 548
2.4 *Data analysis* 548
3 Results and discussion 550
3.1 *Overview of articles, authors and their contributions* 550
3.2 *Changing paradigms and thematic evolution* 552
3.3 *Climate change knowledge production* 556
4 Conclusion 558
References 558

1 Introduction

The knowledge of climate change impacts on the environment, including on the various social and economic sectors that depend on environmental resources, is now more widely known than it was at the turn of the century. For example, in a survey deployed across six countries (Canada, China, Germany, Switzerland, United Kingdom, and the United States) with different cultural and political characteristics, Shi et al. (2016) found that knowledge about the causes of climate change was linked to higher concerns about the phenomenon. The foregoing is similar to the findings made by Tobler et al. (2012) as well, who found that people's knowledge about the causes of climate change was most strongly related to their

https://doi.org/10.1016/B978-0-323-91910-4.00031-5

attitudes toward climate change compared with the other knowledge subsystems that were interrogated. Also, the authors found that respondents' knowledge of climate change seemed to have increased compared with findings from previous years. These studies suggest that educating non-scientists regarding climate change is paying off and raises the hope for better environmental stewardship.

Despite these successes, anthropogenic climate change is yet to be on the decline, with varying trends across the globe. In the Lake Chad basin area, analysis between 1951 and 2015 shows strong signals of increasing temperature trends and decreasing precipitation. Given these trends, the temperature is predicted to increase by 0.65–1.6°C and precipitation to decrease by 13%–11% between 2016–15 and 2026–35 compared with 1991–90 (Mahmood et al., 2019). Due to increasing sea surface temperature, the Greenland ice sheet is reported to have been losing mass at the rate of ~215 Gt/year between 2002 and 2011 (Comiso and Hall, 2014). Meanwhile, using the PRECIS regional climate model under A2 and B2 scenarios for 1960–2049, Lacombe et al. (2012) found that in continental South-East Asia, the temperature is increasing across the region with steeper trends at higher latitudes while precipitation increased or decreased based on landform/land cover (gulf, sea, land). In the Prairie Pothole Region of North America, Millett et al. (2009) report that the climate is getting warmer and wetter with decreased diurnal temperature range. These trend analyses highlight the type, form and degree of changes that are taking place within the climate system.

In addition to the documented trends of climate change, there has been a push within the scientific community to understand the relationship between climate change and extreme weather events. Significant advances are being made in this area. In the United States, Armal et al. (2018) conducted a systematic analysis to identify and attribute the trends of extreme rainfall events to climate change across the contiguous United States. The authors utilized a Bayesian Multilevel Model for their analysis and found that more than half of the climate trends obtained from 1244 climate stations were statistically significant. Such study has only recently become acceptable (Solow, 2015).

According to the Fifth Assessment Report of the Intergovernmental Panel on Climate Change, anthropogenic climate change leads to an increase in the frequency and intensity of extreme daily temperature/precipitation (Bindoff et al., 2013). Although this knowledge is generally accepted by climate change scientists, the attribution of specific individual events to anthropogenic climate change used to be refuted (Solow, 2015). The assessment by the National Academies of Science suggests that the long-held opinion that individual events cannot be attributed to anthropogenic climate change may not be true (National Academies Press, 2016). Meanwhile, Stott (2016) suggests that the use of event attribution to quantify risks can help rebuild efforts after (climate-related) disasters and insurance pricing. Before these conclusions, Otto et al. (2012) and Lewis and Karoly (2013) already attempted event attribution to climate change while focusing on the 2010 Russian heat wave and 2013 Australian heat wave, respectively.

The increasing concern about the effect of more intense extreme events on human health and wellbeing has closely followed the progress made in event attribution to climate change. Concerning research exploring the link between extreme events and health, Deschênes and Moretti (2009) found that extreme heat and cold events in the United States were resulting in immediate increasing mortality. In their sample of all-white deaths occurring in the continental United States between 1972 and 1988, the authors estimate that 0.8% of average annual deaths are attributable to cold events with 4%–7% gains in life expectancy experienced by people who moved from the cold Northeast to the warm Southwest between 1970 and 2000. McMichael (2015) further exposed the effect of extreme events on infectious disease risks and outbreaks. In the aftermath of Hurricane Katrina in New Orleans, gastrointestinal tract and respiratory infections were found to be predominant in the area from late August to September 2005. A review of the 2010 floods in Pakistan—the worst flooding event in the history of the country at the time—shows that this event was accompanied by acute respiratory infections, acute diarrhea, skin diseases and malaria (McMichael, 2015). Furthermore, Hashim and Hashim (2016) show that between 1993 and 2012, over 530,000 people died directly from about 15,000 extreme events in the Asia Pacific region. In Europe, De Sario et al. (2013) assert that heat wave studies suggest a synergistic interaction between high temperatures and air pollution. The authors found that allergen patterns are changing in response to climate change and air pollution.

Apart from mortality, there are also mental health impacts of extreme weather events. Shukla (2013) reviews the effects of climate change-related extreme events on mental health. These effects were found to include the worsening of mental health problems due to health care service disruptions, an increase in admission for mental health problems in the form of posttraumatic stress disorder (PTSD) among others, and even an increase in suicide rates after extreme events. Given the varied effects of climate change presented, a global review of climate change trends, the development of new methods of analysis and the use of multiple scenarios are important. In this chapter, we assess the growth, focus, thematic evolution and contributors to the study of climate change to identify key lessons and gaps that can guide climate change research in the next decade and beyond.

2 Methodology

2.1 Data source

This study is approached from a global point of view. We systematically reviewed publications on climate change using the SCOPUS database. The choice of the database is based on its global acceptance, reputation, and accessibility. Therefore, publications not indexed in SCOPUS are not captured. The period of study is from 2011 to 2021. The search syntax used was *"climate change, trends and scenarios"* or *"climate change and trends and analysis,"* or *"changing climate, trends scenarios."* After excluding some documents based on relevance, the study made use of 2000 (two thousand) publications on climate change, trends, and scenarios across all scales (Fig. 1). The dataset was downloaded and stored in a *.bibtex* format.

2.2 Data retrieval

The syntax used for this study was appraised within the Google Scholar search engine for stability. After careful sorting and to the satisfaction of the authors, we applied the same search protocol ([*climate change, trends and scenarios*] or [*climate change and trends and analysis*] *or* [*changing climate, trends scenarios*]) within the SCOPUS database. This was essential to ensure that studies on the subject were not left out due to the choice of words. A total number of 2000 (two thousand) publications were retrieved based on the syntax between 2011 and 2021 (Fig. 2).

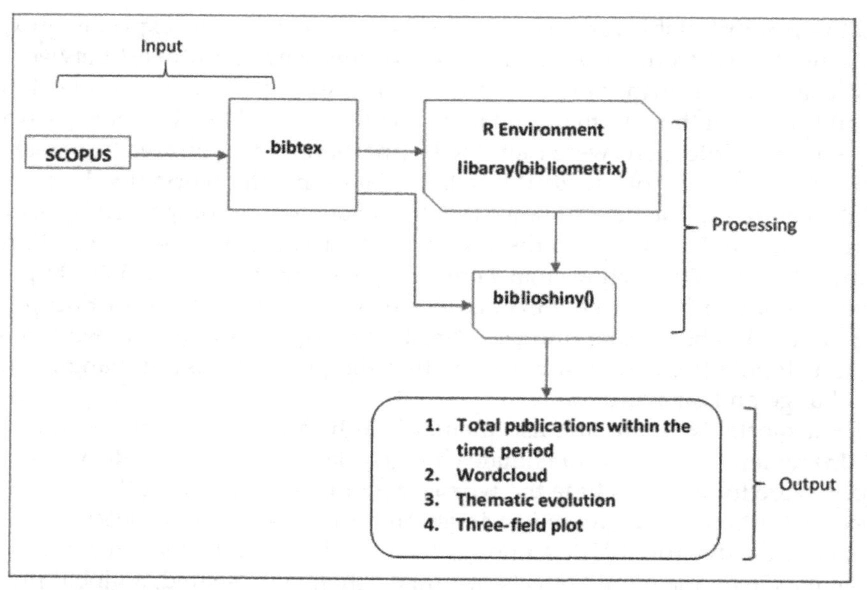

FIG. 1 Flow chart showing the methodology and some selected outputs.

2.3 Data processing

The 2000 (two thousand) publications were downloaded in a *.bibtex* format from the SCOPUS database (Fig. 2) and processed in the R environment (R Development Core Team, 2013) using the bibliometrix package (Aria and Cuccurullo, 2017). The bibliometrix package converts the *.bibtex* format into a data frame-like document using the *convert2df* () function. The *convert2df* function converts the *.bibtex* format into a data frame with variables such as but not limited to authors, countries, affiliations, references, keywords, co-authors, and citations. After the conversion, the data were used to initiate the biblioshiny environment using the function *biblioshiny*(). The biblioshiny environment is a shiny app that provides a web interface for bibliometrix. From the biblioshiny GUI, document information, authors' publication list, word cloud, three-field plot and thematic evolution of keywords were generated (Fig. 2).

2.4 Data analysis

The converted data frame using the *convert2df* function was initially analyzed using the *biblioAnalysis*() function. The function provides descriptive statistics on the dataset, which include information on but not limited to:

1. total number of documents,
2. annual scientific production,
3. country production,
4. most-cited country,

Description	Results
MAIN INFORMATION ABOUT DATA	
Timespan	2011:2021
Sources (Journals, Books, etc)	636
Documents	2000
Average years from publication	4.4
Average citations per documents	17.58
Average citations per year per doc	2.799
References	106176
DOCUMENT TYPES	
article	1733
article in press	1
book	3
book chapter	31
conference paper	156
conference review	4
letter	1
note	4
review	66
short survey	1
DOCUMENT CONTENTS	
Keywords Plus (ID)	9253
Author's Keywords (DE)	5074
AUTHORS	
Authors	7245
Author Appearances	9280
Authors of single-authored documents	57
Authors of multi-authored documents	7188
AUTHORS COLLABORATION	
Single-authored documents	66
Documents per Author	0.276
Authors per Document	3.62
Co-Authors per Documents	4.64
Collaboration Index	3.72

FIG. 2 Document information.

5. most-cited document,
6. most relevant authors and sources,
7. three-field plot.

After initiating the *biblioAnalysis*(), the shiny app, in this case, biblioshiny was launched. Based on the 2000 documents retrieved, various indices were established from the biblioshiny interface. The derived indices were made possible based on the coupling, co-citation and collaboration network

a. Coupling network, where,

$$B = A \times A^T \tag{1}$$

where A is a bipartite network and B is a symmetrical matrix. The strength of the two articles is defined simply by the number of references that the articles have in common. The biblioNetwork from the established dataframe established couplings such as authors, sources and countries (Aria and Cuccurullo, 2017).

b. Co-citation network, where,

$$C = A^T \times A \tag{2}$$

where A is a bipartite network and C is a symmetrical matrix. Here, C (diagonal) consists of the number of cases in which a reference is cited in the dataframe (Aria and Cuccurullo, 2017).

c. Collaboration, where,

$$AC = A^T \times A \tag{3}$$

where A is a bipartite network (Manuscripts × Authors). The matrix ensures the calculation of an author's/country collaboration network (Aria and Cuccurullo, 2017).

3 Results and discussion

3.1 Overview of articles, authors and their contributions

From this study, the Year 2020 witnessed the highest number of publications (300) across the entire period (Fig. 3) at the time of writing. Most of the studies carried out within the study period were published in Science of the Total Environment, the International Journal of Climatology, Theoretical and Applied Climatology, Climate Change, and Climate Dynamics (Table 1). Based on the aim and scope of these journals, it is fair to conclude that most of these

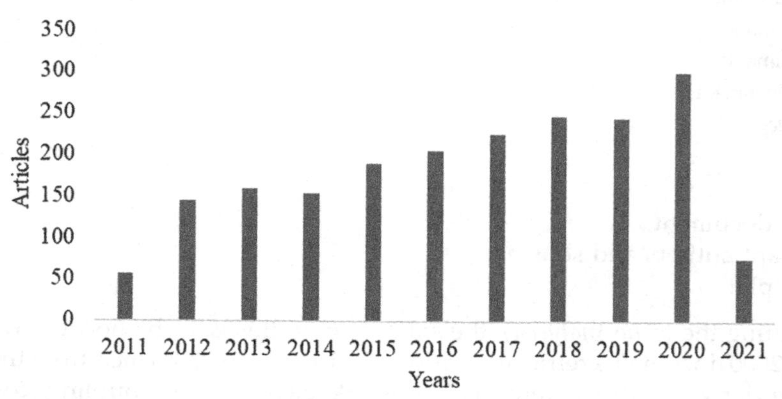

FIG. 3 Distribution of articles between 2011 and 2021 using the search syntax.

TABLE 1 Top publication sources of subject matter from 2011 to 2021.

Journals	Number
Science of the Total Environment	80
International Journal of Climatology	63
Theoretical and Applied Climatology	55
Climatic Change	52
Climate Dynamics	48
Journal of Hydrology	48
Journal of Climate	39
Sustainability (Switzerland)	35
Global and Planetary Change	21
Ecological Indicators	19
Global Change Biology	19
Geophysical Research Letters	18
Hydrology and Earth System Sciences	18
Atmospheric Research	17
Stochastic Environmental Research and Risk Assessment	17
Water Resources Management	16
Advances in Meteorology	14
Environmental Earth Sciences	14
Global Environmental Change	14
Journal of Hydrologic Engineering	13
Shengtai Xuebao/Acta Ecologica Sinica	13
Agricultural and Forest Meteorology	12
Atmosphere	12
Journal of Geophysical Research: Atmospheres	12
Journal of Water And Climate Change	12
Modeling Earth Systems and Environment	12
Atmospheric Environment	11
Modsim 2011—19th International Congress On Modeling And Simulation—Sustaining Our Future: Understanding and Living with Uncertainty	11
Water Resources Research	11

studies are collaborative and interdisciplinary (Deser et al., 2012; Rosenzweig et al., 2013; Newbold et al., 2015). Most of these studies are concerned with climate change, modeling, trends and the climatology of precipitation (Fig. 4) and most of the relevant contributors are to a large extent from China and the United States of America.

China, USA and some European countries are at the cutting edge of research on climate trends, analysis and scenarios. The dominance of these countries stems from the presence of the skilled personnel and state-of-the-art infrastructure needed in climatic studies, within their borders. In addition, the centrality of Europe within collaborations can be tied to the high number of climate-related research institutes that are spread across European countries such as the Centre National de Recherches Meteorologiques (France), Danmarks Meteorologiske Institut (Denmark), Abdus Salam International Centre for Theoretical Physics (Italy), Climate Limited-Area Modeling Community (Germany), Koninklijk Nederlands Meteorologisch Instituut (Netherlands), Max Planck Institute (Germany), and Sveriges Meteorologiska och Hydrologiska Institut (Sweden), among others (Nikulin et al., 2012). Overall, there has been a steady growth in the number of studies from 2011 up to 2020.

3.2 Changing paradigms and thematic evolution

Since the release of the Fifth Assessment Report, climate change impacts have been studied using the representative concentration pathways (RCPs) (Collins et al., 2013). As against previous reports, the Fifth Assessment Report provides the basis for improved and more comprehensive data analysis using a new RCP. This introduction has resulted in more studies that use hybrid models to understand changing climate. Furthermore, coupled with the tremendous growth in Earth Observation Satellites (EOSs) that produce big data for enhanced trends and analysis, post-2011 has been very significant in climate change narratives. The RCPs, which are used for making projections based on these factors, describe four different 21st-century pathways of GHG emissions and atmospheric concentrations, air pollutant emissions and land use. The RCPs include a stringent mitigation scenario (RCP2.6), two intermediate scenarios (RCP4.5 and RCP6.0) and one scenario with very high GHG emissions (RCP8.5). Scenarios without additional efforts to constrain emissions ("baseline scenarios") lead to pathways ranging between RCP6.0 and RCP8.5. RCP2.6 is representative of a scenario that aims to keep global warming likely below 2 °C above preindustrial temperatures.

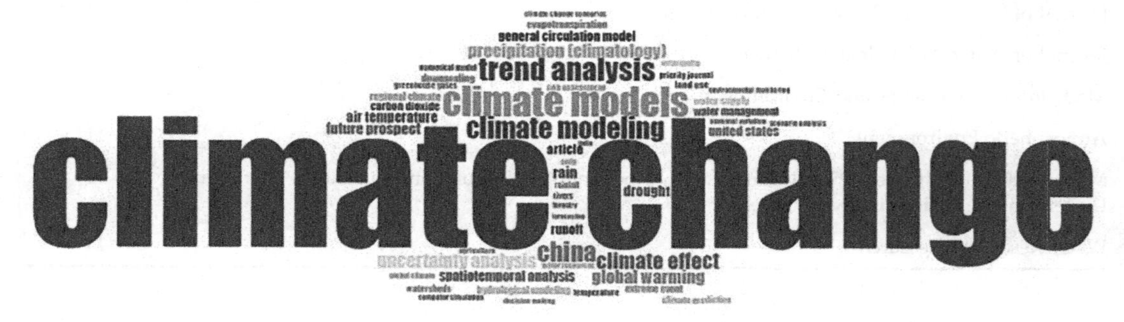

FIG. 4 Most relevant words between 2011 and 2021.

In their study on the past, present and future climate trends under varied RCPs in Uganda, Egeru et al. (2019) posited that in Uganda, the minimum temperature is projected to increase by 1.8°C (RCP4.5) and 2.1°C (RCP8.5) in mid-century, and by 2.2°C (RCP4.5) and 4.0°C (RCP8.5) in end-century. In another study on future drought conditions over South Korea based on the latest and most advanced sets of regional climate model simulations under the RCP4.5 and RCP8.5 scenarios, it was observed that the number of drought months over South Korea is projected to increase (decrease) for the period 2041–70 in the RCP8.5 (RCP4.5) scenario and increase (decrease) for the period 2071–00 in the RCP4.5 (RCP8.5) scenario (Choi et al., 2016). Furthermore, Fix et al. (2018) stated that precipitation extremes are expected to increase in a warming climate, which may have serious societal impacts. The authors claimed that under RCP8.5 between 2005 and 2080, the 1% annual exceedance probability level is projected to increase by 17% on average across the United States, and up to 36% for some grid cells.

To add to this, the paper by Newbold et al. (2015) published in Nature on "Global effects of land use on local terrestrial biodiversity" has been a reference point in understanding the impact of anthropogenic activities on biodiversity. The work posited that the role of humans - measured in terms of land use puts enough pressure on biodiversity and affects their distribution, survival and resilience (Newbold et al., 2015). Another key paper by Deser et al. (2012) on uncertainty in climate change projections where the authors argued that despite the increase in data availability and models, the problem of uncertainty poses a great challenge [especially post-2011] for adaptation planning. In line with this thought on uncertainty, Rosenzweig et al. (2013) in their study on Agricultural Model Intercomparison and Improvement Project (AgMIP) argued that there is the need to address uncertainty, aggregation and scaling to ensure improvements in the understanding of agriculture futures.

From the foregoing, it is clear that the state of knowledge production since 2011 has witnessed improvement in multidisciplinary studies focusing more on interconnections between changing climate and various aspects of the environment (Fig. 5). The use of various global and downscaled models to understand the driving forces behind the changing climate has become more pronounced (Gupta et al., 2013; Arora et al., 2020; Maurya et al., 2021; Mahoney et al., 2021; Akinsanola and Zhou, 2019) albeit without caution to uncertainties in global and regional estimates. Some key terms asides from climate change such as downscaling, ENSO, trend and land-use span the entire study period (Fig. 5).

Temperature and rainfall are often the two climatic variables assessed in climatic studies. However, we found precipitation, drought, and evapotranspiration to be the leading climatic concepts in the quantification of climatic terms (Fig. 6). These three terms have to do with moisture, pointing to a prioritizing of studies involving precipitation. This preference could be due to the hitherto known focus on "warming"—the second word in *global* warming and which has to do with temperature. In terms of centrality and impact, droughts turn out to be one of the major issues in climate change studies, especially within tropical environments. Furthermore, the incidence of multiyear drought in some parts of the United States and sub-Saharan Africa reinforces the importance of droughts and its persistence in both space and time (Szejner et al., 2020; Ayodele-Olajire, 2021; Adedeji et al., 2020; Orimoloye et al., 2021a,b).

Within the last four to 5 years, we found the following terms to be emerging: CMIP5, CMIP6, RCP8.5, and machine learning. The emergence of these terms is a result of the

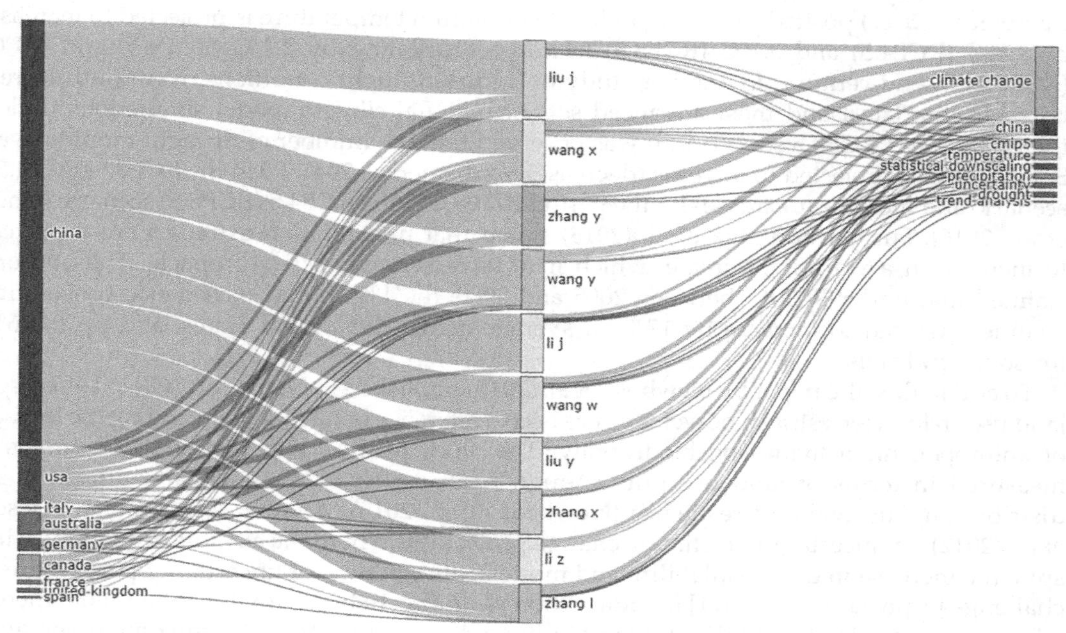

FIG. 5 Most relevant authors, countries, and contributions between 2011–2021.

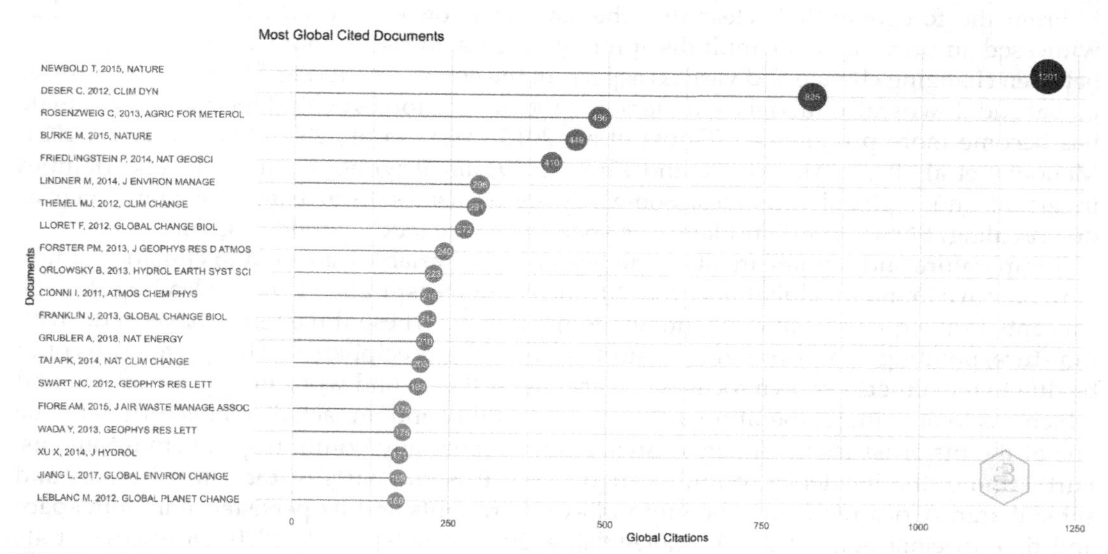

FIG. 6 Globally cited documents.

improvements in the development of climate models such as the coordinated efforts put into the development of the Coupled Model Intercomparison Projects (CMIP). The sensitivity and the performance of the CMIP6 models as against the CMIP5 have been shown in the recently-released Sixth Assessment Report (Physical Basis) of the Intergovernmental Panel on Climate Change (IPCC). This higher sensitivity contributes to projections of about 0.4°C increased warming in this century compared with similar scenarios run in CMIP5.

The CMIP5 protocol was endorsed in 2008 by the Working Group on Coupled Modelling (WGCM), which defined a set of 35 climate model experiments designed to be useful for model differences in poorly understood feedbacks associated with the carbon cycle and with clouds, predictability in climate and determining why similarly forced models to produce a range of responses (https://www.wcrp-climate.org/wgcm-cmip/wgcm-cmip5). In their study on historical Antarctic precipitation and temperature trend using CMIP5 models and reanalysis datasets, Tang et al. (2018), concluded that 37 CMIP5 models show an increasing trend, 18 of which are significant. They however observed that one of the many reasons for the large differences in precipitation is the CMIP5 models' resolution. In another study on heavy precipitation events in a warmer climate using CMIP5 models, using the results of a set of 20 climate models taking part in phase 5 of Coupled Model Intercomparison Project (CMIP5), the authors concluded that CMIP5 models show a projected increase for the end of the 21st century of the width of the right tail of the precipitation distribution, particularly pronounced over India, Southeast Asia, Indonesia, and central Africa during boreal summer, as well as over South America and southern Africa during boreal winter.

Despite the progress made in the last few decades in model evaluation, the CMIP community saw the need for improvement in model evaluation to perform much more efficiently to enable a systematic and rapid performance assessment of the large number of models participating in CMIP. Hence, the implementation of CMIP6 (https://www.wcrp-climate.org/wgcm-cmip/wgcm-cmip6). The use of CMIP6 is mostly aimed at but not limited to improving the understanding of the climate system, providing estimates of future climate change and related uncertainties, providing input data for the adaptation to the climate change, and examining climate predictability and exploring the ability of models to predict climate on decadal time scales (https://cds.climate.copernicus.eu/cdsapp#!/dataset/projections-cmip6?tab=overview). In their study on the comparison of CMIP5 and CMIP6 multimodel ensemble for precipitation downscaling results and observational data in the Hanjiang River Basin, the authors posited that the downscaled CMIP6-MME better-simulated precipitation for most stations compared to the downscaled CMIP5-MME in all seasons except for summer (Wang et al., 2021). In another study conducted in Africa, Babaousmail et al. (2021) posited that CMIP6 models satisfactorily reproduce the mean annual climatology of dry/wet months.

To summarize, factorial analysis was performed on the publications between 2011 and 2021. All the papers analyzed in this study are placed on a two-dimensional (dim) plane to show clusters of research between 2011 and 2021 based on keywords (Fig. 7). The first group of words (red) has to do with applications/applied climatology, while the second group (blue) has to do with models and projections. As expected, in terms of spatial spread, publications on applications are more than those that have to do with science.

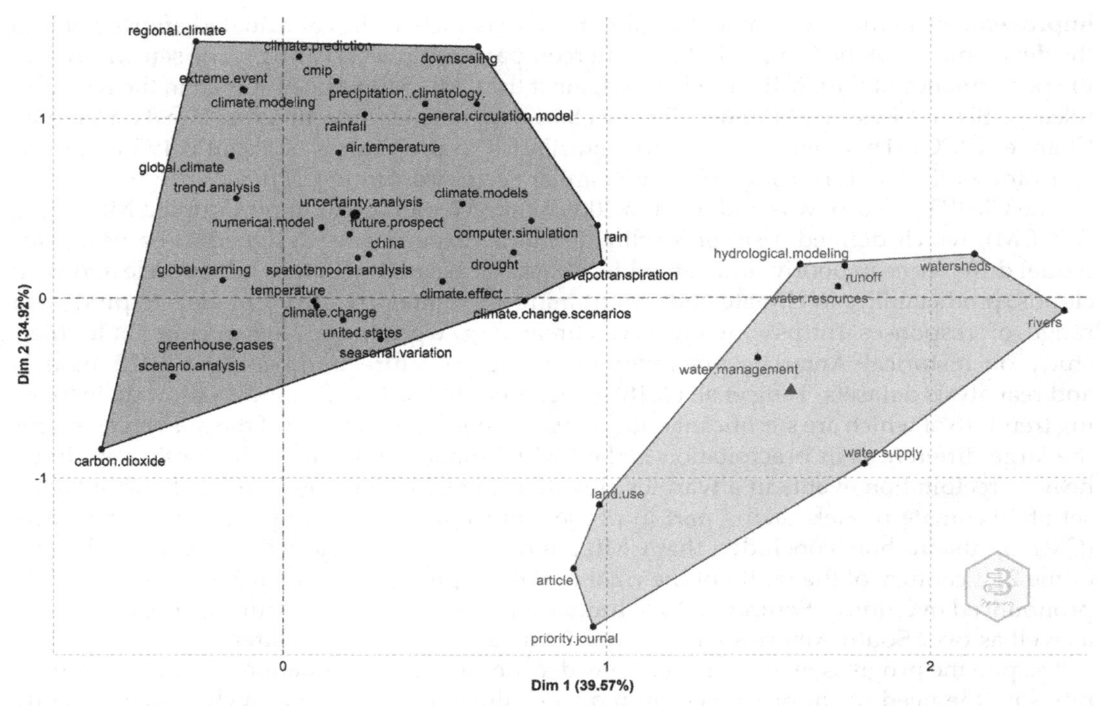

FIG. 7 Classification of keywords used in the search syntax between 2011 and 2021.

3.3 Climate change knowledge production

Contributions have not been on the same scale globally. While Asia (especially China), the Americas and Europe seem to be at the forefront, other continents, especially Africa, are still not contributing as much to climate change knowledge (Fig. 8). Ironically, sub-Saharan Africa and Asia are expected to witness the imprints of changing climate more than some of the places making substantial contributions to the subject matter. The inadequateness of the contributions from sub-Saharan Africa is also seen in the lower rates of interconnections and collaborative works being witnessed across the globe. Countries that are producing more knowledge are seen to be collaborating better on climate change research (Fig. 8).

The predominance of climate studies from the United States, Europe and China highlights the need to move climate financing into research institutions within lagging countries and continents. Financial sources could be local, national (Griffith-Jones et al., 2020) or international (*World Bank Group*, 2020) but with emphasis placed on increasing the research output of climate studies. Such an endeavor will provide finer-scale data and information that can be used in subsequent modelling and policy formulations. Closely related to financing for climate-related research is climate research capacity building for regions of the world that requires such. This is another area in which climate financing can promote equitable climate-knowledge production across the globe. Across Africa, one of such move is the Coordinated Regional Climate Downscaling Experiment (CORDEX)—Africa. CORDEX-Africa

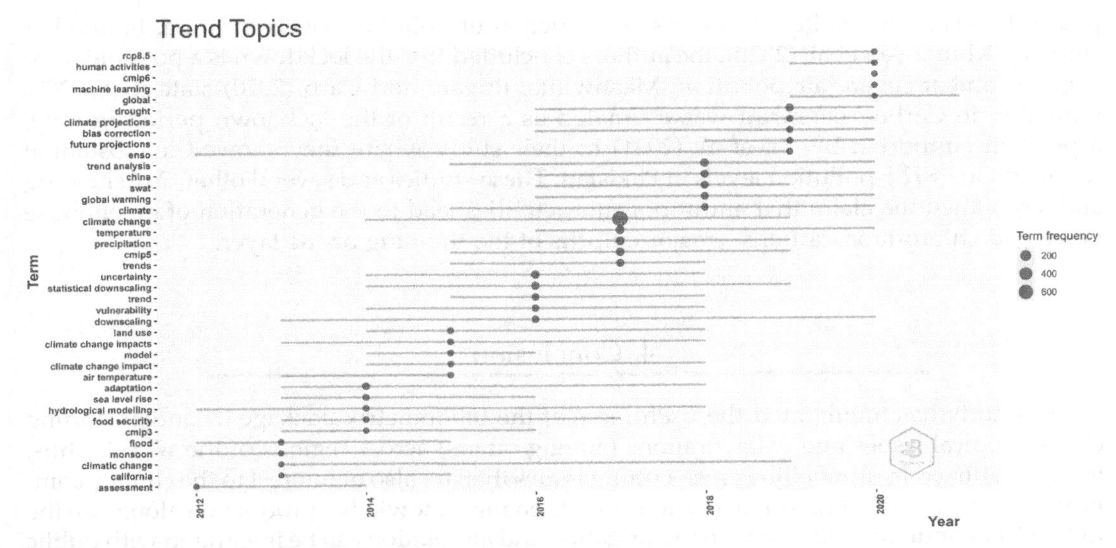

FIG. 8 Trending topics since 2011.

consists of young scientists who are working with downscaled data over the CORDEX-Africa domain. The CORDEX-Africa domain is for Analysis, Foci, Regional Messages, Integrated Approach, Capacity Development and Application. Here, four regional teams across the continent are saddled with the responsibility of addressing climate-based questions covering but not limited to health, biodiversity, agriculture and hydrology. Funded by the International Project Office for CORDEX and the Swedish Meteorological and Hydrological Institute and coordinated by the Climate System Analysis Group of the University of Cape Town, South Africa, the CORDEX-Africa is bridging the gap in knowledge production and putting ensembles from the Africa domain on the global sphere. However, there is the need for more efforts to be geared into the modelling of regional climate especially in regions where the imprints of the changing climate are likely to be more pronounced.

The steady growth in the publication of climate change trends to date is a welcome observation given the scope and magnitude of climate change effects. It is also impressive that 2020 records the highest number of publications in this research area despite the outbreak of the COVID-19 pandemic. The continuous work in publishing climate change findings during the pandemic seems to be related to the continuous link being made between climate change and the spread of diseases (Briz-Redón and Serrano-Aroca, 2020). Another reason for this steady increase, especially in 2020, could be the fact that climate change and COVID-19 affect the same areas of society such as food security, nutrition, public health among others. The pandemic, in its own way, has also brought about the stretching of statistical and modelling tools in climate studies (Briz-Redón and Serrano-Aroca, 2020). Furthermore, researchers are beginning to interrogate the role of anthropogenic activities during the lockdown on the burden of greenhouse gases in the lower atmosphere (Kumar et al., 2020; Khursheed et al., 2020; Rugani and Caro, 2020; Ali et al., 2021). For example, in their study of the environmental impacts of COVID-19 and Nationwide Lockdown in India, Kumar et al. (2020) argue that the phase-I and phase-II lockdown resulted in a massive reduction of pollution levels in India. In another study by Khursheed et al. (2020), the authors concluded that the lockdown is a possible measure to combat severe air pollution. Meanwhile, Rugani and Caro (2020) state that ~20% reduction in Carbon Footprint was estimated as a result of the lockdown period in Italy, a position supported by Ali et al. (2021) in their study where they showed a substantial reduction in PM2.5 pollution levels in Pakistan. These studies and several others are showing and strengthen the claim that anthropogenic activities lead to the generation of greenhouse gases and chlorofluorocarbons—major culprits of the thinning ozone layer.

4 Conclusion

This study has highlighted the usefulness of the bibliometrix package in understanding climate topical issues and collaborations (among others) across regions of the world. Thus, it can be utilized in unravelling cross-cutting issues that are also of interest to the climate community. Given the findings of increasing climate change knowledge production alongside the COVID-19 pandemic, climate change mitigation and adaptation can be integrated with public health to build more resilient communities. This is an area in which regional governments can take leadership and responsibility in ensuring sustainable climate and city goals.

Asides from precipitation, drought and evapotranspiration, which were the leading climate terms during the study period; CMIP5, CMIP6, RCP8.5 and machine learning are rising in use, especially in the past 5 years. This rise reflects the increasing prominence of modelling within the climate change community. Beneficially, the climate change literature has witnessed the refinement of methods within the past decade, through the introduction of the RCPs and the use of EOSs. These advances have led to appreciable knowledge production mainly from China and the United States of America. However, it will not be sufficient to simply maintain the current momentum.

Climate change knowledge production needs to ramp up in other regions of the world as well, especially in sub-Saharan Africa where a significant portion of climate change impacts is expected. In addition, as evidenced by this study, there is a need for more studies on model evaluation across Africa. More importantly, studies should also focus on extreme events such as heat waves, droughts, floods, and dust. In addition, these studies should also x-ray the physics and the dynamics behind these events and their implications on the environment.

References

Adedeji, O., Olusola, A., James, G., Shaba, H.A., Orimoloye, I.R., Singh, S.K., Adelabu, S., 2020. Early warning systems development for agricultural drought assessment in Nigeria. Environ. Monit. Assess. 192 (12), 1–21.

Akinsanola, A.A., Zhou, W., 2019. Projections of West African summer monsoon rainfall extremes from two CORDEX models. Clim. Dyn. 52 (3), 2017–2028.

Ali, S.M., Malik, F., Anjum, M.S., Siddiqui, G.F., Anwar, M.N., Lam, S.S., et al., 2021. Exploring the linkage between PM2. 5 levels and COVID-19 spread and its implications for socio-economic circles. Environ. Res. 193, 110421.

Anon., 2020. Transformative Climate Finance: A New Approach for Climate Finance to Achieve Low-Carbon Resilient Development in Developing Countries. World Bank Group, p. 173. Available from: https://elibrary.worldbank.org/doi/abs/10.1596/33917.

Aria, M., Cuccurullo, C., 2017. Bibliometrix: an R-tool for comprehensive science mapping analysis. J. Informetr. 11 (4), 959–975.

Armal, S., Devineni, N., Khanbilvardi, R., 2018. Trends in extreme rainfall frequency in the contiguous United States: attribution to climate change and climate variability modes. J. Clim. 31 (1), 369–385. https://doi.org/10.1175/JCLI-D-17-0106.1.

Arora, V., Katavouta, A., Williams, R., et al., 2020. Carbon-concentration and carbon-climate feedbacks in CMIP6 models and their comparison to CMIP5 models. Biogeosciences 17, 4173–4222.

Ayodele-Olajire, D.O., 2021. Power in water governance: the case of Prescott active management area, Arizona. GeoJournal, (Springer Science and Business Media Deutschland GmbH). https://doi.org/10.1007/S10708-020-10344-8.

Babaousmail, H., Hou, R., Ayugi, B., Ojara, M., Ngoma, H., Karim, R., et al., 2021. Evaluation of the performance of CMIP6 models in reproducing rainfall patterns over North Africa. Atmosphere 12 (4), 475.

Bindoff, N., et al., 2013. Detection and attribution of climate change: from global to regional. In: Climate Change 2013 the Physical Science Basis: Working Group I Contribution to the Fifth Assessment Report of the Intergovernmental Panel on Climate Change. IPCC, pp. 867–952, https://doi.org/10.1017/CBO9781107415324.022. 9781107057.

Briz-Redón, Á., Serrano-Aroca, Á., 2020. The effect of climate on the spread of the COVID-19 pandemic: a review of findings, and statistical and modelling techniques. Prog. Phys. Geogr. 44 (5), 591–604. https://doi.org/10.1177/0309133320946302 (SAGE Publications Ltd).

Choi, Y.W., Ahn, J.B., Suh, M.S., Cha, D.H., Lee, D.K., Hong, S.Y., et al., 2016. Future changes in drought characteristics over South Korea using multi regional climate models with the standardized precipitation index. Asia-Pac. J. Atmos. Sci. 52 (2), 209–222.

Collins, M., Knutti, R., Arblaster, J., et al., 2013. Long-term climate change: projections, commitments and irreversibility. In: Climate Change 2013: The Physical Science Basis. Contribution of Working Group I to the Fifth Assessment Report of the Intergovernmental Panel on Climate Change. Cambridge University Press, Cambridge. https://www.ipcc.ch/site/assets/uploads/2018/02/WG1AR5_Chapter12_FINAL.pdf.

Comiso, J.C., Hall, D.K., 2014. Climate trends in the Arctic as observed from space. Wiley Interdiscip. Rev. Clim. Chang. 5 (3), 389–409. https://doi.org/10.1002/wcc.277 (Wiley-Blackwell).

De Sario, M., Katsouyanni, K., Michelozzi, P., 2013. Climate change, extreme weather events, air pollution and respiratory health in Europe. Eur. Respir. J. 42 (3), 826–843. https://doi.org/10.1183/09031936.00074712.

Deschênes, O., Moretti, E., 2009. Extreme weather events, mortality, and migration. Rev. Econ. Stat. 91 (4), 659–681. https://doi.org/10.1162/rest.91.4.659.

Deser, C., et al., 2012. Uncertainty in climate change projections: the role of internal variability. Clim. Dyn. 38, 527–546.

Egeru, A., Barasa, B., Nampijja, J., Siya, A., Makooma, M.T., Majaliwa, M.G.J., 2019. Past, present and future climate trends under varied representative concentration pathways for a sub-humid region in Uganda. Climate 7 (3), 35.

Fix, M.J., Cooley, D., Sain, S.R., Tebaldi, C., 2018. A comparison of US precipitation extremes under RCP8.5 and RCP4.5 with an application of pattern scaling. Clim. Chang. 146 (3), 335–347.

Griffith-Jones, S., Attridge, S., Gouett, M., 2020. Securing Climate Finance through National Development Banks. Available from: https://www.econstor.eu/handle/10419/216988. (Accessed 21 September 2021).

Gupta, A., Jourdain, N., Brown, J., Monselesan, D., 2013. Climate drift in the CMIP5 models. J. Clim. 26, 8597–8615.

Hashim, J., Hashim, Z., 2016. Climate change, extreme weather events, and human health implications in the Asia Pacific region. Asia Pac. J. Public Health 28, 8S–14S.

Khursheed, A., Alam, S., Tyagi, V.K., Nagpure, A.S., Khan, A.A., Gaur, R.Z., et al., 2020. Future liasing of the lockdown during COVID-19 pandemic: The dawn is expected at hand from the darkest hour. Groundw. Sustain. Dev. 11, 100433.

Kumar, S., Bhardwaj, S., Singh, A., Singh, H.K., Singh, P., Sharma, U.K., 2020. Environmental Impact of Corona Virus (COVID-19) and Nationwide Lockdown in India: An Alarm to Future Lockdown Strategies.

Lacombe, G., Hoanh, C.T., Smakhtin, V., 2012. Multi-year variability or unidirectional trends? Mapping long-term precipitation and temperature changes in continental Southeast Asia using PRECIS regional climate model. Clim. Chang. 113 (2), 285–299. https://doi.org/10.1007/s10584-011-0359-3.

Lewis, S.C., Karoly, D.J., 2013. Anthropogenic contributions to Australia's record summer temperatures of 2013. Geophys. Res. Lett. 40 (14), 3708–3709. https://doi.org/10.1002/GRL.50673.

Mahmood, R., Jia, S., Zhu, W., 2019. Analysis of climate variability, trends, and prediction in the most active parts of the Lake Chad basin, Africa. Sci. Rep. 9 (1). https://doi.org/10.1038/s41598-019-42811-9.

Mahoney, K., Scott, J., Alexander, M., et al., 2021. Cool season precipitation projections for California and the Western United States in NA-CORDEX models. Clim. Dyn. 56, 3081–3102.

Maurya, S., Srivastava, P., Yaduvanshi, A., et al., 2021. Soil erosion in future scenario using CMIP5 models and earth observation datasets. J. Hydrol. 594, 125851.

McMichael, A.J., 2015. Extreme weather events and infectious disease outbreaks. Virulence 6 (6), 543–547. https://doi.org/10.4161/21505594.2014.975022.

Millett, B., Johnson, W.C., Guntenspergen, G., 2009. Climate trends of the North American prairie pothole region 1906–2000. Clim. Chang. 93 (1–2), 243–267. https://doi.org/10.1007/s10584-008-9543-5.

National Academies Press, 2016. Attribution of Extreme Weather Events in the Context of Climate Change., https://doi.org/10.17226/21852.

Newbold, T., et al., 2015. Global effects of land use on local terrestrial biodiversity. Nature 520, 45–50.

Nikulin, G., Jones, C., Giorgi, F., Asrar, G., Büchner, M., Cerezo-Mota, R., et al., 2012. Precipitation climatology in an ensemble of CORDEX-Africa regional climate simulations. J. Clim. 25 (18), 6057–6078.

Orimoloye, I.R., Belle, J.A., Olusola, A.O., Busayo, E.T., Ololade, O.O., 2021a. Spatial assessment of drought disasters, vulnerability, severity and water shortages: a potential drought disaster mitigation strategy. Nat. Hazards 105 (3), 2735–2754.

Orimoloye, I.R., Olusola, A.O., Ololade, O., Adelabu, S., 2021b. A persistent fact: reflections on drought severity evaluation over Nigerian Sahel using MOD13Q1. Arab. J. Geosci. 14 (19), 1–18.

Otto, F.E.L., et al., 2012. Reconciling two approaches to attribution of the 2010 Russian heat wave. Geophys. Res. Lett. 39 (4). https://doi.org/10.1029/2011GL050422.

R Development Core Team, 2013. R: A Language and Environment for Statistical Computing. R Foundation for Statistical Computing, Vienna, Austria. http://lib.stat.cmu.edu/R/CRAN/doc/manuals/r-devel/fullrefman.pdf.

Rosenzweig, C., et al., 2013. The agricultural model intercomparison and improvement project (AgMIP): protocols and pilot studies. Agric. For. Meteorol. 170, 166–182.

Rugani, B., Caro, D., 2020. Impact of COVID-19 outbreak measures of lockdown on the Italian carbon footprint. Sci. Total Environ. 737, 139806.

Shi, J., et al., 2016. Knowledge as a driver of public perceptions about climate change reassessed. Nat. Clim. Chang. 6 (8), 759–762. https://doi.org/10.1038/nclimate2997.

Shukla, J., 2013. Extreme weather events and mental health: tackling the psychosocial challenge. Int. Scholarly Res. Notices 2013, 1–7. https://doi.org/10.1155/2013/127365.

Solow, A.R., 2015. Extreme weather, made by us? Science 349 (6255), 1444–1445 (American Association for the Advancement of Science).

Stott, P., 2016. How climate change affects extreme weather events. Science 352 (6293), 1517–1518. https://doi.org/10.1126/science.aaf7271.

Szejner, P., et al., 2020. Recent increases in drought frequency cause observed multi-year drought legacies in the tree rings of semi-arid forests. Oecologia 192 (1), 241–259. https://doi.org/10.1007/S00442-019-04550-6 (Springer).

Tang, M.S., Chenoli, S.N., Samah, A.A., Hai, O.S., 2018. An assessment of historical Antarctic precipitation and temperature trend using CMIP5 models and reanalysis datasets. Polar Sci. 15, 1–12.

Tobler, C., Visschers, V.H.M., Siegrist, M., 2012. Consumers' knowledge about climate change. Clim. Chang. 114 (2), 189–209. https://doi.org/10.1007/s10584-011-0393-1.

Wang, D., Liu, J., Shao, W., Mei, C., Su, X., Wang, H., 2021. Comparison of CMIP5 and CMIP6 multi-model ensemble for precipitation downscaling results and observational data: the case of Hanjiang River basin. Atmosphere 12 (7), 867.

CHAPTER

32

Climate change and trend analysis of precipitation and temperature: A case study of Gilan, Iran

Mohammad Reza Goodarzi[a], *Mohammad Javad Abedi*[b], *and Mahnaz Heydari Pour*[b]

[a]Department of Civil Engineering, Yazd University, Yazd, Iran [b]Department of Civil Engineering, Water Resources Management Engineering, Yazd University, Yazd, Iran

OUTLINE

1 Introduction 562

2 Materials and methods 564

2.1 *Study area* 564

2.2 *GCMs and used data* 564

2.3 *Exponential downscale of data and LARS-WG model* 566

2.4 *Change factor (CF) method* 567

2.5 *Lars-WG* 567

3 Performance evaluation of model used 568

4 Trend analysis 570

4.1 *Graphic method of Mann-Kendall test* 570

5 Results and discussion 570

5.1 *Evaluation of LARS-WG model* 570

5.2 *Prediction of GCM models under used scenarios* 571

5.3 *Number of rainy days* 580

5.4 *Determine the type and timing of changes using Mann-Kendall test analysis* 582

6 Conclusions 582

References 585

Current Directions in Water Scarcity Research, Volume 7
https://doi.org/10.1016/B978-0-323-91910-4.00032-7

1 Introduction

The industrialization of societies has led to increased greenhouse gases in recent decades. This increase has caused the Earth's atmosphere to become warmer. The warming of the Earth's atmosphere, in turn, has influenced the state of other components of the climate system and caused climate that is fundamentally different from its natural trend. This phenomenon, which indicates changes in the average long-term climatic parameters, is called climate change.

The phenomenon of global warming and the climate change induced by it has significant effects on various systems such as water resources, agriculture, and the environment (IPCC, 2014). Therefore, continuous increase in greenhouse gas emission will intensify these effects make the weather warmer, change rainfall patterns, accelerate snowmelt, and consequently affect access to drinking water and irrigation. Global warming and its impact on rainfall in different parts of the world are of great significance as the most important manifestations of climate change. In climate change studies, the first step is to produce climate data in accordance with this phenomenon and in the intended time horizon. Currently, the most reliable tool for generating climate scenarios and quantifying the effect of climate change on meteorological parameters are three-dimensional atmospheric general circulation models; however, spatial magnification (cells about 2.5 degrees) is considered as one of their weaknesses (Taylor et al., 2012). In recent years, various groups of scientists around the world have used general circulation models to predict possible changes resulting in climate changes. GCM Models are capable of predicting expected changes in weather conditions for the future (Chen et al., 2012; Pichuka et al., 2017; Rajan, 2014). However, the GCM models in the case of a very large network system, may increase the network reach sizes between 200 and 650 km (Tan et al., 2017). According to the large size of the network, the obtained results are not accurate enough to be used directly to study changes due to various hydrological impacts at the local scale (Huang et al., 2011). To overcome these scale parameters, scalable minimization methods can be used with the potential to bridge the gap between local scaled climatic inputs and the global scaled parameters (Wilby et al., 2002; Wilby et al., 2014). The method of micro-scaling of data can be conducted in both statistical and dynamic methods. Among the statistical micro-scale methods, meteorological generators are very important. These generators statistical models are used to generate a combination of long-term and lost data and to generate different data (Wilks and Wilby, 1999). Micro-scale statistical methods are on the one hand very suitable for simulating precipitation events (Wilby et al., 2002). On the other hand, however, the frequency and intensity of rainfall in heavy rainfall events are probably influenced by prediction of climate changes in most parts of the world; thus the risk of increasing drought and flood in the basin is simulated by statistical micro-scale tools. Moreover, statistical micro-scale tools have the ability to simulate severe climatic events, especially rainfall at the basin scale (IPCC, 2007). With regard to the increasing effects of climate change, LARS-WG model as one of the microscale models is used to simulate weather information. This model has many advantages than other models such as SDSM model (Adnan et al., 2019; Rajabi et al., 2010; Semenov et al., 2002). In recent years, many studies have been conducted in this regard. Some of which are summarized as follows:

In a research, hydrological effects of climate change were studied by general atmosphere circulation models and a small analog scaling model in south of England by Elshamy et al., 2006. In this study, conversion of monthly data into daily data using data production technique was investigated. The results showed that this method scaled daily meteorological data better than small seasonal data (Elshamy et al., 2006; Mourato et al., 2015). The effects of different climate change scenarios were investigated under scenario A2 in 17 Mediterranean basins obtained from two major basins of the Guadiana and Sado rivers in Portugal which showed a precipitation change from +1.5% to −65% using two atmosphere general circulation models and three climate-regional models for the future (Periods 2100–2071). In a study, Semenov studied data simulation using the LARS-WG model in England. In addition, in another study, Semenov examined climate change scenarios in the United Kingdom. In this research, the LARS-WG model was used to construct daily climatic scenarios (Semenov, 2007; Semenov et al., 2002; Singh et al., 2015). Using two HadCM3 and CGCM3 climate models and a small scale SDSM model under scenarios A2, A1B, and B2, they examined the effects of climate change in the Sutlej River Basin in northwestern Himalayas. In their study, an overall increase has been predicted for future periods in the average annual temperature and precipitation in both models.

(Shiferaw et al., 2018) used the output of five general atmosphere circulation models to simulate future climatic conditions in the Ilala watershed basin in northern Ethiopia which they indicated an increasing trend in minimum and maximum temperatures in 1.7°C and 4.7°C respectively and lack of significant precipitation changes as well. In another study, (Goodarzi et al., 2020a, b), investigated the effect of climate change on precipitation and temperature in Seymareh region. The results of this study showed that climate changes influence hydrological variables of the region and shows an increase in temperature and a decrease in precipitation in the future 30 years (2040–2069). In another study, (Goodarzi et al., 2020a, b), they examined the consequences of climate change on the volume of floods in the Azarshahr basin. In this study, they used the Canadian CanESM2 ground system model under three RCP scenarios along with the statistical down scaling method (SDSM) to evaluate baseline and future courses. The results of this study showed that the return period of severe floods will be reduced in the future. Various studies have been conducted to study the trend of climate changes (Ntegeka and Willems, 2008). In a Belgian study using frequency analysis, they investigated severe rainfall with duration varying between 10 min and 1 month. Their study results showed a statistically significant increase in heavy rainfall, with persistence in general atmospheric circulation patterns over the North Atlantic duration periods of 10–15 years.

Van den Besselaar et al. (2003) were commissioned to investigate 478 stations in Europe with maximum rainfall of 1 day and 5 days using the linear regression test. The results of this study showed an increase in heavy rainfall in northern Europe in autumn, winter and spring, and in southern Europe, a slight increase in heavy rainfall in all seasons. It should be noted that the results of the Climate Change Sixth Phase (CMIP6) models have recently been published and are now available. In this phase, the combination of greenhouse gas emissions and socio-economic trajectories has been used for the scenarios (Riahi et al., 2017). Yet, few studies have examined the results of these models on the temperature and precipitation status of the watershed basin. Since assessment of environmental effects of climate change on water resources and biological components is an integral part of hydrological and ecological research in the 21st century, it is important to study the effects of climate change to adopt appropriate policies to mitigate the effects of climate change. The purpose of this study, is to investigate the trend of climate changes in the decades (1980–2010) and to predict these changes in the coming decades (2010–2100) using LARS-WG downscale model in stations of Gilan province. Since IPCC Fifth Reports have been used in climate change studies, in this study attempted to investigate these effects under the influence of the sixth report. In addition, annual rainfall changes and changes created in dry and wet periods for the base and future periods are also compared with respect to the output of CMIP6 model. In the first step, download the AOGCMs data from the database https://esgf-node.llnl.gov/search/cmip6/ and select the parameters of minimum and maximum temperature and precipitation, and then downscaling with the LARS-WG model, in the next step, calibrate Modeling and creating a scenario for precipitation and maximum and minimum temperature for the next period and finally analyzing the process and comparing changes.

2 Materials and methods

2.1 Study area

The case study region of Gilan province, with an area of 14,600 km^2 in northern Iran, is located on the southwestern shores of the Caspian Sea, the northern slopes of Alborz, and the eastern slopes of Talesh. There are 11 synoptic stations in Gilan province and in this study, the required meteorological data have been obtained from two synoptic stations in Rasht and Bandar Anzali. Fig. 1 shows the position of these two synoptic stations. The data used include precipitation, minimum temperature and maximum temperature. The specifications of the studied stations are also given in Table 1. According to the latest meteorological report in 2018, the total summer rainfall for Rasht and Bandar Anzali synoptic stations is 96.7 and 346.6 mm, respectively, and the average summer temperature in 2018 is 26.8°C for Rasht station and 27.6°C for Bandar Anzali station.

2.2 GCMs and used data

In order to estimate and generate future period data (2015–2100), monthly data of above stations including precipitation, minimum temperature and maximum temperature as a base period (1980–2010). The output data of three general circulation models (CanESM5, IPSL-CM6A-LR, and Miroc6) with three scenarios SSP585, SSP434, and SSP119 were used. The choice of the base period duration is due to the limitation of the historical period of CMIP6 models up until 2014 and the existence of the measured data in this statistical period. The process of climate change, especially temperature and precipitation, is the most important issue in the field of environmental sciences. Currently, the most reliable tools for generating climate scenarios are paired three-dimensional models of general atmospheric-ocean circulation (AOGCM) (Pervez and Henebry, 2014).

In this research, the models of the sixth report CanESM5, IPSL-CM6A-LR, and Miroc6, which has been applied using three scenarios shared socio-economic pathways (SSP) in the years (1980–2100). Emission scenarios are used to predict greenhouse gas concentration

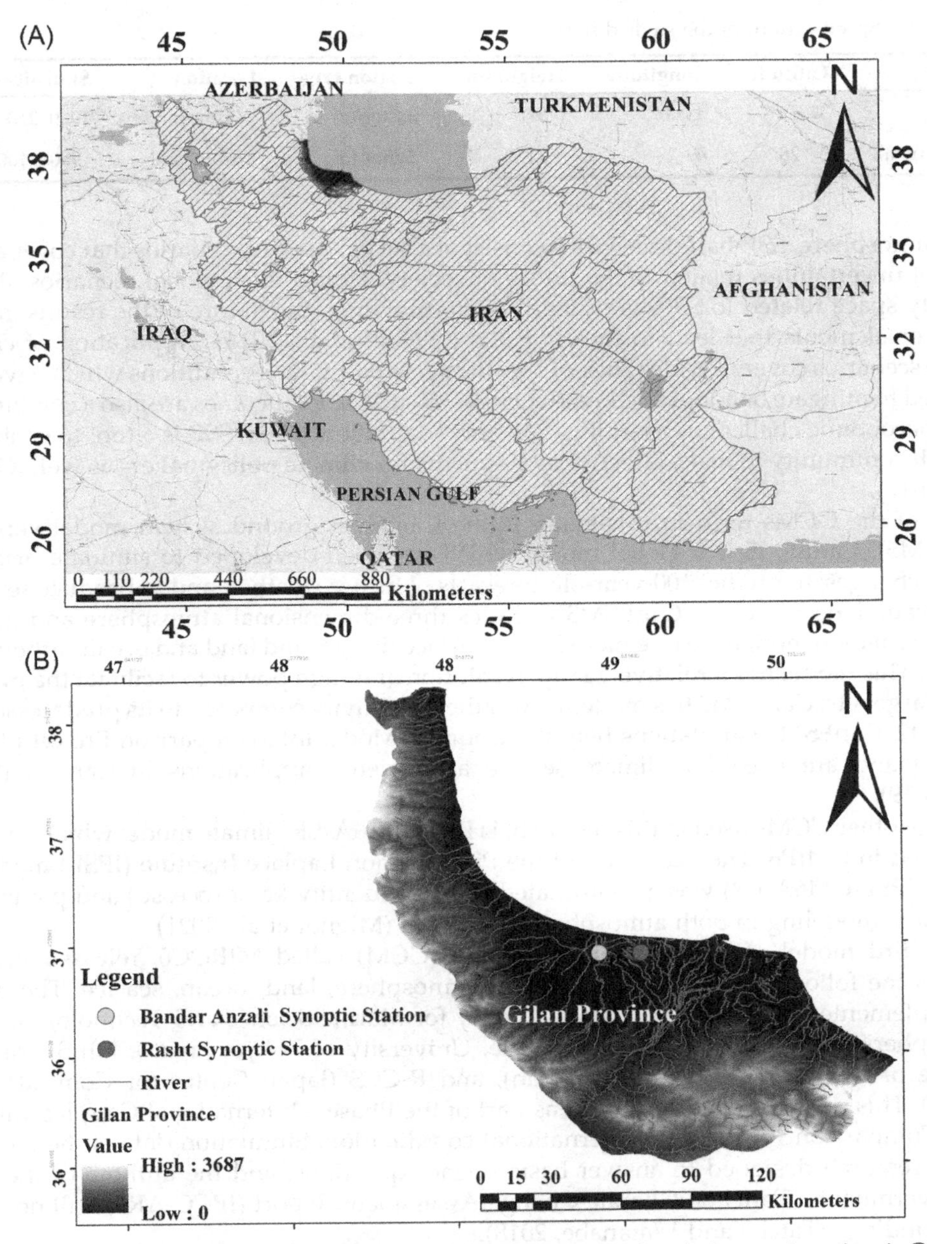

FIG. 1 Geographical location of Gilan province in Iran (A) and location of studied synoptic stations in Gilan province (B).

TABLE 1 Specifications of the studied stations.

Station	Latitude	Longitude	Height (m)	Station type	Location	Statistical course
Rasht	37°19	49°37	−8.6	Synoptic	Within range	1980–2100
Bandar Anzali	37°28	49°28	−24	Synoptic	Within range	1980–2100

in the atmosphere. SSP-based scenarios consist of a set of baseline scenarios that cover a wide range of uncertainties related to the future of society. Unlike most global scenarios, the uncertainty space related to SSPs is primarily determined by the nature of the results, not the inputs or elements that lead to those results (O'Neill et al., 2014). Application of climate change scenarios covers a wide range of specific decision-making conditions which have been designed to mitigate or adapt to climate changes. Hence, SSP outcomes are also a combination of socioeconomic challenges for mitigation and adaptation. Thus, SSP is a tool to enable the research community to make effective assessments for climate policymakers as well (O'Neill et al., 2017).

One of the GCMs used in this study is the Canadian ground system model version 5 (CanESM5). CanESM5 is a global model which has been developed to simulate historical climate changes, to predict 100-year-old forecasts of future weather, and to produce seasonal and decade-long forecasts. CanESM5 includes three-dimensional atmosphere and general ocean circulation models, sea ice model, land surface design, and land and ocean carbon cycle models. This model has a relatively large resolution and high power to facilitate the production of large sets. CanESM5 has moderate weather sensitivity compared to its predecessor, the CanESM2. CanESM5 simulations help the Coupled Model Intercomparison Project phase 6 (CMIP6) and are used for climate service and science applications in Canada (Swart et al., 2019).

One of other GCMs used in this research, is IPSL-CM6A-LR climate model which has been developed in CMIP6. The extraction of the Pierre Simon Laplace Institute (IPSL) model for CMIP6 (IPSL-CM6A-LR) was a coordinated effort to identify key processes and parameters for climate modeling in both atmosphere and ocean (Mignot et al., 2021).

The third model of atmospheric circulation (GCM) called MIROC6, released in 2017, includes the following components: aerosol, atmosphere, land, ocean, sea Ice. The model was implemented by JAMSTEC (Japan Agency for Marine Science and Technology), AORI (Atmospheric and Ocean Research Institute, University of Tokyo, Japan), NIES (National Institute of Environmental Studies, Japan), and R-CCS (Japan Center for Computational Science). This data has been produced as part of the Phase 6 International Coupled International Comparison Project with international coordination. Simulation data is the basis for climate research designed to answer basic science questions, and the authors of the Sixth Intergovernmental Panel on Climate Change Assessment Report (IPCC-AR6) will no doubt cite its findings (Tatebe and Watanabe, 2018).

2.3 Exponential downscale of data and LARS-WG model

The computational network of GCMs is very large, and therefore, they cannot professionally modeling subnetwork climatic features such as topography or clouds in the targeted area.

As a result, GCMs are not able to provide reliable information from temperature and precipitation data for hydrological modeling. Therefore, there is a need to minimize from GCM resolution to very good resolution or even station scale (Wilby et al., 2002). In this research, two methods of downscaling, CF and statistical downscaling have been used to downscaling general circulation model data.

2.4 Change factor (CF) method

In the change factor (CF) method, monthly ratios are generated for historical collections and climate change scenarios are generated for temperature and precipitation.

To create a climate change scenario in each GCM model, the ratio of the long-term average monthly rainfall to the base period rainfall and the difference between the long-term monthly average of the future temperature and the base period using Eqs. (1), (2) with the same GCM model, respectively. Each cell of the computational network is simulated (Jones and Hulme, 1996).

$$\Delta T_i = \left(\overline{T}_{GCM.Fut.i} - \overline{T}_{GCM.Base.i}\right) \tag{1}$$

$$\Delta P_i = \left(\overline{P}_{GCM.Fut.i} / \overline{P}_{GCM.Base.i}\right) \tag{2}$$

In the above equations, ΔT_i and ΔP_i are the scenarios of temperature change and precipitation, respectively, for the long-term average of 85 years per month ($1 \leq i \leq 12$). $\overline{T}_{GCM.Fut.i}$ is the average 85-year temperature simulated by AOGCM in future periods per month (2016–2100) and $\overline{T}_{GCM.Base.i}$ is the average 30-year temperature simulated by AOGCM in the observation period (in this study 1980–2010) for each month. The above calculations are also true for rainfall. After calculating climate change scenarios, the CF method is used to microscale the data (Diaz-Nieto and Wilby, 2005).

To obtain the time series of future climate scenarios, climate change scenarios are added to the observational data (1980–2010) (Eq. 3 and 4).

$$T = T_{obs} * \Delta T \tag{3}$$

$$P = P_{obs} * \Delta P \tag{4}$$

In the above relations T; time series of future temperature climate scenarios (2016–2100) and ΔT; micro-scaled scenarios are climate change. Eq. (4) is related to rainfall. It is important to note that the time series generated for the future by CF has the same variance and average mean as the observational data. This means that the daily values of future data are similar to the observed data, but change with increasing temperature (ΔT) and precipitation (ΔP).

2.5 Lars-WG

One of the most popular models of random data generator that is useful for generating daily precipitation, radiation and maximum and minimum daily temperature in a station in current and future weather conditions is LARS-WG5 model. The LARS-WG model was first designed by a scientist at the Rotamstead Center for Agricultural Studies in the UK and then was developed in Long Ashton research station. Therefore, the name of the model is an abbreviation of the phrase Long Ashton Search Station Weather Generator which can be used for modeling the daily data of future periods using historical data and studying the climate behavior of stations in the statistical period as well as daily network daily of future atmosphere general circulation models (Semenov et al., 2002).

This model has the ability of past and future periods modeling and is one of the generators of weather data. This model uses a semiexperimental distribution for the length of wet and dry series of daily rainfall and solar radiation. Temperature modeling is also performed using the Fourier series. The quasiexperimental distribution of this model is as follows:

$$\text{Emp} = \{a_0, a_i; h_i, i = 1, \ldots, 10\} \tag{5}$$

It includes a histogram with 10 intervals $[a_i, a_{i-1}]$ which $a_{i-1} > a_i$ and h_i represents the number of events observed in ith interval (Semenov et al., 2002). The main part of the LARS-WG model is model calibration. Climatic parameters in the LARS-WG model are simulated in two stages, first the precipitation parameter and its intensity and in the next step, other residual parameters including simulated minimum and maximum temperatures, radiation, humidity and wind speed (Johnson et al., 1996). In the case of simulating a rainfall event, the model first determines whether the day is wet or dry and then estimates the amount of rainfall. Therefore, that the wetter day is considered as a day with precipitation above zero millimeters (pr > 0). The length of each series is randomly selected from the wet or dry semiexperimental distribution for the month at the beginning of the series. In determining the distribution, observational series are assigned to the fish from which they were originated. For a wetter day, the amount of precipitation is simulated from a semiexperimental distribution of precipitation for the intended month, independent of the dry series length or the amount of precipitation in the previous day (Goodarzi et al., 2015). Minimum and maximum daily temperature is modeled as random processes with the mean and standard deviation of daily criteria which depends on the intended wet or dry part of the day. The proposed technique for simulating this process is very close to Racsko's (1991) model (Racsko et al., 1991).

After preparing the daily data of the historical period in an acceptable format for the LARS-WG model, these data were introduced to the model and after examining the trend of changes of these variables, the file of the local climate behavior in the previous period has been in produced based on it. For the evaluation of the power of the model in producing climate data of the case- study area, two methods can be use; first method includes the use of LARS-WG model's statistical test option and the second method related to production of meteorological data in the same period as the historical period based on one of the general atmospheric circulation models and relevant scenarios publication, In this study, both methods have been used.

3 Performance evaluation of model used

In order to investigate the performance of LARS-WG model in simulating climatic variables, coefficient of determination (R^2), error absolute value mean (MAE), root mean square of error (RMSE), and Nash-Sutcliffe efficiency coefficient (E_{NS}), have been used for comparative study. The statistics mentioned are calculated using Eqs. (6)–(9). The coefficient of determination is a dimensionless criterion that varies from zero to one and its optimal value is one (Barrett, 1974) (Eq. 6).

$$R^2 = \frac{\left[\sum_{i=1}^{n} (A_i - \overline{A})(F_i - \overline{F})\right]^2}{\sum_{i=1}^{n} (A_i - \overline{A})^2 \sum_{i=1}^{n} (F_i - \overline{F})^2} \quad (6)$$

The mean squared error and the absolute error mean indicate the amount of model error and the closer it is to zero, the better the performance of the model (Chai and Draxler, 2014) (Eqs. 7 and 8).

$$\text{MAE} = \frac{\sum_{i=1}^{n} |A_i - F_i|}{n} \quad (7)$$

$$\text{RMSE} = \sqrt{\frac{\sum_{t=1}^{n} (A_t - F_t)^2}{n}} \quad (8)$$

The Nash-Sutcliffe coefficient indicates how the regression line between estimated and observational data is close to the regression line with a slope of 1 value (McCuen et al., 2006) (Eq. 9).

$$E_{NS} = 1 - \frac{\sum_{i=1}^{n} (A_i - F_i)^2}{\sum_{i=1}^{n} [A_i - \overline{A}]^2} \quad (9)$$

In these equations, A_i is the amount of observational data (observational monthly average), F_i represents the value predicted by the model, (historical monthly average of the model), i index is the months of the year, and n represents the number of months. The value of the errors mentioned is closer to zero, which is an indicator of the correctness of most models. Also $\overline{A}$ is the mean of the observed data and $\overline{F}$ is the mean of the estimated data.

After evaluating the LARS-WG5 model using error measurement indicators and determination coefficient and ensuring the suitability of the model, the LARS-WG5 model is used to micro-scale the data of general atmospheric circulation models and the data of the period 2010–2100 is calculated using climate change scenarios. Then, monthly precipitation changes are calculated from normal monthly differences of the base period 2016–2100 with the normal values of the base period 1980–2010, and the graphs of precipitation changes related to each station are drawn. Fig. 2 shows the flow chart of the methodology of climate change studies.

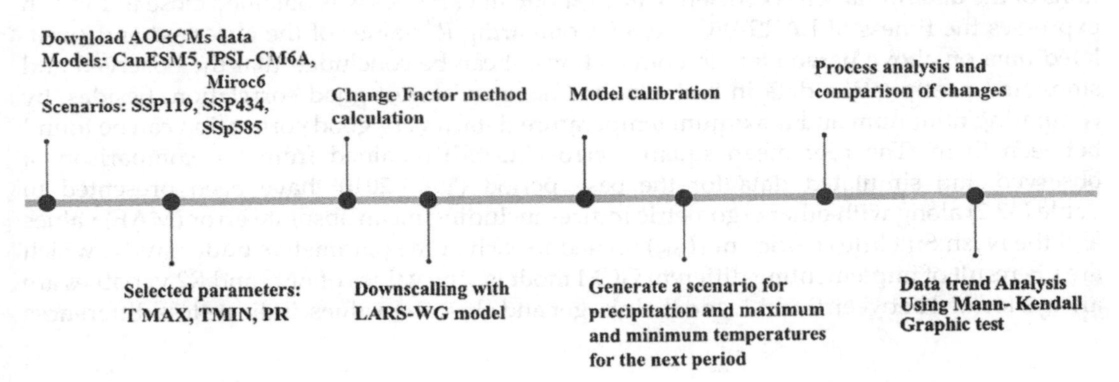

FIG. 2 Flow chart of methodology for climate change studies.

4 Trend analysis

Mann-Kendall test Developed by MANN and KENDALL from 1945 to 1975 (Mann, 1945; Kendall and Gibbons, 1975). This method is one of the most common methods in analyzing the trend of hydrological and meteorological series (Lettenmaier et al., 1994). One of the most important advantages of using this method is that it does not depend on a specific statistical distribution and its partial effectiveness (Partal and Kahya, 2006). Mann-Kendall test has two assumptions of zero and one, the assumption of zero indicates that it is random and there is no trend in the data series and the first assumption indicates the existence of trends in the data series. This test is evaluated both computationally and graphically.

4.1 Graphic method of Mann-Kendall test

Analyzes and interpretations in this test are performed according to the diagrams $U(t_i)$ and $\grave{U}(t_i)$. When the absolute value of U is greater than 1/96, the time series is significant, in other words, if the graph $U(t_i)$ crosses the two lines $Y=1/96$ and $Y=-1/96$, the trend in the series can be considered meaningful. If $U>0$ or the chart $U(t_i)$ has an uptrend overall, a significant uptrend is considered. If $U<0$, the trend is significant and decreasing. If the graphs $U(t_i)$ and $\grave{U}(t_i)$ intersect in the range $-1/96$ and $1/96$, it indicates a sudden change in the mean. In this study, the Mann-Kendall test was used as a sequence based on Sneyers (1990), in which the statistical values of $U(t_i)$ and $\grave{U}(t_i)$ are calculated (Sneyers, 1990).

5 Results and discussion

5.1 Evaluation of LARS-WG model

To calibrate the LARS-WG model, the determination coefficient of the observed and simulated data was calculated using the model for the base period of 1980–2010 in two synoptic stations of Gilan province. In Table 2, the values of R^2 (Determination coefficient) which is used to show the accuracy of the LARS-WG model, the parameters of minimum temperature, maximum temperature and precipitation has been presented that according to the observations of the determination coefficient values about all variables was obtained close to 1 which expresses the fitness of LARS-WG model. Comparing R^2 values of the observed and simulated data on above parameters in both stations, it can be concluded that the observed and simulated precipitation data in both stations have an almost good correlation. Besides, by comparing minimum and maximum temperature data, a very good correlation can be found between them. The root mean squared error (RMSE) obtained from the comparison of observed and simulated data for the base period (1980–2010) have been presented in Table (32.2) along with other ergometric indices including mean absolute error (MAE) values and the Nash-Sutcliffe coefficient (E_{NS}) related to each of the parameters under study; which are the result of implementing different GCM models. The values of ENS and R2 variables are always located between 0 and 1, and their larger and close to 1 values, indicate less differences

TABLE 2 The investigation of the correlation index (R^2) and the values of RMSE, E_{NS}, and MAE obtained from the comparison of observed and simulated data for the base period (1980–2100) related to each of the parameters of minimum temperature, maximum temperature and precipitation in the stations of Gilan province.

	R^2			MAD			RMSE			E_{NS}		
Station	T_{max}	T_{min}	PR	T_{max}	T_{min}	PR	T_{max}	T_{min}	PR	T_{max}	T_{min}	PR
Rasht	0.9988	0.9995	0.9692	0.1907	0.1286	0.2877	0.2598	0.9798	0.3612	0.9986	0.9993	0.9579
Bandar Anzali	0.9993	0.9996	0.9798	0.1361	0.1085	0.3438	0.1779	0.1390	0.4810	0.9993	0.9995	0.9759

(differences between the observed and predicted values during the statistical period) in the data production by the model (Table 2). As it can be seen, the RMSE values obtained from comparison of observed and simulated data for the minimum temperature parameter in the two studied stations varied between 10% and 90%, maximum temperature between 17% and 25%, and precipitation between 36% and 48%. Considering the point that there is no significant difference between simulated and observed values during the base period 1980–2010, it can be inferred that the LARS-WG model had a good performance in simulating and producing climatic data of Gilan province. After evaluating the LARS-WG model and ensuring its fitness, the data generated using the LARS-WG5 model for three climate change scenarios SSP119, SSP434, and SSP585 using three models CanESM5, IPSL-CM6A-LR, and Miroc6 were investigated.

5.2 Prediction of GCM models under used scenarios

5.2.1 CanESM5 model under SSP scenarios

The changes in the values of meteorological parameters in future periods using the CanESM5 model and under the SSP119 scenario are shown in Figs. 3 and 4. Due to these problems and assuming the SSP scenarios are valid, the maximum value of this parameter will occur in the month of August. Also, according to the graph of observational data, we conclude that in the coming years, the maximum monthly minimum temperature will increase according to the CanESM5 model under all three scenarios. Tables 5 and 6 show the comparison of monthly and annual average minimum temperatures in Rasht and Bandar Anzali stations with observational data. In the case of maximum temperature, the process of changes is similar to the minimum temperature. The greatest increase in temperature occurs in July. Also, comparing the observational data of Rasht station, it can be concluded that this parameter will increase relatively much in the coming years under the model and scenarios mentioned (Fig. 3). Also, the annual average of this parameter in both stations using the CanESM5 model under SSP scenarios will increase compared to the observational data of these stations (Tables 3 and 4). Monthly rainfall changes at both stations peak in November. It can be predicted that torrential rains will occur in both stations this month (Tables 7 and 8).

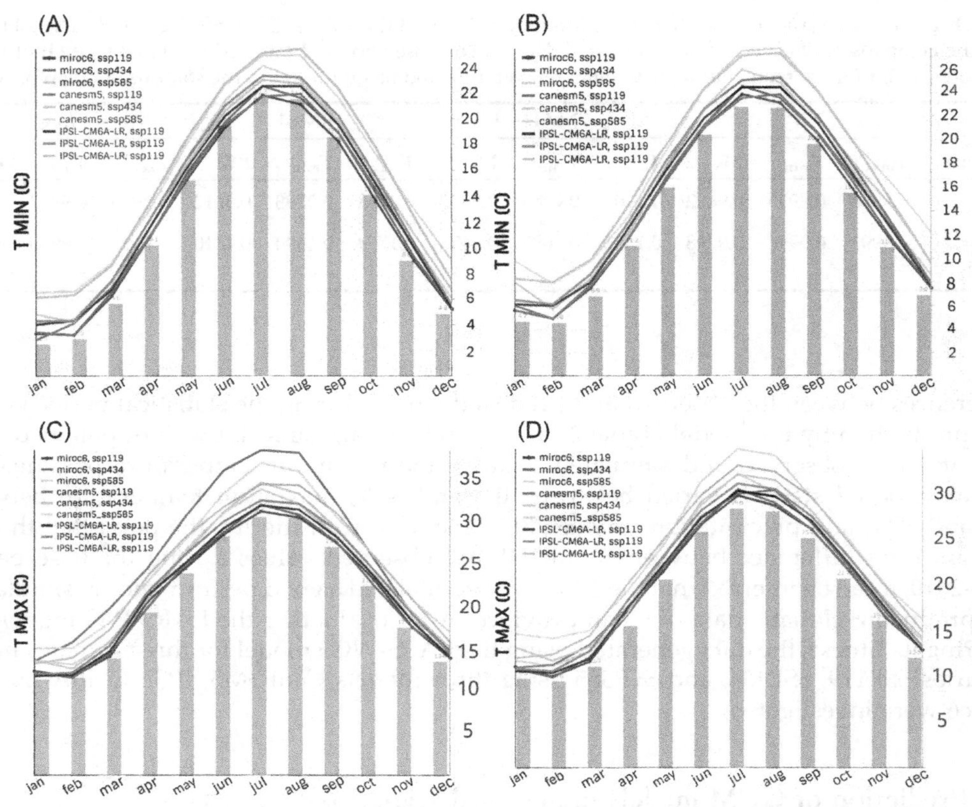

FIG. 3 T_{min} change diagrams in the next 85 years (2016–2100) under three GCM models in Rasht (A) and Bandar Anzali (B) stations using three climate change scenarios and comparing them with observational data diagrams (1980–2010). T_{max} change diagrams in the next 85 years (2016–2100) under three GCM models in Rasht (C) and Bandar Anzali (D) stations using three climate change scenarios and comparing them with observational data diagrams (1980–2010).

5.2.2 Miroc6 model under SSP scenarios

Changes in the values of meteorological parameters in future periods using the Miroc6 model and under the SSP119 scenario are shown in Figs. 3 and 4. Due to these problems and assuming that all three SSP scenarios are valid, the minimum temperature parameter reaches its maximum value in July. Also, according to the observational data graph in Fig. 3, in all scenarios, the minimum temperature will increase in the coming years. Comparison of annual minimum temperature values in both stations under SSP119 scenario with the corresponding observational values indicates that this parameter is constant in the future and under SSP434 and SSP585 scenarios shows an increase in this parameter (Tables 5 and 6).

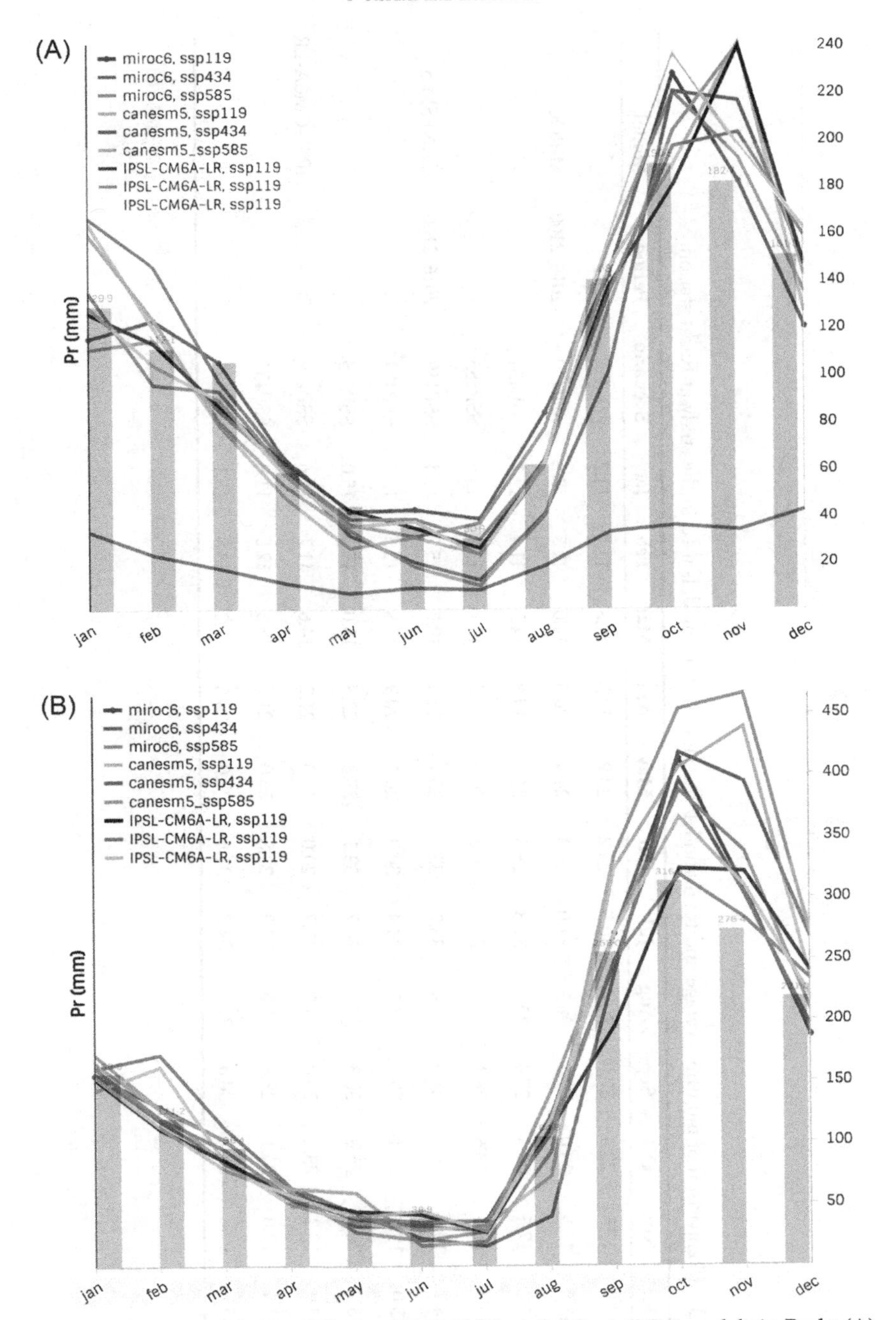

FIG. 4 PR change diagrams in the next 85 years (2016–2100) under three GCM models in Rasht (A) and Bandar Anzali (B) stations using three climate change scenarios and comparing them with observational data diagrams (1920–2010).

TABLE 3 Comparison of monthly average and total annual T_{max} for all three models used in the study at Rasht station.

Year	Dec	Nov	Oct	Sep	Aug	Jul	Jun	May	Apr	Mar	Feb	Jan	Scenario	Period	Model
20.9	13.4	17.5	22.4	27.0	30.4	30.5	28.2	23.9	19.3	13.8	11.4	11.2	_	Base	_
21.2	13.8	17.5	23.0	27.3	30.3	31.0	28.3	24.4	20.5	14.0	11.5	12.2	SSP119	2016_2100	MIROC 6
21.3	14.2	17.5	22.3	27.4	30.6	31.8	28.5	24.9	19.8	14.3	11.4	12.1	SSP434		
22.6	15.2	18.4	23.6	28.1	31.9	33.6	30.8	26.5	20.7	15.7	13.0	13.5	SSP585		
25.6	14.7	19.0	25.4	31.9	37.9	38.2	35.8	30.4	24.4	19.7	15.7	13.4	SSP119	2016_2100	CANESM 5
22.9	15.0	19.3	24.1	29.0	33.1	34.4	30.9	26.1	20.9	15.8	12.5	13.5	SSP434		
24.9	16.9	20.7	25.8	30.3	35.7	36.7	33.1	27.8	22.8	18.0	15.0	15.0	SSP585		
21.7	13.9	18.2	23.4	28.0	31.2	31.7	29.0	25.0	21.2	14.6	11.7	11.6	SSP119	2016_2100	IPSL-CM6A-LR
22.3	14.4	19.0	24.1	28.8	31.9	32.3	29.8	26.0	21.6	15.3	12.0	11.8	SSP434		
24.1	15.7	20.8	25.7	30.8	34.1	34.2	31.5	28.0	23.5	17.2	13.8	13.1	SSP585		

TABLE 4 Comparison of monthly average and total annual T_{max} for all three models used in the study at Bandar Anzali station.

Year	Dec	Nov	Oct	Sep	Aug	Jul	Jun	May	Apr	Mar	Feb	Jan	Scenario	Period	Model
19.1	12.3	16.5	21.2	25.7	28.7	29	26.4	21.1	15.9	11.3	9.6	9.9	_	Base	_
19.4	12.9	16.4	21.6	26.0	28.6	29.7	26.5	21.5	17.0	11.6	9.9	11.0	SSP119	2016_2100	MIROC 6
19.6	12.9	16.5	21.5	26.0	28.9	30.4	26.6	22.0	16.7	12.1	9.8	10.8	SSP434		
20.6	13.9	17.6	22.4	26.7	30.1	32.1	28.6	23.2	16.8	12.8	11.1	11.7	SSP585		
20.2	12.8	17.0	22.0	26.7	30.3	29.9	27.4	22.2	17.5	13.5	11.2	10.9	SSP119	2016_2100	CANESM 5
21.3	14.2	18.3	23.0	28.0	31.4	32.0	29.0	22.9	17.4	13.3	11.7	13.2	SSP434		
23.2	16.2	19.9	24.9	29.4	34.1	34.8	31.2	24.3	19.3	15.1	14.2	14.7	SSP585		
19.9	12.7	17.4	22.3	26.9	29.6	30.2	27.2	22.2	17.4	11.9	10.3	10.5	SSP119	2016_2100	IPSL-CM6A-LR
20.6	13.2	18.0	23.0	27.4	30.4	31.0	28.0	23.1	18.2	12.8	10.5	10.8	SSP434		
22.3	14.9	19.8	24.6	29.5	32.3	33.0	29.7	24.9	20.0	14.5	12.4	12.0	SSP585		

TABLE 5 Comparison of monthly average and total annual T_{min} for all three models used in the study at Rasht station.

Year	Dec	Nov	Oct	Sep	Aug	Jul	Jun	May	Apr	Mar	Feb	Jan	Scenario	Period	Model
12.3	4.8	8.9	13.9	18.3	21.2	21.3	19.2	15.0	10.1	5.6	2.9	2.5	–	Base	–
12.3	5.2	9.0	13.6	18.2	21.1	21.8	19.3	15.3	10.8	6.2	3.2	3.3	SSP119	2016_2100	MIROC 6
12.6	5.3	9.3	14.2	18.9	21.6	22.5	19.7	15.8	10.7	6.5	3.2	3.4	SSP434		
13.8	6.0	10.4	15.3	19.9	23.0	24.1	21.4	16.9	11.4	7.4	4.4	4.2	SSP585		
12.7	5.5	10.1	14.5	18.9	21.9	21.7	19.5	15.5	10.8	6.9	4.1	2.8	SSP119	2016_2100	CANESM 5
13.7	6.7	11.1	15.5	19.7	22.7	23.0	20.6	16.4	11.5	7.3	4.2	4.8	SSP434		
15.6	9.0	13.0	17.2	21.6	24.9	24.9	22.6	18.1	12.9	8.8	6.6	6.5	SSP585		
13.2	5.3	10.0	14.9	19.5	22.5	22.5	20.2	16.2	11.5	6.7	4.3	4.0	SSP119	2016_2100	IPSL-CM6A-LR
13.7	5.5	10.4	15.5	20.2	23.2	23.3	21.0	16.9	12.1	7.4	4.3	4.3	SSP434		
15.5	7.4	12.4	17.7	22.3	25.4	25.1	22.7	18.5	13.4	8.6	6.3	6.1	SSP585		

TABLE 6 Comparison of monthly average and total annual T_{min} for all three models used in the study in Bandar Anzali station.

Year	Dec	Nov	Oct	Sep	Aug	Jul	Jun	May	Apr	Mar	Feb	Jan	Scenario	Period	Model
14.1	7	11.1	15.8	20	23.1	23.2	20.8	16.3	11.2	6.9	4.6	4.7	–	Base	–
14.1	7.5	11.2	15.5	19.9	23.1	23.8	21	16.6	11.7	7.5	4.9	5.6	SSP119	2016_2100	MIROC 6
14.3	7.6	11.4	16.1	20.5	23.5	24.5	21.2	16.8	11.7	7.8	4.9	5.6	SSP434		
15.4	8.6	12.5	17	21.4	24.8	25.9	22.16	17.7	12.2	8.7	6.2	6.4	SSP585		
14.4	7.6	12.4	16.5	20.6	23.8	23.5	21	16.5	11.7	7.7	4.9	6	SSP119	2016_2100	CANESM 5
15.7	9.7	14.2	17.8	21.8	25.3	25	22.5	17.6	12.1	8.3	5.7	8	SSP434		
17.7	12.2	16.1	19.6	24	27.7	27.5	24.3	19.1	13.6	9.8	7.9	9.8	SSP585		
14.94	7.6	12.2	16.7	21.4	24.4	24.4	21.8	17.4	12.6	8.0	6.0	6.1	SSP119	2016_2100	IPSL-CM6A-LR
15.5	7.9	12.7	17.5	22.0	25.2	25.0	22.6	18.2	13.1	8.7	6.1	6.5	SSP434		
17.3	9.6	14.5	19.5	24.1	27.3	27.1	24.3	19.7	14.5	9.9	8.0	8.3	SSP585		

TABLE 7 Comparison of monthly average and total annual PR for all three models used in the study in Rasht station.

Year	Dec	Nov	Oct	Sep	Aug	Jul	Jun	May	Apr	Mar	Feb	Jan	Scenario	Period	Model
1207.7	151.3	182.7	190.5	141.3	62.4	30.89	35.8	44.8	59.3	106.5	112.1	129.9	_	Base	_
1293	120.4	182.3	228.2	142.2	84.2	39.32	43.3	42.5	65.3	105.9	123.5	115.9	SSP119	2016_2100	MIROC 6
278	42.9	34.3	36.31	33.586	18.7	8.504	9.4	7.186	11.5	17.82	23.747	34.1	SSP434		
1258.7	135.1	192.4	220.4	154.44	77.5	37.52	31.6	37.00	60.96	85.436	115.12	111.3	SSP585		
1259.8	126.2	241.1	189.5	145.29	62.1	23.91	31.3	26.6	48.9	79.5	120.5	164.9	SSP119	2016_2100	CANESM 5
1269.2	148.0	216.7	220.7	102	41.5	12.7	20.3	32.3	63.3	98.	146.6	167.18	SSP434		
1237.7	142.0	240.3	204.1	128.6	40.7	10.3	18.3	34.6	53.7	81	124	160.1	SSP585		
1265.3	146.4	239.7	180.9	137.5	64.3	26.6	35.4	43.2	63.5	87.4	113.9	126.5	SSP119	2016_2100	IPSL-CM6A-LR
1252.2	158.8	203.0	197.3	134.5	63.7	30.4	39.2	39.4	60.6	93.6	96.3	135.4	SSP434		
1300.2	162.3	197.1	236.6	160.9	63.5	23.4	38.2	36.1	58.4	89.8	104.6	129.3	SSP585		

TABLE 8 Comparison of monthly average and total annual PR for all three models used in the study in Bandar Anzali station.

Year	Dec	Nov	Oct	Sep	Aug	Jul	Jun	May	Apr	Mar	Feb	Jan	Scenario	Period	Model
1247.9	221	276.4	316.3	258	104.8	37.1	38.9	41.5	52.7	96.4	121.2	158.5	_	Base	_
1855.2	187.2	306.8	412.5	270.6	121.1	36.8	34.4	37.1	61.5	101.6	129.9	155.7	SSP119	2016_2100	MIROC 6
1762	194.6	302.2	395	246.2	115.2	28.6	32.3	32.6	54	85.8	118	157.5	SSP434		
1883.6	208.2	337.4	386.9	270.8	141.1	27.7	20.7	27.8	60.4	102	128.2	172.4	SSP585		
2010.9	233.4	438.7	405.6	323.5	72.4	32.7	29.2	35.7	50.2	83.8	162.4	143.3	SSP119	2016_2100	CANESM 5
1947.5	266.8	393.7	417.6	239	40	16.1	22.5	42.9	64	111.4	172.4	161.1	SSP434		
2151	270.3	466.3	453.6	334.2	92.1	19.6	16.5	59.54	62.8	82.5	127.9	165.6	SSP585		
1709.17	239.6	321	323.46	195.3	110.8	27.4	41.7	44.6	56.22	84.97	111.3	152.8	SSP119	2016_2100	IPSL-CM6A-LR
1733.41	233.2	282.7	318.3	249.3	106.4	36.6	35.7	39.4	55.52	93.99	120.5	161.8	SSP434		
1798.5	218.7	309.2	365.2	271.3	118.6	28.1	44.7	40	57.4	77.96	112.9	154.5	SSP585		

In the case of maximum temperature, the process of changes is similar to the minimum temperature. The highest temperature increase occurs in all three scenarios in July. Studies show that this parameter will increase in the coming years under the mentioned model and SSP scenarios. Monthly precipitation changes at both stations under SSP119 and SSP434 scenarios increase in October. Based on the available evidence and comparing the amount of precipitation in the future period with the baseline period, it can be predicted that under SSP119 scenario, torrential rains will occur in both stations. The point of divergence is the amount of precipitation by Miroc6 model under SSP434 and SSP585 scenarios in Rasht study station, which has a high divergence compared to other scenarios.

5.2.3 *IPSL-CM6A-LR model under SSP scenarios*

Changes in the values of temperature and precipitation parameters for the next period (2016–2100) were also calculated by the IPSL-CM6A-LR model under SSP119, SSP434, and SSP585 scenarios and are shown in Figs. 3 and 4. According to the mentioned figures and assuming the SSP119 scenario prevails, the minimum temperature parameter in both stations in July, under SSP434 in July and August and under SSP585 in August is higher than other months. This parameter will also increase in the coming years under all three scenarios at both stations. The maximum value of the maximum temperature parameter is also under all three scenarios in July. It can also be concluded that this parameter will increase under all three scenarios in the coming years (Tables 3 and 4). The average monthly precipitation changes under SSP119 scenario in November, SSP434 scenario in November and October months, and SSP585 scenario in July and October months. Based on the SSP119 scenario and the available evidence, it can be predicted that torrential rains at Rasht station are likely in the future, but at Bandar Anzali station this probability will be lower. Also, based on the SSP434 and SSP585 scenarios, comparing the observational data of these stations with the predicted data will indicate an increase in the forecasted rainfall in the coming years.

5.3 Number of rainy days

The results of predicating the number of rainy days in the period 2016–2100 AD for the stations of Gilan province showed that the number of rainy days in both stations will decrease compared to the previous period. In some of the models studied in this research, this trend becomes more and more noticeable (Fig. 5A and B). Hence, the highest number of rainy days in Rasht station is related to CanESM5 model under SSP434 scenario with 136 rainy days and the lowest number of rainy days is related to IPSL-CM6A-LR model under SSP585 scenario with 117 days. Moreover, in Bandar Anzali station, the highest number of rainy days is related to CanESM5 models under SSP434 scenario and miroc6 under 119 scenario, with 145 days. The lowest number of rainy days in this station is related to the IPSL-CM6A-LR model under SSP119 and SSP434 scenarios with 115 days which has a significant difference with the number of rainy days observed in this station (150). Although the number of rainy days in each region can be considered as a desirable feature, but proper distribution of these days during the days of the year, especially within the days when water is needed is essential in the field of agriculture. The type of rainfall including showery, heavy, is one of the important characteristics in the analysis of this trait for each region.

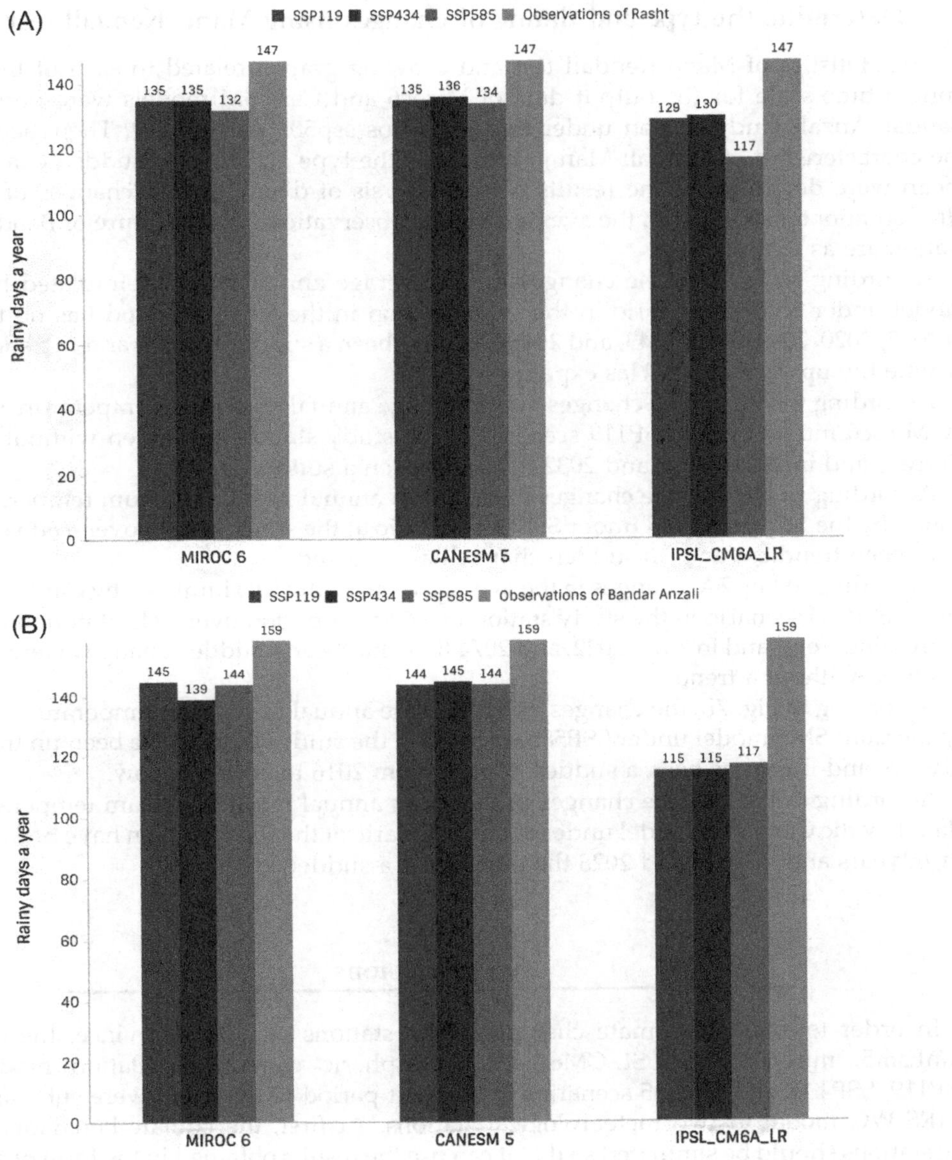

FIG. 5 (A) Number of rainy days observed (1980–2010) and predicted in Rasht station. (B) Number of rainy days observed (1980–2010) and predicted in Bandar Anzali station.

5.4 Determine the type and timing of changes using Mann-Kendall test analysis

The statistics of Mann-Kendall test and drawing graphs related to each of them on an annual time scale for the output data of Miroc6 and CanESM models were performed in Bandar Anzali study station under two scenarios ssp585 and ssp119. Then, according to the characteristics of Kendall Man graphic test, the type and time of sudden change in the mean were determined. The results of the analysis of diagrams and changes of $U(t_i)$ and $\grave{U}(t_i)$ components related to the average annual observational temperature of Bandar Anzali station are as follows:

According to Fig. 6A, the changes in the average annual rainfall simulated by Miroc6 model under SSP119 scenario in the study station in the 60-year period has no trend and in 2017, 2020, 2033, 2034, 2043, and 2044 there has been a sudden change and in a short period of time the upward trend. Has experienced.

According to Fig. 6B, the changes in the average annual maximum temperature simulated by Miroc6 model under SSP119 scenario in the study station have been without trend for 60 years and in 2021, 2029, and 2032 there has been a sudden change.

According to Fig. 6C, the changes in the mean annual mean maximum temperature simulated by the Miroc6 model under SSP119 scenario at the study station over a 60-year period have been trendless and no sudden change has occurred.

According to Fig. 7A, changes in the average annual rainfall simulated by CanESM5 model under SSP585 scenario in the study station in a 60-year period over a short period has seen a decreasing trend and in 2017, 2022, and 2074 there has been a sudden change in terms of Kelly has been without a trend.

According to Fig. 7B, the changes in the average annual maximum temperature simulated by the CanESM5 model under SSP585 scenario in the study station have been un trended for 60 years and there has been a sudden change from 2016 to 2046 in a row.

According to Fig. 7C, the changes in the mean annual mean minimum temperature simulated by the CanESM5 model under SSP585 scenario at the study station have been trendless for 60 years and in 2025 and 2028 there has been a sudden change.

6 Conclusions

In order to evaluate climate changes in the stations of Gilan province, the output of CanEsm5, miroc6 and IPSL-CM6A-LR atmospheric general circulation models with SSP119, SSP434, and SSP585 scenarios in the next period (2016–2100) were subscaled using LARS-WG model in two meteorological stations. At first, the climate behavior model of the stations should be simulated so that it can use the results obtained in the form of equations and statistical distributions during the future climate forecasting process. The evaluation of LARS-WG model was performed through error measurement indicators. Based on the results obtained from the LARS-WG model and according to the good accuracy of the model in simulating the observed data, the good potential of this model in forecasting was determined. Thus the minimum temperature and maximum temperature and precipitation parameters, under the current climatic conditions were well simulated and well predicted for the future. According to the potential of this model to predict climate parameters, this model can be used

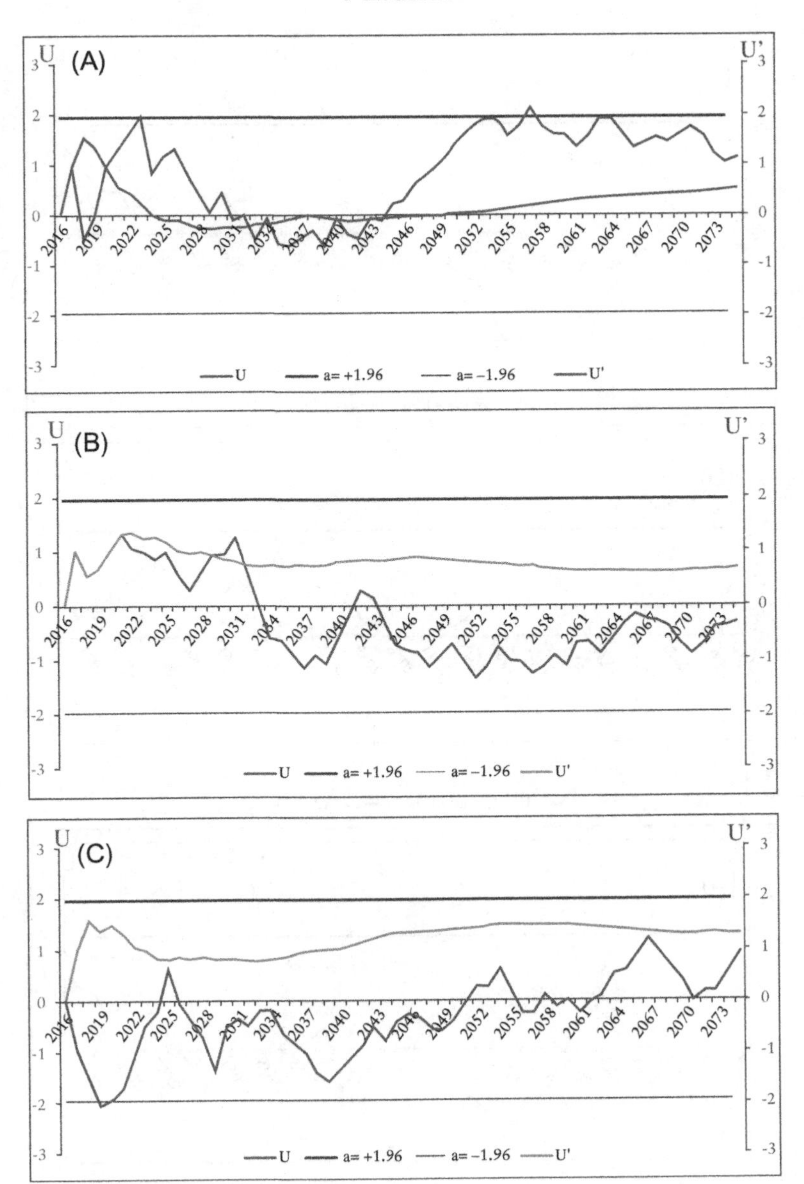

FIG. 6 (A) Changes in u, u statistics of 60-year average precipitation in Miroc6 model data under SSP119 scenario. (B) Changes in u, u statistics of 60-year maximum mean temperature in Miroc6 model data under SSP119 scenario. (C) Changes in u, u statistics of minimum 60-year mean temperature in Miroc6 model data under SSP119 scenario.

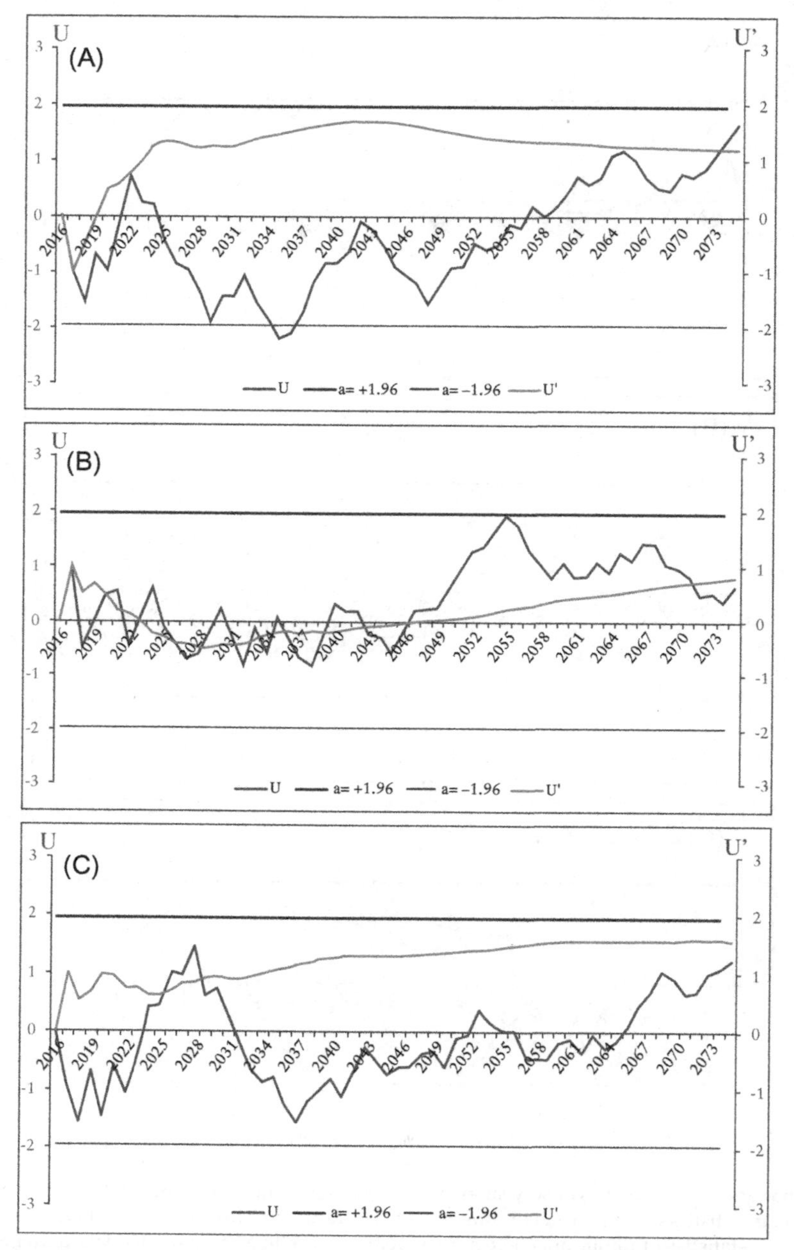

FIG. 7 (A) Changes in u, u statistics of 60-year average precipitation in CanESM5 model data under SSP585 scenario. (B) Changes in u, u statistics of 60-year maximum average temperature in CanESM5 model data under SSP585 scenario. (C) Changes in u, u statistics of minimum 60-year mean temperature in CanESM5 model data under SSP585 scenario.

to assess climate change in this area and areas with similar climates. According to the predicted results, it was observed that most of the maximum temperature values of CanEsm5 and IPSL-CM6A-LR models in all their scenarios, reach the maximum value in July. Similar to the maximum temperature, for the minimum temperature parameter, the CanEsm5 and IPSL-CM6A-LR models will experience their highest values in July, and there is a good correlation between these models at both stations. In general, the results obtained from comparing the maximum and minimum temperatures show that the most critical situation is related to the CanESM5 model under SSP585 scenario with an average increase of 4°C in both Rasht and Bandar Anzali stations. The results obtained for temperature increase in SSP119 scenario were less than other scenarios. Therefore, a temperature increase by about 0.8°C has been seen in all models and scenarios on average, MIROC6 model under this scenario has not a temperature increase and it has been constant. In general, during the studied period, if social and economic pursuits are carried out according to the SSP119 scenario, the temperature will increase on average by 0.8°C, which is a more ideal condition than the SSP585 scenario. Regarding the precipitation, the CanEsm5 and IPSL-CM6A-LR models also have an acceptable and good correlation with each other in all scenarios, and their highest value occurs in the months of October and November. But miroc6 model with SSP434 scenario in Rasht station, has divergence with other models and scenarios and experiences its maximum value in October. In general, for the precipitation parameter a more irregular trend was observed. In low rain months, a decreased rainfall was observed, but still in the highly rainy months, an increased rainfall was observed. However, in some months, the amount of precipitation decrease has been observed, but still the annual averages have increased in all scenarios. Furthermore, in CanEsm5 model under SSP585 scenario, the highest annual precipitation increase is observed but the number of rainy days of this model and scenario has decreased to 8%. Since the number of rainy days predicted by different models and scenarios is fewer than the number of rainy days observed, at the same time, most of the average monthly rainfall in the coming years has an ascending trend, it can be concluded that by reducing the number of rainy days in the coming years, the intensity and volume of rainfall will increase in the long run which will lead to heavy rains. According to the results obtained, it can be said that changes in meteorological parameters in Rasht and Bandar Anzali regions will be under the influence of climate change in the coming decades which can be used in long-term planning to deal with the effects of climate change.

References

Adnan, R.M., Liang, Z., Trajkovic, S., Zounemat-Kermani, M., Li, B., Kisi, O., 2019. Daily streamflow prediction using optimally pruned extreme learning machine. J. Hydrol. 577, 123981.

Barrett, J.P., 1974. The coefficient of determination—some limitations. Am. Stat. 28 (1), 19–20.

Chai, T., Draxler, R.R., 2014. Root mean square error (RMSE) or mean absolute error (MAE)—arguments against avoiding RMSE in the literature. Geosci. Model Dev. 7 (3), 1247–1250.

Chen, H., Xu, C.-Y., Guo, S., 2012. Comparison and evaluation of multiple GCMs, statistical downscaling and hydrological models in the study of climate change impacts on runoff. J. Hydrol. 434, 36–45.

Diaz-Nieto, J., Wilby, R.L., 2005. A comparison of statistical downscaling and climate change factor methods: impacts on low flows in the river Thames, United Kingdom. Clim. Chang. 69 (2), 245–268.

Elshamy, M.E., Wheater, H.S., Gedney, N., Huntingford, C., 2006. Evaluation of the rainfall component of a weather generator for climate impact studies. J. Hydrol. 326 (1–4), 1–24.

Goodarzi, E., Dastorani, M., Talebi, A., 2015. Evaluation of the change-factor and LARS-WG methods of downscaling for simulation of climatic variables in the future (case study: Herat Azam watershed, Yazd-Iran). Ecopersia 3 (1), 833–846.

Goodarzi, M.R., Fatehifar, A., Moradi, A., 2020a. Predicting future flood frequency under climate change using copula function. Water Environ. J. 34 (S1), 710–727.

Goodarzi, M.R., Vagheei, H., Mohtar, R.H., 2020b. The impact of climate change on water and energy security. Water Supply 20 (7), 2530–2546.

Huang, J., Zhang, J., Zhang, Z., Xu, C., Wang, B., Yao, J., 2011. Estimation of future precipitation change in the Yangtze River basin by using statistical downscaling method. Stoch. Env. Res. Risk A. 25 (6), 781–792.

IPCC, 2007. Climate Change 2007. The Fourth Assessment Report (AR4) of the United Nations. Intergovernmental Panel on Climate Change (IPCC), Working Group I, The Physical Science Basis of Climate Change.

IPCC, 2014. Climate Change 2014: Impacts, Adaptation, and Vulnerability. Part B: Regional Aspects, Contribution of Working Group II to the Fifth Assessment Report of the Intergovernmental Panel on Climate Change. Cambridge University Press, Cambridge, United Kingdom/New York, NY, USA. 688 pp.

Johnson, G.L., Hanson, C.L., Hardegree, S.P., Ballard, E.B., 1996. Stochastic weather simulation: Overview and analysis of two commonly used models. J. Appl. Meteorol. Climatol. 35 (10), 1878–1896.

Jones, P.D., Hulme, M., 1996. Calculating regional climatic time series for temperature and precipitation: methods and illustrations. Int. J. Climatol. 16 (4), 361–377.

Kendall, M., Gibbons, J., 1975. Rank Correlation Methods, 1970. Griffin, London.

Lettenmaier, D.P., Wood, E.F., Wallis, J.R., 1994. Hydro-climatological trends in the continental United States, 1948–88. J. Clim. 7 (4), 586–607.

Mann, H.B., 1945. Nonparametric tests against trend. Econometrica 13, 245–259.

McCuen, R.H., Knight, Z., Cutter, A.G., 2006. Evaluation of the Nash-Sutcliffe efficiency index. J. Hydrol. Eng. 11 (6), 597–602.

Mignot, J., Hourdin, F., Deshayes, J., Boucher, O., Gastineau, G., Musat, I., Vancoppenolle, M., Servonnat, J., Caubel, A., Chéruy, F., Denvil, S., Dufresne, J.-L., Ethé, C., Fairhead, L., Foujols, M.-A., Grandpeix, J.-Y., Levavasseur, G., Marti, O., Menary, M., Rio, C., Rousset, C., Silvy, Y., 2021. The tuning strategy of IPSL-CM6A-LR. J. Adv. Model. Earth Syst. 13 (5), e2020MS002340.

Mourato, S., Moreira, M., Corte-Real, J., 2015. Water resources impact assessment under climate change scenarios in Mediterranean watersheds. Water Resour. Manag. 29 (7), 2377–2391.

Ntegeka, V., Willems, P., 2008. Trends and multidecadal oscillations in rainfall extremes, based on a more than 100-year time series of 10 min rainfall intensities at Uccle, Belgium. Water Resour. Res. 44 (7). 15 pp.

O'Neill, B.C., Kriegler, E., Ebi, K.L., Kemp-Benedict, E., Riahi, K., Rothman, D.S., van Ruijven, B.J., van Vuuren, D.P., Birkmann, J., Kok, K., Levy, M., Solecki, W., 2017. The roads ahead: Narratives for shared socioeconomic pathways describing world futures in the 21st century. Glob. Environ. Chang. 42, 169–180.

O'Neill, B.C., Kriegler, E., Riahi, K., Ebi, K.L., Hallegatte, S., Carter, T.R., Mathur, R., van Vuuren, D.P., 2014. A new scenario framework for climate change research: the concept of shared socioeconomic pathways. Clim. Chang. 122 (3), 387–400.

Partal, T., Kahya, E., 2006. Trend analysis in Turkish precipitation data. Hydrol. Process. 20 (9), 2011–2026.

Pervez, M.S., Henebry, G.M., 2014. Projections of the Ganges-Brahmaputra precipitation—Downscaled from GCM predictors. J. Hydrol. 517, 120–134.

Pichuka, S., Prasad, R., Maity, R., Kunstmann, H., 2017. Development of a method to identify change in the pattern of extreme streamflow events in future climate: application on the Bhadra reservoir inflow in India. J. Hydrol. 9, 236–246.

Racsko, P., Szeidl, L., Semenov, M., 1991. A serial approach to local stochastic weather models. Ecol. Model. 57 (1–2), 27–41.

Rajabi, A., Sedghi, H., Eslamian, S., Musavi, H., 2010. Comparision of Lars-WG and SDSM Downscaling Models in Kermanshah (Iran).

Rajan, S., 2014. Statistical Downscaling of Gcm Output, Hydrological Simulation and Generation of Future Scenario Using Variable Infiltration Capacity (Vic) Model for the Ganga Basin, India. Andhra University, Uttarakhand, p. 132.

Riahi, K., Van Vuuren, D.P., Kriegler, E., Edmonds, J., O'neill, B.C., Fujimori, S., Bauer, N., Calvin, K., Dellink, R., Fricko, O., 2017. The Shared Socioeconomic Pathways and their energy, land use, and greenhouse gas emissions implications: an overview. Global Environ. Change 42, 153–168.

Semenov, M.A., 2007. Development of high-resolution UKCIP02-based climate change scenarios in the UK. Agric. For. Meteorol. 144 (1-2), 127–138.

Semenov, M.A., Barrow, E.M., Lars-Wg, A., 2002. A Stochastic Weather Generator for Use in Climate Impact Studies. User Man Herts UK.

Shiferaw, H., Gebremedhin, A., Gebretsadkan, T., Zenebe, A., 2018. Modelling hydrological response under climate change scenarios using SWAT model: The case of Ilala watershed, northern Ethiopia. Model. Earth Syst. Environ. 4 (1), 437–449.

Singh, D., Gupta, R., Jain, S.K., 2015. Assessment of impact of climate change on water resources in a hilly river basin. Arab. J. Geosci. 8 (12), 10625–10646.

Sneyers, R., 1990. Technical Note no 143 on the Statistical Analysis of Series of Observations. World Meteorological Organization, Geneva, Switzerland.

Swart, N.C., Cole, J.N., Kharin, V.V., Lazare, M., Scinocca, J.F., Gillett, N.P., Anstey, J., Arora, V., Christian, J.R., Hanna, S., 2019. The Canadian earth system model version 5 (CanESM5. 0.3). Geosci. Model Dev. 12 (11), 4823–4873.

Tan, M.L., Yusop, Z., Chua, V.P., Chan, N.W., 2017. Climate change impacts under CMIP5 RCP scenarios on water resources of the Kelantan River basin, Malaysia. Atmos. Res. 189, 1–10.

Tatebe, H., Watanabe, M., 2018. MIROC MIROC6 Model Output Prepared for CMIP6 CMIP Abrupt-4xCO2. Earth System Grid Federation.

Taylor, K.E., Stouffer, R.J., Meehl, G.A., 2012. An overview of CMIP5 and the experiment design. Bull. Am. Meteorol. Soc. 93 (4), 485–498.

Van den Besselaar, E., Klein Tank, A., Buishand, T., 2003. Trends in European precipitation extremes over 1951–2010. Int. J. Climatol. 33 (12), 2682–2689.

Wilby, R.L., Dawson, C.W., Barrow, E.M., 2002. SDSM—a decision support tool for the assessment of regional climate change impacts. Environ. Model. Softw. 17 (2), 145–157.

Wilby, R.L., Dawson, C.W., Murphy, C., Connor, P., Hawkins, E., 2014. The statistical downscaling model-decision centric (SDSM-DC): conceptual basis and applications. Clim. Res. 61 (3), 259–276.

Wilks, D.S., Wilby, R.L., 1999. The weather generation game: a review of stochastic weather models. Prog. Phys. Geogr. 23 (3), 329–357.

CHAPTER

33

Innovative triangular trend analysis of monthly precipitation at Shiraz Station, Iran

Mohammad Zakwan[a] and *Majid Niazkar*[b]

[a]School of Technology, MANUU, Hyderabad, India [b]Department of Civil and Environmental Engineering, School of Engineering, Shiraz University, Shiraz, Iran

OUTLINE

1 Introduction 589
2 Materials and methods 591
2.1 Study area 591
2.2 Mann Kendall test 592
3 Innovative trend analysis (ITA) 593
4 Results and discussion 594
5 Conclusion 596
References 597

1 Introduction

Climate changes have triggered the alteration in precipitation patterns all over the world (New et al., 2001; Milliman et al., 2008; Vansteenkiste et al., 2013; Belayneh et al., 2016; Zakwan and Ara, 2019; Abeysingha et al., 2020; Sharafati and Pezeshki, 2020; Zakwan et al., 2022). In essence, rainfall is directly correlated to the runoff process in a water cycle (Elouissi et al., 2017; Ara and Zakwan, 2018; Aiyelokun et al., 2021). Therefore, rainfall pattern changes can contribute to the increase in the frequency of droughts (Myronidis et al., 2018) and floods (Zakwan, 2018; Didovets et al., 2019). Since rainfall is the major component of the hydrologic cycle, changes in rainfall trends will have an important say in shaping the hydrology of the region in question (Huntington, 2006). Thus, hydrologists and geomorphologists have always

Current Directions in Water Scarcity Research, Volume 7
https://doi.org/10.1016/B978-0-323-91910-4.00033-9

remained concerned about spatiotemporal variations in the precipitation and other hydroclimatic variables (Elouissi et al., 2017). As a result, statistical analysis concerned with temporal variations in the hydroclimatic variables, such as various trend analysis techniques, has gained significant attention from hydrologists and geomorphologists (Douglas et al., 2000; Chen and Georgakakos, 2014; Shafiq et al., 2016; Hirca et al., 2022).

The most basic technique used to identify the presence of a trend in time series is the parametric linear regression technique, which has been widely exploited for this purpose (Yin et al., 2010; Zakwan et al., 2019). However, the success of parametric approaches in recognizing the trend is limited to normally distributed time series. With regard to this limitation of parametric tests, they are seldom utilized singly rather than in consultation with nonparametric tests, such as Mann-Kendall, Spearman-Rho, and Sen's slope methods. In fact, Mann-Kendall trend test is the most widely used trend test in hydrology (Hirsch et al., 1982; El-Shaarawi et al., 1983; Hirsch and Slack, 1984; Cailas et al., 1986; Hipel et al., 1988; Alley, 1988; Demaree and Nicolis, 1989; Zetterqvist, 1991; Helsel and Hirsch, 1992; McLeod and Hipel, 1994; Keim et al., 1995; Hamed and Rao, 1998; Douglas et al., 2000; Aziz and Burn, 2006; Brabets and Walvoord, 2009; Haktanir and Citakoglu, 2014; Zhang et al., 2011; Zhao et al., 2018). Moreover, several modifications have been proposed in the Mann-Kendall trend tests over the years. For instance, Hamed and Rao (1998) suggested that in the presence of the significant auto correlation in time series, Mann-Kendall trend tests do not yield reliable results and consequently, the results obtained need to be modified for auto correlation.

In light of improving trend analysis, Şen (2012) proposed an Innovative Trend Analysis (ITA) method. The major advantage of ITA involves visualization of trends and identification of trends of events with various magnitudes. To be more specific, the visual inspection of ITA plots can reveal trend in low magnitude, moderate magnitude and high magnitude events (Güçlü et al., 2020; Zakwan, 2021; Zerouali et al., 2021).

According to the literature, Trenberth (2011) reported declining trend of annual rainfall in tropical regions from the late 19th century. Kampata et al. (2008) analyzed rainfall trends in Zambezi river basin, Zambia and found no significant changes in the rainfall. Jain and Kumar (2012) performed analyzed temporal changes in the precipitation and temperature of various river basins in India. They reported that the rainfall in a majority of basins represent negative trend, while the maximum temperature showed a positive trend. Nnaji et al. (2016) analyzed temporal variations in the rainfall across Nigeria. The results displayed a significant impact of climate change on the rainfall and water resources. Zakwan and Ara (2019) performed the trend analysis of rainfall in Ganga sub-basin using Mann-Kendall and linear regression and reported a significant decline in the precipitation. Şen et al. (2019) applied a novel Innovative Polygon Trend Analysis (IPTA) to analyze trends of monthly or seasonal basis with the help of slope and length of various sides of polygons formed by analysis of monthly mean and monthly standard deviation of variables under consideration. Malik and Kumar (2020) analyzed the rainfall data of 13 districts of Uttarakhand, India, and found mixed results of positive and negative trends. Şişman (2021) identified the power law characteristics of trend of rainfall data in Turkey. Şan et al. (2021) explored the application of IPTA for determining the trends of precipitation in Vietnam. Their study demonstrates that the amount of precipitation has significantly declined in certain months across most of the gauging stations under observation in the country. Hirca et al., 2022 applied IPTA and Mann-Kendall methods to 56-year historical records of precipitation observed at eight measuring stations in Eastern Black Sea Basin, which has the highest annual rainfall with a significant hydroelectric potential in Turkey. Their comparative analysis demonstrated that the numbers of month with an increasing/decreasing trend identified by IPTA and Mann-Kendall were considerably different. This may be due to the fact that IPTA is relatively more sensitive than the Mann-Kendall's test in trend detection. Zakwan and Ahmad (2021) employed Mann-Kendall, Sen's slope, ITA and linear regression to investigate time series of discharge and sediment load of Ganga river. They reported a decline in rainfall patterns, which was the major reason behind a remarkable decline in discharges of Ganga river, as reported by Zakwan and Ara (2019).

In the present chapter, ITTA and Mann-Kendall test are used to detect and compare the results of trend obtained for the annual average and maximum rainfall time series of Shiraz station, Iran.

2 Materials and methods

2.1 Study area

In the current study, the rainfall data of Shiraz synoptic station in Iran was used in the present study. Shiraz, the capital of Fars province, is the largest biological center in southern Iran. It is located at latitude 29 degrees and 48 min north, 29 degrees and 29 min south, longitude 52 degrees and 40 min east and 52 degrees and 23 min west. The location of the study area is depicted in Fig. 1. This city has an average altitude of 1484 m above sea level with an average annual rainfall of 274.7 mm per year during the statistical period of 1923 to 2015. The climate of Shiraz is temperate due to its location in a mountainous region. The maximum and

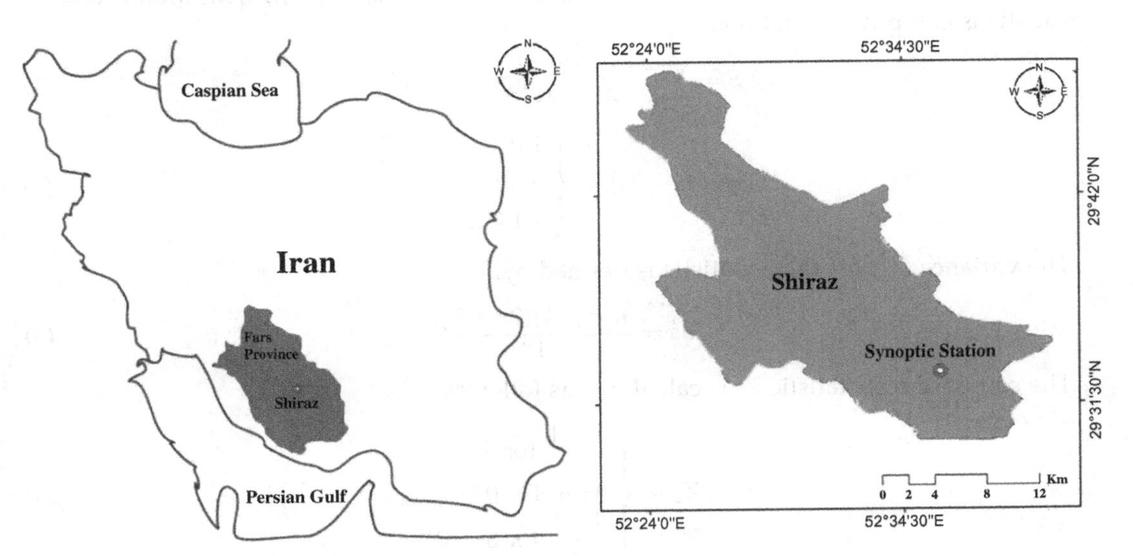

FIG. 1 Location of Shiraz and its synoptic station.

TABLE 1 Characteristics of rainfall data of Shiraz Station, Iran.

Time series	Average	Maximum
90 years (complete time series)	27.35	330
First 30 years (series 1)	29.58	305
Second 30 years (series 2)	25.56	330
Third 30 years (series 3)	26.92	275

minimum temperatures in this region are 40.2 and −14.4 degrees, respectively, while the average temperature is 18°C. Furthermore, the number of frost days is 42 days.

Annual average and annual maximum rainfall time series were prepared from the available rainfall data. Additionally, the two time series were subjected to ITTA and Mann-Kendall trend analysis. In the present chapter, the time series have been divided into three equal parts each containing 30 years data. Table 1 presents the average and maximum precipitation for the complete and three subparts of the data. In this way, a better analysis of changes in the hydroclimatic variable can be investigated as the three time series can be compared with one another.

2.2 Mann Kendall test

Mann Kendall test (Mann, 1945; Kendall, 1975) is a statistical test widely used for the analysis of time series. The null hypothesis H_0 assumes that there is no trend and this is tested against the alternative hypothesis H_1, which assumes that there is a trend. For x_i and x_j as two subsets of data where $i = 1, 2, 3, \ldots, n-1$ and $j = i+1, i+2, i+3, \ldots, n$, the Mann-Kendall S Statistic is computed as follows:

$$S = \sum_{i=1}^{n-1} \sum_{j=i+1}^{n} \text{sign}\ (x_j - x_i) \tag{1}$$

$$\text{Sign}\ (T_j - T_i) = \begin{cases} 1 \text{ if } x_j - x_i > 0 \\ 0 \text{ if } x_j - x_i = 0 \\ -1 \text{ if } x_j - x_i < 0 \end{cases} \tag{2}$$

The variance (σ^2) for the S-statistic is defined by:

$$\sigma^2 = \frac{[n(n-1)(2n+5)]}{18} \tag{3}$$

The standard test statistic Z_s is calculated as follows:

$$Z_S = \begin{cases} \dfrac{S-1}{\sigma} \text{ for } S > 0 \\ 0 \text{ for } S = 0 \\ \dfrac{S+1}{\sigma} \text{ for } S < 0 \end{cases} \tag{4}$$

If $|Z_s|$ is greater than $Z_{\alpha/2}$, where α denotes the chosen significance level (5% with $Z_{0.025} = 1.96$), the null hypothesis is invalid implying that the trend is significant. Similar to some previous studies, in this study, the equations of the Mann Kendall's test (Eqs. 1–4) were implemented in MS Excel, which has an undeniable merit for conducting numerical calculations (Niazkar and Afzali, 2016).

3 Innovative trend analysis (ITA)

In the ITA method, the available time series are divided into two equal halves. The average of both the time series is calculated as $\overline{Y_1}$ and $\overline{Y_2}$. The two parts of the time series are then arranged in an ascending order. Later, a plot is prepared with the first half of time series on the x-axis and the second half series on the y-axis, as shown in Fig. 2. Relative position of scatter point with respect to trendless (1:1) line demarcates the trend, as shown in Fig. 2.

The points above and below the trendless line indicate increasing and decreasing trends, respectively, while the points lying on the 1:1 line represent no trend. The major advantage of this plot is that it can provide the trends of low magnitude, moderate magnitude and high magnitude events. For instance, Fig. 2 illustrates that low magnitude events are trendless whereas the moderate magnitude fall below the trendless line and represent a decline in the magnitude of moderate magnitude events. However, the significance of a trend cannot be determined from the plot, while it can only be delineated based on a confidence limit.

The trend magnitude may be calculated as

$$s = \frac{2(\overline{Y_2} - \overline{Y_1})}{n} \tag{5}$$

The critical trend (Z_s) can be calculated as

$$Z_s = \frac{s}{\sigma_s} \tag{6}$$

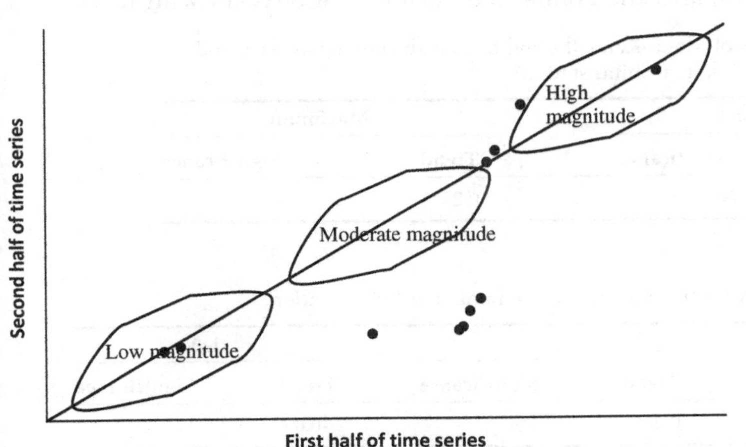

FIG. 2 Presentation of innovative trend analysis (ITA) method.

where σ_s is standard deviation of trend values and can be computed as

$$\sigma_s^2 = \frac{8\sigma^2 \left(1 - \rho_{Y_2Y_1}\right)}{n^3} \tag{7}$$

where $\rho_{Y_2Y1}$is the cross correlation coefficient of the averages of two halves given by

$$\rho_{Y_2Y_1} = \frac{E(xy) - E(x)E(y)}{\sigma_x \sigma_y} \tag{8}$$

ITA was initially proposed by Şen (2012, 2017) presented the methodology to calculate the confidence limits for ITA. Basically, the method allows greater transparency to the trend analysis as compared to the traditional trend analysis techniques, such as Mann-Kendall, Spearman-Rho, and Sen's slope methods. In ITA, the entire time series is divided into two equal parts. ITTA is an extended version of ITA, where the time series can be divided into any number of equal parts.

4 Results and discussion

In the present chapter, the rainfall data of Shiraz station, Iran was subjected to the trend analysis. The analysis of temporal variations of the monthly precipitation was accomplished through Mann Kendall and ITTA. The Mann Kendall considers the entire time series (90 years data) whereas the time series is divided into three equal parts (30 years each) in case of ITTA.

The results of the trend analysis for the annual average and maximum rainfall obtained from Mann Kendall test are reported in Table 2. As shown, both annual average and annual maximum precipitation are declining in Shiraz. Although the trend is insignificant at 5% significance level, the Z values for the annual average rainfall is negative indicating a gradual decline in the precipitation. On the other hand, an insignificant positive trend is observed in Table 2 for the annual maximum rainfall.

The results of trend analysis for the annual average obtained from ITTA are presented in Table 3. As shown, ITTA compares the annual data of the first 30 years with the second

TABLE 2 Results of Mann Kendall trend test for the annual average and annual maximum rainfall at Shiraz station.

Average		Maximum	
Trend	**Significance**	**Trend**	**Significance**
−1.42	No	0.23	No

TABLE 3 Results of ITTA for the annual average rainfall at Shiraz station.

1–2		2–3		1–3	
Trend	**Significance**	**Trend**	**Significance**	**Trend**	**Significance**
−0.13	Yes	0.05	Yes	−0.09	Yes

30 years (represented as 1–2), the second 30 years with the third 30 years (represented as 2–3), the first 30 years with the third 30 years (represented as 1–3). In other words, Table 3 lists the comparative analysis of three time series of the annual average precipitation. It may be observed that the average rainfall declined significantly in the second 30 years in comparison to that of the first 30 years. When the average rainfall for the second and third 30 years is compared, the results demonstrate a significant increase in precipitation. However, the average rainfall in the third time series is still significantly lower than that of the first time series, as represented by a significant negative trend of 1 to 3 in Table 3. Overall, the rainfall has declined as compared to the first time series, which is also in agreement with the negative trend obtained by the Mann-Kendall test. Fig. 3 represent the ITA curve for average and maximum rainfall for time series 1 and time series 2. As most of points plotted below the

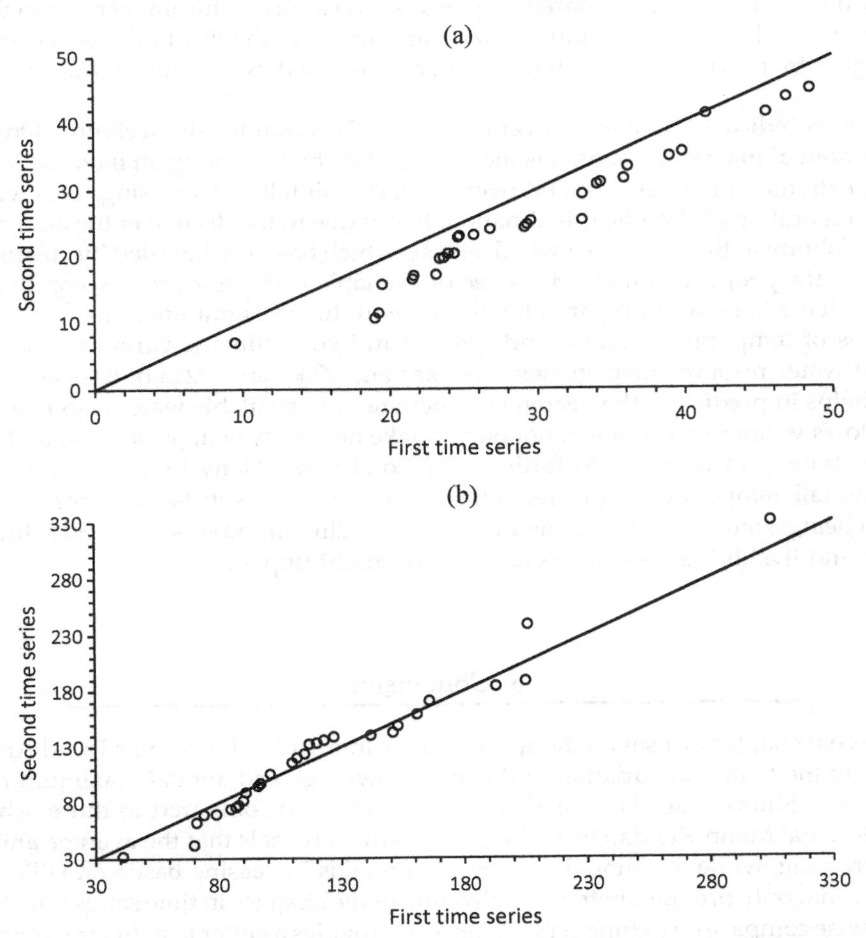

FIG. 3 Presentation of ITA curve for series 1 versus series in case of (A) average rainfall (B) maximum rainfall.

TABLE 4 Results of ITTA for the annual maximum rainfall at Shiraz station.

1–2		2–3		1–3	
Trend	Significance	Trend	Significance	Trend	Significance
−0.004	No	0.23	Yes	0.22	Yes

line in case of average rainfall, it signifies significant decline in average rainfall. On the other, in case of maximum rainfall some of the points also lie above the line, indicating insignificant trend.

Table 4 presents the trends obtained by ITTA for the annual maximum rainfall. As shown, the annual maximum rainfall declined in the second 30 years in comparison to that of the first 30 years. However, the decline was insignificant. When the average precipitation for the second and third 30 years is compared, the results indicate a significant increase in the annual maximum rainfall. Again, the annual maximum rainfall in the third time series was significantly higher than that of the first time series, as presented by a significant positive trend in Table 4.

The results indicate that annual average rainfall is continuously declining. On the other hand, the annual maximum rainfall is increasing, thereby indicating an increase in the magnitude of extreme rainfall events, and overall yearly rainfall is decreasing. This will require twofold preparations to handle this situation: firstly due to the decline in the annual rainfall, water availability in the study area will decrease, which has to be handled by creating awareness among the people about judicious use of available water resources, secondly, the local authorities have to be well prepared for coping with high magnitude rains.

Analysis of temporal variations and forecast in hydro climatic variables are important aspects of water resource management Niazkar and Zakwan, 2021a,b. Basically, the trend analysis helps in predicting the decrement/increment in available water resources. In addition, it allows various stakeholders not only to take necessary actions but also to develop an adequate strategy in response to forthcoming conditions. Many hydro-climatic variables, such as rainfall, temperature and stream flow in a river, have substantial impacts on the life of human beings, fauna, flora, and their ecosystems. Climatic adversities render huge social, economic and livelihood losses especially when caught unprepared.

5 Conclusion

The present chapter focuses on the application of Innovative Triangular Trend Analysis for determining the temporal variation in the annual average and annual maximum rainfall of Shiraz station, Shiraz, Iran. The trends obtained were also compared to those achieved by the conventional Mann-Kendall test. The trend analysis reveals that the average annual rainfall is decreasing, while the annual maximum rainfall is increasing based on ITTA. Furthermore, ITTA not only provides better visualization of the changes in time series but also allows segment-wise comparison of time series. Hence, it provides a better insight into trend analysis and identification of periods with major fluctuations.

References

Abeysingha, N.S., Islam, A., Singh, M., 2020. Assessment of climate change impact on flow regimes over the Gomti River basin under IPCC AR5 climate change scenarios. J. Water Clim. Chang. 11 (1), 303–326.

Aiyelokun, O., Pham, Q.B., Aiyelokun, O., Malik, A., Zakwan, M., 2021. Credibility of design rainfall estimates for drainage infrastructures: extent of disregard in Nigeria and proposed framework for practice. Nat. Hazards. https://doi.org/10.1007/s11069-021-04889-1. Article.

Alley, W.M., 1988. Using exogenous variables in testing for monotonic trends in hydrologic time series. Water Resour. Res. 24 (11), 1955–1961.

Ara, Z., Zakwan, M., 2018. Rainfall runoff modelling for eastern canal basin. Water Energy Int. 61 (6), 63–67.

Aziz, O.I.A., Burn, D.H., 2006. Trends and variability in the hydrological regime of the Mackenzie River Basin. J. Hydrol. 319 (1–4), 282–294.

Belayneh, A., Adamowski, J., Khalil, B., 2016. Short-term SPI drought forecasting in the Awash River basin in Ethiopia using wavelet transforms and machine learning methods. Sustain. Water Resour. Manag. 2 (1), 87–101.

Brabets, T., Walvoord, M.A., 2009. Trends in streamflow in the Yukon River basin from 1944–2005 and the influence of the Pacific decadal oscillation. J. Hydrol. 371, 108–119.

Cailas, M.D., Cavadias, G., Gehr, R., 1986. Application of a nonparametric approach for monitoring and detecting trends in water quality data of the St. Lawrence River. Water Pollut. Res. J. Can. 21 (2), 153–167.

Chen, C.J., Georgakakos, A.P., 2014. Hydro-climatic forecasting using sea surface temperatures: methodology and application for the southeast US. Climate Dynam. 42 (11–12), 2955–2982.

Demaree, G.R., Nicolis, C., 1989. Onset of sahelian drought viewed as a fluctuation-induced transition. Q. J. Roy. Meteorol. Soc. 116 (491), 221–238.

Didovets, I., Krysanova, V., Bürger, G., Snizhko, S., Balabukh, V., Bronstert, A., 2019. Climate change impact on regional floods in the Carpathian region. J. Hydrol. 22, 100590.

Douglas, E.M., Vogel, R.M., Kroll, C.N., 2000. Trends in floods and low flows in the United States: impact of spatial correlation. J. Hydrol. 240, 90–105.

Elouissi, A., Habi, M., Benaricha, B., Boualem, S.A., 2017. Climate change impact on rainfall spatio-temporal variability (Macta watershed case, Algeria). Arab. J. Geosci. 10 (22), 496.

El-Shaarawi, A.H., Esterby, S.R., Kuntz, K.W., 1983. A statistical evaluation of trends in the water quality of the Niagara River. J. Great Lakes Res. 9 (2), 234–240.

Güçlü, Y.S., Şişman, E., Dabanlı, İ., 2020. Innovative triangular trend analysis. Arab. J. Geosci. 13 (1), 1–8.

Haktanir, T., Citakoglu, H., 2014. Trend, independence, stationarity, and homogeneity tests on maximum rainfall series of standard durations recorded in Turkey. J. Hydrol. Eng. 19 (9), 05014009.

Hamed, K.H., Rao, A.R., 1998. A modified Mann-Kendall trend test for autocorrelated data. J. Hydrol. 204 (1–4), 182–196.

Helsel, D.R., Hirsch, R.M., 1992. Statistical Methods in Water Resources. Elsevier Publishers, Amsterdam.

Hipel, K.W., McLeod, A.J., Weller, R.R., 1988. Data analysis of water quality time series in lake Erie. J. Am. Water Resour. Assoc. 24 (3), 533–544.

Hirca, T., Eryılmaz Türkkan, G., Niazkar, M., 2022. Applications of innovative polygonal trend analyses to precipitation series of Eastern Black Sea Basin, Turkey. Theor. Appl. Climatol. 147 (1), 651–667. https://doi.org/10.1007/s00704-021-03837-0.

Hirsch, R.M., Slack, J.R., 1984. A nonparametric trend test for seasonal data with serial dependence. Water Resour. Res. 20 (6), 727–732.

Hirsch, R.M., Slack, J.R., Smith, R.A., 1982. Techniques of trend analysis for monthly water quality data. Water Resour. Res. 18 (1), 107–121.

Huntington, T.G., 2006. Evidence for intensification of the global water cycle: review and synthesis. J. Hydrol. 319 (1), 83–95.

Jain, S.K., Kumar, V., 2012. Trend analysis of rainfall and temperature data for India. Curr. Sci., 37–49.

Kampata, J.M., Parida, B.P., Moalafhi, D.B., 2008. Trend analysis of rainfall in the headstreams of the Zambezi River Basin in Zambia. Phys. Chem. Earth A/B/C 33 (8–13), 621–625.

Keim, B.D., Faiers, G.E., Muller, R.A., Grymes, J.M., Rohli, R.V., 1995. Long-term trends of precipitation and runoff in Louisiana, USA. Int. J. Climatol. 15 (5), 531–541.

Kendall, M.G., 1975. Rank Correlation Methods. Griffin, London, UK.

Malik, A., Kumar, A., 2020. Spatio-temporal trend analysis of rainfall using parametric and non-parametric tests: case study in Uttarakhand, India. Theor. Appl. Climatol. 140 (1), 183–207.

Mann, H.B., 1945. Nonparametric test against trend. Econometrica 13, 245–259.
McLeod, A.I., Hipel, K.W., 1994. Tests for monotonic trend. In: Stochastic and Statistical Methods in Hydrology and Environmental Engineering. Springer, Dordrecht, pp. 245–270.
Milliman, J.D., Farnsworth, K.L., Jones, P.D., Xu, K.H., Smith, L.C., 2008. Climatic and anthropogenic factors affecting river discharge to the global ocean, 1951–2000. Global Planet. Change 62 (3–4), 187–194.
Myronidis, D., Ioannou, K., Fotakis, D., Dörflinger, G., 2018. Streamflow and hydrological drought trend analysis and forecasting in Cyprus. Water Resour. Manag. 32 (5), 1759–1776.
New, M., Todd, M., Hulme, M., Jones, P., 2001. Precipitation measurements and trends in the twentieth century. Int. J. Climatol. 21 (15), 1889–1922.
Niazkar, M., Afzali, S.H., 2016. Streamline performance of excel in stepwise implementation of numerical solutions. Comput. Appl. Eng. Educ. 24 (4), 555–566. https://doi.org/10.1002/cae.21731.
Niazkar, M., Zakwan, M., 2021a. Assessment of artificial intelligence models for developing single-value and loop rating curves. Complexity 2021, 1–21. 6627011. https://doi.org/10.1155/2021/6627011.
Niazkar, M., Zakwan, M., 2021b. Application of MGGP, ANN, MHBMO, GRG, and linear regression for developing daily sediment rating curves. Math. Probl. Eng. 2021, 8574063. https://doi.org/10.1155/2021/8574063.
Nnaji, C.C., Mama, C.N., Ukpabi, O., 2016. Hierarchical analysis of rainfall variability across Nigeria. Theor. Appl. Climatol. 123 (1–2), 171–184.
Şan, M., Akçay, F., Linh, N.T.T., Kankal, M., Pham, Q.B., 2021. Innovative and polygonal trend analyses applications for rainfall data in Vietnam. Theor. Appl. Climatol. 144 (3), 809–822.
Şen, Z., 2012. Innovative trend analysis methodology. J. Hydrol. Eng. 17 (9), 1042–1046.
Şen, Z., 2017. Innovative trend significance test and applications. Theor. Appl. Climatol. 127 (3–4), 939–947.
Şen, Z., Şişman, E., Dabanli, I., 2019. Innovative polygon trend analysis (IPTA) and applications. J. Hydrol. 575, 202–210.
Shafiq, M.U., Bhat, M.S., Rasool, R., Ahmed, P., Singh, H., Hassan, H., 2016. Variability of precipitation regime in Ladakh region of India from 1901–2000. J. Climatol. Weather Forecast 4, 165. https://doi.org/10.4172/2332-2594.10001 65.
Sharafati, A., Pezeshki, E., 2020. A strategy to assess the uncertainty of a climate change impact on extreme hydrological events in the semi-arid Dehbar catchment in Iran. Theor. Appl. Climatol. 139 (1–2), 389–402.
Şişman, E., 2021. Power law characteristics of trend analysis in Turkey. Theor. Appl. Climatol. 143 (3), 1529–1541.
Trenberth, K.E., 2011. Changes in precipitation with climate change. Climate Res. 47 (1/2), 123–138.
Vansteenkiste, T., Tavakoli, M., Ntegeka, V., Willems, P., De Smedt, F., Batelaan, O., 2013. Climate change impact on river flows and catchment hydrology: a comparison of two spatially distributed models. Hydrol. Process. 27 (25), 3649–3662.
Yin, Y.H., Wu, S.H., Chen, G., Dai, E.F., 2010. Attribution analyses of potential evapotranspiration changes in China since the 1960s. Theor. Appl. Climatol. 101 (1–2), 19–28.
Zakwan, M., 2018. Spreadsheet-based modelling of hysteresis-affected curves. Appl. Water Sci. 8 (4), 101–105. https://doi.org/10.1007/s13201-018-0745-3.
Zakwan, M., 2021. Trend analysis of groundwater level using innovative trend analysis. In: Groundwater Resources Development and Planning in the Semi-Arid Region. Springer, Cham, pp. 389–405.
Zakwan, M., Ahmad, Z., 2021. Trend analysis of hydrological parameters of Ganga River. Arab. J. Geosci. 14 (3), 1–15.
Zakwan, M., Ara, Z., 2019. Statistical analysis of rainfall in Bihar. Sustain. Water Resour. Manag. 5 (4), 1781–1789.
Zakwan, M., Khan, I., Ara, Z., Rahim, Z.A., Sharief, S.M.V., 2019. Trend analysis of rainfall in Bihar. In: Water Resources Management (WRM2019), 2019, pp. 79–85.
Zakwan, M., Pham, Q.B., Bonnaci, O., Durin, B., 2022. Application of revised innovative trend analysis in lower Drava River. Accepted, Arab. J. Geosci. https://doi.org/10.1007/s12517-022-09591-5.
Zerouali, B., Al-Ansari, N., Chettih, M., Mohamed, M., Abda, Z., Santos, C.A.G., Elbeltagi, A., 2021. An enhanced innovative triangular trend analysis of rainfall based on a spectral approach. Water 13 (5), 727.
Zetterqvist, L., 1991. Statistical estimation and interpretation of trends in water quality time series. Water Resour. Res. 27 (7), 1637–1648.
Zhang, W., Shousheng, M.U., Zhang, Y., Kaimin, C.H.E.N., 2011. Temporal variation of suspended sediment load in the Pearl River due to human activities. Int. J. Sediment Res. 26 (4), 487–497.
Zhao, G., Mu, X., Strehmel, A., Tian, P., 2018. Temporal variation of streamflow, sediment load and their relationship in the Yellow River Basin, China. PLoS One 9 (3), e91048.

CHAPTER

34

Overview of trend and homogeneity tests and their application to rainfall time series

Deepesh Machiwal, H.M. Meena, and D.V. Singh

Division of Natural Resources, ICAR-Central Arid Zone Research Institute, Jodhpur, Rajasthan, India

OUTLINE

1 Introduction 600

1.1 Examining trend and homogeneity in hydrology: An overview 600

1.2 Rationale of the study 603

2 Materials and methods 604

2.1 Study area and data description 604

2.2 Basic statistical properties 605

2.3 Statistical tests for identifying trend and estimating trend magnitude 605

2.4 Statistical tests for examining homogeneity 605

3 Results and discussion 605

3.1 Rainfall statistics 605

3.2 Spatial distribution and comparison of rainfall trends 607

3.3 Trend magnitudes for annual and seasonal rainfall 610

3.4 Spatial distribution and comparison of homogeneity results 611

4 Limitations and future challenges 616

5 Recommendations 616

6 Concluding remarks 616

Acknowledgments 617

References 617

Current Directions in Water Scarcity Research, Volume 7
https://doi.org/10.1016/B978-0-323-91910-4.00034-0

1 Introduction

Rainfall is an important component of water cycle, which has been widely used by several researchers for diagnosing impact of climate change at different places over the globe (Machiwal et al., 2016). Knowledge about rainfall variability including gradual changes or trends and presence or absence of homogeneity over space and time is imperative to understand many problems that arise in different sectors including agriculture, environment, and water resources, and to find feasible solutions to them. Since the beginning of 21st century, large changes have been observed in rainfall that may have adverse effects on agriculture and reduce food availability especially in developing regions situated in extreme climatic conditions. This may be a probable reason that number of studies analyzing changes or variability in rainfall patterns have been increasing in literature (Meinke et al., 2005; Machiwal and Jha, 2008; Kumar et al., 2010; Jain and Kumar, 2012; Xia et al., 2015; Gupta et al., 2017; Machiwal et al., 2019, 2021; Chargui et al., 2022). A critical review of literature reveals that rainfall variability in most of studies is investigated by identifying trends only. However, homogeneity of rainfall has been examined in a few studies. An overview of several statistical tests available in literature for testing trends and homogeneity in hydrologic time series along with their development and advancement made over the years is provided ahead.

1.1 Examining trend and homogeneity in hydrology: An overview

A trend appears when successive values in a hydrologic time series show affinity toward inclination or declination over time and is considered as a deterministic component (Haan, 2002). Natural and anthropogenic factors, causing changes in hydrologic conditions of a time series, are responsible for introducing both linear and non-linear trends. A linear or non-linear model is needed to mathematically express a trend in a time series. The linear models are widely used to describe a trend in hydrological variables (Shahin et al., 1993). In hydrology, trend may be identified using statistical tests mainly based on two approaches (Machiwal and Jha, 2012): (a) parametric and (b) non-parametric. The parametric tests have limitation that time series should follow a normal probability distribution but they are more powerful than non-parametric tests. On the other hand, non-parametric tests without depending on type of data distribution have a wide applicability in hydrology as most of hydrologic variables do not fulfill normality requirement (Machiwal and Jha, 2012). The simplest test for detecting linear trends is parametric Student's t-test, which requires normality to be present in rainfall time series (Hameed et al., 1997). Hence, before applying Student's t-test for evaluating statistical significance of presence of linear trend, it is necessary to confirm that rainfall series follows a normal probability distribution (Hoel, 1954). This preliminary check has been ignored sometimes in past studies (e.g., Fanta et al., 2001). When a hydrologic time series does not fulfill normality assumption, non-parametric statistical tests may be employed for identifying trends. The widely used non-parametric statistical test for detection of statistically significant trends is Mann-Kendall (M-K) test, which was initially developed by Mann (1945) and subsequently modified by Kendall (1975). The M-K test identifies a monotonic trend without having large changes in mean of a time series. The non-parametric tests are appropriate for identifying trends in censored and non-normal data (Helsel and Hirsch, 1988).

The M-K test has been used for identifying rainfall trends in several studies, e.g., Lettenmaier et al. (1994), Gan (1998), Lins and Slack (1999), Douglas et al. (2000), Zhang et al. (2001), Burn and Elnur (2002), Yue et al. (2003), Machiwal and Jha (2008), Türkeş et al. (2009), Lupikasza (2010), Jain et al. (2013), Asfaw et al. (2018), Meena et al. (2019), Chahal et al. (2021), and Saini et al. (2022). Chahal et al. (2021) examined trend and pattern in 57-year (1961–2017) extreme rainfall of nine stations located in Sahibi basin of Rajasthan, India using M-K test. The results depicted non-significant increasing trends in very wet day rainfall and moderate rainfall days along with the decreasing trend in consecutive dry days and extreme rainfall indices. Saini et al. (2022) examined trends in annual rainfall (1961–2017) of 33 stations in dryland ecosystem of Rajasthan, India by means of M-K test. The mean annual rainfall showed significant rising trend in the station-wise mean annual rainfall at Barmer, Churu, Ganganagar, Jaisalmer, and Pratapgarh stations. There are other non-parametric tests such as Kendall's rank correlation (KRC) test, that is customarily employed for detection of rainfall trends (e.g., Guerreiro et al., 2014). Apart from this, there exist other parametric and non-parametric statistical trend tests in hydrological studies. The parametric trend tests include turning point test, regression test (Haan, 2002), sum of squared lengths test, Wald-Wolfowitz total number of runs test, inversion test, Kendall's phase test, and Kendall's rank test (Shahin et al., 1993). The non-parametric tests are Hotelling-Pabst test (Conover, 1971), Sen's slope estimation test (Sen, 1968), Kendall's rank correlation test (Kendall, 1973), Spearman rank correlation (SRC) test (McGhee, 1985), and Sen's test (Gilbert, 1987), among others.

At the outset of the 21st century, M-K test was criticized for its poor performance under presence of serial correlation in data series (Yue et al., 2002). Positive value of serial correlation coefficient increases variance or M-K test-statistic, and vice versa. A pre-whitening approach was developed by von Storch (1995) for making series trend-free by generating a difference series by multiplying each element of a time series with value of serial correlation coefficient at lag-1 and then taking a difference of each consecutive pair of elements. Subsequently, M-K test is applied on the difference series. The TFPW method has been widely employed in a number of studies reported after the year 2000 (e.g., Douglas et al., 2000; Zhang et al., 2001; Hamilton et al., 2001; Burn and Elnur, 2002). Other than TFPW approach, there are three correction factors or variance-correction approaches: (a) Lettenmaier (1976) approach, (b) Matalas and Sankarasubramanian (2003) approach, and (c) Yue and Wang (2004) approach for modifying the original M-K test-statistic. The Lettenmaier approach is more practical for a long-term series as original observations are not modified (Khaliq et al., 2006). Khaliq et al. (2009) reviewed comparative performance of M-K, Sen's slope estimator, Spearman rank correlation, and regression tests when serial and cross correlations were present in series, and suggested to avoid pre-whitening approach as it affects data slope that is main indicator of trend magnitudes. Nevertheless, it is felt that variance-correction approaches are not being employed in many rainfall trend detection studies, perhaps owing to small change in original M-K test-statistic after variance correction and lack of awareness among researchers. In spite of unraveled limitation of M-K test when serial correlation remains present in series, it is single established and proven method for trend identification for a long time in several subjects. Similar was the case for hydrologic time series as there was no option to identify a trend than M-K test. However, a major breakthrough occurred in year 2012 when innovative trend analysis (ITA) test was proposed by Şen (2012). The ITA test was free from normality assumption and not affected by serial correlation.

Hence, many researchers employed ITA test in addition to M-K test and compared findings of both tests (e.g., Meena et al., 2019; Zakwan, 2021; Zakwan and Ahmad, 2021; Nair and Mirajkar, 2022). Meena et al. (2019) employed M-K and ITA tests to examine trends in 55-year monthly rainfall of 62 stations in arid Rajasthan, India. The M-K test identified significant rising trends in June and September and non-significant trends in July. On the contrary, ITA test indicated all trends significant, and thus, it is over-sensitive in evaluating significance of rainfall trends. Nair and Mirajkar (2022) detected trends in Wainganga basin, Central India, for 1913–2013 period using gridded rainfall data of $0.25° \times 0.25°$ resolution by employing M-K, Spearman's rho and ITA tests. Results indicated existence of decreasing trend in rainfall; however, trend was gradually increasing for extreme rainfall events. Şen (2017) improved reliability of ITA test by adding a statistical significance test application using a synthetically generated time series that included deterministic, ordinary, and stochastic processes, and gamma stochastic process tests. Nourani et al. (2015) combined ITA test with discrete wavelet transformation for improving performance of test in identifying trends. Güçlü (2018) modified original ITA test into double and triple ITA by splitting whole time series into three and four subseries, respectively, which are known as multiple Şen ITA tests. Furthermore, ITA test is revised as innovative triangular trend analysis (ITTA) by separating whole data series into a number of subseries of equal length and comparing all subseries pairwise as triangular array (Güçlü et al., 2020). Zerouali et al. (2021) enhanced efficiency of the ITTA test by coupling it with orthogonal discrete wavelet transformation for partial trend identification. Recently, a new method, i.e., innovative polygon trend analysis (IPTA) method is applied for testing monthly rainfall trends in Susurluk basin, Turkey (Ceribasi and Ceyhunlu, 2021). The IPTA method has emerged from development of ITA test. A few studies reported over-sensitivity of ITA test in deciding statistical significance of rainfall trends (e.g., Machiwal et al., 2019). Further, mathematical formulation of ITA test is reported to be incorrect and its methodology gets affected by the serial correlation present in series (Serinaldi et al., 2020). Reviews of statistical trend tests in environmental variables are presented by Esterby (1996) and Hess et al. (2001). Some additional trend tests are mentioned in Mahé et al. (2001). Kundzewicz and Robson (2004) provided an overview of 4 trend tests and 7 step change tests after classifying tests into four categories: (a) distribution-free tests, (b) rank-based tests, (c) tests based on normal transformations, and (d) tests based on resampling approaches. Theoretical descriptions of a total of 15 trend tests are presented by Machiwal and Jha (2012). Further, Sonali and Kumar (2013) classified trend tests into two categories: (a) slope based tests, and (b) rank based tests. They briefly explained procedures for applying statistical tests with serial correlation effect such as pre-whitening and variation correction approaches, block bootstrap resampling technique with M-K and SRC tests, and sequential M-K test. Mudelsee (2019) emphasized on uncertainty and accuracy of trends identified by linear and non-linear and non-parametric regression tests. Almazroui and Şen (2020) presented methodologies for 7 trend tests, i.e., linear regression, M-K, Sen's slope, Spearman's rho, ITA, partial trend analysis, and crossing trend analysis tests.

Presence of homogeneity in a time series is confirmed when data of series represent same population in such a way that mean of the data series looks invariant (Machiwal and Jha, 2012). A time series becomes non-homogeneous when mean of series does not remain stable over time or a change is appeared in series. Non-homogeneity in a data series is caused by both natural and anthropogenic factors. The natural factors include changes due to climate variability and change whereas anthropogenic factors consist of man-made causes such as method or surroundings of collecting data is changed (Fernando and Jayawardena, 1994). Hence, statistical tests examining homogeneity basically evaluate significance of changes in the mean of a series over different time periods. Three classical methods for testing homogeneity in rainfall series are von Neumann, Bayesian, and cumulative deviations tests (Buishand, 1982). There are some contemporary statistical tests for multiple comparisons, which include Tukey, Dunnett, Link-Wallace, Hartley, and Bartlett tests. However, it is felt that studies employing multiple comparison tests have been increasing in hydrology and climatology over past two decades. In recent years, a growing interest of researchers has been apparently seen in applying tests for examining rainfall homogeneity (Saikranthi et al., 2013; Hänsel et al., 2016; Rahman et al., 2017; Meena et al., 2019; Daba et al., 2020; Kocsis et al., 2020; Machiwal et al., 2021). Daba et al. (2020) analyzed homogeneity of annual rainfall (1980–2017) in the Upper Awash Sab-Basin (UASB) in Oromia, Ethiopia, by applying Pettitt's, Buishand's, standard normal homogeneity, and von Neumann ratio tests. Results indicated presence of homogeneity at 58% stations. Kocsis et al. (2020) applied Pettitt's test to detect change points in monthly, seasonal and annual rainfall of Keszthely (Western Hungary). Results suggested presence of abrupt shifts in spring and winter rainfall. Machiwal et al. (2021) investigated presence/absence of homogeneity in 32-year (1981–2012) rainfall of four months (June–September) and monsoon season (JJAS) for 16 stations in Saraswati River basin of Gujarat, India. Temporal homogeneity was examined by Hartley, Link-Wallace, Bartlett, and Tukey tests, and spatial homogeneity by Levene's and Tukey tests. Results revealed temporal non-homogeneity at Paswadal (June, September, and JJAS), Navawas (August and JJAS), Palanpur (JJAS), and Pilucha (September) stations. Whereas, Levene's test results indicated spatial homogeneity in July, August, September, and JJAS; and non-homogeneity in June.

1.2 Rationale of the study

An extensive literature review shows that studies examining rainfall variability are mostly reported from moderate to high rainfall regions. However, such studies are rare for dry regions (Machiwal et al., 2019) despite the fact that spatial–temporal rainfall analyses are vital for arid regions (Machiwal et al., 2016). It is further revealed that rainfall variability in previous studies is mainly understood by identifying gradual changes (trends) and abrupt changes. However, other properties, for example, homogeneity, stationarity, persistence, etc. are usually not considered and/or ignored while determining rainfall variability (Machiwal and Jha, 2017). This chapter includes a case study of hot arid region of India where homogeneity and trends are tested in seasonal and annual rainfall using multiple statistical tests. It briefly describes methodology of statistical tests employed in case study. Furthermore, limitations of study and future challenges in determining time series characteristics are highlighted. Finally, recommendations are made for future studies and concluding remarks are provided. Furthermore, this study adopts a variance-correction procedure for modifying original M-K test in case significant serial correlation is observed in rainfall data. Also, rainfall homogeneity is investigated, which is tested only in a few literature studies. There are rare studies reported where rainfall homogeneity is tested in Thar Desert region of India. The outcomes of modified M-K test are compared with that obtained from ITA method in identifying significant rainfall trends. This study attempts dealing with diverse results obtained from application of different tests and arriving at a reliable decision by choosing adequate outcome using appropriate logics and reasons.

2 Materials and methods

2.1 Study area and data description

This study is performed in hot arid region of western Rajasthan in India that covers a geographical area of 19.6×10^6 ha (Moharana et al., 2016). The study area is located at latitudes varying from 24°37′00″ to 30°10′48″N and longitude from 69°29′00″ to 76°05′33″E. The study area includes 12 districts of western Rajasthan, i.e., Barmer, Bikaner, Churu, Hanumangarh, Jaisalmer, Jalor, Jhunjhunu, Jodhpur, Nagaur, Pali, Sikar, and Sriganganagar (Fig. 1). The mean annual rainfall is 291 mm for 1901–2010 period and ranges from 150 to 500 mm (Moharana et al., 2016) with relatively large values of coefficient of variation (CV) (35.5% to 105.2%). In summers, air temperature remains high and rises sometimes up to 50°C. Whereas, in winters, temperature may drop up to or below 0°C (Moharana et al., 2016).

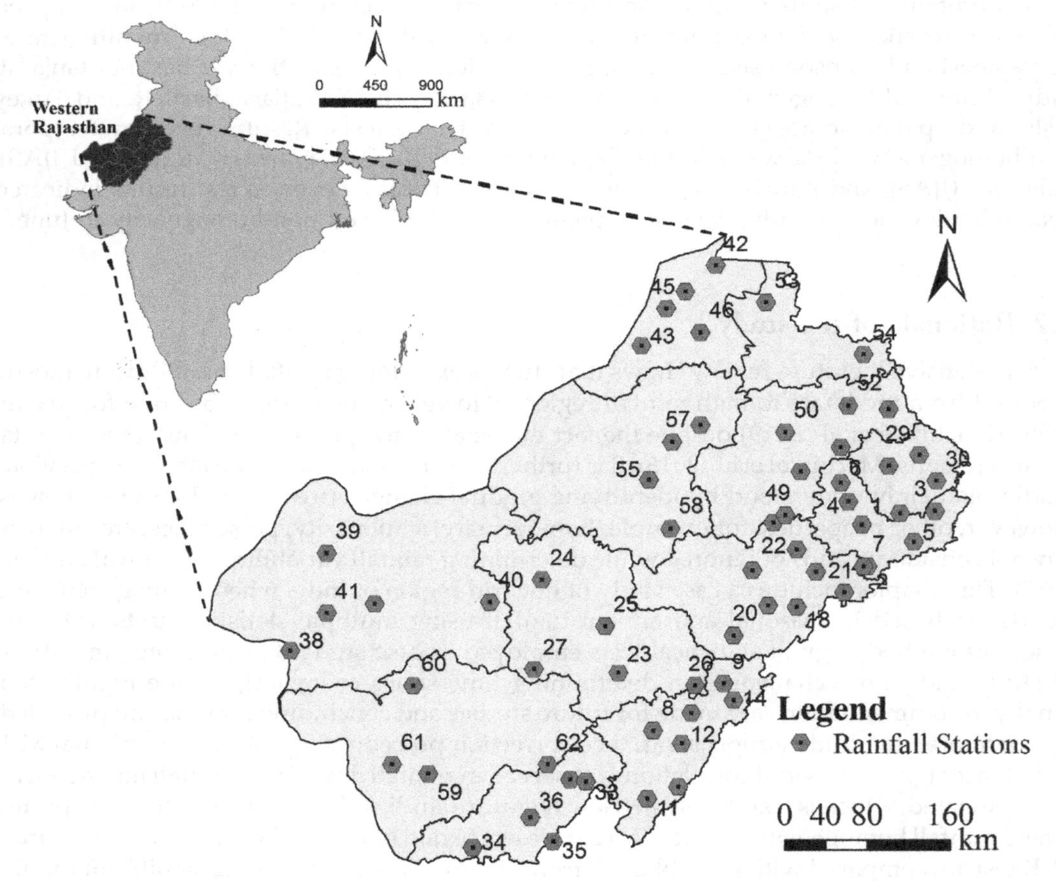

FIG. 1 Map of western Rajasthan with geographical locations of 62 rainfall stations.

In this study, daily rainfall data of 55 years (1957–2011) for 62 stations, collected from Department of Water Resources, Government of Rajasthan, India, were used. Daily data were totaled over years and four periods: (a) monsoon (JJAS), (b) post-monsoon (OND), (c) winter (JF), and (d) pre-monsoon (MAM).

2.2 Basic statistical properties

Basic statistics including average, standard deviation, and CV of annual and seasonal rainfall were computed. Furthermore, annual rainfall data were utilized to draw box-whisker plots across years as well as stations.

2.3 Statistical tests for identifying trend and estimating trend magnitude

In this study, two widely used M-K and ITA tests were employed for identifying rainfall trends. The M-K test has proven its robustness and popularity (Şen, 2017). In the beginning of the 2000s, some studies reported that M-K test may not be trustworthy under presence of serial correlation in data (Yue and Wang, 2002), which led to modification of M-K test-statistic using variance-correction approaches (Machiwal et al., 2019). In this study, original M-K test-statistic modified through variance-correction approach by Lettenmaier (1976) was considered. The ITA test is reported to have no influence if serial correlation remains present in data.

2.4 Statistical tests for examining homogeneity

The rainfall homogeneity (seasonal and annual) was examined by employing four multiple comparison tests, i.e., Tukey, Link-Wallace, Bartlett, and Hartley tests. The Link-Wallace test has applicability only when number of data remains same in all data series. This limitation is overcome by not considering first data of data series before applying test. Details of multiple comparison tests can be found in Machiwal and Jha (2012).

3 Results and discussion

3.1 Rainfall statistics

Range of average, standard deviation, and CV for seasonal and annual rainfall is presented in Table 1. During monsoon season, ranges of average (115–583 mm) and standard deviation

TABLE 1 Basic statistics of seasonal and annual rainfall.

Statistical properties	Monsoon	Post-monsoon	Winter	Pre-monsoon	Annual
Mean (mm)	115–583	4–26	3–20	9–43	135–627
Standard Deviation (mm)	81–283	11–52	6–24	16–44	89–290
Coefficient of Variation (%)	38–121	120–293	99–428	89–255	36–105

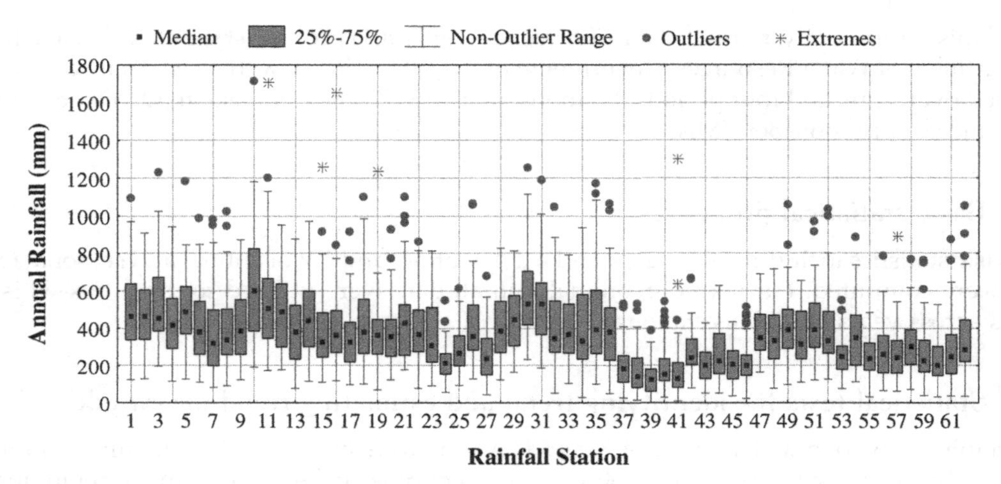

FIG. 2 Box-whisker pots for annual rainfall of 62 stations based on 55-year (1957–2011) data.

(81–283 mm) of rainfall are seen as highest, followed by average (9–43 mm) and standard deviation (16–44 mm) ranges for pre-monsoon (Table 1). However, rainfall variability (CV > 100%) is found large at all stations during post-monsoon and winter seasons having less rainfall. It is inferred that season receiving higher rainfall show a large standard deviation but less rainfall variability and vice versa. Range for average and standard deviation of annual rainfall is from 135 to 627 mm and from 89 to 290 mm, respectively with CV values from 36% to 105%.

Box-whisker plots for annual rainfall are shown in Fig. 2. In arid region, rainfall remains low, and hence, there are little chances that outlier or extreme may occur toward the lower whisker. It is worth mentioning that any data value lying outside region bounded between [upper box value + {outlier coefficient × box length}] and [lower box value − {outlier coefficient × box length}], is considered as an outlier or extreme where box length is computed by subtracting the lower value of box from the upper value. Value of outlier coefficient is considered as 1.5 in case of an outlier and 3 in case of an extreme. Box-whisker plots depict that median of annual rainfall is situated in lower half of box at 25 stations (40%). For most of stations, whisker length on the upper side is longer than that on the lower side, which confirms densely-populated rainfall data over a short range. For similar reasons, all outliers/extremes are lying beyond the upper whisker, which highlights rare occurrence of extreme rainfall in the area but such events are experienced increasingly after the year 2000. Box-whisker plots of annual rainfall were drawn year-wise (Fig. 3). It is apparently visible that median rainfall is elevated up to more than 500 mm during in 9 years (1970, 1975, 1976, 1977, 1983, 1997, 2010, and 2011). Also, outliers/extremes are located beyond upper whisker having a larger length. The unequal lengths of upper and lower whiskers and non-central position of medians in box suggest that annual rainfall is mostly distributed positively-skewed.

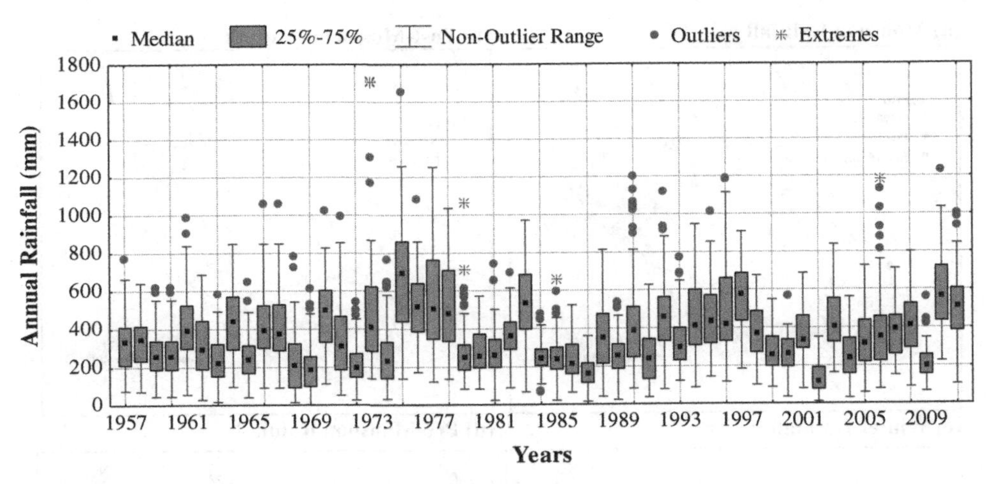

FIG. 3 Box-whisker pots of annual rainfall for 55 years (1957–2011) based on 62 stations.

3.2 Spatial distribution and comparison of rainfall trends

Results of modified M-K and ITA tests are depicted in Figs. 4 and 5, respectively.

According to modified M-K test, rainfall trends are mostly negative in post-monsoon, winter and pre-monsoon when rainfall remains low (Fig. 4). However, modified M-K trends are mostly positive in monsoon and annual rainfall. Interpretation of modified M-K test's results depicts positive (increasing) trends at 56 sites (90% stations) in monsoon season and trends at 4 sites were statistically significant ($P < .05$). Whereas, six sites showed negative (decreasing) and non-significant trends ($P > .05$) in monsoon rainfall. In post-monsoon, declining rainfall trends are found at 48 sites (77% stations) and significantly decreasing trend ($P < .05$) was found at one site. The increasing trends in post-monsoon rainfall are depicted at 14 sites (23% stations) but all are found statistically insignificant ($P > .05$). In winter, decreasing rainfall trends were revealed at 37 sites (60% stations) and increasing trends at 25 sites (40% stations), although all were non-significant ($P > .05$). In pre-monsoon, increasing and decreasing trends are identified at 30 sites (48% stations) and 32 sites (52% stations), respectively. Of them, increasing trend is found significant ($P < .05$) at one site and all decreasing trends are found non-significant ($P > .05$). In annual rainfall, increasing trends are prominent at 54 sites (87% stations) and negative trends at 8 sites (13% stations). Of the increasing annual trends, 4 are found statistically significant; however, decreasing trends are found non-significant ($P > .05$).

It is seen from Fig. 5 that almost all annual rainfall trends, both inclining and declining, detected by ITA test are significant ($P < .05$). On comparing findings of Sen's ITA method with results of modified M-K test, it is apparent that the former is over-sensitive in trend detection. Due to contrast in results of two tests, this study favored outcomes of modified M-K test over ITA test.

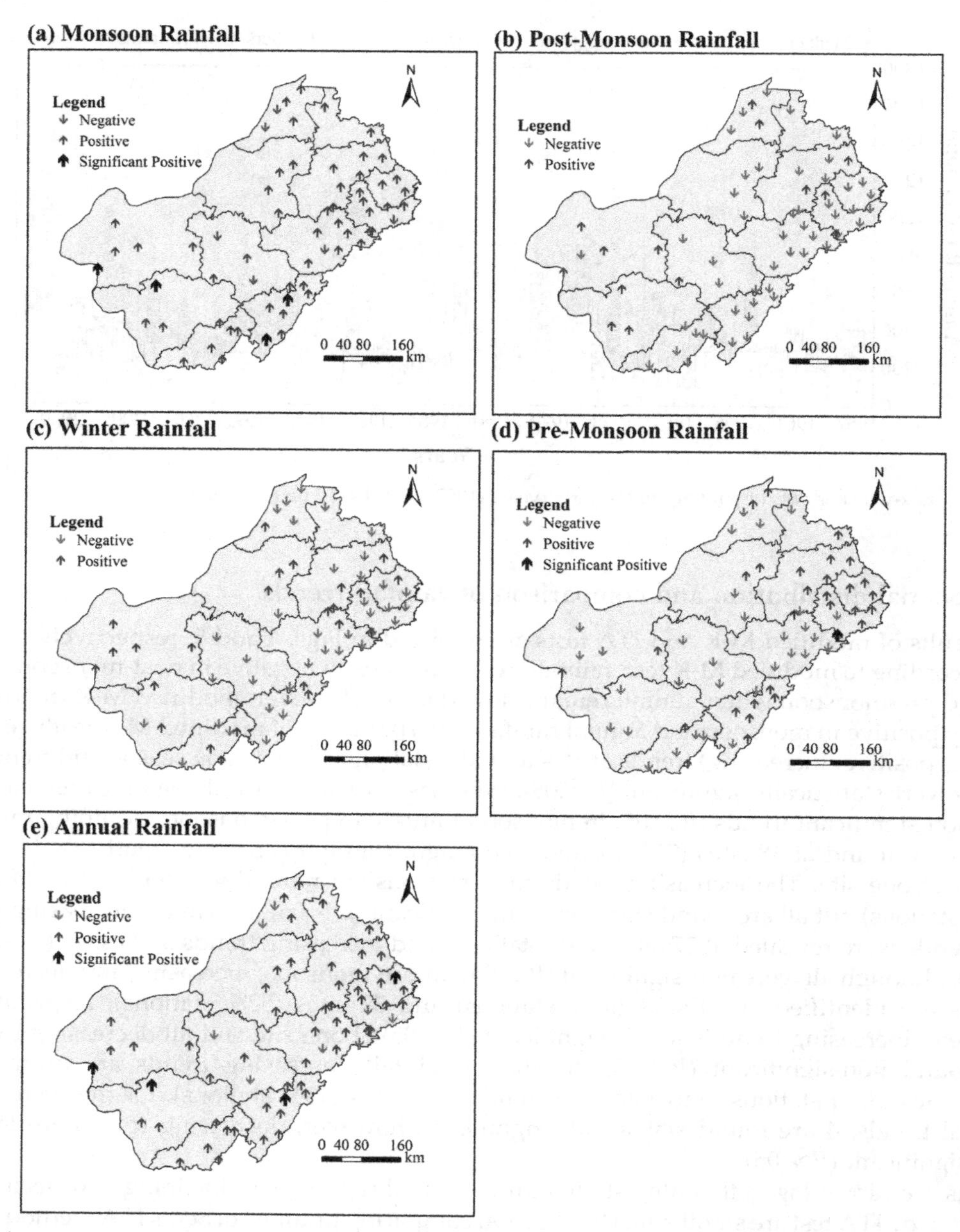

FIG. 4 Spatial distribution of modified Mann-Kendall trends in (A) monsoon, (B) post-monsoon, (C) winter, (D) pre-monsoon, and (E) annual rainfalls over western arid Rajasthan, India.

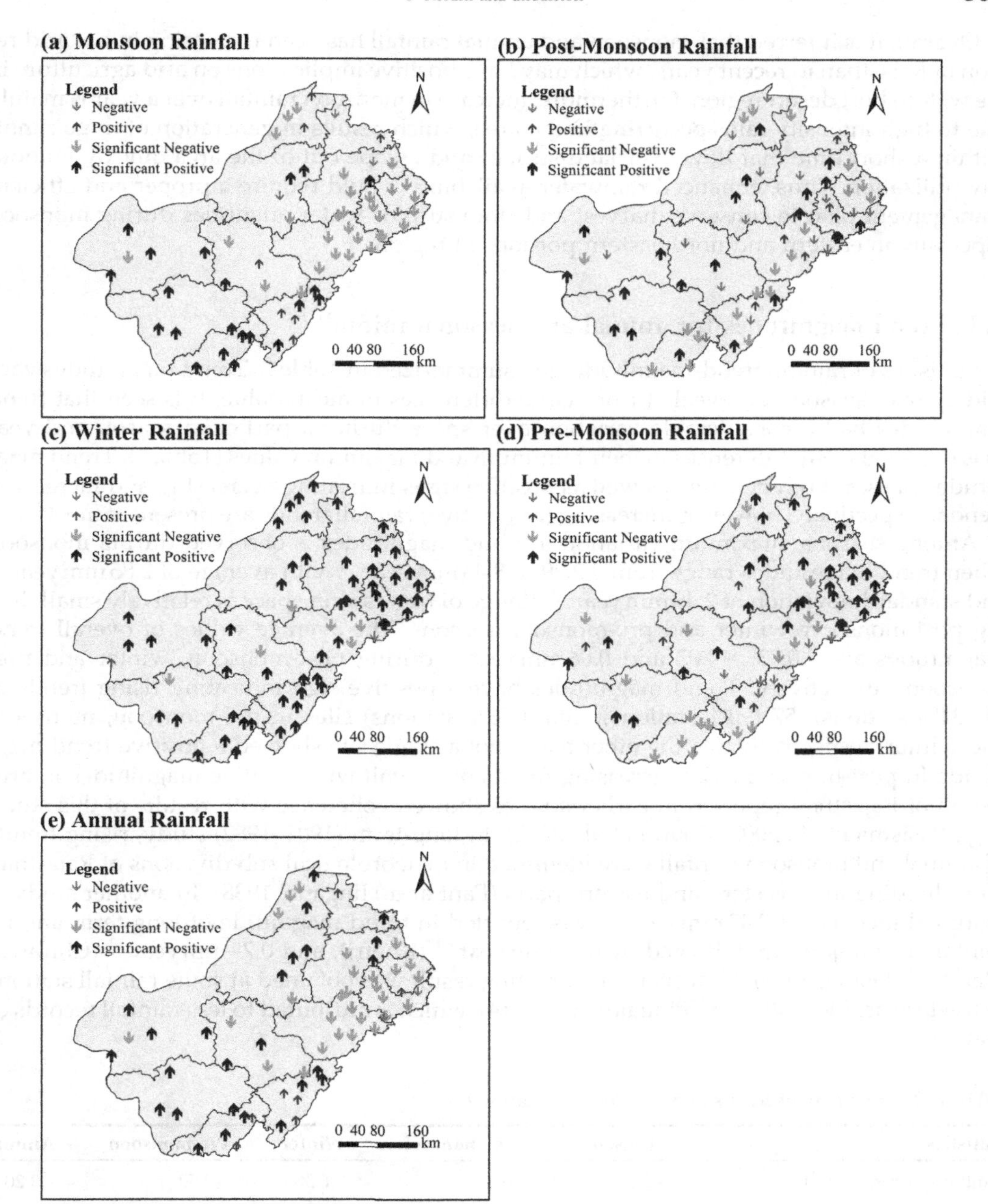

FIG. 5 Spatial distribution of Sen's ITA trends in (A) monsoon, (B) post-monsoon, (C) winter, (D) pre-monsoon, and (E) annual rainfalls over western arid Rajasthan, India.

Overall, it is inferred that monsoon and annual rainfall has been increasing in hot arid region of Rajasthan in recent years, which may have positive implications on arid agriculture in the water-short desert region. Furthermore, increase in monsoon rainfall over a year is mainly due to high-intensity rains occurring in the area, which results in generation of huge runoff within a short time that flow over land surface and escape out of the area quickly without any utilization. Thus, enhanced rainwater availability would require a proper and efficient management plan to conserve, harvest and store surplus water quantities during monsoon especially in eastern and northeastern portions of region.

3.3 Trend magnitudes for annual and seasonal rainfall

Statistics of rainfall trend-magnitudes are summarized in Table 2. Trend magnitudes vary widely over seasons as revealed from large differences in mean value. It is seen that trend magnitudes had a considerable variation over space during a particular season and year as is depicted from differences in their minimum and maximum values (Table 2). Trend magnitudes possess both negative as well as positive signs indicating decreasing and increasing trends, respectively; however, increasing (or positive) rainfall trends are present at most sites.

Among seasons, maximum variation in trend magnitudes is observed during monsoon when trend magnitudes range from −2.29 to 8.43 mm $year^{-1}$ with average of 2.85 mm $year^{-1}$ and standard deviation of 2.49 mm $year^{-1}$. Range of trends over space is relatively small during post-monsoon, winter and pre-monsoon seasons. The average values of overall trend magnitudes are −0.03, −0.03 and 0.06 mm $year^{-1}$ during post-monsoon, winter and pre-monsoon, respectively. Trend magnitudes have a positive sign indicating rising trends at 21 (34% stations), 57 (92% stations), and 4 (6% stations) sites in pre-monsoon, monsoon, and winter, respectively. On the other hand, not a single site showed a positive trend magnitude in post-monsoon. The increasing trends of rainfall with positive magnitudes in arid region of Rajasthan reported in earlier studies showed coherence with results of this study (e.g., Basistha et al., 2007; Kharol et al., 2013). In long-term (1901–1982) study, rising trends of annual and monsoon rainfall were identified in meteorological sub-divisions of Rajasthan after dividing into western and eastern parts (Pant and Hingane, 1988). In another study, a marginal increase of 0.43 mm $year^{-1}$ was reported in trend magnitude of long-term annual rainfall at Jhunjhunun, followed by 0.48 mm $year^{-1}$ (Khetri), and 0.74 mm $year^{-1}$ (Chirawa) (Rao and Poonia, 2013). However, a contrasting result was obtained at some rainfall stations of western arid Rajasthan by Pingale et al. (2014), which is attributed to less rainfall records of that study.

TABLE 2 Summary of results of Sen's slope estimation test.

Statistics	Monsoon	Post-monsoon	Winter	Pre-monsoon	Annual
Minimum (mm $year^{-1}$)	−2.29	−0.43	−0.38	−0.50	−3.20
Maximum (mm $year^{-1}$)	8.43	0	0.17	1.7	10.02
Mean (mm $year^{-1}$)	2.85	−0.03	−0.03	0.06	2.89
Standard Deviation (mm $year^{-1}$)	2.49	0.09	0.08	0.32	2.90

Trend magnitudes in annual rainfall range from −3.2 to 10.02 mm year^{-1} (Table 2). Moreover, positive trend magnitudes are revealed at 54 sites (87% stations) and negative at 8 sites (13% stations). In a recent study, rising rainfall trends are reported for Jodhpur and Jaipur districts based on 1901–2011 data using M-K test (Meena et al., 2015). It is evident that rainfall trends are overall increasing in western Rajasthan. Among probable causes of rising rainfall trends in hot arid region of India are increasing sea-surface temperatures (Goswami et al., 2006), increased atmospheric moisture (Lacombe and McCartney, 2014), and surface latent heat flux over tropical Indian Ocean (Rajeevan et al., 2008), among others.

3.4 Spatial distribution and comparison of homogeneity results

Homogeneity tests' results are presented in Figs. 6–9. In Hartley test results, it is apparent that 39 sites (63% stations) have rainfall homogeneity and 23 sites (37% stations) showed non-homogeneity in monsoon. In post-monsoon, homogenous rainfall is present at 29 sites (47% stations) and non-homogeneous at 33 sites (53% stations). However, homogeneous and non-homogeneous rainfall is depicted at 30 (48% stations) and 32 (52% stations) sites in winter, respectively and at 31 (50% stations) and 31 (50% stations) sites in pre-monsoon, respectively. The annual rainfall is homogeneous at 39 sites (63% stations) and non-homogeneous at 23 sites (37% stations). Results of Link-Wallace test indicated homogeneity and non-homogeneity at 56 (90%) and 6 (10%) sites, respectively in monsoon, 60 (97%) and 2 (3%) sites in post-monsoon, 60 (97%) and 2 (3%) sites in winter, 60 (97%) and 2 (3%) sites in pre-monsoon and 58 (94% stations) and 4 (6% stations) sites in annual rainfall, respectively (Fig. 7). According to Bartlett test (Fig. 8), sites showing rainfall homogeneity are 49 (79%), 36 (58%), 40 (65%), 42 (19%), and 53 (85%) and non-homogeneity are 13 (21%), 26 (42%), 22 (35%), 20 (32%), and 9 (15%) in monsoon, post-monsoon, winter, pre-monsoon and annual rainfall, respectively. Likewise, homogeneous rainfall is found at 58 (94%), 61 (98%), 61 (98%), 62 (100%), and 58 (94%) sites and non-homogeneous rainfall is obtained at 4 (6%), 1 (2%), 1 (2%), 0 (0%), and 4 (6%) sites in monsoon, post-monsoon, winter, pre-monsoon and annual rainfall by Tukey test (Fig. 9). The Hartley and Bartlett test results indicated relatively large number of sites showing non-homogeneity in rainfall of post-monsoon. On the other hand, Tukey and Link-Wallace tests identified more non-homogeneous sites in monsoon.

Sites showing homogeneous and non-homogeneous rainfall are not similar for Hartley and Bartlett tests, whereas similar results of Tukey and Link-Wallace tests are visible over seasons. It is seen that sites showing non-homogeneous rainfall are more in numbers for Hartley and Bartlett tests compared to other two tests over seasons. Thus, Bartlett and Hartley tests seem oversensitive to deviation of rainfall from normal distribution, and if so, it may have affected outcome of tests. Therefore, Tukey and Link-Wallace tests are preferred in this study due to consistency and limitations of Hartley and Bartlett tests. It is stated that Link-Wallace and Tukey tests may perform better than Hartley and Bartlett tests in homogeneity testing of rainfall records. Overall, it is suggested that seasonal and annual rainfall in hot arid region of India are mostly homogeneous and considerable non-homogeneity in rainfall shown at some sites may be attributed to rising rainfall trends in eastern and northeast parts of the area.

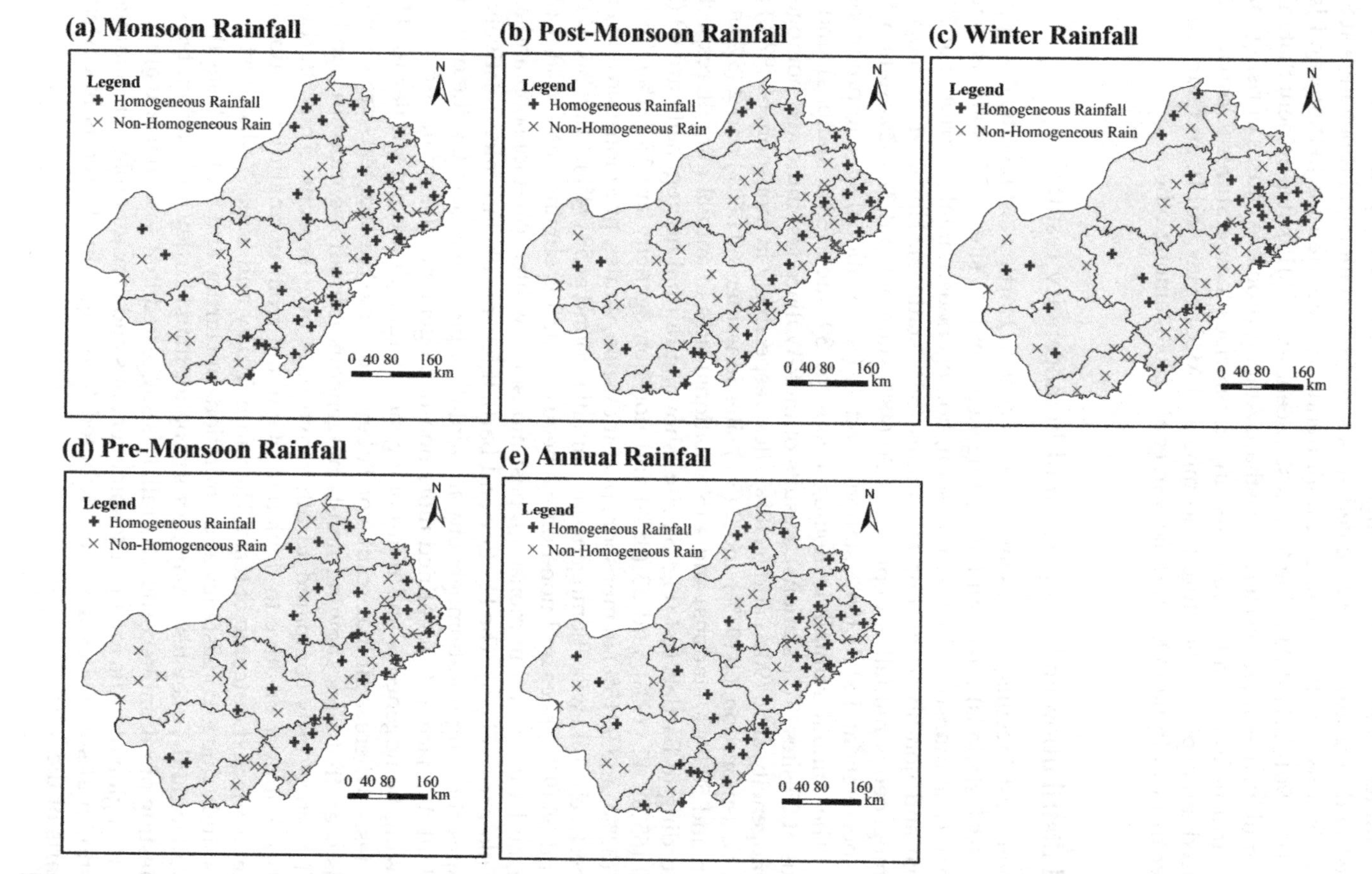

FIG. 6 Sites showing homogenous/non-homogeneous rainfall based on Hartley test during (A) monsoon, (B) post-monsoon, (C) winter, (D) pre-monsoon, and (E) annual period.

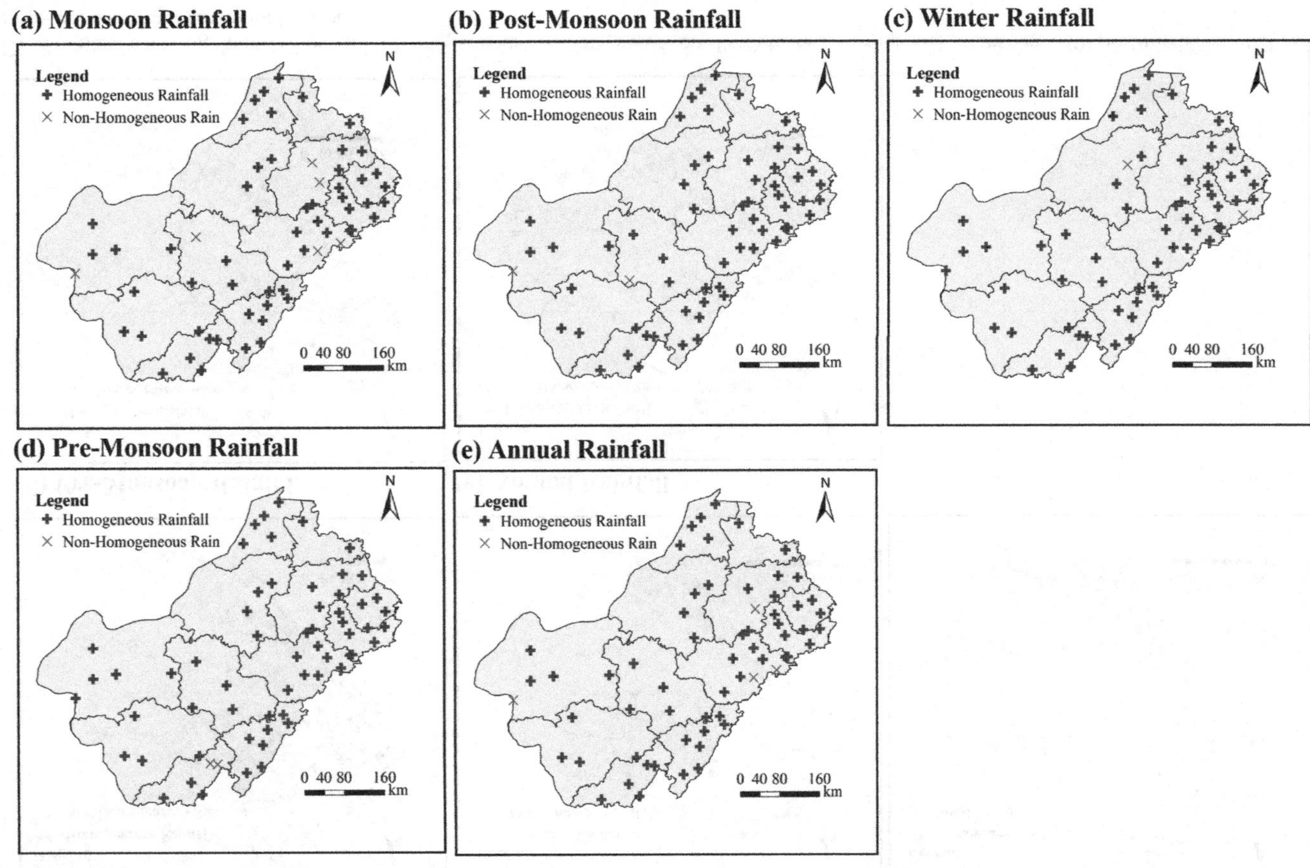

FIG. 7 Sites showing homogenous/non-homogeneous rainfall based on Link-Wallace test during (A) monsoon, (B) post-monsoon, (C) winter, (D) pre-monsoon, and (E) annual period.

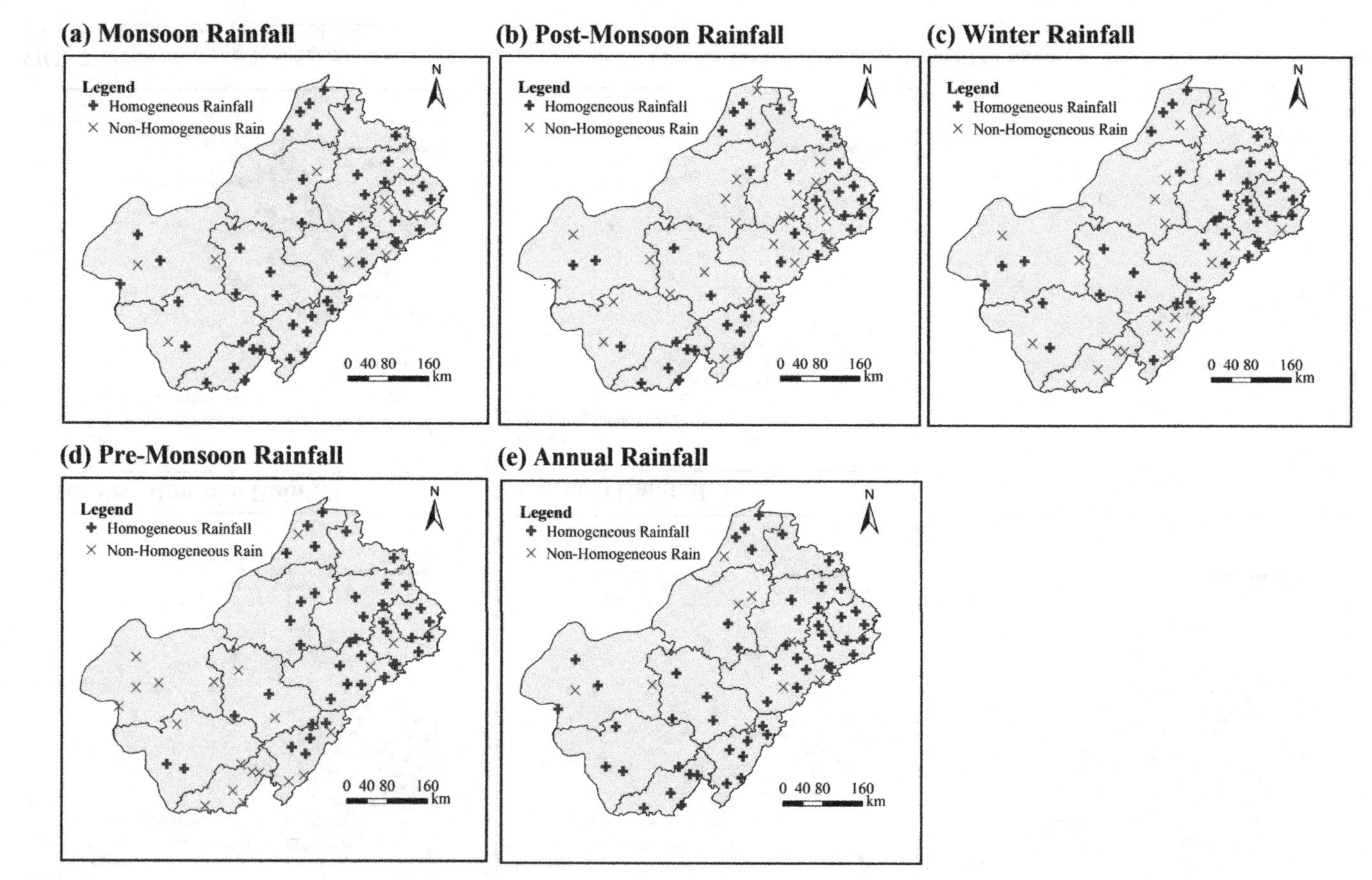

FIG. 8 Sites showing homogenous/non-homogeneous rainfall based on Bartlett test during (A) monsoon, (B) post-monsoon, (C) winter, (D) pre-monsoon, and (E) annual period.

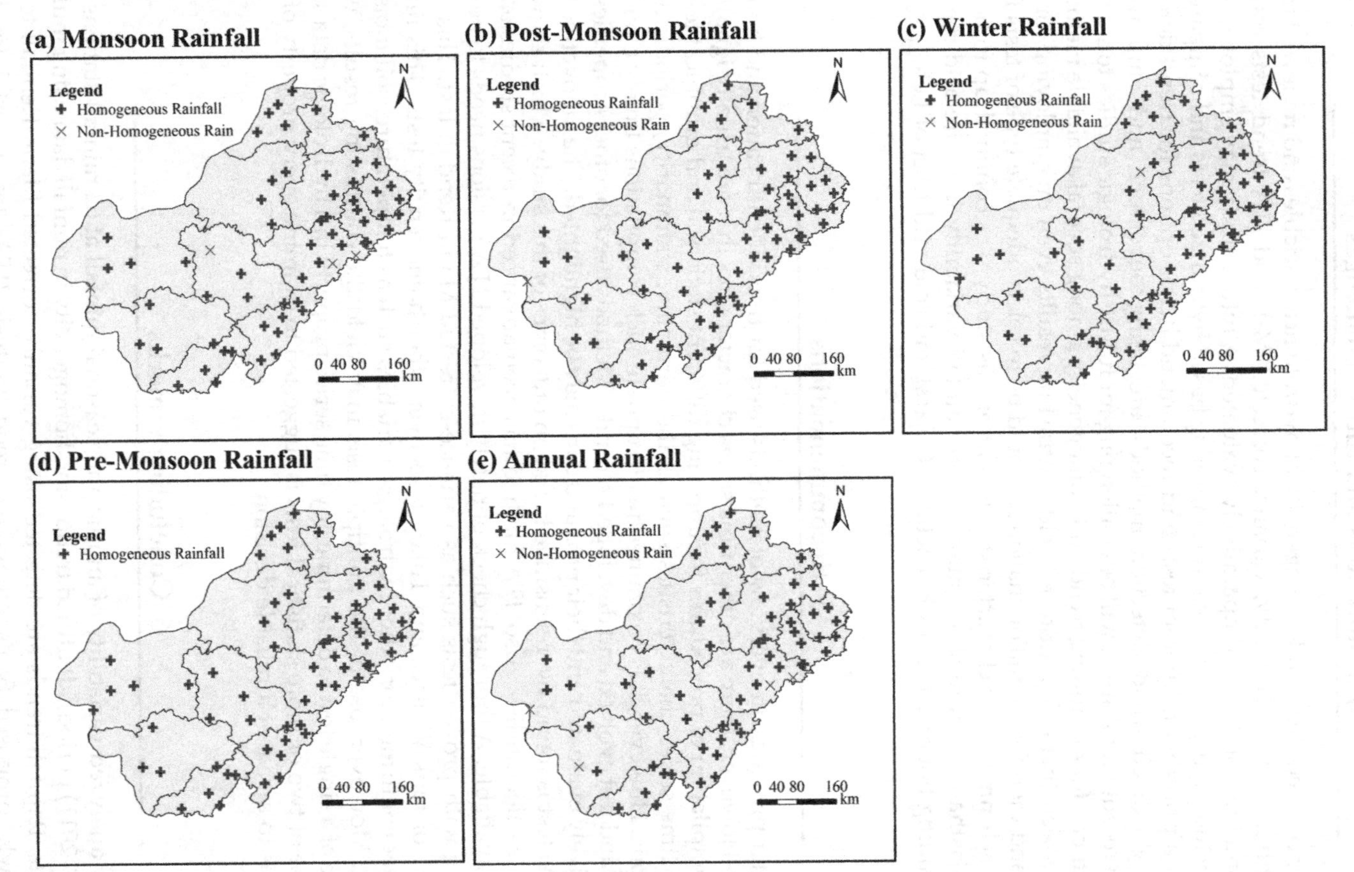

FIG. 9 Sites showing homogenous/non-homogeneous rainfall based on Tukey test during (A) monsoon, (B) post-monsoon, (C) winter, (D) pre-monsoon, and (E) annual period.

4 Limitations and future challenges

This study covered trend and homogeneity of hydrological variables and reviewed the relevant literature. In the case study, variance-corrected M-K and ITA trend tests could be employed, and the results are comparatively evaluated. A future study may employ multiple trend tests including recently proposed resampling based hybridized tests and compare their results to set a priority of different tests and recommend better options for trend detection. Currently, hybridized trend detection methods are being developed and employed as appearing in recent literature. It will be challenging for future hydrologic studies to standardize some methods for examining trend and homogeneity among several available methods. It is apparently seen that different tests yield contrasting results even on employing for same time series, and therefore, a major challenge would be to select adequate type of tests for investigating each time series characteristic. Extent of this study was confined up to hot arid region of Rajasthan; a future study may explore rainfall variability in entire arid region of country including hot arid part of Kachchh in Gujarat and cold arid tract of Leh.

5 Recommendations

This chapter provides an overview of methods used for trend identification and homogeneity testing. However, there are other time series characteristics that are usually ignored. In literature, compilation of procedures for applying different statistical and graphical tests for determining time series characteristics such as the normality, stationarity, periodicity, etc. have been done. However, it is recommended to update such compilations timely by including methods and test evolved and developed recently. Methods used for other properties may be comparatively evaluated and their pros and cons may be highlighted. The innovative trend analysis (ITA) test is found over-sensitive based on outcomes of this study, which shows coherence with earlier studies. Also, ITA method has been advanced in recent studies as new versions are available. A thorough check of newly developed ITA versions needs to be done by comparing with proven tests such as variance-corrected Mann-Kendall test. This study advocates use of Link-Wallace and Tukey tests over other homogeneity tests. This finding may be further confirmed for other hydrologic variables such as temperature, soil moisture, recharge, etc. Moreover, use of the multiple tests may yield the contrasting results, which questions use of a single test to determine any time series property in a hydrologic data series. Therefore, use of two or more statistical tests is suggested to examine every property of time series analysis to arrive at a reliable decision.

6 Concluding remarks

This study analyzed variability of annual and seasonal rainfall at 62 rainfall stations over 55-year (1957–2011) period, identified trends and homogeneity in rainfall data using multiple statistical tests. Rainfall trends are detected using variance-corrected Mann-Kendall test. In addition, newly proposed Sen's innovative trend analysis (ITA) test is used for trend identification. Further, trend magnitudes are quantified using Sen's slope test. Rainfall homogeneity is tested using four multiple comparison tests. Rainfall is found largely variable over time in monsoon ranging from 115 to 583 mm, and largest variability in post-monsoon and winter seasons with more than 100% coefficient of variation. Outliers and extremes in box-whisker plots revealed extreme rainfall occurrences during the 1970s and 2006–2011. Due to the over-sensitivity of ITA test, the outcome of M-K test after variance correction is preferred. The average values of Sen's slope or trend magnitudes are 2.85 and 2.89 mm $year^{-1}$ for monsoon and annual rainfall, respectively, which are considerably high. However, Link-Wallace and Tukey tests showed a fair agreement with each other but deviated largely from findings of Hartley and Bartlett tests. Hence, it is recommended to prefer Link-Wallace and Tukey tests over Hartley and Bartlett tests for examining homogeneity. Overall, it is depicted that seasonal and annual rainfall are homogeneous over most sites in the area and non-homogeneity at some sites is due to increasing rainfall trends. Findings of this study reveal that rainwater quantities received in this traditionally water-short region of arid climate has been rising for last few years. Thus, it would pose a challenge to planners, water managers and decision makers in formulating adequate policies and strategies for managing surplus amount of rainwater efficiently in hot arid region of India.

Acknowledgments

This study is conducted with the necessary facilities and all kind of support provided by the Director, ICAR-Central Arid Zone Research Institute (CAZRI), Jodhpur, India, which is gratefully acknowledged by the authors.

References

Almazroui, M., Şen, Z., 2020. Trend analyses methodologies in hydro-meteorological records. Earth Syst. Environ. 4, 713–738.

Asfaw, A., Simane, B., Hassen, A., Bantidar, A., 2018. Variability and time series trend analysis of rainfall and temperature in northcentral Ethiopia: a case study in Woleka sub-basin. Weather. Clim. Extremes 19, 29–41.

Basistha, A., Goel, N.K., Arya, D.S., Gangwar, S.K., 2007. Spatial pattern of trends in Indian sub-divisional rainfall. Jalvigyan Sameeksha 22, 47–57.

Buishand, T.A., 1982. Some methods for testing the homogeneity of rainfall records. J. Hydrol. 58, 11–27.

Burn, D.H., Elnur, M.A.H., 2002. Detection of hydrologic trends and variability. J. Hydrol. 255 (1–4), 107–122.

Ceribasi, G., Ceyhunlu, A.I., 2021. Analysis of total monthly precipitation of Susurluk Basin in Turkey using innovative polygon trend analysis method. J. Water Clim. Chang. 12 (5), 1532–1543.

Chahal, M., Bhardwaj, P., Singh, O., 2021. Exploring the trends and pattern of rainfall extremes over the semiarid Sahibi basin in Rajasthan, India. Arab. J. Geosci. 14, 966. https://doi.org/10.1007/s12517-021-07320-y.

Chargui, S., Zarrour, R., El Mouaddeb, R., Khelifa, W.B., 2022. Recent trends and variability of extreme rainfall indices over Lebna basin and neighborhood in the last 40 years. Arab. J. Geosci. 15, 203. https://doi.org/10.1007/s12517-021-09334-y.

Conover, W.J., 1971. Practical Non-Parametric Statistics. Wiley, New York.

Daba, M.H., Ayele, G.T., You, S., 2020. Long-term homogeneity and trends of hydroclimatic variables in Upper Awash River Basin, Ethiopia. Adv. Meteorol., 8861959. https://doi.org/10.1155/2020/8861959.

Douglas, E.M., Vogel, R.M., Kroll, C.N., 2000. Trends in floods and low flows in the United States: impact of spatial correlation. J. Hydrol. 240 (1–2), 90–105.

Esterby, S.R., 1996. Review of methods for the detection and estimation of trends with emphasis on water quality applications. Hydrol. Process. 10 (2), 127–149.

Fanta, B., Zaake, B.T., Kachroo, R.K., 2001. A study of variability of annual river flow of the southern African region. Hydrol. Sci. J. 46 (4), 513–524.

Fernando, D.A.K., Jayawardena, A.W., 1994. Generation and forecasting of monsoon rainfall data. In: 20th WEDC Conference on Affordable Water Supply and Sanitation, Colombo, Sri Lanka, pp. 310–313.

Gan, T.Y., 1998. Hydroclimatic trends and possible climatic warming in the Canadian Prairies. Water Resour. Res. 34 (11), 3009–3015.

Gilbert, R.O., 1987. Statistical Methods for Environmental Pollution Monitoring. Van Nostrand Reinhold, New York.

Goswami, B.N., Venugopal, V., Sengupta, D., Madhusoodanan, M.S., Xavier, P.K., 2006. Increasing trend of extreme rain events over India in a warming environment. Science 314, 1442–1445.

Güçlü, Y.S., 2018. Multiple Şen-innovative trend analyses and partial Mann-Kendall test. J. Hydrol. 566, 685–704.

Güçlü, Y.S., Şişman, E., Dabanlı, I., 2020. Innovative triangular trend analysis. Arab. J. Geosci. 13, 27. https://doi.org/10.1007/s12517-019-5048-y.

Guerreiro, S.B., Kilsby, C.G., Serinaldi, F., 2014. Analysis of time variation of rainfall in transnational basins in Iberia: abrupt changes or trends? Int. J. Climatol. 34, 114–133.

Gupta, A., Kamble, T., Machiwal, D., 2017. Comparison of ordinary and Bayesian kriging techniques in depicting rainfall variability in arid and semi-arid regions of Northwest India. Environ. Earth Sci. 76, 512. https://doi.org/10.1007/s12665-017-6814-3.

Haan, C.T., 2002. Statistical Methods in Hydrology, second ed. Iowa State University Press, Ames. 496p.

Hameed, T., Marino, M.A., DeVries, J.J., Tracy, J.C., 1997. Method for trend detection in climatological variables. J. Hydrol. Eng. ASCE 2 (4), 157–160.

Hamilton, J.P., Whitelaw, G.S., Fenech, A., 2001. Mean annual temperature and annual precipitation trends at Canadian biosphere reserves. Environ. Monit. Assess. 67, 239–275.

Hänsel, S., Medeiros, D.M., Matschullat, J., Petta, R.A., de Mendonça Silva, I., 2016. Assessing homogeneity and climate variability of temperature and precipitation series in the capitals of North-Eastern Brazil. Front. Earth Sci. 4, 29. https://doi.org/10.3389/feart.2016.00029.

Helsel, R.D., Hirsch, R.M., 1988. Discussion of "Applicability of the t-test for detecting trends in water quality variables" by Robert H. Montgomery and Jim C. Loftis. Water Resour. Bull. 24 (1), 201–204.

Hess, A., Iyer, H., Malm, W., 2001. Linear trend analysis: a comparison of methods. Atmos. Environ. 35 (30), 5211–5222.

Hoel, P.G., 1954. Introduction to Mathematical Statistics, second ed. John Wiley and Sons, Inc, New York.

Jain, S.K., Kumar, V., 2012. Trend analysis of rainfall and temperature data for India. Curr. Sci. 102 (1), 37–49.

Jain, S.K., Kumar, V., Saharia, M., 2013. Analysis of rainfall and temperature trends in Northeast India. Int. J. Climatol. 33 (4), 968–978.

Kendall, M.G., 1973. Time Series. Charles Griffin and Co. Ltd, London, U.K.

Kendall, M.G., 1975. Rank Correlation Methods. Charles Griffin and Co. Ltd, London, U.K.

Khaliq, M.N., Ouarda, T.B.M.J., Gachon, P., Sushama, L., St-Hilaire, A., 2009. Identification of hydrological trends in the presence of serial and cross correlations: a review of selected methods and their application to annual flow regimes of Canadian rivers. J. Hydrol. 368, 117–130.

Khaliq, M.N., Ouarda, T.B.M.J., Ondo, J.-C., Gachon, P., Bobée, B., 2006. Frequency analysis of a sequence of dependent and/or non-stationary hydro-meteorological observations: a review. J. Hydrol. 329, 534–552.

Kharol, S.K., Kaskaoutis, D.G., Badarinath, K.V.S., Sharma, A.R., Singh, R.P., 2013. Influence of land use/land cover (LULC) changes on atmospheric dynamics over the arid region of Rajasthan state, India. J. Arid Environ. 88, 90–101.

Kocsis, T., Kovács-Székely, I., Anda, A., 2020. Homogeneity tests and non-parametric analyses of tendencies in precipitation time series in Keszthely, Western Hungary. Theor. Appl. Climatol. 139, 849–859.

Kumar, V., Jain, S.K., Singh, Y., 2010. Analysis of long-term rainfall trends in India. Hydrol. Sci. J. 55 (4), 484–496.

Kundzewicz, Z.W., Robson, A.J., 2004. Change detection in hydrological records—a review of the methodology. Hydrol. Sci. J. 49 (1), 7–19.

Lacombe, G., McCartney, M., 2014. Uncovering consistencies in Indian rainfall trends observed over the last half century. Clim. Change 123 (2), 287–299.

Lettenmaier, D.P., 1976. Detection of trend in water quality data from record with dependent observations. Water Resour. Res. 12 (5), 1037–1046.

Lettenmaier, D.P., Wood, E.F., Wallis, J.R., 1994. Hydro-climatological trends in the continental United States, 1948-88. J. Climate 7, 586–607.

Lins, H.F., Slack, J.R., 1999. Streamflow trends in the United States. Geophys. Res. Lett. 26 (2), 227–230.

Lupikasza, E., 2010. Spatial and temporal variability of extreme precipitation in Poland in the period 1951-2006. Int. J. Climatol. 30 (7), 991–1007.
Machiwal, D., Gupta, A., Jha, M.K., Kamble, T., 2019. Analysis of trend in temperature and rainfall time series of an Indian arid region: Comparative evaluation of salient techniques. Theor. Appl. Climatol. 136 (1–2), 301–320.
Machiwal, D., Jha, M.K., 2008. Comparative evaluation of statistical tests for time series analysis: application to hydrological time series. Hydrol. Sci. J. 53 (2), 353–366.
Machiwal, D., Jha, M.K., 2012. Hydrologic Time Series Analysis: Theory and Practice. Springer, Germany and Capital Publishing Company, New Delhi. 303p.
Machiwal, D., Jha, M.K., 2017. Evaluating persistence, and identifying trends and abrupt changes in monthly and annual rainfalls of a semi-arid region in western India. Theor. Appl. Climatol. 128 (3–4), 689–708.
Machiwal, D., Kumar, S., Dayal, D., Mangalassery, S., 2016. Identifying abrupt changes and detecting gradual trends of annual rainfall in an Indian arid region under heightened rainfall rise regime. Int. J. Climatol. 37, 2719–2733.
Machiwal, D., Parmar, B.S., Kumar, S., Meena, H.M., Deora, B.S., 2021. Evaluating homogeneity of monsoon rainfall in Saraswati river basin of Gujarat, India. J. Earth Syst. Sci. 130, 181. https://doi.org/10.1007/s12040-021-01671-6.
Mahé, G., L'Hôte, Y., Olivry, J.-C., Wotling, G., 2001. Trends and discontinuities in regional rainfall of West and Central Africa: 1951-1989. Hydrol. Sci. J. 46 (2), 211–226.
Mann, H.B., 1945. Non parametric tests again trend. Econometrica 13, 245–259.
Matalas, N.C., Sankarasubramanian, A., 2003. Effect of persistence on trend detection via regression. Water Resour. Res. 39 (12), WR002292. https://doi.org/10.1029/2003WR002292.
McGhee, J.W., 1985. Introductory Statistics. West Publishing Co, New York, USA.
Meena, H.M., Machiwal, D., Santra, P., Moharana, P.C., Singh, D.V., 2019. Trends and homogeneity of monthly, seasonal, and annual rainfall over arid region of Rajasthan, India. Theor. Appl. Climatol. 136 (3–4), 795–811.
Meena, P.K., Khare, D., Shukla, R., Mishra, P.K., 2015. Long term trend analysis of mega cities in northern India using rainfall data. Indian J. Sci. Technol. 8 (3), 247–253.
Meinke, H., Devoil, P., Hammer, G.L., Power, S., Allan, R., Stone, R.C., Folland, C., Potgieter, A., 2005. Rainfall variability at decadal and longer time scales: signal or noise? J. Climate 18 (1), 89–96.
Moharana, P.C., Santra, P., Singh, D.V., Kumar, S., Goyal, R.K., Machiwal, D., Yadav, O.P., 2016. ICAR-central arid zone research institute, Jodhpur: erosion processes and desertification in the Thar Desert of India. Proc. Indian Natl. Sci. Acad. 82 (3), 1117–1140.
Mudelsee, M., 2019. Trend analysis of climate time series: a review of methods. Earth Sci. Rev. 190, 310–322.
Nair, S.C., Mirajkar, A.B., 2022. Rainfall trend anomalies over Wainganga Basin, Central India. In: Laishram, B., Tawalare, A. (Eds.), Recent Advancements in Civil Engineering, Select Proceedings of ACE 2020. Springer, Singapore, https://doi.org/10.1007/978-981-16-4396-5_78.
Nourani, V., Nezamdoost, N., Samadi, M., Daneshvar Vousoughi, F., 2015. Wavelet-based trend analysis of hydrological processes at different timescales. J. Water Clim. Chang. 6, 414–435.
Pant, G.B., Hingane, L.S., 1988. Climatic changes in and around the Rajasthan desert during the 20th century. Int. J. Climatol. 8, 391–401.
Pingale, S.M., Khare, D., Jat, M.K., Adamowski, J., 2014. Spatial and temporal trends of mean and extreme rainfall and temperature for the 33 urban centres of the arid and semi-arid state of Rajasthan, India. Atmos. Res. 138, 73–90.
Rahman, M.A., Yunsheng, L., Sultana, N., 2017. Analysis and prediction of rainfall trends over Bangladesh using Mann–Kendall, Spearman's rho tests and ARIMA model. Meteorol. Atmos. Phys. 129 (4), 409–424.
Rajeevan, M., Bhate, J., Jaswal, A.K., 2008. Analysis of variability and trends of extreme rainfall events over India using 104 years of gridded daily rainfall data. Geophys. Res. Lett. 35, L18707. https://doi.org/10.1029/2008GL035143.
Rao, A.S., Poonia, S., 2013. Characteristic features of rainfall and meteorological droughts in Jhunjhunu district of arid Rajasthan. Ann. Arid Zone 52 (1), 1–6.
Saikranthi, K., Rao, T.N., Rajeevan, M., Rao, S.V.B., 2013. Identification and validation of homogeneous rainfall zones in India using correlation analysis. J. Hydrometeorol. 14 (1), 304–317.
Saini, D., Bhardwaj, P., Singh, O., 2022. Recent rainfall variability over Rajasthan, India. Theor. Appl. Climatol. https://doi.org/10.1007/s00704-021-03904-6.
Sen, P.K., 1968. Estimates of the regression coefficient based on Kendall's tau. J. Am. Stat. Assoc. 63 (324), 1379–1389.
Şen, Z., 2012. Innovative trend analysis methodology. J. Hydrol. Eng. ASCE 17 (9), 1042–1046.
Şen, Z., 2017. Innovative trend significance test and applications. Theor. Appl. Climatol. 127 (3–4), 939–947.

Serinaldi, F., Chebana, F., Kilsby, C.G., 2020. Dissecting innovative trend analysis. Stoch. Environ. Res. Risk A. 34, 733–754.
Shahin, M., Van Oorschot, H.J.L., De Lange, S.J., 1993. Statistical Analysis in Water Resources Engineering. A. A. Balkema, Rotterdam. 394p.
Sonali, P., Kumar, D.N., 2013. Review of trend detection methods and their application to detect temperature changes in India. J. Hydrol. 476, 212–227.
Türkeş, M., Koç, T., Sariş, F., 2009. Spatiotemporal variability of precipitation total series over Turkey. Int. J. Climatol. 29 (8), 1056–1074.
von Storch, H., 1995. Misuses of statistical analysis in climate research. In: von Storch, H., Navarra, A. (Eds.), Analysis of Climate Variability: Applications of Statistical Techniques. Springer, Berlin.
Xia, F., Liu, X., Xu, J., Wang, Z., Huang, J., Brookes, P.C., 2015. Trends in the daily and extreme temperatures in the Qiantang River basin, China. Int. J. Climatol. 35 (1), 57–68.
Yue, S., Pilon, P., Phinney, B., 2003. Canadian streamflow trend detection: impacts of serial and cross-correlation. Hydrol. Sci. J. 48 (1), 51–64.
Yue, S., Pilon, P., Phinney, B., Cavadias, G., 2002. The influence of autocorrelation on the ability to detect trend in hydrological series. Hydrol. Process. 16, 1807–1829.
Yue, S., Wang, C.Y., 2002. Applicability of prewhitening to eliminate the influence of serial correlation on the Mann-Kendall test. Water Resour. Res. 38 (6), 1–7. https://doi.org/10.1029/2001WR000861.
Yue, S., Wang, C.Y., 2004. The Mann-Kendall test modified by effective sample size to detect trend in serially correlated hydrological series. Water Resour. Manag. 18, 201–218.
Zakwan, M., 2021. Trend analysis of groundwater level using innovative trend analysis. In: Pande, C.B., Moharir, K.N. (Eds.), Groundwater Resources Development and Planning in the Semi-Arid Region. Springer Nature, Switzerland, pp. 389–405.
Zakwan, M., Ahmad, Z., 2021. Trend analysis of hydrological parameters of Ganga River. Arab. J. Geosci. 14, 163. https://doi.org/10.1007/s12517-021-06453-4.
Zerouali, B., Al-Ansari, N., Chettih, M., Mohamed, M., Abda, Z., Santos, C.A.G., Zerouali, B., Elbeltagi, A., 2021. An enhanced innovative triangular trend analysis of rainfall based on a spectral approach. Water 13, 727. https://doi.org/10.3390/w13050727.
Zhang, X., Harvey, K.D., Hogg, W.D., Yuzyk, T.R., 2001. Trends in Canadian streamflow. Water Resour. Res. 37 (4), 987–998.

CHAPTER

35

Flash floods and their impact on natural life using surface water model and GIS technique at Wadi Degla natural reserve area, Egypt

Sherif A. Abu El-Magd[a], *Ahmed M. Masoud*[b], *Ahmed A. Abdel Moneim*[b], *and Bakr M. Bakr*[c]

[a]Geology Department, Faculty of Science, Suez University, Suez, Egypt [b]Geology Department, Faculty of Science, Sohag University, Sohag, Egypt [c]Egyptian Environmental Affairs Agency (EEAA), Cairo, Egypt

OUTLINE

1 Introduction 622

2 Materials and methods 623
- 2.1 *Study area* 623
- 2.2 *Geology of the area* 625
- 2.3 *Methodology* 625

3 Results and discussions 629
- 3.1 *Land-use changes* 629
- 3.2 *Surface-water model* 631
- 3.3 *Return periods* 632

4 Impact on natural life 635

5 Conclusions 635

Funding 638

Availability of data and materials 638

Ethics approval and consent to participate 638

Consent for publication 638

Competing interests 638

References 638

Current Directions in Water Scarcity Research, Volume 7
https://doi.org/10.1016/B978-0-323-91910-4.00035-2

1 Introduction

The intensity of the precipitation is unexpected due to the climate changes in the arid regions, which will increase the severity of flash floods. However, floods and associated flash floods are probably the most widespread among natural hazards. Floods are expected to occur continuously on earth where the hydrological processes are responsible for flood generation as long as the water cycle runs. Flooding has currently increased property damage and personal loss with the expansion of urbanization and the development of human activities. The same dominance of flood events is observed, striking different areas in Egypt due to insufficient flood defenses and monitoring. However, the rapid deterioration of the environment has elicited the attention of local and international agencies for the formulation of statutes and policies to protect the environment and maintain sustainability. Protected areas are becoming significant and quickly developing mindset, increasing global concern over the need to protect the environment (Langston and Ding, 2001). Rapid and the occasionally unplanned expansions of urbanization in the study site have resulted in environmental degradation.

Various areas in Egypt have been subjected to deterioration due to industrial activities and urbanization. This research focuses on one of the Egyptian protected areas located within the urban context (Wadi Degla) Protectorate (WDP), Egypt. The WDP area is a part of the urban and natural environment, which is not isolated and situated within the urban context. This natural preserve and others all over Egypt are expected to have a significant impact on the national and local levels. Various touristic and relaxation activities, including nature trails and bird watching, mountain climbing, jogging, biking, and night camping, are available in WDP. Furthermore, WDP is important to scientists and researchers for conducting their field research in environmental science, natural life, botany, geology, and zoology. Protected areas are an effective tool for addressing numerous conditions that lead to natural catastrophes, such as severe weather events and major earth movements (Dudley et al., 2015). However, various natural variables, such as hydrological and climatic features, soil types, geological structures, geomorphology, and vegetation, are the most important contributors to floods according to Youssef et al. (2009).

Numerous studies including the traditional approaches, Geographic information system (GIS) techniques, bivariate, multivariate, and machine learning algorithms have been introduced over the past few decades with other investigations to explore, analyze, and predict floods as natural hazards (El-Fakharany, 1998; Abdel Moneim, 2004; Youssef et al., 2005; Mukerji et al., 2009; Moawad, 2013; Moawad et al., 2016; Youssef and Hegab, 2019; Abu El-Magd, 2019, 2022; Abu El-Magd et al., 2020, 2021a,b; Megahed and El Bastawesy, 2020; Mirzaei et al., 2020; Hewaidy et al., 2021). Hagras et al. (2013) and Jin et al. (2015) indicated that the hydrologic models have played an essential role in flood risk management in the recent research, especially in arid and semiarid areas. Hydrological parameters, such as flow discharge, peak discharge, lag time, and volume, can be predicted using such models, allowing for an accurate assessment of flash flood risks. The continuous flow of floodwater during storm events within Wadi Degla toward downstream threatens many areas and needs additional comprehensive management (Fig. 1).

In this context, the hydrological model is mainly based on the rainfall-runoff estimates. Spatial and temporal variations in land use are observed within the catchment area and

FIG. 1 Photography showing the flash flood event in Wadi Degla (Authors, Feb. 2020).

considered for the model. Thus, the surface runoff modeling is used to predict and understand the catchment yields and responses, assess water availability, and evaluate the changes over time (Vaze et al., 2012). This study also provides an insight into the potential impact of flood events in the Wadi Degla natural reserve on the natural life in the area. The authors used the hydrological dataset, land use, and topographical data in this study for the evaluation of the physical and environmental vulnerabilities in WDP.

2 Materials and methods

2.1 Study area

The Wadi Degla area, which is located in the Cairo Governorate, was announced as a natural preserve in 1999. This area lies between latitude 29° 56″ north and 31° 24″ east, covering an area of 260 km^2. It is 30-km long and extends from east to west with a distance of 10 km

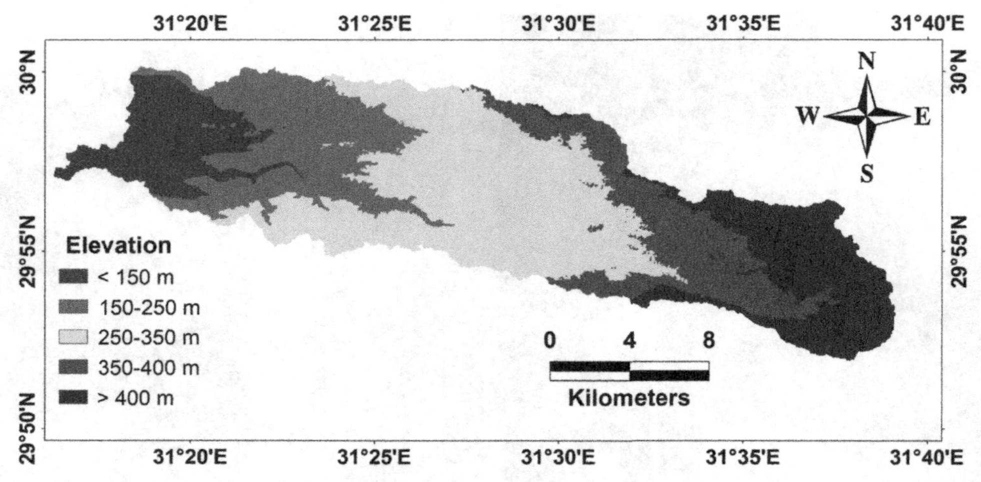

FIG. 2 Digital elevation model of Wadi Degla.

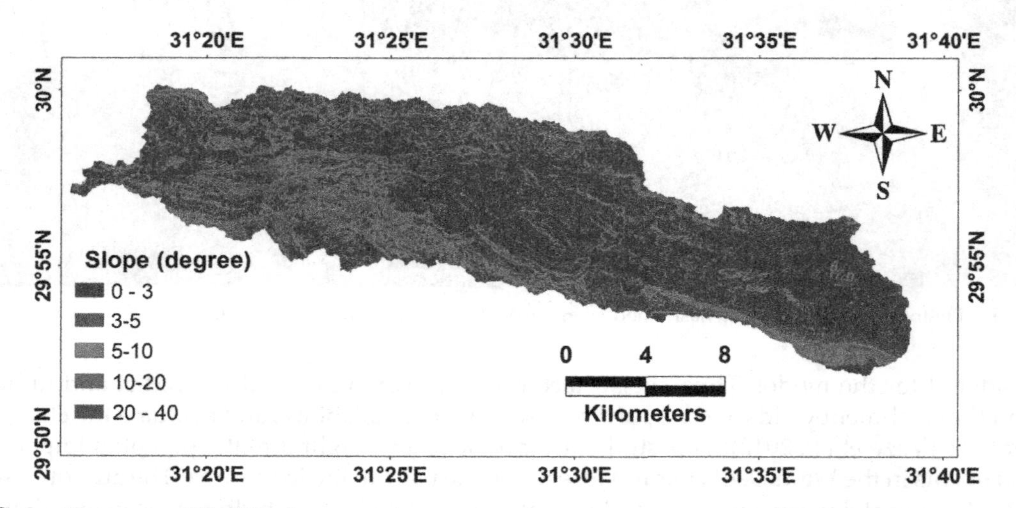

FIG. 3 Slope map of Wadi Degla.

from Cairo. Wadi Degla is millions of years old (upper Eocene period around 50 to 60 million years), and different species of animals, including the Dorcas gazelle, the Nubian ibex, and the red fox, have been traced in the Wadi. WDP travels through the limestone rock units of the Egyptian Eastern Desert, and these limestone units are from the remaining marine environment during the Eocene age (60 million years). Thus, this area is abundant in fossils. These rock units of limestone, which run beside the main Wadi channel, are approximately 50-m high. Fig. 2 shows the digital elevation model of the area, while Fig. 3 shows the slope directions in the area under investigation. The prevalent climatic condition in WDP is presented in Table 1. The average yearly rainfall in the study region is around 12.46 mm according to the Egyptian Metrological Authority. The maximum rainfall (25 mm) was recorded during

TABLE 1 Prevalent climatic condition in the area (EMA, 2019; Megahed and El Bastawesy, 2020).

	Climatic records			
Year 2019	Temperature (°C)	Rainfall (mm)	Evaporation (mm/day)	Humidity (%)
Minimum	7.00	2.00	5.50	40.00
Maximum	34.70	25.00	20.10	56.00
Mean	26.02	12.46	12.74	46.00
Total (12 month)	312.2	149.5	152.9	-

the winter season (EMA, 2019). The mean maximum temperature is approximately 26.01°C in June, while the mean minimum temperature is approximately 7°C in February, with the relative humidity ranging between 40% and 56%.

2.2 Geology of the area

Various geological studies have been conducted in the study site (Moustafa and Abd-Allah, 1991; UNDP et al., 2008; CONOCO, 1987; Hassan and Korin, 2019). The geology of the area is dominated by Tertiary and Quaternary sedimentary rocks. The oldest exposed rock outcrop in Wadi Degla is represented by the Tertiary Eocene limestones (Middle Eocene) of the Mokattam Group, which is found in the southern part of Wadi Degla. The Mokattam Group comprises Observatory Formation, which contains white to yellowish-white, marly, and chalky limestone intercalated with hard, gray dolomitic limestone bands. The clastic shale deposits of the Maadi Formation cover a large portion of Wadi Degla underlying Mokattam limestone. The Maadi Formation is subsequently overlain by the Oligocene Gebel Ahmar Formation, which comprises sandstones and gravels and dominates a large section of New Cairo City. The Gebel Ahmar Formation is capped by basaltic sheet in some places (Moustafa and Abd-Allah, 1991). The Miocene Hagul Formation comprises nonmarine deposits of fluviatile sands and gravels, while the Pliocene Kom El-Shelul Formation contains calcareous sandstone (2–5 m) and represents the Pliocene sediments (Said, 1990). Undifferentiated Pliocene sediments are also discovered in some places underlying the Kom El-Shelul Formation (CONOCO, 1987). Therefore, the floor of the wadis that drains to the Nile is dominated by Quaternary sediments (Fig. 4). The main geomorphologic units in Wadi Degla include the Nile floodplain, Piedmont plain, and Eocene structural plateau. The Piedmont plain is a transition zone between the Eocene plateau and the Nile flood plain. Several sets of normal faults are dissecting the Eocene limestone plateau (Fig. 5). A rose diagram revealed the presence of three main sets of normal faults aligned NE-SW, NW-SE, and NNE-SSW in the study area.

2.3 Methodology

The applied methodology in the current study was summarized in Fig. 6. However, the drainage networks were extracted automatically inside arc GIS (Arc hydro tool) from DEM thematic layer through multiple steps. These steps (Fig. 7A) are as follows. First, DEM fills

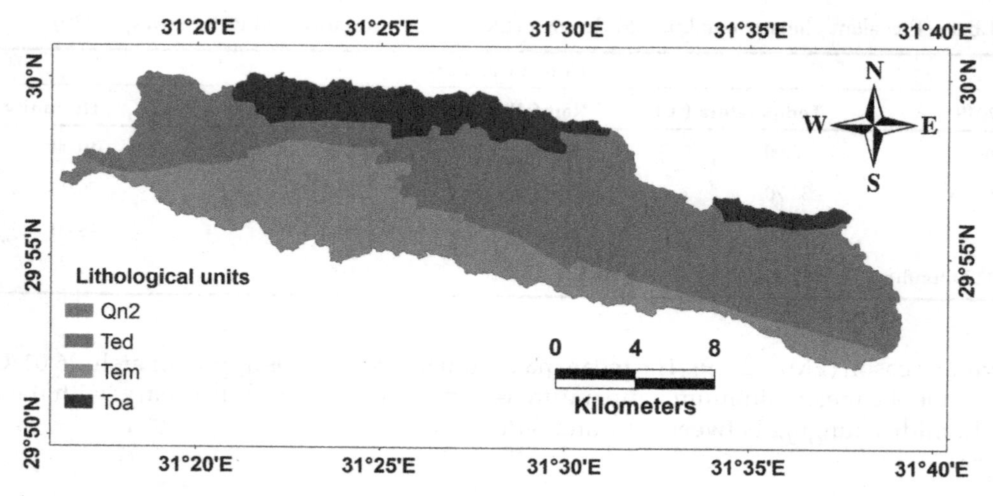

FIG. 4 Lithological units at Wadi Degla.

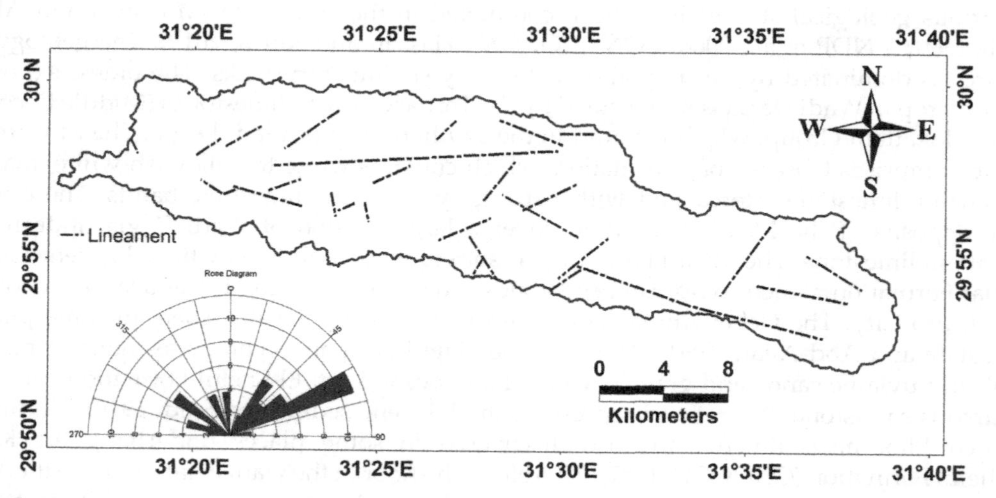

FIG. 5 Structural map of Wadi Degla.

the sinks, followed by the creation of the flow direction (Fig. 7B). Thereafter, the flow accumulation (Fig. 7C) is calculated for each cell in the area based on the flow directions. Finally, the drainage networks (stream orders) and watershed divisions were delineated (Fig. 8). Furthermore, land-use/land-cover changes from 1990 through 2020 were detected using the Landsat Thematic Mapper (TM) and Sentinel-2 acquired from the USGS (https://earthexplorer.usgs.gov/).

The curve number (CN) method was applied in the present study. This method was originally developed by the Soil Conservation Service (SCS) (1964, 1972). However, this method

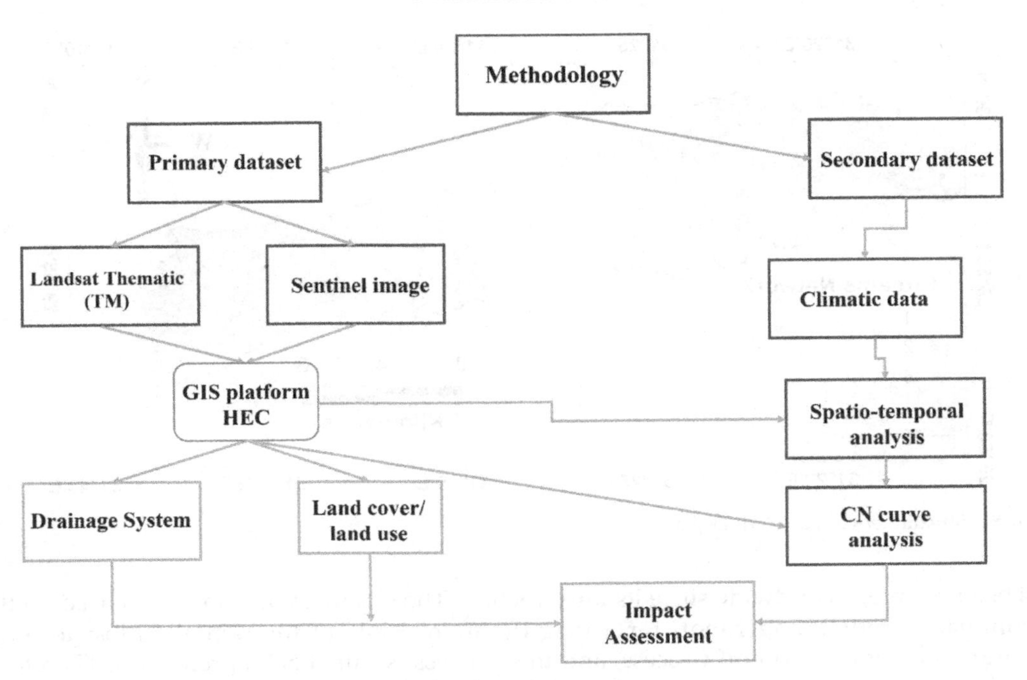

FIG. 6 Flow chart of the applied methodology in the present work.

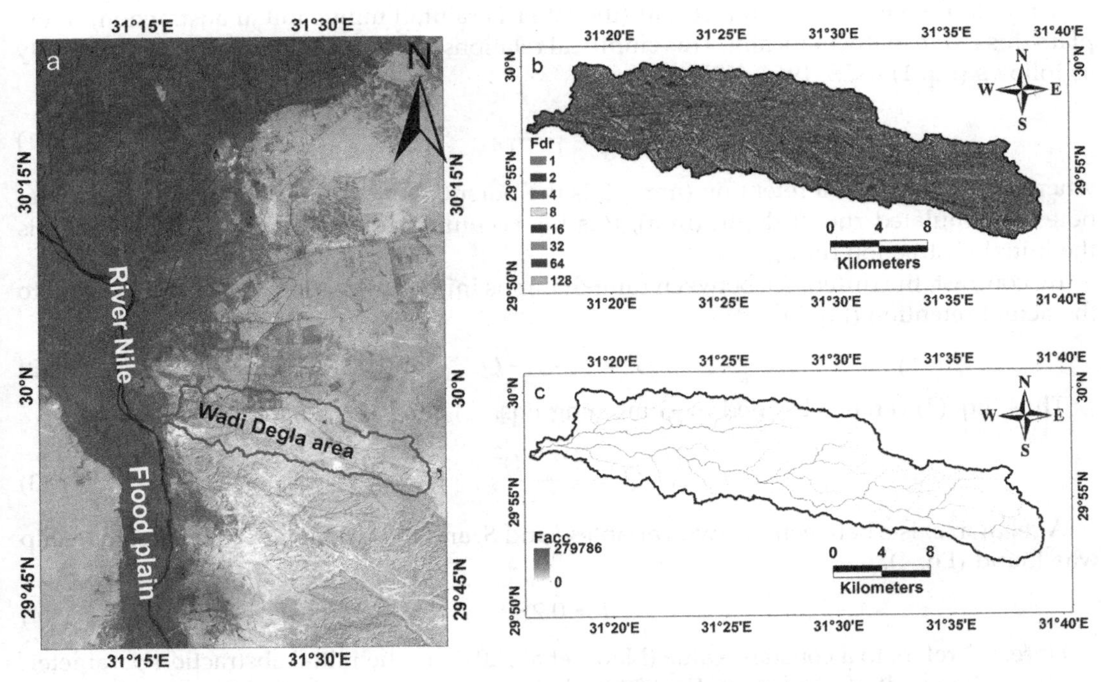

FIG. 7 (A) Sentinel imaginary 2021, (B) flow direction, and (C) flow accumulation.

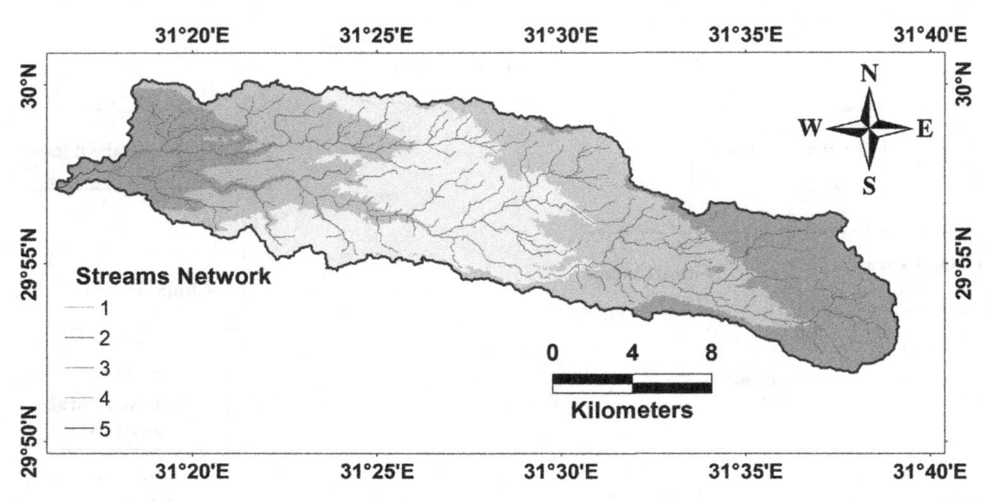

FIG. 8 Stream network at Wadi Degla.

has been adopted worldwide since its introduction. The CN method is initially based on the accumulated rainfall and runoff for a long duration. Some of the rainfall is lost through infiltration during the runoff process, and this process is called actual retention. Therefore, all additional rainfall either becomes runoff or actual retention once runoff has started. The SCS assumed that the ratio of actual to potential maximum retention was equal to that of actual to potential maximum runoff (the latter is rainfall minus initial abstraction) to explain this curve mathematically. This empirical relationship can be expressed mathematically as follows (Eq. 1) (SCS, 1964, 1972):

$$\frac{F}{S} = \frac{Q}{P - I}, \tag{1}$$

where F refers to actual retention (mm), S is the potential maximum retention (mm), Q denotes accumulated runoff depth (mm), P is the accumulated rainfall depth (mm), and I is the initial abstraction (mm).

By contrast, the difference between rainfall minus initial abstraction and runoff is equal to the actual retention (Eq. 2).

$$F = P - I - Q. \tag{2}$$

Thus, Eq. (3) can be obtained by combining Eqs. (1) and (2).

$$Q = \frac{(P - I)^2}{P - I + S}, \tag{3}$$

A relation exists between the two variables I and S, and the following average relationship was found (Eq. 4).

$$I = 0.2S. \tag{4}$$

Here, 0.2 refers to a constant value (Noori et al., 2011) of the initial abstraction parameter. Eq. (5) is the result of combining Eqs. (3) and (4).

$$Q = \frac{(P - 0.2\,S)^2}{P + 0.8\,S} \text{ for } P > 0.2\,S. \quad (5)$$

The possible maximum retention *S* is transformed to the *CN* to provide linear interpolation, averaging, and weighing procedures. This relationship can be presented as follows (Eq. 6) (SCS, 1964, 1972).

$$CN = \frac{25{,}400}{254 + S}. \quad (6)$$

Thus, the potential maximum retention (*S*) can be obtained from Eq. (7):

$$S = \frac{25{,}400}{CN} - 254. \quad (7)$$

Eq. (6) indicates that the *CN* can range from 100 to 0 because the potential maximum retention *S* can possibly vary between 0 and infinity.

3 Results and discussions

3.1 Land-use changes

Rapid urbanization, population growth, biodiversity loss, and encroachment on fertile agricultural land and water resources are all major drivers of global environmental change. Furthermore, scientific studies show that changes in the landscape are strongly linked to the loss of biodiversity and the deterioration of water quality (Sala et al., 2000; Torbick et al., 2006; Batisani and Yarnal, 2009; Abbas et al., 2010; Screenivasulu and Bhaskar, 2010; Rojas et al., 2013; Hassan and Nazem, 2015; Liping et al., 2018). Land-use changes (urbanization, road network construction, deforestation, industrialization, and cultivation) generally play a substantial role in contributing to flash floods, increasing flood impact and frequency. The major hydrological impacts are reflected in the reduction of infiltration rate, decreasing soil porosity, low evapotranspiration, and loss of vegetation. The European Environment Agency (2001) concluded that the main driving processes behind floods include climate change, land changes and land use, population growth, urbanization and increasing settlement, roads and railways, and hydraulic engineering measures.

Overland and Kleeberg (1991) indicated that the relatively small changes in land use can contribute to significant changes in peak flow and runoff depth. Sandstrom (1995) concluded that land degradation and deforestation induced floods. The results of the land-use change of the 1990–2020 images (Fig. 9) show that barren soil and rock are the dominant landscape in the area during the year 1990 (Fig. 9A). However, urbanization started to occur with some vegetation (low coverage of grasslands), which was confined to the Wadi down streams during the year 2010 (Fig. 9B). The comparison of years 1990 to 2020 revealed considerable urbanization, followed by a minor increase in vegetation cover (Fig. 9C). Furthermore, an increase in the urbanization, highways, and roads network in the area (Fig. 10) was observed, leading to high potential risks of low levels of ecological diversity.

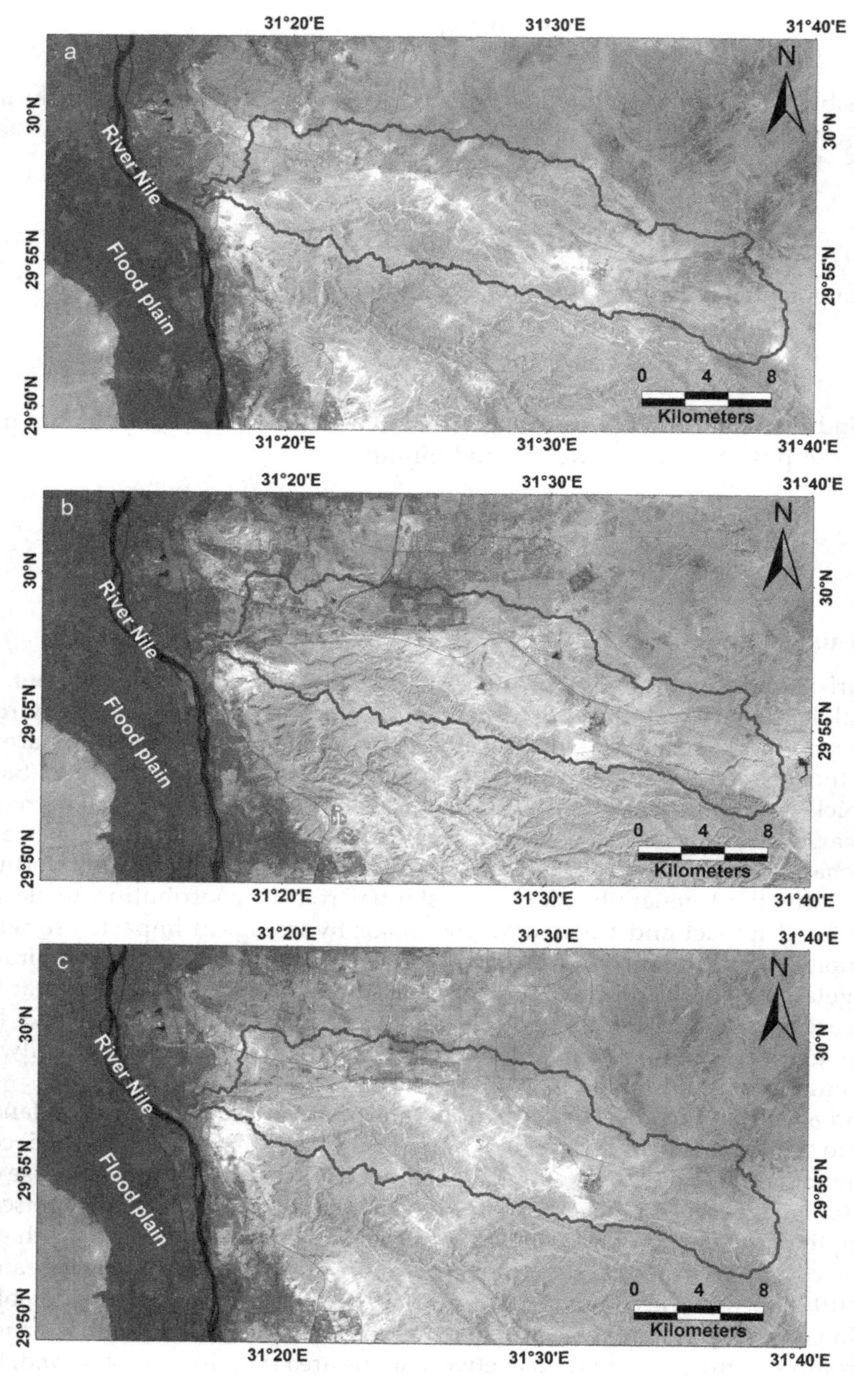

FIG. 9 Landsat 4–5 TM (2 4 7); (A) 1990, (B) 2010, and (C) 2020.

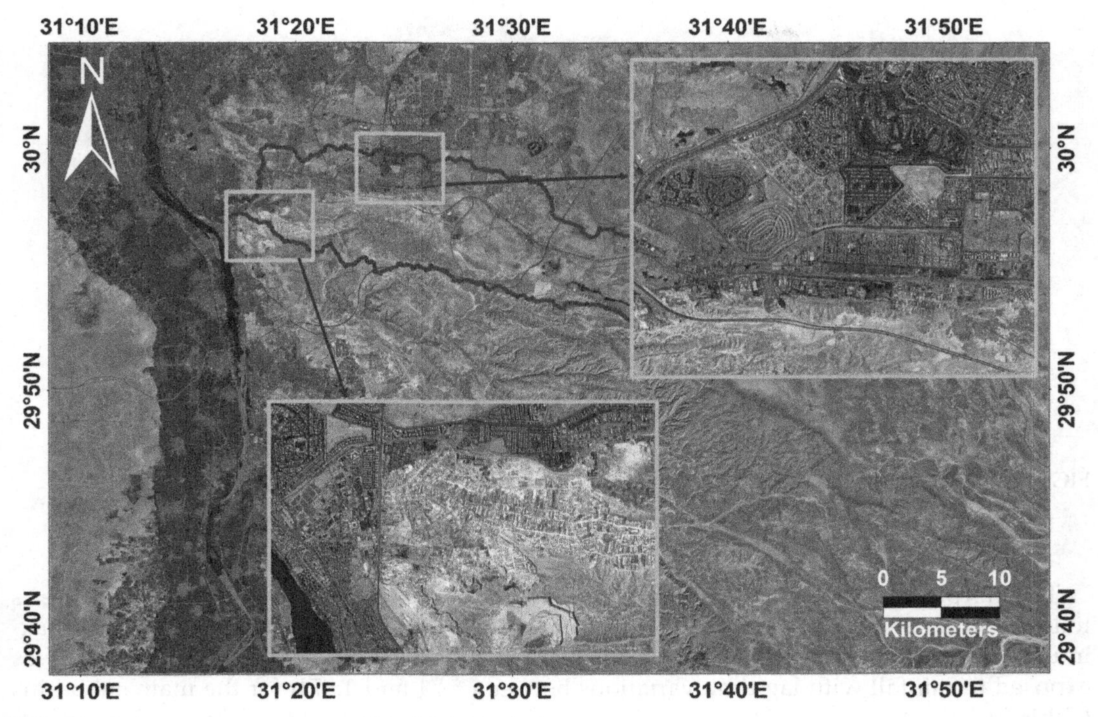

FIG. 10 Urbanization expansion within Wadi Degla downstream area.

3.2 Surface-water model

The hydrographs in the present work were estimated using the flash flood and hydrological parameters of the DEM of the study catchment. However, the accuracy of the hydrograph is generally affected by the lack or absence of rainfall-runoff data for any catchment area. Thus, two scenarios were proposed to evaluate the potential flash flood hazard in the area: (i) a storm with an effective rainfall depth of 25 mm (maximum rainfall during the year 2019) and (ii) return periods of rainfall depths 28.21, 36.88, and 47.53 mm for 25, 50, and 100 years, respectively (after WRRI, 2017). Fig. 11 is a schematic presentation of the three main subbasins in Wadi Degla. The equivalent hydrographs of the area were calculated for these events (Fig. 12A–C). The outlet of the three subbasins (SB_1, SB_2, and SB_3) receives a peak discharge of 19.46, 8.98, and 1.15 m^3/s with a runoff volume equal to 582.09, 283.05, and $211.37 \times 10^3 m^3$, respectively, for the maximum storm event (25 mm) during 2019. The study site experienced growth in urban areas from 1990 to 2020 based on the model findings. Therefore, increasing impermeable areas, altered hydrologic cycles, and decreasing infiltration rates are all consequences of urbanization. Therefore, urbanization in the present work plays an important role in the increase in runoff.

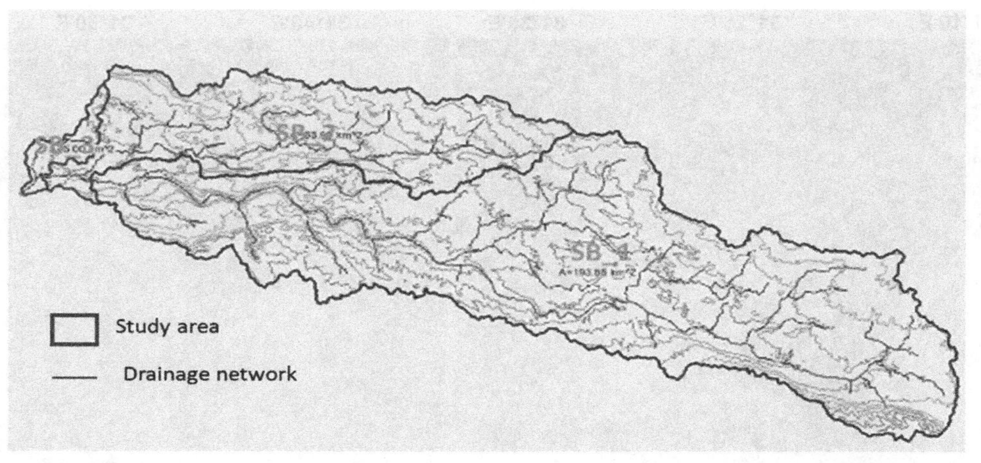

FIG. 11 Schematic presentation of the three main subbasins in Wadi Degla.

The morphometric analysis of the area shows that the characteristics of the subbasins are as follows: surface area of 193.85, 66.94, and 5 km^2; basin slope of 0.069, 0.069, and 0.053; maximum stream slope of 0.0085, 0.013, and 0.007. The model results indicate that the area was exposed to rainfall with lag time variations between 5.71 and 1.57 h for the main subbasins (Table 2).

3.3 Return periods

From a hydrological perspective, a flash flood is an event characterized by high discharges associated with water levels linked to the inundation of adjacent land to the stream network. In October 2019, the study region was hit by a rainstorm that resulted in a flash flood, with rainfall of approximately 25 mm. This event caused extensive damage in the study area, with runoff water flooding most major roadways and submerging numerous urban areas. Thus, the authors of the current study applied the hydrological model in the areas based on maximum rainfall depth (28.21, 36.88, and 47.53) for the return periods of 25, 50, and 100 years, respectively (Fig. 13). The hydrological model was implemented in the entire basin. The geometric characteristics of the basin, including the area, basin slope, and maximum stream slope, are 265 km^2, 0.069, and 0.0082 (Table 3), respectively. The estimated peak discharge from the rainfall depth of 28.21, 36.88, and 47.53 produced approximately a flow of 16.05, 36.7, and 70.27 m^3/s, respectively, according to the model results. However, the lag times computed from the hydrograph are 21 to 21.75 h. Therefore, the resulting volumes from the model are 808,297, 1,770,675, and 3,276,451 m^3 of water for the return periods (Table 4). Meanwhile, Wadi Degla is expected to have high susceptibility to flash flooding possibility according to the morphometric analysis based on the application of GIS techniques.

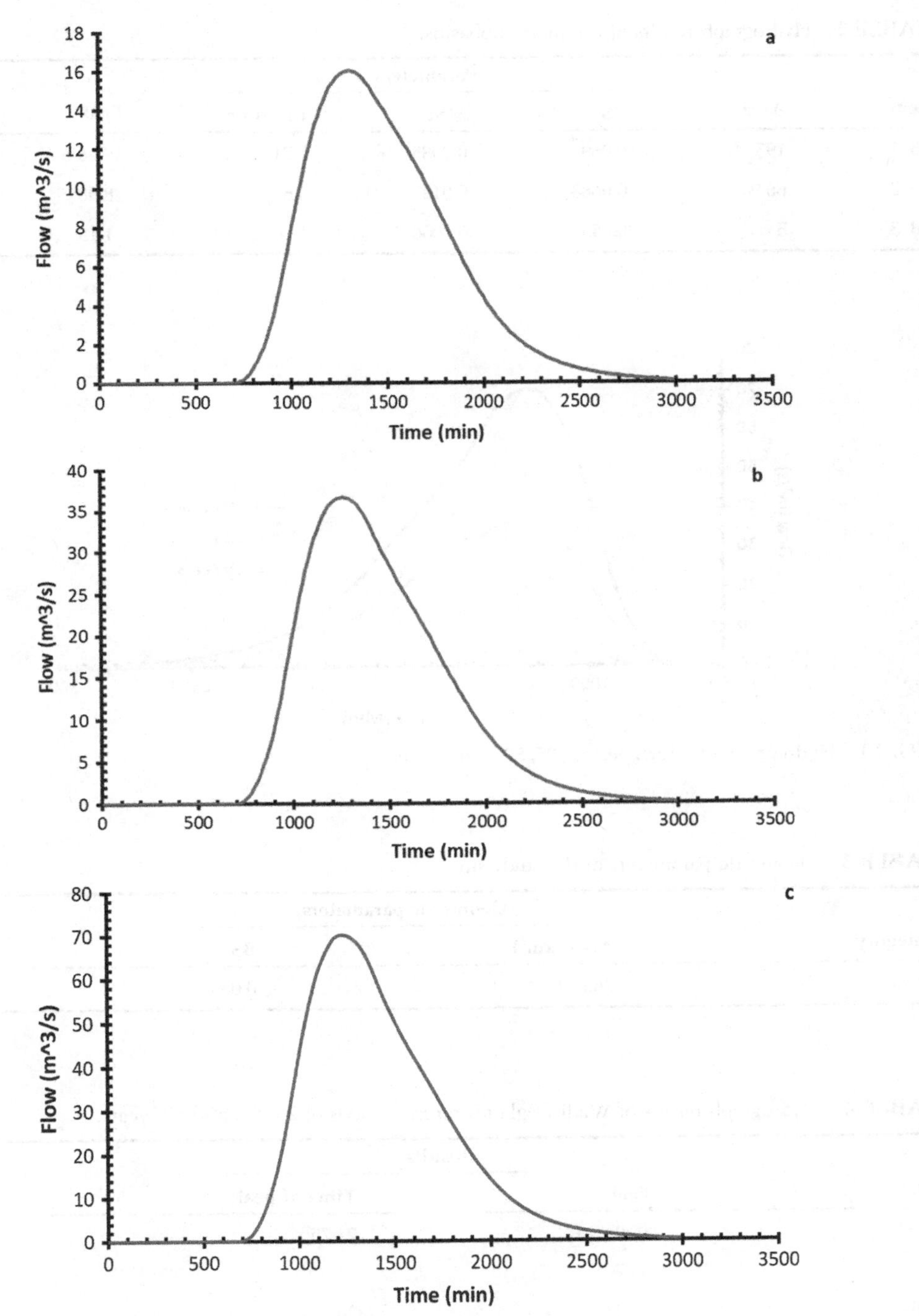

FIG. 12 Hydrograph of the (A) subbasin 1 (SB_1), (B) subbasin 2 (SB_2), and (C) subbasin 3 (SB_3).

TABLE 2 Hydrograph results of the main subbasins.

Item	Area	Parameters			Peak	Volume
		BS	MSS	Lag time		
SB_1	193.85	0.069	0.0085	5.711	19.46	582,086.7
SB_2	66.94	0.0688	0.0127	3.641	8.98	283,050
SB_3	5	0.0534	0.0066	1.567	1.15	21,137.4

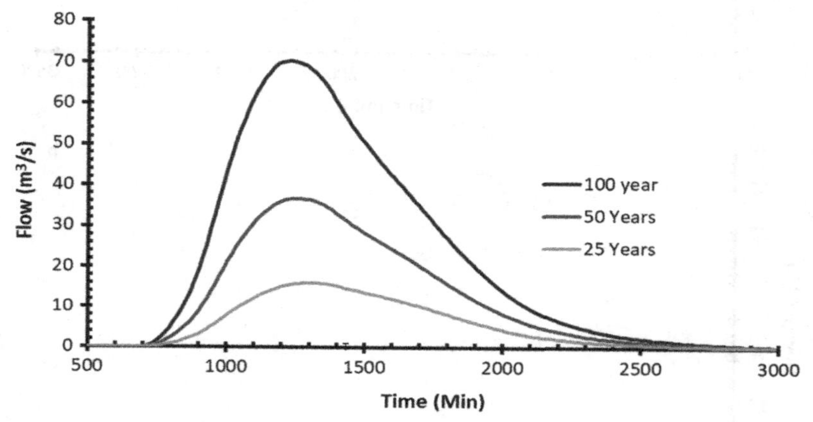

FIG. 13 Hydrograph of return periods 25, 50, and 100 years.

TABLE 3 Geometric parameters of the study site.

Category	Geometric parameters		
	Area (km^2)	BS	MSS
Value	265.79	0.069	0.0082

TABLE 4 Hydrograph results of Wadi Degla for return periods of 25, 50, and 100 years.

Year	Results		Volume
	Peak	Time of peak	
25	16.05	21.75	808,297
50	36.7	21	1,770,675
100	70.27	20.5	3,276,451

4 Impact on natural life

From an environmental perspective, various key natural populations were recorded in Wadi Degla (e.g., Hooded Crow, Spur-winged lapwing, White-throated kingfisher, Green sandpiper, House Sparrow, Spanish Sparrow, Barn Swallow, Pallid Swift, Rock dove, Palm dove, Collared dove, Turtle dove, Brown-necked raven, Eurasian Hoopoe, Sand Partridge, Montague's harrier, Short-toed snake eagle, Pharaoh eagle owl, Spotted flycatcher, and Collared flycatcher) considering biological and ecological systems. However, the accumulation and the availability of a reasonable field database remain an obstacle for effectively understanding the environmental effects of flash floods. By contrast, each system is unique, and generalizations regarding responses to extreme events (Hickey and Salas, 1995) do not translate well between ecotypes. Extreme flooding events tend to remove some paleontology species from Eocene limestone and are recorded on both sides of basins.

Many sinks or pits are recorded in the study site, and these lowlands or holes can store water and help reduce flood peaks. The pond intercepts runoff and holds it until its storage capacity is reached, at which point, the pond overflows and again contributes to runoff (Fig. 14). Rather than being transmitted immediately to its downstream, flood water is partially absorbed within the water body as the water level rises when it reaches sinkholes or anthropogenic activities. However, these ponds prolonged the presence of water, thereby creating favorable conditions for plants. Ponds have ecological biodiversity, which can provide favorable habitats for migratory birds of many species to rest. The ecosystem is an interactive system of plants, microorganisms, and animal communities with their nonliving environment. One of the most crucial impacts of flash floods is reducing soil fertility or contributing to the decline in fertility. Thus, flash floods could significantly alter the severity or the shortages of erroneous vegetation to grow and remain alive (Fig. 15). The flash flood also demonstrated positive impacts in the study site, wherein sediment transport increased along with the organic material, thus providing vital nutrients for downstream species. Unlike dam construction, sediment transport decreased along with the lack of organic materials downstream, which are retained behind the dams. Additionally, instead of fertilizing the main Wadi channel, the fine silts are retained behind the dams.

Disturbances created by extreme flash flood events could change the densities and compositions of populations of natural life. Morgan et al. (1991) concluded that communities are often substantially and frequently altered and inherently unstable with modified hydrologic flow regimes because controlling variables are not predetermined by local natural conditions. This phenomenon allows the dominance of exotic and poorly adapted species in systems with naturally inhospitable conditions. Unlike the disturbance impact of flash flooding, flooding water initiates a return to the natural condition and exposes vegetation cover. This positive impact creates habitat and provides food for some species and increases the population for species covered by vegetation.

5 Conclusions

The land-use/land-cover change occurrence in the study area has dramatically increased from 1990 to 2020. Furthermore, the condition of the soil due to urbanization and road construction is important to determine the possibility of floodwater flowing into the ground or

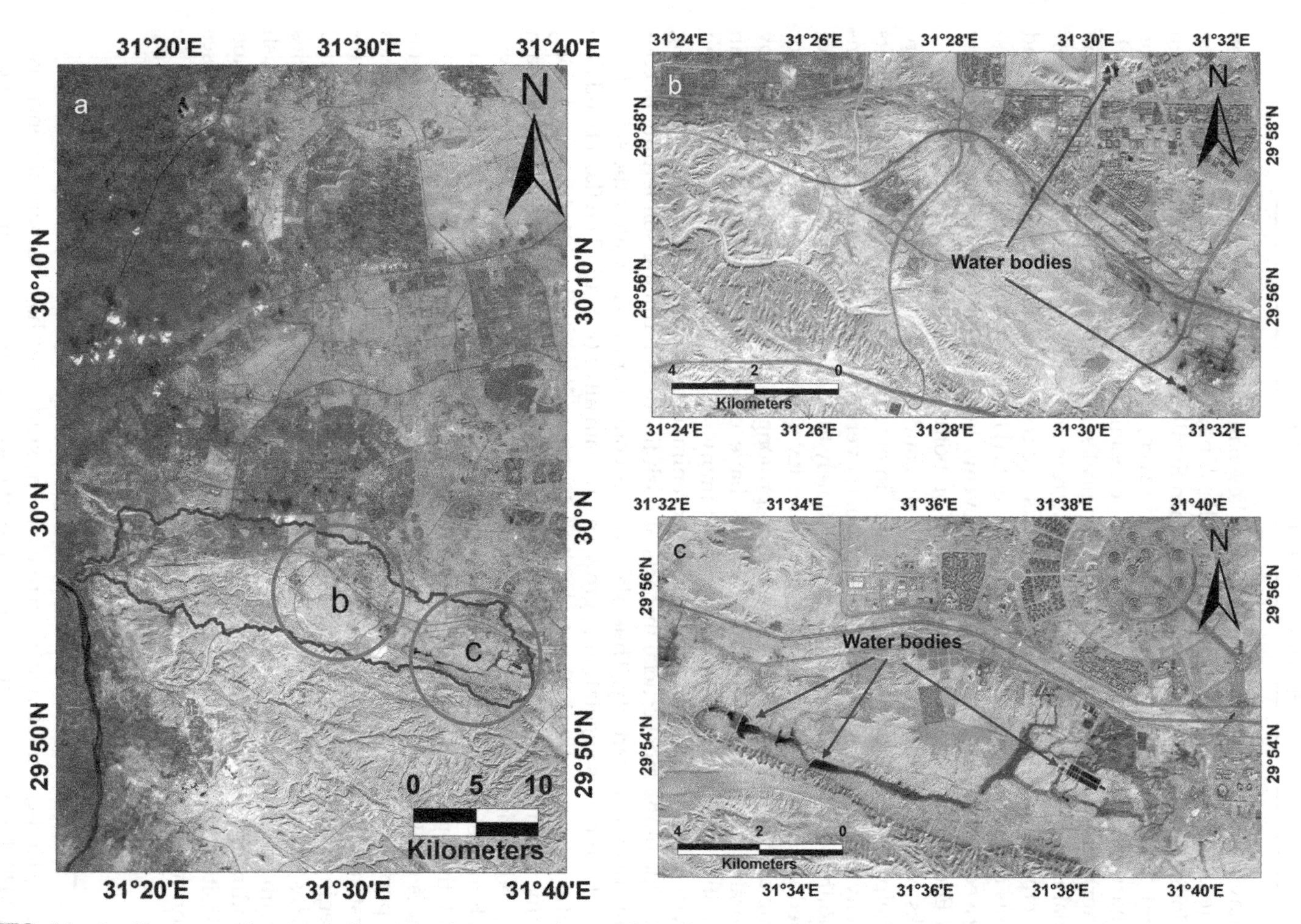

FIG. 14 Satellite images showing the locations of water ponds in Wadi Degla area.

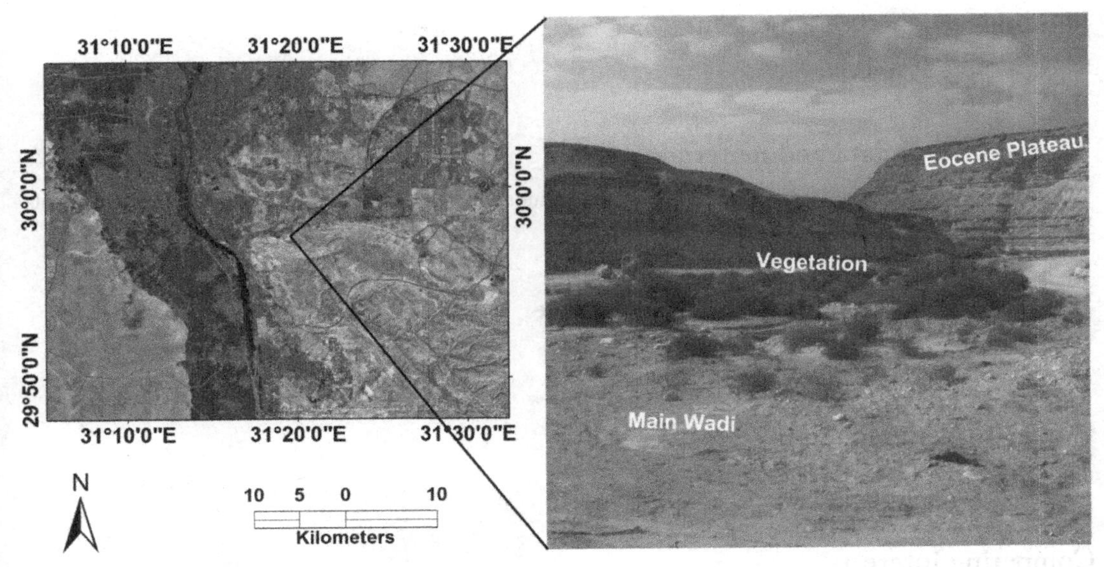

FIG. 15 Low vegetation cover in the study area.

across the land. The overland flow occurs and fills the pits and ground depressions in the area, thus creating depression storage. Various parameters contribute to flash floods; however, the duration and the intensity of rainstorms are crucial in the occurrence and the magnitude of flash floods. Additionally, the soil condition reduces the infiltration rate with the expansion of urbanization in the area. Therefore, the hydrological model was constructed on the basis of the maximum rainfall in the study area for return periods of 25, 50, and 100, and the resulting discharge peaks for the return periods are approximately 16.05, 36.7, and 70.27 m^3/s, respectively.

The necessity to sustain environmental flows should be considered when allocating floodwater management systems (e.g., dams). Existing and future dams in the study site design and operation should also be appropriately altered to minimize environmental impact. However, the environmental impacts can be monitored using various indicators that are consistent with the objectives of the environment. Various components of the environment, including changes in land use, habitat variety, and distribution of water ponds, should also be monitored. Despite the establishment of some protective measures in the area, additional comprehensive studies and protection strategies in the Wadi Degla could be applied to enhance flood risk monitoring and management. Accordingly, environmental, hydrological, and geological studies should be conducted in the area before the development of any new settlements.

To preserve such reserves from overland flow hazards, it is recommended to construct proper dams in the appropriate locations to restore the excess water derived from runoff and could be used as water harvesting systems. Wherever possible, the urbanization and development should be nature-based approaches in the area to flood-risk management. Furthermore, the rainfall-runoff information should be translated into an actionable flood-risk control approach.

Funding

Not applicable.

Availability of data and materials

The data generated or analyzed during this study are included in this manuscript and available upon reasonable request.

Ethics approval and consent to participate

Not applicable.

Consent for publication

Not applicable.

Competing interests

The authors declare that they have no competing interests.

References

Abbas, I.I., Muazu, K.M., Ukoje, J.A., 2010. Mapping land use-land cover change detection in Kafur local government, Katsina, Nigeria (1995–2008) using remote sensing and GIS. Res. J. Environ. Earth Sci. 2010 (2), 6–12.

Abdel Moneim, A., 2004. Overview of the geomorphological and hydrogeological characteristics of the Eastern Desert of Egypt. Hydrogeol. J. 13, 416–425.

Abu El-Magd, S.A., 2019. Flash flood hazard mapping using GIS and bivariate statistical method at Wadi Bada'a, Gulf of Suez, Egypt. J. Geosci. Environ. Prot. 7, 372–385.

Abu El-Magd, S.A., 2022. Random forest and naïve Bayes approaches as tools for fash food hazard susceptibility prediction, South Ras El-Zait, Gulf of Suez Coast, Egypt. Arab. J. Geosci. 15, 217. https://doi.org/10.1007/s12517-022-09531-3.

Abu El-Magd, S.A., Amer, R.A., Embaby, A., 2020. Multi-criteria decision making for the analysis of flash floods: a case study of Awlad Toq Sherq, Southeast Sohag, Egypt. J. Afr. Earth Sci. 162 (2020), 103709.

Abu El-Magd, S.A., Pradhan, B., Alamri, A., 2021a. Machine learning algorithm for flash flood prediction mapping in Wadi El-Laqeita and surroundings, Central Eastern Desert, Egypt. Arab. J. Geosci. 14, 323. https://doi.org/10.1007/s12517-021-06466-z.

Abu El-Magd, S.A., Orabi, H.O., Ali, S.K., Parvin, F., Pham, Q.B., 2021b. An integrated approach for evaluating the flash food risk and potential erosion using the hydrologic indices and morpho-tectonic parameters. Environ. Earth Sci. 80, 694. https://doi.org/10.1007/s12665-021-10013-0.

Batisani, N., Yarnal, B., 2009. Urban expansion in Centre County, Pennsylvania: spatial dynamics and landscape transformations. Appl. Geogr. 29, 235–249.

CONOCO, the Egyptian general petroleum company (EGPC), 1987. Geological maps of Egypt scale 1:500000, NH 36 NW Cairo and NH 36 SW Beni Suef.

Dudley, N., Buyck, C., Furuta, N., Pedrot, C., Renaud, F., Sudmeier-Rieux, K., 2015. Protected Areas as Tools for Disaster Risk Reduction: A Handbook for Practitioners. MOEJ and IUCN, Tokyo and Gland, Switzerland, 44pp.

El-Fakharany, N.A., 1998. Drainage basins and flash floods managements in the area southeast Qena, Eastern Desert, Egypt. Egypt. J. Geol. 42 (2), 737–750.

EMA (2019). Egyptian Meteorological Authority (1949–2019). Cairo, Egypt.www.nwp.gov.eg/.

European Environment Agency, 2001. Sustainable Water Use in Europe. Part 3: Extreme Hydrological Events: Floods and Droughts. European Environment Agency, Copenhagen, Denmark.

Hagras, M.A., Elmoustafa, A.M., Kotb, A., 2013. Flood plain mitigation in arid regions case study: south of Al. Kharj City, Saudi Arabia. Int. J. Res. Rev. Appl. Sci. 16 (1), 147–156.

Hassan, H.F., Korin, A.H. 2019. Contribution to the biostratigraphy of the Middle-Upper Eocene rock units at North Eastern Desert: an integrated micropaleontological approach. Heliyon 5 (2019) e01671. https://doi.org/10.1016/j.heliyon.2019.e01671.

Hassan, M.M., Nazem, M.N.I., 2015. Examination of land use/land cover changes, urban growth dynamics, and environmental sustainability in Chittagong city, Bangladesh. Environ. Dev. Sustain. 18, 9672–9678.

Hewaidy, A.G.A.H., El Hassan, M.A., Salama, A., Ahmed, R., 2021. Hydrology flash flood risk assessment of Wadi Degla basin protected area, East of Maadi, Cairo, Egypt based on morphometric analysis using GIS techniques. Hydrology 9 (3), 66–73. https://doi.org/10.11648/j.hyd.20210903.12.

Hickey, J., Salas, J. 1995. Environmental of effects of extreme floods. U.S.- Italy Research Workshop on the Hydrometeorology, Impacts, and Management of Extreme Floods Perugia (Italy), November 1995.

Jin, H., Liang, R., Wang, Y., Tumula, P., 2015. Flood—runof in semiarid and sub-humid regions, a case study: a simulation of Jianghe watershed in Northern China. Water 7, 5155–5172. https://doi.org/10.3390/w7095155.

Langston, C., Ding, G.K.C., 2001. Sustainable Practices in the Built Environment, second ed. Butterworth- Heinemann, Oxford.

Liping, C., Yujun, S., Saeed, S., 2018. Monitoring and predicting land use and land cover changes using remote sensing and GIS techniques—a case study of a hilly area, Jiangle, China. PLoS One 13, e0200493.

Megahed, H.A., El Bastawesy, A.M., 2020. Hydrological problems of flash foods and the encroachment of wastewater affecting the urban areas in Greater Cairo, Egypt, using remote sensing and GIS techniques. Bull. Natl. Res. Cent. 44, 188. https://doi.org/10.1186/s42269-020-00442-5.

Mirzaei, S., Vafakhah, M., Pradhan, B., Alavi, S., 2020. Flood susceptibility assessment using extreme gradient boosting (EGB), Iran. Earth Sci. Inform. https://doi.org/10.1007/s12145-020-00530-0.

Moawad, B.M., 2013. Analysis of the flash food occurred on 18 January 2010 in Wadi El Arish, Egypt (a case study). Geomat. Nat. Haz. Risk 4 (3), 254–274.

Moawad, B.M., Abdel Aziz, A.O., Mamtimin, B., 2016. Flash floods in the Sahara: a case study for the 28 January 2013 food in Qena, Egypt. Geomat. Nat. Haz. Risk 7 (1), 215–236. https://doi.org/10.1080/19475705.2014.885467.

Morgan, R.P., Jacobsen, R.E., Weisberg, S.B., McDowell, L.A., Wilson, H.T., 1991. Effects of flow alteration on benthic macroinvertebrate communities below the Brighton Hydroelectric Dam. J. Freshw. Ecol. 6 (4), 419–429.

Moustafa, A., Abd-Allah, A., 1991. Structural setting of the central part of the Cairo-Suez district. M.E.R.C. Ain Shams Univ, Earth Sci. 5, 133–145.

Mukerji, A., Chatterjee, C., Raghuwanshi, N.S., 2009. Flood forecasting using ANN, neuro-fuzzy, and neuro-GA models. J. Hydrol. Eng. 14, 647–652. https://doi.org/10.1061/(ASCE)HE.1943-5584.0000040.

Noori, N., Kalin, L., Srivastava, P., Lebleu, C., 2011. Efects of initial abstraction ratio in SCS-CN method on modeling the impacts of urbanization on peak fows. In: World Environmental and Water Resources Congress 2012: Cross Boundaries. Albuquerque, New Mexico, United States, pp. 1–12, https://doi.org/10.1061/97807 84412312.

Overland, H., Kleeberg, H.B., 1991. Influence of land use change on discharge and sediment transport of floods. In: Sediment and Stream Water Quality in a Changing Environment: Trends and Explanation (Proceedings of the Vienna Symposium, August 1991). IAHS Publ. no. 203.

Rojas, C., Pino, J., Basnou, C., Vivanco, M., 2013. Assessing land-use and cover changes in relation to geographic factors and urban planning in the metropolitan area of concepcion' (Chile). Implications for biodiversity conservation. Appl. Geogr. 2013 (39), 93–103.

Said, R., 1990. The Geology of Egypt. Elsevier, Amsterdam.

Sala, O.E., Chapin, F.S., Armesto, J.J., Berlow, E., Bloomfield, J., Dirzo, R., Huber-Sanwald, E., Huenneke, L.F., Jackson, R.B., Kinzig, A., et al., 2000. Global biodiversity scenarios for the year 2100. Science 287, 1770–1774.

Sandstrom, K., 1995. The recent Lake Babati floods in semiarid Tanzania: a response to changes in land cover. Geogr. Ann. A Phys. Geogr. 77A (1), 35–44.

Screenivasulu, V., Bhaskar, P.U., 2010. Change detection in land use and land cover using remote sensing and GIS techniques. Int. J. Eng. Sci. Technol. 2, 7758–7762.

Soil Conservation Service, 1964. National Engineering Handbook, Section 4, Hydrology. Department of Agriculture, Washington. 450 p.

Soil Conservation Service, 1972. National Engineering Handbook, Section 4, Hydrology. Department of Agriculture, Washington. 762 p.

Torbick, N.M., Qi, J., Roloff, G.J., Stevenson, R.J., 2006. Investigating impacts of land-use land cover change on wetlands in the Muskegon river watershed, Michigan, USA. Wetlands 26, 1103–1113.

UNDP, GEF, Cairo University, 2008. Vegetation dynamics assisted hydrological analysis for Wadi Degla. In: Developing Renewable Groundwater Resources in Arid Land, Pilot Case. The Eastern Desert of Egypt.

Vaze, J., Jordan, P., Beecham, R., Frost, A., Summerell, G., 2012. Guidelines for Rainfall Runoff Modelling: Towards Best Practice Model Application., p. 47.

Water Resources Research Institute (WRRI) 2017. Hydrological study for infrastructure protected from flash flood hazrds at wadi Degla downstream, Cairo governorate. Egypt. Internal report. 81 pp.

Youssef, A.M., Hegab, M.A., 2019. Flood-hazard assessment modeling using multicriteria analysis and gis: a case study—Ras Gharib Area, Egypt. In: Spatial Modeling in GIS and R for Earth and Environmental Sciences. Elsevier, New York, pp. 229–257, https://doi.org/10.1016/B978-0-12-815226-3.00010-7.

Youssef, A.M., Abdel Moneim, A.A., Abu El-Maged, S.A., 2005. Flood hazard assessment and its associated problems using geographic information systems, Sohag Governorate, Egypt. In: The Fourth International Conference on the Geology of Africa, Assiut, Egypt, November. vol. 1, pp. 1–17.

Youssef, A.M., Pradhan, B., Gaber, A.F.D., Buchroithner, M.F., 2009. Geomorphological hazards analysis along the Egyptian Red Sea coast between Safaga and Quseir. Nat. Hazards Earth Syst. Sci. 9, 751–766. https://doi.org/10.1007/s12665-014-3661-3.

CHAPTER

36

GLOF Early Warning System: Computational challenges and solutions

Binay Kumar, T.S. Murugesh Prabhu, Anish Sathyan, and Arun Krishnan

Centre for Development of Advanced Computing (C-DAC), Pune, Maharashtra, India

OUTLINE

1 Introduction 641

2 Materials and methods 643

2.1 Study area 643

2.2 Data acquisition 643

2.3 Data interpretation / analysis / computation 649

2.4 GLOF simulation model setup 654

2.5 GLOF early warning 656

2.6 GIS-based spatial decision support system 656

3 Results and discussion 657

4 Salient features of GLOF Early Warning System 657

5 Limitations of the study 659

6 Recommendations 659

7 Conclusions 660

Acknowledgments 660

References 660

1 Introduction

Glaciers are masses of ice which flow out from the snow fields where they originate (Holmes, 1945). Glaciers move by gravitational processes, internal deformation caused by shear stress imposed by overlying ice and snow. There are number of glaciers in India

Current Directions in Water Scarcity Research, Volume 7
https://doi.org/10.1016/B978-0-323-91910-4.00036-4

spanning across States of Jammu and Kashmir, Ladakh, Himachal Pradesh, Uttarakhand, Sikkim, and Arunachal Pradesh.

The Himalayan region is witnessing the impact of climate change which has severely affected the glacier. Himalayan glaciers are considered as the fresh water tower of South and East Asia and are strongly affected by climate change (Vohra, 2007). According to the Fifth Assessment Report of the Intergovernmental Panel on Climate Change (IPCC), climate changes have already adversely affected the glaciated regions globally (IPCC, 2007; UNFCCC, 1992). The Himalayas are the sources of various perennial rivers such as the Ganges, Brahmaputra, Yamuna and many more. Topography, morphology and climate vary significantly (Worni et al., 2012) in this region.

Glaciers originating in valley heads creep slowly downward as tongue-like streams of ice, flowage being maintained by the yearly replenishment of the névé fields (Holmes, 1945). Ultimately the glaciers dwindle away by melting and evaporation, their fronts or snouts reaching a position which may be thousands of feet below the snow line where the forward movement of the ice is balanced by the wastage. Due to deglaciation, increase in the number and size of glacial lakes have been noticed worldwide (Carrivick and Tweed, 2013).

The trend of lake development is spatially variable in response to climate and the evolution of debris-covered glaciers in the Himalayas (Richardson and Reynolds, 2000; Gardelle et al., 2011; Benn et al., 2012; Nie et al., 2013; Bajracharya et al., 2006). The glacial lakes also increased in number and extent in the Himalaya (Wang et al., 2011). A glacier occupying a main valley may obstruct the mouth of a tributary and impound the drainage to form a lake. Glacial lakes are an indirect indicator of glacier change and unstable lakes can pose hazard to downstream locations.

Glacial melt in the Himalayas feed many moraine-dammed and supraglacial lakes (Ives et al., 2010; Nie et al., 2017). Embedded in loose debris and surrounded by sources of falling debris and ice, many of these lakes are prone to GLOF (Clague and Evans, 2000). The moraine dam failure can be due to erosion, water pressure, excessive ablation, avalanche, cloudburst, and earthquake. GLOFs happening in the Himalayan region steps up the degree of impact due to the steep and fragile terrain, causing rapid flow of flood waters causing catastrophic flash floods resulting in huge losses (Bajracharya et al., 2007a,b; Bhambri et al., 2015; Carrivick and Tweed, 2016; Hewitt, 2016; Thakur et al., 2016).

Advances in remote sensing have significantly increased the ability to study the otherwise inaccessible glacial lakes (Aggarwal et al., 2017; Sheng et al., 2016). Due to low accessibility of glacial lakes, remote sensing provides the most feasible technique for monitoring vulnerable glacial lakes in such areas (Buchroithner, 1995; Buchroithner and Bolch, 2015; Huggel et al., 2002; Bhardwaj et al., 2015; Gardelle et al., 2011; Liu and Mayer, 2015; Robson et al., 2015). Studies carried out by the Centre for Development of Advanced Computing (C-DAC), Pune, jointly with the Government of Sikkim, have shown that many glacial lakes in Sikkim Himalayan region have grown over the years. Sikkim has over 250 moraine-dammed lakes, of which around 30 are found to be vulnerable (Kumar and Murugesh Prabhu, 2012). The remote location and harsh climatic conditions pose challenge to undertake regular field investigations and deployment in these regions. Vulnerable glacial lakes can be monitored by deploying all-weather-capable sensors and automatic monitoring systems.

Different types of dam-breach and flood models have been applied to model and asses GLOF impact (Bajracharya et al., 2007b; Huggel et al., 2003; Wang et al., 2008). For dams susceptible to failure, magnitudes of potential GLOFs can be calculated using empirical and physical models (Evans, 1986; Huggel et al., 2004; Kershaw et al., 2005). The application of dynamic models is preferred to empirical models, as the latter represent an over-simplification of complex processes for specific and local-scale scenario modeling (Allen et al., 2009; Worni et al., 2013; Yamada and Sharma, 1993). However, many complex parameters, computational requirements and topographic sensitivity of physically-based models make GLOF modeling challenging. Various hydrodynamic models used for simulation of dam-break scenarios have been used to model GLOF events (Maskey et al., 2020; Aggarwal et al., 2013; Allen et al., 2009). The GLOF simulation model predicts the flood arrival time and inundation depth using the lake discharge and Digital Elevation Model (Fernández et al., 1991; Clarke and Waldron, 1984; Alho and Aaltonen, 2008). In Himalayas, such models have been used to assess GLOF hazards in Imja Tsho and Tsho Rolpa in Nepal and Raphstreng Tsho in Bhutan (Bajracharya et al., 2006). It is imperative that the simulation models must have computational intelligence to simulate different GLOF scenarios with better lead-time.

2 Materials and methods

One of the prime challenges in the development of the system is data availability. A comprehensive glacial lake database was generated and supplemented with field investigations to access the vulnerability of the lakes and prioritize them for monitoring. Base maps such as State boundary, village boundaries, roads, drainage, watershed boundaries, etc. were prepared using Q-GIS software.

The water level data received from the in-situ sensor is stored in PostgreSQL database and analyzed for GLOF event detection in real-time. The GLOF simulation was performed using free and open source hydrodynamic model ANUGA Hydro, which computes the flow over an unstructured triangular mesh by solving Shallow Water Wave Equations using a finite-volume numerical scheme. The model is found to be robust and stable in all the simulations carried out for various scenarios (Mungkasi and Roberts, 2013).

2.1 Study area

The study focuses on vulnerable glacial lakes in the Sikkim, India. Several glacial lakes are present in Sikkim, which have been considered to be vulnerable; however, the present study focuses on Lhonak and South Lhonak glacial lakes (encircled in Fig. 1), located at an elevation of 5362 m, with an area of 1.028 km^2 held by an ice-core moraine dam. As the lake has expanded significantly in the past decades, it is likely to pose GLOF risk.

2.2 Data acquisition

2.2.1 Remote sensing imagery

Multi-temporal multi-resolution satellite imagery, published maps and reports constitute the database necessary for the mapping glacial lakes in Sikkim. IRS-1A/1B/1C/1D/P6, USGS

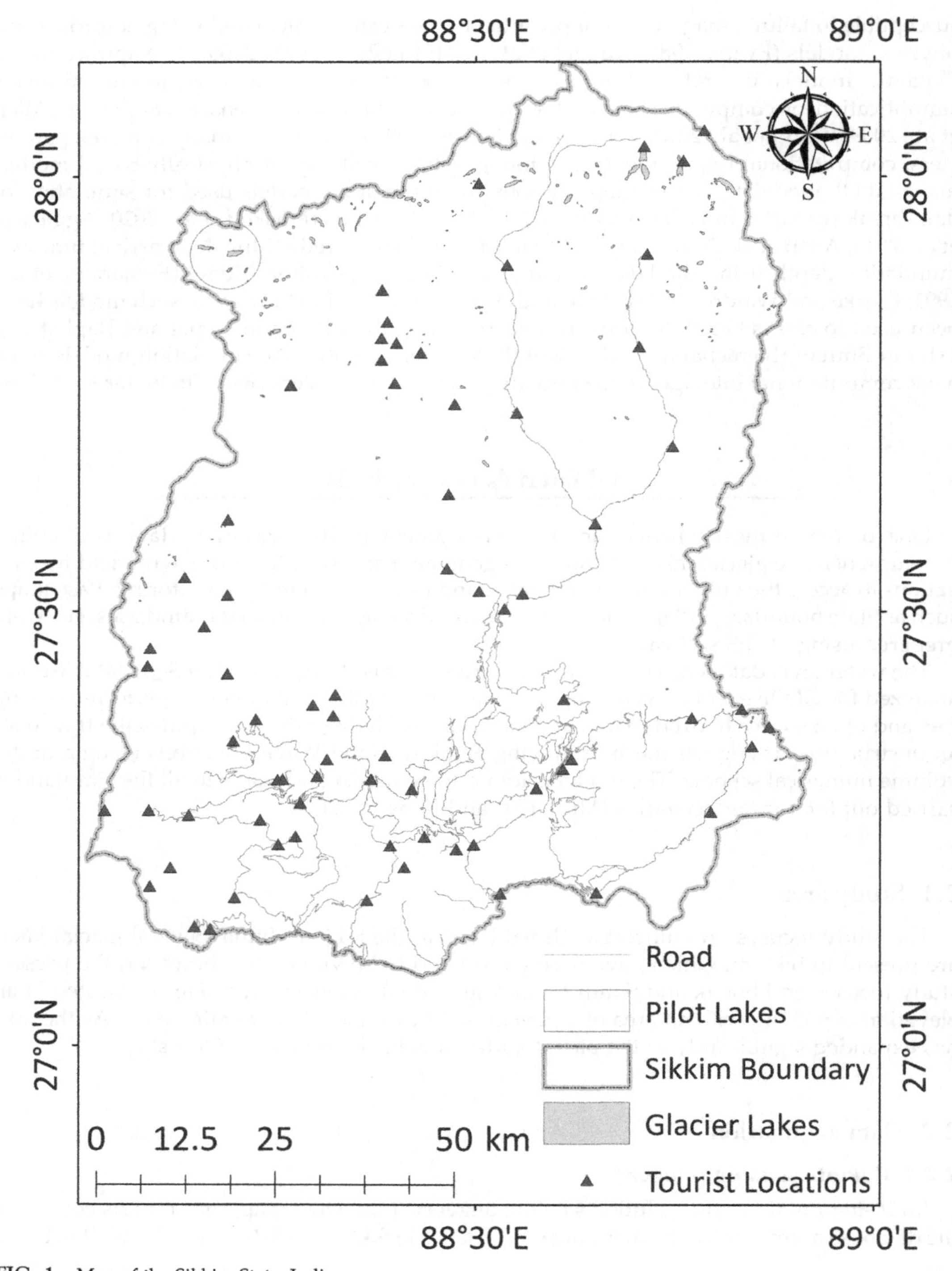

FIG. 1 Map of the Sikkim State, India.

declassified imagery (CORONA, KH-Series), Land Observation Satellites (LANDSAT- MSS, TM, ETM) and Google Earth data were used.

2.2.2 *Topographic data*

ALOS-AW3D30 Digital Terrain Model (2021) was used for the study. "ALOS World 3D (AW3D30)" data set is a global digital surface model (DSM) with horizontal resolution of approximately 30 m.

The elevation profile of the South Lhonak Lake was studied using ALOS DEM where the average depth is about 25 m, which is still a significant depression to hold a large volume of melt water.

2.2.3 *Bathymetry data*

As the DEM data is not sufficient to ascertain the lake depth, which is critical to estimate the lake water volume, bathymetry survey was carried out using an ultrasonic depth-sounding equipment and GPS on an inflatable boat. The survey yielded depth at over 200 points within the lake. The survey revealed that the depth of the lakes to be more than 100 m in the central regions. Only a limited portion of the lake could be surveyed due to the presence of large floating ice blocks (Fig. 2).

FIG. 2 Bathymetry survey using inflatable boat. The floating ice blocks were a major deterrent during the survey and pose danger to the team.

2.2.4 Moraine dam resistivity survey

The geophysical survey data collected at the terminal moraine revealed massive dead ice in the core region, a few meters from the surface, whereas, near the water outlet dead ice was found at deeper levels. The dead ice layer is overlain by ice-debris mixture. The melting of this dead ice may lead to subsidence of the moraine ridge in future. Another important observation is the presence of low resistivity material at deeper level along the survey profile close to water body side. This low resistivity could be either because of saturated/partially-saturated end-moraine or due to piping in the end moraine. Relatively lower resistivity values close to the lake side of the end moraine can be attributed to the partial melting of the buried ice brought up by relatively warmer water body of the lake water in contact.

2D Electrical Resistivity survey was carried out on the eastern flank of the moraine with 5 longitudinal profiles, which on visual interpretation seemed sensitive. Among them, three profiles were drawn along the SW–NE directions with 10 m electrode spacing covering a distance of 240 m and two profiles were drawn from SE-NW directions with electrode 7 m spacing covering a distance of 168 m. The distance between each profile was fixed around 50 to 60 ft.

The apparent resistivity (variation of resistivity represented by different colors) with respect to depth for all profiles are shown in Fig. 3A–D.

Range of resistivity values are shown in Table 1.

The processed image profiles show different zones of very high and low electrical resistivity distribution. Based on the above data, different ranges were assigned to different types of material (Table 2).

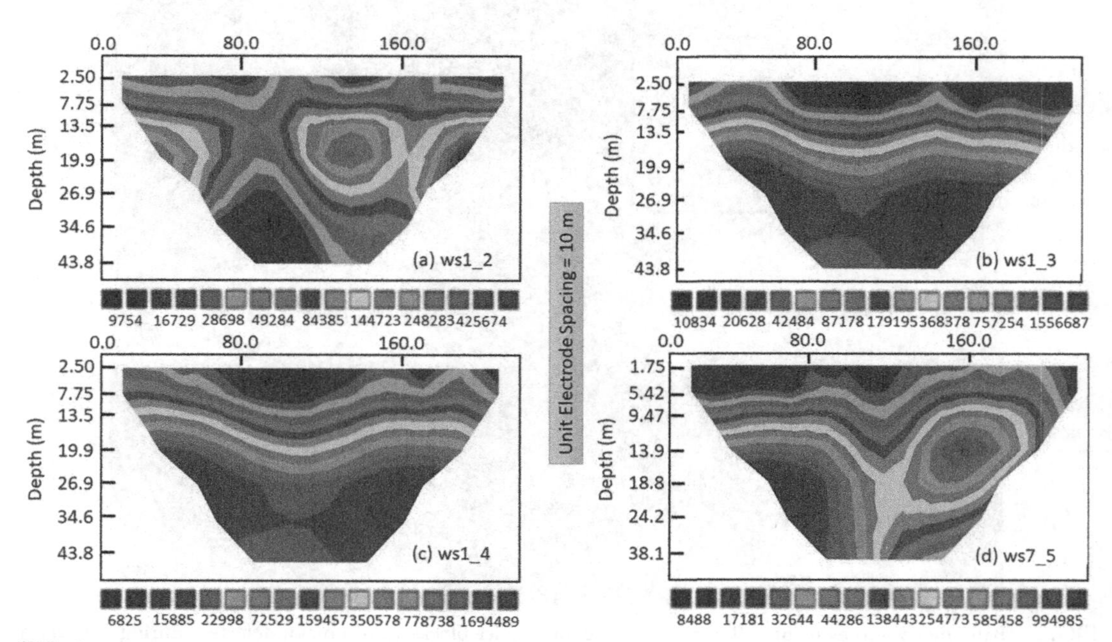

FIG. 3 Resistivity survey profiles of South Lhonak Moraine Dam.

TABLE 1 Resistivity survey data of South Lhonak moraine dam.

Profile name	Range of resistivity (ohm-meter)
ws1_2	9754–425,674
ws1_3	10,034–1,556,687
ws1_4	6825–1,694,489
ws7_1	8688–994,965

TABLE 2 South Lhonak moraine dam composition based on resistivity survey.

Zone	Range of Resistivity (ohm-meter)	Depth (meter)			
		ws1_2	ws1_3	ws1_4	ws7_1
Saturated materials	<2500	–	–	–	–
Partially saturated materials	2000–4000	–	–	–	–
Unsaturated materials	4000–8000	–	–	2.5–3.0	–
Ice bonded permafrost	8000–30,000	2.5–7.0 26.0–43.0	2.5–7.0	2.5–7.0	1.75–8.0
Dead Ice	>30,000	7.0–26.0 7.0–43.0	7.0–43.0	7.0–43.0	8.0–30.1

2.2.5 Glacial lake water level monitoring data

Water level in glacial lakes is monitored by in-house developed all-weather capable in-situ ultrasonic level-sensing system. The data acquisition part consists of ultrasonic sensors and a data logger. The data is transmitted via INSAT satellite communication system to a Central Station. This data is used for monitoring glacial lakes for GLOF events in near-real time and for the prediction of flood inundation (Kumar et al., 2020).

ULTRA-SONIC LEVEL SENSING SYSTEM

The water level monitoring system comprises of three ultrasonic sensors, one an Open-Air Level Sensor (OAS) to measure the distance to the water from the air and two Under-Water Level Sensors (UWS) to measure water level below and above it. The equipment is powered by a solar panel and the sensor data is stored in the data logger.

The sensors operate in two modes to conserve the battery power. In Normal Mode, the sensors are powered every 10 min and in Emergency Mode, the sensors are powered every 2.5 min. The threshold for differentiating normal and emergency mode is programmable and varies depending on the lake characteristics (Kumar et al., 2020).

WATER LEVEL AND ICE THICKNESS CALCULATION

The sensors are mounted on the lower and upper clamps (at known distance labeled "X") for recording the water level and the floating ice thickness (if any). The sensor fixed to the

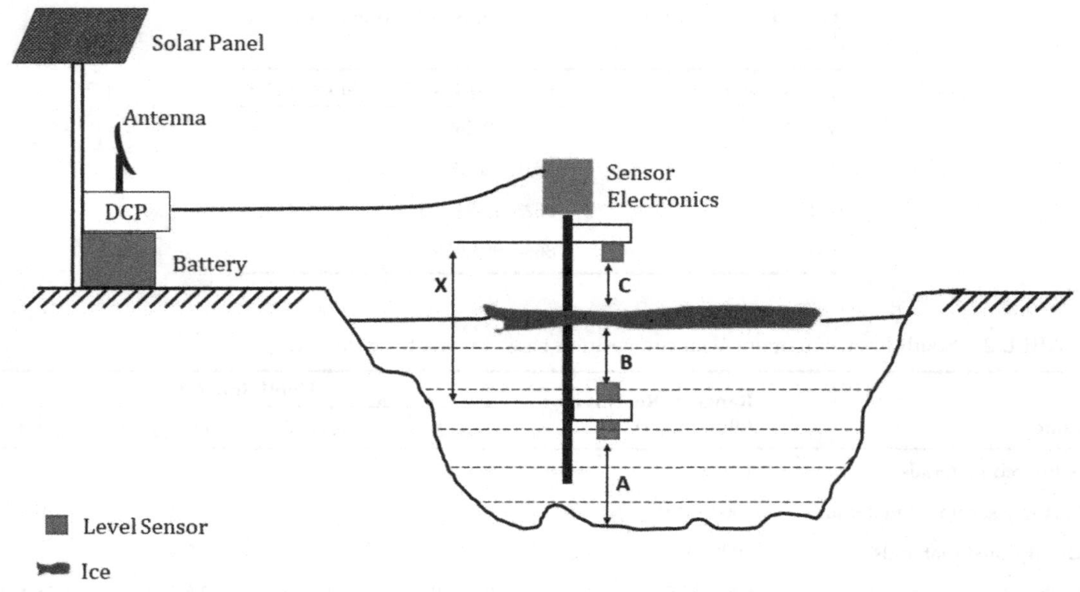

FIG. 4 Sensor mounting arrangement.

lower side of the lower clamp measures the depth of the lake from its location (labeled "A"), whereas the sensor on the upper side of the lower clamp measures the water level to the bottom of the floating ice (labeled "B"). Another sensor (in air) has been fixed to the pole at a known distance "X" from the water sensor. The sensor fixed on the upper clamp measures the depth to the surface of the floating ice (labeled "C") (Fig. 4).

The water level & ice thickness are computed as follows;

$$\frac{Total\ Water\ Level}{Depth} = A + B$$

$$\frac{Level}{Thickness\ of\ floating\ ice} = X - B - C$$

WATER LEVEL CALCULATIONS

The sensors measure the travel time of sound waves to compute water level. The velocity of sound is 347 m/s and 1502 m/s in air and water, respectively.

$$Level = \frac{Sound\ Velocity * Travel\ Time}{2}$$

The signal is sampled at 50 Ksps. The sample count from the transmission trigger to the echo is used for computing travel time.

$$Travel\ Time = Sample\ Count * 20\ \mu sec$$

$$Level = \frac{Sample\ Count * 20 * Sound\ Velocity}{20,000\ cm}$$

Temperature correction for Level:

$$Level\ at\ Temperature,\ T = \frac{Level * Sound\ Velocity\ at\ Temperature\ T}{Sound\ Velocity\ at\ Temperature\ 27^{\circ}C}$$

2.3 Data interpretation / analysis / computation

2.3.1 *Glacial lake area change map*

All the glacial lakes in Sikkim were delineated using remote sensing imagery, published maps, reports and field data, for the period 1965 to 2010. Altogether 771 lakes of both glacial and non-glacial in origin were mapped, 233 lakes were identified as glacial and were studied in detail. The lake area change maps for North and South Lhonak glacial lakes are presented (Fig. 5).

Moraine-dammed lakes are important in monitoring of disaster-prone zones in high-altitude regions. The accumulation of melt water in the lakes has significantly increased the area and volume of the glacial lakes in the past few decades.

The lakes in West and North Sikkim are expanding due to accelerated glacial retreat and melting. The lakes have been increasing in size and volume since 1965. New lakes have also developed in the last few decades (Table 3).

The increase in area of the Lhonak and South Lhonak glacial lakes over the last 45 years is significant. Both these lakes have grown in area by 2 times between 1965 and 1989. The Lhonak has grown nearly 1.5 times and South Lhonak nearly 2.5 times of their initial size in 1989. The areal change of glacial lakes between 1965 and 2010 is illustrated in Fig. 6.

2.3.2 *Estimation of lake volume*

The lake volumes were calculated from the bathymetry data. The bathymetric study carried out in South Lhonak lake showed that the lake area is 126 ha with a measured depth of 260 ft (maximum range of the depth sounding device). The estimated storage volume of the lake is 5,36,38,863.54 m^3.

The volume of the lake was indirectly estimated using Survey of India (SOI) topographical map at 1:50000 and the DEM data. This estimation is approximate and depends highly on the resolution of DEM used, which in the present case is coarse. However, for inferring a worst-case scenario, the volume calculated is used for simulations.

2.3.3 *Heron's method*

The volume of a frustum of a circular cone has been applied by limnologists and biologists to compute lake volume (Taube, 2000);

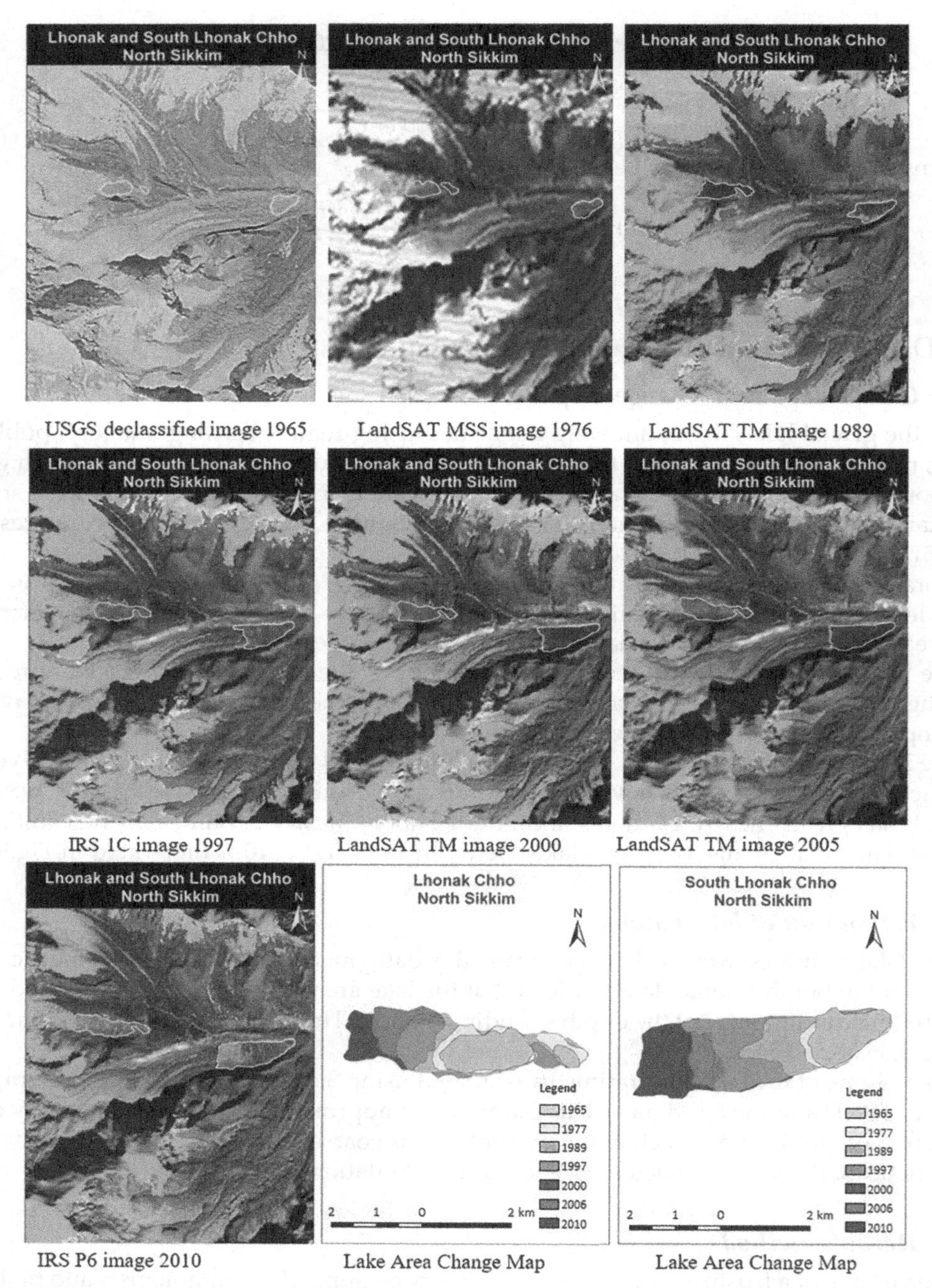

FIG. 5 Time-series analysis maps of North & South Lhonak Lakes, North Sikkim (after Kumar and Murugesh Prabhu, 2012).

TABLE 3 Statistics showing growth in the area of glacial/moraine dammed lakes over the years.

Lake name	Year						
	1965	1976	1989	1997	2000	2005	2010
Gurudongmar Chho A	1.048	1.099	1.099	1.099	1.104	1.115	1.134
Gurudongmar Chho B	0.249	0.322	0.925	1.046	1.046	1.073	1.076
Gurudongmar Chho C	0.480	0.687	0.718	0.728	0.732	0.745	0.745
Chho Lhamo	0.649	0.963	1.031	1.031	1.031	1.031	1.031
Khangchung Chho	1.178	1.261	1.605	1.630	1.661	1.661	1.734
Lachen Khangse Chho	0.360	0.370	0.516	0.523	0.586	0.613	0.613
Glacial Lake feeding river Shako Chhu	0.273	0.409	0.561	0.561	0.561	0.561	0.561
Khora Khang Chho	0.166	0.217	0.269	0.296	0.302	0.342	0.351
South Lhonak Chho	0.242	0.251	0.410	0.633	0.691	0.794	1.028
Lhonak Chho	0.231	0.282	0.418	0.460	0.494	0.652	0.656
Bhale Pokhari	0.090	0.104	0.108	0.114	0.114	0.114	0.114
Glacial Lake feeding river Tikip Chhu	0.069	0.108	0.214	0.257	0.308	0.311	0.311

The lake areas are in sq. km (Kumar and Murugesh Prabhu, 2012).

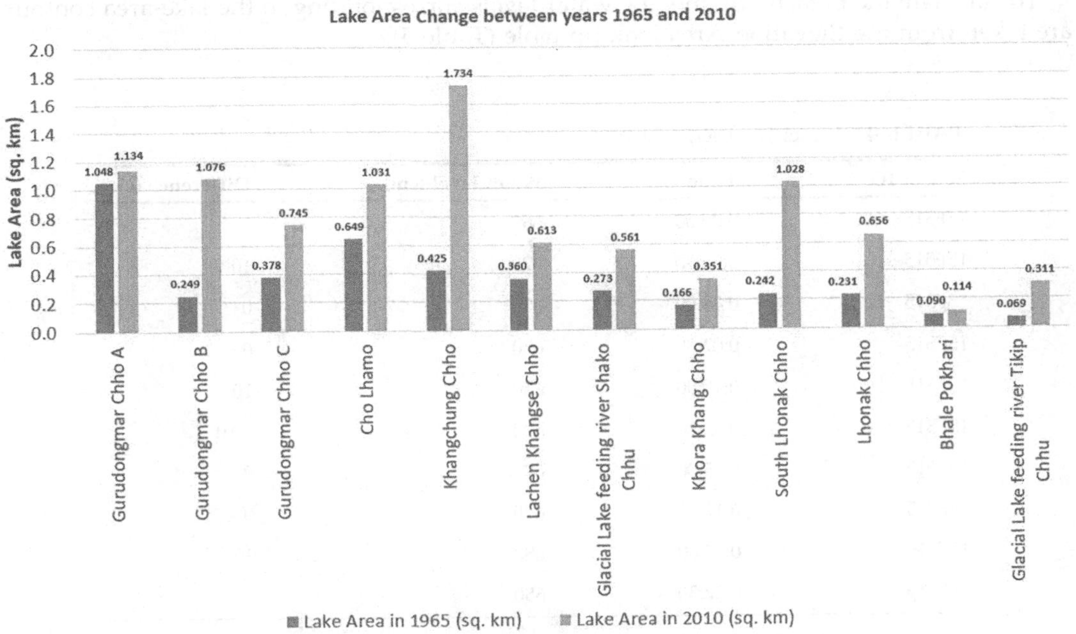

FIG. 6 Glacial lake area change (1965 to 2010).

$$V = \frac{1}{3} * H * \left(A1 + A2 + \sqrt{A1*A2}\right)$$

where

V = volume of water;
H = difference in depth between two successive depth contours;
$A1$ = area of the lake within the outer depth contour being considered;
$A2$ = area of the lake within the inner contour line under consideration;
The procedure consists of determining the volumes of successive layers of water (frustums) to obtain the total volume of lake.

2.3.4 Discharge estimation

The breach discharge is calculated after the water level crosses a preset threshold value. This is based on the "rate-of-change of water level with time" (denoted as dh/dt). If the value drops 5 cm between the last and the current reading and compared with the past half-an-hour readings, the sensor starts recording water level every 2.5 min and the discharge is calculated.

Table 4 shows the fluctuations in lake water level and the possibility for false alarms. In this case, the GLOF happened at 00:52:30, as is evident from the consistent drop in water level.

Since the fluctuations in lake level exceed 5 cm at times, there is a chance for false-alarms. Hence, the water level data is normalized before comparing with the threshold value. The water level time series is checked for consistent drop in values, before sounding the alarm and triggering the GLOF simulation model.

To calculate the breach volume, the water levels corresponding to the lake-area contours are taken from the Elevation-Area look-up table (Table 5);

TABLE 4 Water level sensor data.

Sensor ID	Time	Water level (cm)	Difference (cm)
15F515	0:00:00	660	–
15F515	0:02:30	670	10
15F515	0:05:00	670	0
15F515	0:07:30	670	0
15F515	0:10:00	680	10
15F515	0:12:30	670	−10
15F515	0:15:00	670	0
15F515	0:17:30	670	0
15F515	0:20:00	680	10
15F515	0:22:30	680	0

TABLE 4 Water level sensor data.—Cont'd

Sensor ID	Time	Water level (cm)	Difference (cm)
15F515	0:25:00	670	−10
15F515	0:27:30	670	0
15F515	0:30:00	650	−20
15F515	0:32:30	650	0
15F515	0:35:00	640	−10
15F515	0:37:30	640	0
15F515	0:40:00	620	−20
15F515	0:42:30	660	40
15F515	0:45:00	630	−30
15F515	0:47:30	610	−20
15F515	0:50:00	610	0
15F515	0:52:30	600	−10
15F515	0:55:00	500	−100
15F515	0:57:30	500	0
15F515	1:00:00	500	0

TABLE 5 Elevation-area look up table.

Water level (cm)	Elevation contour corresponding to water level (m)	Contour area (sq. m)
500	5259	5,571,000
600	5260	5,549,000
700	5261	5,505,000
701	5262	5,476,000
702	5263	5,446,000
703	5264	5,438,000
704	5265	5,449,000
705	5266	5,456,000
706	5267	5,465,000
707	5268	5,477,000

Volume = Area difference between two elevation contours (A)* water level difference (h)

Therefore,

$$\text{Volume} = \frac{1^*}{3}(5260 - 5259) * \left(5549000 + 5571000 + \sqrt{5549000 * 5571000}\right)$$
$$= 5559996.373 \text{ m}^3$$
$$\text{Discharge} = 5559996.373 \text{ m}^3 \text{ in } 2.5 \text{ min}$$
$$= 37066.643 \text{ m}^3/\text{s}$$

2.4 GLOF simulation model setup

The model routes the flood waters in the event of GLOF at South Lhonak Chho. The event could be triggered by cloud burst, earthquake, avalanche, dam failure due to piping or ice core melting (Veh et al., 2020). This section illustrates a hypothetical scenario of breach in the lake and simulation carried out to understand the impact of GLOF.

The sequence of GLOF event and early warning can be summarized as follows;

(a) Melt water from parent glacier is stored behind the moraine dams
(b) Water level is monitored every 10 min by in-situ sensors
(c) Sensor data transmitted and received at central station via satellite communication system
(d) Moraine dam breaks, releasing huge amount of water through the dam breach section
(e) The sudden drop in lake level is sensed by the system
(f) The sensor starts recording data every 2.5 min
(g) The GLOF simulation model compares the past hour data with the latest data
(h) Model ascertains the lowering of glacial lake water level
(i) Discharge is calculated with the help of look-up table & the simulation is triggered (Sound Alert as Computer Audio at Central Station)
(j) The simulation model estimates flood level at different locations downstream
(k) Flood inundation map and arrival time at villages downstream are generated
(l) System presents these outputs in map/graph/table formats through the SDSS

2.4.1 Mesh generation

The study area is discretized in to a triangular mesh (47,47,380 cells), with maximum cell size of 900 m^2. The mesh was generated using Sikkim State boundary (vertex locations of the polygon) based on Modified-Delaunay's method. The mesh is stored in NetCDF format (.tsh). To generate such high density of elements a high memory node (HM Node) on the HPC system was used, to avoid memory outflow issues.

2.4.2 Initial conditions: Initial lake level and elevation assignment

The elevation values from the DEM are assigned to the vertices of the mesh. This is represented as the bed elevation in the model, over which the flow is computed.

The current water level in the glacial lakes just before the GLOF event was set as the initial value in the model. After the breach, the flood waters are routed over the terrain from cell-to-cell in the mesh, as the shallow water wave equations are solved using a finite volume scheme. The water level and velocity are tracked over time, until 5h (the simulation duration).

2.4.3 Boundary conditions

A boundary condition defines the state of the cells/elements attached to the computational boundary (domain). Any numerical scheme needs boundary conditions to take care of non-availability of mesh elements outside the domain. In this study, the boundaries are set as reflective condition, as the water is not permitted to flow out of the area. It also satisfies the watershed condition, as the boundary falls on the ridges.

2.4.4 Flow algorithm

The Shallow Water Equations (2022) describe the behavior of a fluid, in particular water, of a certain depth h in a two-dimensional domain, b the elevation of river bed. This scheme neglects the effects of flow in the vertical direction.

The shallow water equations describe the changes of water depth h and horizontal velocities v_x and v_y (in the respective coordinate directions) over time, depending on some initial conditions. The respective changes in time can be described through a system of partial differential equations (https://www5.in.tum.de/SWE/doxy/):

$$\frac{\partial h}{\partial t} + \frac{\partial(v_x h)}{\partial x} + \frac{\partial(v_y h)}{\partial y} = 0$$

$$\frac{\partial(hv_x)}{\partial t} + \frac{\partial(v_x v_x)}{\partial x} + \frac{\partial(v_y v_x)}{\partial y} + \frac{1}{2} * g * \frac{\partial\left(h^2\right)}{\partial x} = -gh\frac{\partial b}{\partial x}$$

$$\frac{\partial(hv_y)}{\partial t} + \frac{\partial(v_x v_y)}{\partial x} + \frac{\partial(v_y v_y)}{\partial y} + \frac{1}{2} * g * \frac{\partial\left(h^2\right)}{\partial y} = -gh\frac{\partial b}{\partial y}$$

The equation for h is obtained, if we examine the conservation of mass in a control volume. The equations for hv_x and hv_y result from conservation of momentum. The two terms involving g model a gravity-induced force (g being the constant for the gravitational acceleration, $g = 9.81\,ms^{-2}$), which results from the hydrostatic pressure. The right-hand-side source terms model the effect of an uneven river bed (b obtained from respective bathymetry data).

ANUGA Hydro (2021) has options to run the numerical scheme with varying polynomial order, viz., DE0, DE1, DE2 (DE stands for Discontinuous Elevation). Based on the recommendation by the model developers and trial simulations, DE1 was used to achieve the required accuracy, optimum compute time and numerical stability. DE1 invokes the numerical solver, Discontinuous Elevation of polynomial order 2 which gives stability when using sudden bed gradients in the model.

2.5 GLOF early warning

2.5.1 Flood inundation map

The GLOF simulation model based on ANUGA Hydro has the flexibility to read and write data during the computation. The simulation model yields the flood progress every one hour after the GLOF event, enabling the disaster manager to track the impact as the flood progresses (Fig. 8).

2.5.2 Flood arrival time and height

The flood arrival time is calculated based on the appearance of water depth value in the simulated output at the village. The flood height changes with time, as the flood progresses downstream from the breach site.

2.5.3 Improved flood forecast lead time with HPC

The simulation was run on PARAM SHAKTI HPC. The system is based on a heterogeneous and hybrid configuration of Intel Xeon Skylake processors and NVIDIA Tesla V100. It consists of 2 Master nodes, 8 Login nodes, 10 Service/Management nodes and 442 (CPU+GPU) nodes with total peak computing capacity of 1.66 (CPU+GPU) PFLOPS.

The 5-h simulation computed over 47,47,380 mesh elements was completed in 31 min on 15 Compute Nodes. This lead time warning is significant to prepare the disaster response plan in the event of GLOF at South Lhonak Lake.

2.6 GIS-based spatial decision support system

Open source tools and packages were used to develop a comprehensive Spatial Decision Support System (SDSS), with sub systems—Glacial Lake Management System and GLOF Early Warning System (EWS). The SDSS helps the user to create, modify and store information (both spatial and non-spatial) related to all the glacial lakes. The GLOF EWS helps the user to monitor water level changes in the glacial lakes (where the sensors are deployed) and automatically generate inundation maps to sound early warning, in case of a GLOF event. Various Free and Open Source Software (FOSS) tools and software packages such as ANUGA Hydro, Q-GIS, PostgreSQL, and PyQt were extensively used to develop the system.

2.6.1 Spatial query

The glacial lakes mapped using multi-temporal data can be queried based on user requirements. For example, the user can query a lake extent by name and the year of mapping.

2.6.2 Sensor settings

The threshold values and breach location of the glacial lakes can be set here. Since the lakes are at remote inaccessible locations, this functionality is indispensable.

2.6.3 System settings

The settings are centralized and customizable at one place under this functionality. The simulation parameters required to run GLOF scenarios can be set under this tab.

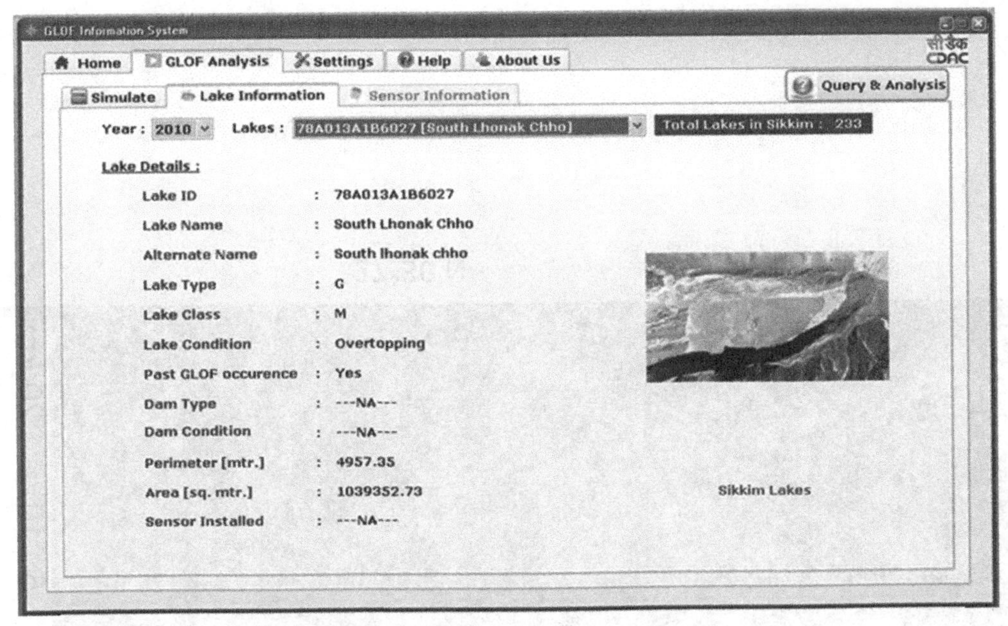

FIG. 7 Lake database management functionality.

2.6.4 *Lake database management*

The glacial lake properties collected based on mapping, field work and other ancillary data are consolidated under this tab (Fig. 7).

3 Results and discussion

The South Lhonak GLOF simulation was carried out to understand the impact of flash flood on the infrastructure and life. The water from the lake flows downstream, causing catastrophic flooding impacting the settlements, industries and tourist locations (Fig. 8A, B, and C).

Fig. 8 shows the impact of probable GLOF scenario at Chungthang, a major town in North Sikkim District, located about 95 km from Gangtok, with a population of around 4000. Along the River Teesta at Chungthang, many hydropower projects and industrial establishments may severely get affected due to GLOF.

The flood arrival time and flood depth at villages downstream are shown in Table 6.

4 Salient features of GLOF Early Warning System

a. First-of-its-kind GIS-based GLOF Early Warning System in India
b. Sensors developed indigenously by C-DAC

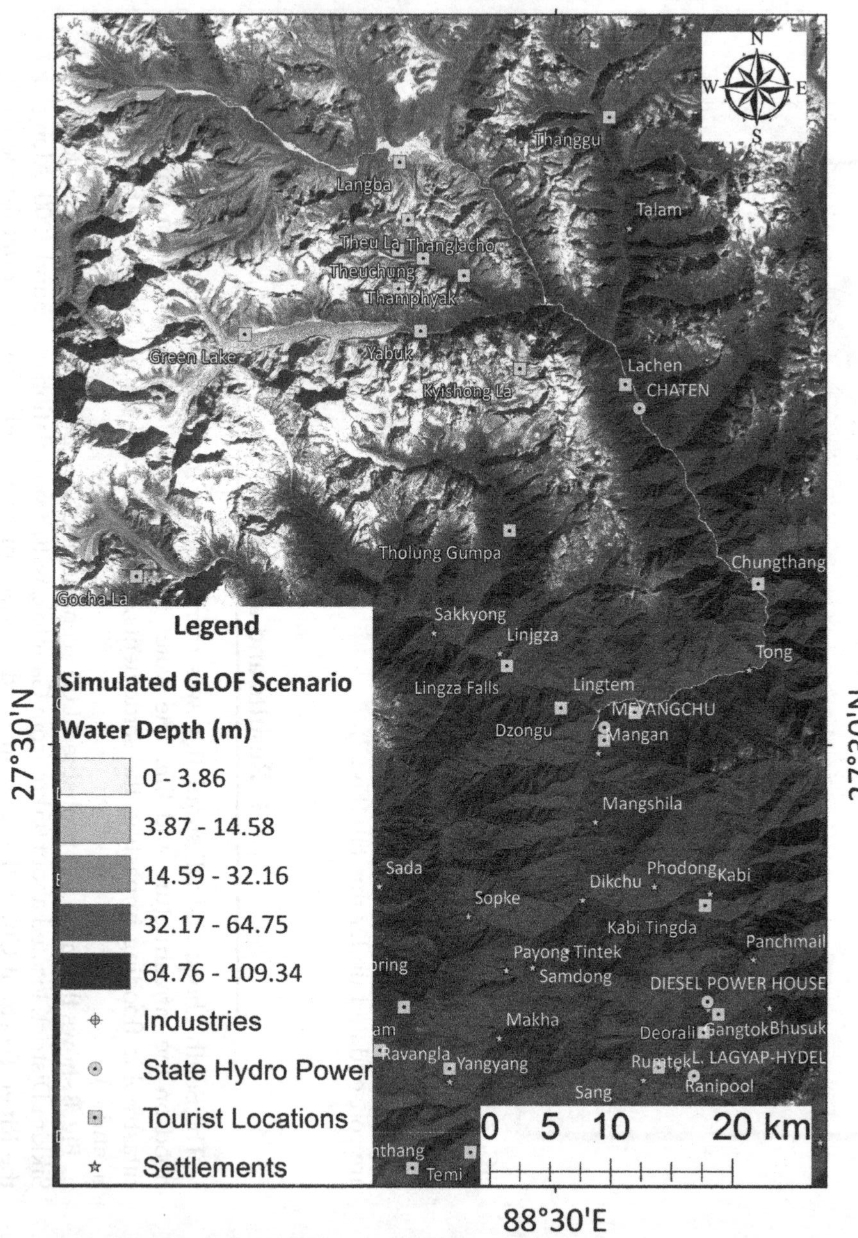

FIG. 8 Simulated flood inundation—South Lhonak GLOF Scenario.

TABLE 6 Flood arrival time and depth at locations downstream.

Station Name	Lachen	Chunthang	Dikchu	Singtam	Mazitar	Rangpo
Flood Arrival Time	57 min	1 h 24 min	2 h 20 min	3 h 26 min	3 h 42 min	3 h 48 min
Water Level (m)	22.58	26.525	21.289	17.531	13.419	10.018

c. Developed on Open-Source technology
d. Real-time monitoring of Glacial Lakes
e. Automated system capable of handling different GLOF scenarios
f. Predicted time of arrival of flood at selected locations
g. Inundation map/Flood Level of selected villages/habitations
h. Exhaustive database of Glacial and Non-glacial lakes
i. Easy-to-use GUI and User-defined Query functionality

5 Limitations of the study

The GLOF Early Warning System deployed in Sikkim runs various GLOF scenarios and generates flood inundation maps. However, the system is not integrated with siren system to alert people living close to river. The ultrasonic sensor system measures the lake water level and requires additional resources to convert level into discharge. In the existing system, debris flow and suspended sediment transport is not computed. The lakes are located in extreme inaccessible restricted areas, due to which field verification of all vulnerable lakes could not be done. Due to very high costing of equipment and logistics, installation of sensors was limited to only 2 vulnerable lakes. Sensor data transmission is possible only through satellite communication system. The sensor system is powered by solar energy and requires being in sleep mode intermittently to conserve battery.

6 Recommendations

The installation of GLOF Early Warning System to monitor all vulnerable glacial lakes is highly recommended. It is recommended to have glacial lake discharge-monitoring sensors along with water level monitoring sensors, Automatic Weather Stations, camera for remote visualization and evaluation of glacial lakes, subject to power and bandwidth availability. Dedicated Data Transmission and Reception System through Satellite Communication System, V-SAT System, RF Systems (erection of Relay Stations/Towers) and Video Transmission System will help round-the-clock monitoring of vulnerable lakes.

The central control station should have fully-automated data reception system, high-end, robust, secure data handling system, uninterrupted high speed connecters, data analyses system, GIS-based Spatial Decision Support System, Mitigation and Response system with rehabilitation location, capacity, shortest route, authority information, other administrative tools, GLOF analyses system, data dissemination system, GLOF prediction and early warning alert system. Automated data dissemination system requires installation of siren systems and integration with existing siren system (if any), SMS based updates/alerts to authorities, web-based data dissemination and alert system and data dissemination in local languages.

7 Conclusions

The Himalayan region is witnessing the impact of climate change, with significant glacier retreat, evident from the formation of numerous glacial lakes. These unstable moraine-dammed lakes are vulnerable to breach, owing to the increase in frequency of geo-hazards. Due to extreme climate and remote location, periodic field investigations and monitoring is not feasible. This calls for the deployment of GLOF Early Warning and Monitoring Systems. The flood inundation forecast is a time-critical application, hence the sooner the results obtained, the better disaster mitigation can be undertaken. GLOF modeling requires many complex parameters, including but not limited to lake volume and discharge, terrain conditions, flow routing, roughness, etc. It is very difficult to model with all the parameters for each glacial lake and no two lakes are similar. When multiple glacial lakes breach, the compute power required would be even more, as the spatial extent of the model would increase manifold with the improved forecast lead time to generate the inundation output. Such time-critical computations can only be achieved using HPC systems as is evident from the results presented for the South Lhonak Lake. There are many other vulnerable lakes in Sikkim which require continuous monitoring and justifies the deployment of GLOF EWS, which can handle multiple dam break scenarios. Further work is in progress to include sediment transport to model suspended debris flow.

Acknowledgments

The authors sincerely acknowledge the Ministry of Electronics and Information Technology, Govt. of India, for funding the project, C-DAC for the support in executing the project, SSCST (end-user), Government of Sikkim for field visit, Department of Forest, Government of Sikkim, for permission to visit protected areas, and Ministry of Defense, Government of India, for providing clearances to visit restricted areas. The authors also extend their gratitude and thanks to Space Applications Centre (ISRO), Government of India, for providing satellite communication facility.

References

Aggarwal, A., Jain, S.K., Lohani, A.K., Jain, N., 2013. Glacial lake outburst flood risk assessment using combined approaches of remote sensing, GIS and dam break modelling. Geomat. Nat. Haz. Risk 7 (1), 18–36. https://doi.org/10.1080/19475705.2013.862573.

Aggarwal, S., Rai, S.C., Thakur, P.K., Emmer, A., 2017. Inventory and recently increasing GLOF susceptibility of glacial lakes in Sikkim, Eastern Himalaya. Geomorphology 295, 39–54. https://doi.org/10.1016/j.geomorph.2017.06.014.

Alho, P., Aaltonen, J., 2008. Comparing a 1D hydraulic model with a 2D hydraulic model for the simulation of extreme glacial outburst floods. Hydrol. Process. 22 (10), 1537–1547.

Allen, S.K., Schneider, D., Owens, I.F., 2009. 2009. First approaches towards modelling glacial hazards in the mount cook region of New Zealand's southern Alps. Nat. Hazards Earth Syst. Sci. 9, 481–499.

ALOS Global Digital Surface Model, 2021. ALOS World 3D—30m (AW3D30). https://www.eorc.jaxa.jp/ALOS/en/aw3d30/index.htm. (Accessed 31 October 2021).

ANUGA Hydro, 2021. https://github.com/GeoscienceAustralia/anuga_core. (Accessed 31 October 2021).

Bajracharya, B., Shrestha, A.B., Rajbhandari, L., 2007a. Glacial lake outburst floods in the Sagarmatha region hazard assessment using GIS and hydrodynamic modeling. Mt. Res. Dev. 27 (4), 336–344.

Bajracharya, R.S., Mool, P.K., Shrestha, B.R., 2006. The impact of global warming on the glaciers of the Himalaya. In: Proceedings of the International Symposium on Geo-disasters, Infrastructure Management and Protection of World Heritage Sites Kathmandu. Nepal Engineering College, National Society for Earthquake Technology Nepal, and Ehime University, Japan, pp. 231–242.

Bajracharya, S.R., Mool, P.K., Shrestha, B.R., 2007b. Impact of Climate Change on Himalayan Glaciers and Glacial Lakes: Case Studies on GLOF and Associated Hazards in Nepal and Bhutan. ICIMOD, Kathmandu.

Benn, D.I., Bolch, T., Hands, K., Gulley, J., Luckman, A., Nicholson, L.I., Quincey, D., Thompson, S., Toumi, R., Wiseman, S., 2012. Response of debris-covered glaciers in the Mount Everest region to recent warming, and implications for outburst flood hazards. Earth Sci. Rev. 114 (1–2), 156–174.

Bhambri, R., Mehta, M., Dobhal, D.P., Gupta, A.K., 2015. Glacier Lake Inventory of Uttarakhand, first ed. Allied Printers, Dehradun, Uttarakhand.

Bhardwaj, A., Singh, M.K., Joshi, P.K., Snehmani, Singh, S., Sam, L., Gupta, R.D., Kumar, R., 2015. A lake detection algorithm (LDA) using Landsat 8 data: a comparative approach in glacial environment. Int. J. Appl. Earth Obs. 38, 150–163. https://doi.org/10.1016/j.jag.2015.01.004.

Buchroithner, M.F., 1995. Jökulhlaup Mapping in the Himalayas by Means of Remote Sensing. Kartographische Bausteine, Institute for Cartography. 12. Dresden University of Technology, Germany, pp. 75–86.

Buchroithner, M.F., Bolch, T., 2015. Glacier lake outburst floods (GLOFs)—mapping the hazard of a threat to high Asia and beyond. In: Velma, et al. (Eds.), Impact of Global Change on Mountains: Responses & Adaptation. CRC Press, USA, pp. 324–345.

Carrivick, J.L., Tweed, F.S., 2013. Proglacial lakes: character, behaviour and geological importance. Quat. Sci. Rev. 78, 34–52.

Carrivick, J.L., Tweed, F.S., 2016. A global assessment of the societal impacts of glacier outburst floods. Global Planet. Change 144, 1–16.

Clague, J.J., Evans, S.G., 2000. A review of catastrophic drainage of moraine-dammed lakes in British Columbia. Quat. Sci. Rev. 19, 1763–1783.

Clarke, G., Waldron, D., 1984. Simulation of the August 1979 sudden discharge of glacier-dammed flood Lake, British Columbia. Can. J. Earth Sci. 21 (4), 502–504.

Evans, S.G., 1986. 1986. The maximum discharge of outburst floods caused by the breaching of man-made and natural dams. Can. Geotech. J. 23, 385–387.

Fernández, P., Fornero, L., Maza, J., Yanez, H., 1991. Simulation of flood waves from outburst of glacier-dammed lake. J. Hydraul. Eng. 117 (1), 42–53.

Gardelle, J., Arnaud, Y., Berthier, E., 2011. Contrasted evolution of glacial lakes along the Hindu Kush Himalaya mountain range between 1990 and 2009. Global Planet. Change 75 (1–2), 47–55.

Hewitt, K., 2016. The human ecology of disaster risk in cold mountainous regions. In: Cutter, S.L. (Ed.), Oxford Research Encyclopedia—Natural Hazard Science. Oxford University Press, Oxford.

Holmes, A., 1945. Principles of Physical Geology. Thomas Nelson and Sons Ltd, pp. 204–252.

Huggel, C., Haeberli, W., Kääb, A., Bieri, D., Richardson, S., 2004. An assessment procedure for glacial hazards in the Swiss Alps. Can Geotech 41, 1068–1083.

Huggel, C., Kääb, A., Haeberli, W., Krummenacher, B., 2003. 2003. Regional-scale GIS-models for assessment of hazards from glacier lake outbursts: evaluation and application in the Swiss Alps. Nat. Hazards Earth Syst. Sci. 3, 647–662.

Huggel, C., Kääb, A., Haeberli, W., Teysseire, P., Paul, F., 2002. Remote sensing based assessment of hazards from glacier lake outbursts: a case study in the Swiss Alps. Can. Geotech. J. 39, 316–330. https://doi.org/10.1139/t,01-099.

IPCC, 2007. Climate Change 2007 Synthesis Report, Intergovernmental Panel on Climate Change [Core Writing Team IPCC]. https://doi.org/10.1256/004316502320517344.

Ives, J.D., Shrestha, R., Mool, P., et al., 2010. Formation of Glacial Lakes in the Hindu Kush-Himalayas and GLOF Risk Assessment. ICIMOD Kathmandu.

Kershaw, J.A., Clague, J.J., Evans, S.G., 2005. Geomorphic and sedimentological signature of a two-phase outburst flood from moraine-dammed Queen Bess Lake, British Columbia, Canada. Earth Surf. Process. Landf. 30, 1–25.

Kumar, B., Murugesh Prabhu, T.S., 2012. In: Arrawatia, M.L., Tambe, S. (Eds.), Impacts of Climate Change: Glacial Lake Outburst Floods (GLOFs). Climate Change in Sikkim: Patterns, Impacts and Initiatives. Published by: Information and Public Relations Department, Govt. of Sikkim, pp. 81–102. ISBN 978-81-920437-0-9.

Kumar, B., Sathyan, A., Prabhu, T.S.M., Krishnan, A., 2020. Design architecture of glacier Lake outburst flood (GLOF) early warning system using ultrasonic sensors. In: 2020 IEEE Recent Advances in Intelligent Computational Systems (RAICS), Thiruvananthapuram, India, pp. 195–200, https://doi.org/10.1109/RAICS51191.2020.9332472.

Liu, Q., Mayer, C., 2015. Distribution and interannual variability of supraglacial lakes on debris-covered glaciers in the khan Tengri-Tumor Mountains, Central Asia. Environ. Res. Lett. 10, 14014.

Maskey, S., Kayastha, R.B., Kayastha, R., 2020. Glacial Lakes Outburst Floods (GLOFs) modelling of Thulagi and lower Barun Glacial Lakes of Nepalese Himalaya. Progress Disaster Sci. 7 (2020).

Mungkasi, S., Roberts, S., 2013. Validation of ANUGA hydraulic model using exact solutions to shallow water wave problems. J. Phys. Conf. Ser. 423, 012029. https://doi.org/10.1088/1742-6596/423/1/012029.

Nie, Y., Liu, Q., Liu, S., 2013. Glacial Lake expansion in the Central Himalayas by landsat images, 1990–2010. PLoS One 8 (12), 1–8.

Nie, Y., Sheng, Y., Liu, Q., Liu, L., Liu, S., Zhang, Y., Song, C., 2017. A regional-scale assessment of Himalayan glacial lake changes using satellite observations from 1990 to 2015. Remote Sens. Environ. 189, 1–13. https://doi.org/10.1016/j.rse.2016.11.008.

Richardson, S.D., Reynolds, J.M., 2000. An overview of glacial hazards in the Himalayas. Quat. Int. 65–66, 31–47.

Robson, B.A., Nuth, C., Dahl, S.O., Hölbling, D., Strozzi, T., Nielsen, P.R., 2015. Automated classification of debris-covered glaciers combining optical, SAR and topographic data in an object-based environment. Remote Sens. Environ. 170, 372–387. https://doi.org/10.1016/j.rse.2015.10.001.

Shallow Water Equations, 2022. https://www5.in.tum.de/SWE/doxy/. (Accessed 31 January 2022).

Sheng, Y., Song, C., Wang, J., Lyons, E.A., Knox, B.R., Cox, J.S., Gao, F., 2016. Representative lake water extent mapping at continental scales using multi-temporal Landsat-8 imagery. Remote Sens. Environ. 185 (2016), 129–141. https://doi.org/10.1016/j.rse.2015.12.041.

Taube, C.M., 2000. Instructions for winter lake mapping. Chapter 12 in Schneider. In: James, C. (Ed.), Manual of Fisheries Survey Methods II: With Periodic Updates. Michigan Department of Natural Resources, Fisheries Special Report 25, Ann Arbor.

Thakur, P.K., Aggarwal, S., Aggarwal, S.P., Jain, S.K., 2016. One-dimensional hydrodynamic modeling of GLOF and impact on hydropower projects in Dhauliganga River using remote sensing and GIS applications. Nat. Hazards 83, 1057–1075. https://doi.org/10.1007/s11069-016-2363-4.

UNFCCC, 1992. United Nations Framework Convention on Climate Change, 1992 FCCC/INFORMAL/84 GE.05-62220 (E) 200705. http://www.managingclimaterisk.org/.

Veh, G., Korup, O., Walz, A., 2020. Hazard from Himalayan glacier lake outburst floods. Proc. Natl. Acad. Sci. U. S. A. 117 (2), 907–912. https://doi.org/10.1073/pnas.1914898117.

Vohra, C.P., 2007. Glaciers of India. In: Richard, S., Williams, J.R., Jane, G.F. (Eds.), Satellite Image Atlas of Glaciers of the World, pp. 259–289. U.S. Geological Survey Professional Paper 1386-F-5.

Wang, X., Liu, S.Y., Mo, H.W., Yao, X.J., Jiang, Z.L., Guo, W.Q., 2011. Expansion of glacial lakes and its implication for climate changes in the Chinese Himalaya. Acta Geograph. Sin. 66 (7), 895–904 (in Chinese, with English Abstract).

Wang, X., Shiyin, L., Wanqin, G., Junli, X., 2008. Assessment and simulation of glacier lake outburst floods for Longbasaba and Pida Lakes, China. Mt Res. Dev. 28 (3/4), 310–317.

Worni, R., Huggel, C., Stoffel, M., 2013. Glacial lakes in the Indian Himalayas—from an area-wide glacial lake inventory to on-site and modeling based risk assessment of critical glacial lakes. Sci. Total Environ. 468–469, S71–S84.

Worni, R., Stoffel, M., Huggel, C., Volz, C., Casteller, A., Luckman, B., 2012. Analysis and dynamic modelling of a moraine failure and glacier lake outburst flood at Ventisquero Negro, Patagonian Andes (Argentina). J. Hydrol. https://doi.org/10.1016/j.jhydrol.2012.04.013.

Yamada, T., Sharma, C.K., 1993. Glacier lakes and outburst floods in the Nepal Himalaya. IAHS-AISH Publ 218, 319–330.

CHAPTER

37

Flood forecasting using novel ANFIS-WOA approach in Mahanadi river basin, India

Sandeep Samantaray[a], *Abinash Sahoo*[b], *and Shaswati S. Mishra*[c]

[a]Department of Civil Engineering, NIT Srinagar, Jammu & Kashmir, India [b]Department of Civil Engineering, NIT Silchar, Silchar, Assam, India [c]Department of Philosophy, Utkal University, Bhubaneswar, Odisha, India

OUTLINE

1 Introduction 663
2 Study area 666
3 Material and methods 668
3.1 ANFIS 668
3.2 FOA 668
3.3 WOA 669
3.4 WOA for parameter calibration of ANFIS 670
3.5 Model performance evaluation 671
4 Results and discussion 672
5 Conclusion 678
References 681

1 Introduction

Floods are the most common natural disaster causing severe harm and damage to life and property. According to estimations of the total economic losses generated by all other disasters, 40% are because of flooding. Real-time flood forecasting provides timely warnings to

https://doi.org/10.1016/B978-0-323-91910-4.00037-6

people living in flood plains and can ease several suffering and damages caused by floods. It also offers valuable information for water management personnel to make best decisions associated with reservoir operations and flood control structures. Flood is a natural phenomenon and is characteristically challenging to model.

In relation to World Resource Institute estimate, India has a more significant proportion of the population open to flood destruction compared to any other nation. Several Indian regions suffer from heavy floods each year because of substantial rainfall, predominantly during June to September (monsoon season). In India, floods occur with significant rain in upper watersheds of main river basins like Ganga, Narmada, Yamuna, Sabarmati, Godavari, Mahanadi, Tista, Brahmaputra, and Krishna. Rivers like Baitarani, Brahmani, and Mahanadi flowing in central India have a common delta. With heavy rainfall during monsoon season, these rivers cause damage to Odisha due to associated floods. Most of the heavy rain in various portions of central India, including Odisha, is due to the low-pressure system/depression movement from Bay of Bengal area to central parts (Pattanaik, 2007; Mohapatra and Mohanty, 2005; Soman and Kumar, 1990). In addition, it is disturbing that change in climatic conditions increases the severity and frequency of extreme events (Li et al., 2016). The fundamental aspect of efficient management and operation of the water resource system is providing an accurate and reliable range of hydrological forecasts. It can be utilized for strategic decisions to tactical adjustments about water storage level in any reservoir, and timely water release specifically to reduce flood occurrences.

Flood forecasting is presently categorized into three main classifications, i.e., the physical, statistical, and data-driven approaches (Khac-Tien Nguyen and Hock-Chye Chua, 2012). Physical process comprises conceptual hydraulic and hydrological models. It has easy operation, reliable accurateness, and a clear physical concept. Yet, it also necessitates an assortment of a substantial data quantity related to river basin characteristics, which is limited in maximum circumstances. Statistical method includes autoregressive and moving average model (ARMA), regression analysis, etc. However, it is not used often for flood forecasting since it doesn't sufficiently illustrate the highly nonlinear nature of flooding process, limiting its forecasting abilities (Hsu et al., 1995). An alternative to physical-based models is the data-driven models. Data-driven models describe a system with restricted assumptions, use generalized relationships amid input and output datasets (Solomatine and Ostfeld, 2008), and show improved performance than physical and statistical-based models. A simple architecture signifies that uncertainty from various sources (e.g., model parameters) can be enumerated and distributed through the model. Among several data-driven approaches, artificial neural network (ANN) is most commonly applied as they can adequately approximate the intricate nonlinear relationship of flood events using self-learning procedure. The major disadvantage of ANN is the unexplained behavior of the network. ANN does not explain why and how while producing a probing solution that reduces trust in the network. To solve this problem, FIS can be utilized where reasoning capabilities of the human mind and language of nature can be interpreted into a mathematical expression. However, selection, adjustment of membership functions (MFs), and formation of fuzzy rules have to be accomplished by artificiality. Adaptive neuro-fuzzy inference system (ANFIS) perfectly incorporates FIS and ANN overcoming the absence of human language manifestation in ANN and the shortage of self-learning in fuzzy theory. Thus, ANFIS can increase work efficacy by providing an enhanced simulation of AI.

Few examples of the application of ANFIS in flood forecasting are discussed in Ghalkhani et al. (2013) and Nguyen et al. (2014). In another study, Lohani et al. (2014) employed ANFIS model for flood forecasting considering water level of different rivers and assessed its performance against other artificial intelligence models. Khan et al. (2018) proposed a fuzzy neural network for predicting peak flow in Bow River, Canada. Jabbari and Bae (2018) evaluated real-time bias adjustment of rainfall dataset using ANN, and assessed improvements in hydrological model for forecasting flood events of Imjin River (North and South Korea). They found that the use of ANN bias correction performance of flood forecasts improved. Cheng et al. (2020) applied LSTM (long short term memory) and ANN for streamflow forecasting at daily and monthly level for a long lead-time period using runoff and rainfall data composed from Ping and Nan River basins, Thailand. These findings suggest that LSTM exhibits superior model performance than ANN model.

The fundamental difficulty faced during the learning process of ANFIS is to obtain a good generalization ability. Effective design of ANFIS-based models needs efficient parameter training for improved accurateness. The parameter learning procedure of ANFIS utilizes derivative-based learning, which has a high possibility of being trapped in local minima. The conventional optimization techniques used in ANFIS are unable to deal with nonlinear objective functions. In recent times, swarm-based algorithms have exhibited their advantages in global optimization and improved the performance of conventional neural network models. In general, there are certain benefits of swarm-based algorithms over evolution-based algorithms. Search space information is preserved over successive iterations in swarm-based algorithms, while in evolution-based algorithms, any information is discarded after forming a new population. Usually, they include fewer operators than evolutionary methods (mutation, crossover, elitism, selection, etc.) and hence are simpler in implementation. Generally, swarm-based are stochastic techniques inspired by biological evolution and natural selection. There are several applications of hybrid machine learning algorithms in modeling various hydrological variables (Samantaray and Sahoo, 2021a,b,c; Aoulmi et al., 2021; Sahoo et al., 2021a, b; Malik et al., 2021; Samantaray et al., 2021; Banadkooki et al., 2021; Sridharam et al., 2021; Adnan et al., 2021; Seifi and Riahi, 2020; Mohamadi et al., 2020; Diop et al., 2020; Alizamir et al., 2020).

Sehgal et al. (2014) developed WANFIS-SD (split data) and WANFIS-MS (modified time series) models to forecast river water levels of Rivers Kosi and Kamla with 1-day lead time. Results demonstrated that for high flood levels, WANFIS-SD performed superiorly to WANFIS-MS. Chen et al. (2015) investigated application of differential evolution (DE), ant colony optimization (ACO), and artificial bee colony (ABC) algorithms in combination with a neural network model for forecasting downstream river flow. Results revealed that ACO and DE algorithms are significantly more capable in optimizing forecasting problems. Tan et al. (2017) proposed ANFIS-FOA and ANFIS models for stock market volatility forecasting. Experimental outcomes revealed that ANFIS-FOA successfully and accurately forecasted stock volatility. Nanda et al. (2016) proposed WNARX (wavelet-based nonlinear autoregressive with exogenous inputs) and assessed it against dynamic NARX, wavelet-based ANN (W-ANN), static ANN, and ARMAX (linear autoregressive moving average with exogenous inputs) models using satellite-based rainfall products for flood forecasting in Basantpur gauge station on River Mahanadi. Results confirmed strength of WNARX in flood forecasting at proposed site. Kumar and Sahay (2018) developed a wavelet based genetic programming (W-GP) model for flood forecasting in River Kosi considering historical daily flow data and assessed its performance against simple AR and GP models. Findings from the study revealed that W-GP models simulated extreme floods superior than GP and AR. Ni et al. (2020) suggested a hybrid model combining Gaussian mixture model with extreme gradient boosting (GMM-XGBoost), for monthly stream flow forecasting and assessed its potential against simple XGBoost and SVM models. It is inferred from results that GMM-XGBoost model performed superiorly than standalone models. Mohammadi et al. (2020) proposed hybrid ANFIS-SFLA (shuffled frog leaping algorithm) model for stream flow time series prediction using historical stream flow data of two different rivers. Performance of proposed model is investigated taking six different input–output scenarios. Results showed that developed hybrid model considerably improved forecasting accurateness and outperformed classical ANFIS model in stream flow forecasting. Samadianfard et al. (2020) used multilayer perceptron model (MLP) and MLP-WOA to predict wind speed at specified stations in northern Iran, with a limited dataset (2004–14). Results obtained from MLP-WOA were compared with MLP-GA (Genetic algorithm) and found that WOA algorithm improved wind speed prediction accuracy. Sattari et al. (2021) aimed at estimating monthly and seasonal rainfall applying SVM, Gaussian Process Regression (GPR), *k*-nearest neighbor (KNN) method and SVM-FOA (fruit fly optimization algorithm) models in Urmia station located in West Azarbaijan province, Iran. Sun et al. (2021) evaluated performance of hybrid SVM-FOA model in scour geometry prediction below ski-jump spillways by comparing prediction results against SVM, ANFIS, ANN, and regression models. Results showed that proposed SVM-FOA method performed well and improved performance of SVM remarkably.

Optimization of artificial intelligence models has always been an open investigation. It is vital for solving shortcomings of conventional machine learning models, such as slow convergence speed, easily plunging into local optima, and poor generalization. Therefore, the major aim of this investigation is to integrate nature-based optimization algorithms (i.e., FOA and WOA) into an ANFIS model and compare their optimization stability, reliability, and capability, and in this manner conclude most adaptive optimization algorithm for flood forecasting problem.

2 Study area

For this study, Mahanadi river basin was selected. It is the fourth largest basin in India, having a catchment area of 141,589 sq. km and accounting for 4.3% of India's overall geographical region. It originates near Nagri town in Raipur district, Chhattisgarh, from around 442 mts above MSL. The river's total length from its origin till converging with Bay of Bengal is 851 km. Average annual precipitation in this river basin is 1572 mm, of which 70% is precipitated from June to October during southwest monsoon season. The river basin is mainly rain-affected with a tropical climate. There is a seasonal fluctuation and considerable variation in water availability.

Because of its location, Mahanadi is one of the flood-prone basins in India (Fig. 1). Most monsoon systems (depressions/lows) form over Bay of Bengal and cross coastal areas of Odisha moving west-north-westward from June to September during southwest monsoon season (Mohapatra and Mohanty, 2006, 2009). This system causes rainfall in the basin, sometimes leading to severe floodings.

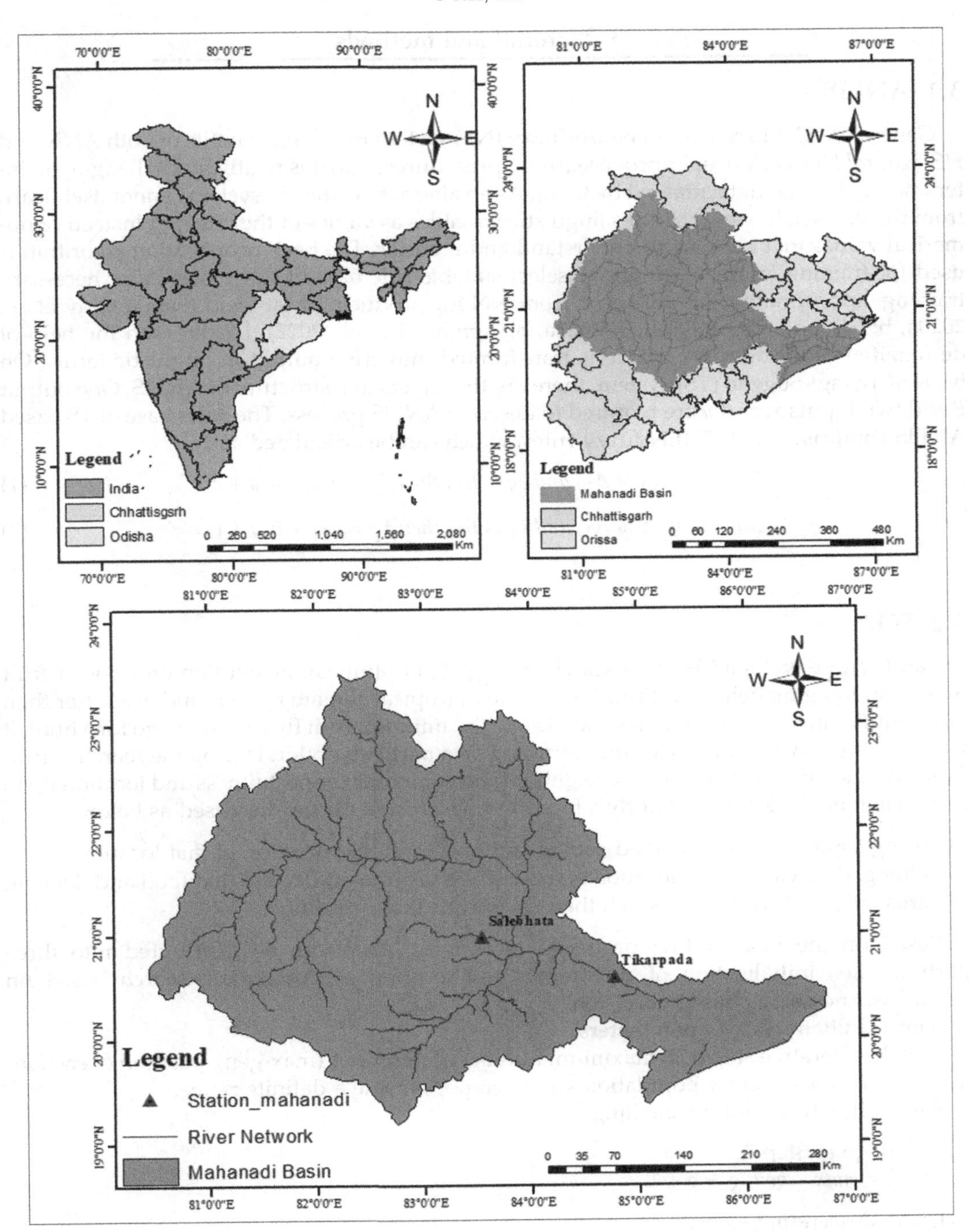

FIG. 1 Location of case study (Johor River Basin).

3 Material and methods

3.1 ANFIS

Generally, ANFIS refers to a neuro-fuzzy (NF) system receiving benefits of both ANN and FIS. Neural Networks can improve learning procedures; but, it is relatively challenging to infer knowledge and data attained by them. As an alternative, the NF system cannot itself learn from the dataset; however, it uses linguistic variables as values of the crisp set instead of numerical values that are easy to understand and monitor. The back-propagation algorithm is used for training inference engine to select suitable rule base in NN block. After necessary training, appropriate rules are shoot from NN for producing best yield (Samantaray et al., 2020a, b; Samantaray and Sahoo, 2021a, b; Agnihotri et al., 2022). Lastly, with the help of de-fuzzifier block output variable is transformed into crisp output in linguistic terms. On basis of Takagi-Sugeno (TS) system, there are five layers in a structure of ANFIS. One output F and two inputs x, and y are required to describe ANFIS process. The rules base of TS based ANFIS comprises of IF-THEN fuzzy rules, which can be articulated as:

$$Rule\,1: if\ (x\ is\ A_1)\ and\ (y\ is\ B_1)\ then\, f_1 = a_1x + b_1y + c_1 \tag{1}$$

$$Rule\,2: if\ (x\ is\ A_2)\ and\ (y\ is\ B_2)\ then\, f_2 = a_2x + b_2y + c_2 \tag{2}$$

3.2 FOA

Pan (2012) developed FOA to search for a global optimization solution on basis of fruit flies' food foraging behavior. Fruit flies reside in tropical climate regions and are better than other entities in terms of osphresis and vision. The minute a fruit fly chooses to go for a hunt, it will randomly fly to search a location directed by a particular odor. During the search, a fruit fly sends and obtains data from its neighbors and compares the best fitness and location up to now (Yuan et al., 2014). A fruit fly's food-finding process can be discussed as below:

(i) using osphresis it smells food source and hovers in the direction of that location;
(ii) after getting close to the location of food, sharp vision is utilized to find food and flocking areas of other fruit flies, which then fly toward that direction.

Based on the food-finding processes of fruit fly, FOA can be distributed into three portions, i.e., initialization of constraints and location of population, search based on osphresis, and search based on vision.

Step 1: Initialization of parameters.

Initialize iterative step (L_0), maximum number of iterations (max-gen), position of each individual (F_i) and fruit fly population size (size-pop) inside a definite range.

Step 2: Iterative hunt by smelling.

(1) $g = 0$, go to Step 3.
(2) $0 < g < max - gen,\ g = g + 1,$

Using descending steps:

$$dL_0 = L_0 - \frac{L_0 \cdot (g-1)}{\max - \text{gen}}$$

Renew each individual position:

$$dF_i = F_{best} - F_i$$

$$F_i = F_i + dF_i \cdot dL_0$$

where F_{best}—optimal location nearest to food source. After that, go to Step 3.

Step 3: Compute flavor's concentration ($Smell_i$).

Input each location (F_i) to ANFIS, and output $E_i = (O_i^5 - d_i)^2$ (where O_i^5—output of ANFIS, d_i—actual value of forecasting).

$$Smell_i = \frac{1}{1 + E_i}$$

Step 4: Position utilizing vision.

$[bestSmell, bestindex] = \max(Smell_i)$,

$$F_{best} = F_{bestindex}$$

Step 5: Iterative optimization.

(1) $g = maxgen$, turn to Step 6.
(2) $0 < g < maxgen$, fly to optimal position, turn to Step 2.

Step 6: Output

Output $F_{bestindex}$.

3.3 WOA

Mirjalili and Lewis (2016) proposed a stochastic optimization algorithm known as the WOA. For optimization problems, it uses a population of search agents for determining global optimum. Like other algorithms based on population, the hunt procedure starts by generating a set of arbitrary solutions (candidate solutions) for a specified problem. Then until satisfying an end criterion, it develops this set. Major difference between other algorithms and WOA is the rules which enhance candidate solutions in every optimization step. Bubble-net feeding behavior is when WOA mimics the hunting behavior of humpback whales to find and attack prey. A humpback whale does the trick by stirring in a spiral track around the prey, generating bubbles along the path. The main inspiration of WOA is this innovative foraging technique. In WOA, the encircling mechanism is another simulated behavior of humpback whales. Using a bubble-net mechanism, the whales loop around preys for starting to hunt them. In this algorithm, the principal mathematical expression proposed is given by:

$$\vec{D} = \left| \vec{C}\,\overrightarrow{X^*}(t) - \vec{X}(t) \right| \tag{3}$$

$$\vec{X}(t+1) = \overrightarrow{X^*}(t) - \vec{A}\vec{D} \tag{4}$$

$$\vec{A} = 2\vec{a} \cdot \vec{r} - \vec{a} \tag{5}$$

$$\vec{C} = 2 \cdot \vec{r} \tag{6}$$

where $\overrightarrow{X^*}$ is the universal best location; $\vec{X}$ is the location of whale; t is the current trial; a is the decreases linearly from 2 to 0; r is an arbitrary number distributed equally amid 0 and 1; $\vec{A}$ and $\vec{C}$ is the coefficient vectors. The stages are explained below:

(a) Exploitation:

The equidistance amid locations of prey and whale is perceived by employing a spiral mathematical approach. After each movement, the location of whale (in a helix environment) has to be adjusted (Kaveh, 2017):

$$\vec{X}(t+1) = D' \cdot e^{bk} \cdot \cos(2\pi l) + \overrightarrow{X^*}(t) \tag{7}$$

where

$$\vec{X}(t+1) = \begin{cases} \overrightarrow{W^*}(t) - \vec{A}\vec{D} \ \ P < 0.5 \\ D' \cdot e^{bl} \cdot \cos(2\pi l) + \overrightarrow{W^*}(t) \ P \geq 0.5 \end{cases} \tag{8}$$

P—random number between 0 and 1

$$D' = \overrightarrow{X^*}(t) - \vec{X}(t) \tag{9}$$

where b and k are the arbitrary and constant numbers. k is equally spread between −1 to 1; b is the logarithmic spiral shape.

(b) Exploration:

When $A < -1$ or $A > 1$, search agent updated at position of an elite agent by an arbitrarily nominated colleague:

In this way, it can be formulated as

$$\vec{X}(t+1) = \overrightarrow{X_{rand}} - \vec{A}\vec{D} \tag{10}$$

$$\vec{D} = \left|\vec{C} \cdot \overrightarrow{X_{rand}} - X\right| \tag{11}$$

in which $\overrightarrow{X_{rand}}$ is the random location of whale in current iteration. More details about WOA can be found in Mirjalili and Lewis (2016). The entire working procedure of WOA is shown in Fig. 2.

3.4 WOA for parameter calibration of ANFIS

First, network data is determined using a matrix, and data consist of training and testing data. Training of ANFIS starts utilizing organized data. Training procedure permits system for adjusting parameters described as input or output of the model.

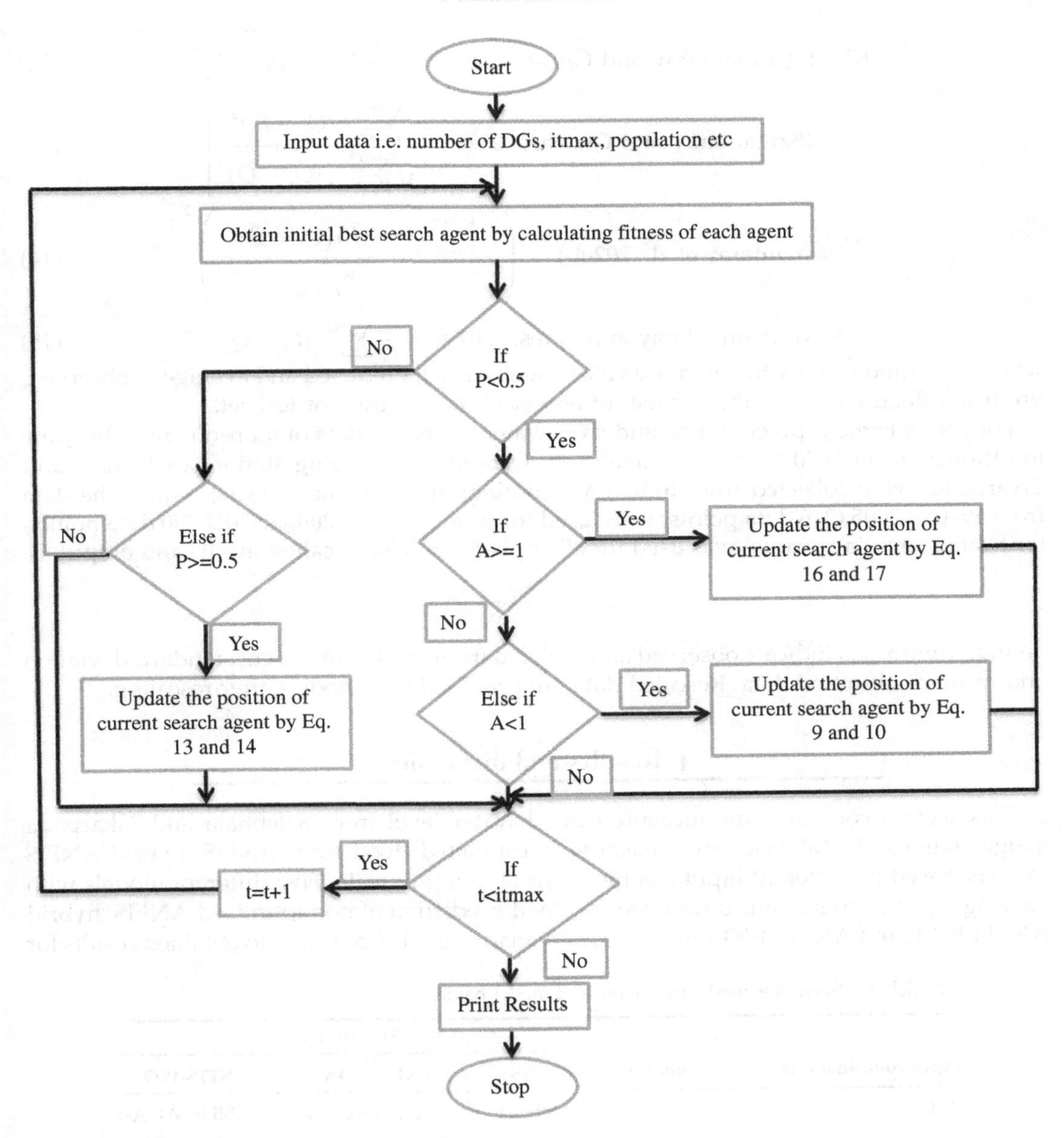

FIG. 2 Flow chart of WOA.

3.5 Model performance evaluation

In present study, RMSE, R^2, MAE, and E_{NS} are applied as quantitative statistical criteria for performance evaluation of proposed methods. The mathematical expressions of applied indices are given by (Samantaray and Ghose, 2019, 2020, 2021a, b; Samantaray et al., 2020b):

$$\text{RMSE (Samantaray and Ghose, 2020)} = \sqrt{\frac{1}{N}\sum_{k=1}^{N}(P_i - Q_i)^2} \tag{12}$$

$$E_{\text{NS}} \text{ (Samantaray and Ghose, 2021)} = \left[\frac{\sum_{k=1}^{N}(Q_i - P_i)^2}{\sqrt{\sum_{k=1}^{N}(Q_i - \overline{Q})^2}}\right] \tag{13}$$

$$R^2 \text{ (Samantaray et al., 2020b)} = \left(\frac{\sum_{i=1}^{n}P_i - \overline{P})(Q_i - \overline{Q})}{\sqrt{\sum_{i=1}^{n}P_i - \overline{P})^2(Q_i - \overline{Q})^2}}\right)^2 \tag{14}$$

$$\text{MAE (Samantaray and Ghose, 2019)} = \frac{1}{N}\sum_{i=1}^{N}|P_i - Q_i| \tag{15}$$

where Q_i, P_i and $\overline{Q}, \overline{P}$ are the monthly values of observed, predicted and average of observed, predicted flood level; n is the number of observed data in train or test set.

For present study, precipitation and river water discharge data of monsoon months (June to October) from 1970 to 2020 are analyzed. Data of two gauging stations (Salebhata and Tikarpada) were collected from Indian Meteorological Department (IMD), Pune. The data from 1970 to 2005 (216 data points) were used for training, from 2006 to 2020 (90 data points) for testing. Standardization was used for all applied models, rescaling inputs and output as:

$$x_{scaled} = \frac{x_{obs} - \mu_x}{\sigma_x}$$

here, x_{obs} and x_{scaled} indicate observed and scaled data, σ_x and μ_x represent standard deviation and mean of observed data. Rescaled data are provided into models after rescaling.

4 Results and discussion

This section compares the measured flood water level from Salebhata and Tikarpada gauge stations of Mahanadi river basin with estimated flood from ANFIS, hybrid ANFIS models, based on different input combinations at monthly scale. Five different models with varying input combinations developed for flood prediction of nonoptimized ANFIS, hybrid ANFIS-FOA, and ANFIS-WOA models are given in Table 1. According to obtained results for

TABLE 1 Scenarios used with inputs and model name.

		Model name		
Input combinations	**Scenario**	**ANFIS**	**ANFIS-FOA**	**ANFIS-WOA**
P_t, Q_t	I	ANFIS-1	ANFIS -FOA-1	ANFIS-WOA-1
P_t, Q_t, Q_{t-1}	II	ANFIS -2	ANFIS -FOA-2	ANFIS-WOA-2
P_t, Q_t, Q_{t-1}, Q_{t-2}	III	ANFIS -3	ANFIS -FOA-3	ANFIS-WOA-3
P_t, Q_t, Q_{t-1}, Q_{t-2}, Q_{t-3}	IV	ANFIS -4	ANFIS -FOA-4	ANFIS-WOA-4

P_t—*Precipitation at "t" month.*
Q_t—*Discharge at "t" month.*
Q_{t-1}—*Discharge at "t−1" month (one month lag).*
Q_{t-2}—*Discharge at "t−2" month (two month lag).*
Q_{t-3}—*Discharge at "t−3" month (three month lag).*

TABLE 2 Performance of ANFIS, ANFIS-FOA, ANFIS-WOA model.

Station name	Model name	RMSE Training	MAE	R^2	E_{NS}	RMSE Testing	MAE	R^2	E_{NS}
Salebhata	ANFIS-1	33.175	36.2458	0.93385	0.9399	35.36	37.9315	0.93049	0.9346
	ANFIS-2	32.687	35.923	0.9347	0.9305	34.821	37.1933	0.93172	0.937
	ANFIS-3	32.28	33.4967	0.93681	0.9421	34.5963	36.825	0.93267	0.9389
	ANFIS-4	31.829	33.234	0.93912	0.9443	32.456	33.8942	0.9354	0.9318
	ANFIS-FOA-1	24.1687	30.012	0.9604	0.964	25.36	30.66	0.9593	0.9628
	ANFIS-FOA-2	20.16	23.034	0.9639	0.9674	23.954	29.985	0.9608	0.9646
	ANFIS-FOA-3	19.7062	22.658	0.96507	0.9687	21.96	28.4192	0.96143	0.9659
	ANFIS-FOA-4	19.2357	22.136	0.96681	0.9693	20.547	26.6571	0.96278	0.967
	ANFIS-WOA-1	13.658	17.8645	0.98764	0.992	15.647	18.768	0.9853	0.9904
	ANFIS-WOA-2	11.24	16.996	0.9891	0.9929	14.98	18.369	0.9874	0.9913
	ANFIS-WOA-3	3.7865	9.5698	0.99142	0.9953	10.492	16.334	0.98965	0.9932
	ANFIS-WOA-4	1.967	8.335	0.9936	0.9968	6.7853	12.001	0.99101	0.9946
Tikarpada	ANFIS-1	35.2279	37.694	0.93108	0.9357	35.8921	38.395	0.9296	0.933
	ANFIS-2	34.226	36.6843	0.9329	0.9394	35.489	38.1642	0.93008	0.9338
	ANFIS-3	32.89	36.114	0.93415	0.9302	35.034	37.48	0.93145	0.9362
	ANFIS-4	32.0036	33.68	0.93765	0.9437	33.961	36.469	0.9333	0.9396
	ANFIS-FOA-1	28.943	31.103	0.9582	0.9619	29.6317	31.5136	0.95554	0.9504
	ANFIS-FOA-2	25.8732	30.853	0.95916	0.962	29.279	31.24	0.95713	0.9511
	ANFIS-FOA-3	21.563	27.649	0.96139	0.9655	25.0067	30.5239	0.9596	0.9632
	ANFIS-FOA-4	21.236	27.1196	0.96214	0.9667	24.5326	30.2381	0.96039	0.9638
	ANFIS-WOA-1	8.054	13.214	0.9905	0.9939	11.985	17.268	0.98875	0.9915
	ANFIS-WOA-2	2.864	6.2165	0.99273	0.996	7.2387	12.52	0.99067	0.9943
	ANFIS-WOA-3	1.584	5.6548	0.99534	0.9976	3.008	9.0541	0.9924	0.9957
	ANFIS-WOA-4	0.894	4.9952	0.99792	0.9984	1.115	5.2167	0.99557	0.9979

conventional ANFIS (Table 2), ANFIS-4 model with RMSE-31.829, MAE-33.234, E_{NS}-0.9443, and R^2-0.93912 during training and RMSE-32.456, MAE-33.8942, E_{NS}-0.9318, and R^2-0.9354 during testing is selected as the best ANFIS model. The best input combination for data indicates that value of flood water level with four-month lag is the most significant factor in monthly flood estimation. However, overall results suggested that ANFIS model did not perform satisfactorily.

Consequently, we tried to improve the performance of ANFIS model by optimizing with FOA and WOA algorithms. In ANFIS-FOA and ANFIS-WOA models, the fruit-fly and whale optimization algorithms (WOAs) were integrated with the nonoptimized ANFIS model. The forecasts by ANFIS-FOA based models based on RMSE, MAE, E_{NS}, and R^2 during training and testing phases are far better than simple ANFIS models. However, compared to ANFIS-FOA models, ANFIS-WOA models produced better flood forecasting results. Among ANFIS-WOA models, ANFIS-WOA4 model exhibited minimum RMSE value of 24.42 m^3/s and MAE value of 16.47 m^3/s) and highest E_{NS} value of 0.652 and R^2 value of 0.864 in testing period. In relation to R^2, N_{SE}, RMSE, and MAE criterion values for training and testing periods, the best architecture had a four-input layer neuronal structure with nine hidden layer neurons and one output layer neuron (4-9-1). The robust ANFIS-WOA model produced the best forecasting results followed by ANFIS-FOA and ANFIS models. For every input combination utilized for flood forecasting, all statistical metrics proved that ANFIS-WOA model performed preeminently.

Performance of ANFIS, ANFIS-FOA, and ANFIS-WOA are also evaluated using graphical representations. For observing the level of agreement amid observed and forecasted flood water levels during the testing period, scatter plots for best ANFIS4, best hybrid ANFIS-FOA (ANFIS-FOA4), and best hybrid ANFIS-WOA (ANFIS-WOA4) models are demonstrated in Fig. 3. The normalized values of the ANFIS-WOA4 model are closer to 45° line ($R^2=0.689$). However, for ANFIS4 and ANFIS-FOA4 models, the values are more scattered from the line, reflecting lower degree accuracy based on the coefficient of determination (R^2).

The results from the table and figures suggest that the forecasting performance for the ANFIS-WOA4 model is superior to ANFIS-FOA4 and ANFIS4 model for long-lead-time monthly forecasting, which the model structure may explain. The ANFIS-WOA4 model processes time-series datasets as an arrangement and a single component as input at one time. Historical sequential information is stored in memory cell that assists the ANFIS-WOA model in capturing trends of datasets for exhibiting more powerful forecasting ability than ANFIS model.

Fig. 4 shows actual and predicted flood discharges together for every collected dataset. Results demonstrate that estimated flood discharges are 5222.854, 5383.541, and 5498.237 m^3/s for ANFIS, ANFIS-FOA, and ANFIS-WOA against the actual peak of 5540.902 m^3/s for the gauge station Salebhata. For Tikarpada station, the estimated peak runoffs are 5306.037, 5463.123, and 5576.293 m^3/s for ANFIS, ANFIS-FOA, and ANFIS-WOA against the actual peak of 5630.345 m^3/s shown in Fig. 4. The figure showed the critical potential of flood and was helpful for flash flood regions with estimated flood indexes.

Fig. 5 illustrates boxplots for different classified datasets for three proposed algorithms obtained by each trainer at the training end. In a boxplot, box represents interquartile range, whiskers relate to farthest flood values, median value is signified by bar in the box, and the small circles represent outliers. Boxplot figures prove and validate superior performance of WOA than the FOA and conventional ANFIS. Note that the boxplot represents the degree of the spread in the predicted data using respective quartile values. The lower end of the plot lies between the lower quartile Q1 (25th percentile) and upper quartile Q3 (75th percentile), with the second quartile Q2 (50th percentile) as the median of the data is represented by a vertical line. Two horizontal lines (known as whisker) are extended from the top and bottom of the box. The bottom whisker extends from Q1 to the smallest nonoutlier in the data set, whereas the other one goes from Q3 to the largest nonoutlier. It is noticeable that the median of the predicted and the observed flood for the ANFIS-WOA models was nearly identical for proposed stations, although the whiskers extend to slightly difference values for each case.

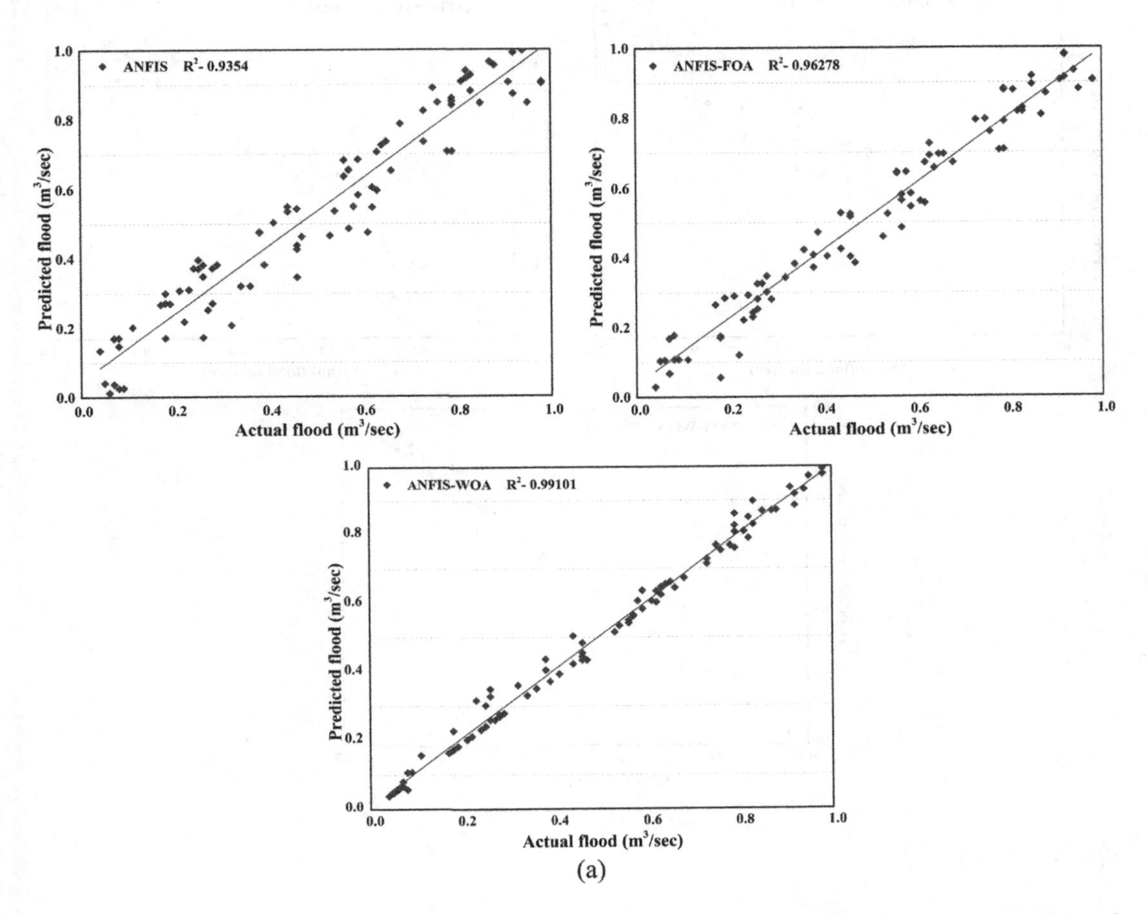

FIG. 3 Scatter plots of observed-predicted flood water level in testing period provided by ANFIS, ANFIS-FOA, and ANFIS-WOA models for best input dataset at (A) Salebhata (B) Tikarpada.

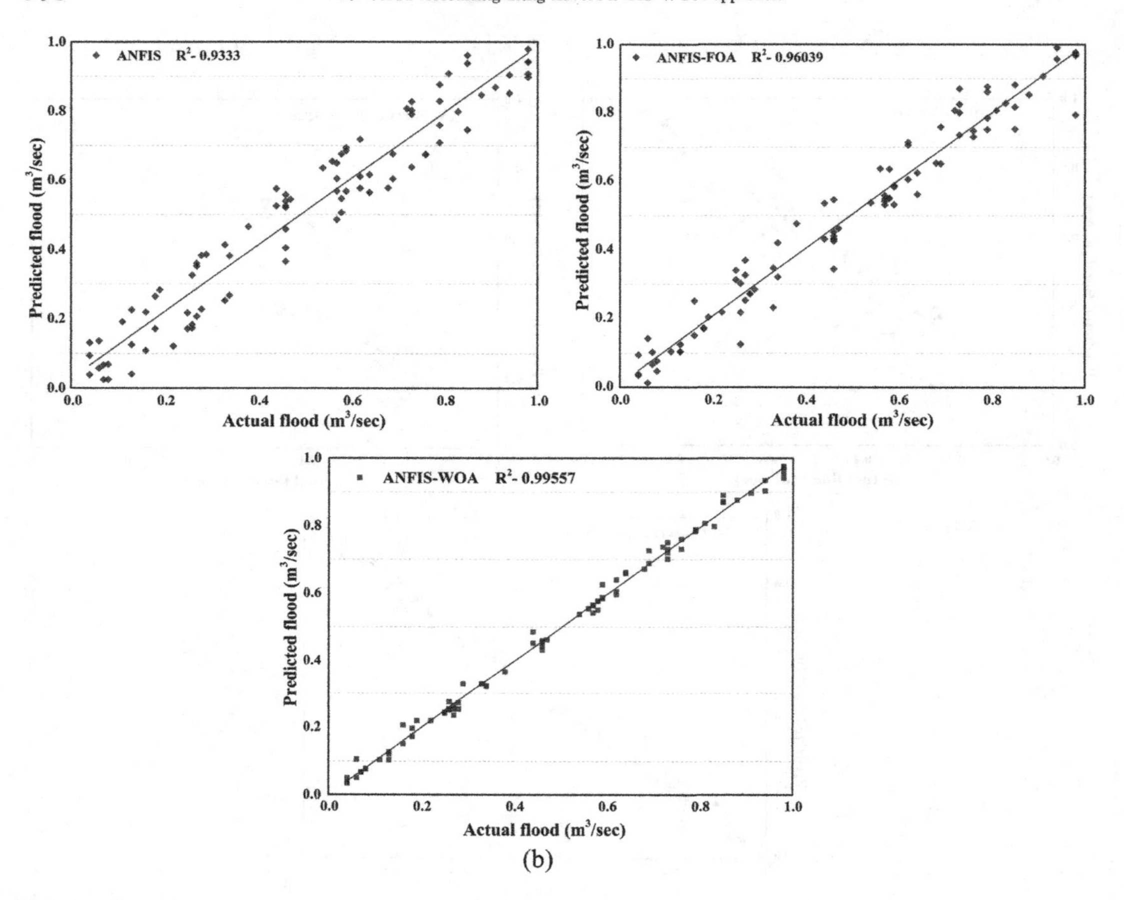
ANFIS R²- 0.9333
ANFIS-FOA R²- 0.96039
ANFIS-WOA R²- 0.99557
Predicted flood (m³/sec)
Actual flood (m³/sec)
0.0
0.2
0.4
0.6
0.8
1.0
(b)

FIG. 3—Cont'd

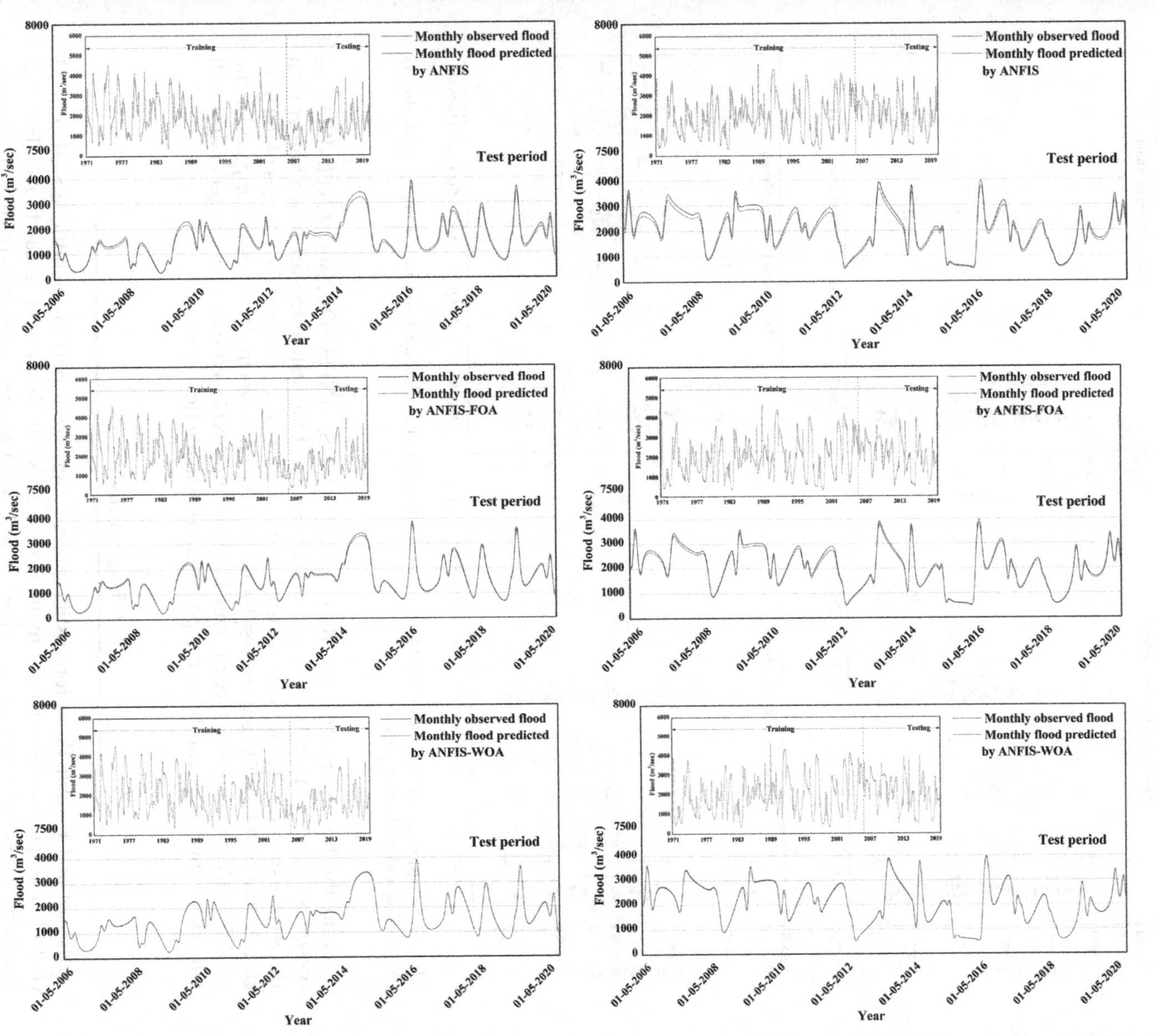

FIG. 4 Comparison between observed and forecasted flood water level using ANFIS, ANFIS-FOA and ANFIS-WOA methods at (A) Salebhata (B) Tikarpada.

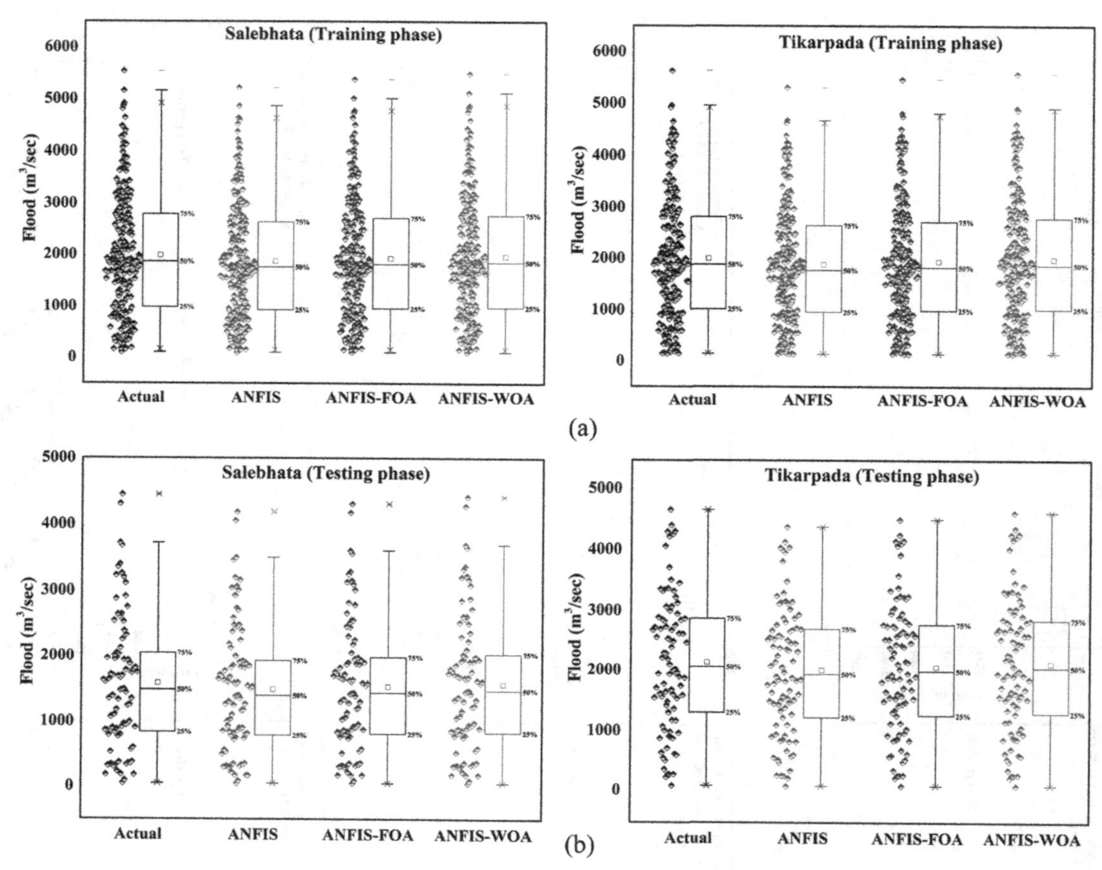

FIG. 5 Boxplots for ANFIS, ANFIS-FOA and ANFIS-WOA-based models for flood forecasting analysis during (A) training and (B) testing periods.

From the histogram of flood flow for the ANFIS, ANFIS-FOA, and ANFIS-WOA as presented in Fig. 6, it can be observed that the peaks of the histograms are much higher than those corresponding to the normal distribution. Also It can be seen that the ranges of the forecasts by the ANFIS-WOA encompass the observed peak flows in all events as compared to the ANFIS-FOA and ANFIS. So we can conclude that the ANFIS-WOA model can generalize well with different inputs and have the potential to manage flood patterns.

5 Conclusion

Robust and reliable flood forecasting plays a significant part in scientific/effective flood control and many water resource management activities. In present research, ANFIS based prediction models incorporated with FOA and WOA as optimizer tools have been adopted for flood prediction. The case study has been performed considering data from two gauging

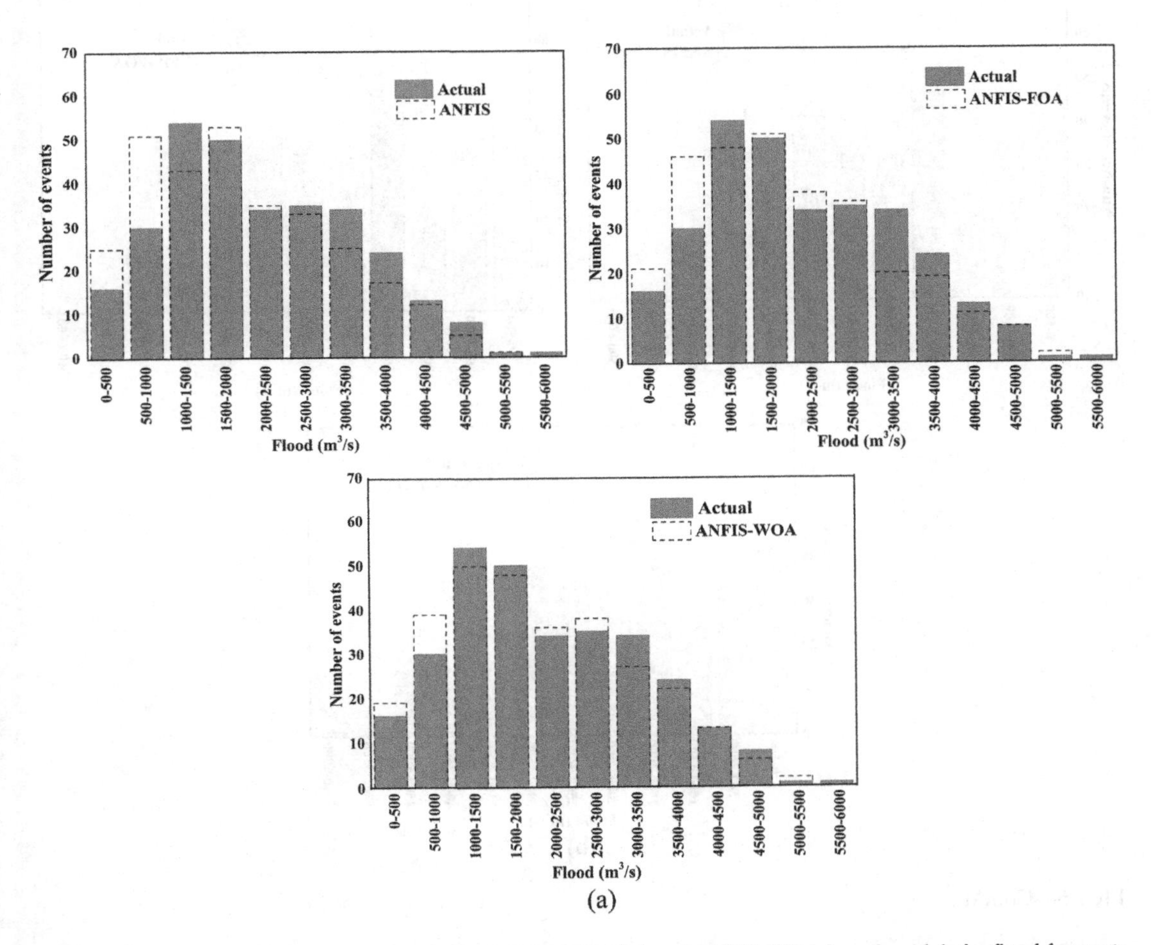

FIG. 6 Histogram plot of best scenario for ANFIS, ANFIS-FOA and ANFIS-WOA-based models for flood forecasting analysis at (A) Salebhata (B) Tikarpada.

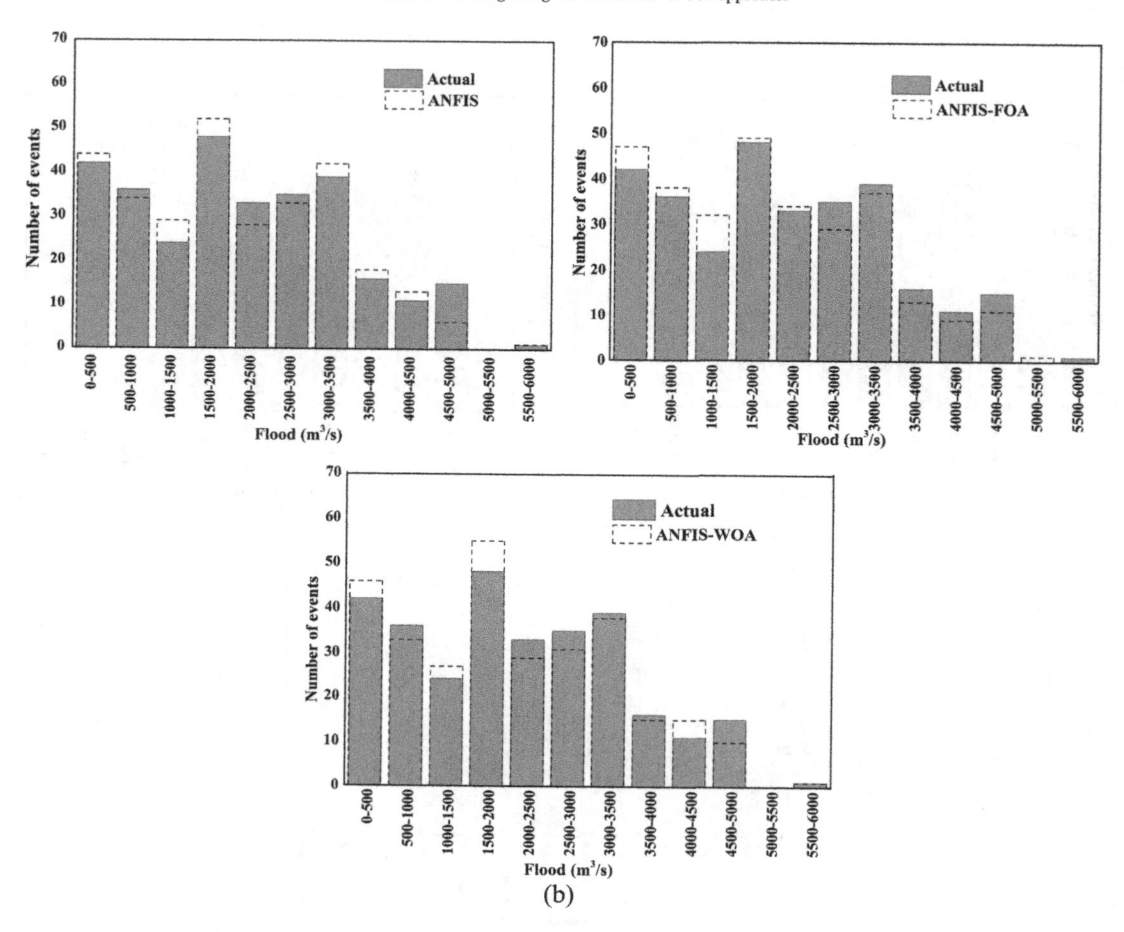

FIG. 6—Cont'd

stations located in lower part of Mahanadi river basin Odisha, India. River flow data of monsoon months (June to October) from 1970 to 2020 were used in this study. Based on obtained results, it was clear that hybrid ANFIS-WOA4 model with three-month lag historical data among input scenarios during training phase was most accurate compared to other equivalents in terms of its RMSE, MAE, R^2, and E_{NS}. This indicates significance of WOA as an optimizer for better accurateness of conventional models. WOA can be effectively utilized to solve nondifferentiable and nonlinear optimization problems in multidimensional space.

During training, the convergence process is a bit slow that is not best for real-time applications. Hence, it is suggested to explore the ability of a related precise preprocessing technique that could identify highly stochastic and nonlinear patterns in the dataset and simultaneously adapt the data quickly. Future research should also focus to improve data availability and quality, extend flood forecasting variables and bridge gap in flood forecast lead time.

References

Adnan, R.M., Mostafa, R.R., Kisi, O., Yaseen, Z.M., Shahid, S., Zounemat-Kermani, M., 2021. Improving streamflow prediction using a new hybrid ELM model combined with hybrid particle swarm optimization and grey wolf optimization. Knowl.-Based Syst. 230, 107379.

Agnihotri, A., Sahoo, A., Diwakar, M.K., 2022. Flood prediction using hybrid ANFIS-ACO model: a case study. In: Inventive Computation and Information Technologies. Springer, Singapore, pp. 169–180.

Alizamir, M., Kisi, O., Muhammad Adnan, R., Kuriqi, A., 2020. Modelling reference evapotranspiration by combining neuro-fuzzy and evolutionary strategies. Acta Geophys. 68, 1113–1126.

Aoulmi, Y., Marouf, N., Amireche, M., Kisi, O., Shubair, R.M., Keshtegar, B., 2021. Highly accurate prediction model for daily runoff in semi-arid basin exploiting Metaheuristic learning algorithms. IEEE Access 9, 92500–92515.

Banadkooki, F.B., Singh, V.P., Ehteram, M., 2021. Multi-timescale drought prediction using new hybrid artificial neural network models. Nat. Hazards 106 (3), 2461–2478.

Chen, X.Y., Chau, K.W., Busari, A.O., 2015. A comparative study of population-based optimization algorithms for downstream river flow forecasting by a hybrid neural network model. Eng. Appl. Artif. Intel. 46, 258–268.

Cheng, M., Fang, F., Kinouchi, T., Navon, I.M., Pain, C.C., 2020. Long lead-time daily and monthly streamflow forecasting using machine learning methods. J. Hydrol. 590, 125376.

Diop, L., Samadianfard, S., Bodian, A., Yaseen, Z.M., Ghorbani, M.A., Salimi, H., 2020. Annual rainfall forecasting using hybrid artificial intelligence model: integration of multilayer perceptron with whale optimization algorithm. Water Resour. Manag. 34 (2), 733–746.

Ghalkhani, H., Golian, S., Saghafian, B., Farokhnia, A., Shamseldin, A., 2013. Application of surrogate artificial intelligent models for real-time flood routing. Water Environ. J. 27 (4), 535–548.

Hsu, K.L., Gupta, H.V., Sorooshian, S., 1995. Artificial neural network modeling of the rainfall-runoff process. Water Resour. Res. 31 (10), 2517–2530.

Jabbari, A., Bae, D.H., 2018. Application of artificial neural networks for accuracy enhancements of real-time flood forecasting in the Imjin basin. Water 10 (11), 1626.

Kaveh, A., 2017. Sizing optimization of skeletal structures using the enhanced whale optimization algorithm. In: Applications of Metaheuristic Optimization Algorithms in Civil Engineering. Springer, Cham, pp. 47–69.

Khac-Tien Nguyen, P., Hock-Chye Chua, L., 2012. The data-driven approach as an operational real-time flood forecasting model. Hydrol. Process. 26 (19), 2878–2893.

Khan, U.T., He, J., Valeo, C., 2018. River flood prediction using fuzzy neural networks: an investigation on automated network architecture. Water Sci. Technol. 2017 (1), 238–247.

Kumar, M., Sahay, R.R., 2018. Wavelet-genetic programming conjunction model for flood forecasting in rivers. Hydrol. Res. 49 (6), 1880–1889.

Li, X., Yao, J., Li, Y., Zhang, Q., Xu, C.Y., 2016. A modeling study of the influences of Yangtze River and local catchment on the development of floods in Poyang Lake, China. Hydrol. Res. 47 (S1), 102–119.

Lohani, A.K., Goel, N.K., Bhatia, K.K.S., 2014. Improving real time flood forecasting using fuzzy inference system. J. Hydrol. 509, 25–41.

Malik, A., Tikhamarine, Y., Sammen, S.S., Abba, S.I., Shahid, S., 2021. Prediction of meteorological drought by using hybrid support vector regression optimized with HHO versus PSO algorithms. Environ. Sci. Pollut. Res., 1–20.

Mirjalili, S., Lewis, A., 2016. The whale optimization algorithm. Adv. Eng. Softw. 95, 51–67.

Mohamadi, S., Sammen, S.S., Panahi, F., Ehteram, M., Kisi, O., Mosavi, A., Ahmed, A.N., El-Shafie, A., Al-Ansari, N., 2020. Zoning map for drought prediction using integrated machine learning models with a nomadic people optimization algorithm. Nat. Hazards 104 (1), 537–579.

Mohammadi, B., Linh, N.T.T., Pham, Q.B., Ahmed, A.N., Vojteková, J., Guan, Y., Abba, S.I., El-Shafie, A., 2020. Adaptive neuro-fuzzy inference system coupled with shuffled frog leaping algorithm for predicting river streamflow time series. Hydrol. Sci. J. 65 (10), 1738–1751.

Mohapatra, M., Mohanty, U.C., 2005. Some characteristics of very heavy rainfall over Orissa during summer monsoon season. J. Earth Syst. Sci. 114 (1), 17–36.

Mohapatra, M., Mohanty, U.C., 2006. Spatio-temporal variability of summer monsoon rainfall over Orissa in relation to low pressure systems. J. Earth Syst. Sci. 115 (2), 203–218.

Mohapatra, M., Mohanty, U.C., 2009. Excess and deficient summer monsoon rainfall over Orissa in relation to low pressure systems. Mausam 60 (1), 25–38.

Nanda, T., Sahoo, B., Beria, H., Chatterjee, C., 2016. A wavelet-based nonlinear autoregressive with exogenous inputs (WNARX) dynamic neural network model for real-time flood forecasting using satellite-based rainfall products. J. Hydrol. 539, 57–73.

Nguyen, P.K.T., Chua, L.H.C., Son, L.H., 2014. Flood forecasting in large rivers with data-driven models. Nat. Hazards 71 (1), 767–784.

Ni, L., Wang, D., Wu, J., Wang, Y., Tao, Y., Zhang, J., Liu, J., 2020. Streamflow forecasting using extreme gradient boosting model coupled with Gaussian mixture model. J. Hydrol. 586, 124901.

Pan, W.T., 2012. A new fruit fly optimization algorithm: taking the financial distress model as an example. Knowl.-Based Syst. 26, 69–74.

Pattanaik, D.R., 2007. Analysis of rainfall over different homogeneous regions of India in relation to variability in westward movement frequency of monsoon depressions. Nat. Hazards 40 (3), 635–646.

Sahoo, A., Samantaray, S., Ghose, D.K., 2021a. Prediction of flood in barak river using hybrid machine learning approaches: a case study. J. Geol. Soc. India 97 (2), 186–198.

Sahoo, A., Samantaray, S., Paul, S., 2021b. Efficacy of ANFIS-GOA technique in flood prediction: a case study of Mahanadi river basin in India. H2Open J. 4 (1), 137–156.

Samadianfard, S., Hashemi, S., Kargar, K., Izadyar, M., Mostafaeipour, A., Mosavi, A., Nabipour, N., Shamshirband, S., 2020. Wind speed prediction using a hybrid model of the multi-layer perceptron and whale optimization algorithm. Energy Rep. 6, 1147–1159.

Samantaray, S., Ghose, D.K., 2019. Dynamic modelling of runoff in a watershed using artificial neural network. In: Satapathy, S., Bhateja, V., Das, S. (Eds.), Smart Intelligent Computing and Applications. Springer, Singapore, pp. 561–568.

Samantaray, S., Ghose, D.K., 2020. Modelling runoff in an arid watershed through integrated support vector machine. H2Open J. 3 (1), 256–275.

Samantaray, S., Sahoo, A., 2021a. Modelling response of infiltration loss toward water table depth using RBFN, RNN, ANFIS techniques. Int. J. Knowl. Based Intell. Eng. Syst. 25 (2), 227–234.

Samantaray, S., Sahoo, A., 2021b. Prediction of suspended sediment concentration using hybrid SVM-WOA approaches. Geocarto Int., 1–27.

Samanataray, S., Sahoo, A., 2021c. A comparative study on prediction of monthly streamflow using hybrid ANFIS-PSO approaches. KSCE J. Civ. Eng. 25 (10), 4032–4043.

Samantaray, S., Sahoo, A., Ghose, D.K., 2020a. Prediction of sedimentation in an arid watershed using BPNN and ANFIS. In: ICT Analysis and Applications. Springer, Singapore, pp. 295–302.

Samantaray, S., Sahoo, A., Ghose, D.K., 2020b. Infiltration loss affects toward groundwater fluctuation through CANFIS in arid watershed: a case study. In: Smart Intelligent Computing and Applications. vol. 159. Smart Springer, Singapore, pp. 781–789.

Samantaray, S., Sahoo, A., Agnihotri, A., 2021. Assessment of flood frequency using statistical and hybrid neural network method: Mahanadi river basin, India. J. Geol. Soc. India 97 (8), 867–880.

Sattari, M.T., Feizi, H., Samadianfard, S., Falsafian, K., Salwana, E., 2021. Estimation of monthly and seasonal precipitation: a comparative study using data-driven methods versus hybrid approach. Measurement 173, 108512.

Sehgal, V., Sahay, R.R., Chatterjee, C., 2014. Effect of utilization of discrete wavelet components on flood forecasting performance of wavelet based ANFIS models. Water Resour. Manag. 28 (6), 1733–1749.

Seifi, A., Riahi, H., 2020. Estimating daily reference evapotranspiration using hybrid gamma test-least square support vector machine, gamma test-ANN, and gamma test-ANFIS models in an arid area of Iran. J. Water Clim. Change 11 (1), 217–240.

Solomatine, D.P., Ostfeld, A., 2008. Data-driven modelling: some past experiences and new approaches. J. Hydroinf. 10 (1), 3–22.

Soman, M.K., Kumar, K., 1990. Some aspects of daily rainfall distributions over India during southwest monsoon season. Int. J. Climatol. 10, 299–311.

Sridharam, S., Sahoo, A., Samantaray, S., Ghose, D.K., 2021. Estimation of water table depth using wavelet-ANFIS: a case study. In: Communication Software and Networks. Springer, Singapore, pp. 747–754.

Sun, X., Bi, Y., Karami, H., Naini, S., Band, S.S., Mosavi, A., 2021. Hybrid model of support vector regression and fruitfly optimization algorithm for predicting ski-jump spillway scour geometry. Eng. Appl. Comput. Fluid Mech. 15 (1), 272–291.

Tan, L., Wang, S., Wang, K., 2017. A new adaptive network-based fuzzy inference system with adaptive adjustment rules for stock market volatility forecasting. Inf. Process. Lett. 127, 32–36.

Yuan, X., Dai, X., Zhao, J., He, Q., 2014. On a novel multi-swarm fruit fly optimization algorithm and its application. Appl. Math Comput. 233, 260–271.

Index

Note: Page numbers followed by "*f*" indicate figures and "*t*" indicate tables.

A

ABM. *See* Agent based modeling (ABM)
Accelerometers, 526
Acoustic Doppler velocimetry (ADV), 485–486
Acoustic transducers. *See* Sonar devices
Actual retention, 626–628
Adaptive neuro-fuzzy inference system (ANFIS), 4, 38–39, 342–343, 468, 664–665, 672–678
ADV. *See* Acoustic Doppler velocimetry (ADV)
Advanced Spaceborne Thermal Emission and Reflection Radiometer (ASTER) DEM, 93
Agent based modeling (ABM), 435, 443
Aggregation function, 61
Agricultural irrigation management, 300
Agricultural Model Intercomparison and Improvement Project (AgMIP), 553
Agriculture
  artificial intelligence in, 22–28
  irrigated, 411
  lifecycle, 19, 19*f*
  productivity, 18, 22–23
Agri-e-calculator, 24–26
AHP. *See* Analytic hierarchy process (AHP)
AI. *See* Artificial intelligence (AI)
Air-glass-water interface, 526
ALOS World 3D (AW3D30), 645
Amazon Web Services, 122
Analytic hierarchy process (AHP), 90, 145–146
AND operator, 270
ANFIS. *See* Adaptive neuro-fuzzy inference system (ANFIS)
ANN. *See* Artificial neural network (ANN)
Annual rainfall trends
  in arid region, trends and homogeneity (*see* Seasonal and annual rainfall, in arid region)
  in Shiraz station, Iran (*see* Shiraz station in Iran, annual average and maximum rainfall)
  in tropical regions, 590–591
Ant colony optimization (ACO), 343–344, 665–666
ANUGA Hydro, 655–656
Arc GIS-Arc Map, 157
Arid region, rainfall trends in. *See* Seasonal and annual rainfall, in arid region
Artificial bee colony (ABC) algorithms, 343–344, 665–666
Artificial drainage, 410
Artificial intelligence (AI), 266
  in agricultural water management
  in agriculture, 26–27
    current and future scope in, 29–30
    lack of irrigation and drainage facilities, 19
    long-term viability and profitability, 22–23
    semiarid and arid areas, 23
    sowing and cultivation, 22–23
  in postagricultural activities, 27–28
  in preagricultural (preparatory) activities, 23–26
Artificial intelligence and machine learning (ML)
  evolution, 7
  scientometric review, 5–7
  shallow learning, 4–5
  Web of Science (WoS) database, 4–5
Artificial neural network (ANN), 4, 20, 20*f*, 202–203, 286, 292–293, 342–343, 664–665
  flood forecasting, 664–665
  side-weir discharge coefficient estimation in trapezoidal and rectangular open channels, 471, 474–476, 476–477*f*
  sour, 77–78
Arvand River
  data and information, 154
  erosion and sedimentation, 154
  future morphologic changes, 160–161
  landsat image, 154–155
  properties, 153–154
  river meanders properties, 157–160
  river width changes, 157
  topographic maps, 155–156
Autoregressive and moving average model (ARMA), 664

B

Bagging, 213–215
Bankfull discharge, 498–499
Bartlett test, 601–602, 605, 611–615, 614*f*
Bat-ELM. *See* Extreme learning machine optimized Bat algorithm (Bat-ELM)
Bathymetry survey, 645, 645*f*
Bayesian network, 10

Bedrock channels
reach level classification, 119–120
using machine learning algorithms, 129
Bedrock reaches, 119
Bernoulli equation, 331–332
Bidirectional long short term memory (BiLSTM), 42–43
Black box concept, 322, 322*f*
Black box error, 322
Blaney-Criddle method, 417–418
Boden watersheds, actual *vs.* predicted groundwater levels, 355–358, 357–361*f*
Box-Jenkins (B-J) models, 246–247
Box-whisker plots, of annual rainfall, 606, 606–607*f*
Bragg wavelength, 522
Brahmani River basin, 204*f*, 220
Brahmaputra River basin, 534–535
Braiding process, 484
Bridge scour, 514–515, 514*f*
Bridges failure, 514–515, 515*t*, 515*f*
Bureau of Indian Standards (BIS), 105, 270
Bursting events
depth-wise distribution of Reynolds stress, 492, 493*f*
temporal distribution, 490–492, 491–492*f*
variation with hole size, 488–490, 489–490*f*

C

Calibrated pile, 524
Calibration, 231
Canadian ground system model version 5 (CanESM5) model, 563–566, 571, 580, 582–585, 584*f*
CAPEX, 23
Cascade reaches, 119
Cauchy particle swarm optimization (CPSO) algorithm, 449
CEEMDAN. *See* Complete ensemble empirical mode decomposition with adaptive noise (CEEMDAN)
Centre for Development of Advanced Computing (C-DAC), 642
Centre of gravity method (COG), 273
Change factor (CF) method, 567
Channel design, 293–295
Channel morphological variables, 123
Channel unit types, 130
Circular channels
design equations, 287–289
optimum design, 286–287
Classification and regression tree (CART), 246–247
Clean water and sanitation, 134
Climate change, 533–534
definition, 562
global warming and, 562
precipitation and temperature data, 553, 582–585
change factor (CF) method, 567
exponential downscale of data, 566–567
flow chart of methodology, 569, 569*f*
general atmospheric circulation (GCM) models, 562–566, 571–580
LARS-WG model, 562–564, 566–571, 582–585
Mann-Kendall test, 570, 582, 583–584*f*
micro-scale statistical methods, 562
number of rainy days, 580–581, 581*f*
study area, 564, 565*f*
trends and scenarios, 558
articles, authors and contributions, 550–552, 550*f*, 551*t*, 552*f*
changing paradigms and thematic evolution, 552–555, 554*f*, 556*f*
data analysis, 548–550
data processing, 548, 549*f*
data retrieval, 547, 549*f*
data source, 547, 548*f*
knowledge production, 556–558, 557*f*
Climatic factors, 18
Cloud computing, 122
CMIP. *See* Coupled Model Intercomparison Projects (CMIP)
Co-citation network, 550
Coefficient of variation (CV), 604–606, 605*t*
Color imaging, 26
Community water model (CWatM), 188
Complete ensemble empirical mode decomposition with adaptive noise (CEEMDAN), 448–449, 455–456
Computational fluid dynamics (CFD), 321, 329
Conditional random field, 11
Conditional sampling technique, 484
Consistency ratio, 206
Continuity equation, 325–326, 412–414
Continuity of mass, 414
Continuity of momentum, 414
Control volume (CV), 323–324
Conventional regression models, 12
Conventional secondary data, 205
Coordinated Regional Climate Downscaling Experiment (CORDEX)—Africa, 556–557
Correlation coefficients, 107, 109–112, 250
Correlation development, 87
Coupled Model Intercomparison Project phase 6 (CMIP6), 566
Coupled Model Intercomparison Projects (CMIP), 553–555
Coupling network, 550
COVID-19 and climate change, 557–558
Crop evapotranspiration (ET_c), 306, 306*t*
Crop monitoring and health assessment, 29
Crop parameters, 417
Curve number (CN), 166, 626–628

D

Darcy-Richards equation-based groundwater models, 369
Darcy's law, 371, 391–393, 398, 414

Data acquisition, 11–12
Data analysis, 123
Davisian cycle, 117–118
Decision-support system (DSS), 434, 436
Decision tree analysis, 202
Decision tree regression (DTR) model, 246–247
Deep learning, 10
Deep percolation, 418, 419*t*
Defuzzification, 269
Delhi-based Sohan Lal Commodity Management (SLCM), 28
Delphi technique, 266
DEM. *See* Digital elevation model (DEM)
Dhapewada dam, 135–137
Differential evolution (DE) algorithms, 665–666
Diffusion wave approximation (DWA), 333–335
Diffusion-wave equation (DWE), 335
Digital bathymetric modeling (DBM), 331
Digital elevation model (DEM), 168, 190, 230, 326–328, 373, 374*f*, 623–625, 624*f*
Direct numerical simulations (DNS), 321, 328
Discharge coefficient, 469
Discontinuous elevation (DE), 655
Discrete wavelet transform (DWT), 449
Dissolved oxygen (DO), in river
  ensemble empirical mode decomposition (EEMD), 455–456, 458*t*, 460–462
  extreme learning machine optimized Bat algorithm (Bat-ELM), 450–459, 454*f*, 458*t*, 461–462
  flowchart of modeling strategy, 457*f*
  performance assessment of models, 450
  relevance vector machine (RVM), 455–459, 458*t*, 461–462
  study site and data, 449–450, 451*f*
Dominant discharge, 498, 503–504, 509–510
Double-ring infiltrometer tests, 374–377, 377*f*
Drainage density, 63, 212–220
DRAINMOD simulation model, 423
DRANSAL model, 421, 424, 426
Driven or buried devices, 518
Drones, 29, 122
DSS. *See* Decision-support system (DSS)
Duhamel-Neumann equations, 396–397
Dynamic linear models (DLMs), 246–247
Dynamic watershed simulation model (DWSM), 187

E

Early warning system (EWS). *See* Glacier lake outburst floods (GLOF)
Earth Observation Satellites (EOSs), 552
Echo sounder, 523, 525
EEMD. *See* Ensemble empirical mode decomposition (EEMD)
Effective discharge, 499–502, 507
  computation, 508
  defined, 499
  in river design and monitoring, 509–510
Effective rainfall, 303–304, 307, 307*t*
Egypt, flash flood in. *See* Flash flood, in Wadi Degla, Egypt
Electrical conductivity (EC), 90–91, 414
Electrical conductivity devices, 523
Embedding coupling, 190
Empirical mode decomposition (EMD), 448–449
Empirical wavelet transform (EWT), 39, 44–45, 48–50, 449
Ensemble empirical mode decomposition (EEMD), 246–247, 448–449, 455–456, 460–462
Environmental Protection Agency (EPA), 226
ENVI software, 168
Equilibrium scour depth, 80
Estuarine water quality models, 10
EU Water Framework Directive, 321
Evapo-transpiration (ET), 417–418
Evolution-based algorithms, 665
EWT. *See* Empirical wavelet transform (EWT)
Expectation maximization (EM), 455
Extreme learning machine (ELM), 468
Extreme learning machine optimized Bat algorithm (Bat-ELM), 450–459, 454*f*, 461–462

F

Farm subsystem, 411, 417–418
FAVOR method, 329
Feedforward neural networks (FFNN) model, 450–453
FFA. *See* Firefly algorithm (FFA)
Fiber-Bragg gratings, 522
Finite element method (FEM), 332
Firefly algorithm (FFA), 349–350, 350*t*
FIS. *See* Fuzzy inference systems (FIS)
Fishing ponds, 140
Fixed scour monitoring instrumentation, 516–519
  driven or buried devices, 518
  magnetic sliding collars, 518, 519*f*
  mercury tip sensors, 518–519
  scuba mouse, 518
  sonar devices, 516–518, 517*f*
  steel rod, 518, 519*f*
  Wallingford tell-tail, 518
Flash flood, in Wadi Degla, Egypt, 622, 623*f*, 635–637
  digital elevation model, 623–625, 624*f*
  geological studies, 625, 626*f*
  hydrological model, 622–623
  land-use change, 629–630, 630–631*f*
  mean maximum/minimum temperature, 623–625
  methodology
    actual retention, 626–628
    curve number (CN) method, 626–628
    flow chart, 625–626, 627*f*
    flow direction and accumulation, 625–626, 627*f*
    land-use/land-cover changes, 625–626

Flash flood, in Wadi Degla, Egypt *(Continued)*
potential maximum retention, 629
sentinel imaginary, 625–626, 627*f*
stream network, 625–626, 628*f*
natural life, impact on, 635, 636–637*f*
prevalent climatic condition in, 623–625, 625*t*
return periods, 632–634, 634*f*, 634*t*
slope map, 623–625, 624*f*
surface runoff modeling, 622–623
surface-water model, 631–632, 632–633*f*, 634*t*
FlexPDE, 400
Float-out devices, 519–520, 520*f*
Flood disaster management, 86–87
Flood forecasting, 4
adaptive neuro-fuzzy inference system (ANFIS), 664–666, 668
adaptive optimization algorithm for, 666
ANFIS-WOA, 670
ant colony optimization (ACO) algorithms, 665–666
artificial intelligence models, optimization of, 666
artificial neural network (ANN), 664–665
data-driven models, 664
evolution-based algorithms, 665
fruit fly optimization (FOA), 668–669
fuzzy neural network, 665
Gaussian mixture model with extreme gradient boosting (GMM-XGBoost) model, 665–666
long short term memory (LSTM), 665
multilayer perceptron model (MLP), 665–666
performance evaluation, 671–672
performance of ANFIS, ANFIS-FOA, and ANFIS-WOA models
boxplot, 674, 678*f*
histogram plot, 678, 679–680*f*
inputs and model name, scenarios used with, 672–673, 672*t*
observed *vs.* forecasted flood water level, 674, 677*f*
RMSE, MAE, E_{NS}, and R^2, 672–674, 673*t*
scatter plots, 674, 675–676*f*
physical-based models, 664
statistical method, 664
study area, 666–667, 667*f*
SVM-FOA model, 665–666
swarm-based algorithms, 665
WANFIS-SD and and WANFIS-MS, 665–666
wavelet based genetic programming (W-GP) model, 665–666
wavelet-based nonlinear autoregressive with exogenous inputs (WNARX), 665–666
whale optimization algorithm (WOA), 669–670, 671*f*
Flood inundation map, 656
Flood risk analysis, 202
Flood simulation model, 202–203
Flow measurement method, 416–417
Flow systems, 368
Fluid flow, in porous media, 391–400
basic assumptions, 391–392
coupled thermo-hydro-mechanical model, 395–397
fluid flow coupled with heat transfer, 394–395
groundwater flow modeling, equations for, 391–392
numerical model, 397–400
poroelasticity model, 392–394
temperature and seepage rates, 400–401, 401*f*
Food and Agriculture Organization (FAO), 228–229, 300
Food production, 300
Free and Open Source Software (FOSS), 656
Frequency ratio (FR), 90, 202
Freshwater fuzzy model, for lake systems-fuzzy lake index (FLI), 272–273
Freshwater quality model, 274
excellent lake, 274–277, 277*f*
Pookode lake, 281*f*
poor lake, 274–277, 278*f*
Sasthamkotta lake, 280*f*
surface viewer, 279, 281*f*
Vellayani lake, 279*f*
Froude number, 468, 471
F-test, 80
Function node, 289
Fuzzification, 269
Fuzzy inference systems (FIS), 78, 274–277
Fuzzy lake index (FLI), 277*t*, 282
Fuzzy logic, 266, 269–270
Fuzzy neural network, 665
Fuzzy set operations, 270
Fuzzy theory, 202

G

Ganga River, 534–535
Gated recurrent unit (GRU), 448–449
Gaussian function, 348
Gaussian mixture model with extreme gradient boosting (GMM-XGBoost) model, 665–666
Gaussian process regression (GPR), 43–44
GCM. *See* General atmospheric circulation (GCM) models
Gene expression programming (GEP), 247–248, 468
General atmospheric circulation (GCM) models, 562–563, 572–573*f*, 574–579*t*
Canadian ground system model version 5 (CanESM5) model, 563–566, 571, 580, 582–585, 584*f*
IPSL-CM6A-LR climate model, 564–566, 580, 582–585
MIROC6 model, 564–566, 572–580, 582–585, 583*f*
General atmospheric ocean circulation (AOGCM) models, 564, 567
Generalized opposition-based learning particle swarm optimization algorithm (GOBLPSO), 448–449
Generalized reduced gradient (GRG), 292–293, 468, 476
Generalized regression neural networks (GRNN), 251–252, 253*f*, 255–257, 257*t*, 258*f*, 260*f*

Genetic algorithm (GA), 266, 343–344, 468, 515–516
Genetic programming (GP), 286, 292–293, 471–472
Geographical information system (GIS), 189–193, 226, 300, 622, 656–657
Geology, 138
Geometric assessment, 87
GIS. *See* Geographical information system (GIS)
GIS-based hydrological models
- calibration, validation and sensitivity analyses, 188–189
- classifications, 185–188
- community water model (CWatM), 188
- digital elevation models (DEM) data, 190
- dynamic watershed simulation model (DWSM), 187
- embedding coupling, 190
- environmental impact study and evaluation, 183–185
- groundwater loading effects of agricultural management systems model (GLEAMS), 188
- groundwater modeling, 194–195
- groundwater potential evaluation, 183–185
- groundwater resources management, 192*t*, 194
- HBV model (hydrological bureau water department model), 187
- HEC-HMS and the Xinanjiang model, 193
- history of, 185
- integration, 183–185
- in Kayu Ara River basin, Malaysia, 193
- loose coupling, 191*f*
- MIKE-SHE model (European hydrological model), 187
- modular groundwater flow model (MODFLOW-2005), 187
- in Sandusky watershed with SWAT, 193
- soil and water assessment tool (SWAT), 187, 192
- tight coupling, 191*f*
- in Zarqa River catchment, Jordan, 193

Glacier lake outburst floods (GLOF)
- dam-breach and flood models, 642–643
- data acquisition
  - bathymetry data, 645, 645*f*
  - glacial lake water level monitoring data, 647–649
  - moraine dam resistivity survey, 646, 646*f*, 647*t*
  - remote sensing imagery, 643–645
  - topographic data, 645
- data interpretation/analysis/computation
  - discharge estimation, 652–654, 652–653*t*
  - glacial lake area change map, 649, 650–651*f*, 651*t*
  - Heron's method, 649–652
  - lake volume, 649
- dynamic models, application of, 642–643
- early warning system, 660
  - flood arrival time and height, 656
  - flood inundation map, 656
  - improved flood forecast lead time with HPC, 656
  - limitations of study, 659
  - recommendations, 659–660
  - salient features, 657–659
- empirical and physical models, application of, 642–643
- flood arrival time and flood depth, 657, 659*t*
- GIS-based spatial decision support system, 656–657
- glacial lake database, 643
- hydrodynamic models, 642–643
- PostgreSQL database, 643
- remote sensing, for vulnerable glacial lake monitoring, 642
- simulated flood inundation, 657, 658*f*
- simulation models, 642–643, 654–655
- study area, 643, 644*f*

Glaciers
- climate change, impact of, 642
- definition, 641–642
- glacier lake outburst floods (GLOF) early warning system (*see* Glacier lake outburst floods (GLOF))
- Himalayan glaciers, 642
- lakes, 642

Global digital elevation model (GDEM), 59–60
Global digital surface model (DSM), 645
Global positioning system (GPS), 92–93, 301–303
GLOF. *See* Glacier lake outburst floods (GLOF)
Gonbad Kavus, 167–168
Google Earth Engine, 122
Google web search, 7
Gorganrood basin, 167–168
Gorganrood watershed, rainfall-runoff modeling. *See* Rainfall-runoff modeling
Gradient boosting decision tree (GBDT), 72
Graphical user interface (GUI), 279
Gray self-memory model (GSM), 342–343
GRG. *See* Generalized reduced gradient (GRG)
GRNN. *See* Generalized regression neural networks (GRNN)
Gross irrigation requirement (GIR), for Kharif and Rabi season crops, 309–315, 309*t*, 313*t*
- Barwala, 301, 302*f*
- computation, procedure of, 303–304, 303*f*
- cotton crop, 309, 310*f*
- crop evapotranspiration (ET_c), 306, 306*t*
- effective rainfall (P_{eff}), 303–304, 307, 307*t*
- meteorological data, 301
- mustard crop, 309–315, 313*f*
- net irrigation requirement (NIR), 303–304, 307–308, 307–308*t*
- pearl millet crop, 309, 311*f*
- reference evapotranspiration (ET_o), 304–305, 305*t*
- rice crop, 309, 312*f*
- spatial database creation and cropping pattern, 304, 305*t*
- wheat crop, 309–315, 314*f*

Ground heat exchangers (GHE), 389–390
Ground penetrating radar (GPR), 522
Groundwater flow modeling, equations for, 391–392

Groundwater level, prediction of, 381–382
Groundwater loading effects of agricultural management systems model (GLEAMS), 188
Groundwater management, 91
Groundwater models, 86–87, 194–195, 370
Groundwater potential
environmental factors, 58
fuzzy logic approach, Marinduque, Philippines
aggregation function, 61–65, 65*f*
annual rainfall, 63
data, 59–60
drainage density, 63
membership function, 61
performance metrics, 66*t*
quantitative and qualitative data, 58–59
weighted aggregation function, 66
zero membership function values, 66
geo-environmental variables, 58
groundwater mapping exploration, 58
hydrological systems, 57–58
statistical and data mining techniques, 58
Groundwater storage-discharge dynamics, 369
Group method of data handling (GMDH), 246–247, 468
GRU-CEEMDAN-GOBLPSO model, 448–449

H

Hagen-Poiseuille law, 391–392
Hargreaves-Samani formula, 303
Hartley test, 601–602, 605, 611–615, 612*f*
Heat transfer, fluid flow coupled with, 394–395
Heavy metal pollution index (HPI), 266, 271–272
membership function, 273*f*
Vellayani, Sasthamkotta and Pookode lakes, 276*t*
Helmhotz free energy, 396
Heron's method, 649–652
High flooding probability, 212
Hillslope-storage Boussinesq (HSB) model, 369–370
calibration, 379, 381*f*
definition, 370–371, 371*f*
derivation, 371–373
field application, 377–379
geomorphic analysis, 379–380, 379–380*f*
groundwater models, 370
hillslope width function (HWF), 380–381, 380*f*
parameterization, 373
performance evaluation statistics, 382, 383*t*
prediction of groundwater level, 381–382
step-by-step procedure, 377–378, 378*f*
subsurface discharge time series, 382–383, 383*f*
validation, 379, 382*f*
Hillslope width function (HWF), 380–381
Himalayan glaciers, 642
Hole size analysis, of bursting events around midchannel bar
bursting events
depth-wise distribution of Reynolds stress, 492, 493*f*
temporal distribution, 490–492, 491–492*f*
variation with hole size, 488–490, 489–490*f*
experimental program, 485–486, 486*t*
Hooghoudt equation, 423
Hydraulic numerical model, definition of, 320
Hydrological cycle, 166
Hydrological engineering center-river analysis system (HEC-RAS) model, 90
Hydrological management, 87
Hydrological response units (HRUs), 230, 379–380
Hydrologic soil groups (HSGs) map, 170–171
Hydrologic submodel, 412–414, 413*f*
Hydrologic time series, 600–603
Hydrology, 140
Hydropower generation, 3–4
Hydrosalinity models, 411, 420
irrigated soil groundwater systems
components, 412
criteria for evaluation, 414–415, 416*t*
deep percolation, 418
drainage investigations, 420
farm subsystem, 411, 417–418
field investigations, 415–420
hydrologic submodel, 412–414, 413*f*
rainfall recharge, 418
recharge from irrigation, 418
salinity submodel, 414
water delivery subsystem, 411, 416–417
water removal subsystem, 411–412, 419–420
in northern India, 421–422
HYDRUS software, 424
Hyporheic-scale, 368

I

IF-THEN fuzzy rules, 668
Image processing technology, 524
IMFs. *See* Intrinsic mode functions (IMFs)
Improved sparrow search algorithm (ISSA), 448–449
Indian Meteorological Department (IMD), 535
Inference rule, 270
Inflow-outflow method, 417
Infrared band acquisition, 120–121
Innovative polygon trend analysis (IPTA), 590–591, 601–602
Innovative trend analysis (ITA), 605
advantages, 590
hydrologic time series, 601–602
seasonal and annual rainfall trends
spatial distribution, 607–610, 609*f*
trend identification and magnitudes, 605
Shiraz station in Iran, annual average and maximum rainfall, 593–596, 593*f*, 595*f*
Innovative triangular trend analysis (ITTA)
hydrologic time series, 601–602

Shiraz station, Iran
annual average rainfall at, 594–596, 594t
annual maximum rainfall at, 596, 596t
Integrated assessment modeling (IAM), 436
Integrated urban water management (IUWM)
agent based modeling (ABM), 435
decision-making, 436
integrated assessment modeling, 436
mathematical optimization approaches, 438
models supporting urban water development planning (UWDP), 436–438
multipattern approach, 435
three steps multibenefit framework, 434–435, 435f
Varanasi City, 440–443, 440f, 441–443t, 443–444f
water resources optimization, 439–440
water sensitive urban design planning support system (WSUD-PSS), 436, 437f
Intergovernmental Panel on Climate Change (IPCC), 642
Intermediate flow system, 368
International Crops Research Institute for the Semi-Arid Tropics (ICRISAT), 24–26
Intrinsic mode functions (IMFs), 246–247, 254–257, 255f, 455–456, 456f, 462
Inverse distance weighted (IDW) technique, 205, 303
IPSL-CM6A-LR climate model, 564–566, 580, 582–585
Irrigated agriculture, 411
Irrigation management, 86–87
Irrigation projects, 142
Irrigation system, 411, 412f
ITA. *See* Innovative trend analysis (ITA)
ITTA. *See* Innovative triangular trend analysis (ITTA)
IUWM. *See* Integrated urban water management (IUWM)

K

Kanhar land, 138
Kanjhari watershed
land use/land cover map of, 374, 376f
soil map of, 374, 375f
Kappa coefficient (Kc), 229–230
Kelantan River watershed, 202–203
Kendall's rank correlation (KRC) test, 600–601
Khariar watersheds, actual *vs.* predicted groundwater levels, 355–358, 357–361f
K-nearest neighbor (KNN) algorithm, 10–11, 123
Knowledge-intensive learning systems, 4
Knowledge production, 123–129
Koppen climate classification, 535

L

Lake database management, 657, 657f
Landsat satellite imagery, 154–155, 168
Land use and land cover (LULC)
changes on water balances
Food and Agriculture Organization, 228–229
groundwater processes, 236
hydrological modeling using SWAT, 230–232
map 1995 to 2020, 227–230, 232–234, 233–234t, 233f
methodology, 229–232
model calibration and validation, 235, 236f, 237t
Palar and river sub-basins, 227, 228f
sensitivity analysis, 234–235, 235t
streamflow, 232
sub-watershed level changes, 235, 238f
topographical divisions, 227–228
rainfall-runoff modeling, 170
Large Eddy simulations (LES), 321
Laser distance sensors, 525
Laser Doppler velocimeter (LDV), 525
Lathe machine, 524
LeachMod, 420
Least absolute shrinkage and selection operator (LASSO), 247
Least squares (LS), 247
Least squares support vector machine (LSSVM), 448–449
Lettenmaier approach, 601–602
Levenberg Marquardt (LM), 342–343
Levene's test, 601–602
Light detection and ranging (LIDAR), 127, 331
Linear regression (LR) model, 246–247, 540, 563–564
Linear trend analysis, 538
Link-Wallace test, 601–602, 605, 611–615, 613f
Local flow system, 368
Local regression (LR), 246–247
Local thermal equilibrium (LTE), 394
Local thermal non-equilibrium (LTNE), 394–395, 403
Logistic regression, 11, 202
Long Ashton Search Station Weather Generator (LARS-WG) model, 562–564, 566–571, 571t, 582–585
Long short term memory (LSTM), 665
Lord-Shulman thermoelasticity, 395
LSSVM-SSA-VMD model, 448–449
LSSVM-WA-CPSO model, 449
LSTM-ISSA-SWT model, 448–449
LTNE. *See* Local thermal non-equilibrium (LTNE)
Lumped hydrosalinity models, 423

M

Machine learning (ML), 4, 21f, 129, 286, 448, 471–473
MAE. *See* Mean absolute error (MAE)
Magnetic sliding collars (MSCs), 518, 519f
Magnitude frequency analysis, of sediment transport
bankfull discharge, 498–499
discharge for specific recurrence interval, 499
dominant discharge, 498
effective discharge, 499–502
computation, 508
in river design and monitoring, 509–510
literature review, 503–507
research findings, 500, 500–503t

Mahanadi River basin, flood forecasting in
adaptive neuro-fuzzy inference system (ANFIS), 668
fruit fly optimization (FOA), 668–669
location, 666–667, 667*f*
performance evaluation, 671–672
performance of ANFIS, ANFIS-FOA, and ANFIS-WOA models
boxplot, 674, 678*f*
histogram plot, 678, 679–680*f*
inputs and model name, scenarios used with, 672–673, 672*t*
observed *vs.* forecasted flood water level, 674, 677*f*
RMSE, MAE, E_{NS}, and R^2, 672–674, 673*t*
scatter plots, 674, 675–676*f*
whale optimization algorithm (WOA), 669–670, 671*f*
Manning formula, 324
Manning's coefficient, 287
Manning's resistance equation, 287
Mann-Kendall (M-K) test, 582, 583–584*f*
arid region, seasonal and annual rainfall trends
spatial distribution and comparison of, 607–610, 608*f*
trend identification, 605
graphic method, 570
hydrologic time series, 600–602
Shiraz station in Iran, annual average and maximum rainfall, 594, 594*t*
trend analysis and variability of rainfall and temperature, 536–537
Mapping flood sensitivity, 202
MARE. *See* Mean absolute relative error (MARE)
Markov random field, 11
Mathematical models, of fluid flow in porous media, 391–400
basic assumptions, 391–392
coupled thermo-hydro-mechanical model, 395–397
fluid flow coupled with heat transfer, 394–395
groundwater flow modeling, equations for, 391–392
numerical model, 397–400
poroelasticity model, 392–394
temperature and seepage rates, 400–401, 401*f*
Mathematical optimization approach, 438
MATLAB, 472–473
Maximum absolute relative error (MXARE), 473–475
Mean absolute error (MAE), 250, 448–450, 461–462, 568–571, 571*t*, 671–678
Mean absolute percentage error (MAPE), 448–449
Mean absolute relative error (MARE), 290, 473–475
Mean shift, 11
Mean square error (MSE), 351–352
Membership functions (MFs), 207, 269, 664
Mercury tip sensors, 518–519
Metaheuristics optimization algorithms (MOA), 448, 454–455
MFs. *See* Membership functions (MFs)
MGGP. *See* Multigene genetic programming (MGGP)
Micro-scale statistical methods, 562
Midchannel bar, hole size analysis of bursting events
depth-wise distribution of Reynolds stress, 492, 493*f*
experimental program, 485–486, 486*t*
temporal distribution, 490–492, 491–492*f*
variation with hole size, 488–490, 489–490*f*
Millennium Development Goals, 433–434
MIROC6 model, 564–566, 572–580, 582–585, 583*f*
ML. *See* Machine learning (ML)
Moderate resolution imaging spectroradiometer (MODIS), 86–87
Modified honey bee mating optimization (MHBMO) algorithm, 287–289
Modular groundwater flow model (MODFLOW-2005), 187
Moisture adequacy index (MAI), 24–26
Momentum conservation, 393, 395–396
Moore-Penrose pseudo-inverse approaches, 250–251
Moraine dam resistivity survey, 646, 646*f*, 647*t*
Multiattribute utility theory (MAUT), 436
Multicriteria decision analysis (MCDA), 436
Multicriteria decision-making (MCDM), 438
Multigene genetic programming (MGGP), 286, 289, 296*f*
application, 289–290
channel design, 293–295
circular channels
design equations of, 287–289
optimum design of, 286–287
funding, 295
overview, 293–295
performance evaluation criteria, 290
relative error (RE), 290–292, 292–295*f*
side-weir discharge coefficient estimation in trapezoidal and rectangular open channels, 471–476, 472*f*, 473*t*, 476–477*f*
Multilayer perceptron model (MLP), 665–666
Multilayer perceptron neural network (MLPNN), 246–247, 448–449
Multiobjective optimization (MOO), 436
Multipattern approach, 435
Multiple criteria decision analysis (MCDA), 90
Multiple linear regression (MLR), 247–248
Multiple nonlinear regression (MNLR) coupled discrete wavelet transforms (MNLR-DWT), 449

N

Nash-Sutcliffe efficiency (NSE), 42, 231–232, 250, 448–449, 456–459
Nash-Sutcliffe efficiency coefficient (E_{NS}), 568–571, 571*t*, 671–674, 673*t*
National Sanitation Foundation (NSF), 266
Navier-Stokes equation, 328
NCHRP Project, 518
Near-infrared band (NIR), 129–130

Net irrigation requirement (NIR), 303–304, 307–308, 307–308t
Neural networks, 266
Neurons, 347
Newton's cooling law, 397
Non-linear models, 600–601
Non-parametric statistical tests, 538–540, 600–601
Normalization, 351
Normalized difference vegetation index (NDVI), 27, 90
NOT operator, 270
Nuapada watershed, 345, 346f, 347t
Numbered bricks, 520, 521f
Numerical model, 397–400, 398–399f

O
One-dimensional (1D) numerical modeling, 322–326, 323f, 325f
Online sequential extreme learning machine (OS-ELM), 246–247
Open-air level sensor (OAS), 647
OPEX, 23
ORELM-EWT-PSOGSA model, 449
OR operator, 270
Outlier robust extreme learning machine (ORELM), 449
Overall accuracy (OA), 229–230

P
Parametric tests, 600–601
Partial least squares regression (PLS), 38–39
Particle image velocimetry (PIV), 525
Particle swarm optimization (PSO), 343–344
Passive sensors, 121
Patna (Bihar), trend analysis and variability of rainfall and temperature
  area of research, 535
  information sources, 535
  Mann-Kendall (MK) test, 536–537
  precipitation, 538–540, 538t, 539f
  Sen's slope method, 537–538
  temperature, 540, 540t, 541f
Pearson's correlation coefficient, 106, 108
Penman-combination method, 417–418
Per capita water availability, 134
Pettitt's test, 601–602
P-factor, 231
Photogrammetry, 526
Physical-based distributed models, 220
Plane-bed reaches, 119
Ponding method, 417
Pool-rifle reaches, 119
Poroelasticity model, 392–394
Posteriori method, 438
Potential evapotranspiation (PET), 93–94
Potential water resources availability (PWRA), 10
Precipitation, 414, 417–418, 533–534, 538–540, 538t, 539f, 589–590
  innovative polygon trend analysis (IPTA), 590–591
  in Shiraz station, Iran (*see* Shiraz station in Iran, annual average and maximum rainfall)
  and temperature, in Gilan, Iran (*see* Climate change, precipitation and temperature data)
Preprocessing signal decomposition (PSD), 448
Pressure-state-response (PSR), 439, 443
Probability density function (PDF), 377–378
Process-response system, 118
Producer accuracy (PA), 229–230
Productive agriculture, 19
Prosopis juliflora, 234
Pulse devices, 521–522
  ground penetrating radar (GPR), 522
  time-domain reflectometers (TDR), 521–522
Pumping tests, 374–377, 377f

Q
Q-learning, 11
Quadrant Reynolds stress, 492
Quadratic rating curves (QRC), 246–247
Qualitative assessment, 87
Quantile regression (QR), 246–247

R
Radial basis function network (RBFN), 342–343, 347–348
  actual *vs.* predicted GWL using, 353f
  comparison results, 359–361, 361t
  data collection and model performance, 351, 351t
  outcomes, 353t
Radial basis function neural network (RBFNN), 252–257, 253f, 257t, 258f, 260f
Radio detection and ranging (RADAR), 121–122
Rainfall infiltration, 418, 419t
Rainfall recharge, 418
Rainfall-runoff modeling
  AHP, 205–206
  ANFIS, 207, 212
  determined CR, 215–219
  distance from river, 212
  drainage density, 212–220
  elevation, 208–209
  flow length, 212
  interrelationship and pairwise comparison, 215
  model performance evaluation, 208
  random forest, 207–208
  sensitivity analysis, 213–215
  slope, 209–210
  thematic map preparation, 205
  using GIS
    computation of runoff, 175–179
    curve number, 175
    data and software, 168

Rainfall-runoff modeling *(Continued)*
hydrologic soil groups (HSGs) map, 170–171
LULC, 170, 173
model validation, 172
SCS-CN method, 168–173, 179
soil map, 173–175
soil penetrability, 172–173
transformation, 166
Rainfall time series, trend and homogeneity tests, 600
arid region, seasonal and annual rainfall in, 603–604, 616–617
average, standard deviation, and CV, range of, 605–606, 605*t*
Box-whisker plots, 606, 606–607*f*
data description, 605
geographical locations, 604, 604*f*
homogeneity tests' results, spatial distribution and comparison of, 611–615, 612–615*f*
innovative trend analysis (ITA), 605, 607–610, 609*f*
limitations and future challenges, 616
recommendations, 616
statistical tests for homogeneity, 605
trend-magnitudes, Sen's slope estimation, 610–611, 610*t*
variance-corrected Mann-Kendall test, 605, 607–610, 608*f*
hydrology, 600–603
Shiraz station in Iran, annual average and maximum rainfall
characteristics, 592, 592*t*
innovative trend analysis (ITA), 593–596, 593*f*, 595*f*
innovative triangular trend analysis (ITTA), 594–596, 594*t*, 596*t*
Mann-Kendall (M-K) test, 594, 594*t*
study area, location of, 591–592, 591*f*
tropical regions, 590–591
Raipur Master Plan 2021, 107
Random forest regression (RFR), 38–39
Random vector functional link (RVFL), 250–251, 252*f*, 255–257, 257*t*, 258*f*, 259, 260*f*
Rank based tests, 601–602
RBFNN. *See* Radial basis function neural network (RBFNN)
RE. *See* Relative error (RE)
Reach-scale, 368
Reference evapotranspiration (ET_o), 304–305, 305*t*
Reflection seismic profilers, 523
Regional flow system, 368
Regional soil salinity model (RSSM), 422
Regression analysis, 600–602, 664
Regularized extreme learning machine (RELM), 448–449
Relative error (RE), 290–292, 292–295*f*, 473–474, 476
Relevance vector machine (RVM), 449, 455–459, 461–462
Remote sensing (RS), 642–645
flood disaster management, 90
groundwater management, 91
hydrological management, 87
hydrological response analysis, 93–94
precision irrigation, 89–90
salinity management, 90–91
watershed management, 87–89
Representative concentration pathways (RCPs), 552–553
Representative elementary volume (REV), 392, 397
Resistivity survey, of South Lhonak moraine dam, 646, 646*f*, 647*t*
Response surface method (RSM), 246–247
REV. *See* Representative elementary volume (REV)
Reynolds-averaged Navier-Stokes equations (RANS), 328–329
Reynolds stress, 488–489, 492, 493*f*
R-factor, 231
Richard's differential equation, 420
River basin-scale, 368
River channel migration, 153–154
River channels and remote sensing
data analysis, 123
electromagnetic energy, 120
knowledge production, 123–129
wavelengths, sensors and river science, 121–122
River discharge, 248, 249*t*, 249*f*, 257*t*
River, dissolved oxygen (DO) in
ensemble empirical mode decomposition (EEMD), 455–456, 460–462
extreme learning machine optimized Bat algorithm (Bat-ELM), 450–459, 454*f*, 461–462
performance assessment of models, 450
relevance vector machine (RVM), 455–459, 461–462
study site and data, 449–450, 451*f*
River water turbidity prediction, 246–247, 249*f*, 257*t*
generalized regression neural networks (GRNN), 251–252, 253*f*, 255–257, 257*t*, 258*f*, 260*f*
modeling strategy, 246, 246*t*, 255–257, 256*f*
performance assessment of models, 250
radial basis function neural network (RBFNN), 252–257, 253*f*, 257*t*, 258*f*, 260*f*
random vector functional link (RVFL), 250–251, 252*f*, 255–257, 257*t*, 258*f*, 260*f*
recommendations, 261
statistics of, 248, 249*t*
United States Geological Survey (USGS), 248, 248*f*
variational mode decomposition (VMD), 254–257
Root mean square error (RMSE), 42, 246–247, 250, 290, 351–352, 379, 448–450, 473–475, 568–571, 571*t*, 671–674, 673*t*
Root node, 289
RVFL. *See* Random vector functional link (RVFL)
RVM. *See* Relevance vector machine (RVM)

S

SAHYSMOD model, 410
Saint-Venan equations, 326
Salinity management, 90–91
Salinity submodel, 414
SaltMod, 420–421
Satellite remote sensing, 86*t*
SCOPUS database, 122–123
Scour holes, 514–515, 514*f*
Scour monitoring, 516–523, 517*f*
  advantages and limitations of devices, 524*t*
  electrical conductivity devices, 523
  Fiber-Bragg gratings, 522
  fixed scour monitoring instrumentation, 516–519
  measurement method in laboratory, 523–525
  numbered bricks, 520, 521*f*
  pulse devices, 521–522
  single-use devices, 519–520
  sound wave devices, 522–523
  using structural dynamic, 525–526
Scuba mouse, 518
Seasonal and annual rainfall, in arid region, 603–604, 616–617
  average, standard deviation, and CV, range of, 605–606, 605*t*
  Box-whisker plots, 606, 606–607*f*
  data description, 605
  geographical locations, 604, 604*f*
  homogeneity tests' results, spatial distribution and comparison of
    Bartlett test, 611, 614*f*
    Hartley test, 611, 612*f*
    Link-Wallace test, 611, 613*f*
    Tukey test, 611, 615*f*
  limitations and future challenges, 616
  recommendations, 616
  statistical tests for homogeneity, 605
  trend identification and magnitudes
    innovative trend analysis (ITA), 605, 607–610, 609*f*
    Sen's slope estimation, 610–611, 610*t*
    variance-corrected Mann-Kendall test, 605, 607–610, 608*f*
Seawater intrusion, 86–87
Sediment transport, magnitude frequency analysis of
  bankfull discharge, 498–499
  discharge for specific recurrence interval, 499
  dominant discharge, 498
  effective discharge, 499–502
    computation, 508
    in river design and monitoring, 509–510
  literature review, 503–507
  research findings, 500, 500–503*t*
Seepage meter method, 417
Sensitivity analysis, 213–215, 232, 234–235
Sen's slope method, 537–538, 600–602, 610–611, 610*t*
Sentinel-2, 625–626
Sequential uncertainty fitting (SUFI-2) algorithm, 231
Shallow water equations (SWE), 326
Shared socio-economic pathways (SSP), 564–566
  CanESM5 model, 564–566, 571
  IPSL-CM6A-LR model, 564–566, 580
  Miroc6 model, 564–566, 572–580
Shiraz station in Iran, annual average and maximum rainfall
  characteristics, 592, 592*t*
  innovative trend analysis (ITA), 593–596, 593*f*, 595*f*
  innovative triangular trend analysis (ITTA), 594–596, 594*t*, 596*t*
  Mann-Kendall (M-K) test, 594, 594*t*
  study area, location of, 591–592, 591*f*
Shuttle radar topographic mission (SRTM), 121–122
Side weir, 467–469, 469*f*
Side-weir discharge coefficient estimation, in trapezoidal and rectangular open channels
  artificial neural network (ANN), 471, 474–476, 476–477*f*
  data, 470–471, 470*t*
  machine learning methods, 471–473
  multigene genetic programming (MGGP), 471–476, 472*f*, 473*t*, 476–477*f*
  performance evaluation metrics, 473–474
  problem statement, 469–470
Sihar soil, 138
Si-Lo Bridge, pier scour measuring systems of, 520, 521*f*
Simulation models, GLOF, 642–643
  boundary condition, 655
  event and early warning, sequence of, 654
  flow algorithm, 655
  initial lake level and elevation assignment, 654–655
  mesh generation, 654
  simulated flood inundation, 657, 658*f*
Single-use devices, 519–520
Singular value decomposition (SVD), 468
Sixth Intergovernmental Panel on Climate Change Assessment Report (IPCC-AR6), 566
Slope based tests, 601–602
Soft-computing methods
  gradient boosting decision tree (GBDT), 72
  range of parameters, 73*t*
  self-adaptive extreme learning machine, 72
  stacked boosting regression tree (SBRT), 72
Soil and water assessment tool (SWAT), 187, 226–227, 230–232
Soil conservation service-curve number (SCS-CN) model, 168–173
  calibration and validation, 179
  features, 166
  with GIS, 166–167
  runoff computation, 171–172
  runoff production potential, 166–167

Soil depth function, 373
Soil moisture
  bidirectional long short term memory (BiLSTM), 42–43
  empirical wavelet transform (EWT), 44–45
  Gaussian process regression (GPR), 43–44
  machines learning (ML) models, 38–39
  multilayer perceptron neural network (MLPNN) model, 38–39
  performance assessment of models, 42
  signal decomposition algorithm, 38–39
  soil moisture content (SM), 38
  study site and data, 39–41
  support vector regression (SVR), 44
Soil texture, 58
Sonar devices, 516–518, 517*f*
Sonic fathometers, 522–523
Sound wave devices, 522–523
  echo sounders, 523
  reflection seismic profilers, 523
  sonic fathometers, 522–523
Sour
  adaptive neuro-fuzzy interference system (ANFIS) model, 78–80
  artificial neural network (ANN) model, 77–78
  maximum scour depth prediction, 72
  parameters on equilibrium depth, 73–74
  predicted scour depth, 75*f*
  prediction equations for, 74–75
  soft-computing methods
    gradient boosting decision tree (GBDT), 72
    range of parameters, 73*t*
    self-adaptive extreme learning machine, 72
    stacked boosting regression tree (SBRT), 72
  statistical error analysis, 76–77
  turbulent two-dimensional jets, 72
Sparrow number, 395
Sparrow search algorithm (SSA), 448–449
Spatial assessment, 108–112
Spatial database creation, and cropping pattern, 304, 305*t*
Spatial decision support system (SDSS), 656–657
Spatial ground techniques, 166
Spearman rank correlation (SRC) test, 600–602
Split-sample procedure, 332–333
Stacked boosting regression tree (SBRT), 72
Statistical analysis, Patna (Bihar)
  precipitation, 538–540, 538*t*, 539*f*
  temperature, 540, 540*t*, 541*f*
Statistical down scaling method (SDSM), 563
Statistical regression methods, 10
Steel rod, 518, 519*f*
Step-pool reaches, 119
Storm water management model, 226
Stream network, at Wadi Degla, 625–626, 628*f*
Student's *t*-test, 600–601
Subbasins (SB), in Wadi Degla, 631, 632–633*f*, 634*t*
Subjective logic (SL), 436
Sub surface drainage (SSD), 423–426
Subsurface hydrologic models, 370
Subtractive clustering (SC), 38–39
Support vector machine (SVM), 11–12, 342–343, 348–349, 455
  actual *vs.* predicted GWL using, 354*f*
  comparison results, 359–361, 361*t*
  data collection and model performance, 351, 351*t*
Support vector regression (SVR), 44, 246–247, 448–449
Surface-water model, 631–632, 632–633*f*, 634*t*
Surface water resources
  boundary and initial conditions, 335
  data retrieval and geometry construction, 331
  hydraulic numerical model, 320
  limitations, 337–338, 338*t*
  mathematical framework, 331, 333–335
  model classification, 322–329
  numerical models, 320–321, 330–335
  one-dimensional (1D) numerical modeling, 322–326, 323*f*, 325*f*
  retrieve of information and geometry building, 333
  three-dimensional (3D) numerical modeling, 322, 328–329, 330*f*
  two-dimensional (2D) numerical modeling, 322, 326–328, 327*f*
  validation and calibration of model, 332–333, 335–336, 336*f*
  water depth deficit analysis, 336–337, 337*f*
Sustainability index, 439
Sustainable Development Goal (SDG)-6, 21
Sustainable urban water development (SUWDP), 434
Sustainable water resource management, 142–143
SVM. *See* Support vector machine (SVM)
SVM with firefly algorithm (SVM-FFA), 349–350, 350*f*
  actual *vs.* predicted GWL using, 356*f*
  comparison results, 359–361, 361*t*
  data collection and model performance, 351, 351*t*
SVR. *See* Support vector regression (SVR)
Swarm-based algorithms, 665
SWAT. *See* Soil and water assessment tool (SWAT)
Symbolic concept-oriented learning, 4
Synchrosqueezed wavelet transform (SWT), 448–449
Synthetic aperture radar (SAR), 122

T

TauDEM toolbox, 377–378
Terminal node, 289
Thematic Mapper (TM), 625–626
Thermal flow
  in local thermal equilibrium (LTE), 394
  in local thermal non-equilibrium (LTNE), 394–395
Thermo-hydro-mechanical (THM) coupling, 389–390, 395–397

Three-dimensional (3D) numerical modeling, 322, 328–329, 330*f*
Tight coupling, 191*f*
Tiltmeters, 525
Time-domain reflectometers (TDR), 521–522
Time-series analysis maps, of Lhonak and South Lhonak lakes, 649, 650*f*
Tirora tehsil, 135, 136*f*
  analytic hierarchy process (AHP), 145–146
  climate and rainfall, 139–140
  geology, 138
  geomorphology and soil types, 138
  hydrology, 140
  irrigation projects, 142
  landforms, 138
  land use, 142
  measurement of water discharge, 141
  priority, 147–148
  slope, 138
  sustainable water resource management, 142–143
  water availability analysis, 140
  water balance study, 141
  water inflow, 140–141
  water quality analysis (WQA), 143–146
  water source sustainability, 142
1:5000 Topographic maps, 155–156
Total dissolved solids (TDS), 414
Total station method, 520
Total water cycle management (TWCM), 434, 436
Trapezoidal and rectangular open channels, side-weir discharge coefficient estimation in
  artificial neural network (ANN), 471, 474–476, 476–477*f*
  data, 470–471, 470*t*
  machine learning methods, 471–473
  multigene genetic programming (MGGP), 471–476, 472*f*, 473*t*, 476–477*f*
  performance evaluation metrics, 473–474
  problem statement, 469–470
Triangular functions, 269
Tukey test, 601–602, 605, 611–615, 615*f*
Turbulent burst, 485, 487
2D electrical resistivity survey, 646
Two-dimensional (2D) numerical modeling, 322, 326–328, 327*f*

U

UAV-based drones, 24
Ultrasonic level-sensing system, 647
Ultraviolet (UV) light, 27
Uncertainty in sequential fitting (SUFI-2) algorithm, 227
Under-water level sensors (UWS), 647
United States Geological Survey (USGS), 123–124, 168, 248, 248*f*
Upper Benue basin, 166–167
Urban distribution systems, 134
Urbanization
  defined, 100
  Indian urban areas, 100
  in Raipur, 100–101
    annual mean temperature, 102–103
    city population, 101*t*
    community toilets, 101
    Köoppen climate classification scheme, 102–103
    location map, 102*f*
    ponds in, 103
    sanitation system, 101
    solid waste and sewage, 101
    storm water drainage networks, 101
    water quality deterioration, 102
Urbanization expansion, within Wadi Degla, 628–629
Urban rain-water management, 202–203
Urban water development planning (UWDP), 436–438
US Bureau of Reclamation, 423
User accuracy (UA), 229–230

V

Varanasi City, integrated urban water management, 440–443, 440*f*, 441–443*t*, 443–444*f*
Variational mode decomposition (VMD), 254–257, 448–449
Vegebot, 28
Vellayani lake, 267, 277
Venn diagram, 20*f*

W

Wadi Degla, digital elevation model of, 623–625, 624*f*
Wadi Degla protectorate (WDP), flash flood in. *See* Flash flood, in Wadi Degla, Egypt
Wallingford tell-tail, 518
Water accounting, 86–87
Water availability, 134, 140
Water balance models, 88
Water balance study, 141
Water contamination, 86
Water delivery subsystem, 411, 416–417
Water development planning index (WDPI), 439, 441–442*t*, 442, 443*f*
Water flux, 410
Water inflow, 140–141
Water level and ice thickness calculation, 647–649
Water quality analysis (WQA), 143–146
Water quality and land use data
  data processing, 104
  data summary, 103–104
  Pearson correlation coefficients, 103, 110–111*t*
  and statistical significance testing, 105–107
Water quality index (WQI), 266, 270–271, 274
  membership functions, 272–273, 272*f*
  ranges, 274, 274*t*
  Vellayani, Sasthamkotta and Pookode lakes, 275*t*

Water quality management
freshwater fuzzy model for lake systems-fuzzy lake index (FLI), 272–273
fuzzy logic, 269–270
defuzzification, 269
fuzzification, 269
fuzzy set operations, 270
membership function, 269
heavy metal pollution index (HPI), 271–272
location map, 267, 268*f*
Pookot lake, 267
Sasthamkotta lake, 267
validation, 280–282
Vellayani Lake, 267
water quality index (WQI), 270–271
Water removal subsystem, 411–412, 419–420
Water resources
engineering, 3–4
management, 95–96, 438
optimization, 439–440
Water sensitive urban design planning support system (WSUD-PSS), 436, 437*f*
Watershed management, 86–87, 134–135
Water source sustainability, 142
Water surface elevation (WSE), 323, 335
Wavelet based genetic programming (W-GP) model, 665–666
Wavelet-based nonlinear autoregressive with exogenous inputs (WNARX), 665–666
WDPI. *See* Water development planning index (WDPI)
Weighted arithmetical index method, 270
Weight of evidence, 202
WGS 84 datum, 135–137
Willmott index (WI), evaluation criteria, 351–352
World Health Organization (WHO), 27, 105
Wroclaw University of Science and Technology (WrUST), 329

Printed in the United States
by Baker & Taylor Publisher Services

Printed in the United States
by Baker & Taylor Publisher Services